Animals, plants, water, wind, materials, and people move through spatial patterns characteristic of virtually all landscapes and regions. This up-to-date synthesis explores the ecology of heterogeneous land areas, where natural processes and human activities spatially interact to produce an ever-changing mosaic.

The subject is of enormous importance to today's society, and indeed for molding the future of areas surrounding each of us. The book bulges with ideas, principles, and applications for planning, conservation, design, management, sustainability, and policy. Spatial solutions are provided for many of society's land-use objectives.

This is an appealing and highly readable text on this major emerging field. Students and professionals alike will be drawn by the attractive and informative illustrations, the conceptual synthesis, the wide international perspective, and the range of topics and research covered.

LAND MOSAICS
The ecology of landscapes and regions

LAND
MOSAICS

THE ECOLOGY OF
LANDSCAPES AND REGIONS

Richard T. T. Forman

Harvard University

CAMBRIDGE
UNIVERSITY PRESS

Published by the Press Syndicate of the University of Cambridge
The Pitt Building, Trumpington Street, Cambridge CB2 1RP
40 West 20th Street, New York, NY 10011–4211, USA
10 Stamford Road, Oakleigh, Melbourne 3166, Australia

First published 1995

Printed in Great Britain at the University Press, Cambridge

A catalogue record for this book is available from the British Library

Library of Congress cataloguing in publication data
Forman, Richard T. T.
Land mosaics : the ecology of landscapes and regions / Richard T.T. Forman
p.   cm.
Includes bibliographical references and index.
ISBN 0 521 47462 0
1. Environmental policy.   2. Landscape ecology.   3. Human ecology.
I. Title
GE170.F57   1995
304.2'3–dc20   94–39092 CIP

ISBN 0 521 47462 0 hardback
ISBN 0 521 47980 0 paperback

Dedicated to the memory of
Henry Chandlee Forman
and
Caroline Lippincott Forman

Contents

Foreword   *page* xiii
Preface   xv
Acknowledgements   xx

PART I. LANDSCAPES AND REGIONS
1 Foundations
Mosaics and the patch–corridor–matrix model   3
Spatial scale   7
Natural process and human activity   14
Ecology of landscapes   19
Ecology of regions   22
Historical development of the field   28
Research methods   29
Conservation, planning, and other fields   36
Appendix: Glossary of common terms   38

PART II. PATCHES
2 Patch size and number
Patch dynamics: origin and persistence   44
Size effect on ecosystems   45
Size effect on biodiversity   54
Genetics in a patch   66
Minimum viable populations   68
Number of large patches   72
Appendix A. Ecosystem/community structure and function 75
Appendix B. Population ecology   78
Appendix C. Equations for species richness   80

3 Boundaries and edges
Types and abruptness   82
Edge development and controls   87
Functions of boundaries   95
Interactions between adjacent ecosystems   101
Width and curvilinearity   104
Moving boundaries   109

4 Patch shape
Common and uncommon shapes   114
Causes   120

ix

CONTENTS

Shape attributes and ecological flows   124
How to measure shape   133
Moving patches   138
Appendix: Equations for measuring patch shape   141

PART III. CORRIDORS
5 Corridor attributes, roads, and powerlines
Internal and external structure   146
Five corridor functions   147
Width and connectivity   153
Changing corridors   157
Linear boundaries, corridors, and troughs   158
Road corridors   159
Trails and powerlines   172
Appendix: Experience with wildlife underpasses and tunnels   176

6 Windbreaks, hedgerows, and woodland corridors
Structural richness   178
Windbreaks and wind   180
Effects on microclimate, soil, snow, and plants   187
Hedgerows as habitats   195
Woodland corridors as wildlife conduits   198
Appendix: Equations for wind and erosion   207

7 Stream and river corridors
Corridor types   209
Key processes   216
Water quality and dissolved substances   228
Fish and aquatic habitats   232
Terrestrial animals and plants   236
Corridor width and connectivity   243
Appendix: Equations for water and erosion   252

8 Networks and the matrix
Network development and structure   254
Dendritic networks   262
Rectilinear networks and wavy nets   265
Functional roles of networks   269
Networks in action   272
Matrix structure   277
Matrix dynamics   279
Appendix: Network equations   282

PART IV. MOSAICS AND FLOWS
9 Mosaic patterns
Ecosystems in a cluster   286
Configuration usage and change   294
Mosaics reflecting geology and culture   300
The third dimension   302
Types of landscape and region   306

CONTENTS

Habitat arrangements and strategic points 310
Measuring mosaics 317
Appendix: Equations for measuring a mosaic 319

10 Wind and water flows in mosaics
A global framework 323
Energy interconnecting mosaic elements 327
Materials and the giant conveyor belt 337
Flows in rough terrain 345
Disturbances in mosaics 351
Spatial pattern and disturbance spread 359

11 Species movement in mosaics
Movement, mosaic, and model characteristics 365
Metapopulation dynamics 372
Gene flow 383
Movement among non-adjacent elements 386
Exotics and invaders 393
Livestock 398

PART V. CHANGING MOSAICS
12 Land transformation and fragmentation
Spatial processes in land transformation 406
Habitat fragmentation 412
Land change patterns 415
Modeling mosaic sequences 423
Toward an ecologically optimum land transformation 426
Reforestation and other patterns 432

13 Land planning and management
Aggregate-with-outliers principle 436
Planning 440
Principles in a generic plan 449
Forest, dry area, and agriculture 456
Planning a suburb 459
Ecological design of house lots 469
Management 474

14 Creating sustainable environments
Concepts of sustainability 481
Time and space 486
Human dimensions 492
Assays 497
Adaptability and stability 502
Learning from history 506
Regional ecology revisited 512
Making a landscape or region sustainable 514

References and author index 525
Index 605

xi

Foreword

In the cloistered tradition of scientific specialization, most ecologists think of the world narrowly, as a system of natural environments beleaguered by human activity. They live, as Aldo Leopold put it, in a world of wounds. They have reason to think this way. Today, less than 10% of the land surface remains in a mostly unchanged state, and only 4% has been set aside in natural reserves. In their own specialty, planners and landscape designers tend to stay in their larger and wholly different world. For them, the bulk of the land has been given over to humanity; and now, they say, people must redesign it to their liking.

Neither view, taken to exclusion of the other, is viable any longer. Ecologists increasingly recognize that they must extend their science to embrace artifactual environments. More reflective planners and landscape designers understand that they cannot afford to ignore organisms – the *right* organisms – needed to sustain a healthy, pleasing environment. A few scholars, of which Richard Forman is a notable example, have come to focus on the intellectually rich domain of overlap between the two fields. The result of their deliberate intermediacy is the new discipline of landscape ecology.

In *Land Mosaics*, Forman joyously embraces the human-altered environment and proposes that its living part can be improved by use of the best principles of environmental biology. He recognizes, as the book's title implies, that the real world consists of finely fragmented habitats. The pieces range from radically altered urban parks and gardens to remnant pockets of the original environment. Across periods as short as a few years, living species arrive, impinge, dominate, yield, and disappear in this kaleidoscope. The vast majority of the inhabitants we never see, because they are too small and obscure: creepy-crawlies, immense in diversity, from insects to fungi and bacteria. All together, they are as important as the towering trees and the birds on which our attention is ordinarily focused. The species that can survive in the habitat fragments come together to determine the health and beauty of the places in which we live.

Forman and other scientifically educated landscape ecologists address the question: 'Given that humanity is now in charge of most of the land surface, and responsible for it, what designs of the terrestrial landscape are best for the rest of life, and therefore for us?' The answers cannot come wholly from unaided esthetic intuition or short-term econometrics. They must come also from a science that adds the living world effectively to a larger picture.

Edward O. Wilson
Museum of Comparative Zoology
Harvard University

Preface

Virtually all the great questions of our time involve land. I rank the following near the top. Can we reduce the number of mushrooming megacities, environmental refugees, and regional wars? As human linkages to the land are severed, can we discover a way to rewrap humans and nature? Can biodiversity be protected? What is lost when civilization converts the soft curves of nature to the hard lines of geometry? Can we shift our attention from the local site or ecosystem to the whole landscape or region soon enough to reverse cascading degradation? Since space on the planet is finite and few will be global managers, is regional decision-making the key to global protection? Is there a way to accelerate sustainable thinking at the time scale of human generations. All wise answers have roots in landscape and regional ecology.

The perspective in this book is captured by adding the following two middle lines to words in a classical Chinese verse (*At Heron Lodge*, by Wang Chih-huan, T'ang Dynasty, 618–906; Bynner & Kiang 1929).

> But you widen your view three hundred miles,
> Then discover a mosaic,
> And see it change,
> By going up one flight of stairs.

The land as seen from an airplane window or on an aerial photograph is the subject of this volume. It is a specific object, with attributes amenable to analysis, modeling, and manipulation. Its ecology is the interaction between organisms and the environment. Productivity, biodiversity, soil, and water are central. Landscapes, from forested to dry and from suburban to cultivated, are explored, and regions composed of many landscapes are introduced.

The ecology of spatial configuration at the human scale burst upon the scene barely a decade ago. Both theory and evidence were thin. But the potential of the conceptual framework and its application was obvious and far reaching. Theories, observations, models, and experiments crowd onto the pages of this book. The rapid transformation in thinking and the weight of accumulated knowledge are amazing. Applications are evident in all land-use areas. Yet the field is young, the uncertainties daunting, and the challenges exciting. Society urgently awaits the waiting discoveries.

Landscape and regional ecology has its feet firmly in science. But science is but a slice of human understanding. Wisdom requires other slices. Much of the core of this subject has come from specialists in ecology and several related disciplines who call themselves landscape ecologists. Yet theories, concepts, and evidence that developed in any context, and that enhance our understanding of the ecology of landscapes and regions, are freely welcomed and incorporated.

Mosaic pattern is the central feature of land, and ecological structure, function, and change of the mosaic is the central paradigm of the book. Spatial arrangement matters. It is the structure of a landscape or region. It determines the movements and flows between local ecosystems, and across the mosaic. It changes in form over time. Spatial arrangement is also a useful handle for decision-makers in planning, conservation, design, management, and policy.

The patch–corridor–matrix model, where all land is composed only of these three types of spatial element, has opened doors to analysis and application. The rapid coalescence of concepts results from a richness of methods, from remote sensing to links with first principles, modeling to field measurements, and landscape-scale experiments to microexperiments. The theories cut across geography, land use, and population density, and are robust from virgin to devastated mosaics, from tropical rainforest to temperate suburbia.

Unlike the landscape, regional ecology is little known. A region has a single macroclimate, but contains ecologically dissimilar landscapes. Human activities tie a region together, as in the idea of regionalism. Natural divisions such as biomes and drainage basins would theoretically be better for organizing the world, but it is unwise to wait for society to redraw the land. Policy can be made and implemented at the landscape and regional scales, where reasonable stability for long-term planning exists.

Regional ecology is pinpointed for three reasons. First, some have used the concepts of landscapes and regions synonymously. Second, assuming the concepts do differ significantly, many of the principles developed at the landscape scale also apply to regions. Third, regional ecology is a little-understood research frontier that will noticeably strengthen conservation, planning, sustainability, and land-use policy. We had better learn the ecology of regions.

Building from diverse origins and ongoing work by pioneering scholars, R. Forman and M. Godron presented in a 1986 book, *Landscape Ecology*, an initial weaving of the field into a land mosaic paradigm. That book remains a good introduction to the subject, and contains a fuller discussion of terminology, landscape typology, natural processes and human roles in the origin of landscapes, and information-theory analyses of landscape heterogeneity.

However, much research and a raft of recent edited books have spurred the field. The pages that follow show how it has mushroomed, coalesced, and reached a new plane. Quantitative and spatial models have proliferated. Geographic-information-system analyses have spread widely. Hypotheses and tests have become commonplace. Empirical evidence has accumulated in most conceptual areas. The present book is a state-of-the-art synthesis. In addition to topics in the preceding book, it features more emphasis on: (a) worldwide coverage; (b) a broader range of applications in land-use issues; (c) extensive literature citations for evidence; (d) integrating human culture and activities with natural phenomena; (e) boundaries and edges; (f) patch shape; (g) stream and river corridors; (h) road corridors; (i) windbreaks; (j) movements and flows across a mosaic; (k) land transformation and fragmentation; (l) land planning and management; (m) creating sustainability on land; and (n) the ecology of regions. The text highlights both the familiar (e.g., patch size and stream corridors) and the frontier (e.g., patch shape and sustainable environments).

Chapters progressively build upon one another. The foundations of landscape and regional ecology, and its simple but important language, are introduced at the outset. Early chapters then explore the array of patch, corridor, and matrix types, plus movements and flows between adjacent elements, as the essential building blocks of land mosaics. Middle chapters fit these spatial elements together in distinctive arrangements, and explore movements and flows through the resulting mosaics. Later chapters then explore the mosaics changing over time, including alternative pattern sequences in planning or managing landscapes and regions.

Fertile soil washes and blows away before our eyes. Biodiversity plummets. Stream corridors are bathed with nutrients from adjacent land. Food and wood production provides less and less to rural families. Houses sprout on the best agricultural soils. And humanity is increasingly divorced from nature. Does anyone see the motif? An ecology of whole landscapes and regions provides a framework to bridge chasms between disciplines, and address questions of our time.

Landscape and regional ecology is analysis for understanding, not advocacy for an objective (other than its use in all land decisions). It makes no distinction between basic and applied principles, and embraces both. It has expanded rapidly from an obscure subfield to a core theme at the heart of ecology; few ecologists now would undertake a study without considering the landscape-level implications. Similarly, it has expanded the content and horizon of several sister disciplines. Its concepts are robust in the traditional natural-resource fields of forestry, conservation biology, wildlife management, soil conservation, agriculture, and water resources. However, the concepts work equally well and increasingly are used in the human fields of regional planning, landscape architecture, transportation, and suburbanization policy.

Yet the separation of these fields, each with a single land-use focus, contributes to degrading the surface of the planet. All land uses compete in large areas, and all require ecological underpinning. Landscape and regional ecology pinpoints mosaic patterns that effectively grow wood, save species, locate houses, graze livestock, save soil, locate roads, protect streams, grow crops, and manage water, either separately or together.

Nothing is immutable and little is irreversible. Thus, we must ask what the appropriate time scale is for considering land use in a landscape or region. Infrequent events, such as major flood, fire, volcanic sky-darkening, hurricane, pathogen, and war, often have long-lasting effects at this spatial scale. Therefore, experience over years or decades is insufficient for wise land-use decisions. What do the documented, generations-long histories of land use from diverse cultures have in common? That answer should figure prominently in a land-use decision.

Indeed, landscapes and regions are exactly the right scales for sustainability. Local ecosystems are commonly transformed in days or years, whereas land mosaic transformations usually occur incrementally over generations. Humans avoid long-term planning and decisions, whereas sustained human–land interactions require them. Large area is a surrogate for long term.

Yet, both the quality and ethics of decision-making are challenged by the landscape and regional perspective. It is simply inept or poor-quality work to consider a patch as isolated from its surroundings in the mosaic. Designs, plans, management proposals, and policies based on drawing an absolute boundary around a piece of the mosaic should be discarded. Moreover, because we know this is wrong, i.e., we know ecological context is as important as content, the practice is unethical. Ethics impel us to consider an area in its broadest spatial and temporal perspectives.

What is the vision? Where do we go? Bits of nature provide key pieces of understanding; broad strokes of nature interacting with humans on land provide different pieces of the pie, and tie the bits into a picture. An ecological understanding of the picture, although incomplete, has emerged. No longer can we tiptoe around because evidence here cannot be applied there. It is time to aim higher, to make those linkages from landscape to landscape and region to region, now based on a foundation of general patterns and principles. Natural-resource reserves are not the goal; rather, they are important pieces in a mosaic where every piece counts. Economic and ecological models have been poor predictors, but now spatially-explicit ecological models at the scale of human land use

xviii

should improve the record. Rather than playing follow-the-leader by waiting until all evidence is in, scholars and decision-makers should aggressively recognize the landscape or region as a keystone.

Indeed, a spatial solution is emerging that will protect the bulk, though not all, of the species, soil particles, mineral nutrients, and clean water in any landscape. This approach incorporates several 'indispensable' spatial patterns, as well as particularly effective patterns, and promises to solve vexing land-use issues.

What is the optimum design for three or more of society's major land-use objectives? In a changing landscape where are the best and worst locations for making the next change? How can we spatially mesh ecological integrity and basic human needs over a time frame of human generations? Which of the principles of landscape and regional ecology will cause the most noticeable improvement in the land? A decade ago such questions were rarely posed. Now I believe we are on the threshold of solutions.

Richard T. T. Forman
Harvard University

Acknowledgements

My recent thinking has been especially influenced by discussions with Michel Godron, Peter J. Grubb, Jerry F. Franklin, Bruce T. Milne, George F. Peterken, Carl Steinitz, Michael W. Binford, Edward O. Wilson, and many students in my graduate classes at Harvard. I warmly thank each of you; your wisdom has guided me to countless discoveries.

For reviewing and strengthening portions of this book, I am deeply grateful to: Per Angelstam, William L. Baker, Jacques Baudry, Michael W. Binford, Ralph E. J. Boerner, James R. Brandle, Jasper Brandt, Françoise Burel, Margot D. Cantwell, Sharon K. Collinge, Thomas R. Crow, Henri Decamps, Kathleen Deagan, Almo Farina, Barbara L. Forman, J. Edward Gates, Wolfgang Haber, Lennart Hansson, Bryant E. Harrell, Kristina Hill, Richard J. Hobbs, Peter Jacobs, James R. Karr, Glenn R. Matlack, Vernon Meentemeyer, David H. Miller, Bruce T. Milne, Robert J. Naiman, Reed F. Noss, Robert V. O'Neill, Paul Opdam, David Pimentel, William H. Romme, Denis A. Saunders, Christine Schonewald-Cox, Karl-Friedrich Schreiber, Frederick Steiner, Fred J. Swanson, Monica G. Turner, Sam Bass Warner, Jr., Thompson Webb III, and Edward O. Wilson. I particularly appreciate Andrew F. Bennett and David R. Foster who reviewed the entire manuscript, and Michael W. Binford who was the problem solver *par excellence*. I thank Joan D. Ferguson and Daniel H. Monahan for major help with the case study in chapter 14. And I especially acknowledge the contribution of George F. Peterken, who worked with me on many portions of chapter 12.

All figures, except a few with separate acknowledgements, were drawn by Amy Bartlett Wright. I am grateful to her for artwork that portrays the irregular patterns of nature, translates general principles from the author, and provides clarity for the reader. I also thank Simran Malhotra, Iram Farooq, and Xiang Yu for carefully checking most of the bibliography. I am delighted and honored with the painting by Elizabeth F. Harrell created for the cover of this book. Finally, Barbara L. Forman's sustained encouragement made this book a joy to write.

Richard T. T. Forman

PART I

Landscapes and regions

1

Foundations

Nature, the earth herself, is the only panacea.

Henry David Thoreau, Journal, *1859*

Five minutes after leaving a tropical city one of the plane's two engines dies. The plane begins dropping. Rich primeval rainforest awaits on one side, and town-dotted farmland on the other. Someone snaps an instant photograph, carries it to the open cockpit, and the pilot explodes in laughter. The passengers remain petrified until we slip over a ridge and bounce onto an unlighted runway.

The object spread out beneath an airplane window, or on an aerial photograph, is the subject of this book. Indeed, an aerial vantage provides a goldmine of information on the ecology of large land areas, such as landscapes and regions. For example in Fig. 1.1, one can predict with some confidence that the forest has many more interior species than the wooded patches on the right. Animal movement funnels through the lobes or peninsulas of the forest. Wind erosion is high in the corridor gaps on the left. Cool-water fish are probably missing from the stream in the far right. The clearing at the top has been expanding, while the clearing below it is contracting. The ecologically optimum location for a cluster of houses is by the field-forest edge in the upper right. The evidence for interpretations such as these and many more is presented in this book. Of course, dropping from the sky to examine the land closely is also essential.

MOSAICS AND THE PATCH–CORRIDOR–MATRIX MODEL

From an airplane, land almost always appears as a *mosaic*. The glorious mosaics of St. Mark's in Venice or the University of Mexico appear as a

3

pattern of colored patches and strips, usually with a background matrix. Tiny stones of different colors are aggregated to create the patches, strips, and matrix. The land appears much the same. Individual trees, rocks, houses, and so forth are the tiny stones. Woods, fields, and housing tracts are conspicuous patches. Roads, hedgerows, rivers, and power-lines are equally striking corridors. Grassland, forest, rice culture, or another land use often forms a background matrix. In short, the individual trees, shrubs, rice plants, and small buildings, analogous to the tiny stones in the artist's mosaic, are aggregated to form the pattern of patches, corridors, and matrix on land.

Mosaic patterns are found at all spatial scales, from submicroscopic to the planet and universe. Land mosaics, however, are at the 'human scale', measured in kilometers to hundreds, even thousands, of kilometers. Thus, landscapes, regions, and continents are three scales of land mosaics.

What causes a mosaic? Much like a child's room with toys, a closed system with no energy input tends toward randomness. Such a lack of organization results in a fairly homogeneous mess throughout, and is expected according to the *second law* of thermodynamics. Without energy input, such as returning toys to their shelves, a system becomes more disorganized (gains entropy). But the land is always *spatially heterogeneous* (an uneven, non-random distribution of objects), that is, always has structure. The key is solar energy. Over geologic time it produces landforms, and today it grows different plants, which provide structure or heterogeneity to the land.[1129,1739,1762,814]

But spatial heterogeneity occurs in two flavors. A *gradient* or series of gradients has gradual variation over space in the objects present. Thus a gradient has no boundaries, no patches and no corridors, but is still heterogeneous. A portion of a moist tropical rainforest is an example where the assemblage of tree species changes gradually over the land. But gradient landscapes are scarce.

The alternative form of spatial heterogeneity is a mosaic, where objects are aggregated, forming distinct boundaries (Fig. 1.2). A land mosaic may contain only patches, or may also contain corridors. No spaghetti-like mosaic of only corridors is known. In short, the land mosaic is directly dependent on thermodynamically open conditions, with solar energy creating and maintaining structure.[501,503]

More specifically, three mechanisms create the pattern. Substrate heterogeneity, such as hills, wet spots, and different soil types, causes vegetation patchiness. Natural disturbance, including fire, tornado, and pest explosions, creates heterogeneity. And human activity, such as plowing fields, cutting woodlots, and building roads, creates patches, corridors, boundaries, and mosaic pattern. Various biological processes commonly modify or enhance the patterns.

4

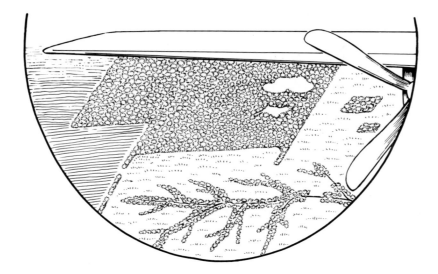

Fig. 1.1. Looking only at pattern, 'landscape detectives' gain extensive information about the ecology of a landscape. Wind is from the left in this cool Russian area, which shows patterns typical of landscapes worldwide.

Understanding heterogeneity is only a first step.[579,646,1253,883,1748] An infinite number of spatial arrangements can produce a particular level of heterogeneity, whether high or low. Specific spatial arrangements or configurations are ecologically much more important.

A more general way to understand form is to relate it to movements and change. One may say that 'Form is the diagram of force'.[1707,14] Form or structure, i.e., what we see today, was produced by flows yesterday. The curving sand dune was shaped by wind, the rectangular vineyards by tractors, and the dendritic stream corridor by water erosion. In addition, a linkage or feedback between structure and function is evident. Not only do flows create structure, but structure determines flows. For example, the arrangement of patches and corridors determines the movements of vertebrates, water, and humans across the land. Finally, movement and flows also change the land mosaic over time, much like turning a kaleidoscope to see different patterns.

Therefore, like all living systems (those containing life), the landscape exhibits structure, function, and change.[501,1745,88,1748] The plant cell has membranes and a nucleus which control the movement of molecules, and over time the cell's anatomy changes. The human body has organs and tubes through which fluids move, and over time the shape of the body changes in interesting ways.

Fig. 1.2. Contrasting regions along the USA–Canada border. Pastureland and native woody vegetation predominate in Alberta and Saskatchewan to the north, and cultivated wheat fields (*Triticum*) in Montana to the south. Differences in land-use policies have differentiated these two regions in the vicinity of the Milk River. Satellite image courtesy of D. H. Knight and ERIM; Image Data Processing by the Environmental Research Institute of Michigan (ERIM), Ann Arbor, Michigan.

Patches and corridors have long been a focus of human activity.[961,646] Ecologists originally focused on patches and patchiness.[1322,612,856] Many became interested in corridors when discussing possible applications of island biogeographic theory on land[400,1224,683], and the ecological roles of hedgerows and windbreaks.[1337,948,91,174,1474]

The patch–corridor–matrix model coalesced when it was realized that a land mosaic is composed only of these three types of *spatial elements*[483,499] (see glossary in chapter 1 appendix). Every point is either within a patch, a corridor, or a background matrix (Fig. 1.2), and this holds in any land mosaic, including forested, dry, cultivated, and suburban. This simple model provides a handle for analysis and comparison, plus the potential for detecting general patterns and principles.[991,683,465,894,1490,1742,1484,661]

Since a mosaic at any scale may be composed of patches, corridors, and matrix, they are the basic spatial elements of any pattern on land. Thus, *landscape elements* are simply spatial elements at the landscape scale. They may be of natural or human origin, and thus apply to the spatial pattern of different ecosystems, community types, successional stages, or land uses.

Because the key spatial attributes are so readily understood, the model has become a spatial language, enhancing communication among several disciplines and decision makers.[484,91,530,1518,678,1472,638] For instance, patches vary from large to small, elongated to round, and convoluted to smooth. Corridors vary from wide to narrow, high to low connectivity, and meandering to straight. And a matrix is extensive to limited, continuous to perforated, and variegated to nearly homogeneous. These scientific descriptors are kept close to dictionary concepts.

Of course, the overall model can be elaborated to recognize additional spatial attributes. For example, *nodes* are patches attached to a corridor. *Boundaries* separate spatial elements and vary widely in structure. *Unusual features* are rare landscape-element types, such as a single major river or two mountains with particular functional significance.

The patch–corridor–matrix model has analogues in other disciplines. Point, line and plane are fundamental concepts in art[870,829] and in architecture.[275,1147] Patch, matrix and mosaic are used in the medical field.[559] The urban planner, Kevin Lynch (1960), recognized five spatial elements, based on what evokes a strong image in a person: district, edge, path, node, and landmark (C. Steinitz & M. W. Shippey, pers. commun.). They are similar to patch, boundary, corridor, node, and unusual feature. Finally, in addition to this foundation in spatial structure, insights are gained by studying spatial variations in movements and flows, different rates of pattern change over space, and scale (K. Hill & M. Roe, pers. commun.).[501,1728,1883]

SPATIAL SCALE

Scale is an underlying motif that sounds in every chapter. However, several useful scale concepts here set the stage.[1102,1883,957]

Time, space and stability

A scarlet-and-black poison-arrow frog sits contentedly on a log in a rainforest, where it is periodically bathed in direct sun penetrating the

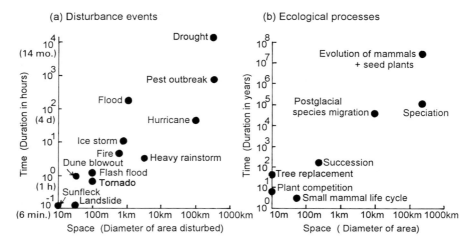

Fig. 1.3. Space–time graphs for environmental and biological changes. Dots represent the approximate centers of ovals within which most examples are found. Adapted from Delcourt et al. (1983), Birks (1986), Forman & Godron (1986), Urban et al. (1987) and McDowell et al. (1990).

canopy. The area of a sunny spot (a sunfleck) is often a few meters across, and the frog, if it did not hop away, would be heated for perhaps 15–30 minutes. A hurricane, in contrast, would alter an area tens of kilometers across, and last for hours. Continental glaciers move over thousands of kilometers, and last tens of thousands of years. These environmental changes or disturbances illustrate a relationship between space and time (Fig. 1.3a).

Most short-duration changes affect a small area, and most long-term changes affect a large area. This generalized *space–time principle* is also observed for many biological responses (Fig. 1.3b) and other ecological attributes.[383,133,1762,384,1084] The principle (like most log–log relationships) is very rough in detail and in predictive ability. The time scales associated with landscapes and regions typically are years to centuries. Exceptions include vertebrate movement patterns over hours or days, and vegetation changes over millennia.

Maps have a *scale*, usually printed in the corner, to show the ratio between distance on the map and actual distance on the ground (see glossary in chapter 1 appendix). For example, a one-to-a-hundred (1%) ratio would be useful for mapping the shrubs and paths in a small garden, and a one-to-a-million (0.0001%) ratio would portray provinces and mountain ranges in a nation. Unfortunately, some literature and disciplines consider the garden map to be large scale, and some consider it to be small scale, relative to the national map. To avoid this confusion

8

and to use terminology generally clear to all, this book only uses the terms, fine-scale and broad or coarse scale.[501,1883] *Fine scale* refers to the pattern of a small area, where the difference between map size and actual size is relatively low (e.g., a garden or neighborhood). In contrast, *broad scale* or *coarse scale* refers to the pattern of a large area, where the difference between map size and actual size is great (e.g., a nation or continent).

The space–time principle implies that phenomena at broad scales are more persistent or stable than those at fine scales. Fine-scale phenomena should be more variable in both time and space. To evaluate such predictions we must briefly consider how an area such as a landscape is regulated or maintained within some stable (or metastable) range. Hierarchy theory and cybernetics are the two major approaches, although a combination of the two may also operate.[1193,638]

Hierarchy theory refers to how a system of discrete functional elements or units linked at two or more scales operates.[1285,605,1250,1762] Thus, a forested landscape might be hierarchically composed of drainage basins, which in turn are composed of local ecosystems or stands, which in turn are composed of individual trees and tree gaps. Each element, from the tree to the forested landscape, functions as a unit, has its own constraints, and exhibits its own degree of stability or variability. The landscape system is a nested hierarchy with each level containing the levels below it.

Flows link the elements at one level, such as movements of animals and heat between nearby local ecosystems or stands. Flows also link elements vertically in the hierarchy, e.g., gap processes affecting stands, or landscape-wide fire regimes affecting nutrient flows within drainage basins, or regional processes controlling local species richness.[1547,878,1488,670] Therefore, to understand the stability of a particular element, a minimum of three linkages must be known. The element is linked to the: (1) encompassing element at the next higher level; (2) nearby elements at the same scale; and (3) component elements at the next lower level. In view of the space–time principle, the encompassing element should provide more stability, whereas the component elements provide more variability. Both the pattern and the processes within a spatial element are dependent on the three hierarchical linkages.

In contrast, *cybernetics* refers to a system in which the elements are linked in feedback loops.[1193] Quite simply, a *feedback loop* has one element affecting a second element, which in turn affects the first. Adult mice produce baby mice that grow into adult mice. Where both linkages are positive or stimulatory, it is a *positive feedback*. The more adult mice present the more babies are produced, and the more babies there are the more adult mice are produced. No stability results from a positive feedback loop; indeed it is a vicious cycle, where soon the world is knee-deep in cuddly mice.

9

In contrast, a *negative feedback* is a marvelous regulatory mechanism to maintain stability. Here, one linkage is positive and the other negative. The more rabbits there are, the more foxes there are; the more foxes, the fewer rabbits; the fewer rabbits, the fewer foxes; the fewer foxes, the more rabbits; the more rabbits, the more foxes; and on and on. The negative feedback does not produce constancy. Rather it keeps both rabbits and foxes varying cyclically in numbers, but within limits. In theory neither species explodes in numbers, as do the mice, and neither species goes extinct.

Grain

A grainy photograph appears full of dots or splotches. The 'resolution' of a photograph, such as an aerial photograph of a landscape, refers to the ability to distinguish or differentiate the objects or spatial elements within it. This relates to *contrast*. As in photography, contrast depends on the amount of difference between adjacent elements, and the relative abruptness of their boundary[501,891] (Fig. 1.2). Thus, a landscape may have high contrast because of sharp boundaries, say, between two similar woodland types. Or high contrast may result from the dissimilarity of woodland and clearing, even though they spatially intergrade along their edges.

Therefore, *grain* refers to the coarseness in texture or granularity of spatial elements composing an area. This is determined by size of the patches that are distinguishable or recognized.[1022,580,501,1883,1223,1884] In landscapes, this is simply measured as the average diameter or area of the patches (and portions of matrix) present. A *fine-grained* landscape has primarily small patches, and a *coarse-grained* landscape is mainly composed of large patches.[39,1474,1888] Thus, scale refers to the spatial proportion of a mapped area, and grain refers to the coarseness of elements within the area.

The resolution of an aerial photograph thus depends not only on the physical grain and contrast present, but also on the ability of the observer to distinguish spatial elements.[458] Similarly animals in the landscape have different *grain responses*, i.e., perceive or are sensitive to different grain sizes.[1022,958,501,1223,1884,39] The deer in Fig. 1.4 responds positively to the clearings provided by wildlife biologists, but does not perceive small spider holes in the ground or the pattern of regions in Texas. The mammoth, an extinct big cousin of the elephant, responded to the mosaic of wetlands and ridges, whereas its fleas responded to individual hairs and varying hair densities on the mammoth's knee. The two animals responded to different grain sizes. Indeed, the mammoth went extinct, but the fate of the mammoth flea is still unknown.

This perception of the grain size of the environment is a key to land-

Fig. 1.4. High contrast and varied grain size in a wildlife management area. Large 250-meter-wide clearings and intervening shrubland strips provide browse and cover for deer, quail, peccary, and wild hogs (*Odocoileus, Colinus, Dicotyles, Sus*). Small fenced rectangles are for birds and other wildlife. Scattered brush piles and lines of brush provide habitat for many small species. Near Encinal, Texas; courtesy of USDA Soil Conservation Service. White-tailed deer (*Odocoileus virginianus*) in Montana; courtesy of US Fish and Wildlife Service.

scape functioning and change. Animals and humans perceive and respond to only a fraction of the multi-scale heterogeneity present. The spatial elements causing a response determine the direction, route, and rate of movement. Therefore, planning and management must focus on both the grain of the landscape and the grain response of its inhabitants.

The hierarchy on land

Suppose you had a giant zoom lens hooked up to your personal spaceship. You begin with a view of the whole planet, and slowly and evenly close the lens until you have a microscopic view of soil particles. At any point you would probably see a mosaic, a heterogeneous pattern of patches and corridors.[501,891] But would the mosaic gradually change in form? Or would it remain for a period and then change suddenly to a new form, like a kaleidoscope that is turned abruptly at intervals? Probably the view through your zoom lens would resemble the kaleidoscope sequence. The quasi-stable mosaics separated by rapid changes would represent *domains of scale*.[590,899,1751,384] Thus, each domain exhibits a certain spatial pattern, which in turn is produced by a certain causative mechanism or group of processes. Overall a mosaic pattern is relatively

11

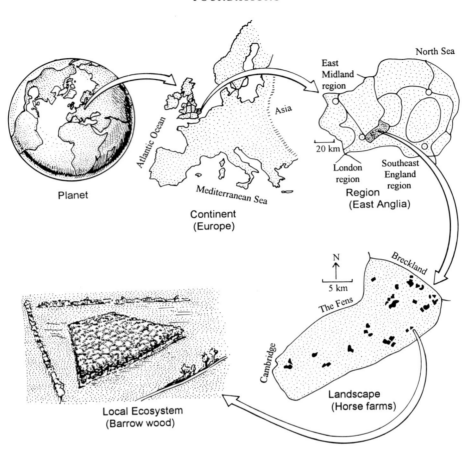

Fig. 1.5. The spatial hierarchy on land. The continent map only includes a few of the regions present, and the landscape map only shows a few of the wooded local ecosystems present. The 'horse farms' or 'woody strips' landscape is distinctive for its wide strips of woody vegetation. Drawn with the aid of G. F. Peterken, Ordnance Survey maps, and Bayliss-Smith & Owens (1990).

stable from the beginning to the end of the scale domain, although minor changes are evident. With such a powerful lens you might wish to test the generality of this idea by zooming over a tropical rainforest, and then a desert.

The planet is spatially subdivided in many ways, including political, economic, climatic, and geographic, depending upon human objectives. The spatial hierarchy used here[1129,1130,1535,72,73,1493] (Fig. 1.5) is selected because of its utility in meshing both human and ecological patterns, processes, and policies. David Miller (1978) further elucidates the follow-

12

ing hierarchy levels in terms of energy and mass, including constraints and flows.

The *biosphere* or planet is subdivided into continents (and oceans). Continents are subdivided into regions, regions into landscapes, and landscapes into local ecosystems or land uses (Fig. 1.5). Local ecosystems, such as woodlots or fields, can, of course, be further subdivided to show their internal patchiness, and so on.

Each level in the hierarchy represents a single scale. Hence, at the scale below the East Anglia region (Fig. 1.5c), landscapes differ in size, shape, and many other attributes, including from fine grained to coarse grained.

Continents usually have distinct boundaries, but in most cases are only loosely tied together by transportation, economics, and culture. Continents encompass extremely dissimilar areas of climate, soil, topography, vegetation, and land uses. Policies and political decision-making at the continental level have been much more often ineffective than effective. At present, an ecology of continents would be mainly an extrapolation from the subunits studied.

A *region* is a broad geographical area with a common macroclimate and sphere of human activity and interest.[888,405,780,781,479] This concept links the physical environment of macroclimate, major soil groups, and biomes, with the human dimensions of politics, social structure, culture, and consciousness, expressed in the idea of regionalism. The southwestern USA, the Loire Valley of France, northern Queensland (Australia), the Andes of Venezuela, the Canadian maritime provinces, and East Anglia in Britain[96] (Fig. 1.5) are regions. Some regions are international (e.g., Scandinavia and Central America), some within a country (Southern California and the Lake Baikal area of Russia), and some are predominantly urban and suburban (New Dehli and New Orleans). Regions often have diffuse boundaries determined by a complex of physiographic, cultural, economic, political, and climatic factors. A region is tied together relatively tightly by transportation, communication, and culture, but often is extremely diverse ecologically.

A *landscape*, in contrast, is a mosaic where the mix of local ecosystems or land uses is repeated in similar form over a kilometers-wide area. Familiar examples are forested, suburban, cultivated, and dry landscapes. Whereas portions of a region ecologically are quite dissimilar, a landscape manifests an ecological unity throughout its area. Within a landscape several attributes tend to be similar and repeated across the whole area, including geologic land forms, soil types, vegetation types, local faunas, natural disturbance regimes, land uses, and human aggregation patterns. Thus, a repeated cluster of spatial elements characterizes a landscape.

Rooted in fields from geography to aesthetics, the term 'landscape' not surprisingly has been variously defined in hundreds of published papers.[1519] The above-described concept, now widely used, integrates a focus on (a) spatial pattern, (b) the area viewed in an aerial photograph or from a high point on the land, and (c) unity provided by repeated pattern.

The *local ecosystem* or land use as a spatial element within a landscape has been described earlier in this chapter. This element is relatively homogeneous and often distinct in its boundary (Fig. 1.5). Nevertheless, all objects are heterogeneous or variegated, composed of patches within patches.[453,482,1410,1441,701,891] The internal heterogeneity, patchiness, and gap dynamic processes within a local ecosystem or land use are elucidated in most ecology textbooks.

Normally it is easy to distinguish individual patches and corridors as local ecosystems in a landscape, and also to distinguish different types of landscape, such as rangeland, suburbia, and rice-paddy land. But in the spatial hierarchy, is there a recognizable and useful intermediate level between the local ecosystem and the landscape? Several lines of evidence presented in chapter 8 suggest the presence of common stable *configurations* or neighborhoods of patches and corridors.[501,1535,1919,251] These should be important in understanding landscape functioning, and be especially useful in design, planning, and management.

Several other spatial units have not been included in this hierarchy of scales. Although oceans (and 'seascapes') are not included, many landscape ecology concepts are probably useful in understanding oceans and their bottoms. Nations are omitted because they vary so widely in size, from thousands of meters to thousands of kilometers across. *Biomes*[198,1096] and *ecoregions*[71] are omitted because, by focusing primarily on the biological dimension, the boundaries usually correlate poorly with human-delimited administrative boundaries; effective political decision-making is essential for land planning and protection. *Drainage basins* (catchments, watersheds) are omitted because they too vary so widely in size, from a tiny stream basin to the whole Nile or Mississippi basin. They are nice for surface-water-driven processes. But watershed divides are often poor boundaries for delimiting animal, human, and wind-driven flows, or protecting home ranges, aquifers, ridges, and view-sheds.

Although boundaries determined by natural processes, such as drainage basins and bioregions, are theoretically optimum, it is not wise to wait for society to redraw the land. To accelerate the use of ecology in design, planning, conservation, management, and policy, we must use regions and landscapes that balance and integrate natural processes and human activities.

NATURAL PROCESS AND HUMAN ACTIVITY

Some years ago an expert in forestry from a developed country was invited to advise in a developing country. A village leader welcomed the expert, and learned that productivity of the local wooded area could be tripled by gradually replacing the rather heterogeneous, scruffy-looking trees with a high-quality eucalypt or pine plantation. Pondering the many ramifications of this profound change, the two leaders strolled through the woods to look more closely. The host observed, 'This tree provides nuts in the dry season; this clearing is where my ancestors won the final battle against invaders; this moist area protects our only clean drinking water; this grove provides the best firewood in the area; this tree is where I was married; this shrub is the only source of fibres for our unique dance; this vine provides the incense for our annual religious festival; this line of decrepit trees provides the children with flutes; this dense bushy area provides at least six major economic products; these virtually unburnable trees on the windward side are essential fire-proofing; and these tall arching trees form the cathedral for reflection and inspiration.' The two leaders embraced warmly, and the visitor returned home to take a closer look at the local tree plantations there.

The overall goal of science or scholarship is understanding, and concepts, models, theories, predictions, experiments, explanations, and so forth are used to gain understanding.[1180,1318] Much of ecology, including landscape and regional ecology, is science. However, some ecological understanding (on interactions of organisms and their environment) comes from studies in social science and the humanities. In this book natural processes, including geomorphic, soil, atmospheric, hydrologic, fire, plant, and animal, receive emphasis. But human activities almost always interact with natural processes to produce the actual patterns, movements, and changes observed. As in the local woods example, these phenomena result from cultural, religious, social, and economic activities overlapping over historical time.[215] Four brief examples from China, Eastern Europe, the USA, and Australia illustrate different ways that human activities and natural processes interact to produce pattern on the land.

Traditional Chinese philosophy focuses on the harmonious relationship among *Tian* (heaven or universe), *Di* (earth or resource), and *Ren* (people or society).[1827] Within this the *Yin* and *Yang* theory emphasizes the duality of natural forces acting upon and within a relationship or ecological system. The *Feng-shui* or wind–water theory expresses the spatial relationship between human settlements and the natural environment.[167,1827,1948,1949] For example, the main function of Feng-shui forests common in rural China is accumulating '*Qi*' (or 'living energy'), the combined flow of energy, material, species, and information. Therefore, Feng-shui forests, as old as the villages, usually are in specific locations (Fig. 1.6). The upper slopes and tops of hills, at water mouths where water enters or leaves a basin, and on steep erodible slopes fulfill the criteria for maintaining forests.

In a village a tortuous channel called an 'ox intestine' brings fresh water by everyone's home, and since everyone cares for it, there is no

Fig. 1.6. Locations where forest should be maintained and temples built based on *Feng-shui* theory. Drawing courtesy and with permission of Yu Kongjian.

tragedy-of-the-commons.[675] At the end is the 'ox belly' a large nutrient-rich pond for fish production, fire control, and growing the symbolic lotus flower. Thus, ecology does not determine the location of forests, nor economics the route of flowing water. Rather, long-term experiments (centuries of 'trial and error') determine the perhaps-optimum locations, which are consistent with culture, religion, social structure, economics, politics, geology, forestry, soils, water resources, and ecology combined.

Ancient Chinese temples and tombs, little constrained by dense populations and utilitarianism, are sometimes said to illustrate an ideal landscape based on Feng-shui.[1948,1949] A temple is placed at the foot of a mountain range, where headwaters converge to form a stream that winds across a flat valley and leaves between protecting hills (Fig. 1.6). Such a location receives 'living breath from all directions', i.e., is the confluence of active flows of energy, material, species, and information, and is therefore ideal for maximizing *Qi*.

In contrast, for two or three centuries in rural Eastern Europe villages were often surrounded by long pastures, and crop fields often separated by hedgerows. Relatively wide stream corridors wound through the farmland. In the twentieth century a major political, social, and economic upheaval changed the face of the land. In parts of Slovakia, for

16

example, equal-sized cooperative communities (collectives or communes) were built at regular points across the agricultural landscape. The distance between cooperatives typically was less than that between former villages, and the total human population was greater. Commonly, each community was surrounded by a large field area dominated by a single crop or land use. Typically, hedgerows were scarce and fields were cultivated right to stream banks. Thus, the political change converted the land from a fine-grained to a coarse-grained mosaic.

Meanwhile, for example in parts of Romania, some rural cooperatives had houses aligned close to and along a road. Behind each home was a large vegetable garden protected by walls. And immediately behind the garden stretched the extensive common agricultural land. In the 1990s a political change quickly resulted in land change. Rural residents began farming small plots, which overnight created a variance in scale. A highly diverse fine-scale mosaic contrasted with the surrounding, still coarse-grained land.

Concurrently a far different pattern occurred outside of cities in the USA, where houses and housing tracts popped up virtually anywhere. Unlike the previous case, fossil fuel was cheap, vehicles plentiful, local administrative units made the decisions, and regional planning was ineffective or absent. Since a city's original location was usually selected because of good agricultural soil, the houses inevitably sprang up on the best soils. Exurban areas became fine-grained mosaics of houses and natural land, which expanded outwards hundreds of times the area of the city. Nature conservation efforts also expanded in the area. The focus was on protecting small patches for rare species or recreation. Corridors were for human transportation, not the fauna. Thus, nature was left fragmented in little pieces, while shreds of nature remained near everyone's home.

Finally, certain Aboriginal peoples in Australia are among the most spatially attuned cultures known. For perhaps fifty thousand years, well back in the ice age, this culture has learned and taught the landscape features from generation to generation. Such accumulated memory and knowledge is far fuller and deeper than that from decades of orbiting satellites; it has a solid core of ecological wisdom. When Australia was connected to Asia, early ancestors apparently walked diagonally across the continent on routes now-called 'songlines'.[269] Major songlines were along watercourses. In today's desert environment those routes are along ancient buried drainage lines often invisible to the eye, or along grassy plains, or from water hole to water hole. The routes are especially rich in wildlife such as red kangaroo (*Megaleia rufa*). Major ceremonial centers are sometimes among the richest wildlife spots. This is partly because hunting there is usually forbidden, interestingly to prevent extinctions where people aggregate. Over the quarter of a million or so human generations every point on the land has been crossed countless

17

times, and sacred sites seem to be nearly everywhere.[269] Many sacred sites, such as rock clefts, valleys, and water holes, not only serve critical religious and cultural functions, but also as rich sources of species for repopulating the land after drought or fire.

Within a local area the few families of Aboriginal people present use fire instead of tilling the land to create crops. A sequence of small fires from several hectares to several square kilometers over a few-year period creates a fine-scale mosaic (D. Carter, pers. commun.)[819,16] (Fig. 1.7a). Here, different food resources are available during each of the perhaps three to six seasons of the year. The mosaic becomes a *cognitive map*, a learned mental picture of landmarks and routes. The residents constantly alter their routes through the mosaic, as concentrated food and water resources change spatially month after month.[819,1825] In droughts the Aboriginal people move farther across the land deriving historical, cultural, and personal meaning from sacred sites and songlines along the way (Fig. 1.7a). Few physical corridors or barriers are present. With such a close link between people and nature, a mosaic of patches makes great sense, but boundary, fence, and road corridors make no sense.

Indeed, Aboriginal people visiting the ordered streets and planned geometric vistas of Australia's capital, Canberra (Fig. 1.7b), have found it highly disordered, as well as boring.[819] It is empty of ecological signifi-

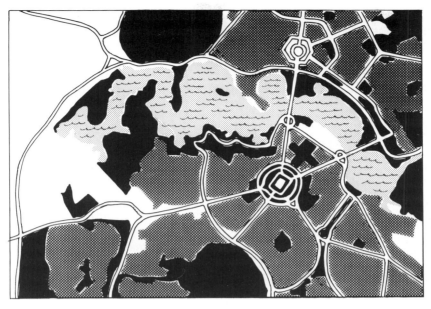

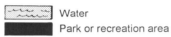 Water    Built area

Park or recreation area   Main road

Fig. 1.7. Two Australian landscapes. (*left*) Painting of desert landscape and storytelling; title '*Mume ku*', artist Mavis Captain (?), 1992, acrylic on board, Mutitjulu Ayers Rock. The following is one of various possible interpretations. Most dots represent trees, shrubs, or rocks. Different-shaded patches of dots represent different vegetation and fire-history areas; the two sandhills on left (each indicated by double curved black lines) separate vegetation patches, whereas the sandhills on the right do not, perhaps because fire crossed over them. The center circles represent an important meeting place; the double and triple straight lines are travel routes; the small circles to the right and left are resting places; and the circles in the corners are campsites (usually with rock, well, or other special feature). U-shaped designs represent women, and the snake designs indicate that they tell 'dreaming tales' of live poisonous snakes. R. Forman photo. (*right*) The nation's capital, Canberra, a glorious planned city designed by W. Burley Griffin and built on pastureland in the twentieth century.

cance, harvestable food resources, wells, religious affiliation, and the immutable links between land and people. It has no real meaning.

ECOLOGY OF LANDSCAPES

Ecology is generally defined as the study of the interactions among organisms and their environment. A landscape was described as a

19

kilometers-wide mosaic over which local ecosystems recur. Thus, *landscape ecology* is simply the ecology of landscapes, and similarly, *regional ecology* is the ecology of regions. The spatial elements within landscapes have been called landscape elements, local ecosystems, ecotopes, biotopes, biogeocoenoses, geocomplexes, sites, and more (see glossary in chapter 1 appendix). The spatial elements within regions are landscapes. This land mosaic or 'ecomosaic'[483,487] paradigm has not only attracted scientists who see rich research opportunities, but also galvanized linkages among disciplines directly solving land use issues.

Nevertheless, various alternative perspectives on landscape ecology have appeared over the years. Carl Troll (1939, 1968)[1519] apparently first used the term when aerial photographs became widely available, and his concept evolved into: 'Landscape ecology [is] the study of the entire complex cause–effect network between the living communities (biocoenoses) and their environmental conditions which prevails in [a] specific section of the landscape . . . [and] becomes apparent in a specific landscape pattern or in a natural space classification of different orders of size'. V. Sukachev & N. Dylis (1964) described biogeocoenology as a similar concept.[1738,1519] A. Vink (1975)[1193] considered landscape ecology to be 'the study of the attributes of the land as objects and variables, including a special study of key variables to be controlled by human intelligence'. I. Zonneveld (1979) indicated that 'Landscape ecology is an aspect of geographical study, which considers the (land) as a holistic entity, made up of different elements, all influencing each other'. Zev Naveh & Arthur Lieberman (1993) viewed landscape ecology as a trans-disciplinary ecosystem–education approach based on general systems theory, biocybernetics, and ecosystemology as a branch of total human ecosystem science. Paul Risser et al. (1984) concluded that 'landscape ecology considers the development and dynamics of spatial heterogen-eity, spatial and temporal interactions and exchanges across hetero-geneous landscapes, influences of spatial heterogeneity on biotic and abiotic processes, and management of spatial heterogeneity'. Geography and geographic information systems also have been considered close to landscape ecology[950,951,647], and other diverse encapsulations of the field are available.[488,271,966,703,639]

Based on the current prevalent concept of landscape ecology, several principles were outlined by Risser et al. (1984), Forman & Godron (1986), Risser (1987), and Turner (1989). These highlighted the distinc-tive nature of the questions being addressed, compared with those in ecosystem science, island biogeography and physical geography, for instance. But the field has moved ahead so rapidly, both empirically and in theory, that the present principles and theories are better absorbed in logical context than in an isolated list (Chapter 14).

Fig. 1.8. Boundary area of two landscapes. The grassland landscape in the foreground has dry ridges, moist valleys, intermittent stream gullies, north slopes, and south slopes repeated throughout its area. The irrigated landscape has individual homes, rectangular fields, canals, and roads repeated throughout. Canals and roads approximately parallel the distinct boundary. California east of San Francisco. R. Forman photo.

The repetition of a few characteristic ecosystems across a landscape means that there is a limit on the variety of habitats available for organisms. A landscape extends in any direction until the recurring cluster of ecosystems or site types significantly changes (Fig. 1.8). For example, one moves from a cultivated landscape to a suburban landscape where the cluster of fields, hedgerows, farmsteads, farm roads, woodlots, and stream corridors changes to a cluster of housing tracts, grassy public spaces, shopping areas, woodlots, and stream corridors.

How sharp is the boundary of a landscape? In mountainous regions where landscape mosaics are usually relatively small, contrasts are great due to sharp boundaries. 'A traveler crossing a mountain range moves into a new mix of ecosystems – that is, a new mosaic – every 20 to 50 km.'[1130]

Where natural geomorphic and disturbance processes predominate in flatter terrain, boundaries of landscapes also tend to be rather sharp. Thus, rock types, soil types, flood regimes, and fire regimes often produce abrupt transitions on land at this scale. The same sharp boundaries of landscapes are observed where human land uses, such as agriculture and forestry, reflect the natural water, soil, and tree species distributions.

However, where human activities and land uses are more independent of the distribution of natural resources, boundaries of landscapes tend to be less distinct. Familiar examples are the boundary between suburbia and forest, agriculture, or dry land. In some areas houses and housing tracts are primarily located based on economic, social, and political criteria, with little regard for natural boundaries.

Often it is useful to observe part of a landscape or region, such as a representative or random sample (e.g., Fig. 1.1). This will be simply referred to as a *portion* of a landscape or region. Other than containing more than one local ecosystem of the landscape (or more than one landscape of the region), no structural unity, extent, or boundary is implied. It is analogous to a portion of a wood, town, or continent captured in a photograph.

ECOLOGY OF REGIONS

Concept of a region

As the planet spins one feels that the atmosphere is circulating around the globe. Atmospheric 'cells' form within this apparent overall circulation, due to differences in solar input and the configuration of continents, mountain ranges, and oceans.[888,72,73,1198] Each cell exhibits its own *macroclimate*, an essentially uniform weather history over a large area, and each contrasts with the surrounding cells. Some regions are the same size as, but many are a subset of, a macroclimatic cell. Thus, a region normally has a single macroclimate, which provides a region-wide control over the soils, ecosystems, and natural processes.

This climatic control of a region is illustrated in the southwestern United States, where scattered mountain ranges are surrounded by desert landscapes. Major fires generally occur in the same year in many ranges. That is, though few fires can spread from one range to another, a drought affects all the mountain landscapes, and results in fires scattered throughout the region.

Some regions develop their own atmosphere though heat energy reflection (albedo), soil heat radiation, evapotranspiration of water, and air pollution.[719,1130,534] The Altiplano of Chile and Bolivia and the tropical rainforest of southern China are examples. Other regions are bathed more or less continually with outside air, and have energy and material flows determined mainly by other parts of the world. The mountains downwind of Los Angeles, the Gobi Desert east of the Himalayas, and the long diverse western half of Chile all have macroclimates determined by conditions to the west. Some regions, such as the North American Midwest and monsoonal areas of India, Indochina and China, are characterized by abrupt climatic changes, when one major air stream seasonally replaces another. Finally, human-caused climate change is increasingly considered at the regional scale.[227,1505]

In addition to macroclimate, human activities determine a region.[780,781] Regions usually contain one major city, or occasionally a few. Transportation and communication usually tie a region together. In some cases, regional boundaries are relatively distinct, and act as strong filters of many inputs and outputs. In short, models of regional change must have an effective balance between nature and humans, or at least ecology and economics.[705,1743,1744,1415,1280,1234]

An example differentiating region and landscape is instructive.[490] New England in the northeastern USA is a relatively distinct, widely recognized region[1634] (Figure 1.9a). It is surrounded by two Canadian regions, two American regions, and an ocean. Except in the southwestern corner, its boundaries are quite distinct physically or culturally. New England is tied together by a cool climate, a tradition of governing by town meeting, a transportation network, and cultural nuances including architecture, religion, and language. However, different portions of the region differ markedly in their ecology, for instance, from wild spruce–fir (boreal) forests in high mountains to the housing tracts and exotic species of suburbia.

This region is composed of at least ten landscape types (Fig. 1.9b). Five cover large areas (oak forest, pitch pine, northern hardwoods, spruce–fir, and agricultural landscapes), and five more (suburban, urban, salt marsh, barrens, and industrial landscapes) are scattered within these. Some landscape types can be subdivided for special purposes, such as cultivated and pasture landscapes instead of agriculture, or oak, transition hardwood, and northern hardwood in lieu of simply oak and northern hardwood landscapes (Fig. 1.9b). Two important alpine areas (Mt. Washington and Mt. Katadin) do not extend for kilometers in width, and therefore are considered as unusual features within their respective spruce–fir landscapes. Overall, the New England region is composed of approximately 100 landscapes.

Spatial arrangement of landscapes

Just as in any level of the spatial hierarchy, the region is composed of patches, corridors and a matrix that vary widely in size and shape. In this case the spatial elements are whole landscapes. Unlike the recurring landscape elements in a landscape, a region does not exhibit a pattern of repeated landscapes.

Usually the distribution of landscapes simply mirrors the typically coarse-grained, geomorphic land surface. Thus, most regions are coarse-grained, or variable-grained with groups of small landscapes (Fig. 1.9b).

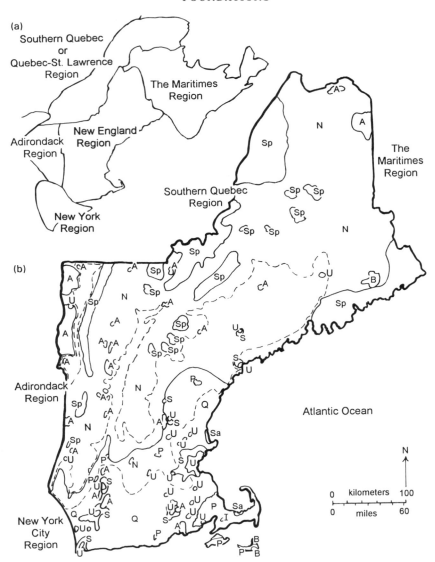

Fig. 1.9. The New England (USA) region, showing (a) surrounding regions and (b) landscapes within. Sp = spruce-fir (*Picea–Abies*); N = northern hard-woods (*Acer–Fagus–Betula–Pinus–Tsuga*); A = agricultural; Q = oak forest (*Quercus*); P = pitch pine–oak (*Pinus rigida–Quercus*); U = urban; S = sub-urban; Sa = salt marsh; B = barrens; I = industrial. Area surrounded by dashed line is sometimes differentiated as a 'transition hardwoods land-scape'. Synthesized from many sources, several courtesy of W. H. Rivers.

24

Some common landscapes have distinctive shapes. For instance, 'corridor landscapes' as kilometers-wide strips include major mountain ridges, wide river valleys, coastal strips, and suburbs along a major transportation route. A suburban ring or band, such as that around Paris, Sao Paolo, or Denver (USA), is also a distinctive landscape shape.

Not surprisingly, the arrangement of landscapes has a major effect on regional flows, and flows have an important feedback effect in producing the arrangement.[535] A major urban center acts as a source of people, vehicles, information, and products. These disperse on radiating transportation and communication routes in all directions across the region. These routes help tie the landscapes together. Some objects, such as air and water pollutants, and people headed for recreation, move out in specific directions. Reverse flows of forest and agricultural products, water supply, and rural people head for the city.

The physical flows linking landscapes are conspicuous in a mountainous region.[1130,1661,535,947] Here, landscapes, such as alpine, coniferous forest, basin grassland, and so forth, tend to be small and have sharp boundaries. It is a fine-grained region, or a portion of a variable-grained region that includes flatter landscapes. In the mountains gravity carries water overland, in streams, and in the ground to lower landscapes. Soil creeps overland or dashes down streams. Wind carries seeds and spores to higher landscapes, as well as downward and along ranges. Animals, often transporting seeds, move upward, downward, and along mountains. Water supplies rush downward in pipes and canals to agricultural and built landscapes.

In short, the spatial pattern or arrangement of landscapes in a region is just as important functionally as the pattern of continents on the globe, local ecosystems in a landscape, or gaps within a woods. Some flows are concentrated, such as water, silt, and industrial pollutants in rivers. Some flows are dispersed, including erosion, seeds, and vehicular pollutants. Some move rapidly and some slowly.[593,1409] And boundaries of landscapes are often filters or places where rates of movement change markedly, a distinctive pattern for fire or dispersing animals.

Wide-ranging species like caribou, tigers, black bears, and vultures are especially sensitive to the arrangement of landscapes. Such species commonly use, and perhaps require, two or more landscapes which cannot be too far apart. As suggested by island biogeography theory, landscapes of a particular type cannot be too far apart for dispersal of their species.[1023,196] For example, in the Southwest, a relatively distinct region in the USA, ski area development is damaging alpine landscapes, which in turn are limited to widely separated high mountain tops. Some alpine plants, invertebrates, and vertebrates may be significantly affected by this loss of stepping-stone landscapes.

25

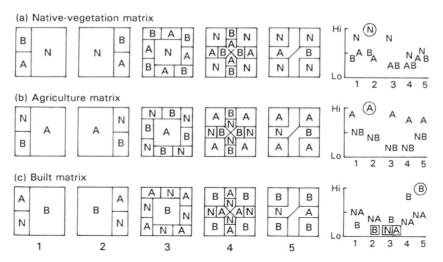

Fig. 1.10. Comparison of spatial arrangements of three landscape types, native vegetation (N), agriculture (A), and built area (B). Each square is a region composed of three to 12 landscapes. The graphs on the right have a horizontal axis representing the sequence of landscape patterns diagrammed on the left. The vertical axis is an overall measure of suitability (low to high) for maintaining natural ecosystems, agricultural productivity, and the built human community, respectively. Points circled on the graphs are the highest, and points enclosed in boxes are the lowest, comparing all 15 regions. See text. Assumptions in these simple geometric models are: (1) the key variables examined are sensitive at the scale of these landscapes; (2) wind blows from the west, and upwind air is the same as that along the western edge; (3) native vegetation produces the cleanest air; (4) for the built area, only large landscapes produce air pollution; (5) for native vegetation, fragmentation into small landscapes is detrimental. The models could be modified with additional variables or different assumptions, and produce different results.

Understanding the key flows and movements among landscapes permits us to search for an optimum spatial arrangement in a region. Simple geometric models readily understandable by decision-makers can be used for direct comparisons. For example, countless spatial arrangements of three common landscape types in a region, native vegetation, agriculture, and built area, are possible. Fifteen contrasting designs or arrangements are compared in Fig. 1.10 using a few simple assumptions. In these models the three key variables are the type of matrix, the sizes of landscapes, and their spatial arrangement.

Where native vegetation is the matrix (Fig. 1.10a), design number 2 exhibits the highest suitability for natural processes. Similarly, with this matrix (top graph) agricultural production appears best in design 1, and the built human community best in design 2. However, where agriculture is the matrix (Fig. 1.10b with middle graph), agriculture is best in arrangement 2, whereas both nature and the built area are best in design 1. Where the built area is the matrix (Fig. 1.10c), arrangement 5 appears optimum for the built human community, and arrangement 1 for natural process and agricultural production. Comparing all 15 models permits an estimate of the hypothesized, overall best designs and worst designs for natural process, agriculture, and the human community.

With such a simple modeling approach one could propose an optimum spatial design for sustainability of a region that integrates the three land-use objectives. Indeed, on average a region is a better bet for attaining sustainability than a landscape. This is because of its larger area, greater complementarity of resources, and a slower rate of change. The space – time relationship (Fig. 1.3) also supports this conclusion.

Linkages with other regions

The arrangement of landscapes in a region not only affects the region, but also neighboring regions. To illustrate, a century ago there was much discussion of how deforestation in the Pine Barrens landscape of New Jersey (USA) would affect climate change in the New York and Philadelphia metropolitan regions, some 50–100 km distant.[1329,1792] The half-million-hectare forested landscape had been subjected to two and a half centuries of intense resource extraction and lack of care.[484] Exposed soil was extensive, shrublands were widespread, tree height had been halved, smoke from rampant fires persisted, and atmospheric heat and water balances were altered. During the ensuing few decades several events, technological developments, and policy changes led to a significant increase in land protection of the Pine Barrens.[496] Fires decreased, the forest regrew, and threatened climate change evaporated.

This scenario also emphasizes that linkages between regions are generally both institutional and physical.[1237] Public policies plus wind determine flow rates of nitrogen and sulfur oxides between regions. Agricultural policies at a supraregional level mean that the delta of the Rhone River receives chemicals from most of central France, and the Mississippi River delta receives chemicals from half of the USA. A boundary as sharp as a linear coastline formed across the plains between Canada and the USA (north of Montana) in five decades (Fig. 1.2). This was caused by big land-use policy change in far-off Ottawa, and little policy change in far-off Washington.[879] The political and institutional change has effectively created a new region, where previously a single more-extensive region had existed.

The remaining sections of this chapter introduce the history and research methods of landscape ecology, and its use in other fields. To provide both the appropriate breadth of vision and an entree into the rich literature available, the concepts and information will be densely packed. The reader is encouraged to focus on the overarching concepts or main points in each section. Of course, one may dip into important details in areas of special interest.

HISTORICAL DEVELOPMENT OF THE FIELD

Several recent histories have explored the roots and rootlets of landscape ecology, especially as it developed in northern Europe[1194,1196,1192,637,1519,1956,1193], and in other regions.[501,1365,491] Rather than summarizing these detailed sequences of concepts, people and time, three broad overlapping phases are presented, with the source disciplines or areas of thought in each.

The initial phase, extending to about 1950, was the *natural history and physical environment phase*. The focus was on understanding nature over large areas.[1931,1089] Among the many contributors Herodotus, A. von Humboldt, and C. Darwin helped develop the origins of science in this area. A. W. F. Schimper, A. P. deCandolle, A. Grisebach, J. Braun-Blanquet, and F. E. Clements contributed more specifically to the origins of a broad-scale ecology. G. White, H. D. Thoreau, and A. Leopold linked nature and ethics. A. Losch, W. Christaller, and C. O. Sauer from geography, and indirectly many climatologists and soil scientists, played important roles.

The second phase, from about 1950 to 1980, was the key *weaving phase*, when diverse threads were meshed to set the stage for today's discipline. The Troll concept of landscape ecology[1734,1735,1737,1738] was further developed and enhanced by geographers and plant geographers, including E. Neef, J. Schmithusen, G. Haase, W. Haber, K-F. Schreiber, T. Bartkowski, M. Ruzicka, and I. S. Zonneveld, several of whom also demonstrated its importance in planning.

But several other disciplines also were effectively incorporated. A reinvigoration in ecology occurred with the biogeocoenosis or ecosystem concept, spurred by the work of A. G. Tansley, V. N. Sukachev, N. Dylis, H. T. Odum, E. P. Odum, G. M. Woodwell, R. H. Whittaker, F. H. Bormann, G. E. Likens, and others. Animal and plant geography syntheses by C. Elton, P. J. Darlington, P. Dansereau, P. Ozenda, H. Walter, and others contributed. R. H. MacArthur and E. O. Wilson's island biogeographic theory[1023] was especially useful. Vegetational and spatial methodologies of P. Greig-Smith, K. A. Kershaw, J. T. Curtis, R. H. Whittaker, M. Godron, and H. Ellenberg, hedgerow studies by E. Pollard and colleagues and a group in Rennes, France,[948] plus agronomic work of A. P. A. Vink were incorporated. Quantitative geography by P. Haggett, A. D. Cliff, E. J. Taaffe, H. L. Gauthier, Jr., and others became an important component. Regional studies by R. E. Dickinson, A. D. Cliff, J. K. Ord, W. Isard, and others began to coalesce. The role of human culture and aesthetics was highlighted by G. A. Hills, J. B. Jackson and A. Rapoport. Landscape architecture and planning work by I. McHarg, G. Eckbo, E. H. Zube, J. G. Fabos, C. Steinitz, H. Kiemstedt, and W. Kehm, plus land evaluation techniques of C. S. Christian, G. A. Stewart, J. Thie, G. Ironside, and I. S. Zonneveld were linked to landscape ecological thought. The interweaving of pairs of these disciplines produced a mound of fascinating swatches. But no clear form of the overall tapestry was evident.

Since about 1980 we have been in a *land mosaic or coalescence phase* of fitting the puzzle pieces together, and seeing the overall conceptual design of landscape and regional ecology emerge. A wide range of people and ideas has been brought together in a series of books during this period. Several edited volumes include a focus on the breadth of landscape ecology concepts[1722,1470,177,1957], habitat fragmentation and conservation[225,1490,672,670], corridors and connectivity[1518,174,1492,1582], quantitative methodology[114,1748], and heterogeneity, boundaries, and restoration.[1742,661,1808,1494] The major authored syntheses focus on land evaluation and planning[1953,1672], soil and agriculture[1800], logging and conservation[683], total human ecosystem[1193], hierarchy theory[1250], statistical methodology[820], river corridors[1038], and land mosaics.[501] But the most important contributors are those in the pages ahead.

RESEARCH METHODS

A rich palette of research techniques is used by landscape and regional ecologists. Some, such as remote sensing, geographical information systems, landscape-scale macroexperiments, and spatial modeling, have rapidly expanded our knowledge and horizons.

Link to first principles

Geologists, astronomers, and physicists are often eyeball-to-eyeball with large complex objects where experimentation is difficult. Yet these detectives of nature have made great advances by linking observable phenomena to existing first principles or basic theory. *First principles* refer to fine-scale or reductionist statements of human knowledge that are highly robust. The laws of thermodynamics, movement of matter, and evolution are examples.

This linkage approach is promising as a foundation for the development of principles at the landscape and regional scales. Before presenting examples, two caveats are in order. First, some basic theory is really a second or intermediate-level principle, being based on finer-scale or more-robust theory. Second, we usually learn from analogy, that is, gain insight by comparison with structural or functional analogues. Yet, differences between a landscape pattern or process and its analogue always exist. A deeper relationship, where a landscape phenomenon is directly related to or derived from its analogue, may or may not exist.

The following are examples where first principles provide insight into landscape pattern or process.[493,503] (a) A thermodynamically open system leads to heterogeneity as well as patchiness. (b) Active or passive movement of matter (or molecular movement) underlies flows in corridors. (c) Membrane physiology is a useful analogy to link structure and function of boundaries in a landscape.[1514,1886,503] (d) Neutral or random spatial models[543,1252] (such as percolation theory) provide insight into movement probabilities of animals across a landscape mosaic. (e) Network theory, including connectivity, circuitry, and linkage density, is important for understanding the movement of animals along intersecting corridors.[995,646,91,223] (f) Central place theory is a useful foundation for learning which patches and corridors are suitable habitats, as well as the dispersal routes of animals.[278,321,1393] (g) Form and function principles underlie the shapes of patches for conservation, and interactions with the matrix.[1707,1340,686] (h) Mass-flow principles, including the Venturi effect, are essential to understand the roles of gaps in a corridor[174] (Fig. 1.11). (i) The counter-current principle is useful in explaining the efficiency of predators foraging along a corridor.[493] (j) Gravity and rheotaxis (upstream movement) provide bases for understanding the one-way or two-way flows of water, mineral nutrients, fish and terrestrial animals in stream corridors.[685] Closer attention to first principles will noticeably strengthen the emerging principles of landscape and regional ecology.

Experiments and observations

Experiments with controlled conditions study the effect of changing a common or normal phenomenon, which helps lead to understanding cause-and-effect and a more exact science.[627] The change is an addition or removal of energy or material. Controls in an experiment are needed to measure the system before alteration, and also to compare an unaltered system with the altered one. Because some landscape-scale experiments require a large area, a long time, and suitable replicates or pseudo-replicates that are scarce[677], a variety of observational and experimental approaches has emerged. Observational and measurement studies of natural patterns and of human-caused patterns are first illustrated. Then landscape-scale and micro-scale experiments are introduced.

Fig. 1.11. Gap in a hedgerow corridor. Note the rabbit browse line at the bottom of hedgerow shrubs, suggesting a rich food source for certain predators. 'Horse farm' or 'woody strips' landscape of East Anglia, UK (Fig. 1.5). R. Forman photo.

(1) *Measurements of natural patterns.* A panoply of measurements, observations, and surveys falls into this category. These include studies of: pollen deposits to understand post-Ice-Age or recent vegetation movement[368,1847,767,382]; mineral nutrient flux between ecosystems[596,1543,1544]; spatial susceptibility to fires and hurricane blowdowns[496,1441,510]; stream corridor structure and change[1503,1661]; movement of mountain zonations[1580,46]; and dynamics of desert pattern.[312] Such studies occasionally include controls, and commonly overlap with the following category.

(2) *Measurements of Anthropogenic Patterns.* Since much of the land has a conspicuous human imprint, studies of anthropogenic (human-caused) patterns and processes are common, even where the vegetation looks relatively 'natural'. The following examples illustrate the breadth of such studies: mineral nutrient fluxes from field to stream[34,999]; effects of boundary curvilinearity on mine revegetation[676]; avian richness affected by hedgerow ditches and woodlots in the matrix[50,1775]; movements of species along hedgerows[1337,718,224] (Fig. 1.11); invertebrates dependent on location in a network[223]; woody perches attracting birds and seeds that change succession[1631,376,1082,722,630]; road underpasses for

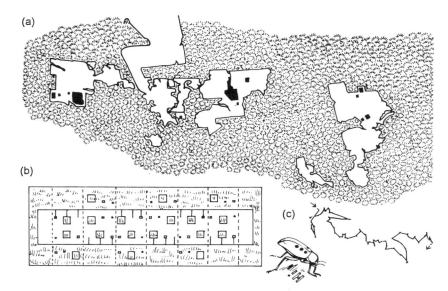

Fig. 1.12. Landscape-scale and micro-scale experiments. (a) Habitat fragmentation effects have been studied in two 100 hectare (250 acre), four
10 ha, and five 1 ha remnants of Amazon rainforest, plus controls in the
surrounding forest (Lovejoy et al. 1984, 1986). Black indicates remnant
wooded patches within white cleared areas within a forest matrix. Courtesy
and with permission of T. E. Lovejoy. (b) Remnant grassland patches of 1,
10, and 100 m², with and without corridors, are created by mowing the
surroundings in the white central rectangle. Courtesy and with permission
of S. K. Collinge. (c) Marked beetle (Tenebrionidae, *Eleodes*), and the route
of a marked beetle through a grass and soil mosaic (4.5 m straight-line
distance) (Wiens & Milne 1989, Johnson, Milne & Wiens 1992). Route
diagram provided courtesy and with permission of B. T. Milne.

wildlife crossing[685]; habitat fragmentation[1309,428,1595]; changing landscape
patterns by water control[1576,315,316], logging[518,662], and other land
uses.[1490,217]

(3) *Landscape-scale experiments*. Studies evaluating the ecological
effects of forest patch size, using both pretreatment and untreated controls, are the best known and most ambitious to date. These experimentally produced remnant fragments include a tropical rainforest project
near Manaus, Brazil[991,990] (Fig. 1.12a), eucalypt forest in southern New
South Wales, Australia[1048], and coniferous forest in Sweden.[672]

Other landscape-scale experiments use different techniques or a
smaller portion of the landscape. Thus, local extinctions of mice in
woodlots were caused to determine recolonization patterns (and metapopulation dynamics) in an agricultural landscape.[1124] Lines of camouflage material have been used to create new corridors.[1114] The structure
of windbreaks has been altered to determine effects on wind pattern, soil

erosion, productivity, and species richness.[174] Invertebrates have been introduced at various locations to evaluate the barrier effect of road corridors, that is, the avoidance or squashing rate.[1032]

Surprisingly few landscape-scale experiments have been done considering the widespread opportunities. For example, one can add to a landscape: mineral nutrients to evaluate stream corridor width; shrubs and trees to block gaps in corridors, or create stepping stones; vertebrates and invertebrates to evaluate movement rates and routes in different configurations; food or cover in different locations for the same purpose; or water to compare erosion rates. One can remove: shrubs and trees to interrupt corridor networks, or curvilinearize boundaries; animals to measure recolonization routes and rates; water to evaluate the effects downwind or downstream; or fruits or flowers to evaluate animal and seed movements among ecosystems.

(4) *Micro-scale experiments*. Experiments at a fine spatial scale are especially useful, because replicates are better controlled, and time, energy, and money budgets are more feasible. Experimental results are then carefully extrapolated to the landscape or regional scale, where some results apply but others do not.[864] Similarly, the medical effects of substances given to mice are extrapolated to humans, and bacteria and fruit flies are used to understand human genetics.

John Wiens, Bruce Milne, Alan Johnson, and colleagues studied 5×5 m fine-scale mosaics or micro-landscapes composed of bare ground and grass clumps in desert grassland[1887,803,804,326] (Fig. 1.12c). By measuring the routes and rates of movement of marked beetles in different mosaic spatial patterns, they evaluated the effect of pattern on process. H-J. Mader experimentally evaluated the movement of spiders across roads.[1032,1033,1034] Analogous fine-scale experiments have explored the movement of insects according to the distribution of flowering plants as micropatches[833], small mammals in experimental model systems[777], and the connections among laboratory bottles simulating micropatches.[504] Plant species richness has been compared in tiny unmown plots surrounded by mown grassland.[1366] Plant and insect richness, in addition to insect movement, has been measured in unmown grassland plots, as well as in changing configurations as grassland is progressively mowed (Fig. 1.12b) (S. K. Collinge, pers. commun.). Wind tunnel studies comparing the effects of different arrangements of buildings are common. Finally, the development of and creativity possible in model systems[1676] hold promise for understanding the ecology of landscapes and regions.

Modeling, analytic, and statistical approaches

Several recent reviews of spatial modeling related to landscape pattern are available.[374,1251,77,1575,1748,17] The following model types simply illustrate the breadth of techniques used: (a) hexagonal packing models[278,646]; (b) general neutral models[263,543,1142,544,1140]; (c) percolation theory[1617,1253,545]; (d) hierarchy theory[1250,1251]; (e) fractal analysis[230,231,1039,1137,1138]; (f) geometric models[1292,518,662]; (g) neighborhood models[1550,1742]; (h) entropy information models[579,1315,501]; (i) various simulation models[1576,1744,1887,678]; (j) geographical-information-systems models[231,807,228]; (k) cellular automata[1951,317]; (1) network models[1670,646,501,894,493]; (m) patch and corridor simulation models[465,1114]; (n) patch dynamic models[954,1320]; (o) graph theory models[251]; (p) flow-based spatial models[857,1576,315,316,1476,1116]; and (q) economic land-use models.[1744,1415,1280,1234,1279]

Some models focus on heterogeneity, some on distance from points, and some on the arrangement of spatial elements (patches, corridors, and matrix) present on the land. Other models focus on fluxes and movement relative to heterogeneity, distance, or spatial elements. And some focus on change in pattern over time. Specific models with realistic variables are useful to minimize oversimplification. At the same time generic models that can link structure, function, and change will be useful in developing general theory.

Geographic information systems (GIS) have rapidly expanded in use, flexibility, and affordability. These are computerized systems that overlay diverse sorts of spatially explicit information about an area, often a portion of a landscape.[231,1726,647] The information is stored digitally, presented visually or graphically, and available for efficient comparisons and correlations. The information may be stored in 'rastor' form, i.e., cell by cell (pixel by pixel) in a grid, or may be in 'vector' form, i.e., at varying distances and directions from points. Since maps of different types of information are commonly at different scales, a GIS with information inserted at a single scale is a real advantage.

The flexibility of the GIS offers promise in a number of areas.[783,1268,815,228,1283,320] For example, modeling may be done to predict ecologically-optimum spatial patterns. The multicolor computer images integrating mountains of data are highly appealing in the decision-making policy arena, and for enhancing communication among diverse disciplines. The GIS is an excellent way to store and retrieve diverse, accumulating land-use information.

Cautions for GIS use have inevitably emerged.[232] Inserting the layers of data is time-consuming and expensive. The additive effect of errors in each information level (e.g., soil type, vegetation type, flood frequency) limits the accuracy and interpretation of a final integrated overlay image. And when presented with a sequence of multi-color images of a changing two- or three-dimensional mosaic, the human mind cannot follow and understand most of the changes; thus a presenter's message is highlighted and the evidence is obfuscated.

Statistical analyses are a particular challenge in landscape and regional studies. Parametric statistics, in which randomness or spatial independence of observations is assumed, have been a mainstay of ecology despite frequent difficulties in fulfilling their assumptions. *Spatial statistics or geostatistics* have emerged as a set of techniques for analyzing non-random, spatially explicit sampled data.[1591,1425,410,364,820] These techniques include spatial autocorrelation, spectral analysis, semivariograms, trend surface analysis, textural analysis, and kriging. The techniques are useful in soil science[1851], geology[822,364], geography[288,673] and ecology.[612,1174,1751] This analytic approach is especially promising since it is difficult to demonstrate any arrangements, movements, or changes in a landscape or region to be random. In effect, both natural processes and human activities contain considerable non-randomness, and hence predictive ability.

Remote sensing, radiotracking, and other techniques

Images from satellites have revolutionized our perception and approaches to understanding landscapes and regions. For the first time a whole region can be examined in a single image. Images showing clear patterns of interdigitating landscapes, as well as ecosystems and land uses within landscapes, are widely available.[1459,1739,650,814,1200] Infrared images underscore the contrasting patterns of vegetated and non-vegetated surfaces, and a variety of spectral wavelengths and color enhancement techniques are used to separate subtle nuances, such as of vegetation type.[1008,801,1011,1364] Computerized techniques have reduced image distortion and increased precision in location finding and orientation.[1477] Frequent passes of a satellite over a spot have provided time-lapse 'photography' and permitted study of rapid landscape change.[158,650,1384,657]

Aerial photography from aircraft became common a half-century earlier. For example, in 1937 practically every canopy tree and little shed in eastern England was clearly visible in German aerial photographs. The long history of black-and-white aerial photography in certain locations provides insight into landscape change over two generations, a useful resource for sustainability studies. Low-altitude flying, including oblique-angle photography, permits an efficient focus on a single site at different times of day or season.[1541,1742,817]

The attachment of tiny radiotransmitters to animals such as mice, geese, and lions has opened up rich opportunities for studying movement patterns. Receivers may be hand-held, placed on vehicles, or used in aircraft. Much of the *radiotracking* data is on population dynamics or dispersal distances of a species, but an increased interest in measuring routes and rates of movement relative to spatial patterns in landscapes or regions offers great promise. Such movements include foraging within home ranges, dispersal out of home ranges, and migration between habitats. Numerous other marking techniques, including dyes, radioactivity, and color bands, are available for the same objectives.

Population-genetic techniques are used to understand how spatial pattern at the landscape scale affects the gene pools, and consequent evolution, of populations over time. Nucleic acid sequences, enzymes, and other chemicals are analyzed from species in patches of different sizes or distributions.[1595,1114,1116,481]

Gases emanating from or absorbed by vegetation patches of different sizes have been measured over kilometers-wide distances in a desert–grassland area using a laser and infrared-gas-analyzer system.[597] This permits instantaneous comparison of fluxes for patches at varying hierarchical scales. In addition, rapid diurnal and several-day changes associated with meteorological conditions are measured. In view of the space–time principle (Fig. 1.3), this provides a rare glimpse into short time-scale changes of a large land area.

Finally, most advances in the discipline have resulted from individual studies by one or a few investigators. Yet three team projects deserve special mention, because of the significance of results from several researchers addressing a series of related questions in a single area over several years. These projects have been led by Paul Opdam in The Netherlands[1258,1257,1788,1260], Gray Merriam in Ontario[1112,1114,1116], and Denis A. Saunders in Western Australia.[1490,1492,1493,732]

CONSERVATION, PLANNING, AND OTHER FIELDS

Researchers, planners, designers, and managers in several fields related to ecology have become major contributors in landscape ecology. Inevitably, they have also integrated the basic theory and thinking of landscape ecology in their several fields (Fig. 1.13). Foresters have incorporated bits into basic forestry planning and management, as have wildlife biologists.[685,334,402] Landscape architects and regional planners have fit pieces into their repertoire of design techniques for parks, suburban development, and river corridors.[11,1582] Geographers have absorbed portions into their work in physical, biological, and historical geography. Park and recreation managers, and efforts in land restoration, use parts of it.[1808,1494] Nature conservationists, conservation biologists, biological conservationists, and soil conservationists have found many portions to be useful.[672,740,324,1103] Landscape ecology should also be useful in range science, agronomy, urban planning, water management, climatology, industrial planning, transportation, and indeed in all fields concerned with land use. And of course ecology has incorporated major portions into its discipline.

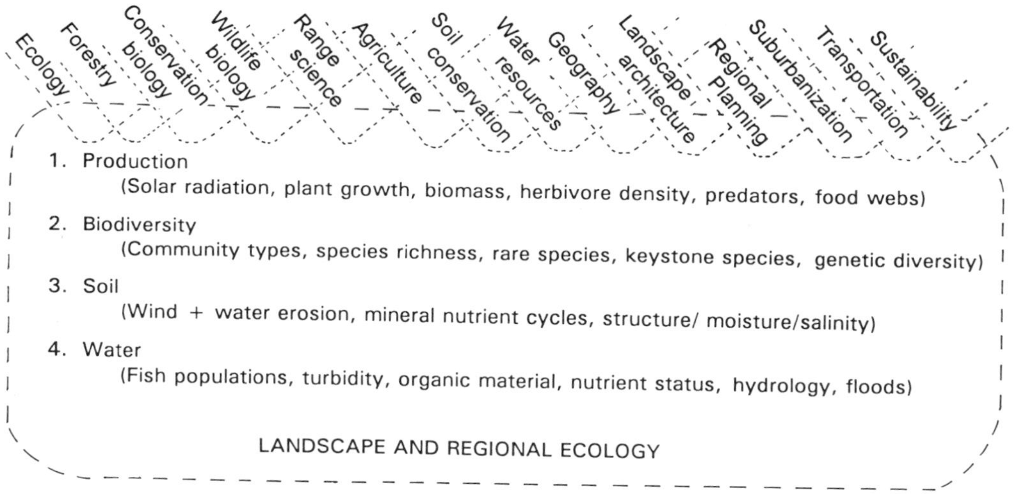

Ecology · Forestry · Conservation biology · Wildlife biology · Range science · Agriculture · Soil conservation · Water resources · Geography · Landscape architecture · Regional Planning · Suburbanization · Transportation · Sustainability

1. Production
   (Solar radiation, plant growth, biomass, herbivore density, predators, food webs)
2. Biodiversity
   (Community types, species richness, rare species, keystone species, genetic diversity)
3. Soil
   (Wind + water erosion, mineral nutrient cycles, structure/ moisture/salinity)
4. Water
   (Fish populations, turbidity, organic material, nutrient status, hydrology, floods)

LANDSCAPE AND REGIONAL ECOLOGY

Fig. 1.13. Four major ecological categories, and important disciplines using landscape and regional ecology. Representative ecological assays in this volume are listed for each category.

The most obvious reason for the rapid expansion of landscape and regional ecology is the subject. It is at the human scale, where nature and people are seen to interact daily, and where land planning, design, conservation, management, and policy must take place. Society craves ecological understanding at this scale. A second reason is its analytic focus. It provides understanding and predictive ability useful for more wood products, species, game, clean water, housing, recreation, or other often-conflicting societal objectives. Advocacy focuses on the intelligent use of landscape and regional ecology in all land-use issues. A third reason is holistic; the mosaic emerges as much more than the sum of its parts.

The fourth reason is the assays or areas of ecological interest. The full meaning of ecology as interactions among organisms and the environment is included (Fig. 1.13), rather than only current interests within ecology. Thus, four categories of ecological assays are recognized throughout, specifically production, biodiversity, soil, and water characteristics. *Ecological integrity*, discussed in chapter 14, refers to near-natural levels present in all four categories.

Solutions to environmental ills such as wind erosion, species extinction, water pollution, septic leaching, aquifer pollution, and suburban sprawl have their roots in this field. Yet perhaps most important is its potential role in sustainability[492], and therefore a full chapter is devoted to this application. Designing a land that effectively meshes ecological

37

integrity with basic human needs over human generations will only be accomplished with a healthy dose of landscape and regional ecology at the core.

APPENDIX: GLOSSARY OF COMMON TERMS

Biodiversity: the variety of life forms, especially number of species, but including number of ecosystem types and genetic variation within species.

Biotope: see *ecotope* and *landscape element*.

Boundary: a zone composed of the edges of adjacent ecosystems.

Configuration: a specific arrangement of spatial elements that is found in different places.

Connectivity: a measure of how connected or spatially continuous a corridor, network, or matrix is. (The fewer gaps, the higher the connectivity. Related to the structural connectivity concept; *functional or behavioral connectivity* refers to how connected an area is for a process, such as an animal moving through different types of landscape elements.)

Corridor: a strip of a particular type that differs from the adjacent land on both sides. (Corridors have several important functions, including conduit, barrier, and habitat.)

Disturbance: an event that significantly alters the pattern of variation in the structure or function of a system. (Usually this refers to natural phenomena, and *human activity* is used instead of human disturbance.)

Ecosystem: a relatively homogeneous area of organisms interacting with their environment. (Although this can apply at any scale, unless otherwise described, ecosystem or local ecosystem refers to a patch, corridor, or area of matrix within a landscape. Also see *patch*.)

Ecotope: 'Tope' refers to place, and thus ecotope and biotope are spatial concepts. Many definitions from many authors and nationalities exist; this author is still learning the shades of meaning. An *ecotope* commonly includes the following characteristics: (a) smallest homogeneous, mapable unit of land; (b) homogeneous in general substrate conditions, potential natural vegetation, and potential ecosystem functioning; and (c) may be composed of patches in different successional stages or land uses. On the other hand a *biotope* commonly: (a) focuses on the living area required for a particular assemblage of plants and animals; and (b) often is congruous with an ecotope, but may be a portion of, or more than one, ecotope. Some examples may help. According to these concepts a biotope for large birds of prey includes several ecotopes. A roadside is a biotope, while the tarmac asphalt road is a landscape element but not a biotope. A hedgerow is a biotope but not an ecotope. A logged clearing within a forest typically is a biotope or a landscape element within a forested ecotope. If the clearing is covered with gravel and sand for a large parking area, it becomes an ecotope. A forest may cover part of an ecotope with homogeneous substrate conditions, or cover one or several ecotopes. Each road corridor, hedgerow, logged clearing, large parking area, and individual forest of one type is a landscape element. For a particular study one may designate the roadside and road as different landscape elements. Landscape elements are actual, not potential, land cover. However, the preceding concepts of ecotope, biotope, and landscape element are not utilized in this volume. Instead, we use *spatial element* or *landscape element* as defined below. Also see 'landscape element'.

Edge: the portion of an ecosystem near its perimeter, where influences of the surroundings prevent development of interior environmental conditions. (*Edge*

38

effect refers to the distinctive species composition or abundance in this outer portion. Also see *boundary*.)

Element: see *spatial element*.

Fragmentation: the breaking up of a habitat, ecosystem, or land-use type into smaller parcels. (Considered to be one of several spatial processes in land transformation.)

Game: animals sought by hunters. (Also see *wildlife*.)

Grain: the coarseness in texture of an area, as determined by the size of patches recognized. (*Fine grain* has mostly small patches, and *coarse grain* mainly large patches).

Habitat: the ecosystem where a species lives, or the conditions within that ecosystem. (*Multihabitat* animals use or require more than one habitat type.)

Hedgerow: a narrow corridor of woody vegetation and associated organisms that separates open areas. (A generic term that includes woody strips of different origins, such as hedges, fencerows, shelterbelts, and windbreaks.)

Heterogeneity: the uneven, non-random distribution of objects. (Contrasts with *homogeneity*, and also with *arrangement* where objects are spatially configured in a particular way.)

Landscape: a mosaic where a cluster of local ecosystems is repeated in similar form over a kilometers-wide area. (A specific object with recognizable boundaries. *Landscape portion* refers to any internal area of two or more local ecosystems, as commonly seen in an aerial photograph.)

Landscape element: each of the relatively homogeneous units, or spatial elements recognized at the scale of a landscape mosaic. (This refers to each patch, corridor, and area of matrix in the landscape. Also see *spatial element* and *ecotope*.

Local ecosystem: see *ecosystem*.

Matrix: the background ecosystem or land-use type in a mosaic, characterized by extensive cover, high connectivity, and/or major control over dynamics.

Mosaic: a pattern of patches, corridors, and matrix, each composed of small, similar aggregated objects.

Multihabitat species: animals that use or require more than one habitat type.

Natural vegetation: plant species composition and cover of an area not planted by humans. (Human impacts and exotic species are often present, but native species usually predominate.)

Network: an interconnected system of corridors.

Patch: a relatively homogeneous nonlinear area that differs from its surroundings. (The internal microheterogeneity present is repeated in similar form throughout the area of a patch. Also see *corridor*.)

Region: an area composed of landscapes with the same macroclimate and tied together by human activities.

Road corridor: a linear surface used by vehicles plus any associated, usually vegetated parallel strips.

Scale: spatial proportion, as the ratio of length on a map to actual length; also, the level or degree of spatial resolution perceived or considered. (*Fine scale* refers to pattern in a small area, where the difference between map size and actual size is relatively low, whereas *broad* or *coarse scale* refers to a large area, where the difference is great. A garden map is fine scale, compared with a broad-scale continent map.)

Sink: see *source*.

Source: an area or *reservoir* where output exceeds input. (Contrasts with *sink* where input is greater than output.)

Spatial element: each of the relatively homogeneous units recognized in a mosaic at any scale. (See *landscape element* referring to a patch, corridor, or matrix unit at the landscape scale.)

39

Species richness: the number of species. (Also see *biodiversity*.)

Stepping stone: an ecologically suitable patch where an object such as an animal temporarily stops while moving along a heterogeneous route.

Stream (or river) corridor: a strip of vegetation enclosing, or on one side of, a water channel. (This may be wider or narrower than a floodplain.)

Stream order: the number assigned to a particular stretch in a dendritic river system and determined by the numbers for upstream tributaries. (For example, two 3rd-order streams coalesce to form a 4th-order stream.)

Sustainability: the condition of maintaining ecological integrity and basic human needs over human generations.

Wildlife: non-domesticated and non-aquatic animals. (Usually the focus is on mammals and birds. Plants and fish are considered separately. Also see *game*.)

PART II

Patches

2

Patch size and number

... nature thus keeps a supply of these plants in her nursery (i.e., under the larger wood), always ready for casualties, as fires, windfall, and clearings by man.

Henry David Thoreau, Journal, *1860*

The landscape mosaic was presented as a whole in chapter 1. Now we begin dissecting it into its building blocks, beginning with patches. A *patch* is a wide relatively homogeneous area that differs from its surroundings (see glossary in chapter 1 appendix). Patches have familiar attributes, such as large or small, rounded or elongated, and straight or convoluted boundaries. These attributes in turn have widespread ecological implications for productivity, biodiversity, soil, and water.

Why are these patch characteristics so important? In forestry the size of woods affects road construction and erosion, how many trees are wasted in edge effects, and the success of tree reproduction. In suburbanization planning the size of housing developments and nature reserves are common controversial issues. In aquifer and lake protection water quality depends on a large patch of natural vegetation. In managing stream systems for fish and fishermen, the size and number of natural patches is critical. In nature reserve design, is it better to have one large patch or several small patches? How many large patches are required to maintain the biodiversity of a landscape? In managing game, small patches are frequently created. And in agriculture an optimum field size has both ecologic and economic implications.

Several of the topics included are reviewed by Forman & Godron (1986). Other useful reviews include: patch dynamics[1320]; the traditional species–area concept[1795]; island biogeographic theory[1023,1323,302,564,1901,198,1763,1536]; remnant patches in landscapes[683,1490,1536]; and genetics in patches[1595].

This chapter begins with the origins and half-life of dynamic patches. Patch size effects are examined for ecosystem processes. Species–area curves and island biogeographic theory lead to evaluating species numbers in remnant and disturbance patches in a terrestrial mosaic. The roles of genetics and minimum viable populations are introduced. Finally we evaluate the importance of number of large patches present.

PATCH DYNAMICS: ORIGIN AND PERSISTENCE

The fortuneteller who predicts change is always right. We may dream of constancy or stability, but change, the universal law, is a key to understanding. Indeed, because change always extends over a limited area, it causes or affects a patch.

A large area in equilibrium that contains many patches in various successional stages has been called a *shifting mosaic*.[154,710] Although the total area remains in a steady state, over time patches in different places appear and disappear. In addition to considering shifting mosaic change, *patch dynamics* focuses on the event or agent causing a patch, and the species changes within it over time.[1320,1719,1750] A near-instantaneous disturbance typically is followed by a successional sequence.[1029,1719] Each patch exhibits directionality, proceeding from initiation toward 'climax'. The balance between the rate of initiation of patches by disturbance, and the rate of succession within them, determines both the rate and direction of change of the whole mosaic. Hence, the mosaic may be degrading or aggrading, slowly or rapidly, or may be in steady state.

The patch dynamic mosaic is part of a broader *landscape change* or *land transformation* process, where corridors and the matrix are also dynamic, and both species and ecosystem processes are dynamic. Indeed, succession is only one of many, mainly human-caused (anthropogenic), processes determining the rate and direction of change in a patch. An old-field on a mountain slope may become a wood, a cultivated field, a swimming pool, or bare rock, depending on whether the predominant force is succession, a farm tractor, excavation equipment, or erosion. Thus patches change in many directions. A landscape may not only degrade, aggrade, or remain in steady state, but in so doing, change in many directions to a different form.

On land five basic causes or origins of vegetation patches are evident and widespread.[409,501] A *disturbance patch* results from alteration or disturbance of a small area, whereas the inverse, a *remnant patch*, appears when a small area escapes disturbance surrounding it (Fig. 2.1). In contrast, an *environmental patch* (of vegetation) is caused by the patchiness of the environment, such as a rock or soil type. A *regenerated patch* resembles a remnant, but instead has regrown on a previously disturbed

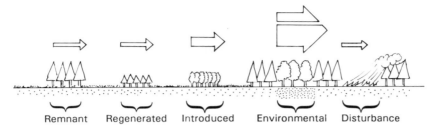

Fig. 2.1. Patch types, their origins and persistence. The five types illustrated assume the original landscape was mature coniferous forest. Thickness of arrow over each patch type is roughly proportional to its persistence or half-life.

site. Finally, *introduced patches* are created by people planting trees or grain, erecting buildings, and so forth.

The diverse assemblage of patch types composing a landscape has several ecological implications. Most important is that the rate of change varies widely depending on the cause or origin of a patch.[501] Environmental patches change slowly, reflecting the stability of the substrate (Fig. 2.1). In contrast, remnant and disturbance patches change relatively rapidly, reflecting the rate of succession, and disappear when they converge in similarity with the adjacent vegetation. In addition, disturbance may be a single or repeated event. Repeated or chronic disturbance, such as annual tractor plowing or daily local pollution, maintains the patch over time. When the disturbance stops, succession takes over. In short, the persistence or half-life of patches varies over many orders of magnitude, depending on the cause of the patch and whether disturbance is single or repeated.

SIZE EFFECT ON ECOSYSTEMS
'LOS' and 'SLOSS'

Which is ecologically better, a large or a small patch (LOS)? Alternatively, which is better, a single large or several small patches (SLOSS)? To be more precise, what is the relationship (shape of the curve) for a variable, such as erosion or species number, plotted against patch size. In the LOS case the horizontal axis is area of a patch. In contrast, in the SLOSS case the axis is the number of (progressively-smaller) patches, while keeping total area constant.[1536,764,1257] Here we focus on patch size from an ecological perspective, but it is important as well in the human domain, as for harvest efficiency, management costs of long perimeters, housing development, and city size.[45,1619,362,80]

45

Fig. 2.2. Pine patches that differ in size, and vineyards that escaped a plague. A single umbrella or nut pine (*Pinus pinea*) contrasts with pine stands of several hectares each in the background. Apparently, due to a high water table, roots of grapes in these fields survived a fungus that essentially eliminated the wine industry across France in the late nineteenth century. Petite Camargue, southern France. R. Forman photo.

Ecological comparison of a large versus a small patch is not simple (Fig. 2.2). The habitat type or environment must be the same; otherwise size is confounded with habitat type, habitat diversity, or both.[207] Environmental heterogeneity in most landscapes is such that, at random, a small patch is likely to be entirely within a habitat type, whereas a large patch is likely to include several habitats. Furthermore, several small patches at random will normally include more habitat types than one large patch of the same total area. Similarly, widely separated patches are normally in areas with somewhat different species pools, and hence represent different habitats. When habitat type is constant, the differences in a variable measured are attributed to area or 'area-per-se'.

Sampling is also a problem, because if sampling intensity is not proportional to area, pertinent information will be missed in large patches.[1013] In patch size studies it seems that the small patch only gets a glance, while monotonous sampling in the large patch goes on forever (Fig. 2.2). The alternative of equal sampling effort in large and small patches, however, is basically a measure of density, the amount per unit area, rather than a comparison of total amount in different sized patches.

SLOSS has generated lively debate. Some say a single large patch is better to protect more species.[1913,1900,1069,397,498,1692,501,137] Others claim several small patches are better.[1558,540,571,1309] One experiment with flies in bottles supported the single-large-patch hypothesis[504], and one with plants in a disturbed grassland supported the several-small-patch hypothesis.[1366,1436,1173] Some studies focus on habitat type or diversity differences, or on various sampling issues, and others focus on area-per-se.

Several major ecological values of large patches are known (see Box 2.1). Some advantages are known for small patches, but few if any for mid-sized patches.[1000] Disadvantages are mainly described for small patches.[1490,671,1536] Interestingly, the values in the box transcend the LOS and SLOSS questions. The advantages hold no matter how many small patches are present, one, ten, or a million.

Additional values can be pinpointed in specific areas. For instance, large natural vegetation patches in suburban landscapes may have a limited number of interior species.[225,1717] Yet these patches often play key roles in ameliorating microclimate, so the downwind neighborhood is cooler and moister.[755,1604,565] They are also hydrologic sponges, absorbing rainfall and reducing floods.

Box 2.1: Ecological values of large patches and small patches

Such a list of values for natural vegetation patches should be second nature to all interested in land use.

Large patches
1. Water quality protection for aquifer and lake.
2. Connectivity of a low-order stream network. For fish and overland movement.
3. Habitat to sustain populations of patch interior species.
4. Core habitat and escape cover for large-home-range vertebrates.
5. Source of species dispersing through the matrix.
6. Microhabitat proximities for multihabitat species.
7. Near-natural disturbance regimes. Many species evolved with and require disturbance.
8. Buffer against extinction during environmental change.

Small Patches
1. Habitat and stepping stones for species dispersal, and for recolonization after local extinction of interior species.
2. High species densities and high population sizes of edge species.
3. Matrix heterogeneity that decreases fetch (run) and erosion, and provides escape cover from predators.
4. Habitat for small-patch-restricted species. Occasional examples are known of species that do not persist in larger patches.
5. Protect scattered small habitats and rare species.

The bottom line: large patches, large benefits, and small patches, small supplemental benefits.

A landscape without large patches is eviscerated, picked to the bone. A landscape with only large patches misses few values. In essence, small patches provide different benefits than large patches, and should be thought of as a supplement to, but not a replacement for, large patches. We may hypothesize that an optimum landscape has large patches, supplemented with small patches scattered throughout the matrix (discussed in later chapters).

No matter what the land-use policy is, almost any landscape will have small patches. But large patches can easily disappear into small ones before our eyes. Maintaining the benefits to society of large patches of natural vegetation is only possible with vigilant planning and protection.

Ecosystem processes

Most ecological emphasis on patch size concerns species. Yet, Richard Hobbs concludes that the effect of habitat fragmentation on ecosystem processes (see appendix A, this chapter) is just as great.[730] Working in a severely fragmented landscape with scattered, remnant natural-vegetation patches, he and his colleagues note major alterations in hydrologic regimes, mineral nutrient cycles, radiation balance, wind patterns, and soil movement. Furthermore, these altered processes change species patterns.

Is there an optimum, a minimum, or a maximum field size for farmers, or forest for foresters? Probably thresholds exist upon which to base decisions, but the subject remains shrouded. Answers may depend on whether a large patch is compared against one or several small patches. Ecological effects for agricultural fields and forest patches will be considered separately in the following sections.

Productivity and biomass

Woody vegetation surrounding a field serves as a windbreak affecting wind velocity in the edge portions of the field. In a field wind speed is low for a considerable distance on the upwind side, and for a shorter distance on the downwind side (Fig. 2.3; see also chapter 6). Plant production is often enhanced in these low wind speed locations. However, a zone of decreased production is commonly present approximately 6–8h (h = height of vegetation surrounding the field) downwind of a windbreak. Here an accelerated wind speed causes desiccation and low temperature[1056,174] (Fig. 2.3; chapter 6).

Reduced production in field edges results from concentrated trampling and grazing by livestock, by shade and root competition from surrounding woodland, and from woodland herbivores that move and munch in the field. In addition, field edges somewhat parallel to the wind direction frequently have accelerated wind speed and probably

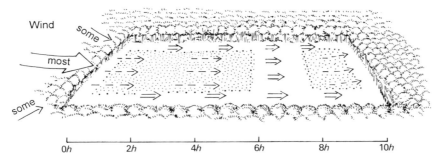

Fig. 2.3. Generalized patterns of wind speed and productivity in a field. Relative to average wind speed over forest, thick arrows in field indicate accelerated, and dashed arrows reduced, wind speed. Unshaded areas of field indicate low plant productivity (explained in chapter 6). h = height of surrounding trees.

lower production (Fig. 2.3). In short, the interior-to-edge ratio of a field is a key limiting factor determining the minimum field size for productivity.

The maximum effective field size is determined by similar factors. As field size increases, *fetch* or *run* increases (the distance without obstructions), permitting wind speed and surface water runoff to accelerate to maximum levels.[1570] This may cause sometimes-severe soil erosion, and consequently decreased productivity. In windy climates wind fetch is a similar problem for livestock production.[241,174] Avian, mammalian, and insect predators that consume agricultural pest species and live in the surrounding woody vegetation, may also help determine maximum field size. If the radius of a field exceeds the effective distance moved by the predators, a pest species may escape predation, and undergo a population explosion.

For productivity perhaps a mid-sized field is ecologically optimum. Its actual diameter depends primarily on climate and the adjacent wooded ecosystem (vegetation height, herbivore and predator densities). Of course, many human or economic considerations also affect field size.[704,1030,1498,236] These range from the ability to turn farm equipment in a small field, to the effort of locating livestock in extensive pastureland. But overall, the ecology of field size remains a question mark.

In the case of forests, the productivity of both natural vegetation and wood products is related to patch size (Fig. 2.4). In general, trees blow down (windthrow) and fires start most along wooded boundaries. A small woods may also lack predators that control herbivore pest outbreaks.[1001,1000] Therefore other things being equal, decreasing the size of a woods means proportionally less wood production per unit area.

49

Fig. 2.4. A remnant patch of eucalypt forest showing exposed edge and corner. As part of a landscape-scale experiment the matrix was cleared three years before to leave this 3 ha (7.5 acre) patch (Margules 1992). Southeastern New South Wales, Australia. R. Forman photo, courtesy of C. R. Margules.

However, where the boundary between woods and surrounding field has been stable for a long time, the many-meter-wide forest edge is normally composed of trees with crooked or leaning trunks, and with large branches along the trunk.[1814,1379] Due to minimal root and light competition on the outer side, these trees generally exhibit higher density and greater diameter growth than trees of the forest interior. Although productivity of wood and natural vegetation is greater in forest edges, the crooked knot-infested trunks are usually of little use for lumber. A tiny woods has only edge conditions. Thus, the interior-to-edge ratio is important in determining the minimum forest size for productivity.

In a clearing, natural reproduction by seeds and seedlings may be important for forest regeneration. Here both minimum and maximum clearing-size considerations are important. Too small a cut means poor reproduction due to shading and root competition by surrounding trees.[1144,1442,971,502,1320,98] Too large a cut often means the center of a clearing receives limited seed dispersal by wind or animals, and a variable microclimate for seedling growth. The minimum dynamic area that includes disturbance, discussed later in this chapter, may be important.[1319,1750] In the southern Appalachian Mountains plant productivity, microclimatic conditions, and arthropod populations are highly sensitive to the size of clearings from 0.016 to 2.0 ha.[1314,1552] Herbivores and hence herbivory may correlate with woods area and/or the size of clearings within woods.[33] In summary, seed dispersal, microclimate, and disturbance appear to be key ecological factors limiting the maximum size of woods for production.

A different approach is to consider the effect of woodland size on forest economics, since many major factors have ecological implications, e.g., transportation and equipment usage, planting and harvesting operations, loss to pests, and so on. Overall, the economics of forestry decreases curvilinearly with decreasing tract size.[1658,1220,1455,118,342] Several studies suggest that often 'diseconomies' accelerate below approximately 20–30 ha (50–75 acre) tracts. Despite a number of advantages of small forest tracts, economics dictate locating new tracts next to large, rather than small, forests.[1658] The two 'ecos', ecology and economics, differ but overlap for forest size.

Erosion and mineral nutrients

As just described, large fields with long fetches for wind and surface water flow increase the probability of soil particulate erosion. Some of the heavier particulates moved, e.g., sand and silt, are commonly deposited in the edges of fields where velocity drops.[174] Larger fields are also more apt to include a tiny zero-order or first-order stream (chapter 7), or be adjacent to a stream. Here, sedimentation and nutrient runoff into adjacent streams doubtless increases with field size. The density of tributaries, or scale of valleys and ridges, will put a maximum limit on optimum field size. Channelization and erosion are accentuated by cultivation, implying that the maximum ecologically effective field size for cultivated fields is smaller than for pastures.

Where wind or water flow accelerates (due to the 'Venturi effect') along edges (Fig. 2.3), erosion typically increases. This is accentuated where livestock trample and heavily graze the vegetation along edges and entrances to fields. Livestock also transport mineral nutrients from grazing to waste elimination areas, often on average from field center to edge.[1429,1882,1501,1534]

Though nutrients from adjacent fields commonly accumulate in forest edges[407], this does not appear to affect decisions concerning forest size. Much more important are logging clearcut size, and the construction and maintenance of roads and the 'landings' for accumulating and loading logs. As in the case of agricultural fields, clearcut size affects erosion, soil organic matter, and nutrient runoff.[1101,1802,1314] But commonly the construction and maintenance of roads apparently causes more erosion than the logging cuts (J. F. Franklin, pers. commun.).

Presumably a tiny woods requires no internal road construction. The smallest woods with road construction may have the highest road-to-area ratio, and hence erosion. Assuming harvested logs are sorted in landings outside small woods, then mid-sized woods would be the smallest to have landings with associated machinery and erosion within them.

Water and aquatic systems

Large fields can be expected to have greater evapotranspiration to the air per unit area than comparable small fields (ignoring the effects of surrounding vegetation), due to exposure to wind and sun. This means

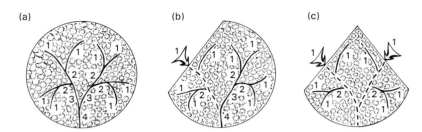

Fig. 2.5. Effect of forest removal on connectivity of a stream system. (a) Connected 4th-order natural stream system completely covered by forest. (b) Forest removal over one first-order stream produces inputs (indicated by arrow) that flow into the 2nd- and 3rd-order streams below, leaving a disconnected network of natural tributaries. (c) Forest removal over another 1st-order stream further disconnects the network.

that in large fields proportionally less water reaches the groundwater. Just as for productivity above, water flow is controlled by microclimate and height of the surrounding vegetation. The diameter of the field is scaled to the height of the surrounding vegetation. As field size, below about 12–16h diameter, shrinks, the proportion of incoming precipitation going to groundwater apparently increases markedly. In fields larger than this diameter, surface water is increasingly likely to be evaporated to the atmosphere.

A large forest can completely cover an aquifer and protect its water quality. Presumably decreasing forest size over an aquifer leads progressively to less water-quality protection, though no threshold seems evident. In the same way, ponds and lakes are completely or partially enclosed and protected.

A large forest also can enclose a complete stream network, such as the dendritic linkages of first-, 2nd-, and 3rd-order streams. When forest removal exposes one 1st-order stream to surrounding influences, the quality of the network decreases markedly (Fig. 2.5). Erosion, nutrients, high temperature, pollution, or other influences are readily transported down into the 2nd- and 3rd-order streams. This effectively breaks the connectivity of the network. Fish that can locomote up and down streams, and move among several separated streams of the same order, are particularly sensitive to the loss of network connectivity.[256]

In principle, progressively decreasing forest size would cross a series of these minimum size thresholds, one for each stream order. Thus, the network system of a 3rd-order stream would become disconnected at one size, and the network of a 2nd-order stream would be disconnected at a smaller size (Fig. 2.5).

The preceding pattern of minimum thresholds is dependent on location of the small patches.[683,1393] In this case, quite different ecological

52

effects result from removing vegetation covering a 1st-order stream, versus vegetation straddling a 3rd-order stream, or vegetation between streams.

The *hydrologic* (water quantity) effect on an aquifer or lake is linearly related to forest size. Like a sponge, a forest absorbs rain water and releases it slowly, thus minimizing flooding. A small woods also functions much like an oasis, where wind from the surroundings arrives horizontally, and dries out the vegetation and soil.[831] In the stream network case, decreasing forest size to expose a first-order stream produces a pulse of water after rain, that is, a flood of short duration that extends into the 2nd-order stream, and less so into the 3rd-order stream (Fig. 2.5). Each successive exposing of first-order streams as a forest shrinks produces a larger pulse into 2nd-order streams, which extends further into higher-order streams.

Most of the same principles apply to 'water quality', including turbidity, temperature, oxygen level, algal production, and toxics. They also apply to stream structure, e.g., large woody debris (logs and branches), bottom sediment accumulation, and fish habitats. The evaluation of patch size refers to a two-dimensional area. In lakes and other water bodies ecosystem processes often correlate with three-dimensional volume.[1720]

Hydrology and erosion coalesce in most reservoirs. Decreasing forest size normally means higher water velocity in streams and rivers, and greater sediment flows. A dam for a reservoir acts as an effective filter, catching most heavy sediment inputs carried by the flowing water.

'SLOSS' and ecosystem processes

Most of the preceding results apply to LOS, but the SLOSS question is equally important. *Risk spreading*, i.e., spatially dispersing resources, is a key consideration. More than one patch for agricultural production or wood products provides stability in the face of disturbance (Fig. 2.2). Fires, tornados, and pests may cover a whole patch, but are less likely to devastate several patches.

Arrayed against this is the ecological disadvantage of movements between patches. Thus, to harvest wood or grow crops in several patches, machinery or farm animals typically must be moved between patches, with many negative ecological consequences.[91]

Human and economic considerations parallel the preceding (assuming the size and specialization of farm or logging operations are optimum and constant). Moving equipment between sites, building and maintaining fences, and so on are economic losses. Having wood or agricultural products from some patches that escape a plague is an economic gain.

Finally, it is evident that we know little about the ecological effects of patch size on ecosystem processes. Productivity, water flow, mineral

53

nutrient cycling, soil erosion, and sedimentation, in relation to field or forest size, are worthy research frontiers.

SIZE EFFECT ON BIODIVERSITY

Biological diversity or *biodiversity* refers to the variety of life forms, especially the number of species, but including the number of ecosystem types and the genetic variation within species (see appendix B, this chapter). Numerous important values of biodiversity have been described, and may be summarized in three categories: (1) food, fibre, and chemical products; (2) ecosystem services of controlling water, recycling, and cleaning the environment; and (3) inspirational, aesthetic, and existence values.[1910,444,445,1912,93,1103]

Species in samples and islands

In contrast to ecosystem processes the effect of patch size on species number has been much studied. We begin with the extensive study of species–area curves since about 1900, followed by a theory on species richness on oceanic islands plus some human colonization patterns, and end with species number in patches embedded in a heterogeneous landscape.

Species–area curves

Supermarkets sell more types of food than little markets. Big areas have more species than little areas. Such generalizations are trivial, but the shape of the curve showing the relationship, e.g., between species number and area, a *species–area curve*, is informative.[786] For most species groups in most habitats, the curve initially increases steeply, and then rather abruptly tends to level off and remain with a slightly increasing slope (Fig. 2.6a). The area at which the abrupt change in slope takes place is called the *minimum area point*.

Species–area curves were traditionally based on sampling within a community (see Hopkins 1955, and Greig-Smith 1983 for methodology issues). The goals were to determine the minimum area for an adequate sampling protocol, and to discover characteristic features or principles governing biotic communities. The minimum area point has been considered an index of how large a community must be to express its essential structural character, or to be representative of a community type.[1293,589,44,1795] Different methods have been used to determine minimum area points from curves.[1293,243,1795] For complex reasons curves for some tropical rainforest and Mediterranean-type communities do not show clear minimum area points.[8,589]

54

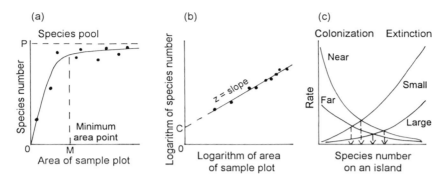

Fig. 2.6. Species–area curves and the core of island biogeographic theory. Species number on nine islands of an archipelago is plotted against area on (a) linear, and (b) log–log, coordinates. (c) Colonization and extinction rates are given for islands near and far from a species source, and for small and large islands. The equilibrium number of species is indicated on the horizontal axis, beneath the point where curves for colonization and extinction rates on an island cross.

In the search for community principles various equations were fit to species–area curves (see appendix C, this chapter). Species number was plotted against the logarithm of area, and suggested an exponential or log-normal distribution or relationship.[576,577,1347,1348] Using a log–log graph, a technique that often straightens out great scatters of points[1901,303], a power function relationship was suggested[55,56] (Fig. 2.6b). In addition to the minimum area point, three characteristics of the curves were proposed as basic features of communities: (1) the slope of the whole curve; (2) the slope of the latter portion of the curve; and (3) the point where the line, if extrapolated, would cross the vertical axis. However, assuming correct sampling, these are basically a product of total species pools and the regularity or aggregation within communities. The minimum area point remains both useful and an enigma.

Species–area curves based on sampling within a terrestrial community were eventually applied to islands in the sea.[358,1347,1348,908] Two new major factors came into play. Islands are often isolated by long distance and inhospitable salt water. And islands are whole ecosystems or patches, where the edge, buffeted by strong maritime influences, differs markedly from the interior.[1584]

Island biogeographic theory

Into the subject waded two young scholars and perceptive observers of nature. One had mathematics as a second language, and lines etched in his blackboard for plotting graphs. The other viewed the world through a tropical rainforest filter, and believed ants were heavyweight champions of the world.

Robert MacArthur & Edward Wilson (1967) observed that in Pacific and other island archipelagoes, (1) large islands have more species than small islands, and (2) islands near a mainland have more species than those more isolated. They developed a theory that explained these patterns in terms of colonization (immigration) and extinction. The number of species on an island represents the balance between the rate of colonization and the rate of extinction[1559,1691,3,1895] (Fig. 2.6c). Islands near a mainland have a greater colonization rate than more-distant islands (the mainland serves as a *species source*, a usually large area from which objects disperse). Colonization rate is higher on larger than smaller islands, due to the bigger target for potential colonizers. Extinction is greater on small than large islands, where population sizes are larger. *Species turnover*, the rate of colonization and extinction, is observed on an island. New (young) islands have low species richness (see glossary in chapter 1 appendix), which gradually increases until the colonization and extinction processes reach an equilibrium number. Island area, isolation, and age are, in order, the major controls on colonization and extinction, and hence species number. This is called the *equilibrium theory of island biogeography*.

Island biogeographic theory was an important historical step in ecology. It contained graphical and mathematical elegance, and provided a simple mechanistic explanation of species richness patterns in an archipelago (Fig. 2.6; see appendix C this chapter). Its predictions were modest[1023,1901,1417,1536,1310], and early empirical data lent support to several aspects.[1559,337,338,1507,1691,397]

Additional island theory concepts emerged. *Stepping stones*, i.e., suitable intervening habitats, can mitigate the effect of isolation. 'Land-bridge islands' (those recently connected to a mainland) contained fewer species than on the mainland, but more species than on older islands.[1561] A land-bridge island may be 'supersaturated' with species (more than a supposed equilibrium number). Following the formation of an island, a period of extinctions until potential equilibrium is reached is referred to as a 'relaxation period'.[196,197,395,1904,1599,1694,1286,982] This is part of an 'adjustment period' characterized by an elevated rate of species change. During the adjustment period a new island or patch attracts species from the surroundings, loses species (relaxation), and is colonized by species.[501] Species–area curves for islands were often plotted logarithmically[302], which smoothed out much of the variation present in nature. Using such a technique, a doubling in species number was sometimes noted when area increased ten times.[358,683]

Critiques and empirical data at variance with the theory emerged early.[1323,302,564,1418] Some resulted from comparing islands with different levels of habitat diversity, such as a small flat versus a large mountainous island, and some reflected problems in methodology and data interpretation.[1023, 303, 1087] Others resulted from attempts to extrapolate the theory to patches in a land mosaic. For example, island theory has been called upon to explain richness patterns on peninsulas[1683,233,1141], enhance management[1052], and design nature reserves.[1904,1913,1691,397,571,1077,683,1536]

However, the many identified shortcomings of the theory, and the wholly different setting of patches in a landscape mosaic make these applications problematic (see Box 2.2). Few people today would use island biogeography as a primary model for studying, planning, or managing the mosaic on land.

Box 2.2: Why not plan landscapes based on island biogeography?

Mosaic characteristics

A mosaic is the central spatial characteristic of a landscape. The mosaic is highly heterogeneous, with habitats ranging widely in type and suitability. Although isolation is a major problem for certain key species[685,1353], most species can cross the mosaic at least at low rates.[877] Thus, for species richness, isolation, a primary characteristic of island biogeography, is generally a minor variable on land. The diverse habitats in a mosaic are sources for many species pools. Species sources for a patch, rather than being strongly unidirectional, are diffuse and multidirectional. Adjacent landscape elements are often major species sources. Species have evolved mechanisms to move through a heterogeneous environment.[1124]

Patch characteristics

Since most patches contain internal habitat diversity, the area-per-se effects present are often difficult to demonstrate, and affect a minority of species.[1321,869,24,1558,524,1207] Disturbance within a patch is a major determinant of, and may either increase or decrease, species number.[501,1319] The minimum dynamic area is better than the minimum area point for sustaining species.[1319] Patch origins are diverse and patch persistence ranges widely. Succession mainly determines species number, rather than land bridge phenomena.[1719,1320] Relaxation is only a portion of a dynamic species adjustment period following patch formation.[1032,991,143,501] The evidence for island size effects on colonization rate is limited.[1556] The ability of early colonists to repel later arrivals may be more important than the area effect.[907,908] Rather than assuming primary control of species number by immigration and extinction rates, species–area patterns seem more directly explained by the importance of the edge effect and the interior-to-edge ratio, also present on islands.[924,1584] Some proposed area-related extinctions may be due to boundary processes.[1514,924,1584] Patch shape varies much more widely on land, and warrants more emphasis than patch size.[540,1905,501,1514]

Methodological problems

When the range of patch sizes is wide, keeping sampling intensity proportional to area is difficult. Species are not distributed randomly within patches, or between patches. Various null models are possible.[263] The ecological meaning of the constants, z and especially c, in the species–area equation is unclear.[1136,604,302,564,1901] Apparent species turnover is often simply behavioral movements or successional changes.[1014,1866,3,564,854] Most species in small patches are generalists[1904,1692,501], whereas probably most in large patches are specialists. A randomly placed small patch will often be on a single substrate, whereas large patches will usually include several substrates. Similarly, small patches commonly differ widely from one another. Small patches often contain many species that utilize parts of the surrounding matrix.[136,623] After three decades well-controlled experiments supporting the island biogeography theory are scarce.

Application of results

Since the equilibrium theory was developed, ecologists have generally altered their focus and now view communities as non-equilibrium, with species populations and number normally in wide fluctuation.[1317,706] Plotting the data logarithmically, while convenient to smooth variations and deal with areas varying over orders of magnitude, obfuscates pattern that may be ecologically significant. The island theory does not fit well with most studies on plants and area.[953,1841,1901,797,1309,541,427] The theory deals with species number, implying that each species is equal. Species composition, and particularly key target species, are of equal interest.

Human colonization of islands

Human colonization of islands over time often leaves archaeologists scratching their heads. Sometimes distant islands are colonized before close islands, small before large islands, and by technologically less-advanced before more-advanced peoples.[847] Examples are Easter Island before Iceland, Samoa before Madagascar, and Hawaii by Southeast Asians rather than North or South Americans. In attempting to understand these puzzles, several concepts emerged that may be useful in studying patches in a landscape.

The ability to live on an island, too small by itself for support, is made suitable by a species also using resources of nearby islands.[199,847] The perpendicular array or screen of islands, e.g., the Hawaiian chain as viewed from the west, is easier to encounter than the same array parallel to the direction of movement. Because people have learned that islands tend to be clustered, the discovery of one island leads to the expectation of discovering more nearby islands.[399] Hence the fifteenth-century European discovery of some Caribbean islands stimulated a flurry of island searches and discoveries.

These and other considerations lead to the conclusion that the spatial arrangements of islands, plus the human expectation of clustered islands, primarily determine island colonization patterns. The abundance of human-introduced species that cause diverse impacts requires caution in interpreting species–area patterns on islands.[139] In summary, the use of several islands by a species, plus the perpendicular array and clustered nature of islands affecting chances of colonization, are useful concepts applicable on land.

Species in a Landscape Mosaic

Remnant and environmental patches

Early attempts to apply island theory on land found species-area correlations (with considerable variability) for mountains surrounded by desert, alpine mountain tops, caves, and marine substrates of different sizes.[1811,343,196,1507] The first statistically-designed terrestrial study to minimize or eliminate habitat diversity as a variable, i.e., to test the area-per-se hypothesis, compared birds and other species in old oak woodlots in a flat agricultural area in New Jersey (USA).[448,498,539] Thirty woods representing ten sizes from 0.01 ha (the size of a large old tree) to 24 ha (the largest remnants meeting the criteria of location, shape, soil, predominant woody species, and vegetation structure) were selected in a statistical design. Sampling intensity was nearly proportional to patch area.

For all species groups studied large patches had more species than small patches, but the shapes of the curves differed markedly. The minimum area point for insect-feeding birds was estimated to be at least 40 ha (100 acres), in contrast to 2 ha (5 acres) for trees, as well as for seed-eating birds (Fig. 2.7a). Mosses and mushrooms were intermediate. This indicated that, of the various groups studied, the avian community was the most sensitive to a decrease in the forest size, and would be the first to unravel.

Approximately half the 35 bird species were *patch-size independent*, i.e., were found in large as well as small patches. But the other half were *patch-size-restricted* species, in this case found only in large, or in mid-size and large, patches. Furthermore, the patch-size-restricted birds differed markedly in the actual size required. As woods became progressively smaller, population sizes dropped, and one-by-one these size-restricted species disappeared (Fig. 2.7b). Small patches contained only size-independent species (a non-random subset of the total), which in large patches were essentially limited to the edges. Surveys of hundreds of woods across Britain also showed a strong correlation between avian richness and area, as well as a progressive loss of species as woods size diminished.[1161]

59

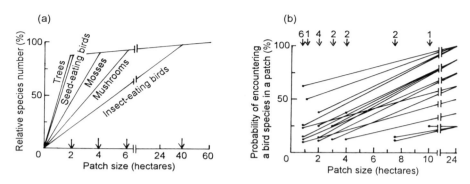

Fig. 2.7. Sensitivity of different species groups and of particular species to patch size in a landscape mosaic. (a) Generalized curves for various groups of organisms in old oak woods (*Quercus* spp.) in New Jersey (USA). Vertical axis is species number in a patch relative to number in a large patch (e.g., 60 ha). Arrows indicate minimum area points for each species group. (b) Changing relative population sizes for each of the 18 area-sensitive bird species present. To simplify various curve shapes, a straight line compares population size in the largest patch (24 ha) with that in the smallest patch (indicated by a dot) in which a species was found. Arrows at the top indicate the number of species that disappeared as patch size decreased. Bird data from Forman *et al.* (1976) and Galli *et al.* (1976); tree data from Elfstrom (1974). Unpublished moss (bryophyte) and mushroom ('morphospecies') data; for mushrooms spring and fall samples during both a wet and a dry year had the same minimum area point.

An additional pattern was illustrated by observations that squirrels (*Sciurus carolinensis*) in New Jersey were common in small patches, but scarce or absent in large patches. Large patches contained owl predators (E. W. Stiles, pers. commun.). This suggests a patch-size restriction, in this case for small patches (see Box 2.1). Other examples of 'small-patch-restricted species' are mistletoe (Loranthaceae) due to herbivory by possums (*Trichosurus*) in large patches[1241,1240], land snails (*Paryphanta*) due to wild pig predation[1240], and several island bird species due to competition with other birds.[396] Small-patch-restricted species may be exceptions compared with the prevalence of large-patch-restricted species, but the subject warrants study.

In contrast to the avian community, none of the 65 tree species encountered in the New Jersey study exhibited patch-size dependence; apparently all could live in large or small patches.[448] The species–area curve indicates that above *c.* 2 ha few tree species are gained by increasing patch size, even by an order of magnitude (Fig. 2.7a). Minimum area points for mosses and mushrooms (Fig. 2.7a) appeared dependent on the amount of fallen dead wood of different tree species (plus the edge effect described in chapter 3).

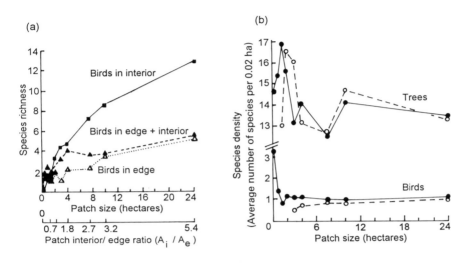

Fig. 2.8. Patch size effects on edge species, interior species, and species density. (a) Comparison for species primarily found in patch interiors, in patch edges, or approximately equal in both.[539,498] Edge width is 30 m. From Forman (1981), with permission of Centre for Agricultural Publishing and Documentation (PUDOC). (b) Solid circles are for whole patches including the edge; open circles are for the patch interior only. Tree data based on 11–31 plots per patch size; bird data based on 32–960 plots per patch size. From studies of Elfstrom (1974) and Forman et al. (1976); see Fig. 2.7.

Thus, an area-per-se effect was found, but immigration and extinction rates were unnecessary to interpret the results. Instead, the pattern correlated with the differentiation of edge species from interior species. *Edge species* are those primarily near the perimeter of a landscape element, and *interior species* are those primarily away from the perimeter, as in the interior of a patch. The pattern also reflects the gradient of minimum patch sizes for the interior species (Fig. 2.7b).

The ecological effect of being near the perimeter, plus the changing ratio of interior-to-edge area as patch size changes, means that large patches are mostly interior habitat, and small patches are all edge habitat[485] (Fig. 2.8a). The patch size where interior area disappears represents a threshold which depends only on the effective width of an edge. Above *c.* 2 ha the number of edge species barely increased, whereas interior bird species increased markedly (Fig. 2.8a).

The importance of the edge effect in patches was pinpointed when *species density* or packing (in this case average species number per 200 m²) was plotted against patch size. Species density was high, i.e., species were packed together, for both birds and trees in small patches (Fig. 2.8b). But above a *c.* 2 ha threshold, species density was low and constant. This pattern of high species density in small patches appears to be general.[623,24,1031,1845,1844,991,1213,672] Therefore, although a small patch contains few or no uncommon species, it is a good place to identify many species in a short time. This can be useful in park design and management.

Hundreds of empirical studies of species number and patch size on land have ensued.[1490,1536] These include plants[953,1841,1558,1309,541,427,1846], arthropods[344,462,1646,1548,1845,1844,1343,871], fish[81,1036], amphibians and reptiles[867], birds[1867,931,757,1013,1000,1240,1775,1433,1434,522], small mammals[602,578,868,107,982,981], and large mammals.[196,1136,1321,1793,107]

Overwhelmingly, larger patches have more species than smaller patches, and area is more important than isolation, patch age, and many other variables in predicting species number. Exceptions often result from the presence of another variable covarying with area[717], or where no specialist interior species are present.[1067] The prevalence of edge species in small patches, and interior species in the patch interior or core, is commonly emphasized.[539,24,1373,1685,1688,671,672,924] For birds, insectivores (insect-feeders) are more patch-size sensitive than seed-eaters or omnivores[498,136,683], and long-distance migrants are suggested to be more patch-size sensitive than residents or short-distance migrants.[24,1016,523,1434,643] Isolation is often important for certain individual species (some of conservation importance), but is usually of minor or no significance for species richness. Species numbers relative to patch age have rarely been compared[1596,107,1000], but presumably reflect successional and adjustment (relaxation) processes in patches of different origins.

A few experimental studies supplement the many correlation studies.[1418] In an important project highlighting the effects of tropical deforestation, Thomas Lovejoy and colleagues[990,991,989] compared species in Amazon rainforest patches of 1, 10, and 100 ha against pretreatment samples, as well as controls within extensive forest. Large patches were richest in species, an edge effect (chapter 3) was prominent, and small patches exhibited only edge conditions. Patch-size effects on trees, insects, birds, and mammals were conspicuous in a short period. Analogous projects in eucalypt forest of New South Wales, Australia (Fig. 2.4) and coniferous forest in Sweden are ongoing.[1048,672] Experiments on grassland[1366,1436] and fruit flies in bottles[504] were described earlier in the chapter.

In short, species–area curves for terrestrial patches are similar to those originally analyzed for samples within a community. Species-richness patterns in patches relate to edge and interior phenomena. On the other hand, in community samples they relate to microhabitat diversity, disturbance, and population aggregation. Where island theory focused on area and isolation[1906] as the leading variables in order, species number in terrestrial patches appears to be controlled by the following variables in order[501] (see appendix C this chapter): habitat diversity, disturbance, area of patch interior, age, matrix heterogeneity, and isolation.

We conclude this section with the *species cascade* or ripple effect, because it appears so sensitive to patch size.[514] In the Amazon rainforest experiment small patches lost their army ants, and consequently lost their ant birds.[990] The same patches lost butterflies adapted to pollinate certain plants, and hence the subsequent demise of the plants can be predicted.[1343]

More dramatic perhaps is the cassowary case. This terrestrial bird, as tall as and able to rip the guts out of a man, is believed to be the only seed disperser for >100 species of woody tropical rainforest plants in Queensland, Australia.[327,1635,329,330,328] The bird normally inhabits large forests. Logging and habitat fragmentation have eliminated the bird from several areas where only small remnants remain. Consequently, a progressive and massive loss of trees and other woody species can now be expected, unless the big bird can adapt or adjust its behavior. In addition to extinctions, the species cascade effect results in population changes throughout the community.

This example emphasizes the importance of protecting *keystone species*, those whose decimation causes widespread ecological effects.[1274] On the average, species higher in a food chain (higher trophic level), such as hawks, snakes, and cats as predators, are more sensitive to patch size than those lower in the food chain, such as plants and herbivores.[498,136,683] (Fig. 2.7). Keystone predators may exert a major control over lower trophic levels, including keeping species richness high.[1274] Therefore, protecting large natural-vegetation patches containing keystone predators is an effective way to sustain species richness throughout the food web.

Disturbance patches

In contrast to the remnant, regenerated, and environmental patches considered in the preceding section, disturbance patches usually exhibit rapid species turnover. For convenience openings or clearings are differentiated from gaps (or 'chablis').[1246] *Openings* produced by major disturbance such as fire, windstorm, and clearcutting are landscape patches (distinct ecosystems at the landscape scale), and undergo the traditional process of succession[1029,1719] (Fig. 2.9). On the other hand *gaps* represent the microheterogeneity or microhabitats within a community.[502,1320] These internal spots are caused by natural processes and disturbances within the community, and their ensuing change is often called *gap dynamics*.[625,1464,728]

Small gaps less than *c.* 5 m diameter (depending on height of the surrounding vegetation) over time are commonly filled by lateral extension from the surrounding plants. Large gaps up to *c.* 20–30 m diameter are

Fig. 2.9. Fire creating a disturbance patch or opening. Washoe County, Nevada (USA). Courtesy of USDA Soil Conservation Service.

mainly covered by growth of existing or colonizing organisms in the gap.[1523,693,389,188,189]

In temperate deciduous forest, plant species richness in the first two years following patch formation is about 10% higher in openings (*c.* 60–140 m diameter) than in gaps (ca. 13–28 m diameter).[1314] Relative to gaps, openings have more light, more light-tolerant species, more plants with chemical defense compounds in leaves, and less herbivory.[895,94,1464,927,1314,1553] Thus, tree reproduction is highly dependent on patch size, a key factor in forestry where growth of natural seedlings is critical.[787,1523,1873,1263,1759] Different-sized disturbance patches are also visually contrasting, due to distinctive plant architectures.[58,653,1246,1872]

In open patches varying from *c.* 25 to 300 m diameter, small mammal populations in the smallest patches differ most from controls (in forest), but the difference is short-lived as gap closure proceeds.[209] Bird populations are also sensitive to the size of openings and gaps.[928,1933] Studies of powerline corridors in temperate deciduous forest suggest that widths exceeding 60–90 m are required for use as a disturbance patch by open-country bird species.[28] Species numbers and population densities for both birds and reptiles are similar in small (<20 ha) versus large (>100 ha) clearcuts in Australian eucalypt forest.[1393] In this forest large clear-cuts have more birds with large home ranges, specialized life history attributes, and social requirements. In Maine (USA) avian species rich-ness increases in clearcuts ranging from 2 to 112 ha.[1460] In addition to disturbance-caused clearings, the area of an ecosystem also affects bird populations in marshes[202] and natural grasslands.[900,1152]

When the height of the surrounding ecosystem is minuscule compared with the diameter of the patch, e.g., in grasslands or an extensive forest burn[987,1882,1152], microclimatic effects imposed by the surroundings are no longer a major factor relating species number to area. Thus, in an extens-ive patch the importance of wind dispersal of propagules from adjacent matrix (typically a negative exponential relationship with distance) is replaced by locomotion. Birds and mammals may carry seeds a consider-able distance into the patch, thus producing a wide patch-edge effect. Similarly, herbivores or predators from the matrix may forage well into an extensive patch.[1601] Analogous phenomena are observed at a fine scale on marine substrates, where the matrix may successfully encroach on a patch until it meets a large well-established colony in the central por-tion.[1600,1601,1507,1276,845,301] Colonies established in the outer portions of a patch are therefore overrun. Erosion in the central part of a large patch, such as after fire, may change conditions significantly for patch interior species, and therefore affect a species–area relationship.

Finally, disturbance within patches must be viewed in broader per-spectives. Species evolved with disturbance regimes, and perhaps all species require and are maintained by disturbance. Some disturbances decrease species number, and some result in higher species numbers. Species–area curves and minimum area points have been used in plan-ning and management (Fig. 2.10a).

But much larger areas are required to sustain a species number over time. A *minimum dynamic area* is the patch space required so that the natural disturbance regime will not eliminate a species[1319] (Fig. 2.10b). For example, a fire or pest outbreak will normally only cover a portion of a large patch, and thus species can repopulate the disturbed area from within the patch. Consequently the effective sizes of large patches should be scaled to the sizes and frequencies of disturb-ances.[986,339,1869,1463,1320,728,78,1750] In conclusion, the effect of patch size on species patterns is one of the most studied foundations in understanding the ecology of landscapes.

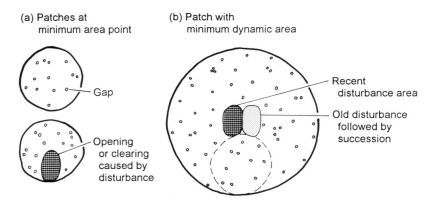

Fig. 2.10. The minimum dynamic area concept. (a) Patch size at minimum area point determined from species–area curve (Fig. 2.6a). (b) This is the smallest possible patch size for maintaining species richness, assuming: that two disturbance patches of the size indicated are always present; that over time these disturbance patches occur randomly across the area; and that a round undisturbed area equal to (a) (e.g., the area within dashes) is always present.

GENETICS IN A PATCH

Males and females within a patch reproduce more among themselves than with individuals in the matrix or other patches. This preferential mixing of genes has important genetic and ecological consequences. In this chapter we focus on a population (see appendix B, this chapter) in a single patch. In chapter 11 the perspective expands to many patches, each containing a subpopulation of a whole population.

Spatial pattern is of central importance in genetics, and many reviews of the subject are available.[1061,1595,916,915,1273,1473] Here we concentrate on the effect of patch size on the genetics of populations.

A *population* is composed of the individual plants or animals of a single species present in a location, in this case a patch, at a particular time. Each cell of an individual organism has several chromosomes, each with numerous genes, and each gene comes in two or more forms (alleles). *Genetic variation* or variability is a measure of how different individuals are in their gene forms. High genetic variation is considered to enhance the ability of a population to adapt to changing environmental conditions.

Low genetic variation, on the other hand, is considered to have three negative genetic consequences (Fig. 2.11). *Inbreeding depression* is the mating among close relatives which produces few offspring, and offspring that are weak or sterile.[1532,1376,266,472] *Outbreeding depression*, in contrast, is the hybridizing of highly dissimilar, nearly genetically-

66

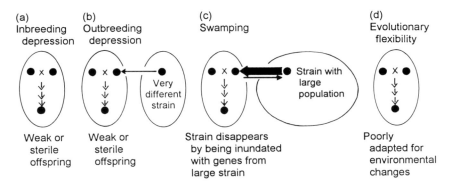

Fig. 2.11. Consequences of low genetic variation in a subpopulation or strain. Vertical oval patches contain small subpopulations. Dots represent interbreeding individuals and offspring over many generations. Horizontal arrows indicate breeding individuals that move between patches.

incompatible individuals, which produces weak or sterile off-spring.[932,1689,1834] And *evolutionary flexibility*, the ability to adapt to changing environmental conditions, is low when genetic variation is limited.[1070,1071,606,514,19] Overall therefore, high genetic variation is considered better than low variation in a population.

Outbreeding depression apparently is uncommon in nature. If genetic variation in a population drops, the chance of outbreeding depression increases when individuals from different populations breed. For example, in California 90% of the rare plant species are reported to grow near a closely related more-abundant species of the same genus.[1352] In cases where different closely–related species can hybridize, the offspring may be vigorous.[933] However, they are more likely to be weak or sterile.[1689] Furthermore, successful hybrids are likely to change the rare species more than the abundant species.[450] The rare species become 'swamped' with genes from the abundant species (producing coadaptation or local adaptation) (Fig. 2.11c). In such a manner, pure or native strains of plants and animals may disappear.

Should a conservationist or manager purposely introduce or move animals and plants around to different areas? How much gene flow is appropriate or optimum?[1595,1536] Should we try to mimic or recreate our perceptions of gene flows in nature? Inbreeding depression and outbreeding depression raise difficult questions. In addition to their ecological, ethical, and other dimensions, the genetic issues are complex. High genetic variation provides for continued evolution, with the probability of future adaptations to changing environments.[606] On the other hand, protecting pure strains by limiting genetic variation saves a wider array of population types in the short run. Yet, maintaining these numerous low-variation strains or populations increases the probability and rate of extinctions over time.

Genetic variation rises and falls in natural populations. For example, *genetic drift* takes place in small populations when the particular forms of genes (alleles) increase or decrease by chance. It is analogous to getting a 5:1 ratio in flipping a coin six times, even though after 6000 flips the ratio is sure to be very close to a 1:1 ratio. Analogously, chance may play a role in decreasing genetic variation in small populations over time.

The first few individuals to colonize a patch represent a tiny population normally with low genetic variation.[1686] The *founder effect* refers to the fact that the particular gene forms in these few individuals will play unusually large roles in the succeeding generations. Later, if the population becomes locally extinct, and a few new individuals recolonize the patch, it is likely that these individuals are quite different genetically than the preceding set of colonizers. In this manner populations following one another on a patch over time may be genetically quite different.

Because management decisions have to be made daily about the purposeful and inadvertant movement of animals and plants among patches, priorities among the genetic issues may be set operationally. Some knowledgeable people consider providing for long-term evolution to be a luxury in the face of accelerating human impact. Some consider outbreeding depression to be a minor localized issue. If these conclusions are valid, the primary genetic issue in land planning is to minimize population extinction due to inbreeding depression. Others disagree.

MINIMUM VIABLE POPULATIONS

Several other characteristics have been correlated with the probability of extinction of a population (see appendix B, this chapter). Of course, exceptions exist. Nevertheless, large animals are more likely to disappear than small animals. Vertebrates with large home ranges (Fig. 2.12) are more liable to extinction than those with small home ranges. Small populations have higher extinction rates than large populations. Population size in turn is often linked to patch size: smaller patches generally contain smaller populations.

A major recent emphasis has been attempting to determine the *minimum viable population size* (MVP), that is, the smallest number of individuals required to maintain a population over the long term.[515,1537,1595,865,1108] Three types of reason are usually given to explain why a small population declines in numbers and/or becomes extinct.[1328,1352] First, a loss of genetic variation may lead to inbreeding, genetic drift, and other genetic changes.[916] Second, demographic fluctu-

Fig. 2.12. Large mammals sensitive to landscape change affecting minimum viable population size. (a) Lowland gorilla (*Gorilla gorilla gorilla*). (b) Florida panther (*Felis concolor*) in the Everglades National Park of southern Florida. (c) Alaska brown bear (*Ursus middendorffi*) at Kamishak Bay, Alaska, a relative of the grizzly (*U. horribilis*). Photos courtesy of US Fish and Wildlife Service.

ations in population size result from often-little-understood changes in birth and/or death rates. Third, environmental fluctuations include fire, floods, changes in predation, competition, disease and food supply, plus human activities, especially habitat alteration or loss. These genetic, demographic, and environmental changes are often interlinked in positive feedbacks leading to accelerated population declines (vortices).[572,1109]

Of course, these linkages may also lead a small population to grow, in which case MVP may no longer be of concern.

Not surprisingly, in view of the diverse species and types of study done, a variety of numbers has been proposed as a general minimum viable population size. For example, experience in animal breeding suggests that with 50 individuals, approximately 1% of the genetic variation is lost each generation.[515] Experiments with fruit flies suggests that with 500 individuals the loss of genetic variation might be balanced by new variation due to mutation.[515] Seal (1985) concluded, based on various vertebrate species in captivity, that this balance might be attained with 250–500 individuals. It is unknown how applicable these results are to natural populations, but the numbers are quite relevant to vertebrates of conservation importance. For example, estimates suggest that only about 100 panthers (*Felis concolor*) live in South Florida[1035], and perhaps 200 grizzly bears (*Ursus horribilis*) in the Yellowstone Park area of the Rocky Mountains[940] (Fig. 2.12b and c).

Surveys of plant species indicate that in general widespread species are more genetically variable than those with limited ranges.[1822] A conifer species in New Zealand appears to lose genetic variation when its population is <8000.[127] A study of 120 populations of bighorn sheep (*Ovis canadensis*) in the southwestern USA desert region indicates that all populations of <50 sheep go extinct within 50 years.[116] Furthermore, in this case almost all populations of >100 animals persist at least up to 70 y. The MVP for different species probably varies widely, and caution in choosing a single number for planning is warranted.[918,1595,915,1536,1103] Some overall value that maintains most or almost all species, however, is a promising objective.

Using the much-cited value of 500 individuals as a base, three major factors warrant corrections or adjustments in the figure. First, when engineers determine the specifications for a bridge or building, their calculations include a 'safety factor'. This in turn is dependent on the expected life span of the structure, plus the environment of the surrounding region. For example, houses and short-span bridges in the USA are commonly designed for 50 y, the interval before major repair or removal is expected (D. Schodek, pers. commun.). If designed for 100 y, the calculated values could be 5–10% higher. Similarly, if designed for major-flood areas, steep mountains, and earthquake zones, the calculations could be considerably higher.

Analogously, a *conservation safety factor* is appropriately added to the MVP calculation. This accounts for the fact that whereas the empirical evidence available is over weeks to decades, minimum viable population size is defined in the long term, say, centuries to millennia or more. In these time periods the occurrence of severe disturbance is virtually certain. Using a hypothesized 20–50% safety factor would produce a MVP of 1000–2500 individuals.

Second, models used to calculate MVP may assume random breeding among all individuals of a population. This is an oversimplification since many factors prevent random mating. Indeed, only a subset of the population reproduces. This subset, often called the *effective breeding population*, is a key to estimating MVP.[863,1894,85] Often the effective breeding population is quite small compared with the total population size, as in many cases of shading by canopy trees, territoriality, and pecking-order hierarchies. Some MVP models are based on an effective breeding population of, say, 25% of the total population.[1894] However, the effective breeding population varies widely from <1% to >50% among different plant and animal species, and it changes over time. MVP will vary just as widely.

Third, while a particular MVP may be appropriate for one population, or an 'average' population, planning must focus on conservation of species aggregations containing numerous populations. A common species today may become rare tomorrow, and must be included in calculations. For example, the hypothetical 1000–2500 MVP range might maintain only half the fauna and flora over time. With so little empirical evidence available, the appropriate amount of upward adjustment is unknown. Nevertheless, in view of the three correction factors, conservation safety factor, effective breeding population, and conserving species aggregations, a target MVP of a few thousand individuals may be a useful working hypothesis.

These estimates are based on balancing the loss of genetic variation in a small population on a patch, against the new variation added from natural mutations. However, if new genetic variation could come from outside the patch, the starting number of 50 might be considered instead of 500.[514] Indeed, this is the case for a subpopulation on a patch which receives its new genetic variation from immigrants (when a population is distributed on a number of patches, a *subpopulation* or local population refers to the individuals on one patch) (chapter 11). Perhaps only one or two individuals arriving per generation is sufficient to balance the loss of genetic variation in a small subpopulation.[910] Thus, adding the three correction factors to 50 produces a possible MVP of a few hundred individuals for a subpopulation experiencing immigration. Therefore, instead of 50 individuals to maintain a subpopulation and 500 for a population, we probably should consider a 'few hundred/few thousand' as a current overall hypothesis of MVP for natural populations.

The conservation goals for a subpopulation and a population differ. For the population as a whole, extinction is forever. Consequently, a goal is to prevent most or almost all populations from going extinct.

71

In contrast, local extinction of separated subpopulations within a whole population is common in natural populations. Among the subpopulations some are smaller than average, and quite subject to local extinctions. But the larger subpopulations provide individuals for recolonizing patches after local extinction, and help maintain the population as a whole. Whereas preventing extinction of whole populations is a common goal, preventing extinctions of subpopulations on patches may or may not be a priority in management and conservation.

Although MVP numbers are rough and will change as evidence accumulates, the concept helps in estimating the size of nature reserves.[1599,143,1894,1483] Extinctions of whole populations occur naturally and at varying rates. However, population extinction due to human-caused habitat loss and other activities is considered to be much higher than recent natural extinction rates. Therefore, the goal is to prevent most, or almost all, extinctions of the diverse populations present.[516] Further research on MVP may help reduce future human-caused collapses of faunas and floras.

NUMBER OF LARGE PATCHES

The preceding discussion has compared large patches versus one or more small patches, and has identified a range of major ecological values of large patches (see Box 2.1), plus a few supplementary values of small patches. Now we move on to consider how many large patches are needed to accomplish these varied ecological values in a landscape. Is one large patch sufficient? Two? Three? Ten? Surprisingly, we have little evidence to answer this question of obvious land planning significance.

First it must be clear that the question refers to large patches of the same habitat type, in the same area, and with minimal internal microhabitat heterogeneity other than differences between patch edge and interior. Otherwise, species richness differences result from little more than habitat or environmental differences.[539,541,1049] In addition, the question of patch number assumes that at least some of the species are sensitive to the large patch sizes involved.[1694,400,764]

The study of birds in New Jersey woods presented earlier in the chapter provides useful evidence on the question of patch number. For the largest woods (24 ha) the curve is still rising at three patches (Fig. 2.13a), though at a diminishing rate. Indeed, the curves for all patch sizes are still rising at three patches (only three replicates of each patch size are included in the study[539]). Thus, more than three 24 ha woods are required to maximize avian richness here at the landscape scale.[498] Indeed, in this study five woods is the smallest number that included all bird species.

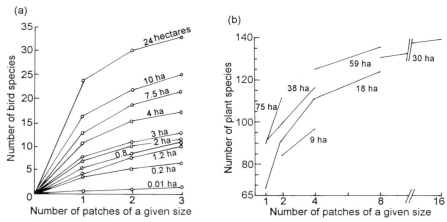

Fig. 2.13. Effect of patch number on species richness. (a) From Forman *et al.* (1976), with permission of Springer-Verlag New York, Inc. (b) Adapted from Fig. 7 in Game & Peterken (1984).

The question of patch number and species richness does not appear to have been directly addressed elsewhere, at least fulfilling the above assumptions, and data are rarely published in a form that permits the appropriate calculations to be made.

A study of mammals in Wisconsin, northcentral USA, included six woodlots of 2.2–2.5 ha in size.[1067] A plot of species richness versus patch number shows little species gain above three woods, in this study essentially limited to generalist species in small woods.

Margaret Game and George Peterken (1984) studied woodland herbs in patches that have never been plowed (ancient woods) in Britain, and found an increase in species richness as the number of woods increases (Fig. 2.13b). However, the rate of species gain for all woodland herbs is slight beyond *c.* 4 to 8 patches. Similar curves limited to woodland herbs of special conservation interest rise only slightly beyond ca. 2 to 4 patches.

Another approach is to consider the *species pool*, that is, the total number of species that live in a habitat type, e.g., within a landscape. If a large patch contains the entire species pool, adding one or more patches will add no further species. If one patch contains 95% of the pool, it is possible that a second patch would add the final 5%, but more likely that three or more patches are required to pick up those final rare species. Thus the rise in richness by adding a second patch normally is not statistically significant (<5% of the species pool is gained). Analogously, if the first patch has 90% of the species pool, a second patch is likely to add a significant number of species, but probably a third patch does not. In this manner decreasing proportions of the species pool can be compared.

A *minimum number point* (analogous to a minimum area point, Fig. 2.6a) appears to be an important conservation target. This may be operationally estimated as the number of patches beyond which the species richness – patch number curve gains <5% of the species pool with the addition of another patch. The minimum number point therefore includes the bulk of, but not all, the species characteristic of the patch type in a landscape.

The lower the percentage of the species pool in the first patch, the more patches typically are required. In areas where the species pool is relatively well documented, it is easy to estimate the proportion of the species pool in a patch of a given size. Then we may roughly estimate the minimum number point by plotting a series of smooth trajectories extending from zero and one to various likely numbers of patches that encompass the pool size.

A quite different approach has been used to identify which nature reserves of a large group are required to protect most of the plant species in a species pool.[1385,591,866,1763,1049] In the Macleay Valley floodplain of southeastern Australia 98 native plant species are recorded in all 432 wetlands present.[1051] To include every native plant species, at least 5% of the wetlands (i.e., 20) are required. More than 10 wetland reserves apparently are required before reaching the minimum number point. Also of conservation interest is that the 20 wetlands constitute 45% of the total wetland area. Indeed, to include all plant species plus all nine of the wetland types present requires protecting at least 75% of the total wetland area.

Similarly, up to 135 wooded patches ranging from 35 to >10 000 hectares in southcentral Australia were evaluated for the plant communities (or alliances) present.[1050,1047] Models with different assumptions suggest that at least 12–18 patches are required to have a 95% chance of including all the plant communities.

These Australian examples combine large, medium, and small patches in selecting the minimum set of nature reserves, and hence are not comparable to determining the number of large reserves required. Probably if several replicates of the largest reserve had been available, the number of patches required would have been considerably less than the 12–20 identified. Furthermore if, instead of choosing patches that include all species, one chose the minimum number point, the patch number required would be still lower.

The several lines of evidence presented do not provide a precise minimum patch number, but they do intersect to provide rough numbers. Apparently when species richness in a large patch is near the species pool, the rate of species increase is slight beyond one or two patches.

74

When species richness of the patch is well below the species pool, the rate of species gain is slight beyond a few large patches.

Other factors should also be considered. *Risk spreading* indicates that two or more patches are required to protect against environmental change or surprise over time. The greatest gain in risk spreading is going from one to two patches. Usually rather little is gained after a few patches.

The preceding analysis has mainly considered species richness in determining patch number. Yet patch size relative to the home range size of wide-ranging species is also informative. If home range is con-siderably larger than patch size, two or even several patches may be required to protect a wide-ranging species.[917,685] The number of aquifers, lakes, and headwater stream networks will also be important in determining the minimum and optimum number of large natural-vegetation patches in a landscape (see Box 2.1).

Finally, this chapter has largely focused on patch size by itself. Later chapters will examine spatial arrangement, but here it bears emphasis that the location of large patches is also important. The movement of species to and from a large patch affects all patches, corridors, and matrix area in the vicinity.

The key role of location is further illustrated by natural vegetation that covers, say, half of an aquifer. If only small patches of vegetation are present, the whole aquifer can be polluted by scattered inputs from agriculture, industry, transportation, and housing. If a large vegetation patch covers the downslope half of the aquifer, the upslope half can be polluted directly, and the pollution then slowly moves through the downslope half. However, if the large patch protects the upslope half of the aquifer, only the downslope half is subject to pollution. In short, both the number and location of large patches of natural vegetation are of central concern in landscapes.

APPENDIX A. ECOSYSTEM/COMMUNITY STRUCTURE AND FUNCTION

An *ecosystem* is an area (or volume) where species interact with the physical environment, and a *community* is the assemblage of interacting species in an ecosystem. The structure of an ecosystem/community refers to the distribution of energy, materials, and species. Functioning refers to the flows of energy and materials in food chains and cycles.

Structure

Biomass in an ecosystem is the mass or weight of living tissue, which is com-monly subdivided into five categories: (1) *producers* are green photosynthesizing plants; (2) *herbivores* eat producers; (3) *predators* eat herbivores; (4) *top predators* eat predators; and (5) *decomposers* break down dead tissue of all five groups. In terrestrial ecosystems, the biomass distribution resembles a *pyramid*, where mass progressively decreases from producer to top predator. Aquatic systems that are greenish exhibit the biomass pyramid, but some clear waters have a *top*-shaped biomass distribution, narrow at the bottom and top, and wide in the middle. Biomass reflects the distribution of both energy and materials in the community.

Vertical stratification refers to the arrangement of plants and animals in the vertical dimension above the soil, in the soil, or in the water. Light intensity is the overriding controlling variable on the stratification of vegetation. In some communities distinct layers of foliage are evident, e.g., from canopy to understory to shrub layer to ground layer. Often each layer reduces light intensity by about 90%. Thus, the soil in a four-layer forest receives about 0.01% of the light striking the top of the canopy. Animal species are sorted out among the layers according to their food and cover requirements. Hence, for example, in a tropical rainforest very different vertebrates live in the canopy, understory, and ground layers.

The horizontal distribution of species in the community is more complex. A *gradient* describes a species assemblage that gradually changes across the area. Alternatively, species may be clustered in somewhat distinctive groups, forming a *patchy* distribution. The internal heterogeneity of a community generally refers to this patchiness, which varies from high to low. *Gaps* form as tiny openings in the canopy due to disturbances and diseases, and the appearance and disappearance of these openings is called *gap dynamics*.

Function

Photosynthesis in chlorophyll-containing plants converts light energy into chemical energy in the form of organic compounds such as glucose. Virtually all life and all food chains are based on these organic compounds. The amount of producer or plant tissue produced in an area is called *primary production*. Photosynthesis also changes some very familiar molecules: $CO_2 + H_2O \rightarrow C_6H_{12}O_6 + O_2$. That is, carbon dioxide and water are consumed, with their atoms ending up in organic compounds. Oxygen is produced along with the organics. Thus, photosynthesis not only makes organic compounds for energy and for growth, it also decreases atmospheric CO_2 and enriches the air with O_2.

Producer $\rightarrow$ Herbivore $\rightarrow$ Predator $\rightarrow$ Top predator is a *food chain*. Often, about 10% of the energy of one level or stage ends up in the next level, though this *energy efficiency* varies widely. Hence, not much energy remains for top predators, whose biomass is usually very small. Also, the length of the food chain is limited, commonly to about four or five, and rarely to seven, levels. The other 90% of the energy goes along two routes. First, plants and animals metabolize, including a process of cellular respiration that liberates CO_2 and heat to the air. Second, decomposers metabolize dead tissue and also liberate CO_2 and heat. Energy therefore *flows one way* through an ecosystem, from light to producer to herbivore to predator to top predator to decomposer to heat in the atmosphere. At any level the two alternate routes are short circuits that transfer energy to the air. Aquatic ecosystems tend to have a *grazing food chain*, where a relatively large proportion of the producer energy flows through herbivores. In contrast, terrestrial ecosystems are often *detritus food chains*, where the bulk of the energy goes directly from producers to decomposers.

76

A *food web* combines the food chains involving each species of an ecosystem. Food web diagrams with arrows showing who eats whom permit evaluation of the stability or vulnerability of each species (and indeed of the whole community). For example, a top predator is vulnerable because changes in different lower levels affect it. Yet a top predator also exerts a major (e.g., keystone) control over the various species at lower levels, and hence over the community as a whole. Energy flows through a food web, as do materials such as mineral nutrients, toxins, radionuclides, and pesticides. *Biological magnification* refers to increased concentration of a material at higher levels of a food chain. Top predators therefore become subjected to a heavy dose of a material that is otherwise dispersed at low concentration in the environment.

Unlike the one-way flow of energy, materials either cycle within or flow one way through an ecosystem. *Hydrology* or water flow is most important. Precipitation drops water that runs downward in *surface runoff* or *subsurface flow* to a stream, river, lake, or aquifer (underground lake). Some water cycles back to the atmosphere in *evapotranspiration*, basically the evaporation from soil plus the (usually greater) transpiration from leaf surfaces. Water that reaches lakes and oceans can also be evaporated back to the atmosphere. Usually, water dropped on a local ecosystem either flows downhill to another ecosystem, or when evapotranspired, the water molecules are blown by regional winds to another ecosystem.

The carbon cycle, an *atmosphere–organism cycle*, has rapid exchanges between organisms and the air in an ecosystem. Carbon in the CO_2 of the atmosphere is incorporated into tissue by photosynthesis, moves through the food chain, and is returned to the atmosphere by cellular respiration. Carbon is stored in the atmosphere, the sea, cool soils, and forest vegetation. Limestone and coal–oil–gas deposits are two additional *reservoirs* of carbon, but with slow, geologic-time interactions with the atmosphere – organism cycle. The human combustion of coal–oil–gas deposits pumps huge amounts of carbon into the atmosphere, with significant consequences for living organisms.

The phosphorus and nitrogen cycles are especially important because these mineral nutrients generally limit primary production. Fertilizers with N and P are widely used to increase agricultural production. Where N and P are plentiful, oceans and lakes are productive of algae and fish. Alas, too much N and P, as in runoff from septic/sewage or from fertilized fields, causes *eutrophication* (algal blooms due to nutrient enrichment) of streams and lakes.

Phosphorus in the soil is absorbed by plant roots, moves through the food chain, and is returned to the soil by decomposition. These are active exchanges of an *organism–soil cycle* without an atmosphere component. Some soils store phosphorus, though in tropical rainforest phosphorus is mainly in the biomass. Additional reservoirs with slow exchanges are rocks and ocean sediments.

Nitrogen similarly flows between soil and organisms. However, specialized bacteria play key roles in this cycle. *Nitrifying bacteria* convert ammonia (NH_3) (from amino acids and proteins) to nitrite (NO_2) to nitrate (NO_3), which is readily absorbed by plant roots. *Denitrifying bacteria* especially in wetland soils lead to the conversion of nitrate (NO_3) to nitrogen gas (N_2). Nitrogen gas composes 80% of the atmosphere, but can only enter the biological cycle by *nitrogen-fixing bacteria* (and blue-green algae), which convert N_2 to NH_3. In this ammonia form nitrogen is readily incorporated into amino acids and proteins in organisms. The nitrogen cycle therefore is an *atmosphere – organism – soil cycle*. Nitrogen is stored in the atmosphere, some soils, and tropical rainforest vegetation. High combustion engines used, e.g., in transportation, industry, and fertilizer produc-

77

tion liberate large amounts of polluting nitrogen oxide (NOX) gases to the atmosphere.

APPENDIX B. POPULATION ECOLOGY

A *population* is composed of the individuals of a species at a place and time. We usually assume that interactions occur among the individuals.

Population structure and growth

A population has a *geographical range* with distributional limits determined by the physical environment or interaction with other populations. At one place within the range, the spatial *dispersion* of individuals is either random, regular, or aggregated. Random patterns are rare, and most species have individuals mainly in aggregations or clusters, separated by areas of low density. When a single aggregation is magnified, it is found to also be composed of smaller aggregations separated by less-dense areas. In this manner interactions among individuals of a population take place over an increasingly fine *series of scales*.

Populations experience *exponential growth* when they increase in proportion to their size, analogous to a bank account increasing faster when more principal is present. A population undergoing exponential growth may be described by the equation $dN/dt = rN$. This nicely illustrates two principles. First, the *growth rate* (dN/dt = change in number of individuals present, N, per unit time, t) is directly proportional to the population size (N). Second, growth rate is also proportional to the *intrinsic rate of increase*, r, an inherent genetically determined characteristic of an individual. For the population, $r = b-d$, that is, the difference between *birth rate* (natality) and *death rate* (mortality). In short, exponential growth simply depends on how many more births than deaths occur, and how large the population is. No other constraints on growth are present.

Population size (N) of course obscures the fact that individuals differ in age and ability to reproduce. The *age structure* is the proportion of individuals in each age class. A stable age distribution with a zero growth rate has about the same number of individuals in each age class, except for fewer in post-reproductive classes. In contrast, a rapidly growing population has a pyramid-shaped age structure, with large numbers of pre-reproductive individuals. A *life table*, like an actuarial table, shows the number of individuals expected in each age class based on fecundity and survival rates. A convenient way to visualize growth rate is *doubling time*, how long it takes to add another N to the population. Today, human populations with low growth rates double in 30–50 years, whereas those with high growth rates double in 15–20 years.

Constraints on exponential growth are important, so we add a component called *environmental resistance* to the equation, as follows: $dN/dt = rN(1-N/K)$. The *carrying capacity*, K, is the maximum number of individuals an environment or place can support. Thus, when the population size (N) is small relative to the carrying capacity (K), the environmental resistance to growth is minimal. The population grows nearly exponentially. But when N is large and approaching K, the environmental resistance is great. The population barely grows. And if N exceeds K, negative growth occurs. This equation for growth rate is called a *logistic equation*, and is illustrated by an *S-shaped curve* where population size levels off at K, the carrying capacity.

Population growth rate is strongly affected by factors both outside and within the population. *Density-independent factors*, such as weather, pests, and human impacts, decrease or increase growth rate no matter how dense the population is. If the external factor is repeated or chronic it may regulate population size, but most such factors cause wide fluctuations in *N*. *Density-dependent factors* have an increased effect with crowding, and therefore tend to regulate, stabilize, or dampen fluctuations in population size. For animals, such factors include increased mortality, reduced birth rate due to physiological changes, and reduced life span. For plants, higher density often causes higher mortality and reduced growth rate or stunting.

Natural selection

Individuals in a population also differ genetically. *Natural selection* refers to the genetically based changes in these individuals over time. The process includes four essential components: (1) *overpopulation*, i.e., more individuals are produced than can survive in the next generation; (2) *variation*, i.e., individuals produced differ genetically, and hence in their use of resources and response to environmental conditions; (3) *competition*, i.e., individuals compete for the limiting resources; and (4) *survival of the fittest*, i.e., only the most genetically fit or best-adapted individuals survive to reproduce and pass their genes on to the subsequent generation. Natural selection, with its changes in gene frequencies from generation to generation, is a key mechanism in speciation and evolution.

Species interactions: competition and predation

The preceding patterns illustrate *intraspecific competition*, where individuals within a species affect the well-being of each other. *Interspecific competition* considered below involves two or more interacting species affecting each other. Species compete for *resources*, the factors that are used by an organism, and increase growth rate when supply increases. Food and space are common resources. On the other hand, *environmental conditions*, such as temperature and pH, influence an individual's ability to use resources, but are not depleted by the process. Thus, a population typically grows until a resource, i.e., a *limiting resource*, is no longer sufficient to permit further growth.

Competition between species thus is for limiting resources. The *competitive exclusion principle* indicates that two species cannot coexist on the same limiting resource. In this case, one species will outcompete the second, which goes extinct. Yet in nature, *species coexistence* is the rule, due to a number of mechanisms. For example, environmental heterogeneity such as hiding places and varied-size food patches permits coexistence. And typically a species uses a range of resources rather than one. Animals have varying food preferences and frequently switch diets. Thus, *diffuse competition* is widespread, i.e., a species competes for many resources with many species, each resource being a small portion of the total used. This mechanism provides stability against fluctuations in population size of the many competitors, and permits species coexistence.

Predators and prey represent two levels in the food chain, unlike competition, which involves species within a single level. *Predators* are animals that consume *prey*, which are organisms consumed by predators. The basic principles apply to herbivores/producers, predators/herbivores, top predators/predators, and parasites/living hosts. Both competition and predation can be incorporated into the environmental resistance component of the equation above for growth rate.

79

When prey density changes, the *functional response* of a predator is to change its rate of food consumption. The predator may: (1) consume a constant proportion of the prey population irrespective of prey density; (2) consume less as satiation sets an upper limit to consumption; or (3) barely increase when prey density is low, e.g., due to low hunting efficiency.

Predator–prey cycles illustrate a *negative feedback system*, where one component stimulates a second component, which in turn inhibits the first. More herbivores result in more predators; more predators → fewer herbivores; fewer herbivores → fewer predators; fewer predators → more herbivores; more herbivores → more predators; and on and on. Both herbivores and predators coexist and persist. Both populations rise and fall cyclically, with peaks for predators slightly after peaks for prey. The predator–prey cycle is a form of population regulation. Of course, population growth of both predator and prey is affected by many other resources and environmental conditions in the ecosystem.

APPENDIX C. EQUATIONS FOR SPECIES RICHNESS

1. *Exponential or log-normal distribution*

$$S_s = c + z \log A$$

S_s = species number in a sample from within a community; A = area of sample; c = intercept on vertical axis (often considered to be a scaling constant for a particular species pool); z = slope of line. On semi-log paper S_s is plotted against the logarithm of A.[577,1795,1347,1348] A logarithmic series concept is related to the exponential distribution.[786,1898,612]

2. *Power function*

$$\log S_s = \log c + z \log A$$

On log–log paper the logarithm of S_s is plotted against the logarithm of A.[55,56]

3. *Island biogeographic theory: area effect*

$$S_i = c A^z$$

This rephrases the previous equation; S_i = species number on an island.[1023,604,1627,302,1536]

4. *Island biogeographic theory: the two key variables*

$$S_i = f [\text{area } (+), \text{ isolation } (-)]$$

S_i = species number on an island; f = a function of; + and − signs indicate whether the variable increases or decreases species number.[1023,501]

5. *Patch in a landscape mosaic*

$S_p = f$ [habitat diversity (+), disturbance (− or +), area of patch interior (+), age (+ or −), matrix heterogeneity (+), isolation (−)]

S_p = species number in landscape patch. The effect of matrix heterogeneity on a patch is as yet poorly understood; less important factors such as area of patch edge and boundary abruptness (chapter 3) could be added.[501]

3

Boundaries and edges

There is something indescribably inspiriting and beautiful in the aspect of
the forest skirting and occasionally jutting into the midst of new towns . . .
Our lives need the relief of such a background, where the pine flourishes
and the jay still screams.

Henry David Thoreau, A Week on the Concord and Merrimack Rivers, *1849*

Artists and photographers use contrast to create stunning images. High
contrast is attained by accentuating the difference between adjacent
areas, and by sharpening the boundary. This chapter focuses on the
myriad types of boundaries that separate ecosystems or land uses in a
landscape (Fig. 3.1).

Of what use are these boundaries within a landscape? In wildlife management edges typically have an abundance of game and other animals
that move along or across boundaries. In questions of biodiversity and
land fragmentation, the ratio of edge to interior habitat is critical. In
aesthetics, views are often dominated by edges. In forestry, edges are
commonly characterized by crooked trunks and blowdowns. In agriculture, edges are often the source of pests, as well as predators controlling
pests. In soil erosion control, boundaries cause wind speed changes and
turbulence. In park and nature reserve management, boundaries may be
barriers to human overuse. In detecting global climatic change, boundaries are often proposed as a sensitive indicator, the 'canary in the mine'.
And in sustainability issues, humans are edge species, but by carving up
the land and increasing edges enormously, we eliminate the key values
of large patches, thus degrading our landscapes.

Useful literature reviews are available for boundary ecology as a whole
(Holland et al. 1991, Hansen & di Castri 1992), as well as for
ecotones[360,1769,403] and the edge effect as related to biodiversity and
wildlife.[1700,684,1941] Boundaries between land and water are analyzed in
chapter 7.

81

Fig. 3.1. High contrast landscape boundaries. The vertical boundary (c. 10 km long) separates snow-covered agricultural fields on left from a protected forest area on right. Woodlots and stream corridors on the left and white frozen lakes on right also have high-contrast boundaries. Boundaries between gray deciduous forest and black coniferous-evergreen forest show less contrast due to less abruptness. Southern Wisconsin (USA). R. Forman photo.

The present chapter begins with the structure of boundaries, and the developmental mechanisms controlling them. Next, functions are presented, including the edge effect and movements and flows across boundaries. Width and curvilinearity are examined as major variables affecting both anatomy and flows. Finally we explore boundaries moving on the land.

TYPES AND ABRUPTNESS

Boundary types and theory

In chapter 1 we noted that introducing energy causes a system, such as a landscape, to become spatially heterogeneous in one of two ways.[503] First, as a *gradient*, gradual differences in concentration of existing, com-

(a)　　　(b)　　　(c)　　　(d)　　　(e)

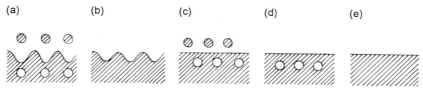

Fig. 3.2. Boundary patterns between two ecosystems. Border may be curvilinear or straight, with tiny nearby patches of one or both ecosystem types.

ponents make a system heterogeneous, but not patchy. Thus, a vertebrate population may be dispersing from a point, or tree species' distributions may be linearly correlated with distance along a moist, tropical mountain slope. Second, as a *mosaic*, a system develops patches (and/ or corridors) with abrupt discontinuities or boundaries. In this sense, with boundaries present in a mosaic but not in a gradient, boundary and gradient are mutually exclusive patterns or concepts.

Three mechanisms produce vegetation boundaries in the landscape[501] (Fig. 3.1): (1) a patchy physical environment, such as a mosaic of soil types or landforms; (2) natural disturbances, including wildfire and tornado; and (3) human activities, such as clearcutting and development for housing. In many cases natural disturbances and human activities sharpen an existing boundary, thus increasing contrast in the landscape.

What types of boundaries result from the above processes? In natural areas *curvilinear boundaries* containing concave and convex surfaces predominate (Fig. 3.2a and b). Here the adjacent ecosystems or land uses interdigitate, suggesting considerable interaction or movements between them. Potentially of special management interest is the *tiny-patch boundary* or ecotone.[618,360] Tiny patches of one ecosystem type may be surrounded by the second type on both sides, or on one side of the border (Fig. 3.2a, c, and d). We hypothesize that the 'curvilinear double-sided' boundary (Fig. 3.2a) is the most common result of natural processes. Especially frequent in human-imprinted areas are linear or *straight boundaries* without coves, lobes, or tiny patches (Fig. 3.2e).

Boundaries are important in all areas of knowledge, and are sometimes differentiated as being *hard* or *soft*.[870] The hard boundary is best illustrated by a straight border with high contrast, as between forest and cultivated field. The soft boundary has varying degrees of softness, from single- and double-sided patchiness to curvaceous[1781,1769,1045,1374,901] (Fig. 3.3).

Few studies have directly compared the ecological roles of hard and soft boundaries. To evaluate the relative penetration of species from the farmland matrix into Great Smoky Mountains National Park (southeastern USA), one study compared two hard boundaries versus two soft boundaries.[23] The latter are perhaps best described as single-

Fig. 3.3. Soft boundary resulting from wildfire. Note curvilinearity and tiny unburned and burned patches. Idaho–Montana border (USA). R. Forman photo.

sided mosaics (Fig. 3.2d). Both vertebrate and exotic plant species crossed the soft boundaries in greater numbers than the straight abrupt boundaries. With soft boundaries the species also penetrated further into the park. Observations in other landscapes suggest similar results.[687,1933,1613,1182,268,503]

Compared with a gradient between ecosystems, an abrupt edge is expected to have a greater density or concentration of edge species present. Hard edges may or may not have more of such species than a soft edge.[1700,683,1227] Edge species may penetrate adjacent ecosystems, or affect the movement of other species between spatial elements.[1350,667,1387]

No net movement exists in a homogeneous system. Hence, the heterogeneity produced by an open system (with incoming solar energy) is a requirement for movements or flows. For example, adjacent high-contrast ecosystems have heat fluxes across their boundary. Or the hoofed animal in an edge capitalizes on the proximity of resources by moving frequently between ecosystems. On the other hand, home-range movements of some vertebrates are completely within a patch, and may be constrained by a boundary.

Thus, the presence of boundaries appears to increase flows, but hard boundaries appear to decrease flows between spatial elements. The effect of boundary abruptness on flow rates between spatial elements may be a curvilinear relationship, but apparently this has not been studied in landscapes. The effect of various soft and hard boundaries on rates of movement could be quite important, for instance, in park and wildlife management.

Gradients are much studied ecologically for their patterns of species, especially plant species, distribution.[350,348,1876,1769,857,65,1693] However, except for the moist tropics, gradients appear to be uncommon in the spatial patterns at a landscape scale. Rather, vegetation and land-use discontinuities across landscapes predominate, and the basic boundary types presented (Fig. 3.2) only hint at the richness of variations found.

Edges in a boundary

Hunters love edges because game species are often abundant. The *edge effect* refers to the high population density and diversity of species in the outer portion or edge of a patch or other spatial element. It is much studied and often considered a principle in wildlife biology and ecology.[943,566,1702,1224,1941] However, not all edges exhibit the edge effect, as we will see in this chapter. The edge effect has been measured in many ways.[1287,1680,1644,555,1779,898,1940] As a minimum, the density and richness of species should be differentiated.[922,1769,715]

An 'ecotone', also much studied ecologically[1781,360,1693], traditionally refers to the overlap or transition zone between two plant or animal communities. Often contrasted with a gradient, the ecotone concept emphasizes a sharp change in species distributions, or a congruity in the distributional limits of species.[1233,350,1876,360,952] Species present in an ecotone are intermixed subsets of the adjacent communities.

The concept of boundary, however, has a strong functional connotation. For example, the cellular membrane is a differentially permeable boundary to materials moving in and out of a cell. A national boundary differentiates insiders from outsiders, and their crossings are controlled at the border, a specific line.[366] A boundary or boundary zone also may exhibit distinctive characteristics unlike the adjacent ecosystems, and thus be considered a system itself.[595,1749] The basic concepts of edge, border, boundary, and ecotone are all useful, and will be used in the following manner.[1377,595,503]

Each landscape element contains an *edge*, the outer area exhibiting the edge effect, i.e., dominated by species found only or predominantly near the border (Fig. 3.4). The inner area of a landscape element is considered the interior or core (chapter 4), and is dominated by species that only or predominantly live away from the border. A *border* is the line separating the edges of adjacent landscape elements. The two edges combined compose the *boundary* or boundary zone (see glossary in chapter 1 appendix). When species distributions within the boundary zone change

85

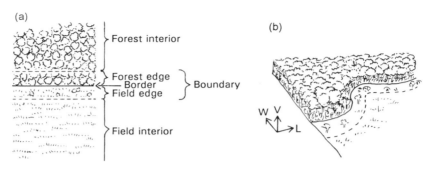

Fig. 3.4. Spatial relationships of boundary, border, and edges. Adapted from Forman & Moore (1992). W = width dimension of edge; V = vertical; L = length.

progressively or evenly from side to side, analogous to a compressed gradient, this describes an ecotone. However, commonly species distribution patterns are more complex within a boundary, as suggested in Fig. 3.2.

Three dimensions of an edge, as well as a boundary, are important for understanding both its anatomy and functioning.[566,1941,503] The edge dimensions are 'width' between the border and interior, 'vertical' including height and stratification, and 'length' along the boundary. Length includes the overall curvilinearity, as well as the arrangement of lobes, coves, and other boundary surfaces along the border (Fig. 3.4b). The internal three-dimensional anatomy of edges varies widely and affects fluxes. Therefore it is considered in detail later in this chapter.

Boundaries at different scales

Landscape phenomena mainly operate in time scales roughly up to a few thousand years, and spatial scales from $c.$ 10 m to 1000 km across (see Fig. 1.3). A specific landscape study, e.g., of hedgerows in farmland or towns in suburbia, focuses on a much narrower range of scales. The space-time principle suggests that the larger an area studied is, the longer the relevant time scale (chapter 1). In addition, we learn from patterns at one scale how another scale may work. Thus, in studying the ecology of boundaries in landscapes, we learn from cell membranes, as well as from political boundaries of large nations.

For understanding boundaries between local ecosystems we recognize at least three important spatial scales.[1879,1880,1102,1762,1883,1748,46] Fine-scale boundaries separate, e.g., blowdown clearings from adjacent forest, or a cultivated field from an adjacent grassland or road. Intermediate-scale boundaries are often demarcated by landforms or long-term land uses, such as where a ridge intersects a valley bottom. Finally, in certain places, boundaries of landscapes are congruous with broad-scale boundaries separating climatic regions, biomes, or tectonic plates.[1654,1839,261,921,600,534]

Despite the huge temporal and spatial differences involved, many boundary characteristics are similar. Boundaries are produced by environmental discontinuities, natural disturbance, and frequently human activity. Most are ephemeral in a century or millennium scale, but some persist. They are differentially permeable. They change internally. And they often move.

EDGE DEVELOPMENT AND CONTROLS

Controlling mechanisms

Is the internal structure of an edge mainly determined by microclimate, soil, or herbivory and predation? The answer, which is not intuitively obvious nor clearly documented, provides considerable insight into boundary anatomy, embryology, and management. We explore the question with a forest-and-field boundary because it is the most studied[943,1814,1700,1379,1908,1941], and because of high contrast between the adjacent ecosystems.

Edge microclimate

Sun and wind are overriding controls on edge microclimate.[273] Both desiccate leaves and increase evapotranspiration (the evaporation of water molecules from plant, soil, and other surfaces). Therefore, sun and wind determine which plants survive and thrive in an edge, as well as having a major impact on soil, insects, and other animals in the edge. These ecological effects increase with a greater difference in vegetation height between the adjacent ecosystems.

In addition to causing higher evapotranspiration, the sun provides light energy used in photosynthesis for plant growth. This determines which shade tolerant or intolerant species are in the edge. The sun also affects the persistence or melting of snow that accumulates in some edges.

On forest edges facing east, north, and west in Germany, for instance, solar radiation peaks in summer.[556] However, south-facing edges have maximum sunlight in spring and fall, because of shading in summer when the sun is high in the sky (Fig. 3.5a). West and southwest, as well as east and southeast, edges receive maximum sunlight in summer. The total annual solar radiation decreases progressively from south, southeast and southwest, east and west, northeast and northwest, to north, with north-facing edges receiving little solar radiation for half the year.

87

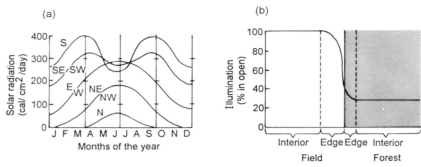

Fig. 3.5. Sunlight on edges. (a) The amount of energy received daily by forest edges facing south, southeast and southwest, etc. (b) Generalized curve showing relative widths of field edge and forest edge based on light. Adapted from Geiger (1965). Originally published in the German language by Friedr. Vieweg & Sohn, Braunschweig 1961.

Diffuse radiation from the atmosphere (sky) is about equal for all edge exposures.

The forest darkens the field edge[531], and the field lightens the forest edge (Fig. 3.5b). How far from the border do these modifications extend? Measurements in coniferous evergreen and deciduous forest adjacent to meadows in Germany suggest that elevated light levels extend only a few meters into both forest types.[556]

In contrast, the field edge width based on illumination was approximately equal to the height of the trees (Fig. 3.5b). Although the curves varied, this result was the same for evergreen and deciduous forest, full sun and overcast days, and winter and summer. The field edge also reflected more energy (higher albedo) than the field interior, despite receiving less solar radiation than the field interior. This resulted from the forest shade shifting some entering solar radiation to longer (red) wave lengths, that in turn are reflected more than the shorter (blue) wave lengths. Finally, the forest protects the field edge at night, decreasing the upward radiation of heat. This sometimes leaves the field edge as a frost-free strip.

In contrast, wind increases evapotranspiration by carrying heat from the matrix, which desiccates leaves. Blowing particles, including sand, silt, snow, seeds, and spiders, often accumulate in forest edges, partly due to the sudden drop in wind speed there.[556] *Impaction*, the accumulation of material on surfaces from moving air, is also important because of the dense foliage in the edge. Impaction of water in blowing fog and mist, various aerosols (droplets), mineral nutrients from fertilizer, pesticides, and toxins can accumulate in edges.

Air flowing over a field is typically reduced in velocity upwind of a forest to a distance of c. 8h (h = height of trees). Downwind of the forest

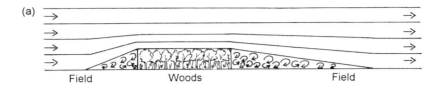

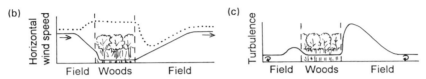

Fig. 3.6. Field and forest edges based on generalized wind patterns. Vertical dimension is magnified to illustrate patterns. (a) Laminar airflow streams and two turbulence zones. (b) Relative wind speeds at 1.5 m above ground (solid line), and at 1.5 m above height of woods (dotted line). (c) Relative amount of turbulence at 1.5 m above ground. Based on many sources.

windspeed is reduced for a much greater distance (Fig. 3.6a and b), about $25h$ and sometimes more.[556,711] Turbulence is present in the upwind field edge. But because the permeability of a woods is low, turbulence is high in the downwind field edge (Fig. 3.6c). This turbulence site may be subject to erosion and a source of dust.

Wind penetration into the forest shows quite different patterns.[556,273] On the upwind side forest edges have elevated wind speed (Fig. 3.6b), penetrating only to about $1h$, compared with the forest center. However, this distance can vary considerably depending on the structure of shrubs and small trees in the edge. The downwind forest edge has elevated wind speed extending in less than $0.5h$.

Temperature and moisture levels of air, soil, and leaves also mainly depend on sun and wind.[388,1452,1453,831] Yet, the distances from a forest–field border in which these are significantly elevated or reduced are generally considerably less than the distances for wind speed or evapotranspiration.[1056,1337]

The 'sea breeze without a sea' is a special case where cool air flows horizontally from woods to field, replacing warm air rising from the field at night. Such horizontal flows carry seeds, aerosols, and other objects from forest edge to field.

A forest clearing with a diameter less than about 6–10h will have slightly less extreme or variable microclimatic conditions than in a large open area.[1405] The width of internal or *inner edges*, i.e. those encircling an enclosed spatial element such as a clearing, is also less than for the outer edges of a forest. For example, microclimatically determined inner edges of a deciduous forest in Czechoslovakia and an evergreen tropical

forest in Costa Rica are apparently well under 0.5h in width.[556,927] In the
1h diameter clearing studied in Czechoslovakia, solar radiation is highest
on the north side due to solar angle. Rainfall is highest on the east side
due to wind; evapotranspiration on the northwest side due to morning
and midday sun; summer soil moisture in the center due to sparse low
vegetation in the clearing; and dew in the center due to night temperature and upward radiation of heat.

In short, the microclimate of field edges differs markedly from that of
field interiors, and forest edges from forest interiors. Sun and wind are
the key determinants, and both can be significantly modified using vegetation management.

Soil in edges

Many vegetation boundaries, both natural and human-created, are
congruous with a boundary between soil types.[1879,1880,740,1886] The contrasting structure and chemistry of adjacent soils provide a range of conditions compressed in the narrow space of a boundary, which are suitable for a richness of plants and soil animals. If the soil boundary is soft,
a heterogeneous area of intermediate or even different soil conditions
is likely. This may coincide spatially with the width of the vegetation
boundary.

Alternatively, the boundary zone between soils may be abrupt, and
coincide with a sharp border between vegetation patches. In this case,
within a patch the differentiation of vegetation edge from interior is
probably caused by some mechanism other than soil.

Nevertheless, in agricultural landscapes many studies show elevated
levels of nitrogen, phosphorus, and other mineral nutrients in forest
edges.[449] Furthermore, some of the most typical edge species in Europe,
such as elderberry (*Sambucus* spp.) in the shrub-small tree layer, and
nettle and bedstraw (*Urtica dioica, Galium apparine*) in the herbaceous
layer, are sometimes called 'nitrophiles' or 'phosphatiles'. These have
been shown to either require high nutrient levels for growth, or to outcompete other species in the presence of abundant nutrients. Such species often produce dense edges with nutrient-rich foliage.[887,449,626,1377,1378]
This is a distinctive and delicious combination for herbivores.

In these landscapes microclimate and soil conditions are clearly interrelated mechanisms controlling edges.[407] In addition to tractors turning
at the ends of fields, wind and water provide transport, carrying fertilizer
and other nutrients from agricultural field to forest edge (Fig. 3.7a). Particles from wind and water erosion are also deposited in edges. The
decrease in wind velocity upwind of a forest, the progressive drop in
wind velocity within the forest edge vegetation, and the impaction process together indicate that abundant nutrient deposits may occur. These
should vary in amount across the boundary, from the beginning of the

Fig. 3.7. Edges affected by mineral nutrients and herbivory. (a) West-facing forest edge subject to high inputs of nitrogen, phosphorus, and other material from large adjacent cultivated field. Ancient woodland, East Anglia, UK. (b) Shrub and herb layer reduced by deer herd on right and protected by fence on left. Nara, Japan. R. Forman photos.

field edge to the end of the forest edge (Fig. 3.6). Snow is not only deposited in edges, but its persistence depends on shade and exposure. Fire, also under microclimatic control, may repeatedly move into edges, such as from grassy to wooded areas in savannas.

Herbivory and predation

Herbivores, including many game species, are commonly at higher densities in edges than in patch interiors. This means that edge plants are the survivors, the species less palatable or less sensitive to trampling.[105,296] In almost any ecosystem the impressive vegetation difference between an exclosure that experimentally eliminates herbivores and its surroundings underlines the point[1638] (Fig. 3.7b). We must hypothesize that the vertical and horizontal structure of edges, as well as their species composition, commonly reflect edge herbivore density. Heavy browsing of the shrub layer by wildlife or livestock permits wind to penetrate further into a woods, effectively widening the edge.

The herbivore density in edges also provides a feast for predators. Hawks, cats, canines, and other predators often focus their foraging on edges. Indeed edges have been called 'ecological traps', because of the concentrated predation on game and other populations of an

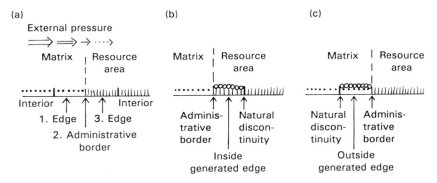

Fig. 3.8. Boundary arrangement for protecting a natural resource area against overuse. (a) Three filters against external pressure by people. (b) and (c) Generated edge inside and outside an administrative border. Adapted from Schonewald-Cox & Bayless (1986).

edge.[123,553,32,38,1387] Since predator species are generally considered more likely to become rare than herbivore species, the provision of edges, including soft edges, may be promising for certain predator conservation efforts. Finally, to the extent that predators sometimes control herbivore populations, we recognize the indirect role of predators in maintaining edge vegetation.

Humans

Clearly humans play overriding roles in maintaining some edges. Vegetation is trimmed or mowed. Berries and firewood are harvested. Edges created by cultivation may be regularly bathed with pesticides. Livestock often concentrate in edges to avoid wind or sun. Human created or 'induced' boundaries in many areas are numerous, diverse, and often differ structurally from natural (or inherent) boundaries produced by substrate patchiness and natural disturbance.[1700,1941,1064]

Management of parks, nature reserves, and other protected areas often focuses primarily on controlling and channeling human access, to prevent overuse of the protected resource.[794] Thus we 'treat the boundary as a skin, whose condition can indicate the health of the entire system'.[1513] Administrative borders superimposed on the land commonly coincide with natural boundaries or discontinuities in some portions, but usually do not coincide along much of their length.[1206,1699]

In controlling the number of people entering a natural resource area, three filters in the boundary are useful. The edge portion (or buffer) of the surrounding land is a first filter against movement into, for example, a park (Fig. 3.8a). The administrative border itself is a second filter. The edge portion of the park is a third filter (Fig. 3.8a). Numerous options for managing the three filters are available[488] (also see Fig. 3.11).

In the typical case where an administrative border and a natural boundary do not coincide, the area between the two has been termed the 'generated edge'.[1514,1513] This is because it normally changes ecologically after the administrative border is established. The generated edge may be either inside or outside the park, depending on where the administrative border is relative to a natural discontinuity (Fig. 3.8b and c). Generated edges offer valuable management opportunities.[1513]

In addition, the 'segmentation' of boundaries is useful in planning, conservation, and management. This requires dividing the boundary or perimeter into many unequal-length segments, one for each adjacent land-use, ownership, vegetation, or boundary type.[1514,1513] In this manner boundary management can be tailored to each segment or segment type.

In short, the structure of edges is significantly molded by all four mechanisms, microclimate, soil, animals, and humans. The interactions among these factors determine the width, verticality, and length dimensions of boundaries. One or two processes of course may predominate in certain locations and for certain periods. Furthermore, the primary controls may alternate between mechanisms within an edge, and those external to the edge. Boundaries are particularly amenable to design, conservation, planning, and management.

Anatomy and development

Suppose a friend bought a nearby tropical forest, cleared a rectangular patch for a few cows, and then sat in a hammock watching the forest–field boundary for decades. At first the border is straight and hard, with foliage mainly limited to ground level in the clearing and canopy level in the forest. Gradually she sees a dense mass of woody plants and tall herbaceous plants develop, mainly under the outermost trees. A thin layer of vines and perhaps sprouts along some tree trunks forms between the canopy and the new vegetation below. In the field edge large numbers of seeds deposited by wind and animals germinate. Some field-edge species tolerate the shade, tree root competition, and livestock grazing. Over time a few saplings in the forest edge become canopy-level trees. These have leaning trunks, and large limbs along the trunks on the clearing side.

This generalized sequence produces the following basic anatomy repeated widely in forest edges. [1754, 1169, 407, 1379, 1908, 1377, 1378] A perennial herbaceous layer or *saum* may be present in the outermost portion of the forest edge. Just inside the saum typically is a dense layer of shrubs and/ or small trees called a *mantel* (Fig. 3.9a). The mantel is sometimes div-

93

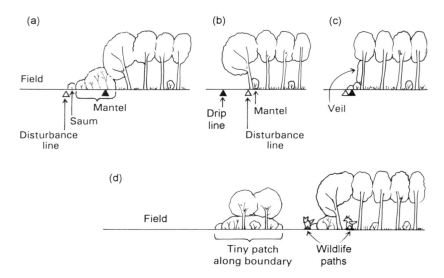

Fig. 3.9. Saum, mantel, and veil of forest edge. (a) and (b) Contrasting saum and mantel development according to juxtaposition of drip line and disturbance line. (c) Veil following recent boundary formation. (d) Mosaic boundary. Disturbance line = outermost extent of disturbance in field, e.g., by plowing or livestock grazing; drip line = outermost extent of canopy-level branches.

ided into a low 'pre-mantel' usually of shrubs, and a higher 'main mantel' of tree saplings. Above these is the *veil*, the usually thin film of foliage connecting the mantel and canopy (Fig. 3.9c).

A young forest edge such as initially created by the observer here is relatively open to the wind and weather. Indeed, development of the saum, mantel, and veil (terms meaning hem, overcoat, and curtain) protects the forest interior from the wind. Similarly, only when the mantel develops are cover and food sufficient to support the high populations of birds and game characteristic of the edge effect.[1247,1162]

Chronic disturbance in the field, say by plowing or livestock grazing, may stop outside the trees, or extend all the way to the outermost tree trunks. The juxtaposition of disturbance line and outer tree architecture determines the degree of saum and mantel development, and their species composition[407,1379,501] (Fig. 3.9). If the 'disturbance line' corresponds to the tree trunk line, the saum is often absent and the mantel is minimal. If the disturbance line is outside the tree 'drip line', typically both the saum and mantel are well developed. Where disturbance line and drip line coincide, both saum and mantel develop to medium size.

The relative heights of the vegetation in adjacent ecosystems is significant in determining the anatomy of their edges, though this is little stud-

94

ied. Forest and field differ greatly in height, which has major implications for airflow, solar radiation levels, deposition of particles and nutrients, and visibility by vertebrates. The anatomy of edges where two forest types, or two grassland types, intersect needs study.

Mammals commonly move along the boundary both inside and outside the mantel[667], sometimes forming paths (Fig. 3.9d). An outside path along the field edge may be accentuated by concentrated grazing or browsing of the forest edge by herbivores.[84] The anatomy and development of the length dimension, e.g., of curvilinear, mosaic, and straight boundaries, is considered in chapter 4 on patch shape.

Hartmut Diershke (1974) studied forest edges adjacent to former pastures in north central Germany, not only documenting the saum and mantel communities, but relating them to soil and especially microclimatic patterns. On forest edges facing southeast, south and southwest a saum community characterized by clover and geranium (*Trifolium, Geranium*) receives both high light intensity and shade during the day. The plants only require a fraction of the light necessary for nearby open-meadow species. The saum sites are warm, half-shaded, relatively dry, and often nutrient poor. Open meadow plants cannot outcompete the saum due to low light and nutrient conditions. In contrast, on northeast, north, and northwest exposures a bedstraw saum community (*Galium, Calystegia*) dominates, and has a cool, half-shaded, nitrate-rich habitat with a well-balanced water budget. Limited light prevents invasion by open meadow plants.

In the mantel few plants grow in the herb layer.[407] The shrubs dominating the mantel exhibit a wide ecological amplitude, and little correlation with microenvironmental variations. Warm dry conditions favor a barberry-dominated (*Berberis*) over a bramble-dominated (*Rubus*) mantel. Herbivory was not measured in the study.

Finally vegetation complexity within an edge doubtless has a major effect on animals of the edge.[1907,268,1644,1387,1584,1177] The distribution of plant species, density, biomass, and cover can vary widely at different points along a single boundary. The options seem rich for creative design and management of edge anatomy, to enhance animal communities and many other societal objectives.

FUNCTIONS OF BOUNDARIES

Membrane theory and the five functions

The cellular membrane is an intriguing and useful model to understand landscape boundaries. Its structure, function, and changes are well known, especially in a medical context.[1855,689,477,1620] It integrates thermodynamic and molecular phenomena and may be more than a simple analogy.

A double layer of fat-like molecules (phospholipids) composes the inner and outer surfaces of a membrane. Cholesterol molecules and large proteins, some with dangling carbohydrates, are scattered through the membrane in relatively distinct patterns. The inner and outer surfaces differ markedly.

These basic structural characteristics largely determine membrane function or physiology. The key function described is being a differentially permeable filter, that is, letting some materials cross but not others. Different materials cross in different places, times, and via different mechanisms: (a) small molecules cross in passive diffusion; (b) active transport (using energy) drives molecules across (and maintains ion gradients); and (c) large particles cross in 'bulk transfers'. In addition, the membrane contains receptor molecules where other objects adhere and interact.

The landscape boundary exhibits these functions and more. It is convenient to recognize five basic functions:[501,109,111,1582] (1) habitat; (2) filter; (3) conduit; (4) source; and (5) sink. Wildlife moving along a grassland–woodland boundary discussed later in this chapter illustrates the conduit function. Forest edge animals moving to feed in an adjacent field illustrate the source function,[1350,531] and eating or otherwise killing animals moving across a boundary[447] illustrates the sink function. Conduits, sources, and sinks are analyzed in detail for corridors in chapter 5. Here we explore the habitat and filter functions.

Edge effect or habitat

Edges are often biological cornucopias. High species richness and density or biomass are documented for many groups, including plants, mammals, birds, and invertebrates[539,716,1941,1721] High levels of flower and fruit production, pollinator and frugivore (fruit-eating animal) density, and seed dispersal are often present in edges.[1708,501] As patches increase in size, so does the number of edge species. However, apparently the curve for edge species tends to level off well before the curve for interior species (see Fig. 2.8a).[1700,485] Edges may also play an important role in species evolution.[1071]

Several categories or types of edge species are usefully recognized, independent of taxonomic group. Most species are generalists that tolerate frequent disturbance, as well as the contrasting environmental conditions on opposite sides of a border. Most species in the edge are common in the landscape. Few are expected to be present only in edges (except perhaps in environmental patches, chapter 2). Typically, no rare species in the landscape live in edges.

In addition to edge residents, boundaries often contain *multihabitat species*, those requiring or frequently using two or more habitat types.[396,501,1941] These organisms capitalize on the complementarity of resources provided by the boundary between two ecosystems or land uses. Indeed, humans building their cities and towns mainly at boundaries are good examples. Though subject to disturbance regimes of both ecosystems, these locations offer ready access to diverse resources, plus stability during stress periods.

Edge species are usefully differentiated from patch interior species. Because a species often behaves differently through the year, and in different parts of its geographic range[1941], edge and interior species are best categorized locally. Thus, *edge species* are those primarily or only near the border of a spatial element, and *interior species* are primarily or only distant from a border.[853,501,1015] Although many, but not all, edge and interior species may survive almost anywhere within a patch, their densities differ according to distance from the perimeter. In addition, 'indifferent species', with similar abundance in both the edge and interior, are probably also common. An indifferent species is even more of a generalist than an edge species, because it persists in both the environmentally diverse edge and in the patch interior.

Tiny gaps, clearings, clumps, and objects creating microheterogeneity within a spatial element also contain edges, called interior or inner edges as in mathematics.[1749,546] These are structurally similar to exterior edges, except that they tend to be less developed and more ephemeral, as emphasized in patch dynamics.[1320] Of course, these 'circular' edges are very short in length, and exposed to the environmental conditions of a small spot or hole, rather than to the conditions of an extensive external matrix.

The promixity of interior and exterior edges may have ecological importance.[33] For example, fewer bird species are reported in exterior edges, where there are edge habitats of tree gaps nearby.[922,1227]

The cellular membrane analogy suggests other functions of boundaries that warrant direct study. Just as large proteins in membranes are sites of crossing of large molecules, so certain plants in landscape boundaries may be important to movements of other species across a boundary. For instance, dense thorny plants like hawthorn, bramble, and cactus (*Crataegus, Rubus, Opuntia*) act as inhibitors to crossings of most large mammals. Certainly they inhibit timid hikers. Similarly, other species that are palatable or otherwise attractive serve as facilitators of movement. Such boundary inhibitors and facilitators could be readily used in wildlife management, or in windbreak construction, to channel or deflect movements and flows (Fig. 3.10).

The receptor function of membranes, where external molecules temporarily attach and interact with other molecules, points to another

Fig. 3.10. A tropical forest edge. The saum is composed mainly of grass, the mantel of tree ferns and palms, and the veil of vines. Note heterogeneous edge anatomy, and its potential effect on movements and flows across the edge. Near Atherton, northern Queensland, Australia. R. Forman photo.

little-studied role in the landscape. Particular species in the edge produce beautiful flowers that attract pollinators, or produce delicious fruits or seeds relished by herbivores[1631,40] (Fig. 3.10). Other plants have herbivore colonies, insect concentrations, or perches where predators are attracted and concentrate their feeding on herbivores. Again, the boundary receptor is probably an important dimension of the edge effect function.

Wildlife managers have frequently increased the amount of edge in a land area to increase game populations.[943] Thus, edge or *boundary length*, or 'boundary density', the total length of boundary per unit area, can be compared for different portions of a landscape.[1287,1680,566,814,547]

Boundary length can also be measured as a 'fractal dimension', with D ranging from 1.0 to nearly 2.0. The former is a straight line boundary crossing an area, and the latter an enormously convoluted boundary where nearly the whole area is boundary (see Box 3.1). In summary,

Box 3.1: Fractals and landscape boundaries

With fractals we have linked art and science, explored unexpected frontiers, and produced dazzling T-shirts. Fractal applications to landscapes were recognized early.[230,1039,899] Fractals in patch perimeter-to-area relationships, mosaic structure, and trajectories of animal, wind, and water movement are introduced in later chapters.

If you ask, 'How long is the coastline of Britain?', the answer can be, 'Its length depends on the length of the ruler.' (or the scale of measurement).[1039] A 1 km long ruler measures major coves and peninsulas, a 1 m stick measuring each shoreline rock results in a longer coastline, and a 1 mm ruler around each sand grain produces a still longer coastline. In effect the coastline length varies, or can approach infinity. *Fractal geometry* is essentially a collection of models, in which fractal dimensions, as scaling parameters, relate some quantities of interest to a resolution or length scale.[1137,1138,379,1887]

Some fractals exhibit a 'self-similarity property'. In this the pattern (or relationship of parts to the whole) is the same or similar at different scales, like the common form of a fern leaf or snowflake. The fractal dimension simply relates the number of parts composing the basic form to a 'similarity ratio' between two scales. The scale at which pattern changes is the level at which a different process is probably operating. In a general sense the British coastline is composed of lobes and coves at any scale. Although self similar, they are not exact copies at each scale. Somewhat different processes cause the patterns at different scales. Perhaps the most self-similar landscape boundaries are those related to a dendritic stream–corridor system.

The 'space-filling property' is a related fractal dimension calculated for the curvilinearity of any boundary. A straight line has a dimension $D = 1$, and D increases progressively with the 'squiggliness' of the boundary. When the boundary is so convoluted that it virtually covers the area, D approaches 2, the fractal dimension of a surface. In an analysis of landscape pattern across the eastern USA (D based on the slope of a perimeter-to-area regression line), Robert O'Neill and collegues (1988a) found $D = 1.23$ for eastern Iowa, where simple rectangular patches in farmland predominate. In eastern Maine $D = 1.44$, where topographic and coastline heterogeneity cause complex shapes of land use. Natural processes, such as erosion, meandering streams, and pest outbreaks, tend to produce boundaries with high space-filling fractal dimensions.

Such soft boundaries may provide stability to the adjacent systems. Consider a 'crazy' drum with a fractal rather than a circular boundary.[1485] The vibration and noise of a drum beat is quickly dampened, because the vibration disappears in its own neighborhood of self-similar forms. This prevents the vibration from spreading over the entire drum surface to produce a loud vibrant sound.

the edge effect reflects the habitat function of a boundary, and, despite important research frontiers, is one of the best known ecological subjects at the landscape scale.

The Filter Function

No absolute barriers or boundaries exist in nature, only filters. In preceding sections we have referred to the cellular membrane and its differential permeability as a useful analogy. A major recent development in studying landscape boundaries is to understand their filter function, that is, how boundaries affect the rates of movements and flows between ecosystems.[1886,1514,211,212,1885,595,503]

Objects such as seeds, silt, wood, and even heat, are carried across landscape boundaries by six *vectors*: wind, water, flying animals, terrestrial animals, humans, and machines. The last, using fossil fuel energy, is not considered further here. Vectors are usefully grouped into two categories or processes according to their energetics, i.e., mass flow and locomotion.[485,1886,501,503] Wind and water movement, as *mass flow* or transport, depends on external heat energy gradients. Animal and human *locomotion* depends directly on expending internal energy. The results of mass flow and locomotion usually differ sharply, and indeed may be opposite. Consider a seed crossing from field to forest through a narrow gap in the edge mantel. If wind transported, it speeds up due to the *Venturi effect* (acceleration due to increased pressure moving through a narrows). If animal transported, it slows down due to behavioral caution.

This effect on rate of exchange is a key role of boundaries. Rates change diurnally, for example, reflecting the ever-changing foraging routes of animals. They change seasonally, e.g., with heat energy flows affected by leaf fall. And they change successionally, as vegetation structure develops. Differential permeability of a boundary also implies that some objects pass readily, and others scarcely or not at all. For the abiotic vectors of wind and water, flow rate depends on their kinetic energy (that associated with motion).[595] The relationship between rate and amount moved is curvilinear. This means that higher velocities have much higher transport capacities. High winds carry huge amounts of particles from dry cropland to riparian forest, whereas medium winds only carry slightly more than low winds. Similarly, mineral nutrients and pesticides move between ecosystems at variable rates, which change with time as well as wind and water velocity.[816,313]

What determines permeability? Membrane theory indicates that a history of abundant fluxes produces a richly textured or heterogeneously structured membrane.[1855,250,477,1620,23,503] Present flows of course respond to that texture. This principle suggests the hypothesis that an anatomically diverse landscape boundary is more permeable to more objects,

100

and that disturbance within a boundary will strongly affect permeability.

A handful of models, measurements, and discussions of landscape boundaries pinpoints a variety of possible factors controlling permeability. In addition to vectors mentioned above, permeability has been related to population density in the edge or in a whole patch, the vertical vegetation structure of the edge, the abruptness of a boundary, the contrast present or relative size of dispersal sink and source pools, the suitability of the patch being entered, and the location of a park boundary relative to the location of a natural environmental discontinuity.[1886,211,212,1513,23,1613,1614,313] Based on this diverse evidence, which boundary types exert the greatest control over flows? Clearly the answer will focus on boundary anatomy. But the chemistry of boundary species, including rich nutrient sources and toxins, also plays a role. The filter function of boundaries offers opportunity for study.

To conclude, the filter function of boundaries has direct application to the design and management of natural resource areas, such as parks and nature reserves. Here preventing overuse by arriving people is critical. Numerous management techniques can be applied to the border itself (center of Fig. 3.11). A similar range of techniques can be applied to the edge portion of the matrix, and to the edge portion of the natural resource area. Most management patterns will have their axis parallel to the border. Management in the matrix edge and along the border focuses on controlling the number of people entering the protected area. Management in the edge portion of the protected area focuses on concentrating human activity in the edge, thus protecting the interior of the natural resource area.

INTERACTIONS BETWEEN ADJACENT ECOSYSTEMS

Movements and flows between ecosystems are at the core of landscape and regional ecology.[485,501] This is a motif that appears in almost every chapter. Nevertheless, introducing some key patterns here provides a valuable foundation.

As just described, six vectors transport objects between ecosystems or land uses. The processes of locomotion and mass flow provide the energy basis for the vectors ('diffusion' at the landscape scale in the usual scientific sense is minor, and in the social science sense includes all processes[501]). Animals and people locomote between ecosystems (chapter 11). Wind carries heat, sound, gases, aerosols, and particles between the spatial elements (chapters 6 and 10). The aerosols and particles are deposited by fallout or impaction.

Water flows between ecosystems either as surface flow or subsurface flow. All water flow carries mineral nutrients in solution. Surface flow

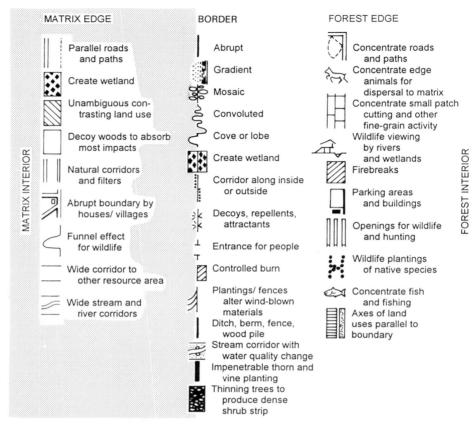

Fig. 3.11. Management examples for edges and border of a natural resource area. Interior and edge of matrix to left, and edge and interior of protected area to right, are delimited by a border in the center. Adapted from Forman (1989). Reprinted from the proceedings of the 1988 Society of American Foresters National Convention, published by the Society of American Foresters, 5400 Grosvenor Lane, Bethesda, MD 20814–2198. Not for further reproduction.

additionally carries particles between sites of erosion and deposition. Water, particles, and mineral nutrients moving from land to stream normally pass through a vegetated stream corridor that acts as a filter (chapter 7).

Flows between spatial elements of course do not depend only on boundaries. The relative abundance of objects on opposite sides of the boundary may be critical. Thus, net movement is often from high to low concentration, though social behavior or prevailing wind direction may reverse the direction of movement. The 'source' and 'sink' concepts, where certain areas provide objects and other areas absorb objects, are also closely related.[212,1357,1117]

Finally, we should view boundary flows in the broader context of the landscape. Systems modeling[1235] and spatial pattern are usefully linked, e.g., by relation theory, energy flow models, or graph theory.[1781,1236,251] For example, in 'relation theory'[1781,1782,1769], patch–matrix interactions are analyzed in terms of four processes: (1) supply to patch; (2) resistance of patch; (3) retention by patch; and (4) disposal from patch. An excess in any of the four rates can damage the patch system. Sources and sinks in the matrix have a major effect on supply and disposal flows, while boundaries and their management control resistance and retention.

To understand interaction between ecosystems further, consider the context of a patch (chapter 9). The adjacent area is composed of one or more *adjacencies*, i.e., the elements in contact with the patch or site of interest. Heat moves between an asphalt tarmac parking lot and an adjacent oak woods.[1133] Building density affects bird species richness in an adjacent suburban woods.[1717] Remnant tree patches receive numerous fires from the surrounding grassland in a tropical savanna.[630] Context also refers to the 'neighborhood'. This includes not only adjacencies, but also nearby elements of the local mosaic linked by active interactions. Principles from a growing literature on the movements and flows among landscape elements in a neighborhood are explored in chapters 9 to 11.

Because each spatial element adjacent to a patch has its own species pool, we hypothesize that the characteristics of a patch are in part controlled by the number of adjacent patches. Although this apparently has not been studied, several patterns seem logical as hypotheses for testing.[1427] Species richness of the patch initially increases with number of adjacent patches, but then drops as the surroundings become finely divided. The curve rises more steeply if each patch is a different type. In similar fashion, modeling combined with field tests can be done evaluating the effect of varying adjacent patch sizes, patch shapes, or number of corridor connections.

Finally, in considering patch–matrix interactions it is important to recognize that the matrix not only is heterogeneous, but changes. For example, patches of *Calluna* heath in southern England harbor different invertebrate populations according to the neighborhood pattern.[1844,1845] However, the neighborhood mosaics change rapidly, so the invertebrates in the patch also remain in a dynamic, rather than an equilibrium, state. Similarly, in the 1930s an area of southern Sweden had species-rich dry meadows mainly surrounded by wet pasture (K. E. Hill and M. Ihse, pers. commun.). By the 1990s cropland surrounded most of the dry meadows, thus significantly altering their species composition and richness.

WIDTH AND CURVILINEARITY

We now focus on two of the three edge dimensions, width and curvilinearity (Fig. 3.4b). Edge width is analyzed from the perspective of what determines it. The curvilinearity of boundaries is primarily examined to see how it affects movements and flows.

Edge width

A small patch with a small interior is mostly edge. But different sides of a patch have different edge widths. For example, based on microclimatic and vegetational measurements, deciduous woods in North America and Europe have wider southern than northern edges, with differing species present.[1814,1941,1066] Because of solar angle, equator-facing edges are wider than pole-facing ones, an effect that apparently is less evident in the tropics.[1379,501,927] Similarly, windward-facing edges can be expected to be wider than leeward edges. Uphill and downhill edges may differ based on exposure to water flow effects. Edge bird communities also differ on different sides of a patch according to sun and wind exposure.[260,1227] A wooded patch or strip of forest in savanna doubtless has a wider edge on the side facing higher fire frequency, or more livestock grazing pressure. These examples pinpoint a general pattern: directional external forces with higher kinetic energy produce wider edges.

In considering microclimate earlier in this chapter we compared various forest edges with field edges (Figs. 3.5 and 3.6), and found their widths varied microclimatically from $>25h$ to $<0.5h$. From widest to narrowest in order they appear to be: (1) field edge downwind of forest; (2) field edge upwind of forest; (3) forest edge upwind; (4) forest edge downwind; and (5) inner forest edge. Some of these widths can be shortened or lengthened by altering edge structure.

The mantel plays a central role in determining width of a forest edge, because of its filter effect on wind, light, and heat. A forest edge in farmland, where the field and forest have been in equilibrium for decades, typically has a dense mantel and is relatively narrow, extending meters or tens of meters across (Figs. 3.9a and 3.10). In contrast, a recently formed edge from logging or road construction has a sparse mantel and a relatively wide microclimatically-determined edge, estimated at 1.5–3h.[683,518,273,1066] Although the recently-formed edge is quite wide, species depending on a dense mantel are low in density.[922,1700,1247,898,716]

Browsing and human activities strongly affect edge width, because they are usually directed at the mantel. 'Browse lines' of extensive stem and twig removal up to a constant height are caused by rabbits, deer, kangaroos, and other herbivores (see Fig. 1.11). These are commonly used as indicators of excessive wildlife population levels.[1638] Livestock produce the same effect. Firewood cutting, trimming, and brush removal also open up edges to microclimatic penetration.[1064] In contrast, some wooded nature reserves surrounded by farmland in England have been protected by enhancing mantel development with dense thorny shrubs and trees, and piles of cut branches and brush.

Edge width also depends on how it is measured. A variable of interest is chosen and measured at intervals along transects from the border toward the center of a patch.[1379,32,1177] The edge width (or penetration distance) extends from the patch border to the point on a transect at which there is no significant change in the variable, in proceeding further toward the center. Connecting these points determined along several transects defines an 'interior – edge borderline', which differentiates the edge from the interior of a patch[924] (also see chapter 4 appendix). An edge width based in this manner on microclimatic modification may be considered a useful, but minimum, value.[1814,924]

The same approach can be taken to determine edge width based on vegetation, e.g., cover, density, biomass, stratification, species richness, and/or species composition of trees, shrubs, vines, and/or herbs.[1814,1379,273] Similar approaches are taken for various animal groups, such as mammals, birds, and insects.[123,555,667,898,119,716]

How do the results of these approaches for measuring edge width differ? Edges defined by insect density and richness, and by vegetation, appear to be meters to tens of meters in width.[1814,1379,555,716,1378] Human effects on suburban woods are concentrated in the outermost tens of meters.[1064] Edges determined by microclimate and insectivorous birds extend tens to hundreds of meters.[928,555,668,1685,1227,273] Those determined by some edge butterflies, mammals, and edge-related nest predation extend hundreds of meters.[1891,568,990] And edge width based on populations of some large mammals, including white-tailed deer (*Odocoileus virginiana*) as an edge species, and tigers and grizzly bears (*Ursus horribilis*) as species sensitive to remoteness from humans, extend hundreds to thousands of meters.[660,794,22]

With such a wide range, from meters to thousands of meters, some narrowing of the edge width concept seems warranted. One approach is to define the 'edge width' of a patch by the distribution of the species, or group of species, that is most sensitive to patch size (hence, width ranges up to hundreds or thousands of meters). Perhaps more useful is to consider the 'edge width' of a patch or other spatial element to be the distance between the border and the point where microclimate and vegetation do not significantly differ from conditions in the interior. Such an edge width commonly appears to be tens of meters (variations are discussed further in chapter 4). The widths measured for animals are, of course, affected by microclimate and vegetation. Nevertheless, an edge width determined by microclimate (and vegetation) is fundamentally different from one determined by the most sensitive species, which are often vertebrates of major conservation importance. The latter is a behaviorally determined or home-range-determined edge width. Yet, using this edge-width concept means that such sensitive interior species are restricted to interior areas much smaller than the patch as a whole. These species are of central conservation importance, and require much larger patches. More work is needed in this area.

The effect of mammalian and avian predators decreasing nesting success of tropical migrants in edges of temperate forest has generated much concern.[186,1891,38,32,643] The problem may be more severe in old edges of agricultural landscapes than in recent edges resulting from logging in forested landscapes.[1461] Finally, edges in grassland are commonly much wider than forest edges.[556,668,1774]

The variability of edge width progressively along a boundary apparently has not been measured in an ecological context. Yet wide and narrow portions of an edge are probably of considerable ecological importance. A dense patch of mantel or a gap in the mantel, e.g., from a path or tree fall, should produce high variability in microclimatic conditions and edge width.[668] In a patch with little interior a gap in the mantel could be particularly important. Also, in such a patch, where the diameter is slightly greater than twice the normal edge width, the interior habitat could easily disappear altogether.

Curvilinearity of boundaries

Overall curvilinearity

The history of civilization is a saga of linearization or geometrization of the land. The soft curves of nature have been replaced by the hard lines of humans. What are the ecological gains and losses from this seemingly inevitable process?

Curiously hardly any studies exist on the subject. To illustrate some dimensions of the pattern, bits of an ongoing project on boundaries between grassland (*Bouteloua–Artemisia*) and woodland (*Pinus–Juniperus*) in northern New Mexico (USA) will be introduced[503] (R. Forman, D. Smith, & S. Collinge in progress). The goals are to see whether curvilinearity has an effect on wildlife usage in an edge, movement along a boundary, or movement across a boundary. Tracks and scats (fecal droppings) of two large herbivores, elk and deer (*Cervus canadensis, Odocoileus hemionus*), are measured in the grassland edge along north-facing woodland boundaries.[660] Eight 150 m boundaries range in curvilinearity from 1.1 to 3.0 (nearly straight to 3× a straight line).

The following results emerged. (1) Usage by both species, based on both track and scat densities, increases with curvilinearity (Fig. 3.12).

106

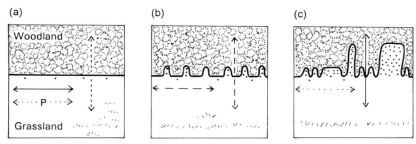

Fig. 3.12. Wildlife usage and movement relative to boundary curvilinearity. Woodland is pinyon–juniper and grassland is grama-sagebrush. Scattered dots represent usage by elk (*Cervus*) and mule deer (*Odocoileus*), based on track and scat densities. Solid arrows indicate much movement, dashed arrows intermediate, and dotted arrows little movement. P = predator movement (coyote, *Canis*). Summary patterns based on unpublished data of R. Forman, D. Smith, and S. Collinge from near Taos, New Mexico (USA).

Large coves in the convoluted boundaries are locations of high usage (usage here probably mainly reflects cover for favorable microclimate and predator avoidance). Movement along the boundaries decreases with curvilinearity. Animals move readily along straight boundaries. In convoluted boundaries animals neither move along the convolutions, nor from tip to tip of the woodland lobes. (3) Movement across the boundaries, in contrast, increases with curvilinearity. Straight boundaries act as partial barriers or filters. (4) Although too few predator tracks (coyote, *Canis latrans*) are encountered for statistical significance, the bulk of those found were proceeding along straight boundaries. These results suggest that boundary curvilinearity is important for wildlife usage and movement, and that further ecological analysis of the nature of curvilinearity is warranted.

Boundary surfaces

Examining the border of a patch usually reveals a sequence of relatively distinct 'boundary surfaces' or forms. These may be categorized in varying detail, combined in various ways, and at varying scales for a particular study. Nevertheless, eight surfaces seem especially common and usefully recognized at the landscape-scale:[503] cove, concave, straight, convex, lobe, corner, finely wavy, and coarsely wavy (Fig. 3.13). The first six surfaces are nearly smooth, and the corner is a specialized form of the lobe.

The sequence of these surfaces along a boundary determines overall curvilinearity. A regular curvilinear boundary is found where only one surface type is present, or two or more types are repeated, e.g., in scalloped, corrugated, crenellated, and dentate boundaries. A self-similar

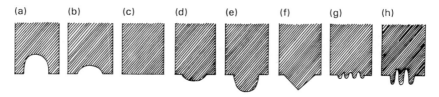

Fig. 3.13. Eight common boundary surfaces.

fractal border is also a regular boundary with repeated surfaces. In theory, boundary surfaces may also be randomly distributed.

However, probably most common in nature are clustered surfaces along a boundary. Thus, a series of coves and lobes may be followed by straight and corner surfaces, which are followed by a series of concavities. The many boundary surface types, and their different-sized aggregations, produce the irregularity in spatial pattern so prevalent in nature (chapter 13).

Lobe and cove functions

We illustrate the ecological roles of two key boundary surfaces, lobes and coves. A decrease in species richness from base to tip of lobes, usually called the *peninsula effect*, has been related to several possible mechanisms.[1562,1683,569,1141,501,1682] These studies show that species richness under certain conditions may also increase or be unchanged from base to tip of a lobe. The peninsula effect is probably common. However, before assuming the peninsula effect is present in a particular lobe, one should examine edge to interior patterns, habitat heterogeneity along the lobe (environment–distance hypothesis[1141]), and a suitable control. Despite the abundance of lobes at the scale of ecosystems in a landscape, we still have little evidence of the peninsula effect at this scale.

The functional significance of lobes promises to be much greater. A *funnel effect* is likely, whereby objects are concentrated and channeled through a narrow place such as a lobe. Thus some animals and humans may enter a patch preferentially at a lobe, while others may avoid a lobe in entering. The same hypotheses hold for leaving a patch. Note again that the patterns may differ for locomotion- versus mass-flow-driven movements. For example, in entering a patch, wind-transported particles are often funneled into coves, whereas some vertebrates head for lobes.

If you were to enter a spatial element such as a forest, meadow, or large building, which of the eight boundary surfaces of Fig. 3.13 would you head for, and which would you avoid? Apparently no studies exist, but presumably the results depend on the relative suitability of the systems entered and left. Perhaps herbivores and predators would respond differently. The question is relevant to all boundary crossing movements.

The tips of lobes are also likely to be of ecological significance. Their exposure on three sides to the surrounding matrix implies a distinct microenvironment and species composition.[1141] In addition, predators such as hawks and certain mammals may preferentially use these special locations for foraging.

Coves are also distinct microenvironments. A comparison of woody plant colonization opposite concave, convex, and straight boundaries on reclaimed surface mines finds much more revegetation opposite concave boundaries.[676] The highest stem densities and stems furthest from the forest border are recorded opposite coves. This is interpreted as a 'cove concentration effect', whereby the rate of change due to colonization is highest in coves.

The grassland-woodland study described earlier in this section found large coves to have many scats and tracks, whereas small and medium-sized coves have relatively little evidence of the herbivores. Large coves are both wide and deep, although potentially cove width and depth (length) could have quite different effects. Several aspects of coves may be important to herbivores. Forage is somewhat different in coves than elsewhere. Hiding or predator avoidance is typically enhanced. Finding a sunny or wind-free spot at a cold time is likely. And finding a shady or windy spot at a hot time is probable.

The functional roles of the other boundary-surface types deserve study. For instance, we may hypothesize movement along a straight surface[123], and a linkage between acute-angle boundaries and substrate.[620] The structure and function of boundary surfaces is useful in understanding patch shape in the following chapter 4.

A nice feature of boundaries is their ease in sculpting. It is relatively simple and cheap with cutting and planting to produce any of the boundary surfaces, or some overall level of curvilinearity desired. The wildlife movement results suggest that one could easily decrease movement to a location if overbrowsing were a problem. Alternatively, one can increase movement for wildlife viewing or other management objectives. The mine recolonization results suggest that one could accelerate revegetation in ecological restoration sites by managing boundary conditions.

MOVING BOUNDARIES

Boundaries are continuously battered by forces on opposing sides. In such a situation movement is the rule, and stasis the exception. In cellular physiology, membranes tend to move in the direction from higher to lower system heterogeneity, a principle that has analogues on land.[1044,1886,23,250,1620,503]

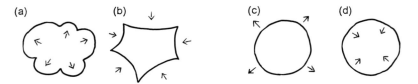

Fig. 3.14. Expanding and contracting boundaries. (a) and (b) Patch shapes that indicate probable direction of movement. (c) Remnant patch. (d) Disturbance patch.

Two types of relatively fixed-location boundaries are those coinciding with a substate discontinuity, and those maintained by humans. Yet in both cases pressure to expand in one direction is exerted by contrasting intensities of both abiotic and biotic forces, such as wind, fire, herbivory, and plant dispersal. Maintaining a fixed boundary, such as along a straight farm woodlot or forest road, against the tendency to move, ordinarily requires human or fossil fuel energy. In contrast, rapidly moving boundaries, such as a spreading pest outbreak or invasion of humans into a rainforest, emphasize the wide variation in rate of boundary movement.

Patterns of movement are equally diverse. In a general sense, boundaries could move like a series of concentric or parallel lines.[142,1606,312] However, since opposing forces are never exerted evenly along a boundary, it tends to change form. Lobes, coves, and other boundary surfaces appear and disappear.[1290,903] Typically, an expanding boundary forms lobes and convex surfaces, while a contracting boundary forms coves and concave surfaces[618,501] (Fig. 3.14 a and b). Changes in boundary form, and their mechanisms, are examined in the following chapter on patch shape.

Paleoecologists, rather than being old timers, are scientific detectives who work out the ecological patterns of pre-history. They have studied more than 20 millennia of vegetation and species movements (including *c*. 10 since the last ice age) based on pollen deposits and other evidence. Several generalizations are pinpointed relative to moving boundaries.[117,367,767,1848,1432,384] Species move 'individualistically' at different rates, rather than together as a vegetation unit of several predominant species. Vegetation units or communities exhibit identifiable boundaries, although the predominant species within them change over time. Both species and vegetation boundaries move in spurts rather than evenly. Where physiographic and vegetation boundaries coincide, vegetation continues to move, albeit more slowly. Species and vegetation boundaries move at different rates along different portions of a vegetation boundary.

The mechanims causing boundaries to move are essentially infinite and must therefore be grouped. The four types of patch based on origin (chapter 2), remnant, disturbance, environmental resource, and introduced, are useful here. The remnant patch tends to expand, and the disturbance patch contract, in both cases due to colonization of the disturbed area (Fig. 3.14c and d). The environmental resource patch has relatively stable boundaries. And boundaries of the introduced patch have varied trajectories, depending on human maintenance regimes.

Russian scientists have had a long interest in boundaries, and have studied boundary movement between forest and steppe, tundra, and bog.[1654,1674,1221,1330 cited in 46] Boundaries commonly alternate advancing and retreating (sometimes described as hysteresis).[261,1839,600,46,1580,47,46] Boundary movement has also been related to catastrophe theory[1341] and chaos theory.[1182,600] Studies of complex coastal boundaries indicate that at a minimum one must evaluate physical properties (e.g., wind and water), species biologies, and ecosystem processes as categories of mechanisms.[1392] As yet, no clear roadmap or synthesis exists for moving boundaries (see Box 3.2).

In summary, boundaries on land are strips of activity where interactions and flows are concentrated. Straight boundaries are the most biologically impoverished, whereas the lobes and coves in curvilinear boundaries exhibit several important functional attributes. Boundary width is usefully defined by differences in microclimate and vegetation, though animals exhibit a much wider range of sensitivity to edge conditions. Boundaries move at different rates and change form. They play important roles in environmental issues from forestry to aesthetics, soil erosion to global climate change, and park management to sustainability. Rich and relatively inexpensive opportunities exist for boundary sculpting and management.

111

Box 3.2: Climatic change: boundaries as 'canaries in the mine'?

The global atmosphere during the human industrial age has changed rapidly, including a 25% increase in CO_2, and a 0.5 °C increase in temperature. The temperature rise is one-third of that believed responsible for the spread and retreat of continental glaciers. If temperature continues to rise a few more degrees, some experts expect an acceleration in the spread of deserts, contraction of forests, translocation of agricultural land, extinction of species now isolated in fragmented patches, and inundation of coastal areas as coastlines move inland.

Boundaries are frequently suggested as sensitive indicators of climatic change.[1191,1593,403,742,1749] Which boundaries would you observe to detect human-caused climatic change quickly and surely?

(1) Fine-scale boundaries including woodlots and property lines? But changes in these mainly reflect human management rather than climate.

(2) Natural boundaries such as treelines on high mountains? But treelines fluctuate, gradually advancing upslope and rapidly retreating downslope due to periodic disturbance.[619,1923,1580]

(3) Biome boundaries primarily determined by climate? Yet if the rate of climatic change is years to decades, and the life span of predominant trees and clonal herbs is decades to centuries, a lag in boundary response is likely.[368,1847,1301,368,1198]

(4) Boundaries in hot arid landscapes where increasing temperature could be lethal to the relatively few dominants? But species in arid areas are there because thay have survived long droughts and stress periods.

Other candidates exist, each with pros and cons. (5) Boundaries containing the (physiological) range limits of many species, such as the northernmost or southernmost limits?[620,384] (6) Less abrupt boundaries perhaps indicating little integration or stability of the adjacent communities?[46] (7) Boundaries in hot arid landscapes where communities have little wood or soil organic matter as a buffer? (8) Vegetation boundaries coincident with steep, or alternatively with minimal, environmental gradients? (9) Land-water boundaries with significant ground-water–surface water exchange?[1183,741]

After choosing a boundary type what should be measured as most sensitive to climatic change? Would you look at attributes at several scales, plant or animal metabolism, herbivory, herbivore densities, predators, internal boundary anatomy, flows across a boundary, expansion or retreat, or changes in boundary form?

4

Patch shape

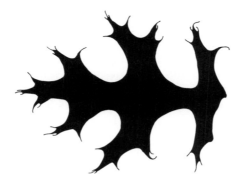

... this leaf reminds me of some fair wild island in the ocean, whose extensive coast, alternate rounded bays with smooth strands, and sharp-pointed rocky capes, mark it as fitted for the habitation of man, and destined to become a centre of civilization at last ... At sight of this leaf we are all mariners, – if not vikings, buccaneers, and filibusters. Both our love of repose and our spirit of adventure are addressed.

Henry David Thoreau, Autumnal Tints, *1862*

The shape of a political district in New York was once described by a critic as 'The Camel Biting the Tail of the Buffalo Which is Stepping on the Tail of the Dachshund'.[1756] While such a convoluted shape is exceptional in landscapes, some familiar simple forms like circles, diamonds, spirals, and stars are also scarce in most landscapes. Yet a set of common patch shapes does result from nature's curves and people's lines.

Patch size is of major ecological importance (chapter 2). However, patch shape is a much richer concept than size because shape varies in so many ways. *Shape* refers to the form of an area (two-dimensional),

113

as determined by variation in its margin or border. The objective of this chapter is to explore how the numerous shapes in a landscape affect ecological movements and flows.

Patch shape has many applications to society. Wildlife can hide in the convoluted margins of patches. The turning radius of a fire truck may determine the shape of a housing development (estate). The shape of logged clearcuts controls the reproductive success of forest trees. The optimum form of a natural area reserve may be round, or alternatively have major lobes. Principles of form and function permit us to predict the functioning of a patch by knowing only its shape. Indeed some shapes can indicate expanding, contracting, or stable patches.

The literature on the ecological effects of patch shape is sparse; this is a frontier area for research. A few reviews address common shapes and their causes[1381,1324], shape effects on ecological flows[501], and shape measurement.[529,1761,1153,1155,67,364] Three-dimensional shapes are discussed further in chapters 10 and 11.

This chapter begins by identifying the common shapes and their causes. Ecological movements and flows are then related to the key attributes of shapes. Next, we consider the ways of measuring shape. Finally, moving patches and three-dimensional patches are introduced.

COMMON AND UNCOMMON SHAPES

Common patch shapes

Holding the rope atop a cliff while a friend rappels to the bottom is a chance cautiously to examine the terrain below. Clearly the ecosystems on steep and moderate slopes differ from those in valleys, and on rolling or flat plains. One sees that the intensity of human activity differs as well. But what are the shapes in natural areas? How does human activity alter patch shape? Does shape vary independently of ecosystem or land-use type?

Partial answers emerge from a heterogeneous, glaciated 23 km² area near Poznan, Poland, where all patches (excluding a few built areas present) were measured and compared.[1324] Terrain ranged from plains (0–7°) to slopes (7–>15°); soils from permeable (poor and fertile) to peat; and vegetation and land uses from open wetland and agriculture to forest. Area, perimeter, long axis, and short axis were measured for each patch (geocomplex), and several indices of shape were calculated[785,1202,1421] (shape measures are presented later in this chapter).

Plains have the most compact-shaped patches (averaging all types), valleys are intermediate, and slopes have the most elongated patches[1324] (Fig. 4.1a). Patches in plains have the smoothest boundaries, and those on slopes are the most convoluted. Variability in shape from patch to

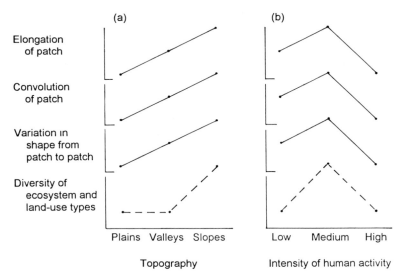

Fig. 4.1. Patch shape attributes relative to (a) topography and (b) human activity. Each curve is on a separate graph, with the vertical axis from low to high. Average values for patches of all types present are plotted. Based on Pietrzak (1989).

patch is least on plains, and greatest on slopes. Finer-scale differences within these topographic areas are also present. Hollows with inhibited drainage have mainly compact patches. Within valleys the higher unflooded portions have more elongated patches than in the floodplain bottoms. On slopes patches oriented up and down are the most elongated, whereas patches extending across the slope and on moderate slopes are less elongated.

Patch shape also correlates with the intensity of human activity in an interesting way.[386,37,654] In this Polish area the most natural patches (i.e., those with lowest human modification) are intermediate in elongation, convolution, and shape variance (Fig. 4.1b). A moderate increase in human activity correlates with maximum levels of elongation, convolution, and variance. But intense human activity correlates with the least elongated, convoluted, and variable patches.

The investigator suggests that the low shape values for the most natural, as well as the most human-impacted patches, represent equilibria. In both the natural and agricultural areas, compact shapes have resulted from diverse processes over a long period.[1324] In contrast, the high shape values for intermediate human disturbance result from the lack of an equilibrium. Land uses here have continually changed over time (in addition to the linearizing effect of slopes and streams). The high ecosys-

115

tem diversity at the intermediate human-disturbance level (Fig. 4.1b) is consistent with this interpretation.

Let us broaden the view to patches in all landscapes, from dry to forest and agriculture to urban. The infinite number of shapes on earth is reducible to a limited number of common ones. Three variables usefully differentiate the shapes: (1) natural versus human created (curvilinear or amoeboid versus geometric); (2) compact versus elongated (length to width ratio); and (3) rounded versus convoluted (number of major lobes present).

Differentiating natural from human-created patches is usually simple because of distinctive boundary forms (Fig. 4.2a and b). Overwhelmingly, boundaries of natural patches are curved, while human-created patches contain one or more straight lines. Although straight lines in natural land exist, e.g., some geologic faults, avalanche tracks, and vertebrate trails, the major processes of wind, water, and glacial flow mainly produce curved landscape patterns. Traditionally, humans often created curvilinear patterns.[654] Modern technology occasionally produces these forms, as by contour plowing, central-pivot irrigation, and cable logging from a point. But most human-created patches have predominantly straight lines, e.g., associated with plowing, drainage ditches, roads, and railroads. Indeed, 90° corners are especially clear evidence of human activity.

Patch *elongation* is effectively measured by the length-to-width ratio ($L{:}W$) (Fig. 4.2), i.e., the dimensions of the longest rectangle that inscribes or just encloses a patch. Elongation results from a unidirectional process. If a major controlling process, such as meandering of a stream or tractor movement in flat terrain, is not strongly unidirectional, or if several processes from different directions predominate, patches tend to be compact. Strong glacial movement or streamline airflow produces elongated patches.

The *convolution* of patches is unlimited in variety. However, an operationally simple measure of convolution is to count the number of major lobes present (Fig. 4.2). Convolution probably results from spatial or temporal heterogeneity in the controlling process(es). Thus, a many-lobed patch may result from wind speed and direction changes during a large fire[709,506] (number 13 in Fig. 4.2), or from leaving woods in the intersection area of farms with different farming objectives (number 24 in Fig. 4.2). A homogeneous or evenly expressed process generally produces a rounded or smooth patch, such as a circular village around a water hole or an elongated glacial lake.

Fig. 4.2 illustrates the common types of patch shape in all landscapes. Except in the lower right of each graph, intermediate shapes exist to fill all combinations of elongation and convolution. Several shapes warrant special comment or emphasis.

116

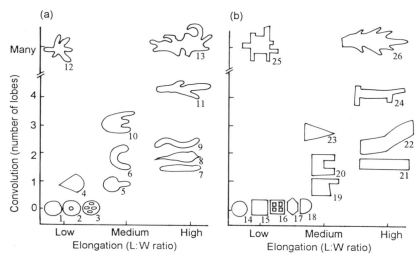

Fig. 4.2. Patch shapes in landscapes arranged by origin and form. (a) Natu-ral patches (curvilinear or amoeboid). (b) Human-created patches (geometric). Elongation and convolution are considered the first two major components of shape. Examples for each numbered shape are: (1) Bog, crater, cirque lake, raised spot in wetland; (2) Slope surrounding dry hilltop, wetland in karst area; (3) Gap dynamics or disturbance within patch; (4) Delta, alluvial fan; (5) Landslide, avalanche, woods that extend along stream; (6) Oxbow, barchan dune; (7) Glacial wetland or lake; (8) Dune, island in river; (9) Woods along stream or river; (10) Mountaintop with lava flows, headlands around fiord; (11) Woods along stream with tributaries; (12) Mountaintop vegetation extending along ridges, disturbance by mammal trampling around waterhole; (13) Fire disturbance, pest outbreak disturbance; (14) Village around well or fort, central pivot irrigation; (15) City block, logged clearcut in checkerboard forestry; (16) Woods with internal clearcuts; (17) Geographic pattern of land use surrounding a cen-tral village; (18) Farm pond with dam; (19) Woodlot in agricultural area, cemetery in city; (20) Woods in suburbia indented by housing development (estate); (21) Cultivated field; (22) Golf fairway, ski slope; (23) Field cut diagonally by later road; (24) Woodlot in the intersection area of several farms; (25) Town or city with development along transportation corridors; (26) Dammed up reservoir.

Round is essentially the only shape produced both naturally and by humans (Fig. 4.2; see numbers 1, 2, 3, 14). 'Amoeboid' may refer to any natural patch since an amoeba takes any of these external forms (Fig. 4.2a). 'Irregular' or 'amorphic' best refers to number 13 where no symmetry or organized structure is evident. Changing clusters in an animal herd are often amorphic or irregular (Fig. 4.3a). 'Stream-lined patches' with smooth outlines (numbers 7, 8, 9 in Fig. 4.2; Fig.

117

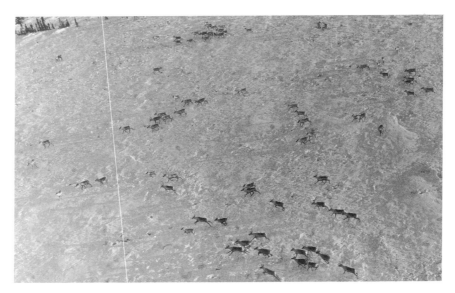

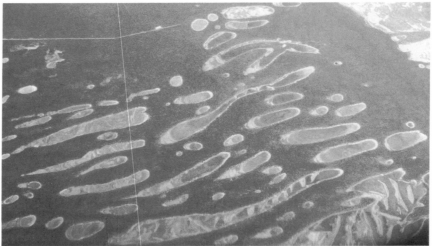

Fig. 4.3. Irregular ephemeral patches and streamlined persistent patches. (a) Caribou (*Rangifer arcticus*) in the Lake Clark area, Alaska. Courtesy of US Fish and Wildlife Service. (b) Salt lakes in flat land of western South Australia. R. Forman photo.

4.3b) also result from even horizontal wind or water flow in the absence of blunt or rough turbulence-causing objects (see chapter 6). 'Indented patches' commonly result either from wind or water flow in flat terrain (number 6 in Fig. 4.2; Fig. 4.3b), or from removal of a

118

portion along one side (number 20). Sponge-like or 'perforated patches' (Numbers 2, 3, 16) result from any of the patch-forming processes (chapter 2), and are variants of unperforated patches. Donut-
shaped or 'ring patches' (number 2) have a single perforation.
Sunburst-like or 'star-shaped patches' (number 12) appear most commonly with 3–6 lobes. In Fig. 4.2b a triangle has three lobes, whereas
a square and a hexagon have no lobes (alternatively one might consider a square to have four lobes).

Fractal shapes represent a special case. Self-similar fractal boundaries
are familiar and well-described in landscapes[1039,1138,1139] (chapter 3). Fractals in the general sense of perimeter-to-area ratio are also used for
patches.[1252,1744,1746] In a hierarchical sense a patch is composed of a cluster of smaller patches, each of which in turn is composed of a cluster of
still smaller patches.[453,482,1250] But it is unclear how similar the subsets of
patches are, or indeed have to be, to be called fractal.[1746,1416,1138] Dendritic
forms associated with erosion are the most likely candidates for 'fractal
patches'. An alternative interpretation, however, is that 'fractals are
everywhere'.

Uncommon shapes

Although unlimited shape possibilities exist, landscape patches represent
a limited subset. No randomly produced shapes exist. Patches with 5, 7,
9 . . . straight sides are rare, because geometric patches on land tend to
be composed of different-sized rectangles, but with few triangles. Spirals, such as in a nautilis or cone, are rare, and consequently earth
mounds (see Fig. 14.1) created by humans in this form are especially
striking from the air. Wave-shaped boundaries, like an octagon with concave boundaries, are rare. The manifold tile designs[628] produced by interlocking geometric building blocks are rare. The hourglass or dumbbell
shape, uncommon but present in many human-dominated landscapes,
apparently often results from indentations in the long sides of a
rectangle.

Hexagonal patches of a vegetation or land-use type are rare. An
extensive geography literature explores the pros and cons of interpreting landscape patterns as hexagons.[278, 985, 321, 646] Based on central
place theory and cell packing theory, a landscape may be a 'grid' of
hexagons (or Thiesson polygons). For example, a village is a central
place surrounded by a hexagonal zone of influence that competes or
intersects with hexagonal zones of influence of surrounding villages.
However, the pattern, actually a hierarchy of increasing-sized hexagons, is a model attempting to simulate reality to gain understanding.
Distinct hexagons are normally difficult to find from an airplane
window.

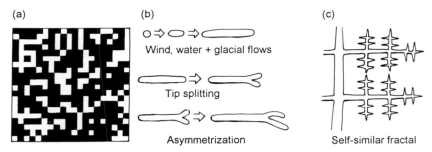

Fig. 4.4. Basic processes and forms produced. (a) Forty per cent of the cells of a 20×20 grid have been randomly selected and whitened.

CAUSES

Origins of shapes

Basic forces and energy-driven processes underlie the creation of shapes.[830,1294,406] Entropy leads to random shapes especially characterized by convoluted forms and rough perimeters (Fig. 4.4a) (also see percolation theory in chapter 8). Movements and flows, such as wind, water, glaciers, transportation, and plowing, tend to be unidirectional, and hence produce elongated forms. A nucleus, e.g., an isolated shrub or settlement, initially tends to form a circular patch around it. Directional forces then cause the patch gradually to elongate (Fig. 4.4b). 'Tip-splitting' tends to take place when a lobe expands in a surrounding medium that contains resistance.[1733,1215,1294,406] Asymmetry (anisotropy) typically develops.[1215] Self-similar fractal patterns[1039] (Fig. 4.4c; chapter 3) and other patterns sometimes develop.[406,1294] The net effect of these forces and patterns is to expect elongated convoluted shapes with rough perimeters. Where such processes are active, a high energy input must be required to maintain circular, lobeless, and smooth perimeter shapes.

Several causes or origins of shapes have been mentioned in the preceding section, including natural versus human mechanisms, wind and water flow on a horizontal surface, and removal of a portion along one side of a patch. A closer look is revealing.

Ephemeral patches, like moving herds of arctic or African mammals (Fig. 4.3a) or shadows from clouds, constantly change in shape. Such patches exhibit a much wider range of shapes than ecosystems or land uses in landscapes. Despite this range, the ephemeral patches lack essentially all the geometric shapes (Fig. 4.2b). What therefore are the major constraints on patch shape in the landscape?

Processes producing round, elongated, tip-split, and asymmetric patterns were just described (Fig. 4.4b) and operate at the landscape scale. Disturbances also normally move, but the rate and direction are uneven,

Fig. 4.5. Shapes of fields and housing developments (estates). Suburban housing organized by a road network is spreading over farmland. Newport News, Virginia (USA). Photo courtesy of US Fish and Wildlife Service.

resulting in elongated patches with convoluted margins. Water meandering across relatively flat terrain produces sinuous patches. Basins with a drainage network tend to have shapes between an oval and an elongated triangle. These are not included in Fig. 4.2 because the ridgeline surrounding a basin hardly ever coincides completely with a patch boundary. Many vegetation patches simply reflect topography produced by water erosion, wind deposition, and glaciation. These natural processes are constraints on patch shape, and doubtless others exist.[1614]

Constraints on the shapes of human-created patches are equally evident. Railroads and powerlines are relatively straight and long, providing a coarse-mesh imprint in which elongated patches predominate. Road networks impose an imprint that is rectilinear and sometimes grid-like in flat areas (Fig. 4.5). These networks in areas of topographic relief are less rectilinear and more variable in mesh size. Patches therefore are

121

more rectangular in flat than hilly areas. Square patches are usually uncommon except where newly formed (e.g., an initial clearcut or cultivated field in an area), or as part of a grid (as in a city).

Rectangular patches that have persisted over time may tend to have an equilibrium $L{:}W$ ratio between 1.5:1 and 4:1 (fields in Fig. 4.5). Longer narrow rectangles often result from longitudinal subdivision of patches perpendicular to access roads or canals, such as land inheritance patterns common in parts of central Europe, Quebec, and New Mexico. Cultivated fields tend to be rectangles because of unidirectional plowing, or composed of different-sized rectangles, resulting from additions or subtractions from an original field. Although hedgerows separate fields, they basically follow the patterns imposed by fields. A distinctive example of the 4:1 L:W ratio is the lakes constructed in the 9th century on a plain near Angkor Wat in Cambodia (Kampuchea).[1630] The lakes were parallel to the contours on gentle slopes, and water flowed by gravity in ditches from lake to lake. Each rectangular lake was 3.8 km (2.5 mi) long, and surrounded by four equal-height dikes. If square lakes of equal area had been constructed, the downslope dikes would have to have been much thicker and higher. If long narrow lakes of equal area had been built, a much greater total length of dikes would have been required for construction and maintenance. The 4:1 L:W ratio was attuned to the terrain and hydrology.

A square or low $L{:}W$ shape, in contrast, often forms at road intersections, as seen for cities, towns, villages, and isolated homes. These patches are organized around a central nucleus, such as a water source, commercial center, or bounded area for protection. Geomorphic forms, such as a ridge or lake margin, commonly interact with the central nucleus to produce elongated built patches. These typically have a $L{:}W$ ratio of 2:1 to 4:1.[702,646] Nineteenth-century New England farmers (USA) often maintained both small square fields to control livestock, and larger rectangular fields for cultivation (in addition to forest and grazing land).[691]

'Mixed shapes' describe cases where a portion of a patch is created by one mechanism, e.g., landform or water flow, and another portion by a second mechanism, such as fire or road construction. Mixed shapes may combine both natural and human mechanisms. The more lobes present (upper part of Fig. 4.2), the more likely that two or more mechanisms are controlling shape in different portions of the patch. Heterogeneity in the substrate or energy flow (e.g., turbulence) leads to convoluted patches.

Obviously no patch stands alone. Each patch boundary is a mirror image of the adjacent ecosystem boundary. An embedded patch is surrounded by a single ecosystem or matrix type. But most patches are surrounded by 2–5 adjacent patches (Fig. 4.5), and a few by more than

20 patches[1202,251]. If a patch expands, it usually does so at different rates along its boundary.[1733,1215,676] Thus, a cove will fill in and make the overall boundary smoother. Or a lobe will expand into the surrounding matrix making the boundary more convoluted. Conversely, when the matrix expands, the patch may lose a lobe or become indented. In effect, shape results from a balance between forces internal to a patch and external to it in the adjacent matrix.[1215]

The world is composed of patches within patches, as considered in scale and hierarchy theory. Sponge-like or perforated patches such as a gap dynamic mosaic (chapter 2) or a golf course are examples (Fig. 4.2; also see Fig. 1.4). In the gap dynamics case a single process, death of old trees, operating at a fine scale perforates the patch. In the golf course case a single process, fairway construction and maintenance, operating at a broader scale, produces similar-shaped patches within the overall patch. Different patch-causing processes at different scales are normal.[476,1250,701] More typical than the golf course example is where internal patches are variable in cause and shape. For example, in mountainous landscapes of New England, USA, large patches of different shapes are created by elevation and direction of exposure.[1410] Patches within these are caused by landslides, impeded drainage, beaver (*Castor*) activity, horizontal lines of conifer (*Abies*) mortality (fir waves)[1606], and less-understood processes.[1410] These diverse processes produce variable shapes, from round to convoluted or highly elongated.

Classification approaches for patch shape

Mentally, we classify shapes into groups, each similar to a familiar regular shape or model. Thus, actual shapes are usually, e.g., somewhat elliptical, approximately square, or nearly star-shaped.

The range of shapes on land has been organized or classified based on natural versus human-created, elongation, and convolution (Fig. 4.2). This is done to combine visible form with causative mechanism, and to be useful in understanding ecological functions.

Alternative organizing principles exist. Drawing on the phylogenetic approach used for plants and animals, we can focus on the developmental sequence of a patch over time. What was the previous shape from which each existing shape developed? But alas, a 2:1 rectangle could come from a square by doubling the area in one direction, from a 4:1 rectangle by halving the area, or even from an ellipse by straightening the boundaries. A star-shaped patch could develop by expansion from a small circle or contraction of a large triangle. This is unsatisfying.

Vegetated patch shapes could also be organized by substrate template, and secondarily by human modification. This is appealing because the bulk of the land is usefully differentiated into three geomorphic categories[610] (chapter 9): (a) erosional (by water), (b) aeolian (by wind), and (c) glacial. *Erosional landscapes*, e.g., in the upper Amazon basin, southeastern USA, and southern Europe, are produced mainly by rapidly flowing water, and have dendritic imprints predominant. (In large dry valleys water-depositional landscapes with irregular, reversed dendritic patterns may be considered a special case.) Wind-depositional or *aeolian landscapes*, e.g., in northern China, northern Africa, and covering a quarter of the USA, typically have flat areas, repeated parallel dune forms and near-vertical (loess) slopes. *Glacial landscapes*, as in Canada, Siberia, and the southern Andes, have smooth, elongated parallel forms with poor drainage patterns. The distribution of soil types results from these same processes but at a finer scale, and vegetation patches typically reflect the landform and soil type distributions. Perhaps the major drawback to a patch classification based primarily on substrate is that round, elongated, or convoluted patches produced in different landscapes would be grouped separately, even though the forms are identical. If ecological function is controlled by form, it seems more useful for similar forms to be grouped together.

Shapes could also be organized more broadly by any causative mechanism. Thus, all shapes produced by logging would be grouped together, and likewise, those by wind deposition, water flow, or cultivation. This places a wide range of dissimilar forms together in most groups, and seems unsatisfying. Therefore, the classification or grouping based on combining visible shape and causative mechanism (Fig. 4.2) appears most useful, at least ecologically.

SHAPE ATTRIBUTES AND ECOLOGICAL FLOWS

Shapes are easy, even fun, to recognize and compare. Art, engineering, and finding a mate often depend on them for success. Yet shapes seem to be equally important ecologically, especially in affecting movements and flows. The ecological functions of patch shapes are conveniently grouped by four major, easily-measured attributes of shape, (a) elongation, (b) convolution, (c) interior, and (d) perimeter.

Three *form-and-function principles* developed for animals and plants and in other disciplines provide a lucid foundation for the subject.[1707,1340,686] (1) *Compact forms are effective in conserving resources*. They protect internal resources against detrimental effects of the surroundings. In compact forms military groups minimize their exposed perimeter to the enemy, and arctic hares (*Lepus arcticus*) with short ears minimize exposure to the surrounding cold and wind. (2) *Convoluted forms are effective in enhancing interactions with the surroundings*. A long common boundary between two objects provides a greater probability per unit area of movements across the boundary, either in one or both directions. A woods with several arms has considerable wildlife movement in and out, and a building with several wings exchanges lots of energy with the surroundings. (3) In addition to active interaction with the surroundings, *network or labyrinthine forms tend to have a conduit system for transport*. A road network, stream system, or telephone system transmits objects or information from one portion of the network to a different portion.

Elongation of a patch

An elongated patch therefore is less effective in conserving internal resources than a round patch. This is widely considered to be true for protecting interior species, as well as species requiring remoteness from human activity (chapter 2). Assuming substrate or habitat homogeneity, a compact patch should contain higher species richness than an elongated patch with its fewer interior species. This is because increasing interior area adds species at a greater rate than increasing edge area[485] (see Fig. 2.8a). For similar reasons a perforated patch (Fig. 4.2) may have little or no interior area. Nevertheless, as yet evidence for these hypotheses is limited[462,540,140,1685,1892,208,213,634,924] (also see shape measurements later in this chapter).

Elongated plots are more efficient than circular plots for sampling a natural community, agricultural field, or forest for timber.[282,824,827,694,1295,150] This is because elongated plots include more environmental heterogeneity or habitat diversity, and are more similar to one another than are round plots. For the same reason sampling efficiency is also greater if the plots are oriented up and down a slope.[694,150]

The orientation of the long axis of the patch relative to flows in the landscape, i.e., the *orientation angle*[501,1742,634], is a key to several ecological phenomena. A compact patch causes turbulence of wind and water flowing past it, whereas elongated forms in the direction of flow tend to be streamlined, causing little turbulence (chapter 6). An area of repeated turbulence adjacent to a patch is apt to differ noticeably in vegetation from the surrounding matrix area.

Wind erosion of cultivated fields is also affected strongly by elongation, depending, however, on field orientation relative to wind direction.[1571,1570] A field oriented at 45° to the wind has only slightly more erosion potential than one perpendicular (90°) to the wind (Fig. 4.6a). However, a field aligned close to (15°) or parallel to (0°) the predominant wind direction has very high erosion potential. Farmers instinctively know this sharply increasing danger in the orientation of fields.

The orientation angle has been hypothesized to be important in species colonization of islands.[540,140,208,1681,1682] On land one study compared birds in north-south versus east-west oriented woods mainly surrounded

(a) Wind erosion potential in fields

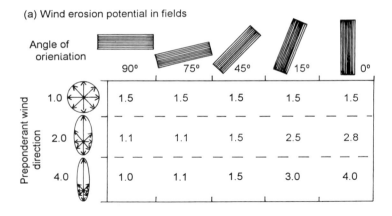

(b) Resident and migrant birds in woods

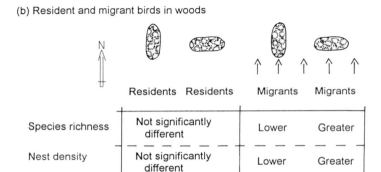

Fig. 4.6. Orientation angle effect on soil erosion and migrant birds. (a) Fields have a 4:1 L:W ratio. Preponderant wind direction is indicated by the longest arrow in a 'wind rose', where arrow length is proportional to amount of wind. Numbers in the table indicate the potential for wind erosion of a field, expressed in relative units that are proportional to the amount and length of airflow on field surface (wind-erosion direction factor). Based on Skidmore (1987). (b) Cavity-nesting birds are sampled in cottonwood (*Populus*) stands (average 7.2 hectare with 3:1 L:W ratio) in Wyoming (USA). Six resident and four migrant species are present. Based on Gutzwiller & Anderson (1987, 1992).

by grassland near the Rocky Mountains.[633,634] Resident bird species used patches of both orientations equally, presumably reflecting their familiarity with the patches (Fig. 4.6b). In contrast, migratory birds exhibit both greater species richness and greater nest abundance in patches oriented perpendicular, rather than parallel, to migration direction. Thus, orientation angle of a patch appears important to these species migrating from afar.

Movement of terrestrial vertebrates, including people, can be considered more broadly relative to orientation angle.[860,1903,1122,75,1852,1508,1032,23,1613] A patch perpendicular to and centered on an animal's route presumably slows movement down, because the animal must cross two boundaries in a short distance. (Exercising caution in approaching a new habitat, animals apparently slow down while crossing most boundaries.) Alternatively, a long detour can be made around the end of the patch. Overall, perpendicularly-oriented patches are more likely to change the direction and lengthen the route of animal movements. In contrast, a patch oriented parallel to an animal's route may enhance movement rate. The animal may enter the patch, be channeled to the far end, and exit. Or it may make a slight detour and be channeled along one side or the other of the elongated patch, without crossing a boundary.

Convolution of a patch

Convoluted patches have a long perimeter and normally abundant exchanges with the matrix.[1202,1514] The number of major lobes is a useful index of the degree of convolution in gross shape of the patch, as distinct from the finer-scale roughness of the perimeter. A convoluted patch causes complex patterns of turbulence in water and wind flow. Most turbulence will be on the open side of a boundary, that is, on the outside of a remnant patch, and the inside of a disturbance patch (chapter 2). The turbulence may create considerable microheterogeneity in soil, water, vegetation, and fauna surrounding a remnant patch (Fig. 4.7a).

A structural effect of a multi-lobed patch is the subdivision of certain populations into partially separated subpopulations. Individuals in the subpopulations are expected to interbreed more within a lobe than between lobes (Fig. 4.7b). This pattern of restricted gene flow is likely to produce more genetic variation for these populations in a convoluted patch than in a round patch where gene flow is less restricted. In an analogous manner, disturbances such as fire and pests are less likely to cover an entire patch if it is lobed rather than round.

Major lobes apparently exhibit both structural and functional ecological characteristics. As described in chapter 3, the peninsula effect refers to the pattern of species richness change proceeding from the 'mainland' to the tip of the lobe.[1683,569,233,1682,1141] If the lobe is wide enough to contain interior conditions in the center, a progressive loss of interior species from base to tip (which is all edge) can be expected. This is probably because either the lobe narrows toward the tip,[766] or certain populations drop out as they become progressively more isolated (Fig. 4.7c). Edge species may also drop out because of conditions associated with isolation. The net effect of these factors is a decrease in species number

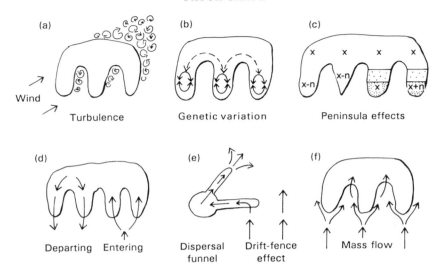

Fig. 4.7. Ecological effects of lobes and patch convolution. (a) Approximate zones of high turbulence in open matrix next to wooded patch. (b) Solid arrow = more gene flow; dashed arrow = less gene flow. (c) x = species number in large upper portion of patch; n = species lost or gained in proceeding progressively to tip of lobe. Shaded area = different microhabitat. (d) and (e) Arrows indicate directions of animal movement, where patch is more suitable than matrix. (f) Arrows indicate directions of objects carried by wind or water from an open matrix to a wooded patch.

from lobe base to tip. Changes in habitat type and/or habitat diversity along the lobe, of course, can accentuate the drop in species richness. Alternatively they may produce a constant species number, or even produce an increasing species number[1141] (Fig. 4.7c).

Animal movement by locomotion must be strongly affected by lobes, though published evidence is scarce.[1903,75,1508,1032,1613,1614] For example, an animal in a patch that must cross a less-suitable matrix is likely to preferentially head out into a lobe, and depart at or near its tip (Fig. 4.7d). Similarly, we may hypothesize that an animal in the matrix heading for a patch will also enter at or near the tip of the lobe, i.e., the nearest point. The channeling of animals from a wide area into a narrow place, such as a lobe, is a *funnel effect*[501] (Fig. 4.7e). A 'dispersal-funnel effect' of channeling animals outward from a patch to the matrix, using lobes, has been incorporated in game reserve planning. Here, the goal of the game reserve is to produce an excess of animals that will disperse into the surroundings for hunters.[943] The same logic applies to nature reserves, where rapid recolonization of surrounding patches, following local extinctions, is important.

The *drift-fence effect* is a variant of this, where a lobe intercepts species moving across a landscape, and channels them inward to the core of the patch[1308] (Fig. 4.7e). In addition to interior species, exotic and pest species may be channeled to the core.[462,540,24] Nevertheless, the drift-fence effect can be incorporated into nature reserve and forestry design to increase the rate of recolonization following local extinctions.[140,1308]

Movements by mass flow presumably are also strongly affected by lobes.[1570,174] Objects in an open matrix carried by wind and water toward a wooded lobe are likely to enter the patch in an adjacent cove, rather than through the tip of the lobe (Fig. 4.7f). This results from the splitting of streamlines by the lobe, combined with acceleration due to the Venturi effect as the cove narrows to its end. All of the ecological effects of lobes may increase with lobe length.

Coves involve the same types of issue and ecological function identified for lobes. As described in chapter 3 for wildlife in New Mexico (USA), coves also offer especially suitable conditions for many animals.[502] Cover on three sides enhances predator avoidance. Forage may differ due to unique microclimatic conditions. During a hot time one side is likely to offer shade. During a windy time one side is usually protected. An hourglass-shaped patch adds the advantage of being able to easily cross from one cove to another on the opposite side. An analogue to these cove effects for animals is the 'hurricane hole' so crucial to Caribbean pirates of past years. They sailed tall ships sometimes overloaded with gold. But during hurricane season pirates rarely strayed far from the few islands with well-known highly protected coves.

Interior of a patch

Large patches contain interior conditions, as well as interior species absent in small patches (chapters 2 and 3). Elongated and convoluted patches should normally have fewer interior species than compact patches of the same size. If edge width is constant in a patch, a line paralleling the patch boundary separates the interior from the edge. However, edge width varies in a patch (chapter 3) according to wind direction, sun direction, and internal edge structure (Fig. 4.8a). In addition, due to external inputs, edges on opposite sides of a lobe may coalesce to become unusually wide. This coalescence eliminates interior conditions in the lobe altogether (Fig. 4.8a). In like manner edges may widen near right-angle corners of geometric patches. In general, an interior area delimited by a line paralleling the patch boundary is usually a good approximation for compact patches. But such a line is a poor representation of the interior area in elongated and convoluted patches (Fig. 4.2) (see chapter 4 appendix for a method of calculating interior area for different shapes).

The core of a patch has sometimes been used in a general sense to indicate the interior or center of a patch.[1685,924] However, a concept different from the patch interior, and close to the dictionary sense of the cen-

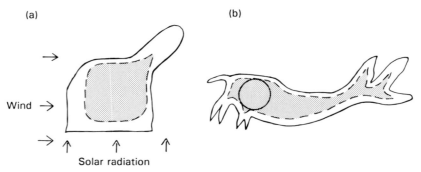

Fig. 4.8. Interior and core of patches. (a) Dashed line encloses patch interior, and (b) circle indicates core.

tral part of an interior or the center-of-gravity area[67], appears ecologically useful. Such a concept can draw on a method long used by geographers. Thus the *core* of a patch is the area of the largest circle that fits within the patch interior (Fig. 4.8b). This largest-circle-fit technique has been used to exclude lobes in comparing the sizes of patches, hence focusing on the undissected portion that retains its integrity.[646]

Ecologically, the core represents the non-convoluted, generally central portion of a patch. This is the critical area for a species requiring remoteness from the surroundings. Home ranges and territories of vertebrates are usually described as ovoid to elliptical.[799,882,321] These forms may be more closely correlated with patch core than interior, which can be highly convoluted. In a round patch, if edge width were hypothetically constant, core and interior areas would be identical.

Perimeter of a patch

Boundary surfaces exhibit infinite variety. Just as major lobes describe patch convolution, little lobes may describe the roughness of the perimeter at a finer scale (Fig. 4.9). The perimeter typically has many more, even orders of magnitude more, little lobes and coves. Replacing the major lobes are single and double teeth, crenellations, corrugations, scalloping, waviness, as well as irregular protuberances and indentations. Instead of whole patch sides being straight, tiny straight sections may compose the sides of teeth.

Probably all the ecological effects of shape described in the preceding sections also apply at this fine scale. But more mechanisms cause the fine-scale patterns. For example, the asexual spread of plants (Fig. 4.9), wind turbulence, and a fallen tree may be more important than bulldozers, farm tractors, and mountain slopes. Yet even the large processes produce complex fine-scale effects on boundary form. Microtopographic

130

Fig. 4.9. Fine-scale roughness of patch perimeter. Palmetto-dominated scrub (*Serenoa repens*) next to sandy margin of ephemeral pond. Armadillos (*Dasypus*) regularly forage into the coves between scrub lobes, whereas bobcat, raccoons, and deer (*Lynx, Procyon, Odocoileus*) normally avoid the fine-scale coves. Archbold Biological Station, southcentral Florida. R. Forman photo.

relief, miniature eddies in flowing wind, and tiny variations in fire intensity are universal. To link perimeter form with ecological function, at a minimum we must examine total perimeter length, plus the proportion composed of straight, smooth convex, smooth concave, and rough boundaries.

Although patch area is often emphasized in landscape ecology, several ecological characteristics correlate better with perimeter length. Populations of edge species, fire ignition, tree blowdown probability, and other factors were considered in chapter 3. In one study the perimeter length of suburban woods is considered a better predictor of avian richness than area, perhaps in part because relatively few interior species were present.[601] And small-mammal density (pika, *Ochotona*) in rockslides in Idaho (USA) mountains best correlates with the perimeter length.[220] The convoluted shape of the habitat benefits these animals that harvest resources in the adjacent matrix, a nice illustration of a form-and-function principle.

Linking ecological characteristics to elongation, convolution, interior, and perimeter attributes is a basis for determining the 'opti-

131

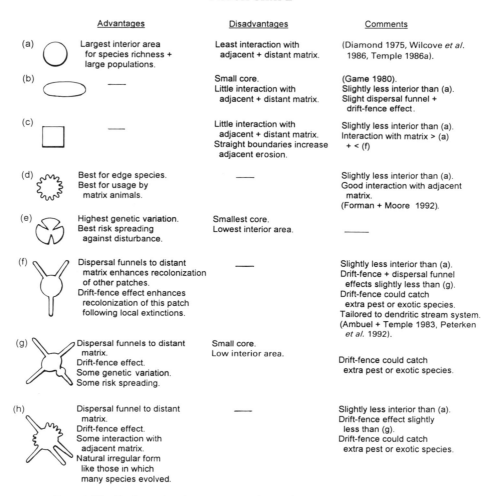

	Advantages	Disadvantages	Comments
(a)	Largest interior area for species richness + large populations.	Least interaction with adjacent + distant matrix.	(Diamond 1975, Wilcove *et al.* 1986, Temple 1986a).
(b)	——	Small core. Little interaction with adjacent + distant matrix.	(Game 1980). Slightly less interior than (a). Slight dispersal funnel + drift-fence effect.
(c)	——	Little interaction with adjacent + distant matrix. Straight boundaries increase adjacent erosion.	Slightly less interior than (a). Interaction with matrix > (a) + < (f)
(d)	Best for edge species. Best for usage by matrix animals.	——	Slightly less interior than (a). Good interaction with adjacent matrix. (Forman + Moore 1992).
(e)	Highest genetic variation. Best risk spreading against disturbance.	Smallest core. Lowest interior area.	——
(f)	Dispersal funnels to distant matrix enhances recolonization of other patches. Drift-fence effect enhances recolonization of this patch following local extinctions.	——	Slightly less interior than (a). Drift-fence + dispersal funnel effects slightly less than (g). Drift-fence could catch extra pest or exotic species. Tailored to dendritic stream system. (Ambuel + Temple 1983, Peterken *et al.* 1992).
(g)	Dispersal funnels to distant matrix. Drift-fence effect. Some genetic variation. Some risk spreading.	Small core. Low interior area.	Drift-fence could catch extra pest or exotic species.
(h)	Dispersal funnel to distant matrix. Drift-fence effect. Some interaction with adjacent matrix. Natural irregular form like those in which many species evolved.	——	Slightly less interior than (a). Drift-fence effect slightly less than (g). Drift-fence could catch extra pest or exotic species.

Fig. 4.10. Ecological advantages and disadvantages of various patch shapes. Characteristics listed under comments are considered to be minor.

mum shape' of a patch (see Box 4.1). The literature currently appears unanimous that a round patch is the ecologically optimum shape. Yet, a patch performs a number of key functions. Consequently, we may hypothesize that a shape something like the last one in Fig. 4.10 (spaceship shape) is ecologically optimum, i.e., the best balance among pros and cons.

Box 4.1: The optimum patch shape

The most common field shape in many landscapes is a rectangle with a L:W ratio between 1.5:1 and 4:1. Is that the optimum shape for cultivation or livestock? What are the optimum shapes for housing developments, parks, logged clearcuts, towns, and nature reserves?

Several options for the optimum shape of a large natural-vegetation patch are listed opposite (Fig. 4.10). The assumptions are: (a) area is constant in all designs; (b) most of the interior species (for this patch type) in the landscape could live in the round patch; (c) the lobes and coves are narrow and composed only of edge effect (i.e., no interior area); and (d) only ecological considerations are given, therefore excluding maintenance costs, aesthetics, recreation, and so forth.

Two of the shapes (Fig. 4.10e and g) exhibit risk spreading and genetic heterogeneity (desirable), but at the expense of having a small core and low interior area (undesirable). Finally, attempting to identify an ecologically optimum shape does not mean that all patches should be that shape. Indeed, attempting to determine the optimum shapes for a group of patches is a different, and much more complex question.

HOW TO MEASURE SHAPE
Key measurements for ecological effects

Elongation

Patch elongation is described by the length to width ratio. Length and width are the dimensions of the maximum-length rectangle that inscribes or just encloses a patch.

Convolution

Patch convolution is described by the number of *major lobes*. A projection or lobe must be scaled relative to patch size. To objectively identify lobes, it may be operationally useful to consider a lobe to be longer than the radius of the largest circle that fits within a patch (Fig. 4.11a). For example, a cross-shaped patch with projections each $\frac{1}{3}$ the length of the patch has four lobes. But if projections are only $\frac{1}{5}$ the length, the patch has no lobes. (Similarly, a rectangle with a *L:W* ratio of 1.5:1 has no lobes, whereas a 3:1 rectangle has two lobes; Fig. 4.2.) It may also be operationally useful to establish a minimum length for major lobes, e.g., 50, 100 or 1000 m, scaled to the study of interest. Otherwise on the occasional huge patch, what normally are major lobes would be considered only perimeter roughness (Fig. 4.11b).

133

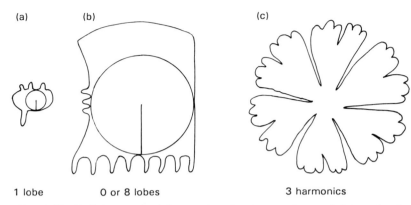

(a) (b) (c)

1 lobe   0 or 8 lobes   3 harmonics

Fig. 4.11. Defining major lobes and perimeter roughness. (a) Length of projection relative to radius. (b) No lobes if length is only relative to radius; 8 lobes if a minimum length is established (here the radius in the previous patch). (c) One harmonic for major lobes, plus two harmonics for perimeter roughness. See text.

For each major lobe it is suggested that at least the maximum straight-line length, plus the average width, should be recorded in ecological studies. The same information for major coves should also be ecologically useful.

Interior

The interior area of a patch is the total patch area minus the area of edge effect. This requires measuring the total patch area, and often an average edge width in a particular landscape can be used for calculation. However, variations in edge width are significant in some studies and applications. As noted above, the core area is measured as the largest circle that fits within the patch interior.

Perimeter

Total perimeter length of a patch is readily measured using a ruler length appropriate to the study or application (chapter 3). In addition, the percentage of the perimeter that is straight, smooth convex, smooth concave, and rough, is probably useful for ecological interpretation (Fig. 4.11b).

These measures are chosen to be simple, widely applicable in landscapes, and directly correlated with, or expressing one or more of, the ecological effects presented in the preceding section. As we understand more about the ecological effects of shape, other measures will doubtless be useful. For example, perimeter roughness could be assayed by the degree of regularity in alternating lobes and coves, number of angular points or inflection points per unit length, or acute (<90°), right (90°), and obtuse (>90°) angles in geometric shapes. In addition, a rich geography literature on measuring patch shape is encapsulated below.

Additional methods

No single measurement or index of shape can unambiguously differentiate all shapes.[935,615,616] In other words, a particular index number can result from many shapes, and it is not possible to reconstruct a shape from this measure alone. Therefore it is useful to choose the components of shape that are of most ecological interest, and select an index that will differentiate the forms for each component. Each component of shape could be linked in a linear equation[442], although that would mask the separate information potentially useful in understanding ecological function.[1823]

A desirable index of patch shape should (a) be simple to calculate, (b) work over the whole domain of interest (e.g., sand grains, cities, or landscape patches), (c) unambiguously and quantitatively differentiate shapes, and (d) permit a shape to be drawn based on the index number.[1636] Myriad indices of shape exist, because it is important in many domains or disciplines. Focusing this discussion around three phenomena, elongation, convolution, and roughness, reflects the desire for ecologically useful measures at this early stage in landscape ecology, rather than attempting to identify a universal measure of shape.[935,156,616] Frolov (1975) came to essentially the same three central characteristics of shape, calling them compactness, dissection, and indentation.[1155]

Before addressing the three central characteristics, let us consider 'compactness' since so many indices focus on it. More than a dozen indices use area and/or perimeter measurements to indicate how much a shape deviates from a circle[364,1287,294,1569] (also see chapter 4 appendix for three examples). Unfortunately none can indicate whether the deviation results from elongation, convolution, roughness, or some combination thereof. Another, the mean radius index, does not work with some convoluted shapes[1761] (see chapter 4 appendix). More importantly, in view of the basic forces producing shapes (discussed earlier in this chapter[799,882], and the predominance of convolution and roughness, using a circle as the control may be less appropriate than using an ellipse, rectangle, or lobed form as the control for landscape patches.

Elongation is the easiest to measure. The $L{:}W$ ratio measures only elongation[364] (three indices in chapter 4 appendix). Elongation indices based on patch length and area are less-precise indices, because area can vary in ways unrelated to elongation (two indices in chapter 4 appendix).

Several shape indices are based on perimeter length, but these cannot differentiate roughness from elongation and convolution[140,364,213] (two examples in chapter 4 appendix).

Convolution and roughness are considered together because one measure, a Fourier series or *harmonic analysis*, may be useful for both.[1154,1155] For this the form of a perimeter is modeled or represented as a sine wave, like the perfectly symmetrical squiggles of a snake. The sine wave has a *frequency* (distance between lobes, or between coves), and an *amplitude* (length of lobes or coves). A 'harmonic' is the frequency or wave length of the sine wave, just as on a radio dial or for light.

To illustrate, for a patch with smooth major lobes like number 12 in Fig. 4.2, count the lobes and determine their average length. Using this information draw the harmonic model of the patch with a symmetrical sine-wave margin.

Similarly, a complex perimeter with major lobes containing small lobes which have tiny lobes and so on can be modeled as a sum of several harmonics. Each harmonic represents a wave length of lobes, and the Fourier series represents the sequence of harmonics measuring increasingly small lobes (Fig. 4.11c). In this manner a perimeter form is expressed (in sines and cosines) as a series of harmonics. The variances between actual lobes and these sine waves are readily determined. For some types of object perimeter differences are extremely subtle, and many harmonics may be required to differentiate them. But probably the first two or three harmonics differentiate virtually all patches of a landscape.

In short, the major lobes of a convoluted landscape patch determine the first harmonic (and occasionally the second when the largest projections are branched into major lobes) (Fig. 4.11). Subsequent harmonics differentiate patches according to perimeter roughness.

For example, the 'giant amoebas' in Fig. 4.12 are clearings logged to provide enhanced wildlife habitat and visual diversity. Elongation (*L:W* ratio) helps to differentiate them, but is insufficient. Convolution (number of major lobes) also helps but is insufficient. A harmonic analysis would measure the wave length and amplitude of major lobes around the boundary of a patch, plus the same attributes for secondary lobes. This simple information would differentiate all the logged clearings from each other. Moreover, it would permit one to draw a symmetrical representation or model shape for each patch. The drawing would have the correct number of major and secondary lobes, plus their average length.

The harmonic of each convoluted patch could also be compared against any regular non-elongated form, such as circle or star. Similarly, the perimeter-roughness harmonics could be compared against any smooth form, such as elliptic or serpentine. Harmonic or Fourier anal-

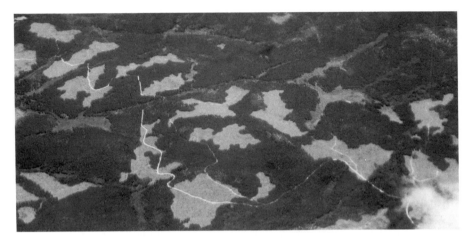

Fig. 4.12. Amoeboid clearcut patches with distinctive lobing patterns differentiable by harmonic analysis. Lodgepole pine forest (*Pinus contorta*), Medicine Bow National Forest, Wyoming (USA). R. Forman photo.

ysis does not solve all shape problems[1072,1821,287,1155,615], but can separate, for example, a one-humped from a two-humped camel, five-pointed forms with concave versus straight sides, a rectangle from an hourglass shape, different cauliflower heads, and serpentine dragons in Beijing's Forbidden City.[1154,1155]

Several additional approaches exist for measuring patch shape.[1636,442,156,323,1761,1168,615,67,364] 'Geometric-figure-fitting' compares various regular polygons (or other defined shapes including circles) with patches to identify the closest fit.[1636,219,67] 'Set-theoretic measures' determine the deviations of shapes in a set or group from a defined shape, such as a circle or ellipse.[1636,935,67] 'Mean-side' and 'variance-of-side' indices define shape by the lengths of sides of polygons.[646,364] The 'sinuosity ratio' is usable for serpentine forms.[610,1761] 'Center-of-gravity' or 'moment measures', e.g., the mean radius index (chapter 4 appendix), determine distances from a patch center as in central place theory.[165,1062,529,67] No indices listed in the chapter appendix differentiate perforated from non-perforated shapes.

Patch shape is of importance in numerous domains and disciplines. Examples are tiles[628], oceanic atolls[1636], lakes[294], sediment grains in petrology[938,617,442], leaves[1707,454,1724], maps[1168], biophysical regions[323], drainage basins[754], wildlife home ranges[799,882,321], wildlife habitat[1287,1680,1043,567], nature reserve design[540,924], human population accessibility[1062], city shape[702,980,616], journey-to-work areas in cities (human 'home ranges')[615], visual perception[1961], and pattern recognition and picture analysis.[1447,1961,424,1289,1288,628]

137

Some studies have directly compared indices for their effectiveness in differentiating shapes. For example, to separate the shapes of oceanic atolls, a best-fit geometric-figure approach is too coarse.[1636] *L:W*-axis ratios and mean-radius indices are unsatisfactory, whereas an index of ellipticity based on area and length (chapter 4 appendix) is useful. However, these results apply to a narrow domain of shapes, mainly differentiating smooth rounded atoll forms with a L:W ratio between 1.5 and 3. On the other hand, a study of the shapes of urban commuter routes found ellipse, non-compact, and concaveness dimensions most important.[615,616] Thus, determining the domain of shapes to differentiate, and which components of shape are of interest, are crucial in selecting the most appropriate indices.

Clearly shape measurement will continue to develop in geography and in other fields such as psychology, petroleum geology, crystallography, and intelligence gathering where the subject is useful. Graph theory[674,1289], fractal analysis[1039] (chapter 3), isomorphism[1724], and other approaches may increase in importance for understanding the ecology of landscape patches. Nevertheless, rather than widening the enormous gap between theory and evidence, the major advances here will come from pioneering empirical studies of the ecological effects of patch shape.

MOVING PATCHES

Like dancers in a nightclub most patches change constantly in shape. Many stay in place but metamorphose as lobes and coves appear and disappear. But some patches move across the surface, often in spurts. To emphasize this dynamic view, consider what can be predicted knowing nothing about a patch except its shape.

Predicting the past, present, and future based on shape

Many causes or mechanisms producing shapes were presented earlier in this chapter. Similarly, many ecological functions are related to distinctive shapes. Therefore, with reasonable probability, some patch shapes indicate their origin[1677] (Fig. 4.13a). Some indicate their present functioning (Fig. 4.13b). And some indicate expected future change[1424] (Fig. 4.13c).

This kind of shape analysis apparently has not been done for a landscape, or for particular types of patch or shape. Studying a time series of aerial photographs would be especially useful for developing predictions of the past and future. Nevertheless, based on qualitatively surveying aerial photographs from a range of landscapes certain hypotheses

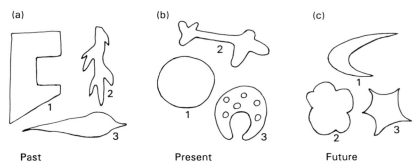

Fig. 4.13. Shapes suggesting past origin, present functioning, and future change. (a) Probable origins or causes. 1 = human-created; diagonal due to a corridor; if patch is wooded and matrix open, indentation is due to invading matrix; if vice versa, patch has spread around indentation. 2 = caused by erosion. 3 = caused by wind or water flow from right to left. (b) Probable functioning. 1 = conserve natural resources, such as water and biodiversity within patch. 2 = abundant movements in and out of patch; abundant animal movements along central strip, a good location for predators and hunters; few disturbances spread over whole patch. 3 = perforations indicate no interior conditions (i.e., only edge conditions) present; abundant animal usage in large cove. (c) Probable changes ahead. 1 = wind-formed patch moves to right. 2 = expanding patch. 3 = shrinking or contracting patch.

are warranted. Past origin can be predicted from many shapes. Present functioning can also. However, seldom does shape indicate future changes. Despite the perceived unpredictability of humans, valid predictions are about equally likely for natural and human-created shapes. Elongation is least indicative, convolution is intermediate, and roughness is most useful in making such predictions.

Advancing and retreating

This chapter began with patch shapes in a diverse Polish area, where patches both on an agricultural plain and on relatively natural slopes appear to be stable.[1324] A proposed equilibrium resulting from a long period of agriculture on the plain is characterized by low levels of elongation, convolution, and roughness. The variability in shape among nearby patches is also low (Fig. 4.1a). Considerable maintenance energy and effort presumably are expended on these geometric shapes. In contrast, on the slopes an equilibrium results from a long period of little human activity. Here, patches exhibit high levels of elongation, convolution, and roughness (Fig. 4.1a). Presumably these natural shapes are controlled by relatively constant or repeated processes.

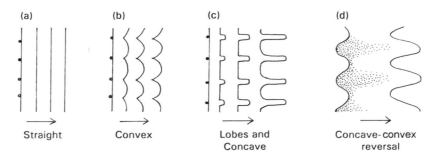

Fig. 4.14. Patterns of advancing or expanding boundaries. (a) Equal-width expansion (uncommon). (b) Convex expansion where certain points grow little or not at all (common). (c) Lobe and concave expansion where certain points grow rapidly (uncommon except for brief periods). (d) First wavy line is border between forest on left and grassland on right; dots indicate woody plants colonizing grassland; second wavy line indicates forest – grassland border resulting from this process. Based in part on Hardt & Forman (1989).

Advancing or expanding patches result from processes that are better understood[619,620,1835,1732,1733,1053,676,312] (see chapter 3). Elongated patches are generally oriented perpendicular to the direction of patch movement, and movement is normally upslope.[1835,141,142,1870,1606,1020,312,1665] This patch movement is apparently due to strong unidirectional wind or water flow, and a resulting feedback between vegetation and substrate.

Normally, expansion is uneven, with some portions expanding to become convex (or lobes), while other portions change little or not at all. The equal expansion in width along the margin of a patch, such as a blowing sand dune or logged strips along a forest edge, is uncommon. Convex boundaries are clues to gradual expansion, as illustrated by a growing delta or resprouting shrubs covering a burned or open area (Figs. 4.9 & 4.13c). Lobes produced by rapid growth at intervals along a margin are also common, such as dune invasion or extension of strip development along a road. But this is temporary, and lobes rarely become very long. Apparently, tip-splitting to form sublobes, often with widening, is common.[1733,1215] This results from lobe expansion into a viscous medium. But the matrix not only provides resistance, it also is heterogeneous. And heterogeneity is likely to split a rapidly growing lobe.

If equal-width expansion of a straight boundary takes place, the surface remains straight (Fig. 4.14a). If surfaces between the marked points in the figure gradually expand, all of the boundary surface will be convex (Fig. 4.14b). This way of forming convexities in expansion apparently is most common in landscapes.

140

But if narrow lobes rapidly expand at the points, most of the surface will be between lobes, and hence concave (Fig. 4.14c). An example of this third case is after wildfire, when rapid regrowth of stream-corridor vegetation extends up a dentritic stream system. Most of the corridor network boundary becomes concave. Gradually, regrowth may tend to fill in the coves. In this example, however, the heterogeneity of the terrain is the reason for the rapid lobe growth. Indeed, the presence of acute angles (<90°) in natural patches has been observed to be an indicator of substrate heterogeneity.[619,620]

The revegetation of Appalachian mine surfaces opposite concave and convex forest boundaries provides further ecological insight into expansion patterns[676,1580] (see chapter 3). The colonization of woody plants onto grass-covered mine sites is greater opposite concave margins than opposite convex margins (Fig. 4.14d). Thus colonization is greatest in coves, basically a 'cove concentration effect'. Surprisingly, this appears to be the only study evaluating the ecological effects of convex, concave, and straight boundaries, yet another promising research area.

If the elevated colonization rate extends only the length of a cove, the net effect is to smooth or straighten the forest-mine boundary. However, the area of elevated plant colonization extends more than twice the length of the cove.[676] The result is to form wooded lobes on the mine surface opposite the original forest coves (Fig. 4.14d). These new lobes in turn cause coves to form opposite the original forest lobes. In effect, a 'concave–convex reversal' took place as the boundary expanded. Mining theory also describes a concave–convex reversal, where equal-width rock surfaces along a corrugated boundary are removed.[1063,501]

Retreating or contracting patches, on the other hand, often have considerable concave boundary length where the adjacent patch has expanded (Fig. 4.13c). Often these patches appear as relicts destined for extinction. The loss of a lobe by logging, windthrow, or fire is a second process characterizing retreat. This normally leads to a smoother boundary. A less-common third process is invasion by lobes from the matrix. This may leave a patch quite convoluted, and at times in danger of being subdivided into smaller patches.

APPENDIX: EQUATIONS FOR MEASURING PATCH SHAPE

A. Measures based on lengths of axes

Form $\qquad\qquad F = \dfrac{\ell}{w}$ $\qquad$ (Davis 1986)

Elongation $\qquad E = \dfrac{w}{\ell}$ $\qquad$ (Davis 1986)

PATCH SHAPE

Circularity $\qquad C_1 = \sqrt{\left|\left(\dfrac{\ell w}{\ell^2}\right)\right|}$ (Davis 1986)

B. *Measures based on both perimeter and area*

Compactness $\qquad K_1 = \dfrac{2\sqrt{\pi A}}{p}$ (Bosch 1978, Davis 1986)

Shoreline development (or Patton's diversity) $\qquad D = \dfrac{p}{2\sqrt{\pi A}}$ (Patton 1975, Taylor 1977, Cole 1983)

Circularity $\qquad C_2 = \dfrac{4A}{p^2}$ (Griffith 1982, Davis 1986)

C. *Measures based on area*

Circularity $\qquad C_3 = \sqrt{\left|\dfrac{A}{A_C}\right|}$ (Unwin 1981, Davis 1986)

Circularity ratio $\qquad C_4 = \dfrac{A}{A_C}$ (Stoddart 1965, Unwin 1981)

D. *Measure based on radii*

Mean radius $\qquad \overline{R} = \dfrac{\Sigma R_j}{n}$ (Boyce & Clark 1964, Lo 1980, Stoddart 1965, Austin 1984)

E. *Measures based on area and length*

Form ratio $\qquad FR = \dfrac{A}{\ell^2}$ (Horton 1945, Stoddart 1965, Austin 1984)

Ellipticity index $\qquad EI = \dfrac{\pi\ell(0.5\ell)}{A}$ (Stoddart 1965, Davis 1986)

F. *Measure based on perimeter*

Shape factor $\qquad SF = \dfrac{p_C}{p}$ (Bosch 1978, Davis 1986)

G. *Measure based on perimeter and length*

Grain shape index $\qquad GSI = \dfrac{p}{\ell}$ (Davis 1986)

Key to variables:
  A = area of patch
  A_C = area of smallest circle enclosing a patch
  l = length of long axis
  n = number of sides, considered as a polygon
  p = perimeter of patch
  p_C = perimeter of circle having same area as patch
  R_j = jth radius of patch, measured from centroid to margin
  w = width of patch perpendicular to long axis

142

PART III

Corridors

5

Corridor attributes, roads, and powerlines

We have an account in the newspapers of every cow and calf that is run over, but not of the various wild creatures . . . It may be many generations before the partridges learn to give the cars a sufficiently wide berth.

Henry David Thoreau, Journal, *1858*

Nature creates corridors in the form of streams, ridges, and animal trails. Humans create roads, powerlines, ditches, and walking trails. How do the two types differ? Nature's are curvy, until humans straighten them. Nature's are continuous until humans superimpose grids on them, and create gaps. Human corridors are generally narrow and costly to maintain.

Corridors, as strips that differ from their surroundings, permeate the land. Vegetation corridors contribute significantly to many goals of society, here lumped into six categories.[491] First, corridors provide biodiversity protection, including key riparian habitats, rare and endangered species, wide-ranging species, and dispersal routes for recolonization following local extinction. Second, corridors enhance water resource management, such as flood control, control of sedimentation, reservoir capacity, clean water, sustainable fish populations, and fishing. Third, linear strips may enhance agroforestry production by acting as windbreaks for crops and livestock, controlling soil erosion, providing wood products, and preventing desertification. Fourth, recreation in corridors includes game management, wildlife conservation for nature enjoyment, and hiking, bicycling, boating, and skiing in suburban greenbelts. Fifth, community and cultural cohesion may be enhanced with greenbelts, which create neighborhood identity, provide wildlife corridors crossing roads that concurrently inhibit strip (ribbon) development along a road, and function as regional topographic barriers that enhance cultural diversity. Sixth, corridors provide dispersal routes for species

145

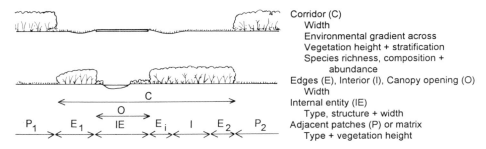

Corridor (C)
    Width
    Environmental gradient across
    Vegetation height + stratification
    Species richness, composition +
        abundance
Edges (E), Interior (I), Canopy opening (O)
    Width
Internal entity (IE)
    Type, structure + width
Adjacent patches (P) or matrix
    Type + vegetation height

Fig. 5.1. Attributes of the internal structure of corridors. Above: road corridor in forested landscape, showing road surface and roadsides (verges). Below: stream corridor of same width in grassland, showing stream, open streamsides, and wooded strips.

isolated in nature reserves, and coastal strips threatened by rising sea level in the event of climatic change.

Useful reviews of corridor attributes are given by Forman & Godron (1986), Schreiber (1988), Brandle *et al.* (1988), Bennett (1990a, 1991), Saunders & Hobbs (1991), Forman (1991), Malanson (1993), and Smith & Hellmund (1993). Reviews of specific topics include the corridor as a conduit[685,688,1228], the corridor as a filter[174,1038,130], road and roadside vegetation[109,111,1492], and trails.[968,1842]

Windbreaks, hedgerows, and wooded strips are presented in chapter 6, and stream and river corridors in chapter 7. Corridors are also commonly interconnected to form networks, as explored in chapter 8.

This chapter begins with general patterns and principles that apply to virtually all corridors. It begins with corridor structure, and then five specific ecological functions of corridors are explored. Width, connectivity, and changing corridors follow in sequence. The final section on trough corridors examines roads, trails, powerlines, and related types of strips.

INTERNAL AND EXTERNAL STRUCTURE

An insect zigzagging from side to side within corridors encounters several variables critical to survival. For instance, corridors vary in width. Vertical stratification and height vary. And the species and density of plants varies the most. Internal structure, as observed or measured from within the corridor, emphasizes the cross-sectional or two-dimensional view, and is reasonably well known for several types of corridor.[1337,948,1938,174,223,1492,1582,377]

The structural attributes affecting the corridor insect are usefully grouped into three categories (Fig. 5.1). (a) Width characteristics include

146

a steep environmental gradient from side to side, two edges that usually differ, and the effects of adjacent ecosystems that may differ. (b) The central portion of a corridor may include a distinctive *internal entity*, such as a stream, river, road, path, ditch, wall, or soil bank, or in a wide corridor, an interior environment may be present. (c) Characteristics of the plant and animal communities include vertical structure, plus species richness, composition, and abundances. Width and the presence of an internal entity or interior environment are key spatial variables controlling corridor function.

Qualitatively, the importance of these attributes is essentially the same, whether the corridor vegetation is higher or lower than the adjacent ecosystems[501] (Fig. 5.1). However, quantitatively, the relative difference in vegetation height between corridor and matrix affects the rates and directions of interactions between them.

If Icarus had discovered how humans could fly, the external forms of corridors would be more familiar to each of us than roadsides. But alas, it is said that he flew too close to the sun, and the wax holding his wings together melted. Instead, views from aircraft emphasize that the external structural attributes of a corridor include not only its form, but its relationship with the adjacent matrix, patches, and environmental conditions.[486,1032, 1229,501,1495,346,720]

In this length or third dimension of a corridor the key structural attributes are length, curvilinearity, alignment relative to the route of an internal entity, and the presence of patchiness or an environmental gradient along the corridor (Fig. 5.2). Width and connectivity are linked from this vantage. Width normally varies along the length of a corridor, which in turn may include narrows and gaps. The presence of gaps (or breaks) determines the connectivity of a corridor. Connectivity and the suitability of areas around gaps are considered keys to functioning.

FIVE CORRIDOR FUNCTIONS

Visualize standing in a large unfamiliar forest or grassland with no corridors. To reach a destination, one takes a convoluted route searching and intruding over a wide area. One could be 'lost' much of the time. Yet if the destination had to be reached repeatedly, a relatively straight route would be learned, and a corridor created to move more efficiently.

This scenario introduces two major theoretical characteristics and advantages of corridors: movement efficiency and protecting the matrix.[493] Channeled movements of matter (or molecules) that are spatially differentiated from an adjacent static area may produce or maintain an observable corridor. Here, movement of objects is greater within a strip than in the surroundings. This pattern results in more-efficient

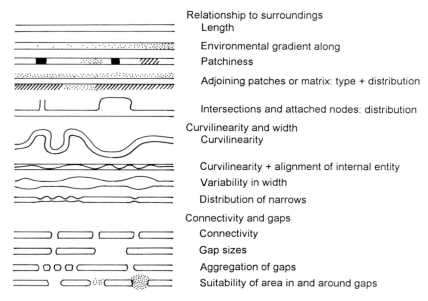

Relationship to surroundings
Length

Environmental gradient along

Patchiness

Adjoining patches or matrix: type + distribution

Intersections and attached nodes: distribution

Curvilinearity and width
Curvilinearity

Curvilinearity + alignment of internal entity

Variability in width

Distribution of narrows

Connectivity and gaps

Connectivity

Gap sizes

Aggregation of gaps

Suitability of area in and around gaps

Fig. 5.2. Attributes of the external structure of corridors.

movement of objects or energy between focal points. It also protects the surrounding matrix from the random or search movements of objects.[1717,1099,187]

Alternatively, active movement may surround a relatively static strip with little movement. This pattern pinpoints the third theoretical characteristic and advantage of corridors. The corridor acts as a barrier or filter to separate patch or matrix subsystems on opposite sides.

Corridors perform five major functions in landscapes: habitat, conduit, filter, source, and sink[501,109,111] (Fig. 5.3). The *habitat* function is well understood, but little documented in most areas. Edge and generalist species predominate.[1337,948,1791,1492,1038] Disturbance-tolerant, riparian, or even some interior species may be present in the central portion of certain corridors.[486,501,1469,1015,130]

A corridor acts as a *conduit* when objects move along it. The corridor is a *filter* or *barrier* when objects are inhibited from crossing between patches on opposite sides (Fig. 5.3). Conduit fluxes may be either inside or alongside a corridor.[123,412,111] For movements either along or across, the corridor serves as a differential filter, that is, movement rates vary both by type of object and over time. The conduit function is much studied ecologically in stream and road corridors[688,111,130,1038], and the filter function in windbreaks.[241,948,174] The movement of wildlife across corridors is increasingly being examined in terms of alleviating road kills,

148

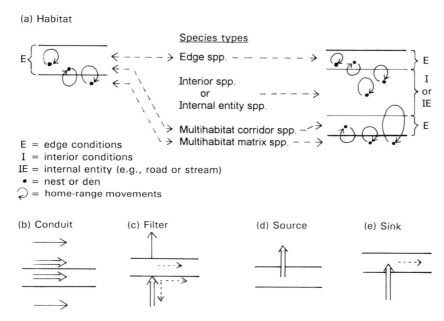

E = edge conditions
I = interior conditions
IE = internal entity (e.g., road or stream)
• = nest or den
↻ = home-range movements

Fig. 5.3. The five corridor functions. (a) Narrow corridor on left, and wide corridor on right; multihabitat species use two or more habitats. (b) Increased probability of movement inside or alongside a corridor. (c) to (e) Movements and flows between matrix and corridor.

and the fragmentation effect of a road network.[1272,1521,103,1397,1033,346,688] Road kills graphically illustrate the filter or barrier function of corridors.

The source and sink functions are understood conceptually[1673,999,501,109] (Fig. 5.3), but as yet the diverse empirical evidence has not been brought together, so the rates and significance of flows are little known. As a *source* (an area or 'reservoir' that gives off objects) of effects on neighboring patches, corridors may, for example, harbor herbivores that feed on crops, predators that control crop pests, trees that disperse seeds, and hunters that hunt in the matrix. As a *sink* (an area or reservoir that absorbs objects) for fluxes from neighboring areas, corridors may, for example, accumulate wind-blown sediment and snow, subsurface mineral nutrients, and animals from the matrix.

Habitat

What lives in corridors? Edge and generalist species predominate. Some multihabitat species and some invasive exotic species may be present (see glossary in chapter 1 appendix). Rare and endangered species are normally absent, unless corridors represent essentially the only remnants of native vegetation in an area.[1337,1400,155,1469,1492]

Line corridors are narrow strips dominated by edge species, whereas *strip corridors* are wide enough to have interior species predominating in the central portion.[499,501] Species in remnant line corridors resemble those in patch edges or gaps, except that fewer interior species are present. Thus, a hedgerow has been called a forest edge without a forest.[1337] Streams, roads, and other entities within a corridor typically have additional species, which vary according to whether the canopy overhead is continuous or open.

The density of edge species is typically high, that is, in a corridor they are packed together.[1211,1337,268,1492,1038] Edges in general tend to have a high species density (chapter 3), but here two edges are in proximity. Furthermore, the two edges usually differ, due to the orientation angle (chapter 4) and to differences in the adjacent patches.

The total richness of edge species in a whole corridor is often high, because the strip crosses various substrate patches.[150] Furthermore, it commonly has an environmental gradient varying from end to end.[233] Some corridors include a winding river or road. Such an internal entity provides a partial barrier to many animal movements within the corridor, and consequently could lead to more overall genetic variation in the resident populations.

Several reasons for these species patterns are evident. Wind and human maintenance regimes commonly mold the form and vertical stratification of corridors. Corridor width may limit the presence of species that have wide home ranges, as predicted by central place theory.[1393,1407] Only generalists can tolerate the steep environmental gradient across a corridor. Only disturbance-tolerant species, such as weeds in a path, survive chronic movements along a corridor. And only disturbance-tolerant species survive being bathed repeatedly by dust, heat, pesticides, and other disturbances from adjacent ecosystems.

Conduit

Form and function principles (chapter 4) focus on the tube or transport structure in corridors. Stream corridors contain movement by the force of gravity. However, virtually all corridors also have movement uphill against gravity using locomotion energy (rheotaxis).[685]

An interesting case of corridor function is illustrated by the 'countercurrent principle'.[493] This principle refers to the high transport rate of energy or material between two adjacent linear systems flowing in opposite directions. Filtration between the blood and urinary systems in our kidney, and heat transfer between adjacent parallel veins and arteries in a bird's leg, are well-known examples. At the landscape-scale the counter current principle is illustrated by the upstream-moving fish, such as the giant 10 m (30 ft) long sturgeon (Acipenseroidea), feeding on downstream-moving microscopic plankton in the Mississippi and Yellow rivers. Similarly, predators learn to move along a hedgerow corridor, in the opposite direction of herbivores moving between resting and feeding areas. The feeding efficiency of the predator (or hunter) is much higher than if either herbivore or predator moved in any other direction. In these cases, the probability of encountering food is increased by the spatial juxtaposition of opposite, parallel flows.

What uses corridors as conduits?[109,685] People and goods move along by walking, and in vehicles on roads. Water, sediments, nutrients, and organic matter move by gravity in stream corridors. Energy, wind, and seeds move in, or adjacent to, many corridors. Animals move along corridors in home range movements, dispersal, mating, and migration. A single animal may move continuously or discontinuously along a corridor.[501,109,688] In the latter case, heterogeneity within the corridor is apt to provide rest stops. Genes may also flow through a corridor as individuals of a population interbreed.[109] And at a broader scale whole faunas and floras may move along a corridor.[688]

Many structural attributes affect the conduit function. Animals can be expected to move more effectively in corridors with few narrows, few gaps, low curvilinearity[1598,503], low patchiness (chapter 4), no environmental gradient, few entrances and exits, little crisscrossing of streams or roads within the corridor, and that are short in length. In stream corridors, if meanders are prominent (high sinuosity ratio[610]), people and animals (e.g., egrets, kingfishers, and river otters) may move directly between meanders, rather than follow the circuitous channel. As considered later in this chapter, the effect of corridor width on the conduit function apparently varies with corridor type, as well as with the object moved.

Corridor length affects the conduit function. A study of hedgerows in Central Bohemia (Czech Republic) found that three quarters of the 41 plant species present are limited to within 200 m of a woods.[695] The remaining species extend along a corridor as far as 250–475 m. Wind disperses 63% of the plant species, but vertebrate dispersal is most important over short distances (< 25 m). Hence, corridor length, and probably also connectivity, have some effect on plant movement, especially of animal-dispersed species.

Finally, a corridor may be located as a 'hedge' against climate change. A 3-km-wide strip of tropical rainforest, the Zona Protectora La Selva, connects the Caribbean lowlands to the high mountains in Costa Rica (a 3000 m rise in only 25 km).[685,688] Although most species are squeezed in narrow elevational levels within the corridor, they may survive by moving up or down slope as climate warms or cools. Latitudinal corridors provide the same benefit, without the narrow elevational limitation.[493,1492]

Filter

Just as for boundaries (chapter 3), the cellular membrane is a useful analogue. A membrane is 'semipermeable', that is, it is a barrier to some moving objects, a partial barrier to others, and quite permeable to still others. The flow rate across a corridor is similarly affected (Fig. 5.3), and indeed permeability can change over time. Objects cross membranes through holes, or by interacting chemically with the structure. In a corridor, gaps are equivalent to holes, and specific corridor plants and animals are sites of interaction with moving objects.[503] As described in chapter 3, wind- and water-driven flows accelerate through gaps, whereas locomotion-driven flows slow down in gaps. The surface of a membrane also has scattered receptors or 'receptive sites', where objects attach and interact. Coves and lobes in a boundary surface, as well as individual plants, analogously may act as receptive sites. Here, herbivores selectively feed or are eaten by predators, or snow accumulates permitting seeds to germinate later.

The overriding result of the filter effect is to separate patches or areas on opposite sides of a corridor. Species composition often differs. Individuals of a population interbreed more on one side than between sides. This may lead to genetic differentiation of subpopulations, as in the case of frogs separated by roads or foxes (*Vulpes*) separated by the Mississippi River.[1639,1403] Differentiation of neighborhoods by a greenbelt or highway is a human analogue.[977,1753,1582] The result may be development of community and cultural cohesion on one side of the corridor. Just as filters can lead to higher total biodiversity, they can lead to higher cultural diversity.

Several structural attributes affect, and can be modified to affect, permeability. Corridor connectivity is a measure of gap frequency, but gaps can be large or small (long or short), and the nature of the area in and around a gap is critical to movement (Fig. 5.2). Corridor width and the presence of narrows are considered key variables affecting the filter function. Streams, rivers, roads, paths, ditches, walls, and other internal barriers generally decrease permeability. A vegetated median strip in a highway, and islands in a river, act as stepping stones to increase permeability.[1639,4] In fact, their location relative to curves in a corridor is probably important. The steepness of the environmental gradient within a corridor doubtless affects flows across it. For instance, a dense windbreak has a much steeper environmental gradient than a porous windbreak. Finally, as discussed later in this chapter, underpasses and tunnels can be constructed to enhance movement across a corridor.

Source

If a new corridor is established across a large field or forest, think how that matrix will change. Some objects such as animals, water, vehicles, and hikers moving along the corridor may also spread out into the matrix. Noise, dust, various chemicals, and some resident corridor species may do likewise. Hence the corridor acts as a source, with diverse effects on the matrix.

Corridor structure controls the source effect. Width determines how many interior species are present. Patchiness and an environmental gradient along the corridor mean the total number of (mainly-edge) species is great, and the actual species available to penetrate the matrix changes along the corridor. Greater corridor curvilinearity doubtless indicates a greater source effect (chapter 3). The nature of an interior entity, such as a stream, road, or wall, stongly affects patterns of movement along the corridor, as well as the resident species composition.

Sink

The corridor just introduced into a grassland or forest also acts as a sink for objects originating in the matrix (Fig. 5.3). Blowing snow, soil, and seeds are trapped in wooded corridors. Water-eroded particulates, pesticides, and animals from the matrix accumulate in stream corridors. Animals are killed crossing a road, or drowned crossing a river. Of course, the sink may be temporary, as snow melts or particulates wash out downstream.

As always structure controls function, in this case as a sink. Width, gaps, and vertical structure determine the proportion of wind-blown materials deposited in a corridor.[174] Curvilinearity, edge structure, and abruptness affect penetration. Internal entities including streams, roads, and walls not only determine the crossing ability of objects, but also, by channeling objects along a corridor, they may affect the sink rate. A predator or hunter situated in a corridor hopes to increase the sink rate of prey or game.

WIDTH AND CONNECTIVITY

Of the structural attributes affecting the preceding five functions of corridors, two stand out: width and connectivity. Both are readily used in planning, conservation, and management. The effect of width of windbreaks is well studied[241,174], but otherwise width and connectivity represent important research frontiers.

153

Fig. 5.4. Narrow and wide road corridors in a giant green network. Created and maintained through agricultural land by ecologists and engineers in highway departments of Western Australia. Photos courtesy of B. M. J. (Penny) Hussey.

Width

Corridor width is usefully measured as an average and a variance. The number of gaps and number of narrows are measured (in various size classes) per unit length of corridor. In addition, the width of internal entities, including streams, roads, trails, and walls, is measured. Because the cross-sectional form of corridors is often smoothed by wind or humans, in some cases it is useful to determine corridor width at different heights above the ground.

The differentiation of strip from line corridors emphasizes the effect of width on the habitat function (Fig. 5.4). Thus interior species are only common in strip corridors.[28,486,412] The presence of interior conditions and an internal structure determines the source pool of corridor species that move out to the matrix.

The effect of corridor width on the conduit function is important for the movement of species among nature reserves, although few empirical data are available.[1162,1393,1118,720] Larry Harris and colleagues[688] recommend corridor widths appropriate for three quite different objectives:

(1) 'When the movement of individual animals is being considered, when much is known about their behavior, and when the corridor is expected to function in terms of weeks or months, the appropriate corridor width might be measured in tens of meters ...' (2) 'When the movement of an entire species is being considered, when much is known about its biology, and when the corridor is expected to function in terms of years, the appropriate corridor width might be measured in hundreds of meters ...' (3) 'When the movement of entire assemblages of species is being considered, when little is known of their biology, and when the faunal dispersal corridor is expected to function over decades or centuries, the appropriate width should be measured in kilometers ...'

The sink function of stream and river corridors controlling erosion, nutrient and water runoff, sedimentation, and flooding is dependent on width. This relationship with width is poorly known (chapter 7). But because the processes are so important in the landscape, often a relatively arbitrary fixed width in the range of tens of meters is recommended. A width that includes the flood plain and hillslopes on both sides of a stream or a river, plus an interior environment beyond at least one hillslope, has been recommended.[501] A variable minimum width tailored to the area drained by each 'stream order' is probably appropriate (chapter 7). Moreover, the appropriate stream-corridor width varies according to soil type, and steep slopes require a wider corridor than modest slopes.

In short, wider corridors probably enhance all five functions, habitat, conduit, filter, source, and sink. The shape of a curve relating a function to corridor width, however, is a research goal. It is likely that the inflection points or thresholds for the different functions will be found at different widths. Probably the most conservative indicator or assay for effective corridor width is, e.g., whether a stream corridor can control chemical and mineral nutrient runoff from reaching a water body. Other conservative assays are introduced in a later section on road corridors acting as a source of impacts on the matrix.

Connectivity

In elementary-school math a system is connected or disconnected (not connected) depending on whether all elements composing the system are linked together. This either/or qualitative concept is quantitatively measured by the degree of connectivity. *Connectivity* varies from 0.0 to 1.0, and is the inverse of the proportion of linkages that must be added to have a connected system.[501,109,493] Simply put, the fewer gaps per unit length along a corridor, the higher the connectivity (Fig. 5.5).

In addition to gap number, the degree of aggregation and the lengths (along the corridor) of each gap are ecologically significant. Aggregated

Fig. 5.5. Corridors of different origin, width, and connectivity in northern England. Road corridors with and without trees, a powerline corridor (top), a stream corridor (upper portion), planted hedges between fields (thin dark lines), tree lines, and wooded strips (separating town from agriculture) are visible. R. Forman photo.

gaps indicate a series of stepping stones in a corridor system. Land use in and around gaps is of particular functional importance for movements, both along and across corridors.[486]

The habitat, conduit, and source functions are all doubtless enhanced by higher connectivity, as species move more readily along a corridor. The filter and sink functions also increase with connectivity, as objects moving across a connected corridor are reflected or absorbed. Hence, just as for width, higher connectivity leads to higher levels of all five functions.

Where a straight corridor is broken, and the area around the gap is difficult to 'repair', amelioration techniques are possible. Underpasses and tunnels are discussed later in this chapter. Andrew Bennett (1990a) proposes maintaining a 'strip of network' between large patches, which is composed of two or three parallel corridors containing interconnections, plus attached tiny patches as stepping stones.

The concept of connectivity focused on structure has long been used in geography and transportation theory[1670,995,646], and more recently in landscape ecology[500,501,465,718,894,1229,1518,1745] (also called connectedness[1111,92]). Thus, one type of object, such as a corridor, can vary in its degree of connectivity. A network (chapter 8) and a matrix (chapters 8 and 12) also usually vary in connectivity.

156

The studies of Gray Merriam and colleagues (chapter 6) emphasize that the concept of *behavioral or functional connectivity* is also important for understanding flows in a landscape or region. This refers to how connected the diverse spatial elements used in the movement of an animal or other object are.[1111,92,493] For example, a vertebrate may forage from woods to hedgerow to beanfield to pond to hedgerow to woods within a home range. Survival may depend on using all of these ecosystems and avoiding others in the mosaic. Hence, the functional or behavioral connectivity of the land for this vertebrate depends on how connected the diverse elements are, as perceived by the animal. Similarly, a mosaic varies in functional connectivity for movement of surface water or wind-blown particles.

CHANGING CORRIDORS

Corridors originate in the same five ways as patches[109,493] (chapter 2). Roads, powerlines, and animal trails illustrate *disturbance corridors* (Fig. 5.5). A wooded strip remaining after land clearing is a *remnant corridor*. A riparian strip and a fault line are *environmental corridors*. A hedgerow or fencerow grown up along a fence crossing a field is a *regenerated corridor*. A planted windbreak or hedge is an *introduced corridor*.

Like patches, corridors also change diurnally, seasonally, and successionally. Many people move once a day between home and work. Many vertebrates move twice a day, morning and evening, between nest (or den) and feeding area. Seasonal changes in foliage density cause differences in permeability to wind and wind-transported objects. Seasonal movements of animals along or across corridors involve migration, e.g., between winter and summer ranges. Corridor usage may be high at one season and minimal at another. Successional changes in corridors presumably parallel those in patches, except that in a narrow exposed situation the development of interior conditions with interior species is limited.[1539,1857,1337,1005,174] Consequently the species composition characteristic of a large old-growth patch of vegetation is unusual in a corridor.

Natural processes changing corridors are usefully differentiated from human activities. Thus, the curvilinearity of natural corridors relates to wind, water, and substrate. Flowing water and wind erode and deposit materials, which produces bulges and indentations, meanders and oxbows. Disturbances such as floods and sandstorms may create or eliminate corridors, and decrease or increase patchiness within them. Native species richness may increase with corridor age.[1539,1337,723,1790,1791] A heavy desert rain may change a surface dendritic imprint to an entirely different imprint, leading at right angles or even in an opposite direction.

157

Analogous changes in corridors result from human activities. External activities such as trimming edges for agriculture, and widening fields right to a stream bank, affect the width, connectivity, and curvilinearity of corridors. Activities originating internally, e.g., from roads, trails, and streams, also mold corridor structure. Both external and internal human activities change the patchiness along a corridor. Human policy and land-use changes cause corridors to appear and disappear. In short, corridors are especially subject to change due to their narrowness, and because the source of change may either be external or internal.

LINEAR BOUNDARIES, CORRIDORS, AND TROUGHS

Corridors and patch boundaries constitute all the linear features in a landscape.[184,503] They differ structurally in that a patch edge separates a patch interior from the matrix, whereas a corridor has matrix on both sides. However, the five corridor functions are exactly the same as the five functions performed by patch boundaries or edges[1886,212,1513,503] (chapter 3). Therefore, although the corridor and patch boundary differ structurally, functionally the concepts merge.

A closer look is informative. A forest–field boundary zone contains a strip of forest edge and a strip of field edge that adjoin along a border[503] (chapter 3). The forest edge, the field edge, and the boundary zone as a whole each exhibits the same five functions as a hedgerow separating pasture from cornfield. A corridor is distinguished from a patch edge by how they differ from their adjacent habitats. The hedgerow corridor differs markedly in structure from both the adjacent pasture and cornfield. In contrast, the forest edge differs only slightly from the forest interior, but is very different from the field on its other side.

Differentiating a corridor from a patch edge is difficult at times, or is done arbitrarily for the purposes of a study. A strip of beach between forest land and sea is the edge of the land. At the same time the beach is a corridor separating forest from sea. In short, functionally, corridors and patch edges are identical, but structurally their difference ranges from none to great.

Most of the preceding concepts and theory in this chapter apply to any corridor.[501] Nevertheless, three types of corridor are so distinct visually that they will now be considered separately. *Trough corridors* are strips with vegetation lower in height than the adjacent matrix. *Wooded strips* are corridors with vegetation higher than the adjacent matrix (chapter 6). *Stream and river corridors* (also see 'riparian corridors'; chapter 7) are vegetated strips containing a channel of flowing water, and may be higher or lower than the adjacent matrix vegetation.

Trough corridors include roads, powerlines for electric transmission, gas lines, railroads, dikes, livestock routes, horseback trails, walking paths, and animal paths. The first six are sometimes called 'rights-of-way'.[1718] All but the last are human-created, and thus often considered unnatural habitats. All depend on maintenance or repeated disturbance. In the case of powerlines, gas lines, and dikes, the maintenance activity or disturbance normally covers the whole corridor width. In the case of roads, railroads, and paths, disturbance is concentrated in a strip within the corridor, leaving distinctly different outer portions of the corridor. Thus, a road corridor refers to the road where vehicles regularly move, plus roadsides or verges on both sides.

Roads in desert, grassland, and agricultural areas, where the surroundings often are also low, are included as trough corridors. Also, all roads and paths are included here, whether the overhead vegetation canopy is open, intermittent, or closed.

Animal paths have existed 'forever', human paths for a few million years, roads for millennia (such as the straight solid Roman roads crossing Europe and the Mediterranean area, an imprint frequently visible today), and hedgerows for a few millennia (they originated with agriculture). Railroads appeared about a century and a half ago, and powerlines and motor-vehicle roads nearly a century ago. In addition to having been present a long time, these corridors are widespread. For example, almost one per cent of mainland Great Britain is road corridor.[1837] The 6.3 million kilometers of roads in the USA cover 8.1 million hectares[6] (equal to the area of Austria or South Carolina), and powerline corridors in the USA cover 2.0 million ha (equal to the area of Sicily or Massachusetts). With such landscape features so long existent and widespread, we may hypothesize that some plants and animals have adapted to these corridors.

The open trough form means that sun and wind penetrate from above, and that their effects on microclimate depend greatly on orientation angle (chapter 4). Solar effects depend on corridor orientation relative to a north–south axis. Wind effects depend on orientation relative to the preponderant wind direction, as well as major storm directions.

ROAD CORRIDORS

Here a *road corridor* (or road reserve) refers to a road, as the surface for vehicle movement, plus any associated, usually vegetated parallel strips.[111] *Roadsides* or verges are the usually open, regularly maintained or disturbed strips within the corridor that adjoin the road. In certain landscapes, especially in parts of Australia and South Africa, parallel strips of natural or native vegetation adjoin the open maintained roadsides. Such strips may be called *roadside natural strips*[109,111,1492,371] (Fig. 5.4). Fences, ditches, soil banks, cliffs, walls, fire lanes, and a median strip of vegetation may be included in a road corridor.

The connectivity of essentially all trough corridors is 1.0. This means no gaps are present and the corridor is completely connected. Almost all these corridors have a strong transport or conduit function, with objects moving between attached nodes. A dike usually is also unbroken, though the reason is a need to maintain its strong barrier or filter function. The five ecological functions of road corridors (habitat, conduit, filter, sink, and source) are prominent in most landscapes.[501]

Habitat, conduit, and filter

Habitat

Just as for all corridors, edge and generalist species predominate in road corridors.[1539,1857,1469,1791,1106,852] A cross section of a road corridor usually indicates considerable heterogeneity, due to the diverse parallel strips included. In addition, patchiness along a road corridor is marked. Maintenance regimes for roadsides often include mowing, trimming, burning, or applying herbicides. Furthermore, material originating from the road, including oil, salt, dust, heavy metals, and debris, commonly bathes the roadside.

As a habitat for species, a 93 to 143 meter wide (*c.* 300–450 ft) vegetated median strip in a highway was found to contain small mammal diversity similar to that of the matrix.[4] In Western Australia wide roadsides have more ant species than narrow roadsides.[846] In the same landscape, roadside natural strips that are narrow (<10 m) have a higher percentage cover of exotic species than wide strips (>25 m) (data of G. W. Arnold, D. Alger, R. J. Hobbs, & L. Atkins cited by 1278). Overall, the number of species in road corridors is usually high, although composed primarily of edge species.

Conduit

Vehicles move large numbers of seeds along roads. For instance, 259 plant species were identified at a car-wash in Canberra, Australia, some of which are believed to have been transported for over 100 km.[1812] Since many seeds are picked up within road corridors, presumably many plants can germinate and become established elsewhere in the road network after being transported (Fig. 5.6b). Occasionally some animals, including frogs and snakes, become established after movement in this manner.[111] Bats also may use road corridor troughs as conduits.[331]

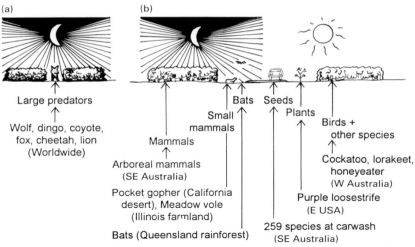

(a)

(b)

Large predators
↑
Wolf, dingo, coyote,
fox, cheetah, lion
(Worldwide)

Bats   Seeds
Small ↑   ↑ Plants
mammals   ↑
Mammals   Birds +
↑   other species
Arboreal mammals   ↑
(SE Australia)   Cockatoo, lorakeet,
honeyeater
Pocket gopher (California   (W Australia)
desert), Meadow vole   Purple loosestrife
(Illinois farmland)   (E USA)

Bats (Queensland rainforest)   259 species at carwash
(SE Australia)

Fig. 5.6. Species conduits within road corridors of different width. (a) Narrow corridor with little vehicular traffic. (b) Wide corridor. Moon indicates mainly nocturnal movement, sun, daytime movement. Sources are: large predators (Bennett 1990b, Saunders & Hobbs 1991); arboreal mammals (Suckling 1984); small mammals (Huey 1941, Getz et al. 1978); bats (Crome & Richards 1988); seeds (Wace 1977); plants (Wilcox & Murphy 1989); birds (Saunders 1980, 1990, Saunders & Ingram 1987, Newbey & Newbey 1987).

The direct movement of animals along roads depends very much on vehicular density.[1774,109] Narrow unpaved roads with few vehicles often are used at night by predators (Figure 5.6a), including fox, dingo, coyote, wolf, cheetah, lion, and perhaps the Tasmanian devil (*Vulpes vulpes, Canis familiaris dingo, C. latrans, C. lupus, Acinomyx jubatus, Panthera leo, Sarcophilus harrissi*).[109,1492] Hikers should be aware of this. Moose (*Alces canadensis*) ambling along a road at night sometimes will charge an unfortunate car. Otherwise, the road surface generally appears inhospitable as a movement conduit for animals.

Open roadsides also are apparently little used as a conduit, except by plants and small mammals[886,1790,1791,493,1278] (Fig. 5.6b). A non-native plant species, purple loosestrife (*Lythrium*), rapidly invaded the eastern USA, apparently facilitated by roadside ditches.[1896] A pocket gopher (*Thomomys*) expanded its range across California desert along roadside vegetation that was irrigated by water runoff from the road.[761] The meadow vole (*Microtus pennsylvanicus*) expanded its range nearly 100 km across Illinois (USA) farmland in 6 years, by moving along the continuous grassy roadside of a new major highway.[561] Despite such examples, for most animals the open road and roadside is an inhospi-

Fig. 5.7. Symbols of hazards in moving along and across corridors. (a) Valley of the revered Ming Tombs near Beijing. (b) Wildlife crossing signs for kangaroo and wombat in eastern Australia, and plant crossing sign in the industrial Midlands of England. R. Forman photos.

table strip, subject to predation and other unexpected dangers (Fig. 5.7a).

Roadside natural strips appear to be important conduits for mammals and birds in Australia[109,111,1492] (Fig. 5.6). A beautiful species of parrot,

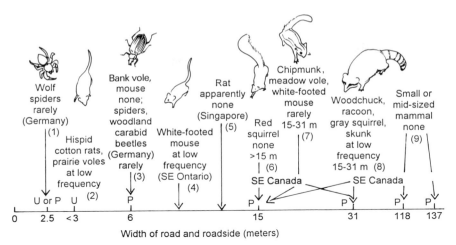

Fig. 5.8. Effect of road corridor width on animal crossings. The maximum corridor width at which animals crossed is given. Sources: (1) Mader (1988); (2) Swihart & Slade (1984); (3) Mader & Pauritsch (1981), Mader (1984); (4) Merriam et al. (1989); (5) Lawley (1957); (6–9) Oxley et al. (1974).

Carnaby's cockatoo, and the western gray kangaroo (*Calyptorhynchus funereus latirostris, Macropus fuliginosus*) regularly use roadside natural strips in home range movement in the Western Australia wheatbelt.[1487,1489,1495,53] Dispersal of arboreal mammals from a home range is also enhanced by such vegetation strips in Victoria.[1650,110] Migration and 'nomadic (wandering, itinerant) movement' of honeyeaters and lorikeets (*Melithriptus, Glossopsitta*) is also reported in roadside natural strips.[1204,111] Presumably, higher traffic density decreases faunal movement in the roadside natural strip, and especially in the open roadside.

Filter

The novelist, John Steinbeck, describes an agonizing tale of a turtle crossing a road, with all the dangers and drama imaginable, leading ultimately to the climax of achieving the far side.[1622] The road corridor as a filter for crossing movement affects animals ranging from terrestrial invertebrates (spiders and carabid beetles) to large mammals including kangaroo, elk, and deer (*Macropus, Cervus, Odocoileus*) (Fig. 5.7b). For example, beetles and spiders in experiments by H-J. Mader almost never crossed a lightly traveled, 6 m wide (19 ft) paved tarmac road in Germany, and also tended to avoid the grassy roadsides[1034,1032,1033] (Fig. 5.8).

163

For wolf spiders (*Lycosa amentata*), a 2.5 m wide paved road was less hospitable than an unpaved road of the same width, though both appear to be barriers to movement.[1033]

Several studies of small mammals show that the probability of crossing lightly traveled roads of 6–15 m width is less than 10% of that for movements within the adjacent habitat[1032,1666,1117,925] (Fig. 5.8). Indeed, half the species studied never crossed the road. In southern Ontario and Quebec, D. Oxley and colleagues[1272] found that small forest mammals, such as chipmunk, red squirrel, meadow vole, and white-footed mouse (*Tamias*, *Tamiasciurus*, *Microtus*, *Peromyscus*), readily crossed road corridors up to *c*. 15 m wide (distance between forest margins). Yet the small mammals rarely crossed road corridors of 15–30 m width.

Mid-sized mammals, including woodchuck, racoon, eastern gray squirrel, and striped skunk (*Marmota*, *Procyon*, *Sciurus*, *Mephitis*), crossed road corridors up to *c*. 30 m wide[1272] (Fig. 5.8). None of the small and mid-sized mammal species was recorded crossing highway corridors of 118 and 137 m width (ca 400 ft). Road surface (gravel versus paved tarmac) did not appear significant, but traffic density did. Large mammals, including mountain goat, caribou, and elk (*Oreamnos americanus*, *Rangifer tarandus*, *Cervus elaphus*), cross most roads[872,1565,1566,346], but the rate of crossing is typically lower than movement rates in the matrix.

Many wetland species, such as amphibians and turtles, exhibit reduced movement across roads.[735,1778,918,1403] The open trough above a road could be a barrier to the movement of some woodland butterflies.[1272,144 cited by 1774] And some nesting birds and large mammals tend to avoid the vicinity of roads altogether.[475,1448,1774,1388,1406] Four-lane highways in The Netherlands are considered least inhibitory to crossing by large mammals, intermediate for butterflies and small mammals, and most inhibiting to birds.[877]

Roads within a home range may cause stress in, and mortality of, individuals.[1565,1567,1515] Roads separating home ranges presumably produce subpopulations with differential gene flow, leading to the expectation of genetic variation.[1032,1033,1117,1042,1403] In summary, the filter effect appears particularly sensitive to (a) width of the inhospitable portion of the road corridor, (b) traffic volume, and (c) mobility and behavior of the species.[1272,1774]

Sink and source

Sink

Locomotion is remarkable because animals can go in directions other than those dictated by the forces of gravity, wind, and sun.[685] Mobile

164

animals regularly forage for food, and avoid predators and disturbances. They also move while feeding and raising young in a territory or home range, in dispersal out of the home range, and in mating activities. Some migrate, thus avoiding cycles of unsuitable weather. Animals move diurnally, seasonally, with successional change, in range expansions and contractions, and as whole faunas over paleoecological time. But mobile animals meet immobile roads (Fig. 5.7b), where vehicular 'mega-projectiles' cause *road kills*.

Road kills of animals by vehicles represent the primary sink in road corridors, and the numbers are staggering (see Box 5.1). Even a field guide to 'flattened fauna' is available.[881] Underpasses (Fig. 5.9) may be the most effective way to reduce this mortality, which is increased by a variety of factors.

Box 5.1: Underpasses and tunnels to alter a collision course

It 'takes guts' to hit the windshield of a moving vehicle, and knowledgeable fly fishermen look carefully at the insects smashed on vehicles. Road kill estimates for insects are limited[519,1837,520,1339], but annual vertebrate numbers based on measurements of short sections of roads are impressive: 159 000 mammals and 653 000 birds in The Netherlands[821 cited in 1774], seven million birds in Bulgaria[1188 cited in 111], five million frogs and reptiles in Australia.[437] One million vertebrates per day are road killed on roads in the USA.[914] The species range from wolves and deer to bats and platypus (*Ornithorhynchus*).[393,736,1757,1479]

Many techniques have been tried to decrease road kills. Reflectors, mirrors, repellents, bait, various fencing types, one-way gates (to escape from fenced roadsides), lighting, wildlife crossing signs, and animated warning signs for motorists generally show modest or no success.[409,1396,1336,467,1397,1828,1497,470]

An alternative approach is to have animals cross over a road, either on a land-surface-level bridge over a sunken road, or on an overpass above the land surface that arches over the road. A land-level grass-covered bridge, 15 m wide in the middle and 50 m wide at the ends, over a sunken highway was studied for faunal movement in The Netherlands (data of W. Nieuwenhuizen & R. C. van Apeldoorn cited by H. D. van Bohemen, pers. commun.). Of 13 mammal species observed, 77% crossed the wildlife bridge, the same percentage that crossed in tunnels under highways. Three species, squirrel, hare, and deer (*Sciurus vulgaris, Lepus europaeus, Capreolus capreolus*), crossed on the land-level bridge, but not in tunnels. Overpasses, designed for wildlife and that arch over a road, apparently have not yet been evaluated for animal crossing.[902]

However, underpasses and tunnels have been established and tested for use by animals in many countries. They are used by badgers in Britain[1386], mountain goats in Montana, USA[1566,1296], toads in The Netherlands[1780], mountain pygmy-possums in Victoria[1042], desert tortoises in California, mule deer in Colorado[1398,1395,1828], and amphibians in Germany, Switzerland, The Netherlands, England, Wales, France, and Massachusetts, USA[1723,1755,918]. Tunnels have been constructed under railroads in Australia[763], and an oil pipeline elevated for caribou and moose movement in the Arctic.[872,346,443] Standardized design, technology, and prefabricated structures are available for amphibian tunnels.[918] So far the most ambitious project is a series of 35 underpasses along 'alligator alley' highway in South Florida, through which panthers, bobcats, bears, alligators (*Felis concolor, Lynx rufus, Ursus americanus, Alligator mississippiensis*) and numerous other species move daily.[685,1035,688]

A tunnel or underpass system includes the essential accompanying fences, walls, rock lines, earth berms, shrubs, and/or trees. The accompanying structures act as drift fences (chapter 4) to funnel animals to an underpass, as sound or visual screens, and as barriers to discourage movement over the road near an underpass. The heights, widths, lengths, angles, and locations of all structures are keyed to target species. It is appropriate to identify at least two types of target species, the largest animal of interest, and the species most sensitive to a road barrier. The underpass or tunnel system is then primarily designed around the behavioral requirements of these species (see chapter 5 appendix for experience from specific projects). Success for these two species types should mean success for most other species. The underpass and tunnel approach will likely enhance movements of most terrestrial mammals, amphibians, and reptiles near roads. This approach also should reduce road kills of birds feeding on road kills. However, reduction of the impressive bird mortality on roads will require other techniques.

Today, mobile animals are threatened by a fixed road grid reinforced with moving vehicles. The solution may lie in porous roadbeds with tunnels and underpasses. In this manner, both natural and human processes are maintained, as 'happy animals cross under oblivious vehicles'. Land-level wildlife bridges are also in the repertoire.

Roadside food is a major cause of road kills. Feeding on spilled grain, grass, seeds on grass, fruits on shrubs, and flying insects leads to road mortality.[734,736,1679,1796,394,417] Large herbivores are often hit, such as kangaroos and deer attracted to grassy roadsides.[102,103,1396,253,318,319,1267] In some areas 'roo bars' are standard equipment on vehicles to protect against kangaroos, and their hopping ability. Collisions with large herbivores are especially common in wooded sections, where a roadside is the only open area present.

Where wildlife corridors and roads intersect, road-killed animals are frequent, and emphasize the need for new policies in highway design and land use. Huge numbers of frogs and toads are killed crossing from

Fig. 5.9. Underpass used by small and mid-sized mammals. Dangerous or impassable during snowmelt or after heavy rain. Possibly Virginia, Courtesy of US Fish and Wildlife Service.

feeding areas to ponds or wetlands during breeding season.[734,735,1778,575,918] A loud crackling sound signals the same process for hermit crabs on Caribbean islands (R. Forman, pers. observ.). In southeastern Canada, mortality of small forest mammals is greatest on road corridors 14–31 m wide, but less on narrower roads.[1272] The death rate of mid-sized mammals peaks at a greater corridor width than that for small mammals.

Predators and scavengers that eat road kills are also killed by vehicles.[1679,1796] Snakes and amphibians basking on warm road surfaces are often killed.[1796,918] Road salt in snowy areas attracts animals and leads to road kills.[872,1271,914] Small mammals and birds living in roadsides are particularly susceptible to road kill during breeding season, when food collecting is critical and young are moving around.[1796,262,548]

Does this sink effect of road corridors have a significant impact on an animal's population? For common small mammals and birds, several studies indicate that only a few per cent or less of a population is killed annually.[575,736,1486,111] This loss is easily recovered by reproduction.

Yet, road kills may have a significant impact on populations of large mammals and some rare species. For example in Florida, vehicles are believed to be responsible for the majority of deaths of many of the large, slowly reproducing animals, including black bear, key deer, and the endangered Florida panther (*Ursus americanus*, a South Florida race of *Odocoileus virginianus*, *Felis concolor*).[685] The same pattern exists for the

167

European badger (*Meles meles*) in Britain.[365] Significant impacts of road kills may affect populations of toads in The Netherlands[1778], hedgehogs (*Erinaceus europaeus*) in Germany[1404 cited by 111], Carnaby's cockatoo (*Calyptorhynchus funereus latirostris*) in Western Australia[1489], and the threatened eastern barred bandicoot (*Perameles gunnii*) in Victoria, Australia.[111]

John Steinbeck tells another story of two people driving in the California hills looking for supper, and by swerving well off the road surface, getting a chicken that was later deliciously roasted.[1621] Road kills also depend on the speed of vehicles. Attempts to change human behavior with road signs warning of animals crossing are apparently ineffective.[1336,318] However, the presence of large carcasses on the road does decrease average vehicle speed. Diverse, sometimes exotic, techniques are used to alter animal behavior to decrease road kills.[102,1396,1336,1397,1828] But underpasses, tunnels, fencing, and possibly flat bridges at land level are generally considered most successful (see Box 5.1).

In conclusion, two-lane highways with high speed limits apparently have the highest road kill rates. But wide multi-lane highways have a much greater ecological effect. They remove more native habitat. More importantly, because many animals do not approach or attempt to cross them, these multi-lane highways are effective barriers that subdivide populations into smaller subpopulations. The subpopulations are more subject to local extinction.

Source

The road corridor also affects the adjacent matrix through the dispersal of particulate and chemical pollutants, water, noise, and roadside plants and animals (Fig. 5.10). Material pollutants include gases (e.g., oxides of carbon, nitrogen, and sulphur), chemicals (e.g., lead and salt), and particles (e.g., carbon, oil, rubber, corroded auto material, litter, and dust from the road).[912, 416, 111] Gaseous pollutants disperse widely and contribute to acid precipitation, but elevated effects on the adjacent matrix are rarely apparent.

Lead is added to petroleum in some regions, where high levels are found in soil, plants, and animals near roads, especially heavily traveled roads. In general, soil has higher lead concentrations than plants.[1363,1830,1865] Therefore, worms and other detritus-feeding animals in contact with soil have higher lead levels than herbivorous insects.[111] Small mammals (e.g., shrews) that feed on invertebrates, generally have in turn higher lead levels than those feeding on plants. The ecological effects of lead levels near roads appears minimal.[1125,560,562,773] However, in view of the severe effects of lead on children, long-term ecological studies are probably warranted.

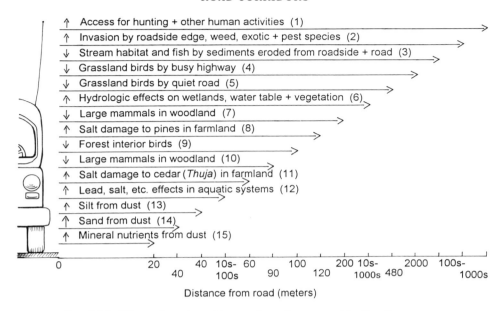

↑ Access for hunting + other human activities (1)
↑ Invasion by roadside edge, weed, exotic + pest species (2)
↓ Stream habitat and fish by sediments eroded from roadside + road (3)
↓ Grassland birds by busy highway (4)
↓ Grassland birds by quiet road (5)
↑ Hydrologic effects on wetlands, water table + vegetation (6)
↓ Large mammals in woodland (7)
↑ Salt damage to pines in farmland (8)
↓ Forest interior birds (9)
↓ Large mammals in woodland (10)
↑ Salt damage to cedar (*Thuja*) in farmland (11)
↑ Lead, salt, etc. effects in aquatic systems (12)
↑ Silt from dust (13)
↑ Sand from dust (14)
↑ Mineral nutrients from dust (15)

0   20   40 10s- 60   100   200 10s-   2000 100s-
          40   100s 90   120   1000s 480   1000s

Distance from road (meters)

Fig. 5.10. Effect of road corridor on the adjacent matrix. Small arrows on left indicate an increase or decrease. Note that the horizontal axis is not linear. Sources: (1) Mech *et al.* (1988), Brocke *et al.* (1990); (4) Lapwing and black-tailed godwit (*Vanellus vanellus*, *Limosa limosa*), 54,000 cars per weekday (van der Zande *et al.* 1980); (5) Ibid, 50 cars per day; (7) Elk and mule deer (*Cervus canadensis*, *Odocoileus hemionus*), Colorado (USA) (Rost and Bailey 1979); (8) White pine (*Pinus strobus*), downwind of snow-plowed highway, southern Ontario (Hofstra & Hall 1971); (9) Maine, (USA) (Ferris 1979); (10) Mule deer, New Mexico (cited by Rost & Bailey 1979); (11) Arborvitae or white cedar (*Thuja occidentalis*), downwind of snow-plowed highway, southern Ontario (Hofstra & Hall 1971); (13–15) Late winter, southern Sweden (Tamm & Troedsson 1955); (2), (3), (6), and (12) Estimates by author.

Salt applied to prevent ice formation on roads alters roadside vegetation[1861,259,416,737], as well as the edge of the matrix where snowplows throw salt into contact with foliage. Detrimental effects on pines and cedar (*Pinus* spp., *Thuja*) extended to 120 m downwind of highway sites in woods in southern Ontario[737] (Fig. 5.10). Salt also dissolves and flows into nearby wetlands, streams, rivers, lakes, and aquifers, sometimes polluting water supplies. Herbicides used in maintaining roadsides similarly flow into water systems.

Dust from roads affects the matrix most in dry climates, and where the chemistry of the road material differs from that of the matrix. Vegetation changes due to this dust are reported to extend 10–20 m from the road[1673]

(Fig. 5.10). Mechanical scraping to maintain roadsides removes the stabilizing effect of plants, and exposes the soil surface to wind and water erosion.

Water runoff from road surfaces in dry climates enhances growth of nearby vegetation and animals.[761] More generally, rain water erodes and washes soil and other particles into lower areas in the matrix[1551] (Fig. 5.9). Streams crossing roads are impacted especially severely, because the uneven stream bottom is smoothed out by sedimentation, sometimes for long distances downstream[699,195,229,290] (Fig. 5.10). This destroys habitat for fish and aquatic invertebrates[281,941], and leads to flooding of wider areas. Sanding of icy roads near bridges commonly leads to the same effect. Until remediation of these damaged habitats is achieved, fishermen would do better going upstream than downstream of a bridge.

Noise due to traffic may be primarily responsible for the absence or decrease of certain matrix species near road corridors.[1448,1774,1388,1406] For instance, forest bird species are inhibited at over 100 m from a major highway corridor in Maine (USA) forest land[475] (Fig. 5.10). In a grassland matrix in The Netherlands, important breeding birds are diminished as far as 2 km from a highway which has an average of 54 000 cars per weekday.[1774] Next to a quiet road with 50 cars passing per day, the birds are inhibited to only 0.5 km. The movement of weeds, pests, and edge species from roadsides into the matrix is also widespread.[26,475,111] This inhibits certain matrix species.[21,186,1891,1941,852]

Road corridors commonly affect, even transform, the matrix through the introduction of people and human activities[1942,1099,187] (Fig. 5.10). The significance of road access to remote areas for hunters and settlers is discussed in chapters 8 and 12. The wide range of source effects presented[1520] (Fig. 5.10) emphasizes that roads are of central ecological importance in most portions of most landscapes. Roads should be core topics in ecology textbooks.

Other road and railroad corridor effects

In landscapes where most natural vegetation has been removed, e.g., for cultivation, pasture, or plantation, the establishment of roadside natural strips on the outer portions of road corridors has a major, positive ecological effect.[109,1490,1492] For example, in extensive portions of western and southeastern Australia, and in the grassland and fynbos areas of South Africa[371], these natural strips are considered of major conservation significance (see Box 5.2).

Road construction is a particularly important cause of erosion, as well as of habitat loss for rare and endangered species. Maintenance regimes for roadsides vary from the intensive use of mowing, herbicides, or fire, to low-intensity use for semi-stable shrublands or natural wildflower protection.[1211,771]

Railroad corridors exhibit most of the same patterns described for roads[1120,1210,872,1615,1167,970], though a few characteristics differ. Railroads

Box 5.2: Roadside natural strips: a giant green network

Roadsides (verges) are typically managed or disturbed regularly to maintain an open, often grassy, strip. But in parts of Australia a striking double strip of natural vegetation is present bordering the open roadsides (Fig. 5.4). These 'roadside natural strips' predominate in landscapes where native vegetation has been largely cleared for intensive use, such as agriculture.

Since c. 1952 in Western Australia, natural strips primarily owe their origin to the public's desire to see native wildflowers[772] (B. Hussey, pers. commun.). Similar strips (sometimes called 'beauty strips') in karri forest of the area and in several regions of the USA are left unlogged to enhance the aesthetic quality of road corridors. But many other benefits accrue from the natural strips, including less soil erosion, less deposition of sand and silt on road, less wind for vehicles, more shade, more habitat for migrant and resident fauna, more chance for gene flow as species move along the strips, and more visually relaxing motoring.

Useful reviews examine roadside natural strips in general[1824,109,111, 1492], as well as geographically from western[1490,1204,52,54,1492,1491] to eastern[1650,108,109,110,962,725] Australia. Possible analogues exist in South Africa[371,372], Britain[1837] and the USA.[1238,1807]

Ecologists and engineers in highway departments at all governmental levels work together in planning and management of the Western Australia roadside natural strips. Management techniques include fencing against livestock, fire management, surveys and monitoring of ecological attributes, and education and active involvement of volunteer naturalists. The spread of weeds and exotic species is reduced through digging, applying herbicides, and planting native species. Managers work with adjacent landowners to minimize inputs of fertilizers, pesticides and fire, and to increase interactions with nearby fragments of native vegetation. Road proximity provides access for management, but management is complex because the natural strips adjoin so many land owners and land uses. Design, planning, and management techniques for these strips of native vegetation are published and widely distributed.[772,109,771,1582]

Roadside natural strips are interconnected to form a giant green network in the landscape, with wide strips on major highways and narrow strips on rural roads. In Western Australia most individual natural strips are 10–70 m wide (road corridors 60–200 m wide), whereas in moister southeastern Australia most natural strips are 5–20 m wide[109,1491] (B. Hussey, pers. commun.). Ignoring patches (chapter 2), the giant green network offers few disadvantages and many ecological and human advantages.[493]

171

Fig. 5.11. Trails varying in barrier, curvilinearity, surface, and edge characteristics. Great Wall of China near Beijing. Route used by potato farmers in the Andes (paramo) above Merida, Venezuela. Path through old-growth coniferous forest in northwestern Montana (USA). R. Forman photos.

tend to be straighter than roads, cause more and larger fires, and spread particles widely over the adjacent matrix.[496] Animal kills per train are perhaps lower than per vehicle on roads, though some spectacular derailings have resulted from large mammals. Some trains carrying grain and other goods doubtless spread seeds, insects, and small mammals widely along the corridor.[1710,1615]

In grassland landscapes often agriculture has replaced native grassland. Here, remnants of native grassland may only persist in places along railroads.[1400,155] These grassland strips are quite equivalent to the roadside natural strips.

TRAILS AND POWERLINES

Trails

Hiking trails, walking paths, horseback trails, rides in European woods, cross-country ski trails, bicycle trails, motorbike trails, snowmobile trails, livestock routes, and animal trails are included. All have a central strip of repeated disturbance for travel, plus distinctive adjacent trail margins affected by travel or maintenance[968,1842] (Fig. 5.11). Relative to road corridors, trails are normally more curvilinear, quieter, less polluted, and more likely to change location. All of the types included result

172

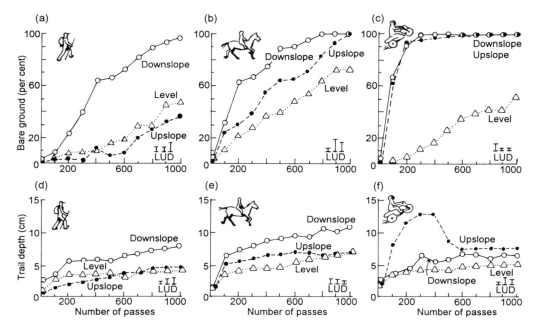

Fig. 5.12. Trail formation by hikers, horses, and motorcycles on different slopes. Experiment on meadows with deep sandy loam soils in Montana (USA). Slopes = 15°; hikers weighed 82 and 91 kg and wore hiking boots with cleated soles; horses weighed 500–579 kg and had uncleated shoes; motorcycle was a Honda 90 running in second gear at speeds of < 20 km per hour. Average standard errors for level, upslope, and downslope points are given in lower right corners. Adapted from Weaver & Dale (1978).

from human activity except animal trails caused by repeated movement of vertebrates along a route.[1818] All are caused by locomotion, except the motorbike and snowmobile trails, which are the most likely to produce noise effects on nearby animal movements.[413,1842] All except the ski and snowmobile trails tend to compact the soil, with consequent problems of wind or water erosion.[178,1766,474] In dry areas where disturbance is periodic but intense, livestock routes tend to lead from water source to water source.[1818,725] 'Greenways' that include trails are discussed in the following chapter 6.

The effect of hikers, horses, and motorcycles was experimentally compared in meadow vegetation (*Festuca, Poa*) in the mountains of Montana (USA).[1842] Several aspects of ecosystem damage were measured, including amount of bare ground and trail depth, which are associated with soil compaction and erosion. Horses and motorcycles create more damage than hikers (Fig. 5.12). Horse and motorcycle damage is greater on slopes than on the level. Hikers cause the greatest damage going

173

downhill. Horses do too. In contrast, motorcycles create deeper trails going uphill.

Fragile habitats such as wetlands, dunes, and alpine areas are frequently damaged by trail effects.[968,1842,774,1079,96,336] Trampling vegetation, and sedimentation from eroded surfaces, may be pronounced. Boardwalks and bridges can be used to protect vegetation and prevent sedimentation of streams and wetlands.[1574]

Trails often change in shape and move in location (Fig. 5.11). Shortcuts, erosion, and adjacent land changes may increase or decrease curvilinearity, or produce braided systems. Introduced weeds are common along trails.[113]

Unlike most roads, existing trails and paths are often used by native mammals as conduits or travel lanes for movement. However, heavy human usage of trails, and probably even occasional usage by domestic dogs that leave scent marks, eliminate almost all conduit use by wild animals. The same factors probably also limit animals from crossing trails. If so, an invisible band of unknown width along trails subdivides a matrix habitat for some animal populations. This pattern would be important in the design of trails in parks and nature reserves. Indeed, native fauna and domestic dogs may be incompatible. Finally, a major ecological importance of trails is the human access to remote areas, where overhunting or other ecological damage may take place.[96,336]

Powerlines

Powerline (electric transmission line), gas line, oil line and dike corridors tend to be long, relatively straight, constant in width, abrupt in their boundaries, and disturbed or maintained relatively evenly over their whole width.[48] Trails or roads may be included within them. Powerline corridors cover about two million hectares in the USA (equal to the area of Israel or Lake Ontario).[421]

Edge and generalist species cover almost all powerline corridors.[1211,1005,28,27,813,723,897] For example, birds were compared in powerline corridors of 12 to 92 m width (c. 40–300 ft) in a forested Tennessee (USA) landscape.[28] Almost all the individual birds are edge species. A third of the 35 species encountered apparently correlate with corridor width. Of these, two are interior species that ventured into the narrow corridors (Fig. 5.13a). However, of most interest are two open-field species that increased sharply in abundance as powerline width increased (Fig. 5.13b).

Except for people and vehicles, evidence for the conduit function of powerline and dike corridors is limited.[1214,1790,1791,493,552] Nevertheless, it seems likely that some species move along within or parallel to these corridors.

174

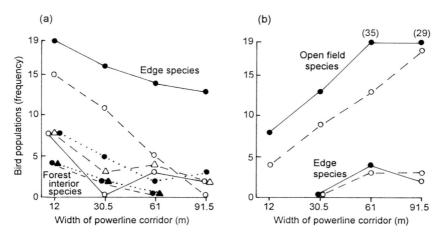

Fig. 5.13. Powerline width effect on bird populations. Open powerline corridors cut through Tennessee (USA) forest land. All 11 species that apparently correlate with width are plotted. (a) Five edge species that decrease with corridor width, plus two forest interior species at bottom, scarlet tanager and red-eyed vireo (*Piranga olivacea*, *Vireo olivaceus*). (b) Two edge species apparently favored by wide corridors, plus two open-country species, field sparrow and yellow-breasted chat (*Spizella pusilla*, *Icteria virens*). Based on Anderson et al. (1977).

The noise emanating from powerlines, especially in wet weather, may be a major factor in the filter function of powerline corridors. Studies by J. Edward Gates and others find that many bird and mammal species, as well as toads, arthropods, earthworms, and snails/slugs, are inhibited in crossing these corridors.[1521, 268, 552] Crossings of shrub- versus grass-dominated corridors (50 m wide) by small and mid-sized mammals have been measured in a wooded area of Maryland (USA).[911] Small mammals (mouse, squirrel, chipmunk; *Peromyscus*, *Sciurus*, *Tamias*) crossed shrubby corridors 10 to 34 times as often as grassy corridors. On the other hand, corridor vegetation type did not affect the crossing frequency of mid-sized mammals (woodchuck, skunk, domestic cat; *Marmota*, *Mephitis*, *Felis*). Insects and mammals may be affected more by powerlines than birds and humans, because of sensitivity to particular wave lengths of the electromagnetic field. Populations of edge species in the nearby matrix may also be affected by the electromagnetic fields.

Migrating birds die from hitting towers and powerlines, especially thin ground-attached wires.[1528,1641,29] The overall sink effect of this appears to be minor, but in places may be significant to certain waterfowl and other wetland species.

Management of powerline corridors usually focuses on preventing tree growth that could interfere with the lines. Herbicides, cutting, and fire are the primary techniques utilized. Studies by William Niering and colleagues indicate that maintaining a cover and high richness of native species offers several benefits, both ecological and for game. In some landscapes this may be accomplished with semi-stable shrublands (requiring minimal tree removal), which surround patches of herbaceous cover.[1211,268,421] In addition, animal movement across powerline corridors can be enhanced by shrub cover, curvilinear soft edges, and natural-vegetation connections between opposite sides.[911,552]

APPENDIX: EXPERIENCE WITH WILDLIFE UNDERPASSES AND TUNNELS

The following lessons about the best locations for underpasses and tunnels have been learned from diverse specific projects. (a) Locate underpasses where cover from woodland or topography comes close to a road on both sides.[102,1356,1566] (b) Keep underpasses and tunnels away from sources of disturbance (e.g., dogs, houses, noise, and pollutants).[109] (c) Lengthen bridges over water or wetlands to include a dry land route for crossing.[685,1228] (d) Include several underpasses or tunnels to provide options and loops for animals to avoid disturbance, and to provide a more-natural distribution of populations.[1035,688] (e) Maintain distances between underpasses such that animals do not have to travel far to find a crossing.[918] (f) In wooded areas establish clearings and salt sources away from the road, so animals are not attracted to roadsides.[470] (g) A line of evergreen shrubs or trees provides a visual barrier, whereas a wall, earth berm, or wide strip of dense woody plants shields noise. (h) Sparsely vegetated approach trails provide good visibility.[1296] (i) Erect fencing along a road near an underpass, and screen traffic by plantings.[1566,918] (j) Animals seem especially disturbed when approaching an underpass while a vehicle is passing, and visibility is obscured.[1296] (k) A funnel-shaped rocky corridor that leads to a tunnel can mimic a scree slope.[1042] (1) A wide underpass provides greater visibility of the other side of the highway.[1296] (m) Concrete walls and floors are suitable, but with tunnels high enough to see the skyline (horizon) at both ends.[1828] (n) Large open underpasses are more successful than tunnels.[1398, 1395] (o) A wide underpass lets a whole assemblage of species through, rather than only the target species. (p) Maintaining underpass or tunnel width greater than height is generally much more effective. (q) Small underpasses may magnify the sound of passing vehicles.[1296] (r) Puddle or small-pond formation and substrate washouts in tunnels suggest that greater effectiveness is attained for terrestrial wildlife if tunnels are not combined with a stream (J. E. Gates conclusion[cited by 688]). (s) Small (20 cm wide × 40 cm high) prefabricated tunnels with road-surface slits for light and air are suitable for amphibians.[918] (t) Design for the most sensitive species.[1228] (u) Mortality by predation or other factors may be high in tunnels.[918] (v) Fencing and tunnels should be durable and easily maintained.[918]

6

Windbreaks, hedgerows, and woodland corridors

These farms which I myself surveyed, these bounds which I have set up, appear dimly still as through a mist; but they have no chemistry to fix them ... The world with which we are commonly acquainted leaves no trace, and it will have no anniversary.

Henry David Thoreau, Walking, *1862*

Since the onset of agriculture, woody strips have spread to become the most conspicuous feature in many landscapes. Thus, the noted British ecologist, Charles Elton, observed (1958) that, 'our own highly managed landscape is still interlaced with a wonderful network of hedgerows and roadside verges. These long winding strips of habitat by the road and land and field margins are the last really big remaining reserve we have.'

Hedgerows were originally created to keep livestock in a field, or to exclude livestock or wild herbivores from a field. Woody strips were used to clearly delineate boundaries. And windbreaks provided protection against wind, as well as defense against marauding humans.

In contrast to trough corridors (chapter 5), *woody strips* have a canopy higher than the surroundings, and include windbreaks, hedgerows, and woodland corridors. *Windbreaks* or shelterbelts are planted to protect against wind. *Hedgerows* are narrow line corridors up to a few tree or shrub diameters in width that separate open areas. The term hedgerow is used in a generic sense to include examples produced by any mechanism, such as fencerows, planted hedges, and remnant lines of trees (see glossary in chapter 1 appendix). *Woodland corridors* are wider strips composed of natural vegetation. All wooded strips are of anthropogenic (human) origin, either directly planted or resulting from human activities on adjacent sides. Most are relatively straight, narrow, and costly to maintain. 'Greenways' that support recreation are included in this chap-

177

ter, but 'roadside natural strips' are described in chapter 5, and stream and river corridors in chapter 7.

Woods may produce wood products, but other economic and ecological values normally are more important. Gifford Pinchot, America's pioneering forester, pointed this out in 1905: 'Not only does [the forest] sustain and regulate the streams, moderate the winds, and beautify the land, but it also supplies wood.' Thus, windbreaks are planted to reduce soil erosion, catch and hold snow, provide soil moisture in steppe areas, protect crops from desiccation, protect livestock, and reduce home energy costs. Hedgerows are widely used to improve crop yields, control livestock, ameliorate drainage and erosion, provide wood products, and enhance aesthetics and a sense of place. Woodland corridors are of particular value as conduits for animal movement, as well as for hiking trails and similar recreation.

Much of the subject is reviewed by Stoeckeler (1962) and Brandle *et al.* (1988). Additional useful reviews of individual topics include: effects on wind[241,711]; effects on microclimate[1776,1446,1093]; erosion and its control[70,274,1950,1012,1714]; hedgerows[1337,948,501]; and woodland corridors.[688,111,1228]

This chapter begins by identifying the key ways in which woody strips differ in structure. Next, windbreaks and their manifold effects on wind, microclimate, soil erosion, snow, and moisture are explored. Hedgerows are then introduced mainly in the context of habitats for plants and animals. Finally, woodland corridors are evaluated as conduits for the movement of animals, an ongoing issue in biodiversity conservation.

STRUCTURAL RICHNESS

In some areas wooded strips are so common the land looks like a football in a net bag. Like strings of the net, the wooded strips are made by humans and constrain the land. Some strips have been directly planted, some have arisen around fences and ditches, and some remain from woodland removal on both sides (Fig. 6.1).

Single-row strips include hedges, roadside treelines, agroforestry lines, some windbreaks, planted hedgerows, and some spontaneous hedgerows[1337,203,174,720] (Fig. 6.2a–d). Shrub lines include trimmed hawthorne (*Crataegus*) hedgerows in northern Europe, and hedges planted for privacy. The presence of a shrub layer, and sometimes an understory layer, under a tree canopy often indicates a hedgerow that grew up naturally over a fence, ditch, or stonewall. A treeline without a shrub layer may result from shrub removal by humans, livestock, or large wild herbivores. Alternatively, treelines are commonly planted to provide shade or decrease wind along roads, or between open areas. Thus Omar Khayyam

178

Fig. 6.1. Woody strips of different origin, structure, and scale. Straight hedgerows formed along fences separating fields (foreground). Probable stream corridor on left and possibly in foreground. Major forested ridge in distance. Pennsylvania (USA). Courtesy of US Fish and Wildlife Service.

observes in the Rubaiyat (*c.* 1100 AD), 'With me along some strip of herbage strown; that just divides the desert from the sown.'

Multiple-row strips are all planted, usually as windbreaks (Fig. 6.2e–h). A double treeline is common along a road, where it is irrigated by runoff from the road surface. Multiple-row strips of equal-height trees and little woody undergrowth usually have been planted for wood production, such as on rich floodplains in Europe. Strips with separate rows of shrubs and trees are generally windbreaks protecting homes and farmsteads, where space is usually available for planting several rows. The highest point of the windbreak may be upwind, downwind, or in the center, an important functional difference as discussed later in this chapter.

Remnant or regenerated woodland corridors differ primarily in time since formation (i.e., age) or disturbance, and consequently in the amount of woody undergrowth and the density of edges (Fig. 6.2i–k). Recent removal of adjacent woodland typically leaves an open corridor with little undergrowth, especially in the edges. Recent disturbance, such

179

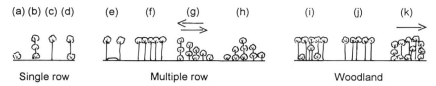

Fig. 6.2. Types of woody strip in cross section. Foliage indicated in canopy, understory, and shrub layers, respectively. Typically, strips (a) to (h) are planted, and (i) to (k) are natural. (e) Includes a road. Arrows indicate predominant wind direction. In (g) the strip differs functionally according to wind direction.

as surface fire, livestock browsing, and overbrowsing by native herbivores, also produces an open corridor. Soon thereafter herbaceous and shrub layers often form. Over a longer period the forest understory and edge mantel (chapter 3) may fully develop.

The major structural differences evident from a side view of wooded strips are the presence of gaps and variability in height, both especially prominent in single-row strips. Vertical stratification of foliage is often conspicuous. Species composition differences and regularity in their distribution also may be marked.

The important attributes evident from a top view are described in the preceding chapter, including length, environmental gradient, patchiness, adjoining ecosystems, intersections, attached nodes, narrows, connectivity, gap sizes, aggregation of gaps, and suitability within gaps (see Fig. 5.2). Three attributes tend to be low in wooded strips: corridor curvilinearity, curvilinearity of an internal entity (chapter 5), and variability in width. Most wooded strips are straight, and where present, most walls, ditches, soil banks, and fences are straight. An important exception is a strip along curvy topographic contours. Variability in width is doubtless important in some woodland corridors.

Not surprisingly, this structural richness has many functional implications especially related to wind and wildlife. It also offers many opportunities for creative design. Although some woody strips may be a challenge to establish, they are easily removed, pruned, sculpted, and enhanced.

WINDBREAKS AND WIND

Wind principles

Wind or air flows in three basic forms. *Streamlines* are parallel horizontal layers of moving air; *turbulence* is irregular air movement typically characterized by up and down currents; and *votices* (or vortexes) are helical or spiral flows, often with a vertical central axis.

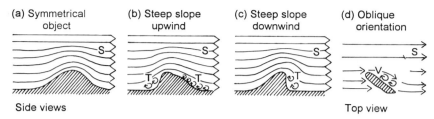

(a) Symmetrical object	(b) Steep slope upwind	(c) Steep slope downwind	(d) Oblique orientation

Side views   Top view

Fig. 6.3. Streamline (S), turbulent (T), and vortex (V) airflows near objects of different shapes. Maximum streamline wind speed is just above the top of the shaded object. Strong turbulence is indicated by large eddies in (b) and (c). Based on several sources.

A *streamlined object*, such as an airplane wing placed in horizontal wind, splits the streamlines and produces little turbulence, because the streamlines come back together almost immediately downwind of the object (Fig. 6.3a). In contrast, a 'bluff object', characterized by a blunt upwind or downwind side, splits the streamlines and causes turbulence (Fig. 6.3b and c). If the blunt side is upwind, turbulence forms in front of the object and usually also on the downwind side. If the upwind side is streamlined and the downwind side blunt, turbulence only forms downwind. If an object is porous, the airflow through the object is called *bleed flow*. This decreases air-pressure differences on up and downwind sides of the windbreak, thus reducing turbulence.

Most natural windbreaks are bluff porous objects. The primary factors determining windspeed patterns near a long thin barrier on a smooth horizontal surface are reasonably well known for air flowing perpendicular to the barrier. Wind speed is a function of (1) barrier height (h), (2) porosity, (3) distance from barrier, (4) height above surface, (5) roughness length in uninterrupted wind, and (6) atmospheric stability[1093] (see chapter 6 appendix). (The Reynolds number relating the viscosity of air and other factors could be added to the wind speed equation.) However, a windbreak is not on an idealized smooth surface, and consequently the roughness of the surrounding area is a major variable affecting wind speed.

Barrier height has been measured as the maximum height, or the average height, of the taller plants in a windbreak. However, wind effects may better correlate with height, if the latter is measured as average height over random points along a windbreak.[711] Distance upwind or downwind of a windbreak is best measured from the margin of the windbreak.

The *porosity* of narrow windbreaks may be assayed optically with visual or photographic measurements. For example, a porosity of 0.2 means that 20% of a vertical plane is unobscured by windbreak vegetation.[711] In wide windbreaks optical measurements are less useful, since air moves through curved three-dimensional spaces. Optical porosity also does not differentiate between large and small pores, or between pores with sharp or smooth edges, e.g., in different designs of snow or sand fences. Therefore, aeronautical engineers prefer using a 'resistance coefficient' to estimate porosity.[74,68,711] Resistance is related to the density of air, windspeed through the barrier, and pressure drop in parallel airflow across a section of barrier (see chapter 6 appendix).

In short, windbreaks alter horizontal wind speed, turbulence, and vortex airflow. Height and porosity of windbreaks are the two major controls on these airflows, and both are amenable to design and management.

Effect on wind

Measurable reductions in airflow typically extend a few h upwind of a windbreak (h = height of windbreak), which creates a partially sheltered area called the *upwind zone*. However, downwind of a windbreak reduced wind speed extends to a distance of a few tens of h.[241,68,174] Two distinct microclimatic zones downwind of a windbreak are distinguished, based on evidence from wind tunnel experiments, mathematical modeling, and field measurements around artificial and natural windbreaks.[1776,241,644,1375,645,1914] These two zones, the quiet and the wake zones, are first described in terms of wind, and later relative to temperature and moisture.

Quiet and wake zones

The *quiet zone* is a triangular area extending from top to bottom of a windbreak, and downwind to ground level at about 8h[1093] (Fig. 6.4). The location of this zone appears independent of porosity, and is characterized by lower horizontal windspeed and higher turbulence than in airflow in the open. From about $8h$ to about $30h$ or more is the *wake zone*, characterized by higher wind speed, greater turbulence, and larger 'eddy' sizes (small circular flows within turbulence) than in the quiet zone. Wind patterns in the wake zone gradually change downwind until they merge with those in the open. Wind speed reductions of about 20% commonly extend some $25h$ downwind of vegetation windbreaks.[1178,711]

Windspeed and turbulence

Horizontal wind speed is usually of interest for one of four purposes: the minimum velocity attained; distance downwind from windbreak to the minimum velocity point; total length of wind reduction; and how the wind reduction changes over that length.[711] Determining these is import-

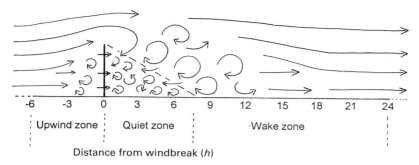

Fig. 6.4. Zones with altered airflow caused by a windbreak. Vertical dimension is magnified for illustration. Vertical line indicates windbreak; h = height of windbreak. Large eddies = strong turbulence. Uninterrupted airflow in the open is to the left of the upwind zone, and to the right of the wake zone. Widths of zones are approximate. Based on several sources.

ant in, for example, maintaining a butterfly population, locating a walkway, and minimizing soil erosion.

Turbulence is of significance primarily because of its effect on exchanges of energy and mass. High turbulence causes a large loss of heat. High turbulence also activates surface particles. This accelerates soil erosion, dislocates insects from their plants, and tears leaves from their stems.

The patterns and principles recognized here are based on idealized studies, with streamlined airflow approaching over an open smooth surface that extends upwind many times the height of the windbreak. However, turbulence in the approaching wind, noticeably alters the windbreak effects.[802,1531,1135,789] This compresses the downwind quiet and wake zones, as well as the upwind zone (Fig. 6.4). In effect, turbulence in approaching wind decreases the effectiveness of a windbreak. Therefore, most windbreaks are in open agricultural and arid land. The turbulence present in suburban and topographically diverse terrain means that windbreaks there are only effective over a narrow zone.

Windbreak porosity and height

In designing windbreaks, height and porosity are the major structural attributes that are altered.[1776,241,711] Height primarily controls the distance downwind that a windbreak effect extends. Porosity primarily affects wind speed, both upwind and downwind of a windbreak. High, medium, and low porosities are usefully differentiated.

Most people believe that dense barriers are better and more effective than porous barriers. But alas, this belief succumbs to Thomas H. Huxley's 'great tragedy of science . . . the slaying of a beautiful hypothesis

183

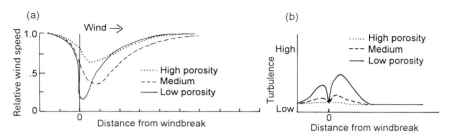

Fig. 6.5. Effect of windbreak porosity on streamline and turbulent airflows. (a) Streamline airflow based on treebelts of different foliage densities; wind measurements at 1.4 m height. From Heisler & DeWalle (1988) with permission of Elsevier Science Publishers. (b) Generalized expected turbulence pattern based on Robinette (1972), Rosenberg et al. (1983), Heisler & DeWalle (1988), McNaughton (1988).

by an ugly fact'. For most human uses a highly or medium-porous windbreak is optimum, though this varies according to the design objective.

A highly porous windbreak, such as a row of planted poplars (*Populus*) (Fig. 6.2c), decreases wind speed only minimally, but has the advantage of also minimizing turbulence (Fig. 6.5a and b). Furthermore, a highly porous windbreak provides a relatively long distance of reduced airflow downwind, although wind speed is only slightly slower than wind in the open.

In contrast, a low-porosity windbreak, like a stonewall or thick dense evergreens, produces a short downwind zone of highly diminished wind speed. However, high levels of turbulence are created both upwind and downwind.

A medium-porous windbreak (e.g., Fig. 6.2b or d) has wind speed reduction nearly as great as the impenetrable barrier. But the shelter extends over a somewhat longer distance downwind (Fig. 6.5a). In addition, the medium-porous windbreak has much less turbulence present (Fig. 6.5b). In many applications a medium-porous windbreak is considered optimum. However, a narrow path may be better protected from wind by an impenetrable barrier. A wide field best accumulates snow and moisture with highly porous windbreaks, but best protects against soil erosion with medium-porous windbreaks.

Turbulence and eddy size in the quiet zone decrease as windbreak porosity increases (Fig. 6.5b). Furthermore, the largest difference in turbulence is between solid and low-porosity barriers. Turbulence differs only slightly between medium- and highly porous windbreaks.[644,1914,711]

The structure of windbreaks is also described in terms of species planted, number of parallel rows planted, distance between rows, and distance between plants within a row, as described later in this chapter.

184

These are useful characteristics that can be manipulated to accomplish different management, production, and conservation goals. Length and width of windbreaks are also important structural attributes (also gaps discussed later in the chapter). The patterns presented here are based on windbreaks that are long relative to their height. The effects of windbreak width are indicated by porosity or resistance coefficient.

Snow accumulates to a greater depth behind medium than highly porous windbreaks.[1512,1545] Snow accumulates over a broader zone behind medium- than low-porosity windbreaks. This is particularly clear when the lower portion of the windbreak is more porous than the upper portion. Such differences are important economically in dry, cold steppe landscapes, where water from snowmelt is critical to plant growth in the spring.

It is unclear whether porosity in the upper or lower portion of a windbreak is more important in determining the width of a wind speed reduction zone, or in causing the overall amount of windspeed reduction. Theoretical reasons support each possibility.[240,1648,1446,629,711] A windbreak of deciduous trees during the leafless period typically exhibits at least 50% of the wind reduction evident during the foliage period.[1178,1648]

Nevertheless, it is clear that an open lower portion of a windbreak acts like a gap in the vertical dimension. For example, a treeline near a road or house, or a heavily browsed hedgerow, typically has little or no shrub layer present (Fig. 6.6). Streamlines are forced through this narrow horizontal opening, causing an acceleration in wind speed in and just downwind of the windbreak.[802,241,690] For example, a typical 20% increase in wind speed causes significant problems to vehicles on a road, for heating a home, and in desiccating wildlife. Under a spreading acacia tree on a hot, tropical savanna afternoon, a person with a friendly lizard enjoys shade, but also experiences wind resembling a blast oven.

Windbreak form in cross section

Wind also molds the external form of windbreaks. Over time a more streamlined form develops, particularly by wide windbreaks in windy climes (Fig. 6.2k). In the midwestern USA windbreaks have often been planted with shrubs on the edges and trees in the center (Fig. 6.2h). By creating a streamlined form on both the upwind and downwind sides, turbulence is nearly eliminated.[241] Changing the cross-sectional form to have tall trees either on the upwind or the downwind edge (Fig. 6.2g) has little effect on wind speed, but adds turbulence[1921] (Fig. 6.3).

Tall windbreaks with vertical sides, and a narrow 0.2–0.3 width-to-height ratio, are frequently recommended for long-distance wind reduction.[629] Such windbreaks of course are more subject to severe alteration by blowdowns and disease than streamlined, multi-species wind-

185

Fig. 6.6. Midday view of blowing soil and treeline without understory. Wind erosion caused by lack of plant cover in surroundings, combined with drought and wind. Dawson County, Texas. Courtesy of USDA Soil Conservation Service.

breaks. It is uncertain how roughness of the upper surface of a windbreak affects the width of the wake zone.[240,629] Experiments with artificial windbreaks suggest that adding horizontal fins atop the windbreak reduces its overall effectiveness.[542] Leaning a windbreak upwind increases, and leaning it downwind decreases, its effectiveness.[826,711,542]

Downwind of a forest (Fig. 6.1), the wind patterns appear similar to those behind an impenetrable windbreak. High turbulence and a relatively short area of reduced wind speed are present.[1178,1776,711] This is probably largely due to the low porosity of a forest (Fig. 6.5). However, turbulence produced by a rough forest-canopy surface, such as of old-growth forest or tropical rainforest, will further decrease the downwind distance of wind speed reduction.

Wind direction

Airflow reaching a windbreak oriented oblique to wind direction is forced over the high point of the plants, and then bends horizontally

186

toward the windbreak[1170] (Fig. 6.3d). This results in a much shorter downwind area of reduced wind than for a perpendicularly oriented windbreak. Some of the effect is due to differences in porosity, as wind intersects obliquely with the plants in a windbreak. However, obliquely oriented stonewalls also result in a change in wind direction, and a small sheltered area. Oblique windbreaks may cause noticeable vortex flows (Fig. 6.3d), and consequent increases in soil erosion.[788] Further research will determine whether these oblique-windbreak problems can be ameliorated with greater porosity, or with intersections in a windbreak network.

The length of an oblique windbreak is also important, because behind short windbreaks much of the area may be affected by wind bending around the upwind tip. Where a windbreak is long relative to h, and oriented more than $c.$ 60° from the wind direction, a significant zone of wind speed reduction can be expected.[788,1531,629]

Gaps in perpendicular windbreaks (where gap width is $<h$) often have wind speed increases of 20% or more.[1179,1776] Horizontally flowing air is 'squeezed' into the gap, causing an acceleration (the Venturi effect). Areas just upwind, within, and just downwind of the gap not only experience higher wind speed, but also less turbulence. This principle has been used, e.g., to accelerate airflow through a house, by planting a windbreak parallel to each side of the house.[690] In a hot or insect infested place that would feel good. In a cold windy site it would feel frigid.

When a windbreak is oriented parallel to airflow, air accelerates around the upwind tip.[1776] However, somewhat reduced airflow with some turbulence is found in a narrow band adjacent to the windbreak, due to drag or friction. Thus, in a hedgerow network, accelerated wind speed through a gap may be ameliorated by placing the gap near the intersection of hedgerows.[1776]

EFFECTS ON MICROCLIMATE, SOIL, SNOW, AND PLANTS

Temperature and moisture

Watching a weather report on television only suggests the many variables composing microclimate. Of these, wind speed is modified furthest downwind of a windbreak (Fig. 6.7a). Evaporation is also affected a considerable distance downwind.

Although the evidence is more fragmentary, the effects on moisture and temperature appear to be limited to the quiet zone, plus the beginning portion of the wake zone[68,1093] (Figs. 6.4 and 6.7a). Air temperature is higher in the quiet zone, and lower in the wake zone (compared with conditions in the open).[644,1375] The maximum temperature appears to be in the quiet zone at about $4h$ downwind.[1093] In the wake zone cooling is

187

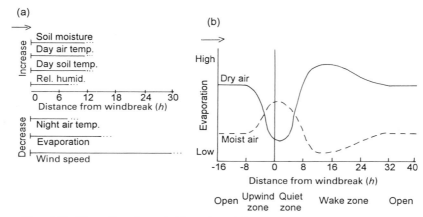

Fig. 6.7. Microclimatic variables altered by windbreaks. (a) Distance is given over which a variable differs significantly from conditions in the open; based on moist air flowing across a field in Britain (Marshall 1967, McNaughton 1988). (b) Generalized expected pattern for evaporation. Adapted from Heisler & DeWalle (1988) and McNaughton (1988).

due to turbulence, not streamline airflow. Actual temperature of course depends on both sun and wind, and the ecological effects of higher temperature near windbreaks varies widely. In a cool climate, productivity depends on warmth, whereas in the tropics more calories may bake the vegetation.

Similarly, relative to open conditions, atmospheric humidity appears to increase in the quiet zone, and decrease in the wake zone.[1573,1093] These temperature and humidity patterns are based on daytime measurements where the vegetation, soil, and slope of upwind and downwind fields are the same. Often of course, fields differ on opposite sides of a windbreak. Differences in heat and water vapor given off by soil and plants will modify overall microclimatic patterns. Also, patterns can be expected to differ at night, and in the early morning and late afternoon, when shadows are greatest.

Evaporation (or evapotranspiration) is altered for a long distance downwind of a windbreak, and may either be higher or lower than in the open.[1134,1092,1093] This depends upon whether the overhead streamline airflow is drier or moister than (saturation-deficit) conditions at soil and plant-surface levels. If dry air flows horizontally, the quiet zone has less evaporation than in the open (Fig. 6.7b). And the wake zone has higher evaporation. If moist air flows overhead, the opposite patterns are expected in the quiet and wake zones.

Soil temperature doubtless follows the evaporation patterns. Dry air overhead should lead to warmer soil in the quiet zone, and cooler soil in the wake zone.[201] Patterns of temperature and other microclimatic variables strongly depend on the orientation angle (chapter 4). East–west windbreaks provide contrasting sunny and shady sides, an effect less pronounced in north–south windbreaks.

In short, these microclimatic variables vary over approximately the same distance that the quiet and wake zones extend downwind. An increase in a variable in the quiet zone (relative to the open) means a decrease in the wake zone, and vice versa. In addition, the magnitude of increase or decrease in the quiet zone appears similar to that in the wake zone.[1093] The greatest difference from the open depends on the microclimatic variable considered (Figs. 6.5 & 6.7). Major differences for most variables extend from the windbreak to the near portion of the wake zone, e.g., to 10–15h. Differences in yield of field crops are also primarily restricted to this area. In the far-wake portion, microclimatic conditions gradually converge with those in the open. Overall, windbreaks modify wind speed and evaporation much further downwind than they affect other microclimatic conditions (Fig. 6.7a).

Soil and erosion control

Soil erosion

A nation that loses its soil self-destructs. Sometimes dramatically, but often invisibly, fertile soil is lost by wind erosion, water erosion, lowered water tables, and salinization. Soil erosion by wind increases where the: (1) soil is loose, dry, and composed of fine particles; (2) surface is smooth with minimal vegetation cover; and (3) susceptible area is large.[1012] Erosion susceptibility is lowest with streamline airflow, intermediate with turbulence, and highest with vortices (see chapter 6 appendix).

Wind erosion, the removal of particles from a surface by airflow, consists of several sequential and overlapping processes. These are particle initiation, transport (suspension, saltation, and surface creep), abrasion, sorting, and deposition.[70,274,1012] Initiation apparently involves oscillation of a particle before it is dislodged, either by another moving particle or directly by wind.

'Black blizzards' or 'dust-bowl days' result from the transport of fine particles (c. 0.002 to 0.1 mm diameter) by suspension.[1930] Such particles may move distances of meters to thousands of kilometers before sedimentation returns them to earth. Although darkened skies and little visibility are dramatic (Fig. 6.6), soil erosion by wind is best represented by slightly dusty afternoons.

Most eroded soil is transported short distances. Mid-sized particles (c. 0.1–0.5 mm) move by saltation or 'hopping'. Typically, such particles rise up to a few decimeters (a foot or two) above the surface, and end up in nearby ditches, windbreaks, and woodland edges (Fig. 6.6). Sand-sized and larger particles or aggregates (c. 0.5–1.0 mm) move short distances along a surface by soil creep, as they are rolled, pushed, or knocked by saltating particles. Abrasion of particles, especially in long fields, is significant in converting larger particles to smaller ones.[1012] This results in greater erosion susceptibility.

Two products of wind erosion are especially noteworthy. First, modest wind erosion selectively removes fine nutrient-rich particles, leaving a less-fertile soil. This loss of soil fertility can be arrested or reversed by windbreaks. These typically provide leaf organic matter, increase soil moisture, accelerate mineral-nutrient recycling rates, enhance the soil fauna, fix nitrogen, increase root decomposition, and minimize the loss of fine particles. Wind erosion commonly reduces soil fertility much more rapidly than soil depth, because of the selective removal of fine particles.

Second, more severe erosion accelerates 'sorting', the separation of soil into deposits of similar particle size. For example, a 'loam' (soil with a relatively even mixture of different particle sizes) in one location is converted into sand, silt, and clay soils deposited in separate downwind locations.

A general equation is often used to predict the potential erosion rate for a site[1920,1012] (chapter 6 appendix). This indicates that the average annual erosion rate increases with: (1) soil erodibility (inherent capacity to erode, based on soil structure and chemistry), (2) soil surface roughness, (3) climate, (4) distance across a field that is unsheltered from wind, and (5) vegetation cover. Control of wind erosion therefore must focus on these variables.

Principles of wind erosion control

Based on long years of experience, five principles for reducing soil erosion by wind are commonly recognized.[1714,1012] (1) Reduce field sizes in the preponderant wind direction, and thus reduce wind speed. (2) Maintain vegetation or plant residues to protect the soil surface (Fig. 6.8a). (3) Maintain soil aggregates or clods large enough (> c. 0.84 mm) to resist the wind force. (4) Roughen the surface, e.g., with furrows, to reduce wind speed and trap moving soil particles (Fig. 6.8b). (5) Cover (or remove) hilltops and other spots susceptible to accelerated streamline airflows, turbulence, or vortices. A combination of practices is always recommended, because of variability in wind direction and velocity, drought intensities, and the temporary nature of some of the control mechanisms.

Windbreaks can contribute to all five principles but the primary role relates to the first principle, namely to reduce the length of erosive surface.[1715] Windbreak height, structure, and porosity effects on wind speed

Fig. 6.8. Soil erosion and a farmstead windbreak. Top: wind erosion in one field without strip-crop plant cover. Center: furrows and soil clods retard wind erosion. Bottom: windbreak of approximately 3–7 rows. Bottom photo looks westward, and cold snow-bearing winds mainly come from the northwest. Note elongated wet spot on left where snowdrift accumulates. (a) = Montana; (b) and (c) = Kansas. (a) and (b) courtesy of USDA Soil Conservation Service; (c) courtesy of US Fish and Wildlife Service.

at varying distances downwind were presented earlier in this chapter. A threshold wind velocity between 19 and 24 km h^{-1} (12–15 mph) is required to start particle movement on most erosive soils.[274,1714] Therefore, the objective is to maintain wind speed over the whole field below this threshold, even during high winds.

Fortunately, barriers reduce wind erosion forces much more than they reduce wind speed.[1714] This is because soil movement by wind varies with the cube of wind velocity. Hence, wind speed reduction can be cubed to indicate roughly the amount of wind-erosion protection by a barrier. For example, at 12h downwind of a barrier in one study, wind speed was reduced to 62% of that in the open. The erosion force was reduced to about 25% (0.62 cubed).[1572] In other words, a particular percentage reduction in wind erosion force extends much further downwind than the same percentage reduction for wind speed.

Herbaceous windbreaks dominated by grass, sunflower (*Helianthus*), or other species are widely used for cultivated areas in cool, dry steppe landscapes.[135,1,134,2,1714,125] Widening the herbaceous windbreaks relative to intervening crop areas becomes a type of 'strip cropping', the alternating of strips of different plantings.[274]

Snow, soil moisture, and plants

Snow accumulation patterns

People create windbreaks to alter the distribution of water molecules, either to accumulate water or to disperse it elsewhere. Snow management is a vivid example, because drifts can block highways, imprison wildlife, and permit children to mimic mice in cozy tunnel networks.

Living and artificial snowfences decrease wind speed, which in turn causes airborne particles to be deposited nearby in a specific pattern[1351,1545] (Fig. 6.8c). This also decreases their deposition further downwind. A well-designed windbreak can keep a highway nearly snowfree, whereas a poorly-designed windbreak can concentrate mammoth drifts directly on the highway. If snow arrives perpendicular to a windbreak, the end portions of the windbreak (several h long) will be relatively ineffective, because wind accelerates and bends around the ends.[1671] If snow arrives obliquely, a huge mound of snow accumulates at a well-defined spot near one end of the windbreak, where a wet soil persists in summer[1889] (Fig. 6.8c). Snow blowing parallel to a windbreak forms deep narrow parallel drifts.[558 cited by 1512] Overall, drifts are smaller if the upwind field is small, heavily vegetated, or sloping up to the windbreak.

Like a bathtub filling up, a windbreak can accumulate a finite amount of snow before reaching equilibrium, where output equals input. This *windbreak capacity* for accumulating particulate matter is the key to designing a windbreak structure that is attuned to winter storm duration and intensity. In addition to snow, windbreak capacity applies to sand, silt, leaves, and even tumbleweeds.

Bathtub capacity depends on tub height and width, drain height, and how you arrange your body in the tub. Analogously, windbreak capacity depends primarily on height and porosity. Other things being equal, a doubling in height is said to increase snow storage capacity four times.[1545] Medium-porous windbreaks that produce smooth, gently sloping, reasonably deep (relative to h) drifts, apparently have the highest windbreak capacity.[1512]

Several windbreak types are normally avoided due to low snow capacities. Short windbreaks, oblique windbreaks, highly-porous windbreaks, and streamlined windbreaks usually trap little snow. A low-porosity windbreak has considerable downwind turbulence (Fig. 6.5b) (this causes some downwind snow to move toward the windbreak, resulting in two parallel snowdrifts).[1776,1512] Minimizing turbulence and vortices is normally a major objective in snow management. Such airflows are highly effective at initiating particle oscillation, which leads to smaller drifts, and hence deposition at another downwind location.

The unresolved issue raised under wind patterns earlier in this chapter, of whether porosity in the lower portion, upper portion, or whole windbreak primarily determines wind patterns, is also important for snow accumulation. A dense windbreak produces a narrow deep zone of snow accumulation. But one may prune away undergrowth. Heavy pruning of undergrowth, e.g., up to 1.8 m (5′8″) height in a line of Siberian elms (*Ulmus*) in Minnesota (USA), produced snow patterns similar to that of an unprotected field.[1512] The open windbreak had little capacity to accumulate snow. Similar studies also found low windbreak capacity with low pruning (only up to 0.8–1.4 m (2′6″–4 ′5″) height). In this case, greater snow capacity resulted from medium-height pruning (to 1.5 m).[558 cited by 1512] These results suggest that snow accumulation is highly sensitive to windbreak porosity.

Alternatively, to accumulate snow one may thin a windbreak by removing alternate trees. A number of experiments has been done on 4.5–6.0 m high elms and ash (*Fraxinus pennsylvanica*) in a single row in the northern Great Plains of North America.[1512,513 cited by 1512] Trees 1.5 m (5 ft) apart produce higher drifts and have higher windbreak snow capacity than those 3.0 or 4.5 m (9 or 14 ft) apart. Thinning alternate trees from a planting 1.5 m apart reduces the drifts and distributes the snow more evenly, and to about $6h$ further downwind. This pattern would be important for recharging soil moisture over a wide area. It also means that these windbreaks can be planted farther apart, and still retain the same amount of snow for soil moisture in spring.

Artificial snowfences with 50–60% porosity have been recommended as optimum for snowpack accumulation.[1355,1351,1545] A bottom gap extending up to 10–15% of the snowfence height increases drift length downwind, and apparently increases windbreak capacity.[1647 cited by 1545] The bottom gap also limits upwind snow buildup, which could convert a non-streamlined to a streamlined windbreak, thus sharply decreasing its effectiveness in catching snow.

Increased snow accumulation in sheltered zones means more infiltration of snowmelt water into the soil, and greater soil water storage. This may reduce water-table lowering. Similarly, snow can be 'harvested' to provide water for livestock ponds, as well as water in wetlands for waterfowl and other wildlife. More soil moisture also decreases soil erosion by linking particles together, which in turn reduces particle oscillation.

Finally, the accumulation of soil particles is essentially the same as that for snowflakes. Therefore, the preceding patterns and principles relating windbreaks and snow also apply to the deposition and accumulation of wind-eroded soil (Figs. 6.6 and 6.8).

Soil moisture, plants, and communities

Overall plant productivity in a field is enhanced by the shelter of windbreaks, but the reasons are complex and the exceptions noteworthy. Growth is highly dependent on leaf temperature. A decrease in wind speed, and especially in turbulence, results in less evapotranspiration and hence warmer leaves. In temperate and cold climates this means more growth, but in hot climates it may be lethal.[607] Reduced plant shaking and mechanical abrasion by blowing particles, especially during droughts, also permit more growth.

These effects of shelter on plants normally result directly in higher agricultural productivity.[1637,948,1445,175] For example, the yield for spring wheat (*Triticum*) during 116 'field-years' in Canadian prairies and northern USA plains was higher between $1h$ and $c.\ 17h$ downwind of windbreaks (relative to production in the open).[174] The maximum crop production was at approximately $3h$ downwind. Similarly, in Ontario, soybean production (*Phaseolus*) was lower from $0–1h$ and higher from $2–9h$, with a maximum at about $4h$ downwind of windbreaks.[174] Similar effects of windbreaks on productivity are found worldwide, though the benefits are less clear in tropical areas.

A stable soil surface and higher soil moisture also can be expected to support a species-rich community in these landscapes. Soil animal communities are enhanced by moist soils, though not by wet spots. The wide range of wind, temperature, and soil-moisture conditions in a gradient downwind of a windbreak permits a large number of plant species to coexist. Consequently, a diversity of foliage, pollen, soil animals,

194

and microhabitats is present, which supports a rich assemblage of herbivores and predators downwind of a windbreak.

HEDGEROWS AS HABITATS

The high density of species and the preponderance of edge species characteristic of corridors (chapter 5) certainly applies to hedgerows and windbreaks.[1937,808,1657,1522] The typical hedgerow is subjected to relatively intense herbivore pressure, and as in forest edges, the plant species present are the survivors (chapter 3). Hedgerows include the wide, regenerated multispecies fencerows between fields typical of eastern North America (Fig. 6.1), the narrow, planted low-diversity hedges typical of Europe, and scattered remnant hedgerows.[1337,750,495,1112]

Internal structure

Essentially all hedgerows and windbreaks are line corridors where edge species predominate throughout. Yet, depending on hedgerow width, forest interior species are probably usually present in low numbers. For example in New Jersey (USA), species richness of forest interior plants increased as hedgerows varied from 4 to 12 m (13–38 ft) wide.[495,501] About 10% of the herbaceous species in hedgerows of 8–12 m width were forest interior species.

In wider strip corridors shade-tolerant plants are enhanced in the central interior environment. For example, American beech and sugar maple (*Fagus grandifolia, Acer saccharum*) may reproduce well in corridors of >30 m and >100 m, respectively.[1379,1228] The modified microclimate in the center of wider hedgerows is considered the primary reason for these results.[501] However, overbrowsing of the shrub layer by livestock or wild herbivores may effectively eliminate the interior environment.[356,1239,1939]

The richness of bird species dependent on remnant vegetation increases significantly in roadside natural strips from 8 to 60 m (25–190 ft) wide in Western Australia.[52,1491] Similarly, in comparing planted windbreaks up to several rows wide in the American Midwest, a few bird species were limited to the widest windbreaks.[808] Width appears to be a good general predictor of bird species richness, nest density, and breeding success in windbreaks.[623,1538,1058,1939]

Vertical structure, which includes complexity, species composition, and trunks for nest holes, also appears to be a major control on avian richness.[808,1939] Windbreaks and probably hedgerows have a high concentration of ground-feeding birds, though the tree canopy is also widely used.[1938,808] A shrub layer is presumably critical for cover against both

195

predators and weather.[356,1015] Hence, hedgerows with shrubs are familiar to hunters for their concentration of game.[435,1312,1337,948,239] Indeed, windbreaks can be designed for wildlife by varying the number of rows, number of woody species, height of vegetation, and the arrangement of plants.[252]

The presence of an internal entity (chapter 5) such as a stonewall, fence, ditch, or soil bank provides an extra microhabitat within the hedgerow.[90] Thus wetland plants, amphibians, and reptiles may be added by a ditch, and desiccation-tolerant species by the equatorward side of a soil bank.[1338] Stonewalls are prized habitats for several small mammals such as chipmunks (*Tamias*) and mice (*Peromyscus*).[720,112] These animals in turn have a significant impact on the remainder of the hedgerow, through seed predation and other activities.[1564,718]

In parts of Europe, the ditch and soil bank are combined in most hedgerows on hill slopes to accomplish several functional objectives.[948,495,223] During wet periods excess water drains away in the ditch network. The soil banks inhibit soil erosion. During dry periods the vegetation atop the soil banks may provide shade and organic matter that reduces soil water loss. And the range of microenvironmental conditions compressed into the narrow space of ditch and soil bank provides habitat for a rich assemblage of plants and animals.

External form

A few external attributes are distinctive and make hedgerows different from other corridors (chapter 5). Hedgerows created early in the development of an agricultural landscape are often along lines separating soil types. In contrast, later hedgerows often subdivide a field and thus tend to be on a single soil type.[258,495,223] The result is that hedgerows probably have less of an environmental gradient along them (chapter 5) than most other corridor types. Of course, an attached or nearby woods will create a distance-from-woods gradient along a hedgerow, which is quite important ecologically.

On the other hand, hedgerows often are perforated with gaps and have low connectivity. In some cultivated landscapes, gaps are frequently found at approximately 30 m from the corner of fields, where a farmer has moved equipment between fields (see Fig. 1.11). Where a hedgerow is composed of several or many tree species, gaps are highly probable due to differential sensitivity to disease or blowdown in severe storms. Where a single tree species forms the hedgerow canopy, the probability is much higher that the hedgerow as a whole will be damaged or destroyed by blowdown or pests.

Windbreaks around small pastures in cool climates sometimes resemble a giant's doodling on the landscape (Fig. 6.9). Several objec-

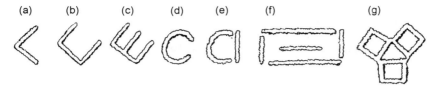

Fig. 6.9. Distinctive windbreak patterns around pastures. (a) to (e) Suggestions for the northern Great Plains of North America (Dronen 1988). (f) Used successfully in northern Asia (Scholten 1988). (g) Used in Scotland (Caborn 1960). Orientation is determined by wind and snowstorm directions.

tives are concurrently addressed in these pastures or paddocks: protection from prevailing winds and cold storm winds, snow accumulation patterns, and access during snow periods.[241,422,1889] Notice the resemblance between these narrow windbreaks for pastures and the multi-row windbreak protecting a farmstead shown in Fig. 6.8c.

Movements to and from surroundings

Almost all animal species in hedgerows move into the adjacent ecosystems. Many species from the adjacent ecosystems move into the hedgerow, and some cross the hedgerow. Observations of these source, sink, and to a lesser extent, filter functions are common. But as yet the literature has not been synthesized to decipher patterns. Therefore, the following discussion illustrates the types of interaction found between hedgerows and their surroundings.

Many hedgerow birds forage in adjacent fields.[1018,1333,1939,182,148,988,1015] Such movements are reported to be more abundant in grain fields containing weed seeds than in pastures.[1939] Movements downwind of a hedgerow, and to old-fields and water sources, also are apparently frequent.[1333,1939] Hedgerow herbivores such as cottontail rabbits (*Sylvilagus*), hares (*Lepus*), woodchucks (*Marmota*), *Apodemus* and insects may also be pests in the adjacent field, especially in the field edge[1777,30,1300,1282] (see Fig. 1.11).

Hedgerow birds and other predators are hypothesized to enhance agricultural productivity, by feeding on potential pest insects, thus preventing population explosions.[1777,964,1350,1311,808,852,809] If this hypothesis is sustained, small fields surrounded by hedgerows, or perhaps large fields peppered with woodlots, would provide a 'natural pest-control' mechanism.[20,30,495,1282] For example, one would predict fewer extensive pest outbreaks and more stable agricultural production in medieval, fine-scale French and English landscapes[1368,1306], than in modern, broad-scale 'macroagricultural' landscapes.

197

Fig. 6.10. Great-horned owl in area of small woods and windbreaks. North Dakota (USA). Courtesy of US Department of Agriculture.

Many birds that live in fields, on the other hand, use nearby hedgerows for singing perches or foraging.[1333,808] Field animals often move to hedgerows as cover against predators[1831], or cover against weather, including sun, wind, and snow.[862,1935,1936,1831,1300]

The concentration of animals in hedgerows also attracts predators from the surroundings[431] (Fig. 6.10), just as hunters are attracted to wooded strips.[239] Where large woods are scarce, certain owls and other predators require two or more smaller woods for their home range. So in a corridor connecting the woods, not much escapes the eyes and silent wing-beat of an owl. Nevertheless, compared with forest edges (chapter 3), predation rates in hedgerows are poorly documented.[1588,1538,1589] Clearly, more study of these conspicuous lines of source-and-sink activity is warranted.

WOODLAND CORRIDORS AS WILDLIFE CONDUITS

Considerable evidence worldwide shows that many animals use wooded strips as conduits in crossing portions of a landscape (chapter

198

5).[685,109,111,688,1492,1228] Species include kangaroos, cockatoos, many small birds, large game birds, nocturnal arboreal mammals, cottontail rabbits, and many other mammals.[808,1492] The concentration of road-killed mammals, where woodlands in open country adjoin or nearly adjoin both sides of a road, is often noted by highway police and wildlife biologists.[734,109,102,87] The evidence is largely based on qualitative and quantitative field observations, and little experimental evidence exists.[1118] Many of the observations are of vertebrate movement along narrow line corridors (chapter 5), rather than woodland corridors (Fig. 6.1).

Only scattered evidence exists that plants move along wooded strips.[1337,495,501,695,493] Some species slowly move short distances along a woody strip by vegetative spread or adjacent seedling establishment. However, most plant movement is 'saltatory.' Seeds are carried by wind or animals some distance to a spot in the corridor, where the species becomes established. Some individuals can reproduce in this environment, and seeds are then dispersed further.

Although some conduit evidence comes from roadside natural strips (chapter 5), these are rather poor examples of woodland corridors because concentrated disturbance lies alongside the entire corridor length. The road and open roadside on one side may bathe a woodland corridor with noise, dust, salt, lead, and heat from the paved tarmac, plus exotic species from the roadside. The road and roadside also significantly affect the source and sink functions of the woodland corridor.

Greenways, the linear conservation areas mainly near built suburban environments, usually have a primary focus on recreation and aesthetics.[1878,961,414,1344,977,700,1753,1582] However, greenways (sometimes called 'greenlining' when networks are formed) are doubtless of considerable importance for species movement, due to the relatively inhospitable and heterogeneous matrix nearby.[362,1597,5,1582] Generalist and edge species presumably predominate in greenways.

Although we know animals use corridors as conduits, the more important question for conservation is whether corridors increase the rate of movement from point A to point B.[1209] For example, a species could move through the adjacent matrix at the same rate as in the corridor. In this case the corridor would be of no advantage to movement of the species. And of course, some species tend to avoid wooded strips in movement across a landscape. A small amount of evidence with marked animals indicates that some species use wooded corridors in preference to moving across open land.[1118,412,1491]

To explore the ecological pros and cons of corridor conduits, and the factors controlling movement, we make certain simplifying assumptions. The two edges of the corridor differ (this is the norm, except where the matrix is the same on opposite sides, or the orientation angle is zero (chapter 4). No internal entity is present (chapter 5). A woodland corridor with interior environmental conditions is the focus, thus essentially ignoring windbreaks and hedgerows. Animals are assumed to be sensitive to the corridor at this scale, and may move inside or alongside the strip.

Enhancing and inhibiting movement

Several types of movement along corridors should be recognized.[111,688,1228] Home range movements, animal dispersal, migration, and wandering (nomadic movement) may take place in corridors. Individual animals move continuously or discontinuously along a corridor. Genes flow by sexual reproduction of resident individuals located progressively along a corridor. Whole communities or faunas may move along a corridor, as in seasonal migration or in response to climatic change.

Earlier it was suggested that movement along a corridor should increase with few narrows, few gaps, low curvilinearity, low patchiness, no environmental gradient, and a short length (chapter 5). Internal and external corridor attributes, and effects of the matrix, are now evaluated in terms of the conduit function of woodland corridors.

The corridor typically offers five parallel routes for an animal to move. A central interior environment is commonly sandwiched between two different corridor edges[499,1598], which in turn are flanked by two different matrix edges. These five microenvironments provide options for a range of species to move. Although apparently no studies have been done, it is highly probable that woodland patch-interior species preferentially move in the central interior environment[93,478,1650], patch edge species move mainly in the two corridor edges[412], and matrix species move mainly in the two matrix edges. Interweaving among the five lines is doubtless common, especially for patch edge species.[557,1145,9,1118] A relatively continuous shrub layer somewhere in the corridor is probably important as cover for many mammals and some birds. Thus, a browsed or burned out understory can be expected to truncate movement of many species.[1239]

Corridor width is considered important for movement[486], though the evidence for this hypothesis is limited.[623,222,720,1103,112] Clearly many animals move along narrow hedgerows.[1337,1216,1123,812,465,808,1114] It seems likely that wider woodland corridors would increase the rate of movement of interior species. Yet most movement in corridors is by edge species, and may not correlate with the width of woody strips.

Moving animals must pass and interact with corridor residents.[976] Such residents are primarily disturbance-tolerant, generalist edge species. Resident vertebrates may have long home ranges. For instance, the white-footed mouse (*Peromyscus*) in southeastern Canada has a home range in hedgerows ten times longer than in nearby woods.[1629,1117,1118]

200

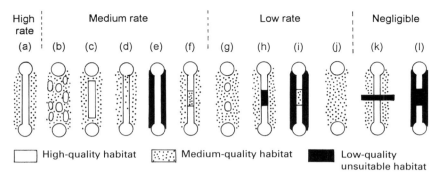

High rate | Medium rate | Low rate | Negligible

(a) | (b) (c) (d) (e) (f) | (g) (h) (i) (j) | (k) (l)

☐ High-quality habitat   ▨ Medium-quality habitat   ■ Low-quality unsuitable habitat

Fig. 6.11. Expected movement rate of animals between large patches, related to habitat connectivity and quality. (a) Connected corridor; (b) cluster of small patches; (d) gradient; (f) patchy corridor; (g) row of stepping stones; (k) barrier intersection. String-of-lights pattern would be at left end.

One may only guess how this distinctive set of residents would affect moving animals.

External corridor attributes are also controls on movement. The effect of corridor length has not been evaluated. But based on general dispersal-distance models[1244], we expect that movement drops exponentially with distance along a corridor. Such a pattern would be modified by an attractant at the end of the corridor, or a deterrent located along the corridor.

Habitat connectivity and quality may be the two primary variables controlling the conduit function. A corridor connecting two patches, and surrounded by medium- or high-quality habitat, presumably supports the highest movement rate of patch species (Fig. 6.11a). Decreasing connectivity of the corridor and decreasing habitat quality in and next to the corridor should reduce the movement rate.[1175] For example, an environmental gradient along a corridor would be expected to decrease movement, since a species would be progressively encountering different environmental conditions and species (Fig. 6.11d). Patchiness along the corridor may decrease movement more, as the species moving encounters not only different environmental conditions and species, but also abrupt transitions and perhaps barriers (Fig. 6.11f, h, i, k and l). On the other hand, patchiness could provide rest stops for a few species, as in the saltatory spread of plants along corridors.[495,501] Overall, a *string-of-lights* combining a corridor and small patches is probably best.[501,1373]

If a high-quality corridor is not present or possible, a second choice is hypothesized to be a *cluster of small patches* (Fig. 6.11b). This is a scatter of stepping stones through which objects such as animals may pass. Unlike a row of stepping stones (Fig. 6.11g), a cluster provides numerous optional routes. It should be effective in permitting moving

(a) (b) (c) (d) (e) (f) (g) (h)

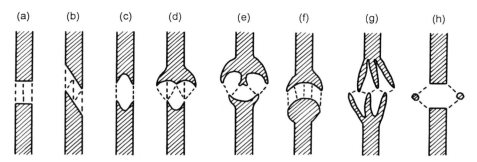

Fig. 6.12. Possible designs to enhance movement across corridor gaps. Dashed lines indicate likely routes.

animals to avoid many predators and disturbances. Experimental studies to determine the optimal arrangement of small patches in a cluster would be welcome. Indeed, no studies exist comparing movement effectiveness in a corridor, a row, and a cluster.

Certainly, the presence of gaps and narrows in a corridor would decrease movement efficiency[1561] (Fig. 6.11c and g), a subject also much in need of study. For example, field observations support the hypothesis that some forest species avoid crossing open areas.[1904,1939,1258,1059,1775] Indeed, arboreal mammals such as squirrels apparently require an essentially continuous line of trees to move between dispersed woods.[93,1216,478,1650,1118,720,412]

If a gap is present in a corridor, ample opportunity exists for design to ameliorate its effect, and enhance faunal movement rate (Fig. 6.12). Corridor ends can be created with more points for crossing (Fig. 6.12e and g) (analogous to a nerve synapse), few points to minimize funneling animals to specific spots (Fig. 6.12b and f), wide tips (Fig. 6.12d and e), or associated small patches (Fig. 6.12h). Until empirical study of such gap designs is done, they must be evaluated using animal behavior and other theory. The size of gaps, the nature of the area in and around them, and gap aggregation presumably are also critical to movement. Indeed aggregated gaps produce a row of stepping stones (Fig. 6.11g). It would be interesting to know the shape of the curve simply relating movement to connectivity.

Unlike windbreaks and hedgerows that are almost always straight, woodland corridors are sometimes curvilinear and variable in width. Using a simulation model M. E. Soulé & M. E. Gilpin (1991) suggested that movement would be more effective along a straight than a curved corridor, because an animal would not have to search or alter direction. This could especially apply to juveniles that have not learned the route. In addition, coves in a curvilinear strip could harbor matrix species (chapter 3) that interfere with movement along a corridor.

202

Finally, the matrix also controls movement in a woodland corridor.[976] Wind whistles through gaps and narrows, across which an animal must move. More significantly, wind may bathe an entire corridor with soil particles, snow, pesticides, fertilizers, or heat. Such a process can be inimical indeed to moving animals, just as it would be to a person. Gap formation and external disturbance are serious intrusions in a corridor compared with a patch of equal area. This difference in disturbance-proneness suggests that corridors themselves are changing more than are patches. Such changes provide windows of opportunity, as well as inaccessible times, for movement of species along a woodland corridor.

Ecological advantages and disadvantages

It is well to recall that woodland corridors are largely remnant or regenerated strips, where the matrix has been changed by human activity. Such corridors have been, are, and doubtless will be widespread in almost all landscapes with a heavy human imprint, except cities. Unlike many other products of human activity, they are inherently neither good nor bad. Faunal corridors are proposed essentially only where an original habitat has been fragmented.

Some controversy lingers as to whether the disadvantages[514,24,1557] outweigh the advantages[1226,1228,1597,1112,1518,685,688,1582] of corridors, or vice versa. The key ecological issues are as follows.

Advantages of woodland corridors

(1) Enhance recolonization to a patch, following frequent local species extinctions therein (i.e., provide for metapopulation dynamics; chapter 11).
(2) Enhance gene flow to a patch, to minimize inbreeding depression in small populations (chapter 2).
(3) Enhance dispersal from a patch, to the matrix and neighboring patches.
(4) Enhance several types or 'needs' for movement in an animal's life history.

Disadvantages of woodland corridors

(1) Act as a mortality sink, by drawing animals to unfavorable conditions in a corridor.
(2) Increase the probability that pests, diseases, exotic species, and disturbances (e.g., fire) will spread to a patch.

The advantages are familiar from chapter 2, and are discussed further in chapter 11. We primarily examine disadvantages here.

A *mortality sink* is a location where considerably more animals enter than leave, and the difference is due to their deaths.[1357,1358] If the collision rate between vehicles and small animals on a road is high, relative to the number of animals crossing (chapter 5), the road is a mortality sink. In studies of hedgerow quality, small mammals moving in low-quality corridors are less successful in reaching the end than those moving in high-quality corridors.[1118,720,112] The animals in the low-quality corridors are exposed to higher predation risk and less-favorable environmental conditions. The low-quality corridors appear to be mortality sinks. Thus, to overcome the disadvantage of corridors listed above, a corridor should be of high quality for the objects moving.

Yet, the effect of a potential mortality sink has to be balanced against the learning process in animal behavior and the adaptability of species. If an animal goes down a corridor alone and is eaten, the animal and the other individuals of its population may have learned nothing. If the animal is eaten, but another individual observes this and escapes, that individual has learned that the corridor may be dangerous. If other individuals learn similar information, either from this individual or their own experience, the population will begin to avoid the corridor. The corridor remains a sink for two types of individual. The young are susceptible because they have learned little. And dispersing animals from afar are susceptible because they also are unaware of local conditions.[112] Therefore, in a low-quality corridor, it seems likely that some individuals will disappear, but the main direct effect on the population is to avoid the corridor.

Of course, if a low-quality corridor is avoided, it is no better or worse than no corridor for moving animals between patches. However, this pinpoints the importance of high-quality corridors, which are used for effective species movement between patches.

A proposed ecological disadvantage of wooded strips is the higher probability that pests and exotic species will spread. Most pests are specialists, and primarily attack one or a few species.[1311,411,1282] Since most corridor species present are disturbance-resistant generalists, a corridor is a relatively improbable location for specialized pests. Damage to the widespread species of corridors normally is of little conservation importance.

However, some pests such as locusts, gypsy moths, blackbirds, and rats are generalists, and when numerous will damage many species. Large inhospitable distances may stop such species. However, the history of spread of such species suggests that they have spread quite effectively across a wide range of landscape patterns.[451,1159,419] A continuous corridor habitat is of little consequence to species that move effectively along a series of stepping stones.[1393] The same pattern can be expected for invasive exotic species. Indeed, the matrix surrounding wooded corridors itself is a major source of weeds, pests, and exotic species.[1278]

The edge of a large patch contains most of the edge species in the landscape for that patch type (Fig. 6.13b and c) (chapter 3). Thus, few generalist species would be expected to reach a large patch in which they are not already present.

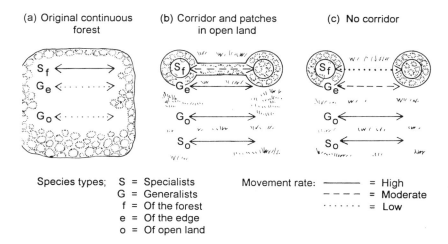

(a) Original continuous forest

(b) Corridor and patches in open land

(c) No corridor

Species types; S = Specialists
G = Generalists
f = Of the forest
e = Of the edge
o = Of open land

Movement rate: ———— = High
– – – – = Moderate
· · · · · · · = Low

Fig. 6.13. Expected movement rates of patch and other species, related to specialists and generalists in the landscape. Patches in (b) and (c) have interior and edge environments.

A series of recent reviews provides a general picture of the few hundred best-known pest species in windbreaks and hedgerows, particularly in North America. Exotic species are included, as well as species moving from windbreak to adjacent fields, and control methods are described. The pests include tree diseases[1311], insects[411,1282], and vertebrates.[1721] The focus is on general life history of pests, not the conduit function of corridors. Nevertheless, no evidence was cited for any species indicating that pests move progressively along a windbreak, or that a continuous windbreak is required to move between two patches or areas. Similarly, the recommended techniques to prevent or control the spread of these pests leads to the same conclusion. Rather, the evidence indicates that the bulk, if not all, of these pest species moves across the field matrix (by wind, flying, or terrestrial locomotion).

In an area of southeastern Australia, Bennett (1990b) found that the six native small-mammal species present were more sensitive to habitat fragmentation than two exotic species. All eight of the small-mammal species were found in narrow roadside natural strips. However, in studying movements, only the native species were recorded moving along corridors between patches.

A few exotic species are reported to have moved along wooded corridors in Australia.[1057,111,962,1278] For example, the spotted grass frog (*Limnodynastes*) expanded 6.7 km along a roadside natural strip <20 m wide in the dry Northwest[1057,111], and a continuous corridor may have been critical to its movement. The 'talkative' common mynah

(*Acridotheres tristis tristis*) represents the pattern for the other exotic species described. This bird is common along roadsides of main highways radiating from major population centers in Australia. These roadsides may have accelerated its spread, but in light of the woods and buildings scattered throughout the areas, it is likely the species would have spread without the corridors.[138] In other words, corridors might increase the rate of spread, even though scattered patches are also suitable for spread.

It has been hypothesized that exotics and pests are favored more by corridors of natural vegetation than are native species.[24,1557,731] At present, the evidence for either exotics or pests is too sketchy to evaluate, but certainly native edge species appear to have no difficulty spreading along corridors, or from patch to patch (Fig. 6.13b). Weeds, for example, are reported to be abundant in some roadside natural strips in Australia.[1278] This is due to wind and especially animal dispersal, a high perimeter-to-area ratio, and abundant disturbance, including fire and nutrient contamination from surrounding agriculture. For each of three woody vegetation types, narrow corridors were found to have c. 25% more exotic species of herbaceous plants than wide corridors.

A corridor as a narrow channel also is an ideal location for human management to arrest the spread of an undesirable species. For example, the aggressive alien tree *Acacia saligna* in South Africa can be successfully controlled by inoculation with its native Australian fungus pathogen.[1083] A corridor would be a convenient location to accomplish this process and monitor the results. Similarly corridors may be convenient locations to control disturbances such as fire. Indeed, wildlife corridors could be monitored to keep a pulse on the health of the land, just as stream corridors are monitored for pollution, hydrology, and so on.

In summary, the proposed ecological disadvantages of corridors appear to be minor or non-existent. Most of the advantages appear to be of major importance (chapter 2). The evidence for all considerations is limited, though growing, so land-use decisions must continue to be made with incomplete evidence. Corridors are no panacea, but they are part of a careful plan for land use and management.

The discussion of trough corridors in chapter 5 shows that for most variables studied, the narrower the corridors are, the better they are ecologically. For wooded corridors, in contrast, most variables indicate the wider the better. Yet a closer look, e.g., at windbreak effects on wind, snow, and other microclimatic variables, shows that narrow, medium, and wide corridors are each optimum for different objectives. In short, conservation and planning decisions must be tailored to the type of corridor, and to the objective to be attained.

Society's decisions on the importance of corridors as conduits, of course, are based on a range of human and economic criteria, such as land ownership patterns, management costs, and recreation, in addition to these ecological criteria. All planning and policy decisions have pros and cons. Just as we now see that the pros of large natural-vegetation patches far outweigh the cons (chapter 2), ecologists and land-use decision makers are moving beyond the pro-or-con controversy phase. We now appear to be well into the overlapping research and implementation phases.[501,1229,1596,688,1492,493,1228,1582,976] The key questions now are: 'Where are the optimum locations? What are the optimum designs? And what can we learn from the corridors in place?'

APPENDIX: EQUATIONS FOR WIND AND EROSION

Horizontal windspeed

$$\frac{U}{U_h} = f\left(\frac{x}{h}, \frac{z}{h}, \frac{h}{z_0}, \frac{h}{L}, \phi\right)$$

where U = average horizontal wind speed for a long thin windbreak on a large flat surface with wind direction perpendicular to windbreak axis; h = barrier height; x = perpendicular distance from windbreak; z = height above the surface; U_h = average horizontal wind speed at barrier-top height in the open; z_o = roughness length taken from the uninterrupted wind profile; L = the Monin–Obukhov stability length (a measure of atmospheric stability); and ϕ = porosity of the barrier. The Reynolds number (hU_h/v, where v is the molecular viscosity of air) also affects the average horizontal windspeed, but with air mixing over a field it is unimportant. After McNaughton (1988).

Wind erosion rate

$$E = f(I, K, C, L, V)$$

where E = potential average annual soil loss (in mass per unit area); I = soil erodibility index; K = soil-ridge-roughness factor; C = climate factor; L = unsheltered travel distance of wind across field; and V = vegetation cover. After Woodruff & Siddoway (1965) and Lyles (1988).

7

Stream and river corridors

Rivers . . . are the natural highways of all nations, not only leveling the ground and removing obstacles from the path of the traveller, quenching his thirst and bearing him on their bosoms, but conducting him through the most interesting scenery, the most populous portions of the globe, and where the animal and vegetable kingdoms attain their greatest perfection.
Henry David Thoreau, A Week on the Concord and Merrimack Rivers, *1849*

A *stream or river corridor* is a strip of vegetation that encloses a channel with flowing water. The corridor may only include the channel and its adjacent banks, or may be wide enough to include a floodplain, hillslopes, and adjacent strips of upland. The focus here is not on the stream or river, nor on the network qualities of a river system, nor on the whole drainage basin, though general aspects of these are included. Rather, the emphasis is on the vegetation corridor, its components, functioning, and dynamics. *Riparian corridor* or vegetation generally refers to the floodplain portion of a stream or river corridor, although the concept sometimes is considered synonymous with stream or river corridor.[130]

Landscape urologists, to use a medical analogy, can fill a bottle with stream or river water, analyze its contents, and tell much about the health of the landscape. Natural processes produce water that may be considered a control. Human activities, either directly in the corridor, or indirectly in the surrounding basin, alter the water in numerous telltale ways. A busy urologist, quietly filling bottles in many channels, can both locate and identify most human activities affecting the landscape.

River corridors are so important to people that every component of society has its hand in the corridor. Its water is extracted for irrigation and drinking supplies. Water flow is altered for flood control, transportation, and hydroelectric dams that generate power. Wastes are carried away in this sewer of society. Fish and fishermen duel in the water. Recreation and aesthetics are enhanced with corridors. Sediments and mineral nutrients are absorbed, and move down the corridor. Beaver, livestock, and other large mammals alter its anatomy. Biodiversity and many rare wetland species are protected here. Agriculture, forestry, roads, and buildings are often rampant in river corridors. Like all corridors, width and connectivity are keys to each of these societal roles.

Useful general reviews are available by Decamps & Naiman (1989), Naiman & Decamps (1990), Brinson (1990), Malanson (1993), and Binford & Buchenau (1993). Detailed reviews include those on: sediment and mineral nutrient flows[610,1503,999,1303]; connectivity and stream processes[1785,345,86,1183]; riparian vegetation[874,1944,1858,1038]; and corridor design.[130,1228]

The chapter begins with a glimpse of the diverse top-view characteristics and cross-sectional structures found in stream and river corridors. Then the dynamic controls by hydrology, erosion and deposition, and animal and human activities are presented. Corridor effects on water quality and fish habitats are next examined. Corridors as habitat and conduit for terrestrial species are described. The chapter builds to a final section on corridor width.

CORRIDOR TYPES

River corridors are by far the most dynamic place in many landscapes. The fierce competition among human uses means rapid corridor changes. Nevertheless, nature's forces almost always remain firmly in control. Floods, droughts, erosion, sedimentation, nutrient flows, fluctuating fish populations, and vegetation succession are pervasive.

Diverse human activities tend to produce two overall effects: (1) decrease the size of a resource, and (2) decrease its variability over time. Thus, the river corridor is narrowed and interrupted, and the river within it is narrowed and straightened. In addition, attempts to decrease fluctuations, and to control the height, extent, and frequency of floods are nearly universal in river corridors.

Before examining the controls on river corridors, we begin with an overview of the types of stream and river corridor, based on their structure or form. The linear or top-view forms presented, as well as the cross-sectional structures, provide a foundation for the remainder of the chapter.

River mosaic concept

The river system is usefully viewed as a series of ecological gradients, a *river continuum*, in which water flow, organic matter, fish populations, and many other factors change somewhat gradually from headwaters to mouth.[1785,345,1616,1146] Gradients also extend across the corridor from upland to river channel. Yet, in most areas the matrix, the floodplain, and the river itself are highly patchy, with relatively distinct boundaries.[760,776,1303,526,816,1182] Thus, a *river mosaic*, superimposed on the underlying gradients, is normally pronounced. For example along a corridor length, relatively distinct patches are evident for water depth, bottom-dwelling animals, fish populations, and rooted aquatic vegetation. The mosaic nature of the corridor lengthwise is evident at many scales from river basin to clay particle.[611]

Similarly, discontinuities are typically present across the corridor, e.g., from upland to channel to upland on the opposite side. These are illustrated by wet spots, exposed gravel areas, oxbows, levees, a hillslope, and upland well-drained soil. In addition, natural disturbance including blowdowns and floods produces patches, and the myriad human activities in corridors result in a patchy mosaic with many boundaries. Significant lateral flows from matrix to stream move across these patches and boundaries in the river mosaic.[1182] At a broader scale, the distribution of ecosystems and land uses in the uplands provides a mosaic through which the corridor passes.

Linear structure

Particles such as gravel, sand, silt, and clay are formed or eroded in one location and deposited elsewhere. Thus, for a typical river system particles are eroded from the upper, steeper straighter portions, and deposited in the lower, flatter meandering portions. Also within most streams a *pool and riffle* structure is prominent. Deep slowly-moving water with sediment deposits on the bottom (pools) alternate with shallow, fast-moving stretches with turbulence (riffles). This is easily visible during *channel flow*, when water between streambanks is moving, and not flooding beyond the banks onto the floodplain.

But stream and river corridors are wider and structurally richer than the stream or river alone. Often one can differentiate five types of linear habitat: the channel (or active channel); *stream or river banks*; floodplain strips; *hillslopes*; and a strip of uplands beyond (Fig. 7.1). A *floodplain* is the relatively flat area that extends from hillslope to hillslope.[610,130] Occasionally we will refer to the floodplain as the portion extending between a hillslope and a stream or river bank.

A stream or river corridor often contains an 'upland strip' of well-drained soil above the hillslope (Fig. 7.1a). The upland strip in turn may include an interior, in addition to an edge, environment (chapter 3). In areas with a heavy human imprint, such as agriculture, forestry, and suburbia, corridor width is commonly highly variable and corridor

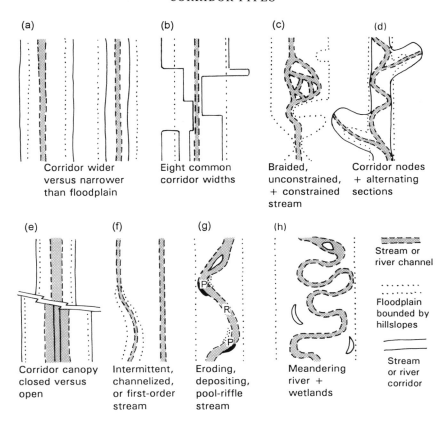

Fig. 7.1. Top view of channel, floodplain, and corridor relationships. (a) Corridor on left includes upland area above hillslopes, whereas corridor on right includes only part of the floodplain. (b) Corridor border includes, top to bottom: (left side) upland, half of floodplain, stream only, an opening on floodplain; (right side) hillslope, an intermittent channel, streambank, and floodplain. (c) Braided portion in upper middle; unconstrained by hillslopes, except constrained near bottom; two alluvial fans on right. (d) Nodes commonly at tributary intersections; corridor is subdivided into alternating sections by stream crisscrossing the floodplain. (e) Upper narrow stream covered by adjacent vegetation; lower wide stream has open trough over it. (f) Intermittent channel on left tends to be straight, but may be curved with tiny floodplain; typical first-order stream or channelized channel on right. (g) During channel flow: dark areas are eroding locations; dotted areas are depositing locations; P = pool; R = riffle; narrow island characterizes high water velocity. (h) Band of meanders within floodplain; wide island characterizes low rate of water flow; two curved oxbow wetlands or lakes left when channel migrated across floodplain.

211

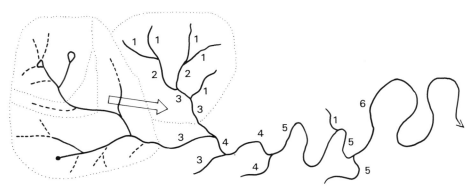

Fig. 7.2. Stream orders and drainage basin patterns in a river system. Dashed lines are intermittent or zero-order streams that only run during part of a year. Black dot = spring; open oval = seepage; . . . = drainage divide; arrow = subsurface drainage between drainage basins. Adapted from Dunne & Leopold (1978).

boundaries highly curvilinear (Fig. 7.1b). External stresses on the corridor, such as the input of dissolved substances, are uneven along its length. Therefore, good design and management require uneven corridor widths, with wider portions opposite matrix areas with higher stress intensity (Fig. 7.1b).

In areas with steep slopes or dry climates heavy rains cause 'alluvial fans' to form in the floodplain. These impinge on the corridor, to create heterogeneity and a curvilinear stream (Fig. 7.1c). Curvilinear streams that extend to the corridor edges subdivide the corridor into alternate segments. This segmentation can be expected to significantly decrease movement of terrestrial animals along the corridor (Fig. 7.1d). The axils of river tributaries are often flooded, and commonly have wet soil that supports vegetation nodes along the river (Fig. 7.1d). Narrow streams are often covered by a continuous canopy of woody plants. In contrast, wider streams and rivers have an opening over the water, which is especially utilized by many flying organisms (Fig. 7.1e).

The 'channels' with actively flowing water (active channels) in a river system are distinguished by *stream order*.[1643,1163] The highest perennially running stream is a 'first-order stream.' Two first-order streams merge to form a '2nd-order stream.' Two 2nd-order streams merge to form a '3rd-order,' and so on (Fig. 7.2). A first-order stream can also flow directly into, e.g., a 3rd or 5th order, but does not change the number of the larger stream.

An *intermittent channel* (or zero-order channel) dries out, i.e., has no flowing surface water, during at least part of the year. Commonly, it is relatively little affected by adjacent vegetation.[1609] On the other hand, a

212

first-order stream is often relatively straight, and is strongly affected by adjacent vegetation (Fig. 7.1f). Streams of about 2nd to 4th order normally exhibit a distinct pool-and-riffle structure (Fig. 7.1g). They erode on the outer convex side, and deposit sediment on the inner concave side.

Larger streams and rivers (from about 5th to 10th order) tend to be highly convoluted, and to have a net deposition rather than erosion of sediment (Fig. 7.1h). The associated wide floodplains create a quite heterogeneous corridor, often with curved wetlands, oxbows, meander scars, vegetated connections between meanders, and intensive human uses. Islands tend to be straight and narrow in small high-velocity rivers, and wide and curved in low-velocity rivers (Fig. 7.1g and h).

In short, these vegetation corridors streaking across a landscape are exceptionally diverse in structure, and can be expected to have numerous functions. As we will see later in this chapter, one major clue to their functioning is the location of the corridor boundary relative to the hillslope crest, and relative to where external stress from the matrix is exerted. A second key is the internal heterogeneity of the corridor, including upland, hillslope, floodplain, bank, and stream or river components.

The stream-order system described best applies to moist climates with dendritic river systems. In dry climates commonly only major streams and rivers flow year round. Much of the land is marked with channels where water flows intermittently, even rarely. Thus, in dry climates intermittent channels, which vary greatly in width and depth, exhibit a richness in structural and functional patterns.

Cross-sectional structure

A cross section of a corridor is almost never seen directly, yet it provides the important third dimension. Large variations in height and depth are present as one crosses the apparently flat corridors illustrated in Fig. 7.1.

Many additional patterns emerge and are outlined as follows. An upland interior environment on both sides of a corridor provides the option of two routes for movement of upland interior species (Fig. 7.3a). If an interior and edge strip are included, most eroded particulate matter from adjacent fields or logged patches is typically arrested and deposited in the vegetation corridor, without reaching the river. Water runoff from uplands is only absorbed by the vegetation corridor in the case of light precipitation. Most water from rainfall runs right through the vegetation and soil (Fig. 7.3a) into the stream, especially in heavy rains. Thus, the corridors generally play a minor role in flood control. Rather, the adjacent, vegetated large patches or matrix are the 'megasponges' that absorb, hold, and slowly disperse water toward streams. On the other hand, dis-

213

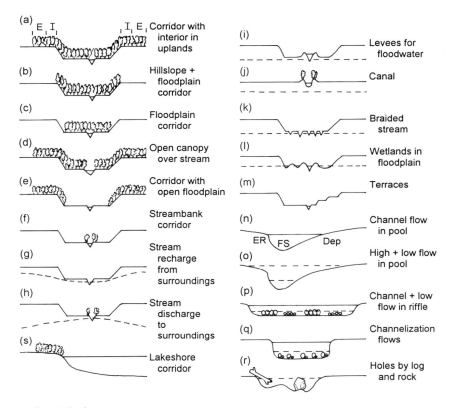

Fig. 7.3. Cross-sectional view of valley, channel, and corridor relationships. E = edge; I = interior; ---- = water table or water surface; ER = eroding bank; Dep = depositing sediment (on point bar); FS = fine-particle sediment.

solved substances such as mineral nutrients and toxics, are significantly affected by corridor width. Such substances often require more than a minimal upland interior width (Fig. 7.3a) to arrest their flow from matrix to stream.

A corridor covering hillslopes eliminates most slope erosion (Fig. 7.3b). Connectivity or continuity along the slope is important, since even a small eroded spot on a slope can alter stream conditions and fish populations a long distance downstream (J. R. Karr, pers. commun).

Vegetation covering the floodplain (Fig. 7.3c) slows water velocity during floods, provides soil organic matter, absorbs some lateral inputs from the surroundings, and provides a route for some terrestrial species that can tolerate wet soils and periodic flooding. Large streams typically have an open space over them, which permits sunlight to raise water

214

temperature, and is a prime feeding trough for some birds, bats, and butterflies (Fig. 7.3d). Human activities, as well as rivers that frequently flood, often create wide open strips down the floodplain (Fig. 7.3e).

A narrow line of vegetation along the stream or river bank cannot control water, dissolved-substance and soil-organic-matter flows, but it does play other important roles (Fig. 7.3f). Shade from bank vegetation of low-order streams helps maintain cool water temperature, decreases diurnal and seasonal variability in temperature, and decreases photosynthesis of plankton and aquatic vegetation. Leaves from the corridor woody plants are a base of the food chain for many aquatic organisms. 'Large woody debris' (the scientists' term for fallen logs and branches[851,1663,1661]) from streambank trees creates fish habitat, as does the cover of overhanging plants. In a dry climate a line of woody plants exposed to the wind and sun (Fig. 7.3f and j) pumps considerable water vertically in evapotranspiration.

The *water table* level (upper surface of saturated soil) relative to the land surface determines much of the internal heterogeneity and dynamics of a stream or river corridor. For example, the water table may be higher in the adjacent matrix than the stream level, in which case water flows into the stream ('stream recharge') (Fig. 7.3g). Alternatively, the adjacent water table may be lower, creating 'stream discharge' (Fig. 7.3h). The former case could be caused by a recent heavy rain on clay or loam soil, while the latter could result from lake water feeding a stream that flows though porous sandy terrain. Human management of water flows often significantly changes water tables and land surface conditions in corridors (Fig. 7.3g–j).

In flat terrain, such as a mountain valley that is dry or is receiving glacial meltwater, the channel of the river is 'full' of sediment and subdivides into several small channels. These in turn interconnect to form a *braided network*, subject to frequent flooding (Fig. 7.3k). All floodplains tend to be heterogeneous in vegetation, due to unevenness in the land surface where a water table is near the average soil surface level (Fig. 7.3l). Floods on a floodplain erode some spots, and deposit levees and accumulations of similar-sized particles in other spots (Fig. 7.3m).

Heterogeneity within the stream or river itself is also important to corridor functioning. The pool portion of a stream typically has an eroding outer side, fine sediments at the bottom, and a shallower inner side (called a 'point bar') where coarse material is deposited (Fig. 7.3n). The riffle alternating with a pool is shallow and relatively even in depth. Coarse gravel, partially sorted into areas of different particle size, covers the bottom (Fig. 7.3p). At low stream flow the pool is relatively deep with a heterogeneous bottom, whereas the riffle is very shallow and somewhat less heterogeneous (Fig. 7.3o and p).

Channelizing of streams by humans tends to eliminate pools, point bars, cutting curves, and slow velocities. Rather, the streams exhibit relatively homogeneous, shallow, and high-velocity conditions, with a bottom of relatively unsorted particle sizes (Fig. 7.3q). Unlike channelized streams, natural streams with large woody debris or large rocks typically have deep holes and big fish (Fig. 7.3r).

Lakeshore corridors exhibit many of the same conditions as river corridors[1212] (Fig. 7.3). Two key differences are evident. First, because crossing of a lake is normally rarer than stream or river crossing, connectivity of the lakeshore vegetation is especially important for the conduit movement of species. Second, the lakeshore is subject to large wave-energy impact. Indeed, it is a small version of a seashore without the associated high salt concentrations. Like the seashore, it changes in form and location when battered with storms.

In summary, stream and river corridors are keyed to the geomorphology and hydrology of valleys. In addition to the stream or river itself, corridors may include upland, hillslope, floodplain, and stream bank. Erosion and deposition patterns create the underlying anatomy of a corridor. The resulting spatial congruencies between water table, land surface, soil type, and slope determine the richness in vegetation and habitats present.

KEY PROCESSES

The distinctive patterns of Figs. 7.2 and 7.3 are what we see in a landscape. They result from four dynamic processes: (1) hydrologic flows, (2) particle flows, (3) animal activities, and (4) human activities. These processes are now presented separately, but in the landscape are highly intertwined.

Indeed, the stream or river corridor should be thought of as a giant tube in dynamic equilibrium.[641] Huge amounts of water and particulates move through it, and extensive animal and human activities affect it. From site to site the internal anatomy of the corridor changes rapidly and markedly. Yet, like a sequence on a movie film, year after year the corridor as a whole looks about the same. Ecologically it may slowly degrade, stay in a steady state, or slowly improve.

The basic stream-order model is used to explore the four key processes. This model is especially applicable in moist climates with dendritic stream and river systems. Modifications may be warranted in other areas. For instance, in dry climates flowing surface water is scarce and episodic, erosion is widespread, and only major streams and rivers flow year round. In sandy areas most water flow is subsurface.[1284,1297] In limestone areas flowing water often follows irregular subsurface and surface channels. In glaciated terrain surface drainage patterns are often partially blocked, producing sluggish streams within extensive wetlands.

The resulting functional effects of these patterns on hydrology, particle flows, animal activities, and human activities are conspicuous indeed.

Hydrologic flows

Relax momentarily on a streambank, and in this fashion watch the stream every day for a year. Lots of water enters your site from upstream, and leaves downstream. A small amount of water enters directly from precipitation. And a small amount exits to the atmosphere in evapotranspiration from the stream and adjoining vegetation. As noted in Fig. 7.3 g and h, water may also either enter or leave your stream through the ground. If you measured these six inputs and outputs of water, you would have an annual 'hydrologic budget' for your stream site. Hydrologic budgets are standard techniques used to understand where water goes, and for managing water.

In a dry climate, stream corridor vegetation evapotranspires a considerable amount of water. This is due to the dry air of the surroundings, as well as the high water table in a floodplain. The result is that humidity over the corridor exceeds that in the open matrix.[1038] Removal of stream corridor vegetation dries the air a bit, but also results in more streamflow in the channel.[426,248]

If stream water always stayed within its (active) channel, the channel could be essentially considered as a pipe, and hydraulic engineering equations could be used to calculate flows and a budget.[708,850] Less precise modifications in the equations might be made for different levels of friction resulting from, say, pools and riffles, curvilinearity, and different particle sizes along the stream bottom. But real hydrologic budgets like yours are still less accurate, because of differences in evapotranspiration rates for various plant species on the streambank, floodplain, hillslope, and upland. Moreover, variations in soil type and water table level across the corridor also affect your budget.

If you really observe every day through the year, it is highly likely that a flood or several floods will wet your leggings or worse. Frequent water-level rises usually cause the stream to overflow its bank every year or two, and cover a narrow adjacent strip. Such a natural process builds the adjacent streambank levee on which you are probably standing, and provides water for nearby depressions in the floodplain.

Floods that go well beyond the streambank change the hydrology of corridors markedly (and hence 'fluvial hydrology' rather than 'hydraulics' is the preferred term[850]). Floodwater of course flows downstream. But the rate depends on distance from the channel, the varying density of corridor vegetation, and the microtopography across the flood-plain.[1859,1581,1503,1609,130] During channel flow your stream trickles or washes by in the active channel. But during floods, water roars past you. The

217

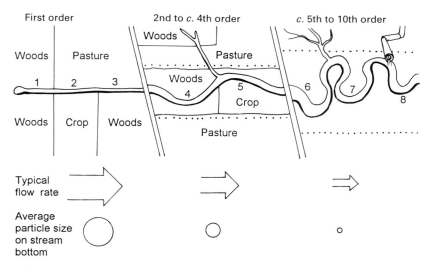

Fig. 7.4. Flow rates and stream-bottom particle sizes along the river mosaic. Numbers 1 to 8 are sections along the river differentiated by a distinct change or boundary. . . . = hillslope. Material entering laterally from a tributary or in subsurface flow also passes through patches and boundaries. Flow rates and particle sizes are represented by large, medium, and small arrows and circles.

depth of flowing water may be several times greater than the channel depth, and the width of flowing water may be orders of magnitude greater. The net effect is that a significant portion of your total annual flow passes in a brief period of flooding.[974,429].

A flood subsides more rapidly if the floodplain upstream is bare, than if it is covered with natural vegetation. Alas though, the flood level is higher if the floodplain is bare. High floods wash out bridges and homes. But floods that periodically cover the floodplain are ecologically desirable to maintain native floodplain plants and ecosystems. Overall, there appears to be no ecological advantage to floods higher than the base of the hillslope.

Average water velocity along a river system is commonly highest in low-order streams, and progressively decreases to a sluggish high-order river (Fig. 7.4). In steep mountain streams, rocks, branches, and logs create a series of mini-dams, to produce a 'stepped-bed' stream structure.[1859,164,851,120] Particles in streams obviously are highly sensitive to flow rate. Consequently, fast-moving low-order streams have rough bottoms mainly of rocks, wood, gravel, and sand (Fig. 7.4). In contrast, river bottoms are typically smooth, mainly composed of silt and other fine material.[946,1785] High velocity upstream and deep turbid water downriver mean that rooted aquatic vegetation is usually limited to intermediate-order

218

channels. In essence, hydrologic flows are a major controller of stream and river corridors, especially though their impact on the erosion and deposition of sediment.[1230,378,1297]

Erosion and deposition

The periodic wide floods have a more lasting effect than simply flushing out heavy precipitation. During a flood, streams and rivers establish new channels and *migrate* laterally across the floodplain. Special curved wet habitats (e.g., oxbow lakes and meander scars) are often seen in floodplains far from the stream (Fig. 7.1h). These habitats are conspicuous evidence of the natural forces causing stream channels to migrate from side to side in a floodplain.

During channel flow, pools have slower water movement than riffles. A small amount of lateral and downstream migration of a channel takes place during channel flow. This results from material being eroded from the outer bank of a pool, and deposited on the inner point bars of downstream pools (Fig. 7.1g). However, during floods, flow rates reverse, so pools have more rapid flows than riffles.[849,76] This scours out fine sediments from the bottom of pools. Floods spread water over the floodplain and saturate the soil. Following a flood the water level in the stream channel drops rather rapidly, leaving water perched above it on the floodplain. The resulting water pressure causes considerable slumping of streambank soil into the pools, which is washed downstream.[850]

In short, the lateral migration of curvilinear streams and rivers produces much of the heterogeneity of habitat and vegetation across a floodplain.[905] The pools of water and the piles of sediment, roughly sorted by particle size, also create corridor heterogeneity.

On your streambank you will see particulate matter pass in the channel. During floods much, much more passes you along the floodplain. However, particles also pass laterally from matrix to stream (in a narrow stream corridor). This sediment transport is especially noticeable in steep terrain, or during high rainfall.

Now let us consider a river corridor in a broader spatial context. The *drainage basin* or basin or catchment (or watershed as used in North America) is the area bounded by topographic divides that drains into a river system[610] (Fig. 7.2). It may be hierarchically subdivided into subbasins, one for each stream and river present. Surface mineral particles in the basin are carried downstream in stream channels. Soil particles are also eroded by water and carried overland by surface flow of water (see chapter 7 appendix for soil-loss equation).

The amount of particulate matter transported to and by streams is dependent on basin area and shape, in addition to steepness, vegetation cover, and land uses.[843] The seasonality and amount of material trans-

219

ported also differs markedly in different climates.[1334] Once in the stream soil particles move in spurts, both in the channel and during floods. Floodplain vegetation is especially effective in trapping sediments from floodwaters.[598,842,429,1503,996,998,307] In this manner, the well-known fertilization and rejuvenation of floodplains are provided free by floods.

Straight channels are most common in steep terrain. Flatter terrain has curvilinear channels, though short straight stretches exist for a variety of reasons. When curvilinearity becomes marked, we call it a *meandering channel*, defined as having a channel length of more than 1.5 times as long as the valley length (the 'sinuosity ratio')[946,610,850,1163] (Fig. 7.5). Sediment deposition exceeds erosion in a meandering channel. Meandering streams and rivers typically appear as a *meander band*, that is, a strip or sequence of similar-sized curves with sinuosity >1.5 within the floodplain (Fig. 7.1h). The sinuosity ratio emphasizes that stream and river corridors are straighter, and often much straighter, than the channel within. The natural boundaries of stream and river corridors are more defined by valleys or hillslopes, than by stream or river banks.

Let us return to observing your eroding pool-and-riffle stream during channel flow (Fig. 7.1g). Pool-and-riffle sequences are most pronounced in eroding streams, though the sequences are also often discernible in meandering streams and rivers. The pool-and-riffle structure provides a range of ecological benefits[850,120]: (1) during low-flow drought conditions some deep water remains in pools, providing habitat heterogeneity (Fig. 7-3o); (2) bubbling water in riffles helps add oxygen to the water; (3) during floods low-velocity pockets are maintained, where fish are protected from high-velocity flows; (4) a sorting and separation of fine and coarse particles provides habitat heterogeneity (Fig. 7.3n and p); (5) heterogeneous streambank vegetation provides cover, shade, and food for fish; and (6) riffles and accumulations of leaves and other organic matter in pools provide a range of habitats for aquatic organisms. These may all be visible from your streambank observation post.

The distance between pools along a stream is surprisingly regular. Measurements of a large number of pool–riffle sequences in temperate areas indicate that the most frequent distance between pools is 5–7 times the width of a channel.[946,850] Eroding streams exhibit a tendency for regularity in pool–riffle sequences, meander sizes, and the spacing of tributary intersections. This suggests that a regularity is also imprinted onto the floodplain portion of a corridor.

Beaver and other animals

Stream and river corridors are also molded or degraded by certain mammals. Beaver (*Castor*) that build dams, elephants that flatten trees, and livestock that cave in banks are conspicuous examples.[159,1095,816]

Fig. 7.5. Meandering stream with stream-corridor vegetation and pasture. Curved wetlands and traces of former channels indicate that the band of meanders migrates across the floodplain. Western Montana, USA. R. Forman photo.

Beaver dams are temporary, which explains much of the remarkable structure and dynamics of the corridors affected. The dam is usually constructed of wood and mud. It causes a pond to form, which covers some or all of the floodplain width. The pond kills most of the previous vegetation. Beaver continue to cut woody plants upstream and downstream from the pond, often including large trees. If appropriate woody plants in the floodplain are scarce, the animals extend their cutting up

221

the hillslope and onto the uplands, thus altering the entire potential width of a stream corridor.[798,1086,816]

If damaged by a small flood, the dam is quickly repaired by these 'busy-as-a-beaver' animals. Rare large floods may remove the dam, and trappers or predators, such as wolves, may remove the beaver. Either way, the pond is replaced by a mud flat, which becomes a (beaver) meadow, which gives way to woody successional stages.[784] Beaver often build a new dam at a new spot, and the cyclic process begins anew with only a spatial displacement. A detailed study of a river system in Minnesota (USA) found that about half of the floodplain area at a given time shows evidence of active beaver impact.[817,1186] But over the time period of a single canopy tree generation, nearly the whole corridor area is affected by beaver.

The sequence of beaver ponds along a corridor has major effects on hydrology, sedimentation, and mineral nutrients.[1184,816,1186] Water from heavy precipitation and snowmelt is held back, thus decreasing downstream floods. Silts and other fine sediments accumulate in the pond rather than being washed far downstream. Wetland areas usually form, as the water table rises upstream from the dam.

The habitats created, however, are unlike those elsewhere in the corridor. The ponds combine slow flow, constant water level, and clear water conditions that support fish and other aquatic populations. The wetlands also have a relatively constant water table unlike the typical fluctuations across a floodplain. Meadow plants colonize a nutrient-rich mudflat. Beaver cutting of woody plants diminishes the abundance of some species, such as elm and ash (*Ulmus, Fraxinus*). But it enhances the abundance of rapidly sprouting species, such as alder, willow, and poplar (*Alnus, Salix, Populus*). Relatively unpalatable species, such as oak, hemlock, and spruce (*Quercus, Tsuga, Picea*) also become more abundant. The diversity of distinctive habitats in the corridor thus results in a high diversity of species.[1412]

Beaver once were doubtless in all rivers, streams, and lakes of north temperate and boreal regions, but have been eliminated by human activity over extensive areas. For example, the first major environmental effect of European colonization of North America was to convert the natural floodplains of moving ponds, meadows, and sprout woodlands into tall floodplain forests, by removing beaver (for fashionable hats and coats in Europe). Except in protected and remote areas with beaver, stream corridors over three centuries later still manifest this profoundly altered and simplified form.

Since most water and sediment enters dendritic river systems in low-order streams, land managers increasingly realize the benefits of beaver (though beaver structures also often conflict with human structures, such as roads and septic systems). Beaver ponds trap mineral nutrients in wetlands upstream, decrease downstream flooding, and decrease turbidity and sedimentation downstream. They provide a striking habitat heterogeneity, and water sources for other animals during droughts.[1412] Beaver dams that are fit to the curves of the landscape provide temporal stability for all of the preceding benefits. Furthermore, dam construction and maintenance require no engineering studies and no bulldozers. Beavers do it free.

Herds of large herbivores also leave an imprint on stream and river corridors. For example, trails of wildebeest, bison, and caribou (*Connochaetes, Bison, Rangifer*) cross river corridors and cause major effects on the downriver environment. The river bank is eroded, and sediments and nutrients cause turbidity, algal blooms, and degraded river-bottom habitat. Until the cut stream bank is covered by plant succession, these effects continue. Elephants, hippopotamuses and livestock at times may move along a corridor, though usually for short distances, causing significant effects on the vegetation, soil, and water.[1581,25,1331,130] Eliminating or carefully managing the use of stream and river corridors by livestock provides significant ecological benefits.[1609]

Some animals in river corridors apparently require continuous, high-quality corridor conditions to survive. The river otter (*Lutra canadensis*) in North America has a linear home range of 5 to 30 km or more along a river.[501] The animal appears to require clear cool water. This normally means that shaded river banks and no major sources of erosion or chemical pollution for long continuous stretches are required to maintain river otters.

Humans altering river corridors

The economic, social, and ecological values of river corridors are well known and critical to society.[1530,377,1038] Reducing flooding, recharging aquifers, supplying irrigation water, supporting fisheries, providing wood products, transportation, recycling nutrients, absorbing wastes, supporting industry, recreation, aesthetics, habitats of rare wetland species, species movement corridors, electric power production, construction materials, agriculture, and livestock illustrate the roles. This richness of competing river system uses creates an exceptional challenge for both planning[122,1582,130] and management.[810,876,378,1181]

Rather than consider the manifold interlocking effects of these, the objective here is to consider how the form and basic ecological functioning of corridors are controlled by four key activities: (1) dams; (2) channelization and diversion; (3) agriculture and forestry; and (4) construction and built objects.

Dams

The preceding section explored the widespread and mammoth effects of beaver and their dams on corridors. Dams associated with *Homo sapiens* are limited or localized in distribution.[584,1829,377] Tiny impoundments provide pleasure and resources to a home, milldams provide water power to support a village or industry, and 'megadams' and reservoirs provide hydroelectric power, water supply, and recreation to a city or region.

For a short distance above a reservoir, the water table is raised, variability in water level reduced, and lateral meandering of the channel reduced. If the flooded area extends from hillslope to hillslope, the native vegetation corridor must extend up the hillslopes and often onto the uplands to provide effective corridor continuity. Downriver from the dam the floodplain is desiccated, and reductions are evident in water flow and channel width, lateral channel migration, water level variation, and habitat heterogeneity.[95,190,1313] These disruptions of natural processes typically extend long distances downstream or downriver, unless the dam is placed just above the intersection with a major undammed tributary. Much sediment moving downstream of course is trapped behind a dam. Where sediment-load transport in the stream is high, the water storage capacity of the reservoir drops rapidly. Many a former pond or reservoir has a channel 'sadly' meandering across a flat sediment surface, and dropping in a narrow waterfall over the face of the dam.

Dams in dry climates may result in more streamside vegetation along intermittent channels.[1609] Water from heavy rains is trapped, raising the local water table, and slowly releasing water downstream. Such vegetation, however, can be expected to differ markedly from the native riparian vegetation. Due to massive erosion by occasional heavy rains in dry areas, sedimentation tends rapidly to fill up the volume behind such dams. A series of such impoundments that trap water resembles a series of oases.

Channelization and diversion

Channelization or straightening a stream accelerates water movement downstream. This drains wetlands and lowers water tables upstream. It reduces the time that floods persist, and reduces the upstream flood height reached. However, floodwater does not disappear into thin air. It accumulates somewhere, and may cause a major flood downstream.

Like the floodplain below a dam, a channelized stream or river causes a reduction in lateral channel migration, habitat heterogeneity, water table level, and variability in water level.[843,838,603] Therefore, few floods cover and rejuvenate the floodplain. Water velocity of course is higher in the straight channel, causing reduced pool-and-riffle dynamics, fine-sediment accumulation, and sorting of deposits by particle size (Fig. 7.3q). Deepening of the channel produces a lower water table in the floodplain. This means that water from heavy rains soaks into the floodplain soil, rather than spreading sediments, nutrients, seeds, and logs over the floodplain during floods.

Diversion of water from a stream to supply drinking water, industry,

Fig. 7.6. Rice culture and livestock on naturally fertilized floodplain sediments. Regular flooding covers floodplain primarily with silt and clay deposits rich in mineral nutrients, all originating from the matrix upriver. Low-input agriculture often produces more net gain (profitability) and stability to the farmer than attempting to increase production with expensive inputs. Probably a floodplain in Vietnam. Courtesy of USDA Soil Conservation Service.

or irrigation produces patterns similar to the upstream effects of channelization. Water table, floods, channel migration, and temporal and spatial heterogeneity are all reduced.[456]

Agriculture and forestry

In the preceding section on animals, livestock were described as important alteration agents for the movement of sediments and nutrients.[713] Much of the effect results from degrading the stream bank or river bank.

Crop fields and pastures also significantly mold corridors (Fig. 7.6). An initial removal of native vegetation from the upland matrix may have the greatest effect.[1586] For example, removing the forest of the Yangtse River basin in central China three millennia ago has resulted in a gigantic raised, widened floodplain with centuries of massive floods.

Often the majority of forestry-related erosion results from road construction and maintenance.[1731,229] Log removal activities, and the resulting soil surface exposed to precipitation, cause the remaining erosion.

Agricultural fields, logged areas, roads, and built areas in the upland matrix, as major erosion surfaces, commonly provide significant amounts of sediment to intermittent channels and low-order streams.[996,998,307] Due to protection by floodplains little sediment is provided directly from upland matrix to higher-order streams and rivers. Instead sediment must first pass through low-order streams.

The width of corridor vegetation on uplands above hillslopes is a key controller of sediment input to low-order streams from forest cutting and agricultural activity in the matrix.[1731,229] In drainage basins with a significant amount of cleared land, enough sediment reaches these streams and is transported downstream for the entire floodplain to be commonly covered by sediment originating from agriculture or forestry practices (Fig. 7.6). This sediment forms almost all the stream and river banks, as well as the surface heterogeneity of the floodplain. Just as sediment from a road-crossing smooths out the stream bottom downstream of a bridge, the sediment from agriculture and forestry tends to smooth out the natural geomorphic heterogeneity of a valley.

If logging activities or agriculture are within the floodplains of a narrow valley with a low-order stream, typically severe sediment deposition into the stream takes place. Particulate matter enters by simply washing down the streambank[1248] (Fig. 7.4). The resulting sediments clog streams, and reduce the number of fish and fishermen.

In contrast, agriculture or forestry in the floodplain of a major meandering stream or river (Fig. 7.6) normally causes little sediment deposition directly into the river. During channel flow periods, little water flows over the floodplain surface to enter the river, except from the riverbank. During flood stages eroded particles from the fields are washed and deposited down the valley. Most of the material deposited is spread across the floodplain. These sediments provided naturally by frequent floods support the high productivity of floodplain crops, pastures, trees, and natural ecosystems.

Construction and built objects

As mentioned in the previous section, road construction and maintenance may have a major impact on a stream corridor through erosion and sedimentation.[130] Associated sand and gravel mining in a river or floodplain usually produces temporary pits. The channel becomes lower with an accompanying increase in slumping and erosion of the riverbank. The nearby water table drops a bit. Constructing a roadbed across low-order streams puts considerable sediment into the stream system, and road maintenance typically keeps the sedimentation process going.

Roads, railroads, and canals are often built in a floodplain due to the flat terrain. These structures generally parallel the valley rather than snaking along next to a meandering channel. As the stream or river migrates laterally across the floodplain over time, in places the stream begins eroding the linear object constructed. Erosion is often severe, especially when sand and gravel are added to attempt to maintain the road location against nature's process. The resulting sedimentation spreads long distances downstream.

Roads crossing a stream or river corridor are potential barriers to water and sediments rushing downstream. Where the roadbed across a floodplain is maintained at floodplain level, the road has little effect on floodwaters. Of course, it is also impassable to vehicles during floods. However, if a solid causeway for the roadbed is built above the floodplain level, water and sediment flows are significantly altered. Water accelerates and deepens the channel, eroding the adjacent roadbed and threatening bridges over the river. Sediment is deposited in the floodplain on the upriver side of the road bed, and in distinctive patterns downriver, resulting from the funneled high water velocity. Any narrowing of a floodplain by a road or series of roads causes high flow velocity, channeling, and erosion. The well-known way around the problem is to elevate a road on pillars across the entire floodplain. Alternatively, swinging vehicles across rivers with cables or cranes, analogous to gondolas at ski areas or the huge trees removed by 'cable logging' in mountains, would be exciting. Careful ecological evaluation, at least, should precede usage.

Buildings on floodplains have a series of familiar problems, including wet and flooded basements, embarrassing malfunctioning of septic systems, and direct physical damage from floods. Building houses on strong stilts or pillars is a traditional way to persist in floodplains. To protect buildings against floods humans sometimes raise riverbanks by building levees upon them (Fig. 7.3i). Thus, a modest increase in water can be contained in the channel without flooding. Unfortunately, this often leads to constructing more buildings in the floodplain. Without failure, a higher flood level eventually arrives, and the destruction is massive. A floodplain is well named.

In conclusion, human activities of diverse sorts appear to purposely or indirectly cause three broad, predictable patterns in stream and river corridors. (1) The stream or river itself is narrowed, resulting in higher water velocity and consequent scouring of material or objects in its path. (2) The stream or river corridor as a whole is narrowed, resulting in more sedimentation, higher floodwater flow velocities, and reduced connectivity of corridor habitat above flood levels. (3) The variability in water level, floodplain habitat heterogeneity, lateral channel migration, and flood periodicity is reduced. These dynamic controllers of stream and river corridor form therefore exert extensive effects on vegetation and animals.

227

WATER QUALITY AND DISSOLVED SUBSTANCES

The preceding focus on water and sediment flows provides the essential foundation for examining how riparian corridors affect water quality. Some detail is important because dissolved substances are sometimes the primary determinant of optimum corridor width, as shown near the end of this chapter.

Water quality is a broad topic that focuses on: (a) physical variables, such as temperature, velocity, and flow regime; (b) chemical variables, including dissolved mineral nutrients, toxic substances, and pH; and (c) biological variables, such as diversity and abundance of organisms, productivity, and pathogens.[972,974,244,130] Some were considered earlier in the chapter and some follow in later sections. Here the focus is on dissolved substances, with nitrogen and phosphorus being the primary examples. Other common examples are salt, pesticides, other organic substances, and heavy metals.

Nitrogen (N) and phosphorus (P) are especially important in the landscape, because they are the two primary mineral nutrients added to soil in fertilizer, as well as in septic effluent. Just as N and P are limiting factors to agricultural production, they are the primary limiting factors in aquatic systems including streams and rivers. Adding N and/or P will usually increase crop production, and also cause algal blooms.[431] The latter effect decreases water quality in many ways, and hence the added N and P is usually considered pollution or eutrophication (over-enrichment). Stream corridors can reduce or eliminate that pollution.

Down the stream corridor

First, consider movement down a river system from headwaters to mouth. In general, water velocity changes progressively from high to low (Fig. 7.4). Particle size on the stream bottom changes from large to small. An eroding stream becomes a depositing river. Shade from overhanging trees goes from complete to minimal. Water temperature progressively rises.

In addition, the base of the aquatic food chain progressively changes. Upstream, dead leaves entering the water are paramount for the food chain. In middle stream-orders with reduced water flow rates, attached algae and rooted aquatic vegetation become the primary base of the food chain. In higher-order rivers phytoplankton (floating algae) and soil organic matter are the primary base of the food chain. These apparently continuous changes along a river system describe the river continuum concept.[1785,345,1146,1616] Various exceptions to these generalized patterns have given rise to additional and alternative concepts such as the river mosaic described at the beginning of this chapter (Fig. 7.4).

Most important to this discussion is the 'nutrient spiraling' concept.[345,1171,1185] In essence, a mineral nutrient upstream is carried and deposited at some lower point along the stream system. Here it may rest in a sediment deposit, or be taken up by vegetation and incorporated into plant tissue. Eventually the deposit or leaf is washed downstream where the nutrient again is temporarily deposited. The process is commonly repeated several times in moving down a river system. In this way the particular mineral element 'spirals' from location to location along the stream. It often combines with different elements during the route, and typically rests for brief periods perhaps of minutes, to long periods of years or decades. The productivity of algae, shellfish, and fish in lakes and estuaries is commonly controlled by hydrologic and mineral nutrient inputs arriving from streams and rivers.[431,1284,294]

Laterally from matrix to channel

Now consider a mineral nutrient flowing laterally from, say, an agricultural field or a logging operation in the matrix toward a 3rd- or 4th-order stream (Fig. 7.7a). It enters the stream corridor vegetation either in *surface water runoff* (over the soil) or in *subsurface flow* (within the soil). Upon entering the corridor, a nutrient typically has four possible destinations.[1076,1659,1303,996,1268]. It may enter the stream and be washed downstream. It may be absorbed by roots and become part of a plant. It may be carried deeper into ground water to pollute an aquifer. Or it may be metabolized by bacteria and liberated to the atmosphere in gaseous form. These are the basic destinations of nitrogen (Fig. 7.7a). Since phosphorus is not liberated in gaseous form, it only goes downstream, to plants, or to an aquifer.

The stream corridor not only determines the destination, but perhaps more importantly determines the rate of flow. Three key corridor mechanisms control rates of nutrient flow.[1457,1581,76,996,998] The first is simply friction: vegetation and soil conditions act as physical barriers or filters to slow down the lateral water flow carrying the nutrient (Fig. 7.7a). The second is the absorptive power of plant roots: plants readily absorb added N and P, incorporate them into tissue, and grow. This is especially so in a nutrient-poor location, usually more typical of uplands and hillslopes than floodplains. And the third mechanism is the absorptive capacity of soil: clay and organic matter (e.g., humus), both composed of fine particles, are particularly effective at absorbing nutrients. Clay and organic matter not only have a large surface area for adsorption, but they also chemically bond with and hold the mineral nutrients.

The passive friction effect of plant stems, fallen branches, and leaf litter slows water and nutrient movement. But these do not hold the

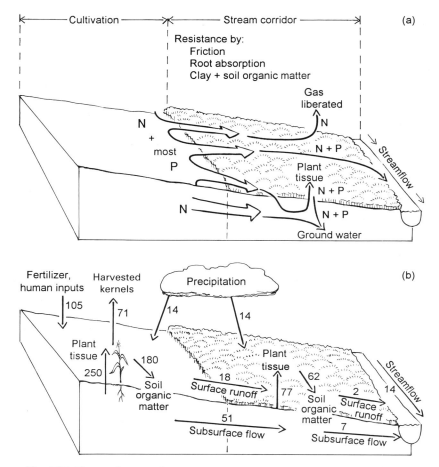

Fig. 7.7. Flows of mineral nutrients in field, stream corridor, and stream. (a) Routes of nitrogen and phosphorus, and resistance by stream corridor. (b) Measured nitrogen flows in kg/ha/year. Rhode River area, Maryland (USA). Based on Peterjohn & Correll (1984), with permission of the Ecological Society of America.

nutrients. Consequently, the next rain washes the nutrients on toward the stream. Absorption by plant roots is a different story, because the nutrient may be incorporated into wood. Here it is retained for years, decades, or even centuries. Or the nutrient may be incorporated into leaves or fine roots, which only last a short period. Leaves may drop directly into the stream. Or the plant tissue may die, decay, and become dead organic matter ('detritus') in the soil.[1121] Soil organic matter often holds the nutrient as long as a few years, but eventually it is either metabolized to the atmosphere, or washed into the stream and flushed downstream.

In essence, to reduce nutrients from the matrix reaching the stream, a corridor can exert a strong frictional force with dense vegetation, absorb with a wide strip of growing plants, and absorb with abundant soil organic matter (and/or clay). These processes temporarily hold the nutrients. Some of the nutrients then can be removed before reaching the stream, by harvesting the plant tissue for wood products or forage.[1943,996] Cutting actively growing and absorbing plants is more effective than removing old-growth.[1804,593,154] A flood often will remove some of the soil organic matter, clay, and vegetation from the floodplain, so those nutrients do not enter and pollute the stream during channel flow. Of course, floods also deposit clay, organic matter, and uprooted or broken plant parts on the floodplain downstream.

A considerable proportion of the dissolved nitrogen, and almost all the phosphorus, entering the stream corridor from the matrix apparently is attached to particles from eroded soil.[842,385,307] Thus phosphorus and some nitrogen basically enter in surface flow of water and soil, particularly attached to clay particles. Consequently, minimizing erosion minimizes mineral nutrient inputs to a stream.[130]

Studies in the eastern USA find that stream corridors filter c. 75–99% of the phosphorus entering from the matrix, leaving only a small portion to reach the stream.[842,1943,1303,996] Filtration of nitrogen is more variable, from c. 10–60% (Fig. 7.7b). Heavy metals such as lead are also readily trapped by stream corridors.[692,996] The effects of a vegetation corridor on the flow of pesticides and other organic substances is less understood.[161,996] High pollutant inputs, however, also may damage the vegetation. This reduces the effectiveness of filtration by a stream corridor.[1248,130]

The effectiveness of different vegetation widths is as yet poorly known.[1303,792,308,270,130] We also know rather little about the friction effect of different vegetation types.

Wetlands sometimes are effective at inhibiting the movement of mineral nutrients from upland to stream.[1794] Unfortunately, the effect is more variable than that for riparian forests.[130] Presumably, the major reason for wetland absorption is the abundance of fine material in the soil, i.e., organic matter and clay. A considerable amount of N and P can be stored in such soils. A nagging question remains, however, since we do not know how long it takes to saturate these soils with nutrients. As soon as they are saturated, no more net absorption takes place. The same question warrants caution in using wetlands for wastewater treatment.

The clay content of soil is determined largely by stream, flood, and wind deposition processes. On the other hand, soil organic matter is enhanced by a complete cover of dense native vegetation[1121], and

231

Fig. 7.8. Intermittent channel and river as key habitats. On left, caiman in narrow strip of woody vegetation in savanna (gallery forest in Llanos) near Calabozo, Venezuela. On right, 'catch-and-return' trout fishing in Snake River with rocky bottom, eastern Idaho (USA). Courtesy of Adrian W. Forman & James Birkenfield. R. Forman photos.

decreased by opening the canopy or eroding the soil surface. Dense vegetation is best produced by natural successional processes that provide for a diversity of species. Livestock grazing, clearing undergrowth, or planting a monoculture doubtless decreases the filtration effectiveness of corridor vegetation.

Corridor width, however, is subject to human control. Wide corridors may be the only defense against especially severe sources of dissolved substances, such as logging on steep slopes, salt or heavy-metal runoff from roads, organic substances from industry, and intensive use of fertilizers and pesticides in fields containing intermittent channels. Wide stream corridors are particularly important where intermittent channels and gullies (Fig. 7.8a) link the matrix to a stream.

FISH AND AQUATIC HABITATS

Where there are fish, there are fishermen. Some fishermen go to catch fish (Fig. 7.8b), and some to tell stories about the big one that got away. But many fishermen go to enjoy, understand, and reaffirm their link with nature.

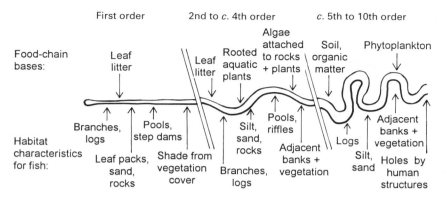

Fig. 7.9. Primary food and habitat characteristics for fish along a river system.

Fish, like any species in a narrow corridor, are highly sensitive to conditions in the matrix and broader landscape. However, in many ways fish populations are an overall assay of landscape quality. Most human activities impact a stream or river corridor, and most fish are high up the food chain in that stream. A 'landscape internist' (to use the analogy of internal medicine) wishing to measure the pulse or condition of the land should simply go fishing.

Food and habitat

To understand how stream and river corridors affect fish we focus on fish habitat. For habitat it is convenient to consider food somewhat separately from spatial location or microsite. Here food will refer to the food chain (for a stream or stream order) upon which the fish depends. Habitat will refer to the particular microsite(s) or physical conditions in a flowing stream or river, where a fish spends much of its time (Fig. 7.9).

In temperate regions it is useful to differentiate cool-water fish, such as trout and salmon (salmonids), from warm-water fish, such as bass, pickerel, and sunfish. Streambank vegetation over streams (Fig. 7.9), especially on the equatorward side, is highly effective in maintaining cool water temperatures. Cool water has high oxygen levels required for cool-water fish.[194,843,471,86,210] In larger rivers bank vegetation typically has little effect on water temperature except with slow flow.[460,130] Many warm-water fish are separated by feeding primarily on plankton, being stream-bottom feeders, or being favored by the low-velocity water in ponds and reservoirs.

The river continuum concept emphasizes the typical shift in the base of the food chain: dead organic matter from leaf fall in low-order streams; attached algae and rooted aquatic vegetation in middle-order streams or rivers; and phytoplankton in high-order rivers[154,973,1735] (Fig. 7.9). In mid- and high-order channels the food chain also depends on soil organic matter carried by floods from floodplains[377,1121], and dead or alive organisms washed downstream (Fig. 7.8). In general, 'detritus' (dead organic matter) is broken down by microscopic bacteria, fungi, and detritus-feeding organisms such as protozoa. The microscopic organisms are consumed by somewhat larger organisms including microcrustacea. These in turn are ingested by aquatic insects, which are eaten by small fish, which are caught by large fish, which form the dreams of fishermen.

Inputs to streams affect different steps in the food chain, any of which in turn affect fish. The microscopic organisms are highly sensitive to pH and heavy metals. Turbidity due to suspended silt especially affects aquatic insects and fish. Turbidity also causes shade that reduces phytoplankton, and hence zooplankton (tiny floating animals) levels. Nitrogen and phosphorus cause algal blooms that reduce light penetration. Many added organic substances cause bacterial blooms that make the stream acid, eliminate oxygen, and kill almost all organisms (except anaerobic microorganisms). The stream corridor therefore must buffer the stream from these and other pollutant sources, to maintain healthy fish populations.

James Karr (1991) has developed an 'Index of Biotic Integrity' as an assay for the ecological health of aquatic systems. The index focuses on the fish community, including species present, relative abundances, and habitat distributions. Such an index recognizes fish as 'canaries in the mine', i.e., as sensitive integrative indicators of activities on the land.

Inputs also affect habitats or microsites within the stream or river. Sediment, such as added sand and silt, wreaks havoc on the pool and riffle structure of eroding streams. It tends to fill in the pools, and then fill in the spaces between coarse gravel in the riffles (Fig. 7.3n to r). Fish depend on heterogeneity within the stream, the deep holes by rocks, the variations in water velocity between pool, riffle, and different sides of a rock, the oxygenation provided by water bubbling through riffles and cascades, the gravels for spawning, the overhangs of steep streambanks, and the accumulations of leaves and wood[842,594,1856,41,106,885] (Fig. 7.9). Aquatic insects and other organisms in the food chain also depend on this habitat heterogeneity.[281,941] The net effect of sedimentation is a smoothing and homogenizing of the stream bottom, i.e., a devastation to most fish populations.

Habitat heterogeneity per unit area tends to be less in higher-order rivers than lower-order streams. Consequently, in rivers, the scattered locations with dense overhanging vegetation, high stream banks by levees, large logs deposited in floods, rock outcrops, tributary intersections, rooted aquatic vegetation, and human structures such as bridge pillars are of particular significance to fish populations.[594]

Fallen logs and branches play an unusually important role for fish (Fig. 7.9). They originate from streambanks, and from adjacent floodplain areas during floods.[1581,76,1664,1623,929] Large woody debris is also a major force in creating the structure and lateral migration of many streams and rivers. In headwater streams, log dams commonly form pools used by fish.[41] A single log can cause a stream to migrate in minutes from one side to the opposite side of a narrow floodplain. Due to water turbulence that scours sediment, large woody debris can overnight form a deep hole in the stream bottom (Figure 7.3r). Big holes have little fish, but are especially important to big fish. Logs also alter the sequential pattern of pools and riffles. Adding logs to a stream or river is a familiar and effective way to enhance fish habitat, fish populations, and fishermen.

In meandering rivers, an occasional log jam can affect up and downstream conditions for many kilometers or tens of kilometers. The logs come from river banks, and from floods that uproot and transport trees. Logs also help cause riverbank undercutting and lateral migration of the river.[851]

River corridor system

We can now pinpoint how each portion of a vegetated stream or river corridor (Fig. 7.3a) affects fish populations. Consider water flowing laterally from the matrix to a stream or river. (1) The upland portion of a corridor helps control the input of dissolved substances, such as N and P in fertilizer, road salt, pesticides, and other organics, and heavy metals. The width of the vegetated upland is typically the major control on these inputs. (2) The hillslope portion is of minor importance in arresting nutrients from the matrix. However, hillslope vegetation is a major control on erosion, nutrients, and sediments originating on the hillslope. (3) The floodplain portion provides soil organic matter to the stream.[1121] Because the channel migrates across the floodplain from hillslope to hillslope, the floodplain is not a dependable buffer against material originating on the hillslope or uplands. (4) The streambank vegetation inhibits nutrient flow, inhibits erosion, and provides soil organic matter, all originating from the streambank itself.[1581,76,69,1248,120] It also provides shade, overhanging vegetation for cover, detritus in the form of leaves, and large woody debris.[194,843,851,460]

In short, vegetation on the stream and river bank is most important to fish. Yet any one of the pollutant inputs will drastically inhibit fish. Therefore, since controlling inputs of dissolved substances requires the widest corridor, the upland portion is normally the critical factor determining corridor width. Future research will have to tell us whether the upland portion is required for the whole length of a corridor, to protect fish habitat and fish.

Finally, most rivers contain *migratory fish*. These move annually from the sea up a river system, and back down the river to the sea. 'Anadromous' species, including salmon, breed in the river system, whereas 'catadromous' species, such as eels, breed in the sea. Dams, agricultural runoff, logging road sediments, and chemical pollution are major impediments and bottlenecks to movement up the river system. Locating suitable gravel-bottom sites for spawning that have not been covered with fine sediment is sometimes a problem.[885] Then movement to the sea is a race against time. The same impediments and bottlenecks must be traversed. But also, around every bend and behind every dam lurk predators, about which juvenile fish know little. And with every summer day, water temperature rises, leaving oxygen levels suffocatingly low.

TERRESTRIAL ANIMALS AND PLANTS

Two of the five major functions of corridors (chapter 5) are emphasized here, habitat and conduit. A third function, the filter or barrier effect, is also of particular importance for stream and river corridors, because of the relative difficulty for some animals to cross a stream or river. For example, radiotagged foxes (*Vulpes vulpes*) in the midwestern USA were found to only cross flowing water in rivers up to a certain width (c. 55–82 m).[1639] In addition, the cranial anatomy of foxes on opposite sides of the >100-meter-wide Mississippi River was found to differ slightly. This suggested that populations on opposite sides of the river had been isolated, or nearly so, from each other for a long period. Rivers and canals in The Netherlands are considered least inhibitory to crossing by large mammals, intermediate for butterflies and small mammals, and most inhibitory for birds.[877]

Corridor as habitat

The steam corridor is exceptionally diverse environmentally, and hence normally supports a high species richness, sometimes the highest in the landscape.[254,1856,1618,1482,825,873,1228] The richness of exotic species is also high.[380] Riparian corridors in dry areas have been called 'linear oases', and contain many rare species.[805] Here, most riparian vegetation (phreatophytes) depends on ground water rather than surface stream water.[1437,249] Strips of species-rich 'gallery forest' snake through grasslands and savannas[926,1394,762] (Figs. 7.5 and 7.8a). Many animals in the surrounding matrix also depend on these corridors for water, food, or shade.[1394]

236

Beyond the stream or river itself, the four portions of corridors, i.e., streambank, floodplain, hillslope, and upland, will be considered sequentially. The types of species using each as a habitat are presented, since the actual species present differ in different regions.

Streambank and riverbank habitats

Species in these habitats must deal with highly fluctuating water levels and soil moisture. These unstable habitats are scoured by water, scoured by ice flows, blanketed by sediment, and damaged by large woody debris. Plants are exposed during droughts, and inundated for different lengths during floods.[1829,584,1530,377,929]

In effect, two types of plant predominate.[249] One is disturbance-tolerant or resistant, with extensive root systems and strong resprouting ability. The other type is opportunistic, a species readily eliminated by disturbances, but which rapidly recolonizes disturbed sites. Such opportunists are usually dispersed by wind (anemochores), but in this habitat often also by water (hydrochores).[1698] They have seeds 'everywhere', that germinate rapidly. Shrubs such as alder, willow, and tamarisk (*Alnus, Salix, Tamarix*) illustrate the resistant species, and herbs such as fireweed (*Epilobium*) illustrate the opportunists. Some common trees, including poplars (*Populus*), exhibit characteristics of both categories. Not surprisingly, bank vegetation in first- and second-order streams differs from that of high-order rivers. With little sediment deposition, mosses and ferns may be common on headwater streambanks. A mix of successional stages is common on riverbanks.

Streambank animals also must deal with the fluctuating environment. On certain rivers beaver (*Castor*) dig burrows deep into the upper parts of riverbanks, and move along the river to feed on bank and other vegetation. River otter (*Lutra canadensis*), weasels, birds, salamanders, snakes, and other predatory vertebrates also live in stream and riverbank burrows. These animals thrive on the adjacent supply of aquatic invertebrates and fish.

Floodplain habitats

The floodplain of a river corridor is especially rich in habitats (Fig. 7.10). This is because the relative height of soil surface and water table varies widely. In other words, different types of 'wetlands' are commonly present, differing in the average number of months per year in which water is at or above the soil surface. With progressively less inundation (or lower water table) the following habitats are often present: marsh or bog; shrub swamp or thicket; forested swamp; and vegetation on poorly-drained, but rarely-inundated, soil of levees, ridges, and domes (hummocks)[810,1230,1684,770,1148,762,185] (Fig. 7.10).

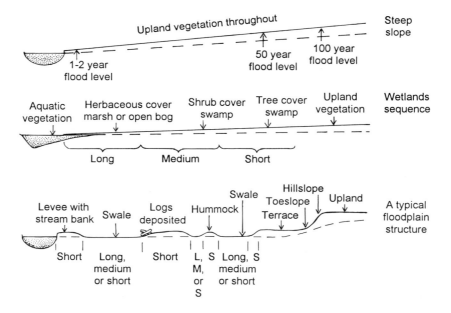

Fig. 7.10. Vegetation diversity reflecting valley structure and length of inundated soil. Long (L), medium (M), and short (S) indicate the relative persistence of water at or above the soil surface. ---- = annual average of a highly
variable water table. Streambanks omitted from two top diagrams.

Though diverse, most floodplain habitats share certain characteristics. They are inundated periodically, have poor soil drainage, are periodically covered with a layer of sediment, and have nutrient-rich soils fertilized by floods. All the habitats are temporary, as the river migrates across the floodplain.

As a whole the floodplain functions as a patch dynamic mosaic (chapter 2) with frequent disturbances and rapid vegetation regrowth.[905] When functioning naturally, the floodplain appears as a mosaic of successional stages with few, large continuous areas of old forest present. Floodplains tend to have many rare species, because of unique conditions, habitat diversity (Fig. 7.10), and habitat loss in the uplands. Seeds of most trees are wind- and water- dispersed. Because elevation of the terrain changes along a river system, corresponding changes in floodplain vegetation are evident as stream-orders change.[926,521,762]

In the short term, average soil conditions are wettest near the river and driest near the hillslope. But because of lateral river migration over the long term, the wettest and driest areas also move across the floodplain. Therefore, periodically the wettest soils are at the base of a hillslope. Few floodplain plants compete well in the fluctuating riverbank environment, or on the drier hillslope soil. Thus, most floodplain species are sandwiched between the hillslope and the advancing and retreating riverbanks. Nevertheless, plant species richness appears to be higher on the floodplain than on either the streambank or the hillslope.[611]

Soil animal communities are severely limited by the high water table of a floodplain. However, on these nutient-rich soils, the vegetation foliage is also rich in nutrients, including N and P. Vertebrate and insect communities feeding on the foliage thus tend to be both dense and diverse.[1703,1618,683,805] Repeated flooding also damages trees, thus producing much dead wood, both standing and fallen. Vertebrates capitalize on the food and nesting spaces offered by this abundance of dead wood. Consequently floodplains are often good locations to find woodpeckers, owls, and arboreal mammals.[1703,1618]

In contrast, few ground-nesting or den-digging vertebrates are expected, because of the high water table and frequent flooding. During floods, non-arboreal animals must move into the hillslope and upland portions of the corridor.

Hillslope habitats

On slopes the plants are usually erosion resistant, having dense root systems that hold soil. While the soil is dryer than on the floodplain, the hillslope is effectively a narrow zone with the water table changing from shallow to deep. Consequently, soil conditions are suitable for a wide range of species, and plant species richness tends to be high. Hillslopes can also be expected to have a high density of animals, including those that nest there and feed in the wetter floodplain habitats.

Upland habitats

Species of the upland portion of the river corridor are characteristic of the upland matrix. They thrive on well-drained soils, and cannot compete on the lower wetter soils, or where regular flooding takes place. Wind speed is often higher at the crest of the hillslope. The air is drier than lower down, where the river and floodplain create moist air.[1038]

If the adjacent matrix is altered by human activity such as agriculture, forestry, or housing, the upland corridor habitats must be tolerant of inputs from the matrix. In addition, the upland portion may be divided into an uphill edge facing the matrix, and a downhill interior above the hillslope crest (Fig. 7.3a). Multihabitat animals (chapter 3) may thrive in the upland portion of a corridor by also feeding in the nearby floodplain or matrix.

Corridor as conduit

Many types of material move down a valley ultimately driven by gravity. However, animals using locomotion may swim, walk, or fly up a stream or river corridor, an anti-gravity process called 'rheotaxis'.[685] This emphasizes the dual roles of these corridors connecting river mouth and headwaters. One direction is a converging system like vertebrate veins, and the other diverges like arteries. The converging system leads to a point, either the exit from a landscape or the mouth of the river. The diverging upstream system leads to the entire surface of a drainage basin.

By looking at the heterogeneity within a river corridor we can visualize probable routes of animal movement along a corridor. The upland interior is a desirable route for many herbivores and predators, because of good visibility at the crest of the hillslope (Fig. 7.11). Also, two adjacent parallel routes with high connectivity, the upland edge and the hillslope, offer detour options and flexibility. The upland edge itself is an important route (chapters 3 and 6). The hillslope may be a route for some animals preferring more cover and humidity than the upland. However, the hillslope is erosion-prone, and an inappropriate route for, e.g., herds of hoofed herbivores. In similar fashion, the hillslope is inappropriate for herds of boots and high-heeled shoes.

Several aspects of floodplains mitigate against their use as conduits for mammals. Frequent floods are most obvious, though in the outer portions near hillslopes, floods are typically infrequent. Streams and rivers migrate laterally across floodplains so routes would have to also shift over time. The high heterogeneity in microhabitats and successional stages in a floodplain suggests that overall movement would be slow. An animal would have to either make a highly convoluted trajectory, or repeatedly cross microhabitats and boundaries. In a similar manner, an animal following a river bank follows a highly convoluted route. Streambanks along low-order streams, however, are roughly parallel to the valley. Although most moving animals presumably prefer

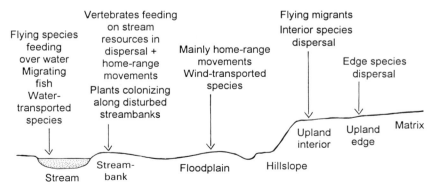

Fig. 7.11. Hypothesized main routes of species using stream corridors as a conduit.

240

routes with visibility, some (e.g., raccoons, *Procyon*) generally follow gullies or streams, where they feed on stream-related organisms.

Home-range, dispersal, migration, and escape from disturbance are different kinds of animal movement.[1225,1228] Movement within a home range, such as the raccoon example, is likely to trace a much more convoluted route than dispersal, migration, or escape. Animals in their home range create a 'cognitive map' of resources present, but also search for food and may defend a territory. Thus, much movement within the streambank and floodplain is home-range movement, whereas the upland interior and edge may be more important for dispersal, migration, and escape along the corridor (Fig. 7.11). Large mammals such as mountain lion, bobcat, grizzly and black bear (*Felix concolor, Lynx rufus, Ursus horribilis, U. americanus*) are known to move many kilometers along stream and river corridors.[1947,940,685,1228] They doubtless move readily among the different portions of a corridor.

Flying animals show somewhat different patterns. The open strip over a river acts as a conduit for species, such as bats feeding on semiaquatic insects or frogs, and butterflies feeding on streambank flowers. Similarly, fish-eating birds including herons, kingfishers, gulls, ospreys, and eagles move along the open river surface. Birds flying within the floodplain vegetation cover have the same constraints as the terrestrial animals just described. However, birds flying over the floodplain presumably are less sensitive to these constraints of internal heterogeneity.

Finally, hawks and related birds of prey are well known to migrate along the windward crests of ridges where updrafts of air (see Fig. 6.3) permit gliding with ease. Such conditions exist above the hillslope and upland interior on the windward-facing hillcrest. Some birds, such as blue jays (*Cyanocitta cristata*), also fly adjacent to forest edges[812], e.g., along the upland edge, where they can quickly duck into cover if a hawk appears.

Some characteristics of stream and river corridors tend to make wind an important dispersal agent. Wind is funneled up mountain valleys, whereas cold air drains down valleys at night (chapter 10). Wind can be bent by forested boundaries or trough corridors to move up or down valleys (chapter 6). Forested river systems in agricultural Nebraska (USA) were found to have a predominance of wind-dispersed plants in the headwater stream areas, but mostly rodent- and bird-dispersed species in lower-river floodplains.[1840] Nevertheless, flying and terrestrial animals probably generally trace a route not much more sinuous than the valley or floodplain. In unusual cases, the animals may follow the twists and turns of streams and banks within the valleys.

Exotic (non-native) plants introduced by humans have spread in stream and river corridors.[682,614] For example in Britain, floods disperse the underground tubers (rhizomes) of Japanese knotweed (*Reynontria*

241

japonica). In dry areas of the western USA tamarisk and Russian olive (*Tamarix* spp., *Eleagnus angustifolia*) have colonized extensively. These species alter sediment flow, thus changing the geomorphic surface, and lower water tables by rapid evapotranspiration. European and American plants predominate along some Australian and New Zealand rivers. Human activities apparently are the primary cause of the spread of these pests.

The spread of pests in continuous stream or river corridors raises a broader issue of planning, management, and conservation. Because pests are so diverse, ranging from extreme specialists to extreme generalists, any land pattern is beneficial to the spread of certain pests

Box 7.1: Conflicting ecological processes in a stream corridor

In New Brunswick, eastern Canada, spruces and firs (*Picea*, *Abies*) are selected for lumber and paper products. In Fig. 7.12, the forest has been harvested, leaving a corridor of spruces and firs along a stream running from left to right. Logging roads and regrowth of hardwoods and conifers are visible in the matrix. The stream corridor performs several key ecological processes, as follows: (1) decreases lateral movement of sediments and nutrients from matrix to stream; (2) provides shade that maintains cool summer water temperature; (3) drops branches and logs in the stream, creating better fish habitat; (4) protects a population of trout; and (5) provides connectivity for vertebrates moving along the corridor.

An explosion of budworm caterpillars (*Choristoneura fumiferana*) that feed mainly on spruce and fir occurred in the landscape. The caterpillars spread rapidly along this continuous corridor for several kilometers to unlogged coniferous forest. The majority of the canopy trees in the corridor died (note gray trees). Corridor processes numbers 1, 2, and 4 became less effective. Process number 3 became more effective. Yet, the effects are temporary. Elevated sediment and nutrient flows usually last no more than a year or two, and elevated tree-fall rates a few years, following this natural insect defoliation.

To many people the caterpillars are a 'pest' that causes a 'disaster'. Yet, insect and other invasions that cause widespread mortality of some tree species are visible most years in different areas of the landscape. Such invasions are considered a normal part of the ecological system.

The budworm caterpillars probably would have spread along a row of stepping stones in lieu of a corridor. Indeed, they probably would have spread without either pattern. Therefore, for the invasion, different land-use patterns may affect somewhat the rate of spread, but have little influence on the probability of crossing the landscape.

Fig. 7.12. Stream corridor as conduit for a caterpillar invasion. New Brunswick, eastern Canada. Photo courtesy of Kevin Johnston.

(see Box 7.1 and Fig. 7.12) (chapter 6). How much weight therefore should be given to pests in land planning?

CORRIDOR WIDTH AND CONNECTIVITY

How wide should a stream or river corridor be? This is perhaps the most frequently asked question by land planners. The preceding sections provide the basic elements for a rather specific answer to this question. Three steps lead to the solution. (1) The first step is to delineate the key ecological processes or functions performed by the corridor. (2) The second is to separate the basic types of stream and river from headwaters to mouth, based on the spatial structure within a corridor. (3) And the third step is to determine the width required for each stream or river type, by linking the most sensitive ecological processes with the spatial structure.

The key ecological processes may be usefully grouped according to their effect on movement along a corridor, movement across, and habitat resources. Movement along a corridor includes the following: terrestrial

243

animals on upland; water in channel and in floods; sediment in channel and on floodplain; dissolved substances in channel; and fish in channel. Movement across includes: water from upland; eroded particles from upland, hillslope, and streambank; and dissolved substances from upland, hillslope, and streambank. Habitat resources are for: fish and other aquatic species in the channel; and the richness of terrestrial species, including rare species, on the floodplain.

As described earlier in the chapter, three types of stream or river from headwaters to mouth are readily recognized (Fig. 7.4): first-order streams; 2nd- to *c.* 4th-order streams; and *c.* 5th- to 10th-order rivers. Intermittent or zero-order channels are also important habitats and conduits (Fig. 7.8a).

Width of first-order streams

A first-order stream with year-round running water may originate from a *spring*, i.e., where an underground channel of water reaches the ground surface. More commonly, the stream originates from one or more *seepages* or seeps, where subsurface runoff originating from precipitation coalesces and reaches the surface.[134] The bulk of the water entering a river system typically enters in first-order streams.[946,610,429] Some enters along the streamside from upland runoff, especially during heavy rain or snowmelt. Most of the other water entering first-order streams comes from 'sources', that is, springs and seepages.

Consequently, the sponge effect of corridors that extend around these sources provides the greatest protection against downriver flooding. In addition, the protection of seepages is particularly important in preventing pollution by dissolved substances carried by water.[130]

'Raindrop energy', the force exerted on a surface by falling water in precipitation, is especially significant where vegetation is removed leaving exposed soil. Raindrops smash into soil particles causing erosion. Tiny channels and larger gullies form, and water runoff accelerates. Flooding and further erosion follow. But vegetation diffuses raindrop energy.

Fig. 7.13. The minimum width of stream and river corridors based on ecological criteria. Five basic situations in a river system are identified, progressing from seepage to river. The key variables determining minimum corridor width are listed under each. The column on the right identifies the outermost location in a corridor required to control a variable. – – – – = seepage, stream or river channel; M = matrix; U = upland; UE = edge portion of corridor in upland; UI = interior portion of corridor in upland; H = hillslope; F = floodplain; MB = meander band; PI = interior of patch of natural floodplain vegetation; PE = edge of patch of natural floodplain vegetation; O = other ecologically-compatible land use.

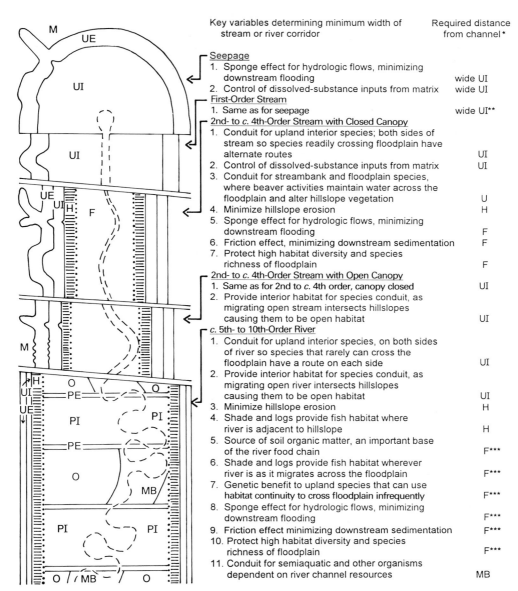

	Key variables determining minimum width of stream or river corridor	Required distance from channel*

Seepage
1. Sponge effect for hydrologic flows, minimizing downstream flooding — wide UI
2. Control of dissolved-substance inputs from matrix — wide UI

First-Order Stream
1. Same as for seepage — wide UI**

2nd- to *c.* 4th-Order Stream with Closed Canopy
1. Conduit for upland interior species; both sides of stream so species readily crossing floodplain have alternate routes — UI
2. Control of dissolved-substance inputs from matrix — UI
3. Conduit for streambank and floodplain species, where beaver activities maintain water across the floodplain and alter hillslope vegetation — U
4. Minimize hillslope erosion — H
5. Sponge effect for hydrologic flows, minimizing downstream flooding — F
6. Friction effect, minimizing downstream sedimentation — F
7. Protect high habitat diversity and species richness of floodplain — F

2nd- to *c.* 4th-Order Stream with Open Canopy
1. Same as for 2nd to *c.* 4th order, canopy closed — UI
2. Provide interior habitat for species conduit, as migrating open stream intersects hillslopes causing them to be open habitat — UI

***c.* 5th- to 10th-Order River**
1. Conduit for upland interior species, on both sides of river so species that rarely can cross the floodplain have a route on each side — UI
2. Provide interior habitat for species conduit, as migrating open river intersects hillslopes causing them to be open habitat — UI
3. Minimize hillslope erosion — H
4. Shade and logs provide fish habitat where river is adjacent to hillslope — H
5. Source of soil organic matter, an important base of the river food chain — F***
6. Shade and logs provide fish habitat wherever river is as it migrates across the floodplain — F***
7. Genetic benefit to upland species that can use habitat continuity to cross floodplain infrequently — F***
8. Sponge effect for hydrologic flows, minimizing downstream flooding — F***
9. Friction effect minimizing downstream sedimentation — F***
10. Protect high habitat diversity and species richness of floodplain — F***
11. Conduit for semiaquatic and other organisms dependent on river channel resources — MB

* Most low-order streams require a markedly variable width of corridors, primarily due to spatial variation of inputs of dissolved substances (e.g., N, Cl, pesticides) from the matrix. This results in convoluted boundaries (as indicated on left side of diagram). For each of the following six conditions the required width is greater than average: (1) an intermittent channel; (2) steep slope; (3) low soil organic matter; (4) a non-loamy soil, either dominated by clay or by sand; (5) reduced cover or density of native vegetation in corridor; and (6) concentrated source of dissolved substances in matrix.

** Minimum width required is normally less than around seepage, and more than around 2nd-order stream.

*** Continuous floodplain vegetation along valley is ecologically optimum. A series of floodplain patches with interior conditions extending from hillslope to hillslope down the valley is an ecological minimum. The distance between patches is determined by the requirement to maintain these six key floodplain functions along the entire length of the river valley.

245

Sources are unusual microhabitats in the basin, normally exhibiting a high water table, slow water movement, and shady conditions. Certain wetland plants especially benefit from these distinctive microenvironmental conditions. Unusual habitats, such as around springs and seepages, are often good spots to find rare species.

In short, buffers that protect seepages and control flooding, are among the widest corridors required in the entire river system (Fig. 7.13). A large amount of water also enters from intermittent channels along the side of first-order streams[307] (Fig. 7.2). Thus, although the required corridor width along the first-order stream is generally less than around a seepage, it is greater than that along other streams.

The actual corridor width in meters required will vary in different climates and topographies, and must be calculated for each area based on empirical data and models. For example, the key variables determining both erosion and water flow rates are relatively well known[429,307,130] (see chapter 7 appendix).

Four physical and chemical variables increase the corridor width required. (1) Steeper surrounding slopes require a wider corridor, due to higher water-runoff velocity. (2) Higher precipitation, and (3) higher dissolved substance input rates require wider corridors to protect the stream. Slope and precipitation values are commonly repeated over the whole drainage basin, so average values often suffice in calculations. In contrast, (4) intermittent channels (considered to be in the matrix) and sources of dissolved substances are typically in specific locations within the basin, and therefore require wider corridors in those locations.

The preceding constraints mean that corridors must vary in width along a stream, rather than being of constant width.[130] Indeed, unlike rivers, considered below, boundaries of ecologically sound corridors will be more curvilinear than the stream. The boundaries will be especially convoluted where entering intermittent channels are frequent.

Three biological and physical variables tend to decrease the corridor width required. (1) Corridors with a dense and complete cover of native vegetation can be narrower, than if the vegetation is reduced or altered. This is due to the sponge effect on water flow, and the absorption effect of plant roots on mineral nutrients. (2) Similarly, soil organic matter is a sponge both for water and dissolved substances. (3) Soil structure effects are more complex, but overall a loamy or silty soil is the optimum filter for water and dissolved substances, and hence requires the narrowest corridor. A soil dominated by clay particles holds water and dissolved substances better than a loam. But because percolation through the soil is slow, much runoff from heavy rains pours over the soil surface into the stream. At the other extreme, a soil dominated by sand particles is highly porous. Therefore, runoff from heavy rains readily percolates into the soil, and pours into the stream from below.

246

To summarize, corridors of first-order streams should be widest around seepages, and wider along the channel than along higher-order streams. These corridor widths are primarily determined by water and dissolved-substance flows, and are linked to the physical, chemical, and biological characteristics of the uplands, rather than to the stream characteristics. Corridor boundaries will be quite sinuous, as widths vary along a first-order stream.

The importance of the river mosaic concept introduced at the beginning of the chapter becomes clearer here. The corridor width required can be readily evaluated for each adjacent patch or corridor in the matrix (Fig. 7.4). For example, a well-managed adjacent forestry patch will typically require a relatively narrow corridor. Cultivation requires a wider corridor, and an entering intermittent channel requires the widest corridor[307] (Fig. 7.13).

Width of 2nd- to c. 4th-order streams

Two major potential sources of erosion and deposition are the streambank and nearby hillslope.[1581,76,1248,69] Dense native vegetation on both is required.[130] But two variables, faunal conduits and dissolved substances, require a corridor wider than simply extending from hillslope to hillslope (Fig. 7.13).

Extensive evidence shows that interior upland species move primarily in interior conditions during home-range activities. We may hypothesize that during dispersal, such species move more readily through upland interior conditions than through other habitats, such as streambanks, floodplains, hillslopes, upland edges, fields, and road corridors. Until evidence shows that these interior species move as efficiently through other habitats, movement corridors that contain interior conditions on the upland should be planned and protected.

In similar fashion, an upland species cannot be expected to move efficiently if it has to repeatedly cross a stream, floodplain, and regular floods. Thus, connectivity of the upland vegetation on at least one side of the river is required. Yet, it seems likely that most upland interior species will move short distances through an edge habitat, or along a hillslope. If so, for short distances the corridor could be narrower and only contain an upland edge. Loss of an upland edge, however, would threaten the hillslope, which could be eroded by runoff from the matrix, or by animals required to move along the hillslope.

However, in certain portions along the stream, the input of soil particles and dissolved substances from the matrix will normally require a corridor wider than the width needed for an animal conduit[307,997] (Fig. 7.13). Just as described for first-order streams, intermittent channels and sources of dissolved substances require bulges in the corridor.

Fig. 7.14. Opening in canopy along river, and edge structure of riverbank vegetation. Blackwater River, south of Perth, Australia. R. Forman photo.

The presence of beaver in a corridor is important because they create a series of ponds often extending from hillslope to hillslope. The same effect is produced by human-made dams along a stream. But beaver keep cutting trees and altering hillslope vegetation and conditions. A faunal corridor therefore must be above the floodplain, and primarily above the hillslope (Fig. 7.13). Providing vegetation continuity on both sides of a stream is important for providing options in a route. Hence, animals that cross the stream can avoid disturbances and danger, and move more efficiently along a valley.

Wider streams tend to have a linear opening in the canopy over them. Here the streambank vegetation exhibits edge conditions (Fig. 7.14) (chapter 3), and where the stream channel reaches the hillslope, the hillslope has edge conditions. Since the channel migrates across the floodplain, over time essentially all portions of the hillslope will be temporary edges. Therefore, maintaining a permanent interior environment requires that the upland area above the hillslope be included in the corridor (Fig. 7.13).

For these 2nd- to *c.* 4th-order streams, a continuous cover of floodplain vegetation is required in the corridor as a sponge for hydrologic control.

248

Unlike the first-order stream corridor, a narrow strip of upland included here provides little water runoff control. Since low-order streams often have narrow floodplains, the sponge effect may require a corridor wider than a minimum-width upland interior. Sediment flow is also reduced by floodplain vegetation, thus reducing downriver sedimentation.[1457,708,843,996,998,307] As described earlier, the floodplain contains a richness of habitats and uncommon species, which are dependent on high water table and regular flooding.

Based on central place theory, some corridors are too thin to support a species with a large-diameter territory or home range.[278,1393,976] For example, a rare cuckoo (*Coccyzus americanus*) in California is essentially limited to riparian corridors >90 m wide.[1407] Beaver (*Castor*) has been described as a central-place forager.[816] However, some animals such as the white-footed mouse (*Peromyscus*) can alter their behavior, from elliptical home ranges in patches to long narrow home ranges in corridors.[1114,1118]

Width of c. 5th- to 10th-order rivers

Most water, sediments, and dissolved substances arrive at a floodplain location in the river either as channel or floodwater flow. Relatively little originates from the adjacent matrix. Therefore, protecting an upland strip has little overall effect on these material flows. Protecting the hillslope with vegetation has a significant local effect on erosion and sedimentation (Fig. 7.13). When the channel is adjacent to the hillslope, the vegetation provides shade and logs for fish habitat. Maintaining vegetation on the riverbanks (Fig. 7.14), and indeed across the meander band, is important for the same reasons. This requires special attention because the channel migrates laterally across the floodplain. Connectivity of the meander band vegetation along the valley will enhance movement of semiaquatic vertebrates, which remain near the channel.

Relatively few upland animals readily cross the floodplain, river, and floodwaters. Therefore, maintaining a continuous strip of interior habitat above the hillslope on both sides of the valley provides parallel conduits (Fig. 7.13) for the two communities separated by the river.

In effect, Fig. 7.13 synthesizes and summarizes the bases for determining the minimum width of any stream or river corridor.

Connectivity of a river or stream corridor

Floodplain vegetation surrounding rivers provides four particularly important ecological functions: (a) minimization of downriver flooding through friction, the sponge effect, and high rates of evapotranspiration[76,1609]; (b) control of downriver sedimentation by trapping sediments during floods; (c) source of soil organic matter important to fish and other river organisms; and (d) habitats for many rare or uncommon species. Hence, a continuous cover of native vegetation will provide optimum ecological conditions, both at a floodplain location and downriver (Fig. 7.13).

River floodplains, however, are usually heavily impacted by many human uses. In view of human population pressures, the design of minimal, in addition to optimal, conditions for an ecologically viable river corridor is important. The two upland strips and the meander band should not be compromised, since significant ecological losses ensue if their native vegetation is lost.

However, the continuous floodplain vegetation might be subdivided into a 'floodplain ladder' without too much ecological loss. Here, patches of natural vegetation alternate with other land uses along the floodplain (Fig. 7.13). Of course, the sponge effect for water will decrease, and sedimentation downstream will increase somewhat.[307] Each vegetation patch should have a significant area of interior, and extend from hillslope to hillslope to provide lateral connectivity. Even infrequent crossings by upland animals provide genetic benefits to populations (chapter 2).

Human activities such as most forestry and agriculture are suitable for the intervening land uses. Sediment and mineral nutrient levels are primarily determined by river flooding, so many common ecological problems with agriculture and forestry are minimized. Human uses, however, should not be sources of toxic organic substances or heavy metals. Floodplains with their high water tables have little buffering capacity to prevent such substances from moving to the river or down the valley.

In short, some ecological degradation takes place where development for other land uses replaces native vegetation in river floodplains. If society chooses to add or maintain compatible land uses here, the floodplain ladder should limit degradation, and maintain most ecological processes.

Connectivity of a stream or river corridor is related to width. But does a corridor have to be continuous? And what is the effect of gap size? Apparently, the only study addressing these questions has examined stream corridor characteristics affecting trout populations in Ontario.[86] Fish diversity increases in warmer streams, but maintaining a trout population requires cool water (Fig. 7.15c). The width of a stream corridor affects water temperature, but the proportion of streambanks covered by forest is more important for keeping the fish cool (Fig. 7.15a).

As the proportion of banks covered by forest increases upstream, several key stream variables decrease, thus benefitting trout.[86] Water temperature, concentration of fine particles, and variability in water flow are

250

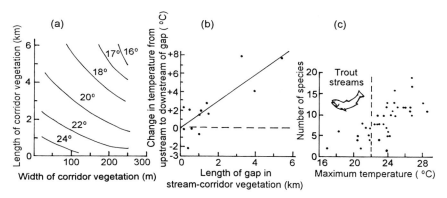

Fig. 7.15. Stream corridor vegetation, water temperature, and fish populations. (a) Effect of upstream vegetation corridors of varying length and width on predicted stream temperature (°C). (b) Change in stream temperature according to the length of a gap in corridor vegetation (i.e., the distance along stream where vegetation has been removed). (c) Fish species richness as related to stream temperature. Of many variables studied, only weekly maximum water temperature clearly distinguished between trout and non-trout streams; streams with temperatures above 22 °C had no or marginal trout populations. All graphs based on weekly maximum temperature (trimean); pool-and-riffle portions of 40 streams studied in southern Ontario, Canada. Based on Barton *et al.* (1985), with permission of the American Fisheries Society.

lower. Of these variables, the trout are most sensitive to water temperature (weekly maximum). An impressive 56% of the variation in water temperature is explained by the proportion of streambank forested within 2.5 km upstream of a site.

Short gaps in a stream corridor probably have little effect on water temperature, except in small, slowly moving streams. Short gaps though may cause erosion and sedimentation that severely impact trout populations and other stream characteristics. However, a gap in stream-corridor vegetation exceeding approximately one kilometer in length causes a significant increase in stream temperature here (Fig. 7.15b). In short, high connectivity of corridor vegetation is critical for trout.

In conclusion, a conceptual framework for determining specific widths of stream and river corridors is presented. By tying it to the spatial components of a valley system, and to the most sensitive ecological processes present, the framework should apply widely in different climatic zones. The answer is not a fixed corridor width, such as 30 m[309,1205,210], 60 m[1164], 100 m[307,997,130], or 300 m[363], on each side of a stream or river. Although technically 'the wider the better' is probably correct, the added ecological benefits become minimal after a certain width. Such an answer is also of limited value operationally, and is no longer adequate based on our knowledge of landscape pattern and process.

Both an optimal and a minimal ecologically determined width for a stream or river corridor varies along a valley. Typically it has asymmetric convoluted margins along opposite sides. Furthermore, the required widths differ for each of the three portions of a river system: the source and first-order stream; eroding streams and rivers with narrow floodplains; and meandering streams and rivers. The conceptual framework (Fig. 7.13) provides a foundation for empirically determining specific widths for different locations. The framework also provides a basis for regulatory actions to protect stream and river corridors.

APPENDIX: EQUATIONS FOR WATER AND EROSION

Soil-loss equation for water erosion

$$A = f(R, K, L, S, C)$$

A = amount of soil eroded by water. R = rainfall. K = soil erodibility (soils high in silt and low in clay or organic matter are most erodible). L = length of a slope. S = angle of a slope. C = vegetation cover.

More soil is eroded as R, K, L, or S increases, and as C decreases.[1917,1126,800]

Surface runoff and subsurface flow

Precipitation lands on a matrix and flows toward a stream in a stream corridor. The equation below lists the key variables determining how much water reaches the stream by surface runoff (SR) and subsurface flow (SF) combined (M. W. Binford, pers. commun.).[610,429,130]

$$SR + SF = f(P, L, S, W, C, M, V)$$

P = Precipitation intensity and duration. L = Land surface slope. S = Soil permeability. W = Water table depth. C = Channel curvilinearity. M = Matrix vegetation cover and density. V = Vegetation width and density in stream corridor. Area of stream basin could be added to the equation.

More water reaches the stream as P, L, or S increases, and as W, C, M, or V decreases. Vegetation in the corridor, matrix vegetation, and channel curvilinearity are the three main variables subject to human design and control.

8

Networks and the matrix

Near the lake, which we were approaching with as much expectation as it had been a university, – for it is not often that the stream of our life opens into such expansions . . .

Henry David Thoreau, The Maine Woods, *1848*

A sandbox is an ideal place. Children are protected from the outside, and become engrossed in creating their own micro-landscapes. They draw networks that crisscross and squiggle. They smooth the surface with their hands forming a surrounding matrix. And together they create a mosaic for the whole sandbox. The mosaic is continually changed and molded, and usually punctuated by big disturbances.

Everyone should have a sandbox, even a tiny one for fingers. The soothing effect, the manifest creativity, and the modeling for landscape and regional ecology are rich benefits. Metaphorically we are the children and the world is our sandbox. To get along and still have a nice sandbox, we need to find land designs that make sustainable sense. Networks and the matrix, analogous to those molded by the children, are the subject at hand.

This chapter moves to a new level. Preceding chapters have explored the attributes of patches and corridors, the building blocks of landscapes. Movements and flows to and from adjacent land have been included as patch–matrix or corridor–matrix interactions. Now we explore the matrix, and hook building blocks together to form mosaics. Patches, corridors, and matrix are combined in distinctive ways to create the finite number of common mosaics on land.

The primary importance of this subject is planning. One may sculpt an individual patch or corridor to attain desired objectives. But sculpting a matrix or network of several spatial elements on a farm, park, forest, refuge, or town requires planning by the owner or manager. And creating a logical mosaic for a whole landscape or region requires regional and government planning.

Perhaps the only overview of the subject is by Forman & Godron (1986). However, useful reviews of portions include: transportation networks[1670,995,646,1518]; hedgerow networks[1337,948,1114,1116]; windbreaks[174]; and the matrix.[501]

The chapter begins with the development of networks, and how to measure them. Dendritic and rectilinear networks are contrasted, and the functioning of networks is explored. Next the structure of a matrix is presented. And we conclude with matrix dynamics.

NETWORK DEVELOPMENT AND STRUCTURE

Corridors of a single type intersect to form a *network*. Roads, streams, walking paths, and hedgerows each form networks. Somewhat less widespread are networks of ski trails, golf fairways, traveling livestock routes, powerlines, gas and oil lines, canals, fence lines, animal trails, and railroads. Networks differ in the objects moving, mesh size, rectilinearity, and so forth. But they all etch the land with a connected net of more-or-less linear elements, and have much in common. They affect essentially all ecological processes at the landscape and regional scales.

Different network types typically overlap and are imprinted upon one another. Surprisingly few dual- or multi-purpose corridors exist, because the objectives and objects moved are often incompatible. Horses, bicycles, vehicles, and trains each do better on their own network. The result is that corridors cover an unexpectedly large total area in most regions.

Networks are equally familiar in other domains. Networks of nerves and blood vessels in vertebrates, xylem and phloem in plants, routes in telecommunications, and wires in a computer exhibit several similar characteristics to those on land. Furthermore, representational or model networks, such as for food webs or in cybernetics, exhibit both similarities and differences.[1274,1275,697,293,1653] No evaluation of such network types for use in landscape ecology has yet been done. Nevertheless, certain analogies will be pinpointed in the following discussion, with the expectation that first principles of network theory do exist, and will become useful at the regional and landscape scales.

Consider a random world (which can be roughly modeled with grid-cell-based percolation theory[1617,543,544]). The land is a dark green grid that you progressively whiten by filling in squares at random (Fig. 8.1a). Suppose a miniature you (*à la* Lewis Carroll's 'Alice in Wonderland') wanted to cross the grid, but can only step on adjacent white squares. At what point can you get across? When the grid reaches a threshold of 59.28 ... % white, you can cross from top to bottom, or left to right (Fig. 8.1b). At first, say from about 60 to 75% white, your route is highly

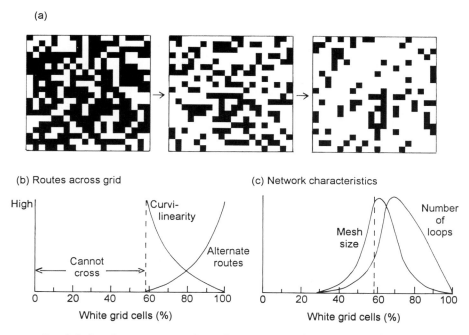

(a)

(b) Routes across grid

(c) Network characteristics

Fig. 8.1. Random patterns with resulting routes and networks. (a) Grid cells are progressively whitened at random to show the 40%, 60%, and 80% coverage stages. Courtesy and with permission of Bruce T. Milne. (b) Routes crossing the grid on adjacent white cells. (c) White network encloses different-sized dark patches. Mesh size = average patch size.

curvilinear or convoluted. Indeed, a network of white cells forms, and you happily have optional routes (Fig. 8.1b and c). Gradually you will find straighter routes across the changing land.

Of course, the real world is non-random, and various landscape patterns can be compared for their degree of non-randomness. Since randomness is a product of entropy (chapter 1), 'non-randomness' is a measure of the amount of energy required to form and maintain a pattern. Thus, a straight road, or two interconnecting roads, of white cells crossing the dark green grid could have been built at the beginning. Or a green barrier could have kept you from crossing until near the end. The more a network pattern diverges from random (or from an underlying substrate pattern), the more it probably costs to create and maintain.

Formation and change

In the broadest sense energy causes networks to form and change. The energy of water flowing over the land creates a dendritic system, as

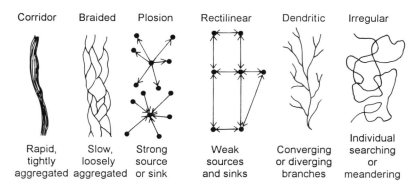

Fig. 8.2. Corridor and networks formed by different forces or types of movement.

streams converge to gouge out progressively larger channels with more flowing water. A mammal herd in less-suitable habitat tends to move together in a single, wide straight route. However, in more-suitable habitat the animals often move in a dispersed, interdigitating network pattern (Fig. 8.2). Engineers have developed flow-capacity equations to describe, for instance, pipeline and road flows. These descriptions are based on the relationship between channel resistance and the viscosity of the material or objects moving.

Nodes also usually play a key role in forming networks. Nodes may be a source of objects dispersing outward, forming an explosion-like pattern[17] (Fig. 8.2). Or the same design results from implosion-like flows toward a node acting as a sink. These 'plosion' patterns are widespread, and result from nodes of functional significance. Dendritic patterns also result from flows, in this case that either converge or diverge. But these branching patterns emphasize that heterogeneity in the matrix also shapes the form of networks. Divergence or splitting of a corridor is commonly located where the matrix has an obstruction or changes in some way. Rectilinear and irregular networks (Fig. 8.2) usually result where horizontal forces are weak or are balanced in different directions, i.e., no strong directionality is present.

Networks change in form through the addition and removal of individual corridors.[425,942,10] For example over a 35 year period in central France, hedgerow numbers, connections, and trees per unit length decreased, while hedgerow length remained constant.[223] Flexibility or adaptability also affects network form. Thus, a system with more rather than fewer optional routes is more stable in withstanding or adjusting to environmental change[939,1116] (chapter 6). Similarly more routes provide more

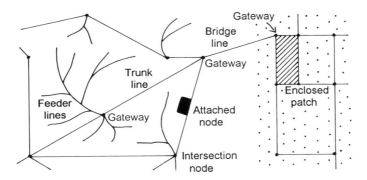

Fig. 8.3. Parts of a network based mainly on transportation theory. Region A on left; region B on right.

options to avoid disturbances, predators, and hunters. A return route or network of routes, analogous to that in a vertebrate blood-circulation system, provides stability for daily foraging, or for migrating animals.

In transportation theory an idealized developmental process for transport is recognized.[1670] A first stage is a scattered number of small nodes, such as small ports along a coast. A second stage is expansion of a few of the nodes, which also become connected to new small nodes in the interior. A third stage interconnects major nodes over the whole area. And a final stage has the emergence of certain high-priority trunk lines. This sequence emphasizes the controlling role of nodes, but also includes the withering of nodes, and even linkages.

Additional structural characteristics from transportation theory appear useful in network ecology.[1670] The major nodes, especially at the periphery of a region, are 'gateways' (Fig. 8.3). 'Trunk lines' are the largest routes that connect gateways. 'Feeder lines' are smaller routes that converge on gateways. And 'bridge lines' connect gateways in adjacent regions. However, the network is dynamic. Any of these can increase or decrease over time, and indeed feeder lines often disappear and reappear.

Structural attributes and measures

Networks are composed of nodes and linkages (corridors) usually surrounded by a matrix. Nodes are located at the intersections of linkages, or are attached to a linkage between intersection nodes (Fig. 8.3). Thus, we start with the structural attributes of linkages and nodes, and gradually expand the view to attributes of a whole network.

Fig. 8.4. Rectilinear hedgerow networks. Upper area is southern England and lower area southeastern Australia. The upper network, composed of managed hawthorn (*Crataegus*) shrubs and scattered trees, may be considered a 'wavy net', due to its irregularly-shaped enclosures and its curvilinear lines, perhaps reflecting the boundaries of medieval woods. R. Forman photos.

Linkages

Linkages are corridors that connect nodes. Linkage width and the curvilinearity of linkages (chapters 5–7) are conspicuous and functionally important characteristics of a network (Fig. 8.4). Each can be measured and expressed as an average and variance for the network as a whole.

258

However, the spatial distribution of wide and narrow, straight and convoluted linkages in the network is just as important as average and variance. High curvilinearity may lead to short-cut movement between lobes, much as mountain hikers sometimes take short-cuts on a zig-zag path. Linkage expansion or expandability is an additional attribute of a network. Just as veins can expand to handle increased blood volume, some streams expand to handle flood water.

Nodes

An *intersection node* is simply the overlap area or junction of intersecting corridors. *Attached nodes* are patches (wider than a corridor) located either at or between intersections (Fig. 8.3). Attached nodes vary by size and distribution, and the node distribution in a network may be regular or aggregated. Woods, towns, and water holes are familiar examples of nodes that have major effects on a network.

The *string of lights* pattern[501], a series of attached nodes connected by corridors, is widespread and conspicuous in aerial photographs and the land, yet is ecologically little studied. Commonly, the pattern is straight or somewhat curvilinear. It is present in both stream and rectilinear networks, and must play important roles in water flow, erosion, nutrient cycles, predation, species movement (chapter 6), and much more.

Intersections

Linkage density, the number of attached corridors per node (or 'degree of a node' in mathematics), is one indication of the importance of a particular intersection node. In transportation theory, the 'beta index' expresses linkages per node for a whole network, or portions thereof. For example, in region A of Fig. 8.3 the gateway nodes (indicated by dots) on the trunkline average 4.7 linkages. In contrast, nodes above the trunkline average 2.7 linkages.

'Intersection angles', formed by adjacent corridors at a node, also provide an indication of directionality of movement. *Rectilinear networks* tend to have intersection angles averaging about 90° (Fig. 8.4). This results from most intersections being 'Ls', 'Ts', and '+s'. However, some intersection nodes have five or more linkages, because of diagonal connections. Region A in Fig. 8.3 has no 90° angles, while region B has only right angles. Dendritic stream networks commonly have intersection angles of about 45°, because of the directionality in water flow.

Hierarchies

The stream-order system from first-order headwater streams to high-order rivers (chapter 7) is a familiar hierarchy in dendritic networks. The analogue in rectilinear networks (those with loops predominant, such as railroad networks) is the feeder, trunk, and bridge line system (Fig. 8.3).

259

In both examples, a directionality in network flows up and down the system is pronounced.

An important attribute of a network hierarchy is the degree of difference of linkages at a particular level. That is, first-order streams could vary from being essentially identical, to each being unique in some characteristic. Thus, to understand a particular linkage in a hierarchy, one must examine at least the next higher level, the next lower level, and the difference at the same level[605,1250] (chapter 1). Although dendritic systems are always hierarchical, rectilinear networks do not always exhibit a hierarchy (Fig. 8.2).

Hierarchies also apply to nodes, such as the different sizes of wooded patches or human population centers. Here a 'gravity model' is sometimes used in geography to evaluate interaction between nodes.[646,995,501] This model indicates that the rate of movements or flows between nodes depends primarily on linkage distance, and secondarily on node size.

Matrix

Just as for an individual corridor, the matrix surrounds and dominates a network and is critical to its function.[1284,526,122] In rectilinear networks the matrix is subdivided into many sections, the size and shape of which also affect the network (Fig. 8.4) (chapters 2 and 4). Patches different from the matrix are also often enclosed within a network (Figs. 8.3 and 8.4). In addition, the topography of the matrix helps determine the actual locations and form of the network, such as hedgerows on the boundary between soil types.[223,224]

Whole networks

Several structural and functional attributes describe the network as a whole. *Corridor density*, the amount or abundance of corridors in an area, is usually measured as the total corridor length per unit area.[893] Thus, a high road density (measured in km/km^2 or mi/mi^2) has been shown to inhibit many large wildlife species in forested landscapes.[1700,1448,1019,187,1099] Similarly, path density for hikers might be a good overall measure of unsuitable habitat for an elusive animal.

In stream systems, 'drainage density', the total stream length per unit area, is analogous to road density.[610,1836,501] A low drainage density (Fig. 8.5a) is usually found on coarse, permeable (sandy) soils. In contrast, high drainage density (fine-textured pattern) (Fig. 8.5b) usually indicates high levels of surface water runoff, impervious bedrock, and (clayey) soils of low permeability. The distance between tributaries along streams and rivers is a measure of the branching pattern in a network.

Fractal geometry could be an alternative measure of corridor density.[230,1137,1138] In this case, a fractal dimension would vary from 1.0 for a

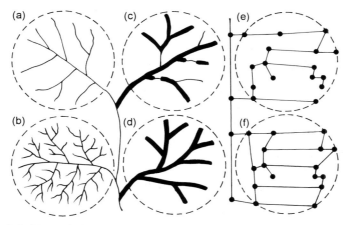

Fig. 8.5. Network density, connectivity, and circuitry. (a) and (b) Low and high drainage density, respectively. (c) and (d) Low and high stream-corridor connectivity, respectively. Thick lines indicate corridor vegetation; thin lines are gaps, i.e., stream sections with some or all corridor vegetation removed. (d) is a connected network (connectivity = 100%); (c) has gaps on five stream sections, and therefore a connectivity of 10/15 = 67%. (e) and (f) Hedgerow networks with low and high values, respectively, for both connectivity and circuitry. Network connectivity (within dashed lines) for (e) = 36%, and for (f) = 48% (see chapter 8 appendix). Network circuitry for (e) = 0%, and for (f) = 19% (see appendix). Adapted from Forman & Godron (1986).

single straight corridor, up to 2.0 in the hypothetical case that the entire surface were covered with corridors.

Network connectivity is the degree to which all nodes are connected. It varies from 0.0 to 1.0, and is the inverse of the proportion of linkages that must be added to have a connected system[493] (Fig. 8.5). To illustrate, a disconnected network in two separate pieces (i.e., missing one linkage) is more connected than the same network in four pieces. Or drawing from geography, connectivity is measured by the 'gamma index', as the actual number of linkages divided by the maximum possible number of linkages in a network (chapter 8 appendix).

Network circuitry is the degree to which loops or circuits are present in the net. The 'alpha index' measures circuitry as the actual number of loops divided by the maximum possible number of loops (Fig. 8.5) (chapter 8 appendix). Loops provide optional routes for moving organisms to avoid disturbances, predators, and hunters. 'Network complexity' may be considered as the combination of connectivity and circuitry[500,501,893], or perhaps also in combination with linkages per node.

Structural connectivity has long been of economic and transportation importance for determining, for instance, the shortest route among a

261

series of cities, the most and least connected areas, the best location for adding a linkage, the linkage most sensitive to disrupting or disconnecting a network, and other time-and-distance optimization problems familiar to traveling salespersons.[1670,646] Such questions may be relevant in understanding the ecology of networks, but at present they are theoretical. Much evidence exists for the movement of animals, and some for plants, along individual corridors (chapters 5 to 7). But very little empirical evidence is yet documented for movements along networks.[1114,1116,688,695] Predators such as wild cats and canines moving at night along lightly used dirt roads and intersections may be an example (see Figs. 5.6 and 8.10). With so little empirical evidence available for ecological movements through rectilinear networks, only the simple, widely accepted alpha, beta, and gamma indices are included.

Finally, the functional significance of networks also depends on where in the network objects interact with the matrix. Do fluxes in and out of the network take place at nodes, along short or distal linkages, or along the length of all linkages?[852,1492,1116] Vehicles in road networks enter and leave at interchange or entrance–exit nodes. Most surface water enters a dendritic stream network along first-order streams, just as CO_2 and O_2 are exhanged in the human body along narrow capillaries. Yet mineral nutrients may enter at any spot along a stream network, and rabbits may head out to feast on a farmer's tender grain at any point along a hedgerow network. In short, the objects moving between network and matrix are highly sensitive to different network structure.

Some network attributes from other domains have analogues in the land. For instance, trees branch in many ways[1246], whereas nerves are connected in a few distinctive patterns.[418] Many properties of connectance in food webs have been described, including chain length, predator/prey ratio, fraction of top, intermediate and bottom species, and 'rigid circuits'.[293,1653,855] Chain length is analogous to stream order, and rigid circuits are present in rectilinear networks.[251] Now we contrast the two major types of network on land, dendritic and rectilinear.

DENDRITIC NETWORKS

Water-caused erosion and deposition create dendritic networks that are often striking in their regularity, or in their variability (Fig. 8.6). In dendritic networks directionality, with converging and diverging branches, is pronounced (Table 8.1). Routes of mammals that seasonally migrate up and down mountain ranges are analogous to a hypothetical vertebrate circulatory system with only one type of vessel. In it blood would alternate going from heart to capillaries, and then from capillaries to heart. Large mammals such as elk, mule deer, and moose (*Cervus canadensis,*

Table 8.1. *Structural attributes of networks*

Dendritic	Rectilinear	Attributes
H	L (–H)	Directionality
H	L (–H)	Converging and diverging corridors
H (–L)	L (–H)	Curvilinearity
H (–L)	L (–H)	Expansion of corridors
H (–L)	L (–H)	Variability in corridor width
H (–L)	L (–H)	Number of hierarchical levels
L	H (–L)	Circuitry
L	H – L	Mesh size
L	H – L	Enclosure size
L	H (–L)	Linkages per node
L (–H)	H (–L)	Linkage angles
L (–H)	H (–L)	Rectilinearity
H (–L)	H (–L)	Regularity of intersection nodes
H – L	H – L	Regularity of attached nodes
H – L	H – L	Variability in corridor context
H – L	H – L	Corridor width
H – L	H – L	Corridor density
H – L	H – L	Connectivity

H = high and L = low level for an attribute. Less common cases in parentheses.

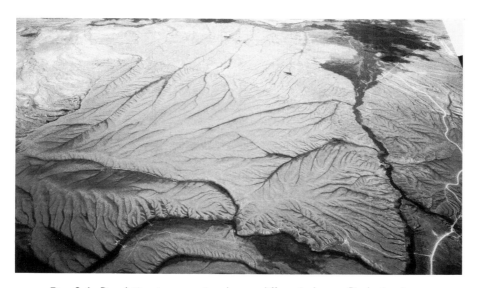

Fig. 8.6. Dendritic stream networks on different slopes. Dark riparian or stream corridor vegetation is visible along major branches, and in large patches at bottom and on right. Dark splotches at top are shadows of clouds. Southern Wyoming (USA); R. Forman photo.

Odocoileus hemionus, Alces americana) in the Grand Tetons of North America go up the network in early summer to scattered subalpine meadows, and return in autumn to winter feeding areas on the plains beneath.[322,1704] Water and sediments only flow down river systems, whereas animals move up and down the dendritic network (chapter 7).

Several structural attributes are characteristically low in dendritic networks (Table 8.1). Circuitry, mesh size, and enclosures are essentially absent. The number of linkages per node is typically three (one branch downward and two upward); rarely is it more than four (Fig. 8.6). The linkage angles tend to be *c.* 45°, with few less than 20°. Angles of more than 60° usually are in areas of geologic faulting or uplifted sedimentary layers[610], or in meandering river floodplains where dendritic pattern becomes less clear. Intersection nodes tend to be regularly distributed in dendritic systems. An exception may be in a dry valley, where descending stream channels diverge and often exhibit aggregated nodes.

The remaining structural attributes of dendritic systems vary from high to low (Table 8.1). Like veins, the expansion of corridors in a converging stream network is important for handling variations in water flow. Also, the range of mineral nutrients and other objects carried downstream will depend very much on how similar or dissimilar the first-order stream areas are.

The distribution of large expanded nodes in dendritic systems is especially significant ecologically and in planning, and holds some predictive ability. Usually, the most probable location of a large node of native vegetation is at an intersection (numbers 3 and 6 in Fig. 8.7). The second most-probable location is on a first-, or first- and 2nd-, order stream (number 1 in Fig. 8.7). Both locations provide important, though

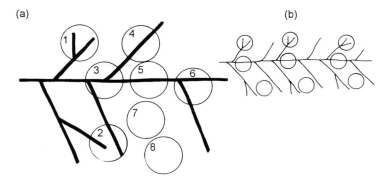

Fig. 8.7. Possible locations of large vegetated nodes in a stream corridor network. (a) Locations relative to main axis and tributaries. (b) Locations 1, 5, and 8 are illustrated upstream, downstream, and in the center of the whole stream-network system.

different, ecological benefits (chapter 7). A large patch (number 2 in Fig. 8.7) that covers first- (and 2nd-) order streams of separate adjacent basins (e.g., which contain 3rd- to 5th-order streams) provides added connectivity benefits as a bridge line (Fig. 8.3). Such a linkage creates a loop or optional route for animal movement. In this way the network as a whole gains some circuitry, enclosure, and mesh size.

The issue of locating large protected nodes in a dendritic system is especially acute for fish, which are restricted to the water and highly sensitive to inputs from land use.[839,1038,130] Fish refuges are often either in the downstream portion of a network, or located just below a headwater area; central locations or specific tributaries are also possible locations.[1006,1054]. Effective movement between refuges makes the stream network a system of refuges.[1037,401,256]

Movement through the system pinpoints another key land-use issue for dendritic networks. Gaps or breaks (chapters 5–7) in a stream corridor are especially significant ecologically. No loops exist, so terrestrial animals and fish cannot use an alternate corridor around a gap. Furthermore, because of gravity and water flow, many effects of the gap are spread widely through the network. A gap in, say, a 3rd-order stream, isolates the upstream tributaries. The main downstream channel is altered. And tributaries just downstream may be isolated. In a dendritic system the importance of corridor connectivity is a network issue, rather than an individual corridor issue.

For planning and conservation, what is the optimum location(s) for natural-vegetation patches in a dendritic network? An answer should emerge by comparing the ecological benefits and shortcomings of each location in Fig. 8.7.

RECTILINEAR NETWORKS AND WAVY NETS
Nodes

Rectilinearity, the high proportion of right angles and straight lines, characterizes hedgerow, road, railroad, powerline, gas line, and oil line corridors, especially in flat terrain (Fig. 8.4). Several spatial attributes are minimal or zero in rectilinear networks (Table 8.1). A hierarchy, if present, has few levels evident, commonly two or three.[1670] Large nodes, either attached or at intersections, tend to be regularly distributed. For example, nodes of livestock routes in New South Wales, Australia are c. 10–13 km apart, the distance a herd is usually moved in a day between water and resting sites.[725] Large open nodes at superhighway entrances–exits in forestland, where sometimes a hawk is visible foraging for small mammals, are usually regularly distributed.

An *intersection effect* may be evident, whereby intersection nodes exhibit ecological characteristics different from those along a corridor.[501] For instance, plant, invertebrate, and bird species richness, including patch interior species, are often higher in the node than along a hedgerow corridor.[305,89,501,909,531] The intersection effect is presumed to result from less-variable microclimatic conditions, plus the higher connectivity provided by additional linkage(s). Also, plants tend to spread from the intersection node into field corners, making the node larger. One may hypothesize an increasing intersection effect with number of linkages, e.g., from 'L' to 'T' to '+' intersections (Fig. 8.4).

The distribution, sizes, and shapes of attached nodes also vary widely. Just as for dendritic networks, large attached nodes usually are of major functional significance (Fig. 8.4). A large node attached to a network is typically a major source of species dispersing widely across the landscape[943], and hence of major conservation and planning significance.[1229,493,688]

Linkages

Both circuitry and connectivity are usually high in rectilinear networks (Table 8.1)[561,893], though both decrease with agricultural intensification.[512,501,680,10,531] In various parts of Europe, reallotment or 'remembrement' programs to aggregate field ownerships can be designed to provide the added goal or benefit of increasing hedgerow connectivity and circuitry.[91] Corridor width and corridor type in any network may vary from high to low.[251]

The number of linkages per node averages about three for hedgerows, and rarely up to four. This is because 'T' intersections are usually more common than '+' intersections.[501,223,251] Road networks typically average *c.* 4 linkages per node, with 3 linkages a bit more common than 5 and 6. In a 'T' intersection often the two opposite linkages are older than the perpendicular linkage, and hence soil and species differences are expected.[91,90,221,224]

Corridor density varies widely, generally from low in remote areas to high in origin or target areas, e.g., for railroad, powerline, and road networks. The effect of *road density* on wildlife populations has been of considerable interest, and research results have been widely used in management and the closing of remote roads.[1700,187,1099,1515] Certain animals avoid the proximity of roads (chapter 5). For example, in Rocky Mountain forestland, when road density is 1 km length per square km (1.6 mi/mi^2), only a half of the total area remains as suitable habitat for elk (*Cervus canadensis*).[1448,1700,1019] When road density reaches 3 km/km^2, only a quarter of the area remains suitable.

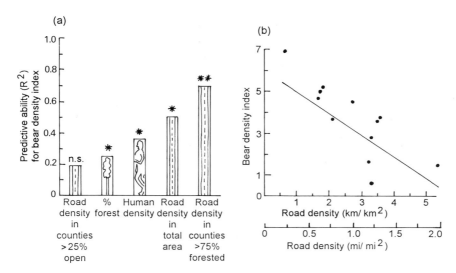

Fig. 8.8. Attributes correlating with black bear density (*Ursus americanus*) in the 20 000 km² Adirondack region of New York. Bear density index = number of legally shot bears/100 mi²/yr. (100 mi² = 260 km²). Road density = km/km² of highways, graveled roads, and other municipally maintained roads in a county. Predictive ability = coefficient of variation (R^2), and ranges from none (0) to complete (1.0 or 100%). A few counties along the perimeter of the region have > 25% open, mainly agricultural, land (where bear harvest is concentrated). n.s. = not a significant relationship with bear density index; * = significant ($p < 0.05$); ** = highly significant ($p < 0.01$). Graph (b) is for the 12 counties where forest cover exceeds 75%. Adapted from Brocke *et al.* (1990).

In the Adirondack Mountain region of New York, black bear populations (*Ursus americanus*) decrease as forest removal, human population density, and road density increase (Fig. 8.8a). However, road density in mainly-forested areas is the best predictor of bear density.[187] No evidence of a threshold in the bear density versus road density curve is present (Fig. 8.8b). Road length *per se* is not inhibiting the bear, as in the elk case. Rather, the roads, especially the tiny 'first-order' roads, provide access to remote areas for hunters. Closing and opening the tiny roads is an effective management action for regulating hunter access and black bear populations. Road density effects have been demonstrated for several other large mammals, including moose (*Alces*).[187,1492,111]

The *mesh size* of networks is readily measured as the average area or diameter of the enclosed patches or sections of matrix. It is essentially the inverse of corridor density. The greater the corridor density, the smaller the mesh size of the enclosed patches. On the other hand, agricultural intensification increases mesh size by removing fence lines and hedgerows.[1337,680,10,1518] Species richness and composition are highly sensitive to size and shape of the enclosures (chapters 2 and 4).

For example, in France a particular predatory beetle (Carabidae) is found in areas where small fields are surrounded by hedgerows. When the mesh size exceeds *c.* 2 ha (5 acres), the beetle disappears[392] (Fig. 8.9). Plants such as *Alliara officinalis, Lapsana communis, Galleopsis dubia,* and *G. tetrahit* are essentially limited to the same fine mesh.[1156,89] An owl species disappears above about a 7 ha mesh size[934], and other local birds also appear sensitive to mesh size.[305] Presumably the species disappear when fields are so large that the microclimate is unsuitable and food density too low.

It has been proposed that mesh (or grain) size would be a useful overall measure or index of the ecological 'health' of a landscape.[501] Any species population can be plotted against mesh (or grain) size (Fig. 8.9). This

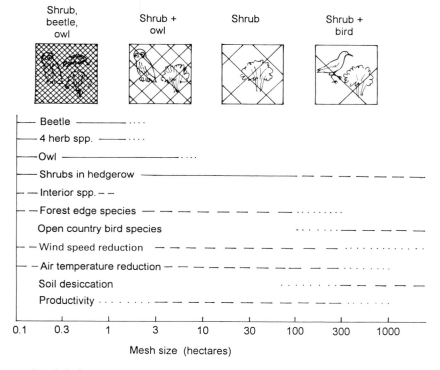

Fig. 8.9. Species and processes related to mesh size of network. Mesh size = average area of fields in locations in northwestern France; logarithmic horizontal axis. Solid line = observed results (based in part on *Les Bocages* 1976); dashed line = hypothesized patterns based on chapters 2, 5, and 6; dotted line = range within which phenomenon ends. See text.

268

clearly separates, for example, the overall requirements for forest versus open country species. Indeed, functional processes and soil and microclimatic conditions are similarly pinpointed along a mesh size gradient.

Wavy nets

Walking paths, animal trails, and braided areas in rivers are usually in the form of *wavy nets*, with variable angles and curvilinearity, especially in rough terrain.[610,1669,35,713,1038,1955] Such nets tend to be linked with topography, and often affect water and soil flows.[223] Wavy nets mimic nature, while rectilinear nets introduced by humans require energy for both construction and maintenance (Fig. 8.4). Green networks for conservation are wavy nets, since they generally follow natural topography.[1229,688,493,1228]

Directionality, convergence, divergence, and corridor expansion appear more pronounced in wavy than rectilinear nets. Otherwise, all the rectilinear attributes (Table 8.1) appear to apply equally to both types. Most wavy nets would be poor models for catching fish. They contain a high variability in mesh size, and frequently change in form.

FUNCTIONAL ROLES OF NETWORKS

In a maze, success in reaching the inner chamber usually depends on taking unlikely routes near the beginning, followed by luck or trial-and-error perseverence in exploring dead ends. The maze is a network designed to confuse. A good maze has countless convoluted, counter-intuitive, cul-de-sac corridors that frustrate and delight.

Though a vertebrate foraging up a stream system can encounter many dead ends, fortunately the functions of other terrestrial networks are more straightforward. We may recognize three broad functions: (1) a habitat; (2) a conduit for flows and movements along the corridors; and (3) a barrier against flows in the enclosed matrix or patches.

Habitat

Corridors are habitats for edge species (chapters 5–7). However, a network has different habitat roles. Because of the connection and proximity of two corridors at a right or acute angle, it is more likely they can serve as a habitat for some patch species, even in the absence of a patch. This appears to be the case for kangaroos using roadside strips in Western Australia.[54]

The location of a corridor within a network also determines what species are present. Thus, in Brittany, France the carabid beetle fauna differs in different parts of a hedgerow network.[223] Some species are near the periphery, some nearer the center, and some in wide corridors within the network. The probability of finding a beetle or forest plant in a hedgerow is much higher if the species is in an attached hedgerow.[89] The same pattern was found for forest plants in New Jersey (USA).[89] Finally, in one study, certain tree species, such as ash (*Fraxinus excelsior*), are essentially limited to corridors near villages.[223]

In the preceding section species were related to mesh size. The habitat implications of this are that total species richness in the network is increased by increasing the variance in mesh size. Just as for fish nets, a fine-mesh network and a coarse-mesh network each has a certain monotony. But a variable-mesh can support more species and ecological processes, and is a useful planning objective if a network imprint is to be present.

Flows along linkages

The flow rate along a single corridor, based on engineering equations, is the inverse of the product of a corridor's resistance times the viscosity of the material or object moving. In landscapes 'corridor resistance or quality' results from habitat heterogeneity and disturbance along a corridor (see Fig. 6.11).[720]

But 'network resistance' adds the effect of having to cross intersections and nodes of various sorts, plus the optional routes offered by circuitry. Nodes function effectively as control points[1235], where predators and hunters wait, stop lights reduce big crashes, zoning reduces corner-cutting, and source and sink actions may predominate. In the case of predators moving along a dirt road network (Fig. 8.10), intersection nodes may be locations of concentrated activity and interactions.

Animals commonly learn two or more alternate routes between intersections (Fig. 8.10). One loop in a network provides two alternate routes, two loops provide four routes, and a few loops provide numerous alternate routes (chapter 8 appendix). Loops and network circuitry provide flexibility and stability, so the animal can avoid disturbances, predators, and hunters.

Variable-width corridors in a network offer additional flexibility to some movements. Indeed, a matrix surrounding differently shaped patches is effectively a network with coarse uneven linkages. Some parts of this network are wide and suitable for animals passing unnoticed, while other parts are narrow bottlenecks.

This leads to the question of reverse flows. In a diverging dendritic stream network, vertebrates could forage up to the headwaters in the morning or in the spring, and return in the evening or autumn. Here, the opposite movements are time separated[1955], as in home-range foraging or

270

Fig. 8.10. Irregular and straight routes of vertebrate movement. Left: bison or buffalo herd (*Bison bison*) follow a wavy net that includes a main (or trunkline) route. Lower right: coyote (*Canis latrans*). Photos courtesy of US Fish and Wildlife Service. Upper right: tracks of coyote, Servietta National Wildlife Refuge, New Mexico. R. Forman photo.

seasonal migration. Alternatively the movements could be concurrent. If the parallel oppositely moving objects interact, a 'counter current' process operates to maintain frequent interactions.[501,493] The predator, for example, could wait for prey at strategic nodes, or optimize a route through the network to fill its belly. But the prey can also optimize a route through the network that starves the predator. For strategic cat-and-mouse games networks never cease to amaze.

Barrier

A study in Denmark indicates that, compared with an open landscape, a hedgerow network may reduce average wind velocity at ground layer by perhaps 15%.[802] This has numerous effects on soil and water characteristics.[802,501,1474] Soil erosion and evapotranspiration are reduced. Energy flows are altered. Streamflow is reported to be reduced in the cool wet season, and enhanced in the warm dry season. Overall, temporal variability through the year is greater in the open landscape. But spatial variability from spot to spot is higher in the hedgerow or 'bocage' landscape.

A single corridor that species of the matrix must cross may reduce their movement rate. Yet a series of corridors in a network can be expected to reduce much more the movement of matrix species[963,964], both because more lines must be crossed, but also some searching in the network is likely. This aspect of landscape resistance[501,877] obviously is dependent on mesh size.

An ecologically important result of this barrier effect is the isolation of subpopulations within the subdivided sections of matrix. Reproduction is greater within than between subpopulations. Such isolation causes a high degree of genetic variation for matrix populations sensitive to the network barrier effect. Clover, butterflies, and mice illustrate such genetic changes in populations.[42,169,681,1871]

NETWORKS IN ACTION

The functions of barriers, flows along linkages, and nodes are nicely combined in designing rangeland for livestock[1612] (I. R. Noble, pers. commun.). If you owned a big ranch or paddock in a dry grassland landscape, how would you arrange the water sources, fencing, and best grazing areas? One solution is to place the wells or water troughs about equidistant apart and in medium-quality grazing areas. Connect the nodes (water sources) with fences to produce a giant network. Thus, in every enclosure, moving livestock that encounter a fence are funneled to water in a corner. Concentrating the heavy livestock trampling in corners of enclosures does relatively little harm. The poorest grazing areas with the most fragile vegetation and soil are thus protected by distance from concentrated trampling and wind erosion. And the best grazing areas with the greatest productivity are similarly protected from intense trampling.

M. Kozova and colleagues (1986) tie several network concepts together in relating ecological stability to connectivity, based on an area in Slovakia that varies in the amount of fragmentation. Fig. 8.11 illustrates the conversion of a mapped portion of the heterogeneous landscape into a network of nodes and linkages for wildlife movement. The network highlights the species sources, loops and alternate routes, problematic routes, and possible locations of subpopulations in a metapopulation (chapter 11). Most important from wildlife management, conservation, and planning perspectives, the network pinpoints the five major gaps in the network.

(a) Map of landscape area   (b) Routes of wildlife movement

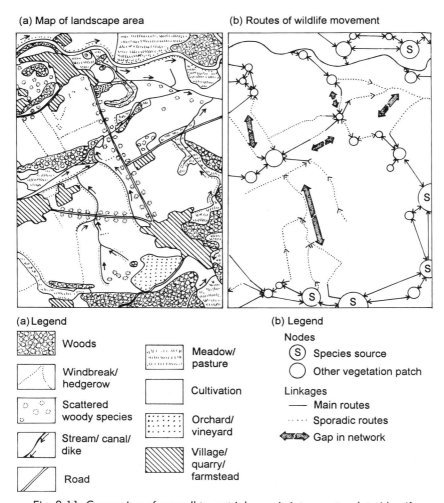

(a) Legend

Woods

Windbreak/ hedgerow

Scattered woody species

Stream/ canal/ dike

Road

Meadow/ pasture

Cultivation

Orchard/ vineyard

Village/ quarry/ farmstead

(b) Legend

Nodes

(S) Species source

◯ Other vegetation patch

Linkages

—— Main routes

····· Sporadic routes

◀/▶ Gap in network

Fig. 8.11. Conversion of a small terrestrial mosaic into a network to identify key gaps. (a) Three villages, five farmsteads, and five roads act as barriers or filters to wildlife movement. Cultivation surrounds woods and meadows as key wildlife habitats. (b) Wildlife habitats are nodes, and movement routes are linkages. Major gaps in the network are indicated by wide broken arrows. Adapted from Kozova *et al.* (1986).

The study uses a mathematical model in which animal populations forage and disperse along a network according to various mobility parameters. The quality of nodes, corridors, and matrix is varied from high to low based on generalized field data: regional nodes are considered to be 55–65 ha (*c.* 150 acres); local nodes 5–10 ha; regional corridors 500 m (*c.* 1600 ft) long and 30–50 m wide; and local corridors 1000–2000 m

273

long and 10–20 m wide. Alpha, beta, and gamma connectivities (chapter 8 appendix), corridor density, mesh size, number of alternate routes, critical points in the network, and distance parameters for mobility dispersal are calculated. Based on modeling and field observations, higher network connectivity appears to: (1) increase dispersal between subpopulations (local populations) at nodes; (2) decrease dispersal mortality, due to corridor protection; (3) increase average subpopulation growth rate, due to recruitment, recolonization, and natality; and (4) decrease variation in (pre-winter) subpopulations.

The authors conclude that the resistance and persistence of mobile (multihabitat) animal species represent a useful measure of ecological stability, and that these attributes are enhanced by higher habitat connectivity.[893] Connectivity measures (alpha, beta, and gamma) effectively evaluate conditions around an individual habitat, or a portion of a landscape. The authors also conclude that landscape connectivity only 'constitutes a territorial precondition and the overall landscape stability depends primarily on economic systems obeying ecological laws'.

In one of the few long-term integrated, even elegant, studies in landscape ecology, Gray Merriam and colleagues have studied small mammal dynamics in hedgerows and woodlots of a southern Ontario agricultural landscape. Using field measurements and experiments with the common white-footed mouse (*Peromyscus leucopus*), they found that the species becomes locally extinct in about 2–5% of the woods each year.[1123] However, recolonization from farmsteads and other woods by movement along hedgerows (fencerows) rapidly repopulates a woodlot.[1123,718,1114] Thus the early equilibrium–island-biogeography model was of little use. Instead a patch–corridor–matrix model, based on combining field measurements, experiments, and mathematical modeling, progressively developed. This seemed to provide a robust understanding of dynamics in the vertebrate species and the landscape as a whole.

Experimental snow or sand fences, camouflage netting corridors, and observations of natural hedgerows indicated that movement is greater where there is vegetation cover both in the ground and overhead layers.[1114] Measurements of movement in hedgerows confirmed that corridor quality was important to movement rate.[720,112] Animals moved best in wide hedgerows with a continuous cover of woody plants. Gaps in the woody plant cover, and to a lesser extent narrows, reduced movement. Radiotracking indicated that mice placed in an unfamiliar landscape move almost entirely in hedgerows, especially favoring wide and wooded corridors.[1118,1114] The home ranges of resident hedgerow mice were about ten times longer than those of the same species in woods.[1114]

A mathematical model was created to mimic movement among four patches.[465] The model predicted that population growth rate is lower and extinction probability greater in isolated patches, than in patches connected by corridors (Fig. 8.12a). Field data supported the prediction. The effect of the arrangement of patches and corridors was then examined with a five-patch model[939] (Fig. 8.12b). The model suggested that population growth and survival in a patch depend on connections with other patches, and increase with the size of the largest geometric figure in which it is connected. Thus, patch populations in a connected pentagon should persist better than those in a triangular or linear arrangement.

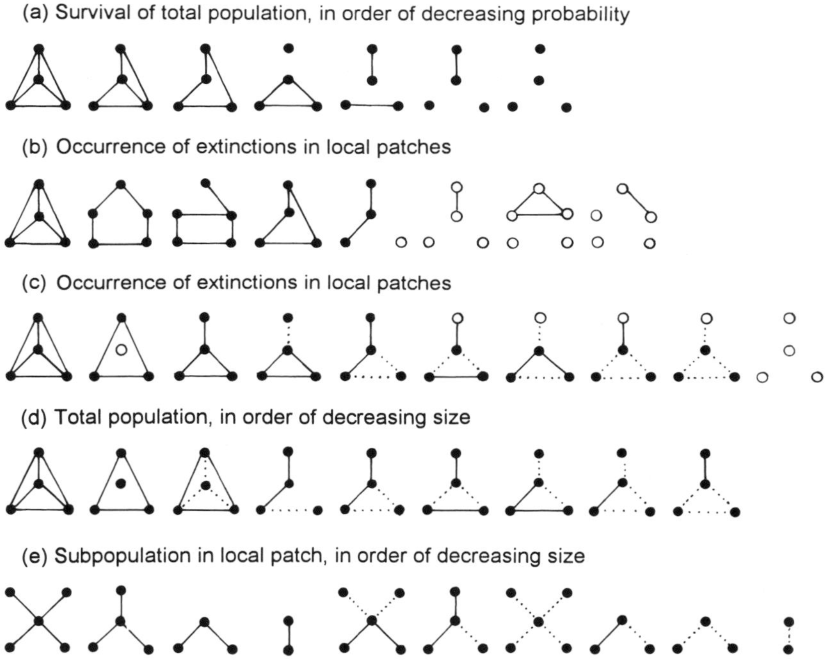

(a) Survival of total population, in order of decreasing probability

(b) Occurrence of extinctions in local patches

(c) Occurrence of extinctions in local patches

(d) Total population, in order of decreasing size

(e) Subpopulation in local patch, in order of decreasing size

Fig. 8.12. Effects of patch and corridor configurations on populations of white-footed mouse (*Peromyscus leucopus*). Predictions based on simulation modeling over time, plus some field data from woodlots and hedgerows in southeastern Ontario. (a) Comparing seven arrangements of corridors among four patches. The initial total population is equally divided into four subpopulations; the final population is largest at the left and smallest at the right. (b) In comparing four versus five patches with various corridor arrangements, open circles indicate patches where subpopulations become locally extinct. (c) Effect of corridor quality on local extinctions. Solid lines = high-quality corridors; dotted lines = low-quality corridors. (d) Effect of corridor quality on final total population size. (e) Local patch affected by number of corridors (with attached patches) and corridor quality. The local patch, in each case on the horizontal mid-line of the diagram, has the largest final subpopulation at the left and smallest at the right. Adapted from Fahrig & Merriam (1985), Henein & Merriam (1990), and Merriam et al. (1991).

275

Models for population size and persistence were then modified to incorporate corridors of high and low quality.[720] These permitted several predictions (Fig. 8.12c–e). Networks with only high-quality corridors are better than those with mixed high- and low-quality corridors. With corridor quality constant, more connected patches are better. Isolated patches connected with a low-quality corridor have the highest local-extinction probability. And adding an extra patch with a low-quality corridor decreases the total population.

These analyses of corridor quality suggest that a low-quality corridor acts as a mortality sink (chapter 6). Further studies are needed, since an alternative expectation is that animals quickly learn to avoid a dangerous low-quality corridor (A. F. Bennett, pers. commun.). Nonetheless, considerable evidence supports the conclusion that, compared with no corridor, even a low-quality corridor or stepping stones increases connectivity and enhances species movement (chapter 6). These in turn help counteract the demographic and genetic effects of isolation.

During the summer period of agricultural production the mice also move into the matrix. Hence, the landscape is functionally connected for the mice, which move through patches, corridors and matrix.[92,1115] Among the various habitat types in the landscape, the hedgerows have the highest immigration rate, probably because they are connected to more patches than almost any other habitat.[1112,1113,1116] Though the species undergoes large population fluctuations and movements each year, some evidence is available suggesting stability of the network population, both genetically and over centuries.[1124,1114]

The empirical results from many years of field studies in this landscape paint a surprisingly full picture of mouse dynamics in a network. The simulation models provide predictions and hypotheses, but the assumptions and parameters of the models are based on these field studies, and hence are unusually realistic. Consequently the overall results, such as having high population stability when several patches are interconnected or when a network only has high-quality corridors, and having low stability in isolated patches or with low-quality corridors, provide reasonably solid guidelines for planning and management.

Finally, we may ask what is the optimum form of a landscape network? First, it bears emphasis that a network is no substitute for large natural-vegetation patches. A giant green network is part of a broader land-use planning and conservation strategy, which accomplishes biodiversity, water, production, recreation, and community objectives.[1229,493,111,1492] An optimum network form is hypothesized to include a high level of circuitry, plus a high variance in corridor width, curvilinearity of linkages, and mesh size.[493]

Yet, it should be obvious that direct research on the ecology of networks is scarce. Considering the ubiquity and size of networks, this is a worthy frontier in which to think big.

276

MATRIX STRUCTURE

How to identify the matrix

When you are in the 'middle of nowhere', you are probably in the *matrix*. It covers an extensive area, is highly connected, and controls landscape or regional dynamics, in this case your every move.[501] In a similar manner it encloses and affects patches, as well as corridors.

Often the matrix in a landscape is obvious. For example, the bulk of an area may be forest, residential, cultivation, wetland, or even glasshouses (greenhouses). But in some areas two or three types are approximately equal in area, and it is uncertain which functions as the matrix. Michel Godron described how the three prime attributes, area, connectivity, and control over dynamics, are used sequentially for identification of a matrix.[501]

Total area is the first and easiest criterion. If one element type covers well over half the area, or is much more extensive than the second-largest element type, it should be considered the matrix (Fig. 8.13a and b).

But if the two most-extensive element types are similar in total area, connectivity should be used to differentiate them. Just as for a network, 'matrix connectivity' varies from 0.0 to 1.0, and may be readily measured as the inverse of the proportion of linkages that must be added to have a (fully) connected system.[493] If connectivity of one leading element-type is significantly higher than that for a second, the former should be designated the matrix (Fig. 8.13c and d) (Fig 8.13d has twice as many shaded as white sections. How many changes would be needed to connect each type?) Or if area and connectivity are both slightly higher for a particular element type, it should be considered the matrix.

If in unusual cases the matrix designation is still unclear (Fig. 8.13d), the third attribute, 'control over dynamics,' becomes the determining factor. Though difficult to measure directly, this is the most important assay. Indeed, area and connectivity are surrogates or indirect measures of the third attribute. Control over dynamics is the reason it is useful to

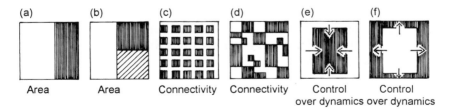

Fig. 8.13. Characteristics determining the matrix of a landscape. White land-use type is the matrix, and covers 60% of (a), 45% of (b), and 50% of the others. Arrows indicate net direction of flows. The key factor determining the matrix is indicated for each grid.

277

identify a matrix (Fig. 8.13e and f). If climate, natural disturbance regimes, or human activities changed markedly, which element type would exert the greatest influence in determining the future landscape? Examples of control over dynamics are an element being the source for: seeds of recolonizing plants, herbivore herds, keystone predators, blowing particulates, gaseous pollutants, vehicles, human masses, flood water, and heat. Pollen from lake and bog deposits tell paleoecologists what the matrix vegetation was at different points in time.[367,767,1850,1848] The matrix has the greatest control over landscape and regional dynamics.

Shape

Rarely the matrix is an uninterrupted area with other element types clustered at its periphery. A remote forested area with settlements at the edge, and an elevated dry plain with peripheral moist slopes, are examples.

The norm, however, is for the matrix to be cut up by corridors or perforated by patches.[1706,683,518,1492] A *subdivided matrix* has sections that are separated by more or less equal-width filters, such as hedgerow, road, and other networks.[948,806] The filters are daily bathed and controlled by inputs from the matrix. Edge species move back and forth between corridors and matrix[1111,1114] (chapter 3). Some isolation of subpopulations and processes takes place in the matrix sections, which tends to increase variation within the matrix.

Alternatively, in the absence of corridors, a *porous* or *perforated matrix* has patches scattered over its area. These also are bathed by the surrounding land[794], though large enclosed patches are buffered somewhat from matrix effects. The isolation of matrix subpopulations is much less than that produced by the subdivision of a matrix.

The proportion of interior and edge in the matrix is greatly affected by its shape. If an uninterrupted matrix is progressively subdivided by corridors, the amount of interior habitat rapidly plummets, and edge habitat skyrockets. At some yet-to-be-determined point (e.g., 25% corridor and 75% matrix area), little if any matrix interior remains. It is all edge habitat dominated by edge species and processes.

On the other hand, if the uninterrupted matrix is progressively perforated by small patches, the interior environment disappears, and edge increases, more slowly. The apparent threshold point when the matrix interior has essentially gone due to perforation is later (say, 40% patches and 60% matrix) than in the preceding case of creeping corridors.

In either case, however, large round or square areas of matrix such as nature reserves disappear near the beginning of the process.[518] Furthermore, human-created networks in landscapes tend to be more regular, symmetrical, and straight-boundaried than patches, with their variations in size, shape, and aggregation. Consequently, a subdivided matrix tends to be less variable, and probably less species rich, than a perforated matrix.

Techniques for measuring a matrix[1426] are suggested in the appendices of chapters 2 and 9. Sections of a subdivided matrix are measured in the same way as for patches. The perforation or porosity of a perforated matrix may be assayed by grouping patch sizes and shapes in classes.[501] For example, the number of large, medium, and small patches of certain shape categories (chapter 4 appendix) would describe the perforation of a matrix, to compare with another matrix. Perforation is of considerable interest for the movement of wind, water, and species through a matrix (chapter 10).

MATRIX DYNAMICS

Resistance and heterogeneity

Landscape resistance was described as the effect of structural characteristics of a landscape impeding the rate of flow of objects (species, energy, and material).[501] Since boundaries separating spatial elements are locations where objects usually accelerate or slow down[1669,174,503], it has been suggested that *boundary-crossing frequency*, i.e, the number of boundaries per unit length of route, is a useful measure of resistance. For example, a herbivore typically slows down, exercising caution, in crossing a boundary into a new ecosystem. Therefore to cross a mosaic we expect an animal would normally choose a longer route with fewer boundaries.

Alternative general measures of resistance could be simply the relative resemblance of a matrix to the optimal habitat for a species.[877] Or the area could be subdivided into more-suitable and less-suitable habitats, so the amount of less-suitable habitat along different routes is a measure of resistance.

In developing ecologically based plans for locating new forests in southern Holland, Bert Harms, Paul Opdam, and colleagues provide additional insight into landscape or regional resistance. Landscape resistance was initially calculated for the movement of a group of woodland birds (e.g., nuthatch, *Sitta europaea*).[678] Built areas, glasshouses, and busy roads were considered to increase resistance, and woody vegetation to decrease resistance. Resistance (calculated for each cell in a raster GIS, and mapped) was a key to the accessibility for forest birds to colonize various proposed sites for establishing new forests.

The analysis was then broadened to calculate resistances for different groups of animals, i.e., large mammals, small mammals, birds, and butterflies.[877] The following resistance components were calculated for areas with building development, rural agricultural land, and open water prominent: percentage cover of woods and/or density of hedgerows; percentage cover of water; and percentage cover of built-up area. In addition, the presence of a broad river or canal and/or four-lane motorway was evaluated as a barrier.

In all landscape areas resistance to species movement is less with more forest cover present (Fig. 8.14). Landscape resistance is most serious when forest cover is below 25%. In built areas butterfly movement is most inhibited, especially when forest cover is less than 5%, and movement of forest birds least inhibited (Fig. 8.14). On the other hand, in rural areas with little forest cover bird movement is most inhibited, and large mammals least affected. A similar pattern is evident in areas with considerable open water, though the degree of landscape resistance is greater. Highway and river/canal corridors are serious barriers to bird movement, and least inhibitory to large mammals (Fig. 8.14). In all areas small mammals are intermediate in response to resistance. A simulation model for movement of species relative to landscape resistance was then used to evaluate the proposed locations of new forests.[877]

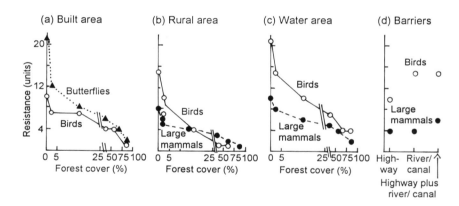

Fig. 8.14. Resistance of different types of matrix or barrier to species movement. Estimated resistance values indicate the relative probability that an individual animal will not cross an area or barrier. Estimates are based on literature survey, consultation with experts, and field observations in The Netherlands of the following representative species in each group. Large mammals (1 species; deer, *Capreolus*); small mammals (2 species; squirrel, *Sciurus*, and noctule, *Nyctalus*); birds (12 species, mainly forest songbirds); and butterflies (3 species). Only the groups with the highest and lowest curves are plotted; other curves are intermediate. Small mammal curves are the same as for butterflies in (a), and as for large mammals in (b). Adapted from Knaapen *et al.* (1992).

These are measures of landscape or regional resistance where clear patches and corridors are evident. But resistance applies equally well to a matrix without distinct enclosed patches and corridors. Indeed, the question is one of heterogeneity and scale (chapter 1). Nothing is homogeneous. A relatively homogeneous matrix is heterogeneous or variegated at a finer scale, where micro-patches and micro-corridors are evident.[1881,1877,834] Fine-scale heterogeneity is important if the rate of a process or distribution of objects varies in relation to it.[1583,1509,956] Therefore, matrix resistance varies for most species, and can be measured to indicate the probability, rate, and route of crossing by an object.

There are many measures of heterogeneity, which ranges from high to low in an area.[1583,1322,1276,698,883,1748] But just as spatial arrangement matters at the landscape and regional scales, the explicit configuration of micro-patches and micro-corridors is often important for flows across a matrix. Therefore, both heterogeneity and spatial arrangement within a matrix matter ecologically.

One further heterogeneity issue is quite helpful in understanding a matrix.[579,501] Suppose you looked carefully in one portion of a matrix and were able to identify 13 microhabitats present. You then explored other portions of the matrix, and found approximately the same 13 at each place. Meanwhile, your friend in a second landscape surprisingly also found about 13 microhabitats in each portion. But at each place in the second landscape about half the microhabitats were previously recorded, and half were new. You found *microheterogeneity*, where the differences in a small area were repeated in similar form throughout. She discovered *macroheterogeneity*, where the main differences were across the broad area. An animal could move across your matrix more easily than across her matrix, where new habitats would be continually encountered. But her matrix would be much richer in biodiversity.

Changing shape

The control over dynamics by the matrix has major implications for change in the landscape mosaic over time. An extensive mosaic batters corridors and patch edges with seeds, dust, people, heat, and so forth. The matrix thus squeezes, invades, and sometimes wipes out enclosed elements. The result is a decrease in landscape heterogeneity. In addition to traces of former patches, evidence for this process is seen where the matrix bulges, i.e., has convex boundaries relative to a corridor or patch[501] (chapters 3 and 4). Where natural processes dominate, straight corridors generally become curvilinear, even disconnected.

Yet often, matrix boundaries are concave, such as around a round or amoeboid patch. This typically means that the patch is expanding at the expense of the matrix. For example, villages expand in agricultural land, and overgrazed patches bulge in desert grassland.

In short, the matrix takes its place alongside the patch and corridor as a fundamental model of land. But the three types of spatial element do not produce a mosaic like that fixed on the floors of a beautiful Roman Bath. At any moment there is open competition among the spatial elements, or an uneasy truce between invasions and retreats. This produces a highly changing pattern reminiscent of a movie going backwards at ten times its normal speed. The following chapter now focuses specifically on mosaics, those small configurations around a spot, as well as whole landscape and regional mosaics.

APPENDIX: NETWORK EQUATIONS

Gamma index for network connectivity

$$\gamma = \frac{\text{number of linkages}}{\text{maximum possible number of linkages}} = \frac{L}{3(V-2)}$$

Alpha index for network circuitry

$$\alpha = \frac{\text{number of circuits or loops}}{\text{maximum possible number of circuits}} = \frac{L-V+1}{2V-5}$$

where L = number of linkages; V = number of nodes.
Adapted from Haggett *et al.* (1977), Taaffe & Gauthier (1973), and Forman & Godron (1986).

PART IV

Mosaics and flows

9

Mosaic patterns

'Do you see,' said Henry, 'anything here that would be likely to attract Indians to this spot?' One boy said, 'Why, here is the river for their fishing;' another pointed to the woodland near by, which could give them game. 'Well, is there anything else?' pointing out a small rivulet that must come, he said, from a spring not far off, which could furnish water cooler than the river in summer; and a hillside above it that would keep off the north and northwest wind in winter.

F. B. Sanborn, The Life of Henry David Thoreau, *1917*

Vertebrates are often marked and followed to determine their habitat requirements, and the locations of their encounters with other animals. Suppose we try this on a suburban family using radiotracking techniques. Miniature radio transmitters are worn as fancy belt buckles or earrings, and you have a high-tech antenna to locate family members at any time. You record the cluster of workplaces, food stores, roads, schools, local parks, night spots, and so forth visited. Each person's 'home range' is then mapped using the spatial arrangement of locations and routes taken. The amazingly different mosaics would help planners design better suburbs, based on the cluster of locations that people use, and perhaps need (S. B. Warner, Jr. pers. commun.).

In previous chapters we considered patches, corridors, and matrix by themselves. Now we begin hooking these building blocks together to form mosaics. Mosaics are evident at all scales. Here, we focus on the neighborhood clusters of ecosystems or land uses, plus whole landscape and regional mosaics. On land there may be a limited number of common mosaics to understand.

In landscapes and regions context is usually more important than content. That is, the surrounding mosaic has a greater effect on patch functioning and change than do the present characteristics within the patch. *Context* here includes three components: adjacency, neighborhood, and

285

location within a landscape. *'Adjacencies'* are the spatial elements in contact with the patch or site of interest (chapter 2). *'Neighborhood'* is the encompassing local mosaic linked by active interactions. Neighborhoods and locations within a landscape will be the emphasis here in understanding context.

Movements and flows in previous chapters have mainly been between a patch or corridor and its adjacent spatial element. In the following chapters 10 and 11, flows are between non-adjacent elements, involving a minimum of three ecosystems: the beginning, the intervening, and the target element. Therefore, in the present chapter we focus on clusters or configurations of ecosystems where spatial arrangement is critical to movement. In addition, we spotlight the structure of whole landscape and regional mosaics, across which wind, water, animals, and people move.

The structure of the mosaic strongly affects crop growth and erosion on a farm, biodiversity and aesthetics in a park, wood production and fish in a forest, wildlife movement and extinction in a refuge, water and livestock production on a ranch (paddock), and all ecological characteristics of a town. As humanity and the planet roar down the track of time, decision makers that will firmly grasp the handle labelled 'spatial planning' are needed at the helm.

No review of this chapter topic exists, so references cited in the text will provide the needed background. However, the volumes by Saunders *et al.* (1987) and Turner & Gardner (1991) contain much information relevant to mosaics.

The chapter begins with a cluster and a configuration of ecosystems. The uses of a configuration by animals are presented, along with its stability. Then the roles of geomorphology and human culture in creating whole regions and landscapes are described. We explore the vertical or third dimension of land. The types of regions and landscapes are introduced. Next, habitat arrangements for biodiversity and wildlife are introduced, along with strategic points in the land. We end with how to measure a mosaic.

ECOSYSTEMS IN A CLUSTER

A first law of geography states that 'everything is interrelated, but near objects are more related than distant objects'. One may object that it is impractical or too idealistic to consider everything in making a land-use decision or action. Animals, seeds, heat, mineral nutrients, genes, information, and much more move between ecosystems. Yet a useful, interaction or spatial-flow principle seems possible by also drawing from ecological theory.

From ecosystem science we learn that 'energy and mineral nutrients flow from one object to another within, or between, ecosystems.' From behavioral science, 'because species find certain habitats more suitable, many locomotion-driven movements are directional, toward patches of the same type.' Combining these principles with the geography law provides the following 'spatial-flow principle' or guideline[489], useful for estimating which ecosystems to consider in planning and management. *All ecosystems are interrelated, with movement or flow rate dropping sharply with distance, but more gradually between ecosystems of the same type.*

Thus, in Fig. 9.1a patch A at the bottom has the greatest interactions with the adjacent ecosystems or land uses, B and C. Interactions between this A patch and the B ecosystems decrease with distance. However, interactions between this A patch and the other A patches decrease more gradually than in the case of the B patches.

Clusters and configurations

The spatial-flow principle implies that an asymmetric cluster of ecosystems will be interlinked by flows or movements that are somewhat predictable in relative rate. Robert Woodmansee (1990) illustrates and further develops the concept in a dry grassland area of Colorado (USA), where an ecosystem or site is commonly congruous with a soil type (or soil polypedon[800,1546]). An *ecosystem cluster* (or site cluster) is recognized as a spatial level of hierarchical organization between the local ecosystem and the landscape.[501] It describes a group of spatial elements connected by a significant exchange of energy or matter.

For example, Gus Shaver and colleagues have carefully documented the changing patterns of flows for nitrogen, phosphorus, and other

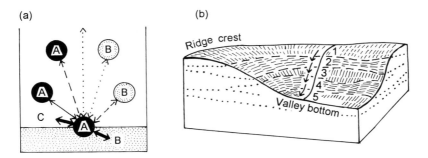

Fig. 9.1. Flows or movements in an ecosystem cluster and a catena. (a) Flows between a patch and its surroundings in relation to distance and ecosystem or land use type. Amounts of flow: ——> ——> - - - -> · · ·> (b) Flows among five soils in a catena. Adapted from Hole & Campbell (1985) and Woodmansee (1990).

material through a sequence of plant communities in the Arctic tundra of Alaska.[1543,1544] The communities interact, indeed are tied together, by these flows.

The ecosystem cluster can include the concept of a *catena*[1919], traditionally used by soil scientists to describe a connected sequence of soils on a slope from ridgetop to valley bottom[800,1546,740] (Fig. 9.1b). The connecting process is typically downslope movement of water and soil, which follows distinct flow paths. Soil development and productivity are generally enhanced at the bottom of the slope, where mineral nutrients and water accumulate.[1501] The soil catena concept conveys an inherent length or area, which is characteristic for a series of ecosystems linked by downward flows.

However, many flows are not simply linear and downward.[147] Animals locomote all over the place, and wind blows up all valleys and across slopes. Thus a concept applicable to a two-dimensional (or 3D) area is more realistic, and is described as a *flowweb*.[1919] This is the network of flows or movements among neighboring ecosystems.

Many flowweb movements are evident in grasslands. Spectacular photographs were taken of the 1930s American 'dust bowl', an area with blowing soil in Colorado–New Mexico–Texas–Oklahoma–Kansas, the size of Thailand or France.[1930] The photos mask the fact that most of the soil material moved short distances from field to gully or fenceline, and was redeposited in another location by subsequent wind eddies and storms. Snowdrifts by hills and windbreaks are also short-distance and rearranged phenomena.[1919]

Wildlife and livestock typically feed in one ecosystem, drink in a second, and rest and leave wastes in a third.[943,1525,1500,1495,1535,1534] This cluster of ecosystems is tied together by animal movement, and is differentially affected by herbivory, fertilization, trampling, and erosion. For instance, a flock of sheep is described as grazing in lush lowland pasture next to a railroad in New Zealand.[1919] When trains pass, the terrified sheep race to the ridgetop, and defecate and urinate. The typically low-nutrient ridgetop has become rich, using this unusual fertilization technique.

The extent and boundaries of landscapes are recognizable by the repeating cluster of ecosystems or land uses present (chapter 1). Several characteristics of the cluster are also repeated in similar form across the landscape. These include the flows or interactions among the ecosystems of a cluster[499,501], geomorphic land forms, microclimatic patterns, and the set of natural disturbance regimes and human activities that characterize a cluster. The relative abundance of ecosystem types within a cluster may vary somewhat more from cluster to cluster.

The cluster and flowweb concepts imply no particular spatial arrangement of the ecosystems. Yet, we imagine the spatial elements are not

288

randomly distributed, and we know that spatial pattern matters ecologically. Thus, the concept of a *configuration* is used to refer to a specific arrangement of spatial elements (patches, corridors, and/or matrix) that is repeated in other locations.[488,251] It is a repeated, spatially explicit cluster of local ecosystems or land uses. Although configuration is used for a neighborhood of ecosystems within a landscape, the concept can apply at any scale.

Each whole landscape certainly has a unique unrepeated structure. In contrast, almost any 'adjacency arrangement' is probably repeated in similar form elsewhere. That is, the pattern of a patch (or corridor) with its adjacent patches and/or corridors is general. For example, a patch surrounded by two other land use types (patch A at bottom of Fig. 9.1a), or surrounded by four corridors (e.g., field with hedgerows), or attached to one corridor and surrounded by matrix (e.g., clearing with road in evergreen forest), is common. Generally, alluvial fans are at the lower end of stream channels, hedgerows are between fields, and successful farmsteads are by busy roads. Adjacency arrangements are poorly known ecologically[1844,633,634,430,1151,1150], but easily studied.

In the spatial hierarchy on land (chapter 1) the configuration represents a level between the landscape and the local ecosystem. Although an adjacency arrangement could be a configuration, typically configurations are 'neighborhoods' of adjacent and non-adjacent ecosystems.[430] Thus, overall, we expect that the patterns of adjacency arrangements are common, whereas the arrangement of ecosystems or land uses in a whole landscape is unique. We hypothesize that a finite number of configurations exists, and more importantly, that a limited number of common configurations exist.

A configuration is spatially recognizable, but depends on being functionally tied together by frequent movements and flows.[488] This functional characteristic means that the boundary of a configuration is identifiable, i.e., where interactions among spatial elements decrease sharply. Flows linking elements in a particular cluster do not have to be of a single type, such as moving sheep or silt. Hence, in an interacting cluster, fire may move uphill, phosphorus wash downhill, seeds blow horizontally, and wildcats forage in all the ecosystems.

A cluster of a few hedgerows, woodlots, and cultivated fields in Ontario, all linked by seasonal small-mammal movement, is an especially well-studied example of a configuration (chapter 8). Rearranging, say the distribution of low-quality corridors, woodlot sizes, or crops grown within the configuration, significantly alters mammal dispersal, growth rate, and persistence.[1124,720,1114,1116]

Nature conservation in The Netherlands offers another example linking function and structure. Success in protecting egret and heron (Ardeidae) populations requires three characteristics: (1) a wooded area

289

for nesting and roosting that is larger than a certain minimum size; (2) a marsh for feeding that is larger than a certain minimum size; and (3) that the wooded area and marsh are less than a certain maximum distance apart. Indeed, movement between non-adjacent habitats is common for many vertebrates.[1690,1701,554,1704,1142] Thus, a configuration can be determined or created to provide for the required wildlife movement and conservation.

The cluster and configuration concepts apply equally well to humans. The leadoff quotation for this chapter, when Henry Thoreau is teaching school after his Harvard graduation, emphasizes *context over content* in selecting a site, i.e., the surroundings are more important than what is in a location. It also emphasizes the cluster of spatial elements used by those Native Americans. Indeed, identifying such clusters or configurations doubtless is a good predictor for locating most archaeological sites.[131,341,132,1610,1611] And for the same reason, the configuration of required resources should be a keystone in the planning and design of human communities.

Familiar spatial patterns

An analogy with nutrition may be helpful in identifying widespread, stable spatial patterns.[251] Suppose a landscape is a large protein or carbohydrate molecule. And let us say that wooded patches, bean fields, stream corridors, hedgerows, and so forth, are analogous to atoms, such as carbon, hydrogen, oxygen, and nitrogen. What therefore are the analogues to intermediate-sized amino acids and glucose sugars? That is, what are the distinctive stable groups of atoms that make up part of a large molecule? Are there widespread, distinctive stable combinations of patches and corridors, independent of landscape type? If so, they should be especially useful for understanding landscape fluxes, and for planning and management.

Indeed, if 'self-organizing principles' exist[920], a configuration is a likely prediction or product. That is, given a mix of land uses plus ample time, they would become arranged in a limited number of ways. Perhaps that is why different villages, farms, or beaver flowages often look so similar within a landscape.

Certain spatial patterns underlie and help us understand a configuration. These are so widespread that their mention would be trivial, except that in most cases their ecology is poorly known and highly important.[501]

Regular

Regular patterns have approximately equidistant ecosystems or land uses surrounded by a matrix or network. Examples are school yards in suburbia, clearings from dispersed-patch logging, fields in a hedgerow

Fig. 9.2. Repeated relatively regular patterns on ancient deposits of silt and volcanic ash. Agricultural production on former native grassland in eastern Washington. Courtesy of USDA Soil Conservation Service.

grid, polygonal meadows in wet tundra, sinkhole wetlands in limestone karst topography, vegetation stripes in a desert[141,142], and ancient dunes on a plain (Fig. 9.2).

Aggregated

Aggregated ecosystems are illustrated by beaver ponds in separated mountain valleys, petrol or gasoline stations in towns of an agricultural landscape, alpine habitats on clustered volcanic peaks, drumlins left in groups by a glacial margin, and clustered orchards or vineyards on hilly soils.

Linear

Linear distributions of patches are widespread, including a row of desert oases, wooded patches scattered along a river, villages on a coastline, clearings along a new Amazon road, and African villages by a long rocky ridge.

Parallel

Parallel corridors are familiar, such as roads in suburbia, mountaintop paths in ridge-and-valley areas of the Appalachian Mountains, rivers cutting across plains, strip clearings in forest for wildlife, and windbreaks in cool dry areas.

Parallel separated windbreaks are often planted perpendicular to the prevailing wind direction to reduce surface wind velocity over a broad area. No cumulative effect takes place to produce more wind reduction downwind of several windbreaks.[711] Indeed, maximum wind reduction often is behind the first windbreak, because the turbulence produced by this windbreak slightly reduces the effectiveness of subsequent windbreaks.

Associated

Each of the preceding examples is for a single ecosystem or land use type repeated over space. Perhaps more important in a configuration of clustered ecosystems is the association or linkage of different types. Thus if one ecosystem is present, there is a high probability that a second or additional ecosystem will be present with it. Examples of ecosystem associations are a sprouting young-hardwood stand near an active beaver pond[1183], north- and south-facing slopes (Fig. 9.2), two vegetation types linked by soil or other factors[1283], a pond near a cranberry bog (*Vaccinium macrocarpon*), and a commercial center near a residential area.

Some 'incompatible' ecosystems, in contrast, are normally distant from one another. Usually, a farmstead is separated from a woodlot, expensive housing is not downwind of a major air-pollution source, cultivated fields are away from refuges with wild herbivore herds, and mountain lion (*Felis concolor*) and human habitations are distant from grizzly bear (*Ursus horribilis*) territory in western North America.

The concept of a *buffer*, as an area that lessens or cushions the effect of one area on another, is common in land-use discussions.[1229,297] It relates to the juxtaposition of two areas with incompatible interactions. A buffer separating two areas is sometimes proposed to minimize negative interactions[999,683,1229,5,592], or to reduce steep gradients in an edge.[1514,1689] The buffer may repel, or may absorb, those flows. These are the same two functions accomplished by the many ways of sculpting and managing boundaries[1514,490] (chapter 2).

Too often little consideration is given to what the land uses will be in a proposed buffer. For instance, an intervening area itself is a source of flows, which affect both of the adjacent areas. An alternative approach is initially to consider what land uses are appropriate adjacent to a patch. Designing a configuration, in this case a neighborhood with compatible adjacencies, therefore becomes the goal. If desired, some portion of the configuration may, of course, be called a buffer.

Detecting patterns common to all landscapes

The bewildering diversity of regions, landscapes, and local ecosystems has shrouded our ability to detect widespread spatial arrangements of ecological importance. More provocatively, what universal spatial patterns exist? How do you compare an Indonesian mountain–rice landscape, a North American suburb, and an African savanna? It is an 'apples and oranges' problem, or more like a 'whales and mastodons' problem. Yet, understanding common spatial patterns at the configuration scale could open up a raft of interesting research questions, and provide opportunity at exactly the convenient scale for much land planning, design, and management.

One approach to reduce both complexity of pattern and diversity of objects is to convert everything to a common language. 'Graph theory' is an attractive possibility.[674,1915] In this case graphic models, composed only of circles and connecting lines, are drawn based on the arrangement of, and presumed interactions between, spatial elements. Here, circles represent spatial elements, and connecting lines represent common boundaries between elements.

In an attempt to detect common and uncommon spatial patterns, Margot Cantwell selected 25 aerial photographs to represent a breadth of climates, vegetation, continents, land uses, and human population densities.[251] They also covered four orders of magnitude in area. Landscape 'graphs' based on graph theory were simply drawn on clear plastic overlay sheets for direct comparison.

Each landscape portion as a whole differs, as expected. But little detective work was required to identify a dozen distinctive configurations representing smaller areas. Seven are common, of which three are in >90% of these highly-diverse landscapes (Fig. 9.3).

The three common graph-theoretic graphs represent the following configurations of patches, corridors, and matrix. One, called a 'spider' for convenience (Fig. 9.3a), represents a matrix area surrounding many patches. Examples are scattered bogs in coniferous forest and schoolyards in suburbia. A second, the 'necklace' (Fig. 9.3b), represents a corridor bisecting a heterogeneous area, such as a stream corridor or powerline passing through a sequence of fields separated by hedgerows. The

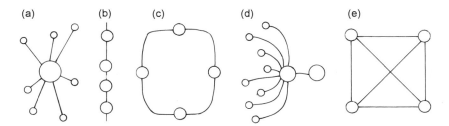

Fig. 9.3. Graph-theory graphs representing five arrangements of spatial elements in landscapes. A circle represents a spatial element; a connecting line represents a common boundary between two elements. See text. Adapted from Cantwell & Forman (1994).

third, a 'graph cell' (Fig. 9.3c), represents a unit in a network of intersecting corridors. Examples are a city block and an area surrounded by roads connecting a few villages. The 'rigid polygon' in Fig. 9.3 is rare (1 of 25 landscapes). The 'candelabra' is of intermediate frequency, and because of the potential roles suggested by its central circle, this graph-theory pattern is of special ecological interest.

While the approach is exploratory, it offers the benefits of focusing on patch–corridor–matrix pattern, including interactions between spatial elements, and direct comparison at any scale and any location. You can compare a vegetable garden with the Earth from a satellite. The results suggest the presence of common configurations worldwide. Generic planning and management strategies could easily be targeted to each configuration.[251]

A different approach to understanding interactions within a configuration is to consider how connected a certain type of spatial element is, no matter where it is located. Do hedgerows or house clearings have some inherent degree of connectivity (e.g., linkage density measured by the beta index), irrespective of context?

Seven element types (fields, clearings, woods, hedgerows, rivers, roads, and house clearings) were common in the 25 landscapes and their graphs.[251] The number of connecting lines was recorded for each example (each circle) of every type in every landscape. Two types, fields and house clearings, exhibited a relatively constant number of connecting lines. Fields almost always had four connections, reflecting their rectangular shape, with each side in contact with a hedgerow or patch. House clearings typically had only two connections, to a road and to the surrounding patch or matrix. Roads and rivers were the most connected elements, while woods had the highest variability in connectivity.

In short, the preceding connectivity analysis suggests good predictive ability for certain spatial elements in a configuration, but not for others. Overall the common-language approach offers promise for identifying common, as well as uncommon, spatial configurations independent of landscape or region.[458,17]

CONFIGURATION USAGE AND CHANGE

Animals in configurations

An alternative approach to delimiting a configuration is to consider what combination of ecosystems is used by a single animal or a population.[666,39,430,969] For example, Graham Arnold measured the amount of woods, fields, roads, farmsteads, and hedgerow ditches at increasing distances from birds' nests in English farmland.[50] The goal is to determine whether nest sites can be predicted based on the amounts of such

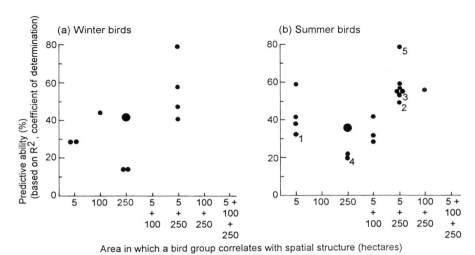

Fig. 9.4. Local versus broad-scale effects on bird density. Each point refers to an individual species or related group of species; large points = species richness. Birds are sampled in 5 ha (12.5 acre) quadrats, each containing hedge, fields, and other spatial attributes such as ditches and woods. Bird samples were then correlated with 43 spatial and habitat variables in the 5 ha quadrats, and in 'concentric' circles of 100 ha (= 1 km^2 = 250 acres) and 250 ha (= 2.5 km^2 = 1 mi^2 = 625 acres). All but two bird groups (not plotted) are significantly correlated with one or more variables. Birds are plotted according to whether they correlate only with local variables (5 ha), only with variables at 100 ha, with variables of both 5 ha and 100 ha areas, and so on. (1 = number of species with territories; 2 = number of passerine territories; 3 = number of woodland species territories; 4 = number of thrush nests; 5 = number of thrush territories). Adapted from Arnold (1983).

elements, and if so, within what distance. As expected, species show different patterns, but overall the best correlations are with the total area of woods and the hedgerow-ditch density (chapter 8).

In evaluating spatial scale for winter birds, almost all the variability in bird densities and species richness can be accounted for at two scales, 5 ha (12.5 acres) and 250 ha (1 mi^2) (Fig. 9.4a). The 100 ha scale adds little to our predictive ability. Essentially the same results are found for summer birds (Fig. 9.4b). In other words, the avifauna appears to be especially sensitive to both the immediate, mainly adjacent, 5 ha area, and the larger neighborhood area of 250 ha. This suggests a possible relevant size of a configuration in this agricultural landscape, plus the key elements required within it. Little spatial arrangement within the 250 ha area is implied. Of course, birds are only one of several types of movements and flows in a mosaic.

Similar information may be available for determining the apparent requirements for nesting of various hawks, eagles, and waterfowl. Nest site location for the northern spotted owl (*Strix occidentalis caurina*) in western Oregon (USA) correlates with the amount of old-growth and mature forest in each of seven concentric bands, from a radius of 0.0–0.9 km to 2.4–3.4 km.[505,255,1426] For this species the connectivity of the forest habitat is important. The animals avoid open areas where they are subject to predation by the larger great-horned owl (*Bubo virginianus*). Spotted owls will cross small cleared strips. In this case, the forested landscape is structurally disconnected into sections or fragments[493] (chapters 5 and 8), while at the same time it is behaviorally or functionally connected[1111,1116] for the spotted owl.

Home ranges for species of large animals may also be useful in delimiting configurations. A monkey troop, family of eagles or deer, or wolf pack requires a number of ecosystems to provide for food, water, escape cover, and raising young. The species also may exert a major effect on those ecosystems. The configuration of elements could reflect the seasonal requirements of species, such as deer and bear that move between habitats.[666,917] Or the cluster of ecosystems could provide the necessary resources for different stages in the life cycle of a species, such as the insects and agricultural pests that sequentially require a water body, forest edge, and field.[1350,1427,20,21]

Most of the preceding examples tell what habitats are required in a cluster. Yet rather little is known quantitatively about the movements among clustered habitats, a key to understanding spatial configurations. John Wegner and Gray Merriam (1979) studied small-mammal and bird movements among adjacent woods, hedgerow (fencerow), and field in southern Ontario farmland. The greatest movement is between woods and hedgerow, with least movement between woods and field (Fig. 9.5). This pattern becomes more accentuated from early to late summer, and is most pronounced for the small mammals studied. Such data quantitatively comparing interactions among a number of spatial elements are scarce.

Nevertheless, general patterns of spatial interaction are consistent with the configuration concept. In the eastern USA Thoreau (1993) found that squirrels (*Tamiasciurus*, *Sciurus*) and jays (*Cyanocitta*) carry oak seeds (*Quercus*) to pine woods (*Pinus*), whereas wind carries pine seeds to oak woods. Thus after logging, the oak woods and pine woods 'switch locations'. A recent study found that acorns are transported by jays an average 1.4 km from the tree.[812] The route is typically from woods, along a hedgerow, to lawns and clearings, where acorns are carefully planted near wooded edges.

Certain keystone species are so dominant that they help create the configuration of ecosystems itself. Humans, termites, and elephants

296

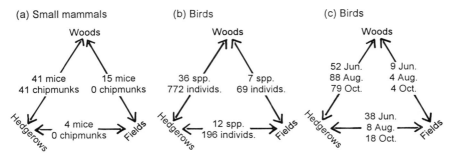

Fig. 9.5. Vertebrate movements among woods, hedgerows, and fields. (a) White-footed mouse and eastern chipmunk (*Peromyscus leucopus, Tamias striatus*). (b) Total number of bird species and individuals. (c) Percentage of total number of individuals in (b) during each of three selected months; nesting peaks in July and migration in September. Monthly measurements May through October in southeastern Ontario; deciduous woods and hedgerows. Adapted from Wegner & Merriam (1979).

spring to mind. But beaver (*Castor canadensis*) is more prominent in some cool moist areas. Beaver build dams, create ponds, raise water tables that produce wet meadows, turn streamside forest into shrubby coppice areas, and open up woods at a distance by toppling trees. The spatial arrangement of the habitats is quite predictable, and the elongated area molded by a beaver family often approaches or exceeds a kilometer in length.[1183]

The beaver is an organizing force, and its activity illustrates an *influence field*, the area significantly affected by flows from a node or patch.[1670,646,501] Usually the amount of influence decreases curvilinearly with the square of the distance (or as a negative exponential function).

The classic example of organizing forces and influence fields is the economic distribution of goods around villages and towns in rural areas.[276,1758,1568,265,216] This is modeled as a hierarchical hexagonal structure stretching across the land.[278,205,1554] Other administrative units such as counties or forest management units also have a distinct boundary, and usually a central node from which people exert control. An individual farm or property may be similarly organized, whereas housing tracts often lack the organizing node.

A second model of influence fields portrays concentric 'Thunen bands' around a population center. In this case different land uses, such as cultivation, pasture, and forest, sort out at increasing distances.[651,1554,535] A third somewhat-different example is the drainage basin or catchment that is controlled by water movement. In each of the three examples the object or unit is typically repeated across a landscape, and has an internal configuration that varies within a narrow range. Pinpointing the distribution of ecological organizing forces and mapping their influence fields should help identify the sizes, shapes, and locations of configurations in a landscape.

Stability

The configuration concept is founded on interaction among neighboring ecosystems, but may also imply interdependence among them. In addition to functioning as a unit with distinct or indistinct boundaries, the internal spatial arrangement of elements is critical to the functioning of a configuration. These statements describe all living systems. All systems containing life, including a configuration, have a structure, function with flows, and change over time.

In an interdependent system what happens if a key element disappears or is destroyed?[1274] Suppose an important central pond or woodlot or stream corridor disappears in an agricultural configuration. Countless animals and plants disappear, and soil may begin seriously to erode. Other spatial elements previously affected by flows from the extinct element will change, with a cascade effect to further elements. Thus, the configuration might continue, but in an altered form.

Alternatively, spatial elements on the periphery may increase interactions with surrounding configurations, leading to break up and disappearance of the original configuration. Such a pattern follows when a wolf pack (*Canis lupus*) moves or breaks up, and the space is subdivided by adjacent packs.[1098] A natural floodplain that becomes developed for farming and buildings alters virtually all flows between river and surrounding fields and woods, as highlighted in the graph-theoretic 'candelabra' graph (Fig. 9.3d).

Stability is used here in the general sense of persistence with low variation (also see chapters 12 and 14). In considering the stability of a configuration[1750], one may ask whether one of its spatial elements is more variable over time than the others. In four diverse areas of North America, sequences of ecosystems on slopes have been compared for a variety of mainly ecosystem parameters.[895] In each sequence the most-variable ecosystem is identified. In all four sequences, water flow across the landscape appears to be the primary driving factor influencing variability in the parameters measured.

A patch receives flows and 'information' from most or all of the spatial elements in a configuration. This includes adjacencies[489,492], nearest neighbor patches or corridors (chapter 9 appendix), and the encircling matrix with its control-over-dynamics feature (chapter 8). It appears likely that the stability of a patch within a configuration is often primarily dependent on conditions in the matrix[466] (S. A. Levin, pers. commun.).

298

Fig. 9.6. A possible configuration of spatial elements. Movements of water, animals, farmers, and tractors may tightly link these elements (or a larger set, or a subset). Regularity in spatial pattern, primarily of human origin, is conspicuous within and among these woods, orchards, vineyards, pastures, hedgerows, and stream corridors. The narrow hedgerows in the foreground contain at least eight woody species providing economic products. Southern Toscana (Tuscany), Italy. R. Forman photo.

Until the structure and function of configurations are better understood, it is premature to say much about their overall stability.[78,1750] It may simply be low, with configurations frequently forming, breaking up, and reforming. If high, some cases may persist because of resistance to outside influence, like a wolf pack that marks boundaries with urine and uses many teeth, or a crusty old farmer that marks boundaries with signs and patrols with an antique gun.

If influence fields are wide, and flows cover much of the configuration rather than just moving between adjacent elements, stability may reside in resilience, the ability to rapidly recover following disturbance of an element.[501,1750] 'Mosaic stability' (chapter 12) may play a role in persistence. However, if a low level of 'redundancy' is present (limited numbers of each element type), and if many individual spatial elements are disturbed or changed, a configuration probably cannot remain in its present form.

In short, it should be evident that configurations (Fig. 9.6), as a hierarchical level between a local ecosystem and a landscape, are poorly known and represent an intriguing research frontier. In fact, their existence itself can be challenged. Nevertheless, if their structure, function, and change can be understood, this will be of considerable importance for design, conservation, and management. The individual local ecosystem is too small for meaningful sustainability, and the whole landscape is too large for most people and organizations to deal with effectively. Both generic and specific designs and management regimes, targeted for clusters of ecosystems, are at the scale that can quickly make a difference on land.

MOSAICS REFLECTING GEOLOGY AND CULTURE

While including configurations, we now focus on *land mosaics*, i.e., whole regions composed of landscapes, and whole landscapes composed of local ecosystems. The structure of these mosaics was conceptually outlined in chapter 1, but here the focus is on geomorphology, human culture, and their interaction in producing the mosaics.[501,271] Ecological understanding of them as 'ecomosaics'[483], of course, remains the overall objective.

Geomorphic processes are the mechanical transport of organic and inorganic material.[408,1836,1663] Landslides (slips), surface erosion, soil creep, and material carried in solution by surface and subsurface water are familiar examples.[1524,1712] Some processes produce *landforms*, the geomorphic features of the Earth's surface, such as deltas, eskers, sand dunes, and eroded gullies. The size and shape of landforms, however, are equally determined by the influence of climate on rock weathering.[501] Hence an exposed hill-sized protruding rock often resembles a giant loaf of bread or haystack in the wet tropics, but in a Mediterranean-type climate it may resemble a butte with jagged edges.

Landforms molded by geomorphic process and climate thus create a heterogeneous template for a region, upon which soil, vegetation, and animal communities develop, and sort out into local ecosystems. The spread of disturbances such as fire and flood reflect landform distribution.[506,1742,1663] Landforms affect temperature, wind, and moisture conditions in an ecosystem, as well as the flows of seeds, heat energy, water, mineral nutrients, and large fierce animals across the region.

In some areas scattered over the globe, landscape mosaic pattern simply mirrors landform pattern. But most of the world is patterned by the activity of humans on landforms (Fig. 9.6). Agriculture[1382,1383,237,1612], deforestation[1706,1899], and altered fire regimes[1465,1466,332], for example, significantly modify the land surface. Thus, geomorphology and culture together produce the views from an airplane window.[1628,1382] In unusual cases, such as certain suburban landscapes, culture actually overwhelms or obfuscates geomorphology. Here, residential areas of similar appearance simply cover over previously distinct landforms. Whole streams 'disappear' into storm-sewer pipes, and patches of prime agricultural soil are 'lost' beneath lawns, houses, and roads.

Four examples of very-different mosaic patterns resulting from four cultures were presented in chapter 1. In parts of rural China the location and protection of woods, fields, and buildings around drainage basins have been based on 'Feng-shui theory.' In parts of Eastern Europe a regular pattern of fine-scale garden and house plots, surrounded by large fields, has been nested within a regular coarse-grain pattern. Outside cities in North America houses and clusters of houses have appeared in seemingly unplanned locations, including streambanks and the best agricultural soils, resulting in a stunning fragmentation of both natural and built areas. And in Australia, Aboriginal people have used fire and an intimate knowledge of the land to create a fine-scale mosaic of resources in lieu of cultivation. The fine-grained areas sit within a coarse-grained network of sacred sites, almost invisible to non-Aboriginal people. Examples from any region illustrate the different spatial effect of culture on regional and landscape structure.

Suppose powerful Indians or Chinese had colonized North America centuries before the Europeans. Suppose Indonesians or Italians had colonized Australia (Fig. 9.6). Or suppose Maya and Inca were today the major countries of Latin America. Think how different the land would look.

Nature is usually aggregated. When objects are dense relative to resources, such as certain bird territories, desert shrubs, and plant cells, they compete and tend toward regularity in distribution. But the norm is to find plants, animals, landforms, patch types, and so forth in clusters, separated by low-density spaces.

Add humans, and the land becomes more regular (Fig. 9.6). Farms, villages, logging patterns, roads, fields, houses, and much more are all more regularly distributed than the previous natural aggregations. Extreme regularity stands out in the grids of Roman roads, the American Midwest, cities, rice paddies of Malaysia and Japan, and hexagonal models of rural land.

Understanding the human process of altering natural patterns is important, because it provides points of entry for altering behavior and implementing policy. Perhaps three general steps summarize the process. People acting as a physical agent make a change.[1842,1369,1362] Then the human perception, especially aesthetic and economic, of the new form takes place, based on receiving and processing the information in light of previous experience.[1382,1960] Finally, further actions and changes based on the perception ensue.[1634,1189] In this fashion both humans and the landscape 'evolve' over time.

301

This process is illustrated in mid-nineteenth-century New England, when many farming families left for richer soils in the American Midwest. Old-field succession followed field abandonment across the region. Folks from cities saw the farmland reverting to weeds and trees, and interpreted this as due to lazy, sloppy farmers up country.[1633] Partly as a result of these perceptions, states advertised and tried to fill the land with industrious farmers. Eventually, city folk bought and fixed old farmhouses, converting portions of the area to vacationland.

In an analogous manner, ecologists study hedgerows and woods of different widths.[495,89] But why do the hedgerows differ in width? In some cases width results from broad-scale economic considerations, where production in 'every square meter' counts[236] (Fig. 9.6). In other cases hedgerow width results from the small decisions of an individual farmer, such as his or her neatness in maintaining straight tidy lines, or progressiveness in accepting change.[235,1190] In northeastern China a row of poplars (*Populus*) commonly separates fields, both because it enhances regional economic production, and because the farmer protects poplars as nest sites for magpies (e.g., *Pica*), which may bring happiness.

Geomorphic process and culture are an uneasy marriage. Big human actions periodically appear that strain or break it. And big natural disturbances, inevitable in occurrence but unpredictable in timing, may be absorbed, or may lead to a new face on the land.

THE THIRD DIMENSION

Air is a volume of molecules in gaseous form, and wind is that volume being transported by energy differences. Hills and mountains, valleys and gorges are also volumes, but fixed in place. Blowing seeds, flying animals, and ambulating animals and people go up and down, and around about, these fixed three-dimensional objects. Regional and landscape ecology has traditionally focused two-dimensionally, but the vertical dimension is also critical to structure and dynamics in many landscapes.

The vertical dimension focuses on land topography and vegetation height, though of course buildings are important in some areas. Topography and vegetation height may be positively, negatively, or not correlated. Consequently, the smoothness or roughness of the top surface of the landscape varies, depending on the nature of this relationship.

Consider a wooded stream corridor passing between two steep hillslopes. The vertical dimension of the slopes suggests a site where eroded particles and mineral nutrients enter the stream in large amounts. The dense hillslope vegetation probably provides a major input

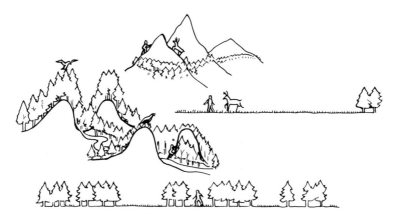

Fig. 9.7. Vegetation height relative to topographic relief. Animals and people perceive and move differently according to the horizontal pattern of relative vegetation and topographic heights. See text.

of seeds and leaves to the stream. Fish and other stream organisms respond markedly to these concentrated inputs.

Birds and bees flying between hilltops respond more to the top of the vegetation. If the streamside has small trees and the hilltop tall forest, the animals either fly horizontally over a chasm where they are subject to predation, or drop down into the chasm and have to fly up and out, or do not cross at all (upper two hills on left in Fig. 9.7). But if streamside trees are tall and hilltops have low vegetation, the animals may fly horizontally just over a flat canopy surface (lower two hills in center of Fig. 9.7). Two-dimensional views minimize such patterns and processes, whereas three-dimensional views, especially in rugged topography, provide key insights.[1709,13]

Common solid-geometric forms may provide useful models for ecological understanding in three-dimensional terrain. For example, a large snow drift in an alpine landscape is an asymmetric lens that shrinks in thickness through the summer (W. D. Billings, pers. commun.). As snow melts, buried vegetation is progressively exposed around the periphery. Consequently, the shrinking snow lens is surrounded by bands of plants that grow, flower, attract pollinators, fruit, and disperse seeds progressively later in the season. Fortunately, plants emerging later from under the lens have faster life cycles, because 'winter is coming'.

Other common forms mimicking the land are half spheres, half cylinders, troughs, tubes, serpents, cones, and so forth. Combined forms mimic complex terrain such as ridges and valleys. It should be possible to delineate a series of predicted ecological patterns and processes

303

Fig. 9.8. Mountains illustrating ruggedness, regularity, scale-independence, verticality, and alternating concave–convex surfaces. Would any pattern differ if the photograph had been printed upside down? Iran. Courtesy and with permission of E. Laporte.

associated with each form. This would enhance understanding of the third dimension of landscapes and regions.

The product of erosion, glaciation, and other geomorphic processes is a rough surface. This may be measured as *ruggedness*, the sum of the slopes, valleys, peaks, exposures, and elevations present[501] (Fig. 9.8). At least some vertebrates, including game species, appear sensitive to ruggedness.[100] In western North America mountain goats and peregrine falcons (*Oreamnos americanus, Falco peregrinus*) use steep rugged terrain, while pronghorn antelope and greater prairie-chicken (*Antilocarpa americana, Tympanuchus cupido*) are nearly restricted to flat or gently rolling terrain. This contrast in habitat may result from the probability of finding food, cover from predators and hunters, or a suitable microclimate during severe conditions. Or it may reflect the availability of a series of spatially separated ecosystems, which are required diurnally, seasonally, or through the life cycle.

Of the various measures of ruggedness available[1642,1523,1107,100], one of the simplest appears to incorporate best the myriad attributes in the concept. Thus, the amount of ruggedness is proportional to the total length of contour lines per unit area on a topographic map.[100]

A general look at topography or topographic maps of a region suggests a scale-independent pattern, where large valleys are dissected by small valleys, small by smaller valleys, and so on (Fig. 9.8). Indeed, this has been modeled with fractal geometry.[1039] While the two-dimensional appearance or form of the topographic dissection is relatively scale-independent over several orders of magnitude, a limit or threshold is present based on downward flows.[1157] Dissection proceeds until hillslope lengths are just shorter than that needed for surface-water flow to cause erosion and channelization.[754,641,1157] In other words, tributaries will continue to appear until the smooth slopes around first-order streams can no longer be eroded to form channels.

The alternating concave and convex slopes above first-order (and intermittent) streams in mountains are striking for their abundance and regularity[642] (Fig. 9.8). Because of concave and convex surface differences, solar radiation input, soil moisture and development, and directions of water flow differ. It is probable that vegetation and faunal differences exist as well. For instance, in parts of the central Appalachian Mountains, pine forest (*Pinus*) tends to be on convex surfaces (noses), oak forest (*Quercus*) on side slopes, and northern hardwoods (*Fagus, Acer, Betula*) on concave sufaces (hollows).[642] In central Massachusetts (USA) concavities often have a thin loam soil with ash (*Fraxinus americana*) being common, whereas convexities have a thick loam with oak (*Quercus rubra* or *borealis*) being abundant.[1640] Concave and convex surfaces may be useful building blocks in studying the ecology of rough-terrain landscapes.[642]

In rugged terrain with steep slopes, the hiker or animal in a narrow valley feels closed in (foothills in Fig. 9.7). Yet hiking in the high alpine zone the landscape appears open, and the series of surfaces at increasing horizontal distance provides a large 'depth-of-view' effect (Fig. 9.7 top). Vegetation height makes little difference to these perceptions at either location.[501] In contrast, to a hiker a flat plain appears open if treeless (Fig. 9.7 middle), or closed if forested. Little depth effect is evident in either case. In all four cases (mountain valley, mountain top, treeless plain, and forested plain) variability from spot to spot is low. The view is typically either all closed or all open.

Variability in 'openness', however, can be increased in the plain (Fig. 9.7 bottom). Equal-sized clearings alternating with forest, as in a dispersed-patch logging regime, produce a regular open–closed pattern along a route. However, high variability for an animal or hiker is produced by different-sized clearings and woods on the plain. Spatial variability, rather than humankind's tendency for regularity, is an attribute to consider in designing a sustainable land (chapter 14).

305

With increasing topographic relief or steepness, vertical flows and movements appear to increase rapidly relative to horizontal flows[408,1663] (Fig. 9.8). Thus, cold air drainage down into valleys on still nights, and wind that carries seeds and pollen up valleys to high elevations are common meteorological events with strong vertical components. Disturbances, including blowdown, fire, and landslide, usually have a strong vertical dimension in mountains. Water, particulate matter, and mineral nutrients move downward in many processes. Many animals migrate vertically between low winter habitat and high summer habitat, and indeed relatively straight stream valleys tend to be a busy up-and-down route.

Yet one factor does tend to produce horizontal movement in mountains, the zonation of vegetation. Horizontal belts of particular vegetation types, or of distinctive mosaics of land uses, are common in mountains. A belt typically varies in width due to differences in exposure and solar radiation. Nevertheless, although apparently little studied, a belt probably acts as a track for foraging animals. Habitat conditions are more constant, and energetics generally more favorable, for horizontal than upward movement.

In rugged mountains the conical form of peaks means that higher zones have less area than lower zones. This, along with climatic differences, may contribute to the reduced richness of birds, mammals, and fish at higher elevations.[766] Some vegetation zones are rings around the mountain (chapter 4), whose shape and consequent ecology are little known.

Mountain slopes are also covered with repeated, vertical wedge-shaped valleys that narrow upward (Fig. 9.8). Ridges therefore form a dendritic network converging upward, whereas valley bottoms converge downward. 'Ridge networks' are doubtless used for vertical movement by animals favoring open dry routes, compared with stream networks used by species favoring protected moist routes. A third more-convoluted route is to follow a particular slope exposure, such as mammals migrating upward in spring on snow-free slopes facing the equator. The movement of species in mountains offers much research opportunity, and is important for evaluating human impacts such as recreation, ski-area development, and logging (M. A. Folger, pers. commun.).

TYPES OF LANDSCAPE AND REGION

Terrain types

What are the basic types of landscape in a region, or indeed regions within a continent?[888,655,1130,72,73,1809] A particularly useful first step is to consider the landforms present in an area and their cause.[610,1836,1165] The bulk of the land has been shaped by three broad processes, namely, moving water, wind, and ice. Each process produces 'landforms', the natural features of the Earth's surface resulting from erosion or deposition. Flat or gently-rolling plains, and lake deposits in low areas, result from all three processes. However, each process produces a distinctive set of (upward-protruding) erosional and depositional landforms. These are somewhat predictable in size and shape. Therefore, an ecologically informative first question about any region or landscape would be, 'Is it water-created terrain, wind-created terrain, or glacial terrain?'

Water-created terrain

Streams and rivers form this (fluvial) land surface. The surface is eroded upstream, and deposits accumulate downriver. Nearly half of the USA, large areas of Australia, and extensive portions of other continents exhibit water-created terrain.

In eroding areas a dendritic stream system is the universal signature. V-shaped valleys are equally characteristic. The depositing portions of a meandering stream or river are also distinctive (Fig. 9.9d). A strip of floodplain, often with terraces and traces of former meanders, adjoins the river. Floodplain soils (with alluvium often predominant) are extremely heterogeneous from spot to spot, ranging from sands to silts, clays, and loams. The river may lead to a fan- or hand-like 'delta' in a lake or sea. In dry areas an alluvial fan is often produced in a valley. A 'coastal plain', submerged by former high sea levels, today experiences stream erosion and deposition.

Wind-created terrain

Wind erodes and deposits material to form this (aeolian) land surface, particularly characteristic of dry regions with sparse vegetation (Fig. 9.2). Large surface areas of Central Europe, Central Asia, Central Australia, Argentina, and the northern third of Africa are formed by wind. In the USA large areas (mainly formed during glacial climates) are found in Washington and Idaho, plus from Texas, Kansas, and Nebraska northeastward to Illinois and Michigan.

Rocks and hills are eroded, rounded, and smoothed by wind. Two types of deposit are present (Fig. 9.9b). 'Sand dunes' are usually near the sources of sand, and may be new and actively moving, or ancient and stabilized by vegetation. Of the many dune shapes two are especially common. Parallel linear dunes form perpendicular to the wind direction. And crescent-shaped 'Barchan dunes' have their points and steep concave side pointing downwind (see Fig. 4.13c).

The other type of wind deposit is 'loess', composed of silt usually blown far from its source. Deposits often extend dozens (even hundreds) of meters deep, and become thinner at greater distances from a source.

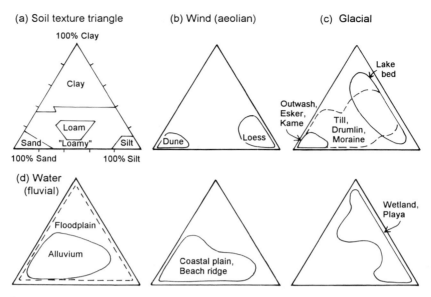

Fig. 9.9. Distinctive particle sizes in three types of region characterized by wind, glacial, or water deposits. Deposits are directly compared with soils in the soil texture triangle (a): clay soil = predominately fine (clay) particles; silty soil = predominately intermediate-sized (silt) particles; sandy soil = predominately coarse (sand) particles; and loam soil = a somewhat-even mixture of sand, silt, and clay. Loams and loamy soils with silt predominating are typically best for agricultural and forest production. In (d) the average for beach ridge is sandier than for coastal plain, and playa sandier than for wetland (swamp, bog, marsh). Adapted from US Department of Agriculture and Way (1978).

Water erodes the loess deposits, forming distinctive fern-like (pinnate) stream systems, with very steep slopes. Therefore, recent loess deposits tend to have smooth parallel hills, whereas old deposits are highly dissected with rugged steep hills.

Glacial terrain

Glaciation gouged out the land and deposited material often far from its source. Virtually all of Canada and the northeastern third of the USA were glaciated a number of times. Glaciation was extensive in the Andes, northern Europe, northern Asia, and the high mountains of Eurasia.

U-shaped valleys, often with small (cirque) lakes at their head, characterize glaciated mountains. In flatter areas elongated, north–south lakes, wetlands, or fiords are often present, as well as nearly round bogs. Glacial deposits include sandy snake-like ridges called 'eskers' (Fig. 9.9c). Low oval hills called 'drumlins' are clustered near some former glacial margins. 'Kames' are sandy knobs interspersed with depressions called 'kettle holes'. And 'terminal moraines' are ridges of mixed-size particles left at the ends of glacial advances.

Water-created, wind-created, and glacial terrain characterizes almost all the land, and forms distinctive regions. A few other finer-scale processes form landscapes or portions thereof. 'Valley-alluvium terrain' in a dry valley results from water running down mountains, depositing coarse sediment, and 'disappearing' in the valley bottom as subsurface flow. 'Karst-limestone terrain' in a moist climate is pock-marked with hills and depressions, and has water flowing in subsurface channels. And 'tidal-flat terrain' in coastal locations usually has saltmarsh or mangrove swamp vegetation covering river deposits subject to marine tides.

Patch–corridor–matrix types

While countless geographic and hierarchical classifications of land have been developed for different purposes, none seems to focus on the most important characteristics for understanding the ecology of regions and landscapes. The spatial arrangement of patches, corridors, and matrix should be at or near the core.

A preliminary and rough grouping of more than 200 aerial photographs from many parts of the world suggests that they can be usefully placed into six piles, based on the predominant spatial pattern present[492] (Fig. 9.10). (1) *Large-patch* landscapes have one or more large patches surrounded by a matrix (e.g., an extensive vineyard surrounded by natural vegetation, or a large forest surrounded by grazing land). (2) *Small-patch* landscapes similarly have small patches surrounded by a matrix (e.g., small bogs in coniferous forest, or woodlots surrounded by cornfield). (3) *Dendritic* landscapes have the major imprint of a dendritic river system (e.g., wooded stream corridors in savanna, or hilly upland forest surrounding forest on floodplains). (4) *Rectilinear* landscapes have a similar imprint by a net of straight corridors (e.g., a road network, or

(a) Large patch   (c) Dendritic   (e) Checkerboard
   (b) Small patch   (d) Rectilinear   (f) Interdigitated

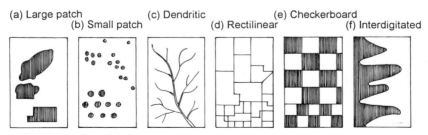

Fig. 9.10. Six types of landscape based on predominant spatial pattern. Adapted from Forman (1990b).

a hedgerow network or 'bocage'). (5) *Checkerboard* landscapes, with no prominent network, have two or three patch types of similar size alternating across the area (e.g., fields of different crops, or forest about half cut by dispersed-patch logging). (6) *Interdigitated* landscapes have alternating major lobes of two types (e.g., uplands and lowlands in mountains, or built and agricultural areas in the ex-urban fringe).

Few specific landscapes would fit only one of these. Rather they should be thought of as the points of a six-pointed solid, and any particular landscape would fall somewhere within the volume. Additional basic types may exist, and each type may be subdivided. Indeed an approach that includes change in spatial pattern (chapter 12) might be more ecologically informative.

A traditional alternative approach is simply to name the landscapes by their predominant vegetation or land use, such as suburban, rice-paddy, grassland, spinifex–mulga, salt marsh, boreal forest, and industrial landscapes. For these to be useful one must be familiar with a spinifex–mulga area (*Triodia–Acacia*) in Australia, and a boreal forest (*Picea–Abies*), say in Canada or Siberia.[1916,1290] In addition, ecological properties vary greatly among the examples within a category, e.g., grassland landscapes differ and industrial landscapes differ.

The basic types of region are less clear ecologically. They may be named geographically, such as the Upper Midwest, Northern Queensland, Western Wales, Midi, Lake Baikal, Llanos, or Maritimes Region. Or perhaps macroclimatic, soil, or ecoregion names could be used.[888,72,73] However, an *urban region*, with a major city and various suburban, intervening and surrounding landscapes, is one rather distinct type. A *mountainous region* with landscapes at different elevations is another. How do other regions sort out in an ecologically informative way?

HABITAT ARRANGEMENTS AND STRATEGIC POINTS

The patch–corridor–matrix approach pinpoints general patterns and principles that cut across the incredible diversity of species, habitats, landscapes, and regions. Nevertheless, understanding is further enhanced by focusing directly on the habitat types present. Thus, we now change gears and focus on biodiversity, and then on wildlife habitats. Efforts to identify the best collection of natural areas to protect biodiversity[440,441,1912] provide considerable insight into the ecology of a landscape. And success is important to decrease the elevated rate of human-caused species extinctions.

Selecting biodiversity areas

How well do the parks in a country protect biological diversity? A map of the national parks of Ecuador was overlaid on a map of the biodivers-

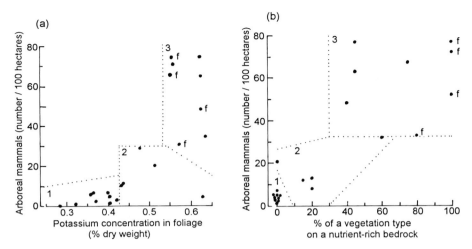

Fig. 9.11. Leaf nutrients and geological substrate as predictors of mammal density. Each point refers to one of the 22 tree communities present. 1 = leading dominant species in the community is *Eucalyptus sieberi*, *E. consideniana*, *E. agglomerata*, or *E. globoidea*; 2 = *E. obliqua*, *E. cypellocarpa*, or *E. muellerana*; 3 = *E. fastigata*, *E. bridgesiana*, *E. dalrympleana*, *E. viminalis*, or *E. radiata*; f = *E. fastigata*. Seven tree-inhabiting mammal species were observed during 337 logging operations in southeastern New South Wales, Australia; most common were greater glider (353 individuals), feathertail glider (328), and sugar glider (120) (*Petauroides volans*, *Acrobates pygmaeus*, *Petaurus breviceps*). (b) The degree to which a particular tree community is found on the most nutrient-rich substrate is plotted. This geological substrate (Devonian granodiorite or tonalite) is one of five present. Adapted from Braithwaite (1991).

ity 'hot spots', i.e., the richest locations for species, and almost no overlap existed.[1695] Similarly in Nepal most parkland is at 500 m and at 4000–5500 m elevation, a pattern totally unlike that for either bird or mammal species richness.[766] Parks generally are located primarily for aesthetic and recreational purposes.

In a forested region of southeastern Australia with scattered pastures and with eucalypts as tall as redwoods (*Sequoia*), Wayne Braithwaite and colleagues have discovered a linkage among biodiversity, soil, and economics. The greatest richness of arboreal vertebrates is in the forest type (*Eucalyptus fastigata* predominant) that has been almost entirely removed for grazing[171,170,172] (Fig. 9.11). The reason is that most of the animals feed and depend on tree foliage that is rich in mineral nutrients such as potassium. The soil in flat lowlands on certain geologic substrates is richest in nutrients. Therefore, trees in these locations have the most nutrient-rich foliage.

But grass also grows best on these soils, and hence sheep that eat the grass have the highest production. In effect, by maintaining these spots of high production that supplement extensive areas of regular sheep production, society unknowingly has caused the richest habitats for native vertebrates to become rare.

In the Smoky Mountains of Tennessee and North Carolina a national park has protected exceptionally species-rich coves or valleys, while around the park these valleys were the first areas cleared for agriculture. This pattern of species-rich habitat being targeted for other uses is probably widespread. It also means that species and habitats of little economic use are greatly overrepresented in a region.[1493]

Park boundaries also do not correlate well with home ranges of large vertebrates. The Florida panther (*Felis concolor*) ranges well beyond the Everglades Park in Florida. Wolves, bears, mountain lions, and eagles have home ranges partly in and partly out of parks in Canada and the USA.[906,1206] The problem is accentuated by the tightening noose of incompatible land uses encroaching on many parks.

Several approaches have emerged to identify the optimum or minimum locations required to protect all or some proportion of the species in a landscape or region. One approach ('gap analysis') superimposes a map of species-rich spots onto a map of existing protected areas. The difference between the maps indicates the areas or gaps needing protection based on species-rich sites.[1229,1527,1090]

A second approach that identifies where new forests or restored habitats should be located in Holland was described in chapter 8. In essence, a group of indicator species is studied in detail, simulation models developed, the land mapped for different levels of accessibility for species colonization plus persistence after colonization, and the sizes, numbers, and locations of proposed new woods identified.[678,877]

Locating nature reserves to protect biodiversity requires knowing the distribution of species, a difficult challenge in most parts of the world. Nevertheless, guidelines are available for cost-effective surveys[66,1049] and various survey methods.[218,1176 cited by 1047] The studies of Chris Margules, Mike Austin, and others indicate that two general approaches exist, both using surrogates for actual species distribution lists.[1052,1046] One uses a more general unit of diversity such as an ecological community, species assemblage type, or some environmental class.[280,1304,1213,1360,1091,1346] The other relates the probability of a species occurrence to some independent predictor, such as one or more environmental variables.[65]

Estimating the species distributions is easy compared with deciding on how many reserves and where to locate them. Quantitative, somewhat-objective approaches usually involve creating an index, say of richness and rarity, and ranking sites.[1360,1046] Indices must be used cautiously because key individual variables are obscured. Evidence from different regions and vegetation in Australia show that some 20 to 90% of the sites, or of the total area, must be protected to encompass all the species

312

present.[1051,1047,1346] This range, almost too wide to be useful, may reflect actual species distributions, but also may result from the different data bases, indices, and statistical analyses used.

A somewhat different approach would center nature reserves on natural boundaries, and be large enough to include a major amount of interior area in each adjacent habitat.[841,758,469,1213] This also enhances multihabitat species (chapter 3).[666,39,430] The administrative boundary[1513] length would be about 20–30% shorter than for two separate reserves of the same total area. The total amount of interior habitat protected by the single reserve could be greater or less, depending on how the boundaries of the two reserves correlate with natural or existing boundaries. Of course, a single large reserve in one habitat best protects that habitat type.

Three conclusions emerge from these analyses. First, except in unusual cases, it is unlikely that all species in a region, or even a landscape, can be protected in nature reserves. Second, nature reserve location will always result from political human decision-making. For instance, in agricultural areas of Britain tiny woods are typically neglected or lost, large woods are reserved for wood production, and medium-sized woods are a focus of nature conservation. Wood production currently appears to have more political support than nature conservation on land (G. F. Peterken, pers. commun.). And third, providing ranked categories of locations (rather than locations ranked individually), which are determined with a relatively-objective method and with easily understood rationales, is most likely to be used by decision makers.[1385,1763]

Wildlife habitat patterns

A different approach to sustaining biological diversity combines habitat distribution analysis with patches, corridors, and matrix. Wildlife biologists have long shown that increasing edge habitat in the landscape increases game and many other wildlife populations.[943,359,1702] But, because species richness and certain uncommon species are enhanced in large patches (chapter 2), maximizing edge is detrimental. Nevertheless, these two objectives of society, protecting species richness in large reserves and increasing game populations in edges, are quite compatible. Large patches of natural vegetation are surrounded by ample edge habitat distributed throughout the matrix.[1224,501] The edge may be produced by many small patches, by corridors, or by convoluting the boundaries of large patches (also see Box 2.1).[501,503] With this approach it is unnecessary to know the population, distribution, name, or even existence of all species. Most or all species will be included in the several large patches (chapter 2).

However, the locations of large patches are important. Thus a collection of nature reserves[1225] effectively incorporates (a) patch context, i.e., surrounding habitats and connectivity, in addition to (b) combinations of habitat types, and (c) disturbance and regeneration regimes. Combining large patches in carefully chosen locations with connecting corridors and stepping stones across the landscape produces a 'nature reserve system' rather than a collection of reserves.[501] In a system the interactions among reserves (objects) are as important as the reserves themselves.[1235]

Yet, natural-resource reserves are not the goal. Rather, they are important pieces in a mosaic where every piece counts. Planning of the whole landscape mosaic is required to enhance wildlife and protect biodiversity.[665,666,906,1213,271,501,1786,1040]

Since many vertebrate and insect species use and require two or three habitats diurnally, seasonally, or in their life cycle, the proximity of the habitats is critical.[665,666,39,777] To enhance habitat conditions for such species, wildlife biologists have used the concept of *interspersion*, the presence of habitats or land uses at intervals in or among other habitats, or simply the spatial mixing of habitats.[943,356,1555,566,1702,764] Boundary length has been used as a measure of interspersion, though this does not capture the idea of intervals or scattering.

In addition, when three or more habitat types are mixed, 'adjacencies' can also be varied. That is, pairs of contiguous habitats are increased or decreased for different management purposes. Interspersion accomplishes two objectives. First, the boundary length between habitats is high, and hence edge species are enhanced. Boundaries between different land uses contain varying concentrations of exotic (or ruderal) species, a possible useful assay for comparing landscapes (H. Rambouskova, pers. commun.). Second, the proximity of habitats throughout an area means that multihabitat species (chapter 3) are enhanced throughout the area. Both edge and multihabitat species are generalists.

These patterns are illustrated in Fig. 9.12. Design *b* has more habitat interspersion than *a*. Designs *b* and *c* have the same boundary length, but *b* has greater interspersion. Designs *e* and *f* have the same interspersion, but adjacency lengths in *f* are greater for speckled–black and speckled–white habitats, and lower for black–white. Compared with *e* design, *g* has more interspersion, more black–white adjacency, and equal adjacencies for speckled–black and speckled–white. Compared with *e*, design *h* is higher for interspersion as well as for all habitat adjacencies.

A result of interspersing more than two habitats is to produce *convergency points* ('junctions' or type of 'covert') where three or more habitats converge.[943,566,764] In Fig. 9.12 designs *e* and *g* each have one convergency point, while designs *d* and *h* each have two. Such locations are of special importance to certain species, such as quail (*Colinus*).[943,683] The arrangement not only provides a diversity of adjacent resources, but

314

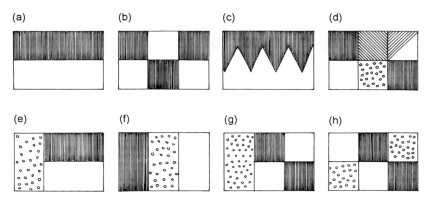

Fig. 9.12. Habitat interspersion, adjacencies, and convergency points. See text.

also stability through stress periods. Three habitats converging at a point are common, but four or more are usually uncommon (squares are rare in landscape graphs[251] (Fig. 9.3)). Convergency points generally appear to have low stability, or high maintenance cost, because habitats in proximity tend to lose their distinctiveness due to colonization by nearby species. Indeed, without maintenance, a convergency point may become the nucleus that expands to become a new patch.

Convergency points are important well beyond wildlife habitat.[501] They are often funnels for flows of water, eroded particulates, and mineral nutrients, as well as for moving animals. Thus, they are ideal locations for predators and hunters. Indeed, for the same reason, they are ideal for wildlife-observing blinds and platforms.

In summary, combining habitat type information with the broader patch–corridor–matrix analyses provides additional fine-scale insight into the structure of landscape mosaics. Enhancing both biodiversity and game species are quite compatible objectives using this approach.

Strategic points in a mosaic

Try covering copies of an aerial photo or map of a landscape, or a region, with clear plastic. Have people mark the ten (or 20) key, ecologically *strategic points* in the mosaic. You may wish to define 'strategic' in terms of importance, particularly in the long term, and 'ecological' to include productivity, biodiversity, soil, and water attributes. How congruent are the points marked? The rationales given for each mark will be just as interesting as comparing the different marks. Two approaches to lowering the variability among marks seem useful.

315

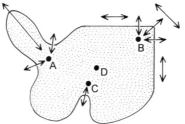

A. Best control over inputs from other landscape.
   Good access to other landscape.
B. Best overall access to other landscape.
   Best control over movements within
       other landscape.
C. Best access to major cove.
   Good access to area of this landscape.
D. Best protection from other landscape.
   Best access to area of this landscape.

Fig. 9.13. Example of strategic points in a landscape. See text for other ways of identifying strategic points.

First, one could identify strategic points based on the theory and empirical results absorbed, say, from the pages of this book. This analysis might identify three categories of points to mark: (1) locations that are strategic because of their contents and their source effect, such as large natural-vegetation areas, cities, and unusual features (e.g., a single major river, or the only two mountains); (2) locations where change is the issue, such as especially sensitive areas, and places that, if damaged or destroyed, have a long recovery time; and (3) locations or 'flux centers' where flows or movements are concentrated. These may include steep slopes, stream and river corridors, major lobes as funnels, stepping stones, certain corridors and intersection nodes in a network, and patches adjacent to (especially upwind or uphill of) one of the preceding. Nothing inherently makes patches, corridors, or matrix more strategic; all three can be in each category. The abundance of flux centers as strategic points, however, is apt to be pronounced.

A second approach to determining strategic points ignores the characteristics of a site, and instead evaluates the position of a location relative to the size, and especially the shape, of a landscape. For example, a mathematical model was used to compare the populations of an uncommon songbird (Bachman's sparrow, *Aimophila aestivalis*) in South Carolina (USA).[979] The model predicts that the population in mature forest would be 12% greater if the forest is situated in the center of a landscape, than if it is located at random. Indeed, sparrow populations are predicted to be 21% greater if the forest is in the center, than if it is in a corner of the landscape.

In the irregularly shaped landscape of Fig. 9.13, four possible strategic points for animals or people are proposed. To evaluate the pros and cons of each, one must at least consider access to, protection from, escape from, and control over certain terrain. Each of these in turn should be related to the landscape area itself, as well as the area surrounding the landscape. Some results are listed (Fig. 9.13), but other factors may be as important. Which point would you choose as most strategic?

316

A strategic point analysis is also applicable to an individual patch (chapter 2). For instance, the spread of fire or a pest outbreak would vary significantly according to location within a patch (Fig. 9.13). Doubtless there are strategic points where the probability of avoiding predation, finding food, or finding a mate is highest.

A more theoretical yet familiar approach is to use gaming or military strategy[1806] (Kristina Hill & Kongjian Yu, pers. commun.). Spatially-explicit games such as checkers and chess are useful models, and Japanese 'Go' or Chinese 'Chess' will be used for illustration. The board is a perfect 19×19 grid, and two opposing players take turns placing black and white 'stones' on the grid intersections. The goal is to control more than half of the board, and one encircles a large or small area with black, or white, stones to control it (usually enclosing some of the opponent's stones at the same time).

Experts know, based on abacus mathematics plus experience, that certain points are more strategic than others, both at the outset, and progressively as the board fills up with stones and conquered territory. As the pattern of stones becomes progressively more complex, mathematical guidelines become largely replaced by empirical experience. Applying gaming and other mathematical modeling approaches to identify strategic points in a mosaic may be useful, especially if a spatially simple landscape is assumed. A region may be more challenging to model in this way, because of its characteristic coarse-grained mosaic of landscapes. Anyway if you don't know 'Go', try it; you'll like it.

MEASURING MOSAICS

The subjects of mapping, remote-sensing, and geographic information systems (GIS) were introduced in chapter 1. Useful recent reviews of each, and a rapidly changing technology mean that the subjects are not developed further in this volume. Mapping or cartography has a long history and considerable theory. An enormous empirical base emphasizes land and geomorphology[280,655,779,1890,1456], as well as vegetation and climate[1727,517,72,901], and includes fine-scale nature mapping.[636,635] Similarly, remote-sensing has grown mainly from air photo interpretation to multi-spectral satellite imagery, and has a strong theoretical, quantitative, and empirical foundation.[1953,1740,573,43] GIS has developed rapidly and become highly usable, useful, and used.[1726,228] Not surprisingly as reviews point out[232], it has also been misused, especially in the well-worked-out cartographic areas of boundary delimitation, spatial aggregation, scale, error propagation, and color mixing for clarity or obfuscation. It is said that 'a man (or a woman) should come with instructions'. Analogously, a GIS can be an important tool in landscape and regional ecology, when the user is well grounded in the theory and practice of both cartography and remote sensing.

A rich array of measurements for heterogeneity and spatial pattern is available. The particular measurements selected are targeted to accomplish particular objectives, such as to understand better, or plan and manage better, a region or landscape. Because methods are so closely tied to objectives as well as to theory, they are presented together throughout the book. For instance, models of spatial structure and pattern are outlined in chapter 1. Many ways to measure patch shape are given in chapter 4, and measures of network attributes are explained in chapter 8. Nevertheless, some general measures of the spatial heterogeneity in a mosaic listed here have been used by a number of researchers, and may be useful in addressing a variety of questions.[579,1745,1748,501]

Four categories of measures or indices are identified (see chapter 9 appendix for specific measures). Two categories, 'diversity measures' and 'boundary (or edge) measures', essentially analyze the heterogeneity of a mosaic. The abundance of patches, boundaries, and so forth is critical, but their locations are of minor or no importance. In contrast, the other two categories, 'patch-centered measures' and 'all-patch pattern measures', depend on both the abundance of objects and their location relative to one another. One challenge emerging is to determine where heterogeneity is ecologically important, versus where spatial pattern is required to understand process (M. G. Turner, pers. commun.).

Diversity, richness, evenness, and dominance measures are various ways of determining the relative numbers of types, sizes, or shapes of patches (or corridors) present in a mosaic (chapter 9 appendix). Are there many or few types? Do one or two types predominate, or is there about the same number of each type present? Some of the measures are analogues of 'Shannon–Weiner information theory' measures of species diversity (other aspects of information theory are also useful in ecology[114,501]). William Romme and Dennis Knight used these measures to understand fire history in Yellowstone Park (USA).[1439,1441] Of the measures, richness, dominance, and evenness are the most informative, since they highlight single important variables, and do not combine and obscure variables.

Boundary (edge) number, boundary length (or density), and fractal dimension (chapter 3) are various measures of the type and predominance of boundaries in a landscape (chapter 9 appendix). They have been used to understand wildlife habitat[566], peninsula sizes on a rocky coast[1141], beetle movement[1887,803], and 50-year patch-shape changes in regions.[1234,1902] Patch density and matrix contiguity or connectivity are other overall essentially spatially-independent measures (chapter 9 appendix).

318

Spatially explicit measures of landscape or regional pattern express the locations of patches relative to one another (chapter 9 appendix). A measure may focus on the neighborhood, e.g., relating a patch to the surrounding spatial elements, or on arrangement of the whole landscape. An isolation-of-a-patch measure focuses on a single patch. Measures such as for isolation-of-patches, dispersion-of-patches, nearest-neighbor, contagion or aggregation, and accessibility-of-a-patch, provide average values and a variance, based on measuring all patches present. In these cases the distance between patches is important.

A host of other possible measures is available to measure mosaics. Grain size[501], adjacency analysis[430,1151,1150], anisotropy, backbones in percolation theory[543], texture[673,1200,1174], matrix contiguity[1426], and the cumulative frequency distribution of patches of various sizes[698] have been used in landscape ecology. Geography and cartography are rich sources of spatial measurement and analysis.[1670,995,646,289] Vegetation methodology is as well.[856,612,1169,1322,590,820] Spatial statistics, including semivariance, autocorrelation, and kriging, offers a newer supply of measures.[288,1425,364]

Usually two or three carefully selected measures are sufficient to provide the necessary breadth for addressing a specific question. In this manner one minimizes the probability of misinterpretation based on a single measure.

To conclude this chapter on mosaic pattern, we have now used the building blocks of the preceding chapters to construct mosaics of many spatial elements. The spatial patterns in these mosaics are numerous and complex, a worthy research frontier. Although most spatial arrangements evident in landscapes are well beyond the equations summarized in the chapter appendix, some patterns are presently understandable and predictable. We are now poised to consider a central and exciting issue in the next two chapters. How do the patterns affect flows and movements across the landscape and region?

APPENDIX: EQUATIONS FOR MEASURING A MOSAIC

Diversity measures

$$\text{Relative richness} \quad R = \frac{s}{s_{\max}} \times 100$$

where s = number of habitat types; $s_{\max}$ = maximum possible number of habitat types. From Romme (1982) and Turner (1989).

$$\text{Relative evenness} \quad E = \frac{H2(j)}{H2(\max)} \times 100 \qquad H2 = -\ln \sum_{i=1}^{s} p_k^2$$

where $H2$ (j) = modified Simpson's dominance index for landscape j; $H2$(max) = maximum possible $H2$ for s habitat types; s = number of habitat types; p_k = proportion of area in habitat k. From Romme (1982) and Turner (1989).

$$\text{Diversity} \quad H = -\sum_{k=1}^{s} p_k \ln p_k$$

where s = number of habitat types; p_k = proportion of area in habitat k. From O'Neill *et al.* (1988a) and Turner (1989).

$$\text{Dominance} \quad D_o = \ln s + \sum_{k=1}^{s} p_k \ln p_k$$

where s = number of habitat types; p_k = proportion of area in habitat k. From O'Neill *et al.* (1988a) and Turner (1989).

Boundaries (edge) measures

$$\text{Edge number} \quad E_{i,j} = \sum e_{i,j} \times \ell$$

where $e_{i,j}$ = number of horizontal and vertical interfaces between grid cells of types i and j; l = the length of the edge of a cell. From Gardner *et al.* (1987), Turner (1989) and S. Turner *et al.* (1991).

$$\text{Fractal dimension} \quad D = \frac{\log P}{\log A}$$

where A = area of a two-dimensional patch; P = perimeter of the patch at a particular length-scale. From O'Neill *et al.* (1988a), Milne (1988, 1991a, 1992) and Turner (1989).

$$\text{Relative patchiness} \quad P = \frac{\sum_{i=1}^{N} D_i}{N} \times 100$$

where N = number of boundaries between adjacent grid cells; D_i = dissimilarity value for the ith boundary between adjacent cells. From Pielou (1977), Romme (1982) and Turner (1989).

Boundary length (total length of boundary per unit area) and Boundary density (number of boundaries per unit area) are also measures.

Patch-centered measures

$$\text{Isolation of a patch} \quad r_i = \frac{1}{n} \sum_{j=1}^{i=n} d_{ij}$$

where n = number of neighboring patches considered; d_{ij} = distance between patch i and any neighboring patch j. From Forman & Godron (1986).

$$\text{Accessibility of a patch} \quad a_i = \sum_{j=1}^{i=n} d_{ij}$$

where d_{ij} = distance along a linkage (e.g., a woodland corridor or hedgerow) between patch i and any neighboring patch j; n = number of neighboring patches considered. From Lowe & Moryadas (1975) and Forman & Godron (1986).

APPENDIX: EQUATIONS FOR MEASURING A MOSAIC

All-patch pattern measures

Dispersion of patches $R_c = 2d_c \left(\dfrac{\lambda}{\pi} \right)$

where d_c = average distance from a patch (its center or centroid) to its nearest neighboring patch; λ = average density of patches. If $R_c = 1$, patches are randomly distributed; if $R_c < 1$, patches are aggregated; if $R_c > 1$ (up to a maximum of 2.149), patches are regularly distributed. The equation is also a measure of aggregation. From Pielou (1977), Forman & Godron (1986), O'Neill *et al.* (1988a) and Ripple *et al.* (1991).

Isolation of patches $D = \Sigma(\sigma_x^2 + \sigma_y^2)$

Patches are located on a grid with x and y coordinates; the average location and the variance for all patches are calculated for the y coordinate. σ_x^2 and σ_y^2 are the variances on the x and y coordinates, respectively. The equation is based on the 'standard distance index'. From Lowe & Moryadas (1975) and Forman & Godron (1986).

Nearest neighbor probabilities $q_{i,j} = \dfrac{n_{i,j}}{n_i}$

where $n_{i,j}$ = number of grid cells of type i adjacent to type j; n_i = number of cells of type i. From Turner (1988, 1989) and Ripple *et al.* (1991).

Contagion $C = 2 s \log s + \sum\limits_{i=1}^{s} \sum\limits_{j=1}^{s} q_{i,j} \log q_{i,j}$

where s = number of habitat types; $q_{i,j}$ = probability of habitat i being adjacent to habitat j. From Godron (1966), O'Neill *et al.* (1988a), Turner (1989) and Gardner & O'Neill (1991).

Patch density (number of patches per unit area) and Contiguity (aggregation or dispersion of patches) (Williamson & Lawton 1991) are also measures.

10

Wind and water flows in mosaics

Thus from generation to generation it goes bounding over lakes and woods and mountains . . . sailing . . . on various tacks until the wind lulls, to plant their race in new localities – who can tell how many miles away? . . . And for this end these silken streamers have been perfecting themselves all summer, snugly packed in this light chest, a perfect adaptation to this end – a prophecy not only of the fall, but of future springs.

Henry David Thoreau, The Dispersion of Seeds, *1862*

Weather always seems to come from somewhere else. Storms, cold fronts, and dry air, produced by energy and water flows in adjacent or distant regions, keep rolling in. We, along with vegetation and animals, get frozen, broiled, desiccated and drenched. But the circulating air brings more than heat and water. Air-borne particles increase rain just downwind of most cities. Sulfur oxides and heavy metals from factories damage vegetation in a downwind plume. Dust from cultivated fields filters through the cracks of far-off houses.

Weather and climate also depend on the internal characteristics of a landscape. The arrangement of land uses and the contrasts between adjacent ecosystems in the landscape help create or sustain flows of energy, water, and air. In short, both flows from outside and pattern within determine the wind and water flows in mosaics.

The objective of this chapter is not to learn meteorology or global cycles, though bits of them are introduced. Rather, after eight chapters primarily exploring spatial pattern and its ecological significance, this is the leadoff chapter with principles focusing on flows, movements, and changes in the land mosaic.

The principles in this chapter are a key to spatial planning, management, and conservation, because of the extensive flows among local ecosystems in a land mosaic. Plans and management schemes must highlight not only adjacencies (chapter 9), but sources and sinks at varying

322

distances. Upwind and upslope sources of heat, water, and pollutants are highly significant in, for example, suburban development, nature reserve protection, crop growth, and forest production. Downwind and downslope sinks at various distances also affect the ecological characteristics of an ecosystem or land use.

Scale is important in every section of this book (chapter 1). In chapter 5 local wind patterns interacting with vegetation, soil, and microclimate were examined down to scales of meters and hours. On the other hand, major vegetation transformations in regions occur over thousands and tens of thousands of years. Nevertheless, in considering wind and water flows in this chapter, most phenomena at landscape and regional scales are significant over years to centuries.

This chapter depends on and builds from preceding concepts, especially those in chapters 1, 6, 8, and 9. No review covers the whole subject, but individual topics are usefully reviewed as follows: flows between adjacent ecosystems[501]; energy flows and general meteorology[1446,1131,1243,1477,12,1162]; material transport in mosaics and topography[556,1132]; and the spread of disturbance.[1742]

We begin by briefly introducing the global framework as background to the patterns of horizontal circulation. Then we focus on the land mosaic where vertical energy flows and contrast among ecosystems determine horizontal heat-energy flows. These flows in turn transport particles and gases from source ecosystems to sink ecosystems in the mosaic. The distinctive airflows in mountainous areas are considered along with 'sheet' flows of water. Finally, we explore how disturbance affects a mosaic, and vice versa.

A GLOBAL FRAMEWORK

Air tends to circulate around the globe within six zones, which are roughly separated by high and low atmospheric pressure areas at certain latitudes (Fig. 10.1). Thus, equatorial zones are evident between c. 0° and 30°, mid-latitude zones between c. 30° and 60°, and polar zones above c. 60°.

Wind or air movement results from differences in air pressure. Thus, air moves from high pressure regions to regions of low pressure. However, because of the Earth's rotation, air movement also has a strong east–west directionality (Fig. 10.1).

Global air circulation also interacts with the arrangement of continents (Fig. 10.1) and sea currents to create distinctive climatic regions. Palms will grow in southwestern Britain where west winds bring heat from warm ocean water. Along the west coast of South America and southern Africa eastward-flowing air crossing cold northward-flowing

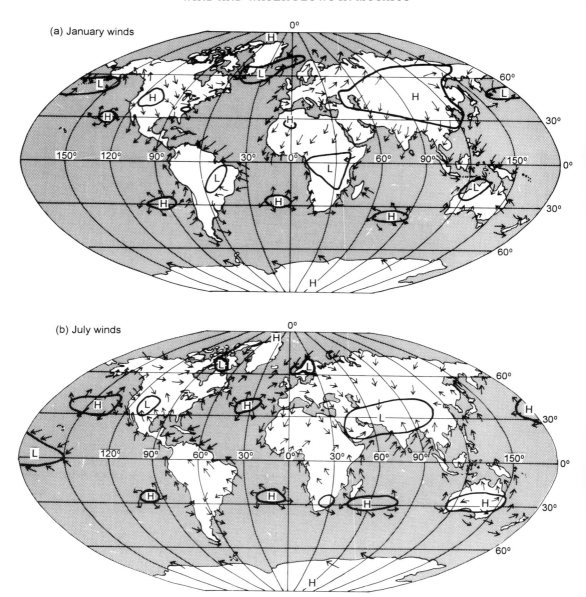

Fig. 10.1. Predominant wind directions relative to continents, atmospheric pressure, and season. Major atmospheric pressure areas: H = high; L = low. Based on Rumney (1968), Sabins (1987), Moran & Morgan (1994), and other sources.

ocean currents produces the deserts of Chile/Peru and Kalahari, respectively. Monsoonal regions in India and Indochina result from moist air seasonally flowing shoreward from the warm Indian Ocean to the south.

The concept of a region is in part defined by 'macroclimate' (chapter 1), the climate or long-term weather of a large area. Large areas or 'cells' with a similar climate are a result of global air circulation that interacts with the distribution of continents, mountain ranges, and sea currents.[1462,1197,1198]

The primary ecological significance of this air circulation is horizontal transport between ecosystems, between landscapes, and even between regions. The atmosphere receives heat and materials from the land or sea in one place and transports them to another location.

Global air circulation not only links regions, but also connects continents. Smoke from forest fires in western Canada has been observed in Europe, dust in the Caribbean from the Sahara, haze in the Arctic probably from Europe, and radioactivity in North America from nuclear explosions in Russia and China.[1007,1925,1401,151,1409]

Below the high-altitude 'stratosphere' and its ozone layer is an atmospheric layer called the *troposphere*, the star of this chapter. This is the lowest 10–20 km (6–12 mi) of the atmosphere, characterized by temperature normally decreasing with height, appreciable water vapor, and the presence of most of our weather.[1243,1162] The troposphere extends much higher in equatorial than polar areas (Fig. 10.2). A weather report is largely about action in the troposphere.

In the stratosphere and upper troposphere air circulates around the Earth essentially unimpeded. Friction from the planet affects airflow below c. 1–1.5 km.[1243] Thus, the lower portion of the troposphere, extending from the Earth surface to the level where the frictional influence is absent, is called the *planetary boundary layer* (or frictional, atmospheric boundary, external boundary, or simply boundary layer).[1446,1131,1132,1243] Topographic roughness, as well as daily change from daytime sun to nighttime heat flow, causes up-and-down airflow (turbulence) within it.[556,1132,1243] At night and in calm cool conditions the boundary layer may only extend upward to a 100 m or so (Fig. 10.2). However, by day it may extend upward a kilometer or two over rough terrain in windy conditions with *thermals*, i.e., columns of rising warm air.

Global circulation and mixing in the troposphere are well illustrated by a remarkable three-decade record of atmospheric CO_2 measured at 3400 m elevation on Mauna Loa, a Hawaiian peak in the north central Pacific.[1413] From 1960 to 1990 average annual CO_2 concentration rose steadily from 320 to 350 ppm, a 10% increase. Similarly, in Antarctica from 1975–90 the rate of CO_2 increase was almost identical, 335–350 ppm.[1413] Of equal interest, however, is the regular sawtoothed shape of

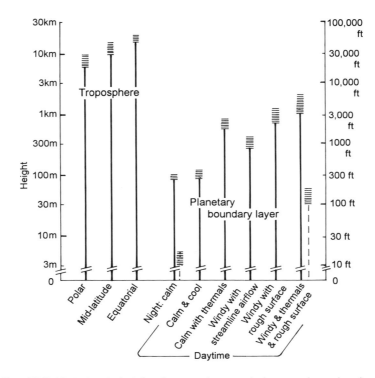

Fig. 10.2. Variation in height of troposphere and planetary boundary layer. Horizontal dashes indicate the approximate range of variation. The two vertical dashed lines refer to a 'turbulent surface layer' at the bottom of the boundary layer.[257] Note logarithmic vertical axis. Based on Oke (1987) and Ahrens (1991).

the curves. Each year in Hawaii the spring–summer peak is *c*. 6 ppm CO_2 higher than the winter minimum. This apparently results from high summer and low winter metabolism (respiration) in terrestrial ecosystems of the northern hemisphere, which are thousands of kilometers from Hawaii. In Antarctica, on the other hand, the summer peak and winter dip of CO_2 are only *c*. 1.5 ppm apart. The southern hemisphere has much less land surface for terrestrial ecosystem metabolism, and a somewhat-separate circulation.

To conclude, flows between local ecosystems or between landscapes are part of, and controlled by, a global circulation pattern. Almost all the flows are in the troposphere. Turbulence associated with local ecosystems links the mosaic land surface with horizontal flows in the planetary boundary layer.

326

ENERGY INTERCONNECTING MOSAIC ELEMENTS

Vertical energy flows

Solar radiation and albedo

The surface of the sun is hotter than that of the Earth, and hence radiates much more energy than from the Earth. Radiant energy moves in different wave lengths, e.g., from very-short gamma radiation to very-long radio waves (Fig. 10.3). The majority of the sun's energy or solar radiation is given off at wave lengths less than 3 micrometers in length, so-called *shortwave radiation*.[1243,12,1162] In contrast, most of the earth's radiation is emitted between *c.* 3 and 25 (or 1000) micrometers as *longwave radiation*.

A space capsule in 'outer space' receives approximately 2 calories per cm^2 per minute (1370 watts/m^2/min.) over its surface, the same amount of incoming solar energy that strikes the Earth's atmosphere. The atmosphere then differentially filters the solar energy (Fig. 10.3).[550,1405,1446,12] Almost all the shortest wave radiation, gamma and ultraviolet (UV) rays, is absorbed by a layer of ozone and oxygen several kilometers high in the atmosphere. The bulk of the solar energy (maximum at *c.* 0.5 micrometers = green light) comes right through the atmosphere to the earth surface. Indeed, most of this comprises the 'visible' wave lengths

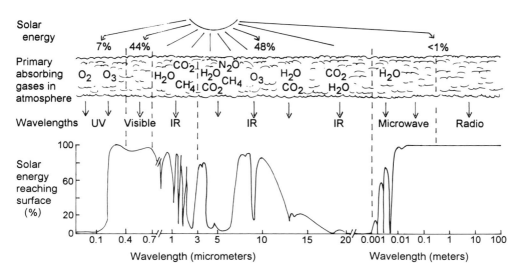

Fig. 10.3. Solar energy of different wave lengths passing through atmospheric gases to reach the Earth's surface. Generalized global averages. Adapted from Sabins (1987) and Ahrens (1991).

327

or rainbow colors from blue to red that we see. This is also used in photosynthesis by plants (see Appendix A of chapter 2).

Infrared radiation (IR) is invisible to our eyes. Its wavelengths are too long, extending from *c.* 0.7 to 1000 micrometers (1 millimeter) (Fig. 10.3).[1477,12] A small amount of shortwave solar radiation reaches the earth between 0.7 and 3 micrometers ('near-infrared radiation'). But most important is the Earth's or terrestrial longwave radiation given off above 3 micrometers. The maximum energy is emitted at *c.* 10 micrometers. Indeed, the band from 8 to 10 is called an 'atmospheric window', because of the abundance of IR radiation (heat energy) given off, and its ready passage through the atmosphere to outer space. Thermal remote-sensing instruments on aircraft use this atmospheric window to produce stunning and informative infrared images of landscapes.

Several atmospheric gases act as absorbers of specific wavelengths of radiant energy.[1477,1243,12] Ozone (O_3) especially absorbs energy below *c.* 0.3, and at 9.5, micrometers. Water vapor (H_2O) and carbon dioxide (CO_2) are important absorbers in several IR bands. Methane (CH_4) and nitrous oxide (N_2O) also absorb at specific bands in the IR. The amount of each absorbing gas in the atmosphere is affected by human and land-use patterns, and in addition, affects ecological processes on land. For example, an increase in atmospheric CO_2 apparently raises temperature by preventing terrestrial heat from escaping to outer space, and a drop in O_3 lets life-killing UV radiation through.

Neighboring local ecosystems receive about the same amount of solar radiation for all wave lengths. Nearby landscapes receive about the same amount of visible energy, but often differ in the amount of longwave radiation, due to differences in broad-scale atmospheric gas content over different landscapes.

Direct solar radiation from the sun is the only strongly directional energy flow in a landscape or region. Indeed, this linear input of solar energy is the main reason that north and south slopes almost always differ ecologically.[179] The equatorward- or sun-facing slope receives more solar energy per unit area than the poleward-facing slope.

The direct input of solar energy is supplemented by 'diffuse sky radiation', largely the solar energy scattered by clouds and large particles, that then reaches the earth.[1243] This is particularly important in winter months, and in mid and high latitudes where the direct solar angle is low.

Direct and diffuse radiation that strikes the land is either reflected or absorbed.[550,551,1131] Some of the shortwave radiation (0.3 to 3.0 micrometers) reaching the Earth is absorbed, and some is directly reflected back to the atmosphere and space. This reflection of shortwave energy is called *albedo*.[904,828,1446,1162] Albedo is higher for smooth, light-colored surfaces, and lower for rough dark sur-

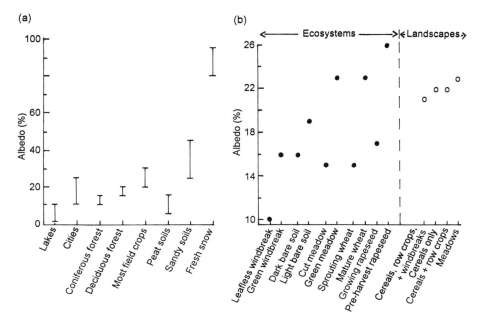

Fig. 10.4. Albedo or reflection of solar energy by different surfaces, ecosystems, and landscapes. (a) General ranges based on Geiger (1965), Reifsnyder & Lull (1965), Rosenberg et al. (1983), Moran & Morgan (1994), and T. Webb III, pers. commun. (b) Measurements during the growing season in specific neighboring ecosystems and landscapes near Poznan, Poland, based on Ryszkowski & Kedziora (1987) and data from Tamulewicsz & Wos (cited by Ryszkowski & Kedziora 1987). Wheat = *Triticum*; rapeseed = *Brassica napus*.

faces.[361,556,1451,1452,1475] Therefore, fresh snow with an albedo of 80–95% of incoming energy, only absorbs 5–20% of the energy (Fig. 10.4a). This barely melts the surface snow. Dry light-colored sandy soils have albedos of 25–45%, in contrast to dark peat soils of 5–15%. Most field crops have albedos of 20–30%, deciduous forest 15–20%, and coniferous forest 10–15%. Lakes are usually the lowest reflectors in a landscape.

The shortwave energy not reflected is absorbed by soil and vegetation. Therefore for example, peat and coniferous forest absorb more shortwave energy than sandy soil and crops (Fig. 10.4a). Lakes and wetlands are major energy sinks. Virtually all the arriving longwave radiation (>3 micrometers) is absorbed by the vegetation and soil.

In effect, incoming energy is about constant over a landscape. But the differences in albedo, due to contrasts among local ecosystems within the landscape, result in different energy budgets. One ecosystem can have most of its incoming solar energy reflected, while the adjacent ecosystem a meter away absorbs most of the sun's energy, and heats up.[1133]

Heat for soil, air, and evapotranspiration

Where does the energy then go? In a clear sky with little of the absorbing atmospheric gases present (Fig. 10.3), energy radiated from a land surface goes directly back to outer space. But normally the atmosphere contains significant amounts of the gases, especially where humans have generated pollution. In this case some radiated energy cannot get through and is absorbed by the atmosphere.

The energy originally absorbed by the soil and vegetation heats them up.[179,1451,1453] This is quite interesting, in part because an object radiates energy according to its surface temperature (a person with a fever radiates a lot, and a hot sun much more). A dry surface that absorbs the most energy (e.g., black soil or paved tarmac) usually radiates the most energy.[527,257,739] This radiated energy then heats up the atmosphere.

But the plot thickens. A 'wave length shift' takes place. Shortwave radiation absorbed by soil and vegetation raises their surface temperature. Consequently, they radiate more, but in the form of longwave infrared radiation. The longwave radiation doesn't penetrate the atmospheric gases as well, and hence is mainly absorbed by the atmosphere. This longwave radiation is basically heat, and the air gets warmer at the landscape, regional, and global levels. This is exactly the process in global warming due to the buildup of 'atmospheric greenhouse gases'. It also partially explains how a glasshouse or greenhouse works, where glass functions in lieu of the gases.

The other key function of the energy absorbed by soil and vegetation is to provide heat for metabolism, growth, decomposition, snowmelt, evapotranspiration, and the like. *Evapotranspiration* (ET) is the vaporizing or evaporation of water mainly from soil and plant surfaces. The amount of water pumped upward from the soil to these surfaces varies widely according to vegetation and land use. For instance, in the southeastern USA converting deciduous forest to pine forest results in 20 cm more water per year being pumped out in ET.[1660]

Evapotranspiration is also of special interest because a large amount of energy is consumed in evaporating water molecules.[180,1133] Deserts seem hot, not because solar inputs are unusually large, but because of little evapotranspiration. Without ET the energy absorbed by the soil is largely radiated as heat ('sensible heat', which thermometers sense and you feel).[608,1131] This raises the air temperature. An irrigated desert area is much cooler. Here, a significant portion of the plant-and-soil-absorbed energy is used in ET from crop and soil surfaces. This evapotranspiration energy ('latent heat') therefore is not available to raise air temperature.

330

Several of these energy flows are illustrated in a mosaic landscape about 40 km south of Poznan, Poland, studied for over two decades as reported by L. Ryszkowski and A. Kedziora (1987). The land has a topographic relief of only a few meters, and contains both sandy and peat soils. Its cover is 69% cultivated fields (half in cereals), 14% in shelterbelts (c. 160 year old windbreaks of black locust, poplar, oak, pine, and spruce (*Robinia, Populus, Quercus, Pinus, Picea*) and small woods, 12% in meadows and pastures, and 5% in villages, roads, small lakes, channels, and wetlands.

Since incoming radiation energy is constant over the mosaic of ecosystems, the first control on energy flow is albedo. The highest reflection is by fully developed crops and meadows (Fig. 10.4b). The lowest albedo (i.e., greatest energy absorption) is by windbreaks, dark soil, meadows after cutting, and sprouting winter wheat. The fully developed crops reflect more than twice as much energy per unit area as the leafless windbreaks. In addition, albedo changes through the seasons.

Four agricultural landscapes that differ markedly in the proportions of cereals, row crops, meadows, and windbreaks are also compared. They exhibit similar albedos (Fig. 10.4b). A landscape dominated by meadows reflects the most energy, and a landscape with many windbreaks absorbs the most.

Compared with the various local ecosystems, the mosaic landscapes are intermediate in albedo. The radiation energy emitted by soil and vegetation is relatively constant from ecosystem to ecosystem in this area, because their surface temperatures are similar. The maximum radiation is only 1.14 times that of the minimum.

Incoming energy that is not reflected is mainly used to vaporize water in ET, heat the air, and heat the soil[257], which together increase plant productivity. Evapotranspiration is highest in the windbreak and lowest on bare soil, a two-fold difference (Fig. 10.5). Among the fields, meadow has the highest ET and wheat (a cereal) the lowest. Bare soil heats the air most, and a windbreak heats the air least, a more than five-fold difference. Much less energy goes into heating soil (only c. 5%) than into ET. Most energy enters the soil by day and is given off at night. The nocturnal heat energy helps maintain frost-free plants and animals, which is especially important at the beginning and the end of the growing season.

Thus the windbreak uses 40% more energy in ET than the wheat field, whereas the wheat field heats the air three times as much as does the windbreak (Fig. 10.5). Consequently, the windbreak can evaporate 17 cm more water than the same area of wheat. Indeed, during the growing season, water evaporated by windbreaks exceeds incoming rainfall by 62%. The windbreaks dry out the adjacent soil. Several reasons for high windbreak ET are evident: more soil moisture (from intercepting rain or snow), longer roots (both horizontally and vertically), less leaf (stomatal)

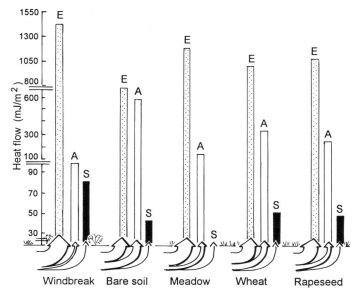

Fig. 10.5. Heat used in evapotranspiration (E), and in warming air (A) and soil (S) of different ecosystems. Average heat flows measured are for the whole 225-day growing season. See Fig. 10.4 caption. Based on Ryszkowski & Kedziora (1987).

resistance to ET, and a greater canopy height and roughness that causes turbulence and local high wind velocities.[1131]

Overall energy budgets thus vary widely from ecosystem to ecosystem in a landscape mosaic. To see how much landscapes may differ in energy flows, models with different 'real world' proportions of cereals, shelterbelts, other crops, meadows, and non-vegetated areas, plus various assumptions of planting times and horizontal wind, were compared.[1475] Despite the big differences in ET, air, and soil heating among ecosystems (Fig. 10.5), and especially in the role of windbreaks, the eight model landscapes compared have very similar values for all the energy flows. Four main reasons are recognized: (1) the various field and soil types have rather similar albedos; (2) no large forests with quite-different energy flows are present; (3) shelterbelts cover only a tiny proportion of the land; and (4) energy moves horizontally between ecosystems to average out differences at the landscape level. Clearly, energy flows could differ more among landscapes within a region if fields and soils had more contrasting albedos, if large forests were present, or if shelterbelts (windbreaks) around small fields were more frequent.

The fourth point, horizontal energy flow, to be considered below, is a key to flows across landscapes and regions. For instance, after harvesting grain in summer, a hot field gives off heat that flows into an adjacent woods and home. This dries the woods and broils the neighbor.

Advection

The horizontal movement of energy or material carried by air has been termed *advection*, and numerous examples are documented world-wide.[1675,1446,79,1131,1011] The contrasts in albedo, air warming, and the like just observed between adjacent local ecosystems create energy gradients or differences. Heat flows from warm to cool locations (in the absence of other pressure differences). Therefore, a temperature contrast between ecosystems causes heat to flow horizontally in the mosaic.[1131,1132] This air movement caused by heat differences is the norm, not the exception. Since heat and air are constantly flowing, gases and particulates are also carried between ecosystems. No terrestrial ecosystem can be isolated from these inputs and outputs.

To visualize these heat patterns, temperature is measured over time along transects through forests with clearings.[556,64] For example, air temperature was measured from a summer mid-afternoon to the following morning along a 7 km (4.5 mi) transect through forest, young stands, and bare areas in Germany[556] (Fig. 10.6). In the middle of the starry night the young stands and bare areas are coldest (<12 °C), whereas the large forest is warmest (>15 °C). The forest doubtless absorbs more solar radiation, but also evapotranspires more, and puts less heat into the air than do the open areas. However, the open areas radiate more heat to the atmosphere, leaving cold spots at night. Where would you sleep the next time you are lost on a clear cold night?

The 3–4 °C (5–7 °F) night temperature differential between forest and adjacent open patches means that on a typical windless night, heat and air are moving horizontally from the forest to the clearing. In fact, this air movement has been called the 'sea breeze without a sea'. The warm forest air moves horizontally to the cool clearing, while the cool-clearing air is radiating heat and moving vertically toward cold outer space. Such horizontal and vertical processes carry gases, aerosols (droplets), spores, seeds, and other particles on calm nights.

Advection is also prominent at the broader scale of regional air pressure differences.[173,1132] Thus a summer daytime sea breeze cools a coastal landscape. Indeed, such winds largely eliminate the local temperature-gradient flows just described. For instance, along the forest-and-clearing gradient (Fig. 10.6) the maximum mid-afternoon temperature is relatively uniform (*c*. 24 °C). This is due to air mixing by normal daytime winds. The temperature in the clearings is *c*. 1 °C higher than in the forest. But without wind it would be several degrees higher due to more sunlight. The horizontal temperature profiles are most uniform across

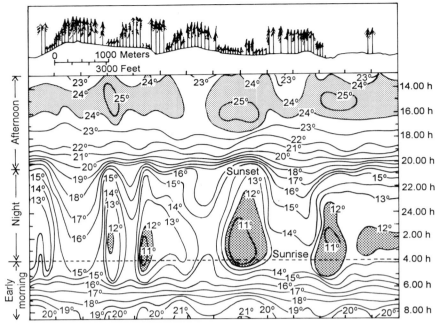

Fig. 10.6. Comparing summer temperature changes in wooded patches and clearings over 20 hours. Along a 7 km line (diagram at top) chest-height temperature measurements were made (in °C) every hour (reading downward). Upper shaded areas = highest afternoon temperatures; lower shaded areas = lowest night temperatures. Maximum topographic relief = 20 m. Adapted from Geiger (1965). Originally published in the German language by Friedr. Vieweg & Sohn Braunschweig 1961.

the landscape just before sunset and just after sunrise. A sharp decrease and a sharp increase in solar heating at these times override local ecosystem contrasts that develop during the day and at night.

Wind is directional across a landscape, whereas local energy gradients[556,64] may create air movement in any direction within a landscape, depending on the juxtaposition of adjacent ecosystems. In fact, materials can be locally transported at night between adjacent elements in the direction opposite to the prevailing daytime wind direction.

In addition, air moves beyond the adjacent ecosystem to affect many spatial elements, sometimes far downwind. Advection by warm air may cause any of the same three effects as for incoming solar energy, namely, increase soil temperature, air temperature, or ET. The increase in soil temperature may melt permafrost, or enhance seed germination and root growth, especially early or late in the growing season. The warming of air by advection may melt snow, stimulate or inhibit animal movement, increase productivity, or minimize frost, thus lengthening the growing season.

The increase in evapotranspiration due to advection is called the *oasis effect*. The acceleration in vertical water movement through oasis plants results from daytime blowing of warm dry wind from the surrounding desert. It is a water pump driven by heat advected from the hot desert soil and rock.[913,608] The oasis effect also occurs in moist areas. For instance, ET and water loss from a moist woods was calculated to be 5% greater due to advection from an upwind shrubby area.[80] Nevertheless, the oasis effect is especially important in savannas, grasslands, and deserts.

Wind-driven advection energy is a major property linking landscapes in a region, and indeed regions in a continent.[1132] The predominant directionality of wind across a region of course provides considerable predictive ability about which landscapes will affect which others. Since advection can move heat (sensible heat) horizontally, the 'source' landscape will be an energy-liberating system such as desert, grassland, or city. The 'sink' will be a landscape such as forest or moist peatland. Wetlands and lakes are energy sources in autumn when air is cool, but they are sinks in spring when air is warm.[1217,747,234,126,1454] Wind of course also advects particles and gases, but here the focus is on energy, and its effect on downwind soil, air, and ET.

In a 43 km (27 mi) cross section through a series of landscapes in Alberta, Canada, vertical plumes of hot air are seen over farmland, non-cultivated terrain, and prairie landscapes[747,1132,1243] (Fig. 10.7). Between these are cooler air pockets over the cooler surfaces of a lake and irrigated land. In the planetary boundary layer, commonly more than *c.* 70 m above the vegetation, differences due to local spatial variations on the ground are averaged out and scarcely detectable. In the Canadian landscapes vertical thermal plumes (created by convection) form beneath the boundary layer (Fig. 10.7). They are tilted downwind by regional wind that is faster aloft. Therefore most energy and material transport in the air column enters the adjacent landscape below the boundary layer. This means that the advected energy will enter the adjacent 5–10 km wide landscape, and affect its air, soil (or lake), or ET.

Thermals, the columns of rising warm air, enhance or support soaring birds such as vultures, gulls, and eagles.[1302] The sensitivity of such birds to thermals is illustrated in India where vertical walls in town absorb the early-morning solar radiation, heat up, and produce heated air that begins to rise. Consequently vultures begin soaring from towns an hour earlier than the birds from open countryside.[298] If you have experienced a bumpy small-plane flight on a sunny summer day, where the plane responds to thermals over fields, forests, and towns, you may reconsider wishing to soar like a big bird.

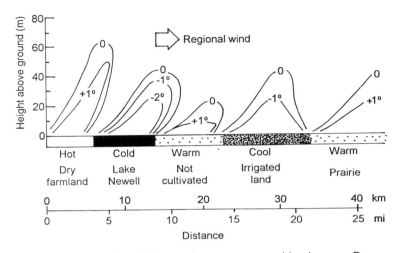

Fig. 10.7. Warm and cool air transfers across several landscapes. Degrees (°C) above or below the temperature of the boundary layer (or interfaces between landscapes) are given for the afternoon of 6 August 1968 in southern Alberta, Canada. Local parcels of warm air were also measured up to a kilometer above warm surfaces. Adapted from Holmes (1970), with permission of the American Society of Agronomy, Inc.

Climate change

Climate change and its effects are not explored in this book, because numerous recent volumes are available on the rapidly developing topic. The subject links a multitude of phenomena, including the greenhouse effect, global warming, increased CO_2 and other gases, more atmospheric dust, jet stream movements, El Niño effects, greater climate variability, shift of vegetation zones, disappearance of familiar vegetation types, appearance of new types, Earth axis tilting, sun spots, ocean warming and surface changes, migration of agriculture, changes in precipitation regimes, increased soil erosion, greater fire frequency, pest outbreaks, melting glaciers, sea-level rise, extensive floods, loss of nature reserves and parks, and massive species extinctions.

Nevertheless, a few issues can be highlighted in the context of land mosaics. First, though much emphasis is currently global, most effects and solutions are to be found at landscape and regional scales.

Second, landscape and regional changes alter climate, especially over decades and centuries.[1792,1505] Converting agricultural landscapes to woodland in southern France, or desert to irrigation in Egypt, or forest to agriculture in the American Midwest, in each case has changed the energy budget, and hence the climate.

336

Third, climate change alters landscapes and regions (the converse of the second point). The primary effects considered are on vegetation[367,1847,1848,382,1466], soil[86], and land use.[131,129,132,204,1741,128] Paleoecology (chapter 14) provides a basis for predicting the effects of climate change. The documentation of ecological changes in landscapes over millennia during and since the last ice age, as well as in shorter time scales[132,1084,1516,1466], explicitly includes correlations with past climate changes.[1848,1084,1516]

Finally, climate change associated with greenhouse gases, such as the sharp CO_2 increases measured in Hawaii and Antarctica, is curious and serious.[1413,1191,1593] Unlike expectations from the space–time principle (see Fig. 1.3), the temporal scale of climate change is short, years to decades, but its spatial effect is broad, continents to the globe. Anthropogenic climate change and weather modification must be addressed at landscape and regional scales, where land use plus wind and advection play central roles.[1505]

MATERIALS AND THE GIANT CONVEYOR BELT

Thus, we can visualize winds in the troposphere (Fig. 10.2) acting as a *giant conveyor belt*. They pick up warm or cold air from one ecosystem and deliver it to the adjacent system (Fig. 10.7), or to many spatial elements downwind. The upper portion of the troposphere is primarily involved in long-distance transport, particularly of gases and light-weight particles. In the lower portion the planetary boundary layer interacts directly with turbulence from local ecosystems. The lower portion is especially involved in short-distance transport, including gases and considerable particulate matter.

The conveyor never stops, though winds at the ground often stop. However, the conveyor moves at different rates depending on regional air-pressure differences. It is a principal linkage between landscapes, as well as between regions.[1131,1409]

But the troposphere wind carries much more than energy. Living structures such as spiders, seeds, spores, and pollen are picked up in one ecosystem and deposited elsewhere. Non-living particles including dust and soot get on and off the conveyor. Gases, including SO_2, NOX, O_3, CO, and CO_2, are transported between spatial elements. And water in different forms, vapor, droplets, and snowflakes, is picked up and dropped off. No patch is an island. Each patch is intricately linked to and dependent on other patches, through this giant conveyor belt.

The conveyor goes over mountains and expands down into large valleys. But tall columns of rain cut through the belt, and almost anything

337

in the conveyor at those points is rapidly washed to the ground. Hail and snow likewise pour through the belt.

Tornadoes, typhoons, and hurricanes meander through the moving belt as if they did not 'see' it. Such windstorms add unusually large particles like leaves, mice, roof material, and chairs into the conveyor. Big fires also cause smoke, particles, and gases to rise right up through the conveyor. Nuclear explosions on the land surface similarly shoot dust and radioactivity up through the conveyor. Most material ends up in the troposphere. Yet some particles and gases reach the stratosphere above most weather activity. These materials may circle the globe many times, but gradually filter downward and enter the conveyor.

Particles and gases

Particles

Experiments in the absence of wind or rising air have been done to determine the fallout rates of various seed types, pollen, and spores. Ash and fir seeds (*Fraxinus, Abies*) fall faster than birch and dandelion (*Betula, Taraxacum*), which fall faster than pollen of pine (*Pinus*), which falls faster than clubmoss and puffball spores (*Lycopodium, Lycoperdon*).[556] Such studies never include tomatoes, coconuts, and watermelons (*Lycopersicon, Cocos, Citrullus*), perhaps because of danger to investigators or the cleanup problem. These studies are useful first steps, but omit a key ecological factor, i.e., the role of the giant conveyor.

Most seeds, pollen, and spores are liberated from the plant in dry, usually moving, and often turbulent air. These are optimal weather conditions for lifting the live particles up into the planetary boundary layer. Thus the distance moved has little to do with the isolated fallout rate. Rather, it has much to do with the rate of conveyor movement, and the attributes of the ecosystems beneath. Seeds of numerous types, pollen, spores, and even spiders with tiny balloons have often been collected high over the mid Pacific and all other oceans. This transporting role of the conveyor helps provide the supposedly-amazing, but actually quite expected, recolonization of remote bare areas. It underlies the essential resilience of ecological systems.

The same processes operate on non-living particles, such as dust from soil and soot from combustion. Seeds, pollen, and spores are liberated above the ground on plants, so they are in the airstream at the outset. Soil particles must be lifted off the ground, which requires dry soil and air plus high wind velocity, as in turbulent and vortex airflows (chapter 6). Sandstorms lift fine sand grains into the conveyor belt, but most soil material carried is silt or clay. Soot particles are carried upward from surface and canopy fires, chimneys, and smokestacks. Their movement

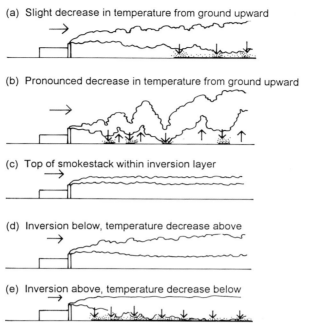

Fig. 10.8. Downwind smoke and pollutant deposits related to temperatures above the ground. Adapted from Geiger (1965). Originally published in the German language by Friedr. Vieweg & Sohn, Braunschweig 1961.

to the giant conveyor is rapid, because the particles are lifted in a column of rising heat from combustion.

Occasionally when the boundary layer is moving slowly, a *temperature inversion* forms over a portion of a landscape, or even an area larger than a region. For instance, an inversion may form at night when heat is given off by a city under high atmospheric pressure conditions. This inversion often 'burns off' by mid morning, and hence is short-term. Instead of the usual progressive decrease in temperature from the ground upward, in an inversion a layer of warm air forms in the conveyor. It is cooler below the layer and colder above. Particles carried upward in rising air then accumulate under the inversion layer, because the (relatively-cool) rising air cannot pass through the warm layer. The sky looks dirty and incoming solar radiation is decreased. Ecosystems beneath may be bathed in the accumulated particles.

The presence, height, accumulation, and pattern of deposition of particles often can be indicated by the plume from a smokestack.[556,1243,12] Under typical conditions of a modest decrease in temperature from the ground upward, a symmetrical widening plume is seen (Fig. 10.8). Where the ground is hot, an up and down zig-zag plume reflects the uneven columns of rising heat from different ecosystems. If the top of the smokestack is within a warm inversion layer (cooler both above and below the plume), a thin horizontal plume is expected. If the inversion layer is below the top of the stack, the bottom of the plume is smooth. Conversely, if the inversion layer is above the smoke, the top of the plume is smooth.

In all cases the particles return to the ground. The first case (Fig. 10.8a) spreads particles widely, both near and far from the smokestack. The second case may drop particles on certain cool nearby ecosystems such as wetlands and lakes, but otherwise deposits them far from the stack. The third case carries nearly all particles to distant locations ('dilution is the pollution solution' = 'dump it on your neighbor'). The fourth case causes a dirty sky for a long distance downwind, since few particles can escape either upward or downward, and deposits them far from the source. The last case deposits most particles over nearby ecosystems. Because a plume also spreads horizontally with distance, the highest-density (number per m^2) particle fallout is in the last case.

Considerable amounts of blowing particulate matter such as soil and snow accumulate by hedgerows and narrow windbreaks (chapter 6). Most accumulation is due to the decrease in wind velocity by a windbreak, rather than by direct 'impaction' on the plant surfaces (chapter 3). A forest, on the other hand, may be quite effective at absorbing moving particles, through a combination of impaction and sedimentation in decreased airflow.[455,1868,57,1075,892,596] The typical, rough upper surface of a forest causes turbulence.[388,1496] This funnels particles into (and out of) the forest interior where airflow is low, and also increases the impaction of particles on foliage.

Measurements of particle densities in boundary layer air moving across a landscape reveal the source and sink roles of local ecosystems in the mosaic. For example, a flat area of fields, forests, and villages in northern Europe illustrates the high levels of dust particles produced in villages, and the absorption of particles by a forest.[556] The cleanest air is over the downwind portion of a forest. Since a forest is nearly as impenetrable to horizontal wind as a stonewall, airflow goes up over the forest. But also strong turbulence is produced immediately downwind of the forest (chapter 6). Thus a sharp rise in particle density was recorded over an open area just downwind of a forest. This dirty air doubtless results from turbulence causing erosion of soil particles from the open area.

Gases

Most of the patterns for particles apply to gases such as SO_2, NOX, O_3, CO, and CO_2. Because the mass of a gas molecule is so small the fallout rate usually can be ignored. Gases are dependent on rising or descending air, wind, and the giant conveyor. The destination of gases is also somewhat different.[1409] Often they are simply diluted to a negligible concentration. Or they may disappear by reacting chemically with another molecule to become a third molecule. Sulfur dioxide produced by combusting fossil fuel, say in a factory, is a good example. In dry air it moves hundreds or thousands of meters in high concentrations, where it may damage plants, animals, and human respiratory systems. But in moist air, such as with high humidity, clouds, and fog, the SO_2 molecule combines with H_2O molecules to form a dilute solution of sulfuric acid, H_2SO_4. When this rains on vegetation, soil, awnings, statues, and marble gravestones, many damaging effects ensue. Some say acid rain is more acid than rain.

Numerous volumes on the causes and effects of gaseous air pollutants are available. Nevertheless, carbon dioxide warrants special mention because of current interest in its role as a greenhouse gas and in global carbon cycling, as well as its relationship to the giant conveyor.[12,1928,1162] Oceans are both sources and sinks for CO_2. However, on land the key sources are respiration by soil and vegetation, burning vegetation and fuelwood, and fossil fuel combustion for transportation, industry, and heating buildings. The key sinks are plant photosynthesis and the atmosphere.

Warm moist soils with organic matter, such as in parts of the tropics and subtropics, and during summer in colder regions, have high respiration rates. Basically, bacteria and fungi thrive on warm moist conditions to decompose dead organic matter and produce CO_2. During the day much of the CO_2 respired from the soil may be absorbed by photosynthesizing, growing vegetation at a site. But at night when plants are not photosynthesizing and wind is low, rising warm air carries most of the soil CO_2 to the conveyor belt.

Carbon dioxide is then carried to other ecosystems where it may or may not be absorbed. Photosynthesis occurs in all systems with green plants, but the rate varies widely. Vegetation with a large green surface (i.e., a high leaf-area index) and rapidly growing (i.e., adding wood or other stem, root, or leaf tissue) has the highest net productivity. Therefore, it is the greatest absorber or sink for CO_2. Thus, much CO_2 carried to a young growing forest or actively growing crop field is absorbed and turned into plant biomass.

But bare, dry, cold, or built areas can remove little CO_2 from the conveyor. Areas of old-growth vegetation in decades-long steady state absorb CO_2, yet liberate an equal amount in respiration. If the conveyor belt does not encounter a sink, CO_2 simply accumulates in the atmosphere. Gases such as CO_2 also accumulate under an inversion along with particles (Fig. 10.8).

As described earlier in this chapter, CO_2 is an 'atmospheric greenhouse gas'. In recent decades both CO_2 concentration and average temperature of the atmosphere worldwide have been significantly increasing. Carbon

341

dioxide buildup prevents heat from escaping to outer space. Some say that just as your temperature rises when you become ill, the temperature of the planet is rising. A more cautious summary of rising atmospheric CO_2 and temperature emphasizes the considerable debate as to whether one is the cause of the other.

Water

Water is transported by the planetary boundary layer in the forms of snowflakes, raindrops, droplets, and vapor.[1131] The first three forms are created locally in clouds and transported short distances. Water vapor entering the giant conveyor is largely transported near the ground, though some is mixed aloft and transported long distances. Immediately downwind of a source, such as a lake, wetland, or forest, is a plume of moist air.[747,234,126] This plume is enriched with water vapor, and in some cases is full of water droplets (i.e., appears as a cloud).

For example, in irrigated land surrounding an oasis, evapotranspiration in the upwind portion causes moister air in the downwind portion.[913] This decreases ET water loss from the downwind area. Consequently, large oases have less average water loss per unit area than small oases. One study found that a 10 km diameter (6.5 mi) oasis annually pumps out an average 42.5 cm (17 in) of water in ET, whereas a 1 km oasis pumps 47.5 cm, and a 0.1 km oasis pumps 53.0 cm of water.[913] Those are big differences in a dry area. Similarly in Russian farmland, ET pumps out 0.5 cm of water per day just upwind of a forest (Rauner 1963, cited by D. H. Miller, pers. commun.). Yet in the moist plume several kilometers downwind of the forest, ET is still reduced by 25%.

Fog droplets, often associated with pollutants, usually move at night and early morning only over a portion of a landscape, before solar radiation 'burns off' the temperature inversion and associated fog. Nevertheless, huge amounts of water may be transported from several ecosystems or a water body, and concentrated on a single spatial element.[74,1281,446] For instance, a mossy or elfin woodland atop a Puerto Rican peak is bathed in clouds, and receives approximately 10% of its water per year from droplet impaction on the plants.[756] In Hokkaido in northern Japan a 100 m wide 'fog-removing forest' collects enough droplets to clear the bottom 300 m of the planetary boundary layer, and permit solar radiation to reach rice paddies downwind.[753,1128]

Raindrops from the upper troposphere move through the boundary layer and, under common windy conditions, approach an ecosystem at a slant. This may be locally important in a dry plain where windbreaks alternate with strips of grass, and trees intercept many of the diagonally falling raindrops.[1579] Some of the water runs down the stems to roots, and thus enhances tree survival in the arid climate.

Snowflakes in mountainous topography are typically transported horizontally a few tenths of a kilometer, whereas on plains they tend to move up to a few kilometers.[1919] Snowflakes approach the vegetation or ground at a low angle. Hence, snow accumulations are highly dependent on vegetation pattern and microtopographic variations, such as concave and convex forms.

In short, water in all its forms, and the hydrologic cycle as a whole, are quite dependent on horizontal and vertical flows in the giant conveyor. In general, water is collected over a broad area and deposited in concentrated form in a small area. Movement of water from mountains to sea may involve alternating phases in streams and the planetary boundary layer. For example, it is estimated that a third to a half of the water entering the Amazon basin from the Andes is recycled before reaching the mouth of the river.[1480,1481] That is, the water is evapotranspired by vegetation and then dropped again in rain. Indeed, a particular water molecule may be recycled several times through trees and rainstorms *en route*.

Material transport from source to sink

The preceding sections have underlined the importance of the planetary boundary layer, the driving forces of energy flow and wind, and the particles and gases transported. Here we will tie these and additional threads together, by focusing on mechanisms, processes, and the overall results, in moving from a source to the conveyor to a sink. Since source or sink areas cover three levels of scale, local ecosystem, landscape, and region, generally we will simply refer to areas.

What do major source areas look like? Topographic or vegetational roughness usually indicates an important source area, because turbulence tends to launch energy and materials upward into the boundary layer (Fig. 10.9). Dry areas nearly bare of vegetation, especially if dark colored, also tend to be important sources. Parking lots, cultivated fields, and deserts are examples. Sun-facing slopes and recently burned areas are good source areas. In these locations soil radiates to heat the night air. The warm air rises and carries materials to the conveyor. For the same reason, in autumn a wetland or lake which is warmer than the surrounding ecosystems is a good source.

The troposphere transports energy, particles, and gases horizontally, and delivers them to a number of other ecosystems, landscapes, or regions. Several delivery processes are possible.[1132] If the vegetation or soil is cooler than the boundary layer, a net downward movement of heat takes place. Particle deposition occurs due to gravity. Washout of particles and gases takes place in rain. And impaction of particles and

Fig. 10.9. A changing mosaic of sources and sinks for materials moving in the boundary layer. Turbulence carries heat, particles, and gases from soil, buildings, and vegetation upward to the boundary layer, and also in the opposite direction. Turbulence changes diurnally and seasonally, and differs in magnitude over forest, field, village, valley, and ridge. Southwestern Germany. R. Forman photo.

droplets (aerosols) takes place in moving air, when they adhere to surfaces such as stems, leaves, and houses (chapter 6).

The upwind or leading edge of a sink tends to receive the greatest input.[1172,1132] This is because air reaching the downwind portion has been partially cleaned of its material. It is like removing the air filter from an automobile. The intake side is always dirtier than the outflow side.

What do major sink areas look like? Again topographically or vegetationally diverse areas extract more particles and droplets than smooth areas, due to turbulence (Fig. 10.9). Impaction is a particularly important absorber. A cold area surrounded by warm areas may be a significant sink, because there is little air rising. Indeed air may move downward under a number of conditions. A wetland or lake in early summer is an example. An actively growing young forest or crop may be a sink for CO_2. Where ET contributes to cloud formation, an area downwind is a sink for all types of material carried in the conveyor.

Finally, several overall effects of the atmospheric transport can be identified. A source area may give off or lose either ecologically desirable or undesirable resources. A sink area may be enhanced by moderate levels of input, but inhibited by high input levels. A concentration effect is common, whereby energy or material is gathered from many areas

344

and deposited in one or a few areas. However, dilution and dispersal, the opposite effect, also occur.

Local ecosystems in a landscape are scoured and bathed by the conveyor. More importantly they are interlinked by transport. Their essential attributes of productivity, biodiversity, mineral nutrient cycling, and the like are determined by and reflect the process. Many interactions among ecosystems in a neighborhood configuration, as described by the 'spatial flow principle' (chapter 9), depend on the conveyor.

Landscapes in a region are interlinked by the giant conveyor belt. Perhaps only human movement rivals in ecological importance this linkage among landscapes.

Regions also are linked by the giant conveyor. However, regions are in part defined and delimited by differences in macroclimate. Adjacent regions tend to have different air pressure and wind patterns. Therefore, flows and linkages across a boundary between regions are probably less predictable than those within a region. In other words, atmospheric linkages between landscapes and between local ecosystems appear to be more dependable than those between regions.

FLOWS IN ROUGH TERRAIN
Ridges and valleys

If you had the opportunity to create a television series for children explaining how mountainous areas work, you might provide a large number of three-dimensional models of individual mountains for the children to fit together in various arrangements. Some models are of elongated ridges and others are of rounded convex peaks or hills. You might provide valley models in the form of troughs, and concave depression models for lakes, wetlands, and internal drainage areas. The models are of different sizes. A warm movable sun is provided, and the model surfaces are heat sensitive, turning reddish when warm and bluish when cool. A fan is provided for wind, and tiny pieces of white plastic or other light-weight material are available to see where windy and sheltered areas are, and where snow might accumulate. You can decide whether the children will make thunder, rain, and floods.

The ruggedness of mountains and its effect on species movement was explored in the previous chapter. Here, we will first consider airflows with ridges and valleys, the elongated forms, and then with hills and hollows, the rounded convex and concave forms. Wind principles given in chapter 6 for windbreaks apply equally to ridges and valleys. Thus, the distance effects upwind and downwind of an isolated ridge are generally predictable for winds arriving perpendicular, diagonal, or parallel to the ridge. Several parallel ridges are like several parallel windbreaks. When they are close together, turbulence is rampant. The following patterns for diverse topographic areas are evident in summer when solar radiation causes the greatest contrasts on land.[64,63 cited by 1132] In winter, and to a lesser extent in cloudy times, incoming solar radiation is limited, air and soil temperature are both low, and wind is high. These all tend to decrease contrasts within a mountainous landscape or region.

Moist air arriving perpendicular to a high ridge is forced up over the ridgetop and continues downwind, producing dramatic differences on land along its route. This may happen on ridges hundreds to thousands of meters high, such as Coral Sea air crossing the mountains of Queensland, Australia, or Pacific air crossing California up over the Sierra Nevada.

The air pushed upward cools, the moisture condenses, and precipitation pours onto the windward slope, even producing rainforests.[1923] The now-drier air continues on over the ridge. As it moves downslope on the downwind side, the air is warmed. Here, it produces dry, even desert conditions, as in western Queensland, Australia and Nevada, USA. The dry area with descending air downwind of a mountain is called a *rain shadow*.

High wind velocity on the ridgetop (chapter 6) desiccates and shapes the vegetation, forming beautiful 'windswept' trees or even alpine tundra. Hawks and other soaring birds often forage or migrate along the windward slopes of ridges. This route takes advantage of the line of uplifting air.[298,1302] Since most migration takes place at night, towers and high structures tend to be lethal to birds, and should be excluded from such established pathways.

Ridges are also major sinks for airborne materials.[1411,994,151] Because trees are somewhat 'stair-stepped' up the windward slope, some air enters from beneath the canopy and moves diagonally up through the foliage. Thus, a large surface area is exposed for impaction of particles. Many pollutants and mineral nutrients accumulate on ridges in this manner, although others accumulate because clouds carrying particles and gases tend to remain and bathe ridges. Snow carried to ridges often accumulates in great depth just downwind, where wind velocity suddenly drops. Snow accumulations varying in depth over several orders of magnitude are typical at local horizontal scales down to meters. This is because turbulence, redeposition, and shading produce complex patterns in mountainous and hilly terrain. Nevertheless, these snow accumulation patterns are exceptionally important in determining spring flooding, and in supporting later summer streamflows often in distant low-elevation locations.

Valleys are V-shaped (due to water erosion) or U-shaped (due to a glacier) troughs that receive water and air from higher elevations. Generally, north- and south-facing slopes receive very different amounts of solar radiation. The sun-facing slope heats up and warms the air. Some of the warm air moves horizontally in the valley to the opposite, cool pole-facing slope. There, it may melt snow or heat vegetation and soil.[1127] But much of the warm air rises vertically. Ridgetop ecosystems may be warmed at night by this rising heat.[1132]

Cold air drainage describes the reverse flow of cool air that moves down slopes, usually at night. This replaces the rising warm air, and means that valley bottoms are often colder than upper slopes. For this reason experienced campers in dry rough terrain let tourists sleep alone in valley bottoms. This moving air, or ventilation, can cause frost in the valley bottom long before frost appears even on pole-facing slopes.[124,1446] Frost also forms in flat areas on calm nights due simply to loss of heat by night radiation. Cold air drainage, just as with any air movement, also transports seeds, spores, dust, particles, gases, and can even spread a forest fire down a valley.[1768,1408]

These widespread, rather familiar patterns can be nicely tied together by visualizing over a 24-hour period the longitudinal airflows along a valley bottom, combined with the vertical flows up and down the side slopes.[556,1243] Beginning shortly after sunrise, solar heating causes flows up the slopes to begin early (Fig. 10.10a). Meanwhile air continues to move longitudinally down the valley bottom, since this 'night air' is still cooler than air outside the valley. The upslope air movement is in thin layers against the slopes. Some air is also sucked downward in the middle of the valley by the longitudinal airflow. Soon the longitudinal airflow disappears, and the valley rapidly heats up. Only the upslope flows remain (plus downward vertical replacement air mid-valley).

Near midday a wind up the valley typically forms (Fig. 10.10b). Together with upslope winds, this carries evapotranspired moisture upward, sometimes contributing to the formation of cumulus clouds. By late afternoon the upslope flows cease, leaving only up-valley wind. By early evening downslope cold air drainage begins (Fig. 10.10c), and by late evening the down-the-valley airflow commences (Fig. 10.10d).

Note the phase difference. Vertical slope airflows change first, reflecting local energy flows mainly on the slopes. Longitudinal valley winds change later, reflecting broader-scale energy flows of the valley in its surrounding plain or landscape area.

These temperature-driven airflows cause a multitude of ecological effects. For example, birds (but probably not hang-gliders) may use the upslope flows throughout the day only in narrow zones close to the slopes. Nocturnal predators should find more temperature-sensitive food, such as snakes and insects, higher on a slope than in the cold valley bottom. The diversity of frost-sensitive plants is also greatest on the upper slopes.[1923] Wind-borne seeds, spores, and pollen liberated during the afternoon are likely to be carried up rather than down a valley.[1840]

347

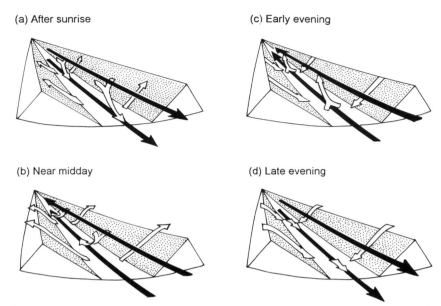

(a) After sunrise

(c) Early evening

(b) Near midday

(d) Late evening

Fig. 10.10. Reversing airflows up and down a valley and its slopes during a summer day. Black arrows = longitudinal valley wind; white arrows = wind on slopes, and mid-valley flows associated with slope winds. Adapted from Geiger (1965). Originally published in the German language by Friedr. Vieweg & Sohn, Braunschweig 1961.

Fires are blown up mountain slopes by day, and may be blown downslope at night.[48] Polluting particles and gases tend to accumulate down in the valley by night. Tree blowdowns are often concentrated in gaps at the heads of narrowing valleys, where wind is funneled and accelerated.[1496] If you build a cabin at the base of a slope, downslope airflow may mean down-chimney airflow and smoke in your cabin, but only at night.

Convex and concave surfaces

Ridges and valleys have elongated cylinder-like convex and concave surfaces. But here we focus on individual spherical mounds, hills, even volcanos, plus hollows and depressions. Convex and concave surfaces tend to have different vegetation.[641,1640] Upward projecting convexities are more common than concavities, because the latter often disappear by filling with water, soil, and wetlands. However, natural disturbance and human cutting of forests commonly leave isolated wooded patches and clearings, which have the same convex and concave airflow characteristics as hills and hollows (Fig. 10.11).

Fig. 10.11. A concavity with high-variability in meteorological conditions affecting animals and plants. White-tailed deer (*Odocoileus virginianus*) in Texas. Courtesy of USDA Soil Conservation Service.

Convex surfaces drain cold air off at night and hence are less apt to experience frost or as low temperatures as surrounding flatland. Solar input of course varies on different exposures, but convexities overall usually do not reach as high a daytime temperature as the surroundings. In short, convex surfaces have a more constant environment than a plain. Of course, like ridges, their exposed position means they are subject to greater wind velocity, advection, and impaction input of fine particulates and droplets that adhere to vegetation. However, coarser particles such as snowflakes and sand grains are swept off the convex surface to lower areas.

Concave surfaces, in contrast, are more variable environments than the surroundings (Fig. 10.11). They get colder at night and hotter by day, often by several degrees centigrade.[1243] In temperate and colder regions concavities are frost pockets, where late spring and early autumn frosts occur, resulting in a short growing season. Low-lying beaver ponds and cranberry bogs often get autumn frosts long before most of the surround-

349

ing landscape does. Yet on cool windy days wildlife can sunbathe in warm concave clearings (Fig. 10.11).

For large depressions, temperature is about the same whether they are shallow or deep.[1607] In small-diameter depressions, deep ones are colder mainly because turbulence does not reach to the bottom.[556] Yet, very-small deep depressions may be warmer than any of the previous pockets or the surrounding plain. This is because the steep side soil-banks are sources of heat, and compensate for much of the potential direct radiation to the sky by radiating some heat horizontally and downward at night. Concave surfaces receive little input from impaction. However, they may receive considerable coarse particulate matter such as snowflakes and sand, due to the reduced wind velocity in a depression.

Gentle convex and concave surfaces maintain streamline airflow (chapter 6). Steeper-sloped surfaces, however, produce turbulence. For example, a recent clearcut patch in a forest has a daytime eddy with downward airflow on one side, and upwind airflow on the opposite side.[556] Where in the clearing would you build your chimney and cabin?

Convex mounds and mountains also produce vortex winds (chapter 6). Horizontally flowing air encounters and bends around a mound, forming vortex eddies with vertical and oblique axes.[1496] This may be particularly significant ecologically, because vortices are even more effective than turbulence in removing particles from a surface. Hence just downwind of these convex forms we may expect processes such as soil erosion, snow removal that subjects vegetation to deep freezes, high ET rates, desiccation inhibiting vertebrate use, and removal of herbivorous insects from plants.

Sheet flows of water

The somewhat-linear flows of water in gullies, streams, and rivers were examined in chapter 7, and the three-dimensional or volumetric flows of water in evapotranspiration and precipitation were outlined earlier in this chapter. Here we link these water flows by briefly considering the nearly two-dimensional *sheet flows* of water in rough terrain. These occur as 'surface runoff' or 'subsurface flow' (chapter 7). Vegetation type and land use of course control the rate of sheet flow, as shown by, e.g., a large decrease found after converting deciduous to coniferous woods.[1660] Because the substrate is highly heterogeneous, flow models are extremely useful.[1576,315,316,1575,128] Principles from other domains such as electricity and heat, where flows through heterogeneous systems are routine, may also be applicable (M. W. Binford, pers. commun).

Water flowing over the surface results largely from heavy rain, rapid snowmelt, or relatively impervious substrates. The rather regular alternating convex and concave surfaces in most rough terrain provide considerable predictive ability. Sheets of water accelerate in flowing off convex slopes, and decelerate on concave surfaces. If the bottom of a depression or valley is relatively impervious, water accumulates until it drains downward, evapotranspires, or overflows the depression. Thus, depressions with puddles, ponds, or lakes act as temporary storage reservoirs for surface water. But most surface runoff runs into intermittent-flow gullies, streams, and rivers, or soaks into the soil to join subsurface flows.

Surface water runoff is also a major cause of erosion, transport of particulate materials, and sedimentation in rough terrain (chapter 7).[1917,1664,1126,1663,1661,130] Erosion typically leaves a network of intermittent stream channels over the surface of slopes.[1524,641,1643,1129,997] Many of the soil particles carried downward are dropped as sediment on lower slopes and in depressions such as wetlands and lakes.[998,181,997]

In contrast, subsurface sheet flow of water causes no erosion. However, it does carry fine particles, mineral nutrients, and other materials in solution.[1804,974,154,593,146,1409,1801,314] Subsurface water may flow through soil and other porous materials below the surface and reach a water body.[1297] Yet much more is usually evapotranspired, because the water is largely in the root zone where plants may absorb it. The more water transpired upward means less water arrives in depressions. Where ET is high, e.g., on sun-facing slopes and highly ventilated locations, often little or no subsurface flow continues downslope.

Subsurface water may also take a deeper route by penetrating into rock crevices and layers to emerge far downslope or downriver. Subsurface flow eventually emerges from the soil at a point, i.e., a 'spring', or diffuses out over a broader area as a 'seepage' (Chapter 7). Springs and seepages are often the concave ends of tiny valleys, where the water forms a first-order stream. As fishermen well know, a spring or seepage may also enter directly into a lake or river underwater. During summer when lake water is warm, these submerged input locations are good spots to catch cool-water fish such as trout.

Finally, with the preceding principles, let us return to the three-dimensional models created for the television series. Using different arrangements of the models, wind directions, and sun positions, we should be able to identify the worst and best locations for vertebrate diversity, game populations, rare plants, particular soil conditions, and water regimes, including massive snow accumulations and floods. Adding information from chapters 13 and 14 we will also be able to pinpoint the best and worst locations for buildings in rough terrain, including your cabin.

DISTURBANCES IN MOSAICS

Disturbance is an event that significantly alters the pattern of variation in the structure or function of a system.[1320,501] Examples of natural dis-

Fig. 10.12. Volcanic explosion as a natural disturbance that flattened a centuries-old forest. Mount St Helens, Washington, USA in 1980. Coniferous forest dominated by Douglas-fir (*Pseudotsuga menziesii*) trees c. 70 m (220 ft) tall. Photos courtesy of USDA Soil Conservation Service.

turbance are a fire that kills wetland vegetation, a severe-air-pollution event that eliminates key insects from a landscape, and a volcanic explosion that flattens a forest (Fig. 10.12). These examples are of brief acute increases in some component of the physical environment. Brief acute decreases, such as drought in a usually moist area or the absence of fire in a fire-prone ecosystem, are also disturbances. The rapid increase or decrease of a species, such as pests, diseases, herbivores, predators, and exotic species, can also be disturbances, but these are included in the following chapter 11. While most disturbance studies are on effects within an ecosystem[986,83,1160,1463,1320], here we focus on disturbance in an existing mosaic.[899,1742,878,487] Some disturbances spread only within an ecosystem, while others spread within and between ecosystems.[1748]

Disturbances are normal, though infrequent, in an ecological system. If acute environmental increases or decreases are frequent, species adapt to them.[1729,658,484,588,1730,1320,1038] That is, over time the individuals of most generations experience the events, and only the resistant individuals survive, reproduce, and pass their genes to the next generation.

Thus, periodic, acute environmental changes become part of the ongoing background pattern of variation for the system. (If changes become chronic rather than periodic and acute, they may be considered to be a 'stress' rather than disturbances.) A fire is not a disturbance in a grass-land, pine forest, or Mediterranean-type ecosystem that experiences fire every few years.[658,893,484,1730,496] Persistence of the plants and animals depends just as much on fire as on soil, water, and CO_2. Fire prevention would be a disturbance, since the species present would succumb to outside invaders.[1009,861,496] On the other hand, fire is an obvious disturb-ance in a system where fire is rare.[984,1410] Similarly, the presence of old-growth forest suggests that volcanic explosions are rare (Fig. 10.12), since tall trees are improbable adaptations to such events.

Adaptation depends not just on frequency, but on the regularity or unpredictability of environmental changes[279,769] (chapter 12). Species may adapt to even occasional changes if they are regular or cyclic. Thus insects that only live for months and only meet winter once, die as indi-viduals, but do fine as a species, because winter comes regularly once a year.

In the following discussion four common categories of disturbance are illustrated: fire, water, windstorm, and air pollution. The examples are listed or described to provide insights for three questions that will be addressed in the last section of this chapter. Does natural disturbance increase mosaic heterogeneity and structure? How does landscape struc-ture affect the spread of disturbance? And how can we design landscapes to minimize disturbance spread?

Fire

(1) William Romme and Dennis Knight studied fire history in coniferous forest of Yellowstone Park (USA) and found[1439,1440,880]: essentially no over-lap in burned areas since 1600; time since last burn (e.g., fuel load) is a good predictor of fire susceptibility; the shapes of burned areas (chapter 4) had an average length-to-width ratio of 2.5:1 (range 1.1–3.7:1), with wavy borders and about three major lobes.

(2) David R. Foster (1983) studied southeastern Labrador and found: white birch (*Betula papyrifera*) areas up to several km² have sharp bor-ders with the adjoining conifer forest, and are restricted to steep slopes that have burned in the previous 110 years; and for 4000 years the total amount of burned area has remained in relative equilibrium[452], though burned areas move around within the limited topographically rough por-tions of the landscape.

(3) Monica Turner and Susan Bratton (1987) studied a coastal barrier island in Georgia (USA) and found: fire starts in open habitats, moves into wooded areas, and is a major determinant of landscape heterogen-eity; wild horses spend most time in open habitats and move into wooded habitats, whereas deer (*Odocoileus*) exhibit the opposite pattern; and movement of animals apparently increases with boundary length and landscape heterogeneity.

(4) R. Forman and Ralph Boerner (1981) studied the Pine Barrens of New Jersey (USA) and found: a coarse-grained mosaic of burned variable-sized patches (average 36 ha) had a fine-scale mosaic of rather constant-sized 6 ha patches superimposed on it (i.e., a fine-scale pattern created since the 1940s when effective fire control was introduced); a genetically-distinct race of a fire-adapted pine species (*Pinus rigida*) dominates an area with the fewest natural stream-corridor firebreaks[483,588]; and a landscape homogenizing effect appears in severe droughts about every 20–25 years, when extensive hot fires kill stream-corridor vegetation.

(5) Jerry Franklin and R. Forman (1987) modeled Pacific Northwest (USA) forest landscapes and concluded that: both fire susceptibility and fire spread are greatest early in the rotation of a dispersed-patch cutting regime, due to drying out of the moist forest with scattered clearings; and several alternative strategies of spatially arranging cuts would markedly reduce this disturbance spread.

(6) Roger Suffling and colleagues (1988) examined fire histories in boreal forest landscapes of northwestern Ontario and found that: intermediate levels of forest fire are associated with highest landscape heterogeneity; and that fire control in fireprone landscapes increases landscape heterogeneity, but decreases heterogeneity in landscapes with less-frequent fire.

(7) Malcolm Hunter, Jr. (1993) examined boreal fire histories in Quebec and Labrador[506,1291] and found that: the average natural-fire sizes were 12 710 ha and 7 764 ha, respectively; yet in areas dissected by lakes and rivers average fire sizes were 322 and 770 ha, respectively.

Many other useful studies on fire patterns at the landscape scale are available.[1916,658,709,1932,978,484,1730,613,279,78] The preceding fire patterns, plus additional disturbance patterns following, are presented as background to address questions of landscape disturbance and design at the end of this chapter.

Water

Giant floods that escape levees or floodplains often extend widely over a landscape area. Nutrient-rich sediment is spread in patchy deposits (chapter 7). Water velocity is often low, but because the land has rarely experienced these exceptional floods, the removal of biomass and damage to human structures may be great. The spread of floodwater depends on topographic relief, especially elongated ridges, channels, and levees. Local soil type is also relevant since water pressure can undermine a barrier at a single point, and open up an additional area to flooding. In dry areas 'stream capture' occurs, whereby during a flood the surface topography is rearranged so water from one stream flows into an adjacent stream. The resulting dendritic-network patterns differ considerably from the pre-flood pattern.

A severe drought may decrease landscape heterogeneity, by selectively killing wetland or high-moisture-dependent species. The loss of these least-drought-resistant species thus favors the spread of widespread upland species. However, where wetlands cover much of the land, a drought may increase heterogeneity.

Hailstorms affect relatively small areas within a landscape, and thus tend to temporarily increase heterogeneity. In contrast, unusual snow events often blanket large portions of a landscape or even regions, thus tending to temporarily decrease heterogeneity. Contrast among local ecosystems however is often increased by snow. This is because large differences in snow accumulation are created by local topographic and vegetation nuances. Heavy snow accumulation translates into spring and summer streamflow and soil moisture. Snowmelt in spring, and especially when rain falls on snow, may cause huge floods in distant landscapes or regions.

Windstorm

'Tornados' are vortex airflows with extreme wind velocities (to 150 m/s) that commonly flatten trees or forests, and damage smaller plants and animals. Generally they last minutes at a spot and affect an area only tens to hundreds of meters wide, and 400 m or longer (Fig. 10.13) (T. Webb III, pers. commun.).

At the other end of the scale of windstorm intensity are moderate winds that topple trees only because of special local topographic or vegetational conditions.[1361,1496,508] Trees in saturated soils, such as from a soaking rain, are susceptible to blowdown. Trees located at the ends of funnel-like spatial patterns are susceptible because airflow accelerates there (chapter 6). Hence, on a mountain ridge, trees at a gap or pass at the head of a valley are susceptible, because wind is funneled up the narrowing valley.[1496] Similarly, trees next to clearcuts are susceptible where wind is channeled or funneled by diagonal forest borders.[15,518] Like the tornado these blowdowns normally only affect a local ecosystem.

More instructive at the regional and landscape levels are hurricanes or cylones that are kilometers to tens of kilometers wide, and whose high winds last for hours or days.[1563] David Foster and colleagues have published a highly informative series of studies covering hundreds of square kilometers from the Harvard Forest to Mt. Pisgah in southern New England.[507,508,510] Based on patterns from a well-documented 1938 hurricane, approximately a quarter of the predominately-forest area had severe tree damage (mainly blowdowns), and another 50% experienced moderate damage.

355

Fig. 10.13. Natural tornado disturbance usually increasing heterogeneity by creating elongated patches at the scale of local ecosystems. Probably short-grass plains in Oklahoma. Courtesy of USDA Soil Conservation Service.

In agricultural and young forest areas of this gently rolling landscape few trees are flattened (Fig. 10.14). With a distinct southeast wind, south- and east-facing slopes suffer heavy forest damage, while northwest slopes are protected.[1203] Trees at the northwest ends of ponds and lakes are heavily damaged, while those on south and east shores are little affected (Fig. 10.14). In this way landscape heterogeneity increases.

The height, composition, age, and density of a forest stand also affects its susceptibility to wind effects. Tall stands and trees are most likely to blow down. Conifer stands are more susceptible than deciduous woods (Fig. 10.14). A canopy of fast-growing pioneer species is more apt to fall than one of slow-growing later-successional species. To a lesser extent old trees and low-density stands correlate with severe damage. In short, three variables provide high predictive ability for susceptibility to hurricane damage across a diverse landscape: exposure direction, height of vegetation, and species composition of the canopy.[507,508,510]

Patches created by tree blowdown varied up to 3.25 km², but most were small, less than 10 ha (25 acres). These blowdown patches, however, contained variability in damage intensity. The largest contiguous area of the same damage intensity was 35 ha, and most such relatively

356

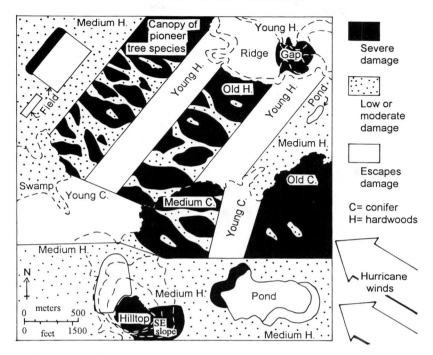

Fig. 10.14. Expected results of a severe hurricane on a landscape. Young, medium-age, and old stands of conifers or hardwoods represent a hypothetical pre-disturbance mosaic in southern New England (USA). Conifer = *Pinus strobus*; hardwoods = *Quercus, Acer, Betula, Fagus, Carya* (and *Tsuga*); pioneers = *Pinus, Populus, Betula papyrifera*. Predicted damage distribution is primarily based on studies of Foster (1988a, 1988b) and Foster & Boose (1992).

homogeneous blowdowns were less than 2 ha. Large blowdown patches were mainly in older stands, conifer stands, and exposed stands. Most tree fall by lakes and existing clearings was within 100 m of their boundary (Fig. 10.14).[1110,15]

In this manner landscape heterogeneity was increased by the disturbance. The particular landscape structure has no effect on the 100 km wide swath of high-velocity airflow, but it strongly affected the amount and spread of blowdowns. If all tall trees, conifer stands, and canopies of pioneer species had been in protected locations, history would read differently.

Air pollution

Chronic air pollution, including its effect on plants and aquatic systems over years and decades, continues to be a key issue.[152,975] Some species apparently adapt[565], and some are inhibited or die. Brief acute-pollutant-buildup events are well documented, such as for atmospheric acidity and lead inputs.[975,1587,1838] Yet the ecological effects of these disturbances are little known.

For instance, during August 7 to 13, 1984 Gene Likens, F. Herbert Bormann, and colleagues recorded an acidic cloud/fog water event that covered whole regions.[1838,152] From the mountains of Virginia and New York to Acadia National Park, Maine, the pH of cloud moisture was measured at an extemely low 2.8 to 3.09 (15–20 × more acid than usual). Very-high average ozone levels of 0.09 ppm were recorded. And the concentrations of sulfate and nitrate were 7 to 43 times greater than the expected averages in precipitation. All of these acute pollutant levels are known to damage many plant species in greenhouse conditions. Nevertheless, it is unknown what effects this disturbance event had on the natural vegetation or animals.

Most conifers are highly sensitive to SO_2 pollution. Lichens have been used for decades as general assays of urban pollution levels in many European cities, which are sometimes described as lichen deserts.[565] Symptoms of SO_2 damage in some higher plants appear within hours of an experimental acute-pollution event, and may be severe within days.[176,352]

An experiment applying 1.0 ppm SO_2 for 4 h to an enclosed natural old-field community at Hutcheson Memorial Forest, New Jersey (USA) in mid-June resulted in a 17% reduction in photosynthetic surface totaled over the remaining growing season.[291] A second acute SO_2 exposure three days later resulted in a total of 24% reduction, whereas a second exposure (on different plots) 15 days after the first produced a 32% drop. In this last most-severe case, sensitive regrowth tissue of the plants was damaged by the second SO_2 application. All three treatments resulted in species composition change, especially an increase in grass cover, for the remainder of the growing season. An acute air pollution event can be expected to have important effects on a natural terrestrial ecosystem, though evidence apparently is scarce.[423]

Earlier in this chapter we saw how specific ecosystems in a mosaic, such as forests, ridges, and frequent-inversion locations, are effective sinks within the mosaic. Such locations are therefore most likely to experience air pollution. Gases especially may be damaging because they bathe organisms and enter through pores. Particulates are least likely to cause damage in brief acute events, because their sunlight-blocking is minimal, and most particulates quickly reach the soil. Radioactive fallout may cause damage to organisms wherever it is deposited. And finally, in 'ozone holes' the ultraviolet radiation penetrating can damage organisms within a landscape or region.

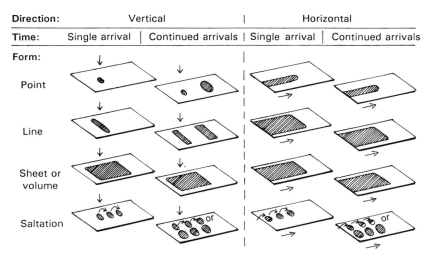

Direction:	Vertical		Horizontal	
Time:	Single arrival	Continued arrivals	Single arrival	Continued arrivals

Fig. 10.15. 'Arrival models' comparing areas altered by inputs that vary in direction, time, and form. Inputs may be energy, material, species, or disturbance. The area affected by continued arrivals depends on the absorptive capacity of the land. Adapted from Kangas (1989).

SPATIAL PATTERN AND DISTURBANCE SPREAD

In tying some of the preceding threads together we can identify hypotheses and some patterns. These are organized around the three questions posed in the preceding section.

Does natural disturbance increase mosaic heterogeneity and structure?

Disturbance is characterized by a large energy change, that may arrive from different directions and in various forms.[830] *Arrival models* define the predominant direction and form of inputs (Fig. 10.15). A tornado in a linear or column form touches down at a spot. A flash flood at the base of mountains arrives 'end first', as a line or cylinder moving along the surface. A line of fire moves along the surface as a front, and passes a spot.[1230,1371,1372] An essentially two-dimensional sheet of water eroding a mountainside or spreading over flatland arrives as a front, but persists in its effect by blanketing the area (Fig. 10.15). A three-dimensional volume of snow or rain pours from above, and also persists by blanketing the area. Alternatively, as in species dispersal, a large energy change such as fire or windstorm can jump from spot to spot ('saltation'), rather than moving smoothly across a surface.[501] And finally, the width of a disturbance almost always changes as the disturbance moves, leaving a distinctive squiggly bordered signature on the land.

An occasional extensive fire tends to homogenize the land by reducing different types of vegetation biomass to a similar-appearing blanket of ash.[710,496] Yet most fires are small, and increase heterogeneity.[49] A high fire frequency, as well as a low fire frequency, may decrease heterogeneity, whereas an intermediate disturbance frequency is associated with a more heterogeneous mosaic.

Several other fire patterns tend to increase landscape heterogeneity. Fire usually makes patches with a few major lobes, produces wavy boundaries, and increases boundary length (chapters 3 and 5). High winds often produce distinctive long, narrow burned patches (Fig. 10.15). Fire produces a high variance in patch size, with many tiny patches and a few huge patches. Variations in fire frequency in different ecosystems lead to vegetation differences based on species composition and fire-adaptations.[484]

Floods, severe droughts, and heavy snows all tend to homogenize by blanketing a mosaic, and reducing extremes in biomass and species composition. However, in each case the opposite effect of increasing heterogeneity may also take place. Floods drop patches of different sediment types in a fine-scale mosaic. Snow accumulates in highly variable depths due to local topographic and vegetational differences. Although severe drought selectively removes wetland species, in wetland landscapes this tends to increase heterogeneity. Hailstorms generally damage small areas. Melting snow, especially with rain-on-snow events, causes floods far downstream. During severe droughts vertebrate populations aggregate around water holes. In short, although some water-related disturbances reduce heterogeneity, most increase it.

A rare acute air-pollution event that blankets a portion of a landscape could decrease heterogeneity by selectively damaging certain dominant species. However, atmospheric disturbances generally increase mosaic heterogeneity. Tornados damage small areas, often leaving a distinctive elongated path. Moderate-windstorm blowdowns resulting from funneling wind also affect small areas. Hurricanes flatten many small patches but few large areas. Their effect is primarily on local exposed sites, such as particular slopes and sides of lakes. They also primarily affect patches of tall trees and only certain species. And air pollution is concentrated on certain ecosystems in a landscape.

The answer to the overall question therefore is that some disturbances decrease heterogeneity. However, the bulk of them create more heterogeneity and structure in a landscape mosaic.

How does landscape structure affect the spread of disturbance?

The preceding patterns of disturbance spread suggest that certain local ecosystems are especially 'susceptible' to disturbance initiation, whereas others are 'resistant'.[151,508] In addition, certain spatial elements 'channel' or enhance disturbance spread, while others are 'barriers' or filters that inhibit its spread.

A patch may be particularly susceptible or particularly resistant to any type of disturbance. Exposed slopes and margins of lakes and clearings are particularly susceptible for the initiation of forest blowdowns. Conversely, the opposite slopes and margins are resistant to hurricanes. Clearings that dry out an area of moist forest create edges that are susceptible to fire. Certain habitats, such as steep dry slopes in a moist landscape or grassland on a barrier island, frequently are sources of fire that may spread to adjacent types. Wetlands, lakes, and rivers of course are fire resistant. Deep water holes resist drought effects. Certain slopes and vegetation structures have little snow accumulation, while others are deeply buried. Recently burned patches tend to be fire resistant.[658,1439] These results suggest that all local ecosystems in a landscape can be mapped for susceptibility and resistance to disturbance.

Barriers and channels affecting disturbance spread are common, but apparently less studied. Wetlands, lakes, and rivers are barriers that prevent extensive fires. Long ridges and levees are barriers to floods, while troughs and valleys are channels. Diverse topography and vegetational structure provide friction inhibiting floods. Straight boundaries channel wind and waterflow, whereas curvilinear boundaries provide turbulence that reduces the overall flow rate. Thus, barriers to and channels for the spread of disturbance can be added to the map of susceptibility and resistance in a landscape.

How can we design landscapes to minimize disturbance spread?

Most disturbance increases heterogeneity of the mosaic, yet occasionally disturbance spreads extensively to create a more homogeneous landscape. In some areas large disturbances are important natural processes, with which certain species evolved, and which are required for maintenance of the species.

In other areas minimizing the spread of disturbance may be an important planning or conservation goal. For this, three approaches are suggested: minimize site conditions for the initiation of disturbance; increase mosaic heterogeneity to reduce spread; and increase the filters or barriers to spread.

Minimizing disturbance initiation begins with a map of the local ecosystems particularly susceptible to the various disturbances. Actions then might include eliminating such sites or, e.g., for fire, establishing a wetland or burned firebreak around or upwind of the site. Boundary length of logged areas can be minimized to reduce fire in moist forest. Keep tall trees and certain susceptible canopy species off hurricane-exposed slopes, as well as the margins of lakes and clearings. Logging should be done to minimize the funneling of wind that causes blowdowns in specific spots. And the main way to prevent floods is to maintain or reestablish the connectivity of vegetation corridors along first-order streams, as well as gullies with intermittent flow, across the drainage basin.

Once initiated, most disturbances are inhibited in their spread by heterogeneity in a mosaic.[1742,489] Fire and blowdowns may expand until encountering a resistant patch. In a fire-prone area, human fire control increases heterogeneity.[484,1651] But in a rarely burned area, fire control decreases heterogeneity, leading to the potential for an extensive fire. Logging as a creator of heterogeneity is more effective in more homogeneous areas away from, e.g., lakes and wetlands as natural firebreaks. In dry landscapes maintaining a wide distribution of water holes buffers the severe drought effect on vertebrate and invertebrate populations. Finally, 'boundary-crossing frequency'[501] has been suggested as an assay of disturbance spread, because boundaries between patches appear to be filters or barriers to many disturbances (chapter 8).[1439,506,508,492]

One may create many types of filters and barriers to disturbance spread. A straight patch boundary perpendicular to the direction of spread is expected to be a more-effective barrier than a curvilinear boundary (chapter 3). In addition to patch boundaries as filters, a resistant corridor perpendicular to disturbance movement is useful. If a disturbance spreads along a corridor connecting patches (chapter 6), the corridor may be an effective bottleneck, where fire or pest spread can be arrested.[490] The same, but greater, opportunity exists for arresting disturbance moving along a series of stepping stones (chapter 6). A straight boundary or corridor could be used to deflect a disturbance diagonally away from a key location. In this manner levees may block and channelize floodwater. Maintaining woody vegetation in floodplains provides friction against waterflow, resulting in downriver floods that last longer but do not rise as high.

While overall heterogeneity is a general inhibitor of disturbance spread, much more effective is landscape structure, the specific arrangement of patches, corridors, and matrix.[1742,1332] Targeting susceptible elements and the arrangement of ecosystems immediately surrounding them is one major step. The other major step is the focus on the arrangement of corridors and patch boundaries as barriers, relative to the directions of disturbance movement. Many of the preceding observations probably apply not only to natural disturbance, but also to the spread of human activities on land. Finally, it should be clear that the subject is in its infancy, and although portions emerge clearly, a complete synthesis awaits the future.

This chapter on wind and water flows in mosaics has emphasized the central importance of horizontal transport in the troposphere, the giant conveyor. This horizontal transport process interacts with the mosaic through turbulence above local source and sink ecosystems. In this way every ecosystem in a landscape, as well as every landscape in a region, is inextricably linked. In rough terrain local ecosystems additionally interact through up and down flows. Disturbances, as acute increases or decreases in energy or material, also spread across the landscape, but are somewhat amenable to control through design and planning. The next chapter now explores the movement of animals and plants across a land mosaic.

11

Species movement in mosaics

In our most trivial walks, we are constantly, though unconsciously, steering like pilots by certain well-known beacons and headlands, and if we go beyond our usual course we still carry in our minds the bearing of some neighboring cape; and not till we are completely lost . . . do we begin to find ourselves, and realize where we are and the infinite extent of our relations.

Henry David Thoreau, Walden, *1854*

Dragons have played a key role in history and now are nearly extinct. This keystone species (or guild of species) apparently had population centers at least in Ethiopia, northern Europe, and China.[738] The large ugly scaly reptile-relatives generally lived in deep caves, lakes, or rivers, and foraged singly across landscapes, both on the ground and in the air. They were attracted to virgins in distress, and repelled by heroes with swords. Chinese dragons were frightened by iron, centipedes, wax, leaves of the lien tree (*Melia*), and other objects. Populations fluctuated, with sighting records in Europe higher during the Middle Ages, in the seventeenth century when classification treatises were published, and in the nineteenth century when Darwin's evolutionary theory suggested relationships with known dinosaurs, such as pterodactyls. With increased human population and technologies including satellite images, remote habitats have decreased and dragon populations crashed. Most recent sighting reports have been of Chinese dragons, in areas where the species has traditionally received generosity and respect. But alas, too little information is available on how dinosaurs and these mythological dragons used pattern in moving across a landscape or region, the subject of this chapter. Therefore we must learn from movements of today's less fearsome animals and plants.

The subject of species movement at this scale is critical in many arenas. Conserving biodiversity depends primarily on landscape and regional pattern. Forest regeneration in most regions depends on seed dispersal. Invasions of exotic species are inhibited or enhanced by landscape pattern. Plagues of insects or herds of herbivores attack crop fields one after another. And migratory fish move long distances through heterogeneous river systems.

This chapter, along with the preceding one, is about movement across mosaics. It assumes familiarity with concepts in chapters 2, 3, 5, 8 and 9. Indeed, chapter 8 considers the resistance to species movement by a landscape, and chapter 9 examines species movement in mountains. No overall reviews of species movement in mosaics exist, but the following are useful for particular topics: animal movement[1669]; movement among patches and corridors[501,1490,1492,1228]; metapopulations[1257]; species invasions[451, 1159, 624]; and gene flow.[1595,1536] Vertebrates and vascular plants are stars of the chapter, since relevant studies on tiny organisms are scarce.[1911,743,1327] Population ecology is outlined in Appendix B of chapter 2.

This chapter begins with general types, models, and principles of species movement. Metapopulations, with species extinction and recolonization, are presented next. Genetic dimensions are introduced in the context of mosaic pattern. Then, species movement between non-adjacent spatial elements is explored. Finally, we consider the introduction of non-native species and livestock.

MOVEMENT, MOSAIC, AND MODEL CHARACTERISTICS

Types of species movement

Animals and plants

Several important types of animal and plant movement are easily recognized.[80,1098,537,1704,1669] Most vertebrates have a *home range*, the area covered in day-to-day movements for food, shelter, etc. (Fig. 11.1). Many of these species also have a 'territory', a smaller area around the den or nest that is defended, especially against other individuals of the same species. *Animal dispersal* refers to essentially permanent movement away from the area of birth or residence. For example, insects commonly fly or are blown varying distances away. Subadult vertebrates leave a home range, and establish a new home range often at distances several times the diameter of the original home range.

Migration usually refers to the cyclic movement of animals from the birth area to another area and back again (other definitions are used in genetics, river dynamics, and biogeography). Thus, migration corresponds to cyclic environmental changes, such as the changing seasons (Fig. 11.1b). Many birds migrate latitudinally between cool and warm

Fig. 11.1. Species with home-range and migration movement. Left: opossum (*Didelphis virginiana*) with young in home range; the only genus of marsupials (pouched mammals) in North America. Maryland. Courtesy of US Fish and Wildlife Service. Right: great blue heron (*Ardea herodias*) c. 120 cm (4 ft) tall; it may migrate hundreds of kilometers between winter and summer ranges. Louisiana. Courtesy of USDA Soil Conservation Service.

regions, while others in a mountainous region migrate altitudinally between high and low landscapes. Movements that do not fit neatly into the preceding categories of animal movement are usually called 'wandering' or nomadic or vagrant behavior, though probably this mainly reflects our lack of understanding. For example, many bird species exhibit wandering behavior through the half dozen or so seasons of the year in Australia.

Each of these movements can be subdivided by type of activity.[1700,743,1669,380] *Foraging* refers to the searching for a resource, usually food. The behavior and routes of foraging for food, water, or nest material may differ markedly. Other movements also occur, as in avoidance, escape, and breeding behaviors.

Scale differences are evident within any of these animal movements. Thus local movement of a wolf (*Canis lupus*) near a fallen tree may differ from broad-scale movement along a frozen stream, though both are home-range movements. Most latitudinally migrating birds fly long straight distances at night, alternating with short convoluted routes while foraging for energy-rich food by day (Fig. 11.1b).

Plant dispersal is simpler, because, as some say, plants do not behave. Plant seeds, spores, and other propagules are moved by 'vectors': wind, water, flying animals, running animals, and humans (chapter 3). *Short-distance dispersal* and *long-distance dispersal* of plants are usually differentiated.[1772,681,1177] Short-distance means dissemination to a site near the parent plant or, more ecologically useful, to a location within the breeding population of the parent. Similarly, long-distance dispersal means to a site beyond the parent plant's population. Hence, long-distance plant dispersal is required for a population to spread.

In dispersal, heavy fruits and seeds rapidly decrease in density with distance from a parent plant.[1708,1635,757] For the few of these seeds carried long distances by vectors, the considerable stored energy in the seed increases the probability of successful plant establishment. In contrast, light-weight seeds, pollen, and spores often move long distance in prodigious numbers, especially if wind dispersed. Such seeds, though, have little stored energy for successful establishment. Thus, seed dispersal patterns are somewhat predictable, but successful establishment and reproduction of a plant depend on conditions where a seed lands.

Navigating and searching

Have you ever walked at random? A statistically-random walk could be done by picking a series of compass directions and distances from a table of random numbers or the rolls of dice, and following those precisely. No one and no animal takes this 'random walk'. Indeed, the highly patchy environment makes such a trajectory extremely difficult. Nevertheless, models assuming random walks have been used to simplify complex movements. When each individual of a population is assumed to move at random the process is called *diffusion*, and the model of it a diffusion model.[1244,115,836,163,1078] The process is analogous to what happens after breaking a perfume bottle in a still classroom: luscious floral molecules passively diffuse across the room.

In foraging, where the location of a target resource is unknown, an animal may cruise along at a continuous steady rate. This search route tends to have relatively few turns, and cover a large-diameter area. More common, though, is a *saltatory movement*, where an object such as the animal alternates moves and pauses.[501] The pauses increase the probability of changes in direction. Therefore, routes are convoluted, and may only cover a small-diameter area. Resting, feeding, ambush of prey, reproduction, and so on usually require pauses. Any object moving across the land, including a soil particle or mineral nutrient, may exhibit saltation[501]; for species it has also been called 'jump dispersal'.[1323]

Some say that women navigate through a landscape primarily using landmarks, such as buildings, road intersections, and hills, whereas men navigate primarily by 'following their noses', i.e., by general directions such as southward, windward, or uphill. Apparently, the success rate in reaching targets or destinations is equal. In either case the navigator has learned some information about the landscape and target in advance. Learning the non-randomly located cues in the environment is a key to efficient movement. Inexperienced young animals learn fast, and do better than random walks within a home range or in migration, yet still experience high mortality (Fig. 11.1a). Follow-the-leader is normally a good game plan in unfamiliar terrain. Thus 'traplining', the following of an individual which has learned the route, is a familiar tactic and an efficient process for, e.g., nectar-feeding butterflies that must locate certain flowers.

Local searches and movements take place within a habitat, but 'multi-habitat species' (chapter 3) moving between habitats provide a focus in mosaics. Such species cross boundaries, where movement rates usually decrease due to behavioral caution (chapter 3). Also, direction often changes at boundaries. Other characteristics of people and animals moving across boundaries are reported, but remain just an interesting research frontier.[212,1613,366] Nevertheless, *boundary-crossing frequency*, the number of borders between ecosystems per unit length along a route, is a simple measure of 'landscape resistance' to movement or flow[501,877] (chapter 8). In designing natural resource reserves for wildlife or wildlife viewing, alternative routes of movement can be evaluated in this manner.

Finally, movement between two points must be considered in terms of distance.[501] A straight line measured in meters or kilometers is the shortest distance. Yet 'time-distance' measured in minutes and days is generally more useful, especially in a mosaic. This is the shortest time required to go from one point to another; the route taken may be straight or convoluted. A convoluted route through suitable habitat may be much faster than a straight route through less suitable habitat, or a route with high boundary-crossing frequency.

'Topological space', as illustrated in many train and airline route maps, is an alternative way of considering movement routes.[1670,995,501] Here, the order of points, lines, and patches is presented precisely, but distance among them is elastic and only approximate. Such a view of space focuses on the origins and destinations, and minimizes the curvilinearity in navigation, search, and learned movement routes.

Population arrivals

Individuals of a population arrive in an area in the same ways as disturbances described in the 'arrival models'[830] (chapter 10). The population may arrive 'end first', as a front line, as a two-dimensional sheet, or as a three-dimensional volume. A population may arrive horizontally across the surface, or vertically from above as illustrated in Fig. 11.2. And it often skips areas when spreading.

The movement or spread of a population strongly depends on where the population arrives in the mosaic structure. For example, a population of light-weight seeds or floating spiders may arrive from above in rain washout to a spot. If it arrives within a large patch of suitable habitat, the population spreads and thrives (Fig. 11.2a). But if arrival occurs

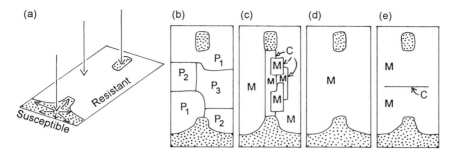

Fig. 11.2. Arrival and spread of a population in a mosaic. (a) Arrival in an area resistant to invasion, and in two spots susceptible to invasion, one of which is surrounded by suitable habitat for subsequent spread. (b) Spread between shaded areas requires crossing a patchy area (no corridors); note the different sequences of patch types (P1, P2, P3) as possible routes. (c) Spread along network corridors (C) in a matrix (M). (d) Spread across a matrix. (e) Spread across a matrix where a corridor is a barrier; only two possible routes.

in a small suitable patch or an unsuitable habitat, little population movement or persistence may ensue. In other words, the spread of a population depends on the arrival form relative to the existing mosaic.

Key mosaic attributes

The preceding section explored characteristics of individuals and populations affecting movement across a landscape or regional mosaic. Here, we focus more directly on certain attributes of the mosaic that control movement, no matter whether it is home-range, dispersal, migration, diffusion, saltatory, or arrival movement.

Spatial elements vary widely in their *suitability* or hospitability for an individual or population.[1669,969] To simplify we often classify a patch, corridor, or matrix as more suitable, less suitable, or unsuitable (e.g., see Fig. 6.11). Hence, the arrangement of suitable and unsuitable ecosystems is a major determinant of the spread of species (Fig. 11.2b). It is quite analogous to the distribution of 'susceptible' and 'resistant' elements affecting the spread of disturbance (chapter 10).

However, unlike disturbance, vertebrates have 'targets' which generally are suitable patches containing food or shelter. A target may be a specific previously encountered location, or any suitable location encountered while searching for a resource. The animal may search for the target along a highly convoluted route, or alternatively, may have learned the shortest time-distance route to the target.

The mosaic, though, is the template across which the search or learned movement must take place. Patchy, network, and matrix are the three

369

commonest templates (Fig. 11.2) (chapter 8). Navigating through a patchy area involves a high boundary-crossing frequency, as well as usually alternating between more-suitable and less-suitable patches. On the other hand, navigating along suitable corridors of a network, or through a suitable matrix, requires little boundary crossing. The matrix generally has wider connections and a more-direct route than a network. Depending on its relative suitability, the matrix therefore may permit more rapid movement.

The spread of populations, like that of disturbance, is especially dependent on the arrangement of corridors. These may act as conduits to channel individuals. Yet, equally pertinent is their function as barriers or filters to movement[1381] (Fig. 11.2e). A medium-sized unsuitable patch usually can be circumvented, but encountering a medium-length unsuitable corridor may require a major detour.

Many other attributes of a mosaic control movements. Narrows, bottlenecks, and stepping stones affect movement rate (chapter 5). The connectivity of a network or matrix affects rate (chapter 8). Perforation of a matrix affects rate (chapter 8). Patch shape, and especially its orientation angle relative to the moving organism, affects rate (chapter 4). Peninsular interdigitation and its orientation angle are critical (chapters 3 and 9). For some vertebrates the existence of escape routes or terrain, available in case of predators or disturbances, is a central factor in habitat selection.[917]

In short, movement of individuals and populations depends on the major spatial attributes of each mosaic.[1484,497] The proportion of species affected is dependent on spatial scale. It appears that some species are sensitive to the arrangement of landscapes in a region, many species to the pattern of local ecosystems or land uses in a landscape, and most species to the arrangement of ecosystems in a neighborhood or configuration (chapter 9).

Population models

The modeling of movement in populations has had a long and rich history. Animal movement, human demography, epidemiology, and gene-flow models have been developed by scholars in many disciplines. Most early models in ecology assume simple random diffusion and a homogeneous surface.[1244] The homogeneity assumption is useful within habitats. Movements of course can be measured at different scales, and may appear random at one scale, though highly non-random at other scales.[832,1003,1667,31] 'Diffusion-reaction' models are a biological variant of diffusion models, where individuals reaching a target change in some way, such as dying or reproducing.[955] Diffusion models have also been combined with other models (including predator-prey models[1143],

370

Markov series, percolation patterns[543,545], and fractal geometry[1652,1138,803]). Overall, the results of diffusion models do not fit well with widely-observed patterns in nature.[648,697,1760,464] Since the assumption of spatial homogeneity is inconsistent with the central mosaic attribute of landscapes and regions, these background models are not considered further here.

The assumption of a heterogeneous environment makes diffusion models more realistic.[954] This is a distinct improvement, yet, as seen in chapter 9, numerous spatial arrangements can produce the same level of heterogeneity. Therefore, models of movement that include explicit spatial arrangement are needed for landscape and regional mosaics.[1670,646,1381,663,463] The following examples illustrate our rather rudimentary knowledge, and emphasize that this is an important and promising frontier.

Central place theory, where a core location is a source of objects that move to the surrounding 'hinterland', has been widely used in geography (chapter 9). When objects from adjacent central places compete, a pattern of hexagons that enclose the area of influence of each central location appears in the landscape.[278,646] Most early work assumed surface homogeneity, but gradually constraints such as rivers, mountains, and highways were added, so model predictions were of different-shaped polygons. Central place theory has also been used to understand animal foraging, though too seldom is it related to the arrangement of habitats.[200,267,104,1508,1393]

Models of connectivity and circuitry are used to understand movements along networks (chapter 8). These permit a range of route optimization problems to be addressed. In addition, node size is varied using a 'gravity model'. In this, the movement between two nodes is dependent on distance, as well as on the population sizes of both nodes (Fig. 11.2c) (chapter 8).[1670,501] Movement increases with node size, but decreases with the square of the distance. Therefore distance exerts more control than node size on interactions between nodes.

To test spatial and fractal models John Wiens, Bruce Milne, and Alan Johnson studied the movement of marked beetles (Tenebrionidae) relative to the size and distribution of grass clumps, bare areas, and beetle home ranges.[1887,803,804,326] They concluded that movements are not significantly different from random at grain sizes (chapter 1) smaller than that of the grass and bare patches present. Near-random movement is also observed in areas smaller than the home range of the beetle. However, beetle movement was highly non-random at the grain size of grass and bare patches, and also at the grain size of home ranges.

Lenore Fahrig used a general model of population dynamics in a landscape composed of patches and corridors to understand movement.[463,464,466] Model results led to the hypotheses that: (1) the most

371

important determinants of average local population size are the probability of dispersers detecting new patches (positive relationship), and the fraction of organisms dispersing from the patches (negative); and (2) the dispersal distance (negative) is the main factor determining whether local populations will be influenced by the exact spatial relationships among patches. The author concludes that simple population-dynamics models are unlikely to be useful in landscapes, and in addition, that complex simulation models for a specific system are unlikely to provide generality.[464] She recommends developing the simplest possible simulation model to study a question, 'running experiments' with the model, and stating the relationships uncovered between input parameters and output responses as hypotheses to be tested.

Finally, Peter Kareiva and colleagues have provided a range of insights by linking a series of population-movement models with tests using insects and experimental plantings in different arrangements.[832,833,835,836] The preceding examples illustrate that spatial arrangement can be included in a population model not just as an added variable, but as a central focus of the model. They also illustrate how little we know about population movement in mosaics. Hardly any of the major mosaic attributes elucidated in the preceding section are as yet considered in population modeling. This is a subject ripe for development.

METAPOPULATION DYNAMICS

In a continuous suitable habitat all the individuals of a plant or animal population may be together and interact (chapter 2). At the opposite extreme, suitable habitat may be widely scattered in patches containing individuals, but with no movement and interaction among the patches. The first case is a single population, and the second case is several populations.

Between these extremes is a *metapopulation*. This is a population consisting of spatially-separate subpopulations that are connected by the dispersal of individuals[959,1909,1869,528,1257,1112,664] (Fig. 11.3a). Each subpopulation (or local population), as well as the metapopulation as a whole, changes in size over time, and may persist for long periods.

However, metapopulation dynamics is of special importance, because subpopulations often drop to zero (a local extinction), such as in small isolated patches.[1328,1352] If each subpopulation dropped to zero, this would mean extinction of the whole metapopulation. However, because individuals move between subpopulations two results occur. First, the subpopulation 'extinction rate' (the number of species disappearing from a patch per unit time) is lowered. Second, when

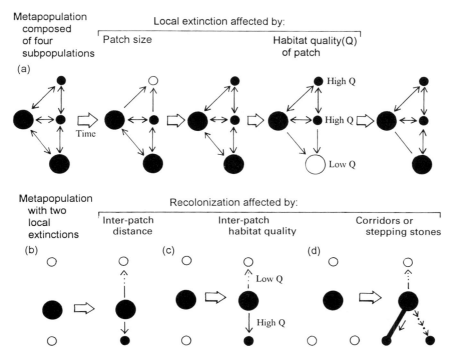

Fig. 11.3. Local extinction and recolonization in a metapopulation. Solid circle indicates subpopulation is present in a patch; open circle indicates local extinction of a subpopulation. Solid arrow = movement between patches; dotted arrow = infrequent movement between patches.

this local extinction does take place, recolonization of individuals reestablishes a new subpopulation at the site. Consequently, with extinctions followed by colonizations the metapopulation as a whole persists (Fig. 11.3a). The local scale is highly unstable, but the broad scale has more stability.

Local extinctions and recolonizations are common in nature. This rate of species turnover varies widely, e.g., from none to about half of the patches each year, based on diverse studies of land snails in ponds, ragwort (*Senecio jacobaea*) in Dutch dunes, birds in woodlots, and beetles (Phyllotreta) in cabbage and grassland patches.[166,1510,1771,833,1258,1783,1784,1436] It is not known whether, or how much, subpopulation extinction and colonization rates are higher in distinct patches than in separated areas of continuous suitable habitat.

The steepest decline in population sizes of birds appears to be when forest cover (of an originally forested landscape) is reduced between 60% and 20%.[537] This is the phase of deforestation when *metapopulation dynamics* are most pronounced, that is, extinctions of subpopulations followed by recolonizations. However, a declining subpopulation in a patch may be rescued by the arrival of individuals in time to halt local disappearance of the species.[199,1536] This 'rescue effect' is particularly amenable to actions of the conservationist or land manager, by changing the neighborhood arrangement around a patch (chapter 9).

The recolonization of patches may be extremely slow in some cases, e.g., at the time scale of centuries. 'Ancient woodlands' are areas scattered about Britain that have never been plowed and presently are wooded.[1367,1369,1306] George Peterken lists five plant species strongly associated with ancient woodland throughout their ranges in Britain[1306]: lime, service, oxlip, lily of the valley, and hay-scented buckler fern (*Tilia cordata*, *Sorbus torminalis*, *Primula elatior*, *Convallaria majalis*, *Dryopteris aemula*). Thirty-four such species are recognized for an agricultural area of central Lincolnshire with scattered ancient woodlands and secondary woodlands (those regrown since medieval time). These species have a 'strong affinity for ancient woods, showing little or no ability to colonise secondary woodland, and rarely found in other habitats.'[1306] The only woody plants included are lime and service; 94% of the species are woodland herbs.

Animal species are also related to ancient woodlands.[1306] The black hairstreak butterfly (*Strymonidia pruni*) is confined to the Midlands, and almost exclusively in ancient (coppice) woods. A few subpopulations survive in hedges, and two thrive in secondary woods close to ancient woods. Some beetles also may be sluggish in colonizing other patches, and hence good indicators of ancient woodland. Other groups characteristic of ancient woodland such as slugs (terrestrial molluscs) are also, not surprisingly, slow recolonizers of patches.

Oliver Rackham's detailed study of Hayley Wood, an ancient woodland in relatively-flat East Anglia (see Fig. 1.5), provides further insight into isolation and colonization rate.[1367] Neighboring secondary woods, including one only 100 m distant, are on similar soils, and originated several decades to a few centuries ago. Woods are separated by large cultivated fields with few hedgerows remaining. Characteristic ancient woodland plants present include oxlip, dog's mercury, Herb Paris, a grass, and woodland hawthorn (*Primula elatior*, *Mercurialis perennis*, *Paris quadrifolia*, *Milium effusum*, *Crataegus laevigata*). These species are absent or rare in the secondary woods. Little colonization across distances of only 100 m to a few kilometers has occurred over a period of decades to centuries.

A small wood adjacent to Hayley Wood began developing five decades before the study.[1367] Oxlip, which varied from *c*. 4 to 2 million plants in Hayley Wood during that period, invaded into this adjacent secondary woodland at an average rate of approximately 1.3 m (4 ft) per year. Dog's mercury has also invaded very slowly. As noted above, oxlip has barely crossed a 100 m field to the nearest secondary woodland. In short, despite a large subpopulation in a patch, the colonization rate of other, even nearest-neighbor, patches may be extremely slow.

Spatial pattern

The subpopulations may be in separated areas of continuous suitable habitat, such as a matrix. They may be in separated corridors of a network, as in a stream system. More commonly though, they are observed in patches separated by less suitable habitat, or connected by corridors or stepping stones.[1116,1257,51,1306] The patches may be natural, such as oases in a desert or beaver clearings in forest. However, metapopulation dynamics is becoming of major interest in planning and management, because of the human-caused fragmentation of once continuous habitat. In this case a continuously-distributed population is converted to a metapopulation, where extinction and colonization become critical.

Extinction rate is believed to largely depend on the habitat quality and the area of a patch[1260] (Fig. 11.3a). A subpopulation is more apt to persist in high than low quality habitat. Since small patches tend to have small subpopulations, the extinction probability is high on a small patch.[398,1510,1417] Although large patches have somewhat more edge habitat than small patches, the much greater area of interior habitat in large patches provides a buffer against local extinction. Furthermore, extinction rate increases with higher environmental variability[387, 837], such that a population widely fluctuating in size is more prone to extinction.

In contrast, recolonization following local extinction depends primarily on the pattern in the surrounding mosaic. Recolonization rate appears to be high when patches are close together (Fig. 11.3b), the matrix habitat is suitable (Fig. 11.3c; chapter 5), or connectivity between patches is provided by corridors or stepping stones (Fig. 11.3d; chapters 5 and 6). In effect, patch characteristics affect extinction, and the surroundings affect recolonization.

To compare the relative effects of area and isolation on interior species richness, eight bird studies are examined[1257] where an original habitat became fragmented (each study excludes edge and matrix species, and sampling effort is roughly proportional to patch area[758,1016,1258,202,137,59,1775,1597]; see chapter 2). Isolation is generally measured as the total amount of suitable habitat within a certain radius of a patch, or the shortest distance to a larger patch that might serve as a species source. Corridor density has also been used as a measure of isolation.[1775] Amount of habitat may be the total area of patches, corridors, or both. The relative importance of patch area and isolation depends on the scale and dispersal capacity of the species.[1257] Nevertheless, all eight studies cited above found area to be the leading predictor of species number. Most also found a correlation with isolation as a secondary predictor.

Avian metapopulations in woods

A series of studies by Paul Opdam, Jana Verboom, Boena van Noorden, Alex Schotman, Rob van Apeldoorn, and colleagues in Dutch landscapes has provided a detailed view of metapopulations and, along with parallel studies on other species, one of the clearest pictures of landscape ecology worldwide.[1261,1258,1259,1775,1254,1784] Hundreds of woods and numerous hedgerows surrounded by a matrix of agricultural fields are studied. Centuries before, this region was continuous forest. Woods are selected to minimize habitat variability, and each correlation study is one to three years in length. The primary independent variables measured (horizontal axis of Fig. 11.4) are hedgerow density (including lanes), road proximity, distance to nearest wood, density of woods, and total area of neighboring woods. The focus is on the presence or absence of forest-dwelling birds, including the nuthatch, marsh tit, hawfinch, golden oriole, and great spotted woodpecker (*Sitta cothraustes, Parus palustris, Coccothraustes coccothraustes, Oriolus oriolus, Dendrocopos major*).

The number of bird species restricted to woods is greater in larger woods. Species richness also is greater with higher corridor (hedgerow) density, and with a higher density of woods (low inter-patch distance), in the surroundings (Fig. 11.4). Small remote woods are most likely to be missing one or more bird species.

Based on models using parameters of the well-studied nuthatch in The Netherlands, the role of a large *source patch* (or core area) in metapopulation dynamics is evaluated.[1788] The source patch contains a large subpopulation from which individuals frequently disperse to the surrounding mosaic. The model suggests that the overall probability of extinction of the metapopulation decreases with more source patches, as well as with more woodlots in the surroundings. This is because of dispersal from woodlots to source-area patches. It also suggests that local extinctions of subpopulations are low near a source area, due to continued dispersal from the source patch.

A study of three consecutive years finds that the frequency of local extinction is strongly correlated with patch size.[1783, 1788] Small patches are most likely to miss a species for one or more breeding seasons, and

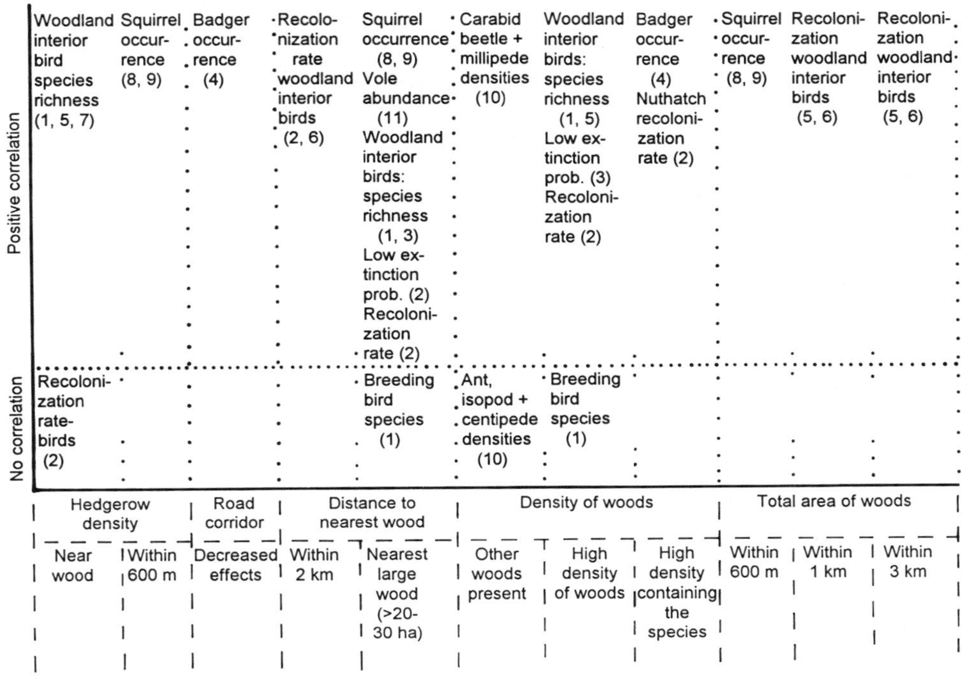

	Hedgerow density		Road corridor	Distance to nearest wood		Density of woods			Total area of woods		
	Near wood	Within 600 m	Decreased effects	Within 2 km	Nearest large wood (>20-30 ha)	Other woods present	High density of woods	High density containing the species	Within 600 m	Within 1 km	Within 3 km
Positive correlation	Woodland interior bird species richness (1, 5, 7)	Squirrel occurrence (8, 9)	Badger occurrence (4)	Recolonization rate woodland interior birds (2, 6)	Squirrel occurrence (8, 9) Vole abundance (11) Woodland interior birds: species richness (1, 3) Low extinction prob. (2) Recolonization rate (2)	Carabid beetle + millipede densities (10)	Woodland interior birds: species richness (1, 5) Low extinction prob. (3) Recolonization rate (2)	Badger occurrence (4) Nuthatch recolonization rate (2)	Squirrel occurrence (8, 9)	Recolonization woodland interior birds (5, 6)	Recolonization woodland interior birds (5, 6)
No correlation	Recolonization rate-birds (2)				Breeding bird species (1)	Ant, isopod + centipede densities (10)	Breeding bird species (1)				

Fig. 11.4. Effect of area surrounding a woods on species in the woods. Based on Dutch agricultural landscapes. Numbered references are: 1 = Opdam & Schotman (1987); 2 = Verboom et al. (1991); 3 = Verboom et al. (1991) (Landschap); 4 = Lankester et al. (1991); 5 = Van Noorden (1986); 6 = Opdam (1991); 7 = Van Dorp & Opdam (1987); 8 = Verboom & Van Apeldoorn (1990); 9 = Opdam et al. (1992); 10 = Mabelis (1990); 11 = Van Apeldoorn et al. (1992).

hence are most dependent on metapopulation dynamics. The area of a wood explains most of the variation in species number for the 16 bird species characteristic of mature forest. Local extinction rate is negatively correlated with the estimated maximum number of territories, as well as the proportion of optimal habitat territories.[1788]

The frequency of recolonization of a vacant patch by a species is partially correlated with density of woods in the surroundings (Fig. 11.4). Therefore, small patches near large patches are most likely to benefit from metapopulation dynamics. Small isolated patches may remain vacant for extended periods. Recolonization rate is not correlated with subpopulation size. Rather it has been correlated with a connectivity index, which combines distance to woodlots and the size of woodlots

377

within 2 km.[1789] In contrast, local extinction is not related to the connectivity index. The conclusion is that local recolonization does not significantly affect the extinction probability in a patch.[1789]

Approximately 40–50% of the suitable 0.3 to 30 ha woods have no nuthatches.[1068,1775,1260] Therefore, a considerable proportion of the suitable patches remains unoccupied.

Higher habitat quality leads to less extinction. Colonizers and young move from lower- to higher-quality habitats. But in small patches there is not much high quality habitat available.

The studies are also expanded to see if a regional trend exists. The presence or absence of forest-dwelling bird species is recorded in 234 mature woodlots clustered in groups of 10–12 in each of 22 landscapes. Two variables, patch size and the location of a landscape within the overall region, explain 66% of the variation in species richness.

The total number of species, as well as of interior species only, in patches also correlates with the density of hedgerows (including lanes) surrounding patches[1775] (Fig. 11.5a). The recolonization rate of vacant patches by the birds studied does not correlate with the density of hedgerows surrounding woods.[1788] However, corridor density probably is quite important for recolonization of other groups, such as squirrels, raccoons, small mammals, amphibians, and beetles.[561,669,812,223]

Total species richness also correlates with the distance to a large nearby wood, as well as the total amount of wooded area within 3 km (Fig. 11.5a). The same is true for the number of interior species. Species richness appears generally unrelated to density of woods in the surroundings, total wooded area within 1 km, and the presence of a hedgerow connected to a woods.

The regional trend detected[1775] is a gradient from near a forested region eastward in Germany to an open area southwestward by the North Sea (Fig. 11.5). About half the forest bird species are inhibited by isolation at the scale of up to c. 10 km (6 mi), whereas the other half are affected at the scale of the regional gradient of c. 100 km. Four species sensitive to patch size show quite different patterns near the eastern forest region than in the more-distant central/southern area (Fig. 11.5b). In addition, nuthatch, marsh tit, and wood warbler (*Phylloscopus sibilatrix*) are more sensitive to isolation than the golden oriole and lesser spotted woodpecker (*Dendrocopos minor*).

These mosaic-and-bird studies illustrate that metapopulation dynamics depend on the location of a landscape in the region. The studies also reemphasize that 'context is as important as content'. Neighborhood, rather than within-patch, attributes are primary considerations in determining the long-term nature of a local ecosystem or land use.

378

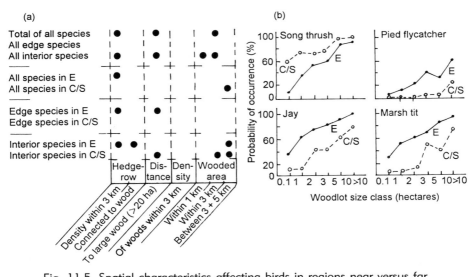

Fig. 11.5. Spatial characteristics affecting birds in regions near versus far from a forested region. The eastern region (E) and central/southern region (C/S) of The Netherlands, each c. 60 km wide, are compared by sampling woods in agricultural landscapes of similar appearance. East of E is a forested region in Germany, and west of C/S is mostly open land. (a) The spatial characteristics plotted are those in statistically significant multiple regressions, after correcting to eliminate effects of area and habitat. (b) Four species sensitive to patch size and differing in occurrence between E and C/S regions: song thrush (*Turdus philomelos*), jay (*Garrulus glandarius*), pied flycatcher (*Ficedula hypoleuca*), marsh tit (*Parus palustris*). In each region 13–29 woods of each size class are sampled. Based on van Dorp & Opdam (1987), with permission of *Landscape Ecology*.

Other metapopulation examples

To provide broader perspectives on metapopulations, several examples of other animals or a different landscape are encapsulated as follows.

Red squirrels (*Sciurus vulgaris*) studied in some of the same landscapes are absent in about half the suitable woods. The probability of occurrence correlates with patch area. It also correlates with three characteristics of the surrounding mosaic (Fig. 11.4): (1) hedgerow density within 600 m; (2) distance to nearest wood of more than 30 ha; and (3) total wooded area within 600 m.[1260] The probability of red squirrel presence is high in patches near forests of more than 30 ha.[1789]

Bank voles (*Clethrionomys glareolus*) move about 500 m across the landscape. Approximately 11% of the suitable patches apparently had experienced local extinctions of voles, but recolonization took place within a few months. Local extinctions are most probable in woods of less than 0.5 ha.[1767,1260] Vole abundance in woods decreases with distance from large woods of more than 25 ha.[1767]

Badgers (*Meles meles*) have dropped sharply in number since 1960 in these landscapes, and fragmentation may be important.[919] Modeling badger populations as metapopulations suggests certain spatial patterns that would increase the probability of the survival of the species.[1255]

Insects also are sensitive to pattern at the landscape scale (based on sampling transects of pitfall traps in 22 woods). If a woods has other woods in the surroundings, it tends to have more carabid beetle and millipede individuals, but not more centipedes, isopods, and ants, in the forest floor[1021] (Fig. 11.4). Flying carabid beetles are found in all woods, whereas wingless carabids are in less than 10% of the woodlots. An ant species dispersing by winged queens is in every woodlot, whereas ants with poor dispersal capabilities are only in 30–50% of the woods. Fertilizers from neighboring farms intrude into and decrease habitat quality more in small than large woodlots. In a similar landscape in Poland with little fertilizer intrusion into woods, wood ants inhabit smaller woodlots than in the Dutch landscape (Mabelis data[1260]).

Almost all similar evidence comes from woods of temperate agricultural landscapes. A small amount of evidence suggests that for marshes the number of bird species is also dependent on area and isolation.[1257,202] This is especially important in certain areas where wetland habitats (and their species) have disappeared at a greater rate than the surrounding upland habitats.

In an agricultural landscape of southeastern Ontario, local extinctions and recolonizations of three migratory birds (wood thrush, ovenbird, scarlet tanager; *Hylocichla mustelina*, *Seiurus aurocapilus*, *Piranga olivacea*) were measured in 71 woodlots.[1797] A median 10% (range 0–25%) of the subpopulations go locally extinct in a year. Recolonization is 22.5% (0–50%). No woodlot-size effect is evident for the tanager. But for two of the three species, extinction and recolonization rates are higher in small woodlots.

The patterns described in chapter 8 for white-footed mice and eastern chipmunks (*Peromyscus*, *Tamias*) in the Ontario landscape represent another good example of metapopulation dynamics.[718,1114,1116] Subpopulations in woodlots become extinct and are reestablished through colonization. The main difference is that distinct hedgerow corridors provide a network that connects most of the patches. Furthermore, at certain times of year the mice move from the patches and corridors into the matrix of fields.

Winking and source–sink habitats

Richard Levins (1970) defined the concept of metapopulations, and E. O. Wilson (1975) made the concept come alive as 'a nexus of patches, each patch winking into life as a population colonizes it, and winking out again as extinction occurs'. With the expansion of empirical studies at the landscape-scale, the role of patch attributes and patch context has become paramount in determining the rate of local extinction followed by recolonization, i.e., the *winking* rate.[1901,633,1797,1257] (Generally, 'turnover' is where species replace one another along a line or over time.) Patch attributes include food, cover, water, or sunlight. Context or isolation is primarily considered as the amount of patch area and/or corridor density surrounding a patch, or the distance to a larger patch.

We assume the winking rate varies over time. Hence, a patch will have brief vacant periods and long vacant periods, separated by periods of occupation by a species. The determinants of winking are of key interest. In the Ontario woodlot study of birds[1797], winking took place in small subpopulations of all three focal species. One species, scarlet tanager, may have been affected by isolation. However, two species, wood thrush and ovenbird, had winking primarily in small patches.

A three-year study of birds in small cottonwood (*Populus*) stands in Wyoming (USA) attempts to determine the best predictors of winking rate, by measuring numerous spatial and habitat characteristics around each patch.[633,634] The best predictor of winking rate is patch area, and the second best is distance to the nearest patch (streamside). Significant correlations are also found with patch boundary-to-area ratio, presence-or-absence of adjacent wetland (palustrine), and area of nearest patch (streamside). Several adjacency attributes (chapter 9) do not correlate with winking rate: number of adjoining land-use types, and presence-or-absence of adjacent riverine wetland, mixed rangeland (pastureland), irrigated cropland, and roads.

Habitat quality and 'source–sink' relationships may be useful in understanding these results. A *source habitat* has more dispersal than immigration of individuals, and a *sink habitat* has immigration exceeding dispersal.[969,1779,748,1357] In population studies a source habitat is commonly considered to be a high-quality habitat where births exceed deaths. The excess individuals disperse and colonize less-suitable sites.[525,1158] Similarly a sink habitat is considered to be a low-quality habitat (with mortality exceeding natality). High- and low-habitat quality normally refers to internal or local characteristics of soil, moisture, food, vegetation structure, and the like.

But the Dutch and Ontario studies indicate that winking rate is strongly affected by landscape structure (Figs. 11.4 and 11.5). Patch size, neighborhood effects, and isolation all correlate with the extinctions and recolonizations of subpopulations on patches in the mosaic. This result is consistent with an increased recognition that, in addition to local resource availability, understanding population dynamics requires a focus on the broader spatial scales relevant to both the organisms and the processes under study.[1745,1883,835,430,1359,969]

In sinks three scenarios occur for subpopulations. A subpopulation may go extinct nearly permanently. Winking may take place where a sequence of temporary subpopulations follows one another. Or a continued high immigration rate enables the subpopulation to persist. A large source patch in the neighborhood can provide this third effect in a small patch.[748,1357,1359] A medium-sized neighboring patch may function similarly, but its edge (and the dispersing edge species) (chapter 3) will play a larger role as a filter, through which the limited-size populations of interior species must pass.[1613,212] Finally, a temporary sink, such as a dry spot becoming wet at the end of a drought, may turn into a source.

Metapopulations in management

Landscape structure or the arrangement of spatial elements is a good handle for the planner, designer, and manager. With this approach one does not have to know every species and its dynamics over time to take effective action. Metapopulation dynamics takes place in continuous suitable habitat such as a matrix, but the number of species converted to metapopulations doubtless increases markedly with the increase of habitat fragmentation into small patches and corridors.[1255,1256,537] Thus, a first overall management step is to decrease habitat fragmentation, by reestablishing viable connectivity among fragments. This should decrease the number of metapopulations, as well as the local extinction rate, by converting some to populations of continuously distributed and interacting individuals.

Identifying the species that undergo winking in existing patches is a next step. Species with low dispersal capabilities, such as some arboreal mammals and vegetatively reproducing plants, are of special concern. The beginning of metapopulation dynamics may or may not be an early symptom of a species heading for trouble.

A third step is to identify sources and sinks. Sources, including source patches, may need to be expanded or established. Local habitat quality is an important determinant of whether an element is a source or a sink for a particular subpopulation. Less variable environments such as many mature forests, peatlands, and oases often require special protection. Species with limited dispersal capability are often concentrated in such habitats.

Several landscape attributes also affect the source-sink relationship.[1255,1114] Matrix activities commonly affect habitat quality in patches and especially corridors, such as by inputs of fertilizers, pesticides, and fire. Buffer strips separating the patch and matrix have been proposed to ameliorate this effect.[1021] Nevertheless, the focus must be on the two key attributes of metapopulations, extinction and recolonization.

Preventing extinction and winking is one objective. High habitat quality in a local patch helps. A large size helps, by normally providing a large ongoing subpopulation that decreases the probability of extinction. And a high immigration or colonization rate decreases the probability of extinction. Of course, the implications for breeding, gene flow, and social behavior will differ markedly between an ongoing subpopulation with little colonization, versus a subpopulation maintained by constant immigration. The spatial characteristics enhancing this immigration are the same as those enhancing recolonization after extinction of a subpopulation.

Recolonization of a patch is enhanced by the proximity of a large source patch. Corridors connecting from a source enhance recolonization. Stepping stones do also. A less-inhospitable matrix does too. Low inter-patch distances, or a high density of patches that decreases isolation, enhance recolonization. Indeed, the metapopulation concept itself implies that all patches and all other spatial elements in the mosaic are important, not simply those containing individuals of the species. Although certain spatial elements are most important, every element in the landscape affects the winking rate.

Managing to minimize winking, i.e., the probability of extinction and recolonization of the various patches, focuses on spatial arrangements within a landscape. However, as seen in the Dutch studies, the regional scale is also important in management (Fig. 11.5). Metapopulation dynamics of the same species in the same type of woods in the same type of agricultural landscape differs according to the location of the landscape in the region. Agricultural landscapes near a forested landscape in Germany apparently benefit from a high regional immigration rate of birds, so winking is less. On the other hand, agricultural landscapes near the North Sea dune area suffer from a low regional immigration rate of songbirds. In conclusion, metapopulation dynamics is a rapidly growing area of research, and its role in planning, conservation, and management will doubtless increase in importance at the landscape and regional scales.

GENE FLOW

Most mating and exchange of genetic material takes place within a population or subpopulation, although some is exchanged between subpopulations. The transfer of new genes (alleles) and genetic combinations between populations (or subpopulations) is here called *gene flow*.

The genetic consequences of patch size were presented in chapter 2, particularly in estimating minimum viable population sizes

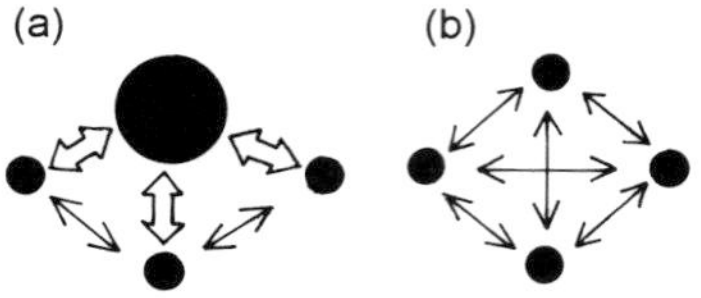

Subpopulation:	(a)	(b)
Winking rate	Low	High
Local extinction rate	Low	High
Recolonization rate	High	Low
Genetic:		
Inbreeding depression	Low	High
Strains disappear due to local extinction	Low	High
Strains disappear due to "swamping" (outbreeding)	High	Low
Differences among subpops.	Low	High
Total variation of metapop.	Low	High
Adaptability to environmental changes	Low?	High?
Evolutionary rate	Low?	High?

Fig. 11.6. Expected patterns from gene flow in two metapopulations. Thickness of arrow represents relative amount of gene flow.

(MVP).[515,572,915,865] These consequences were viewed primarily in the contexts of extinction and evolution. In the present chapter we will also focus on extinction, and then on evolution, relative to subpopulations in a metapopulation.

For small subpopulations local extinction is a normal, frequent, and important process in metapopulation dynamics[570] (Fig. 11.6). In chapter 2, a minimum viable population size (MVP) for a whole population was hypothesized to be a few thousand individuals, while a MVP of a few hundred was estimated for subpopulations (compared with effective population sizes of 500 and 50, respectively, sometimes cited in the literature[515,116,1536,1352]). The figure of a few hundred individuals is an average for the subpopulations present. Therefore, some subpopulations will be small and will experience local extinctions. Meanwhile, some will be larger and rarely or never go extinct. These larger subpopulations on source patches provide individuals that recolonize empty patches, and are a key to maintaining the metapopulation as a whole (Fig. 11.6a).

While mutations occur in the whole metapopulation, new genetic variation in a subpopulation primarily depends on gene flow from other patches. In the metapopulation model gene flow to a subpopulation results from dispersal of individuals from other patches (Fig. 11.6). One or two individuals arriving and mixing genes with a subpopulation per generation is hypothesized to be sufficient to maintain long-term genetic variation.[965,916,910] Through this mechanism the negative effect of inbreeding depression[266,916,127,472] (chapter 2) may be minimized, and the extinction rate of subpopulations reduced.

The evolutionary implication of metapopulation dynamics is equally important.[1811] With a winking rate >0 (chapter 2), some new subpopulations will form by recolonization from other patches. The *founder effect*, in which the genes of the few initial colonizing individuals play an unusually large role, is important here. This implies that the new subpopulation is likely to be genetically quite different from the previous subpopulation in a patch. The same pattern occurs when a subpopulation is reduced to a few individuals ('genetic bottleneck') and then expands.[1273,930]

Furthermore, with a high winking rate, the subpopulations on different patches will be genetically quite different from each other[1061] (Fig. 11.6). Therefore, despite being composed of small subpopulations on separated patches, each threatened by inbreeding depression, the metapopulation as a whole maintains a high level of genetic variation. Indeed, we may hypothesize that genetic variation of a metapopulation typically exceeds that of an equal-sized population on a single large patch.

The large genetic variation means that metapopulations may be expected to adapt well to changing environmental conditions.[514,19,1073] Environmental change should normally increase the winking rate. But also there is a high probability that one or more of the genetically diverse subpopulations will survive even severe environmental change. A high winking rate combined with the founder effect means that adaptations and evolution should occur at a rapid rate (Fig. 11.6).

Of course, a high winking rate also means that the local extinction rate is high.[570] Pure strains disappear. Subpopulations on local patches disappear. If extinction on all patches coincides time-wise, the metapopulation or population as a whole disappears.

The last individual also may effectively disappear by *genetic swamping* (Fig. 11.6). In this, a large well-adapted subpopulation hybridizes ('outbreeding') with a small population low in genetic variation. The genes from the larger subpopulation gradually replace or 'swamp out' those of the original small subpopulation.[1689,450] Consequently the small subpopulation never drops to zero, but instead continues in an altered form. Indeed, such a process applied to many small subpopulations will temporarily remove differences among them, including pure strains of local importance. Therefore, human introductions and reintroductions of subpopulations in the landscape require evaluation and caution for possible outbreeding effects.[1536]

Finally, we note that the rate of recolonizing patches in a metapopulation also has genetic consequences.[1061,916,1255,481] Subpopulations on opposite sides of barriers or filters in the landscape can be expected to differ genetically.[1639,1403,930,1273] In addition, the establishment of corridors connecting land that has been fragmented enhances the conduit function, i.e., the recolonization rate of species between patches (chapters 5 and 6). In this process individuals may move directly from patch to patch, or gene flow may take place sequentially through resident individuals in a corridor.[109,1228] In either case, the higher recolonization rate decreases inbreeding depression on a patch. It also decreases genetic differences among patches. Indeed, ultimately if the winking rate drops to zero, and interbreeding and gene flow become frequent among all patches, the metapopulation effectively disappears by being transformed into a single population without subpopulations.

MOVEMENT AMONG NON-ADJACENT ELEMENTS

Interactions between adjacent spatial elements were emphasized in chapters 2 to 7 on patches and corridors. Here we focus on interactions among elements that are spatially separated by one or more other ecosystems. A few examples are also given in chapter 9 illustrating the concept of a configuration.

Routes

The routes of numerous animal and plant species have been recorded using a range of marking techniques (chapter 1). Most studies focus on home-range sizes, dispersal distances and directions, habitat selection, or population dynamics. To date, the information apparently has not been reviewed or synthesized to determine patterns and principles of movement in mosaics. Therefore, we begin by identifying nine common arrangements of spatial elements in which most movement probably occurs. These represent specific combinations of patches, corridors, and matrix (Fig. 11.7).

The boundary separating spatial elements is a well known 'travel lane' for some animals[123,1484] (Fig. 11.7a) (chapter 3). Wildlife often moves along a sequence of corridors or corridor segments, where a change in an adjacent ecosystem type creates the next corridor segment (Fig. 11.7b) (see necklace in Fig. 9.3). Species move through a sequence of individual corridors in a network surrounded by a matrix[222,223] (Fig. 11.7c) (chapter 8). Alternating corridors and patches provide a route (Fig. 11.7d). A row of stepping stones is a pattern for movement (Fig. 11.7e).

A 'cluster of small patches' or stepping stones is probably much more efficient than a row, because of the many options available in avoiding problems in the matrix[51] (Fig. 11.7f) (chapter 6). A continuous sequence of different patches (i.e., a patchy area) is often observed without corri-

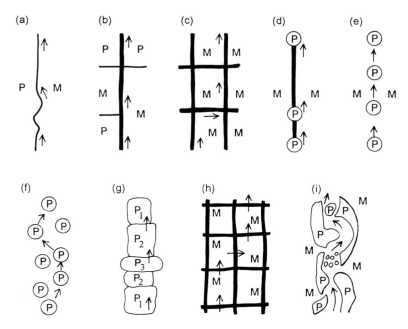

Fig. 11.7. Common spatial arrangements through which most species movement probably occurs. P = patch; M = matrix; thin line = boundary; thick line = corridor.

dors (Fig. 11.7g). This is probably an inefficient route due to high boundary-crossing frequency, plus the high probability of having to cross a low-suitability patch (chapter 8). Where a matrix is subdivided into sections by a network, matrix species must cross corridors (Fig. 11.7h). Finally, a species may move exclusively within a matrix that is continuous, but confront various constraints due to locations of other spatial elements (Fig. 11.7i).

Seed dispersal

Most plants move by dispersing seeds by wind or animals. Wind dispersal of seeds is probably most effective across an open matrix or up a valley corridor (chapter 10). Several of the proposed routes have a high heterogeneity that would produce considerable air turbulence. This should produce a patchy deposition of seeds. Most seed dispersal probably does not extend as far from a source in turbulent air as in a strong streamlined airflow.

In open country the maximum dispersal point is a short distance downwind of a seed source.[1074,62,1245] Beyond this peak dispersal point, seed densities apparently decrease exponentially. In effect, most seeds go a very short distance, yet a few go a very long distance. For example, three species in a Wisconsin (USA) agricultural landscape show the exponential decrease with distance.[811] Basswood or linden (*Tilia americana*) disperses the shortest distance, maple (*Acer saccharum*) intermediate, and ash (*Fraxinus americana*) the farthest. Very few seeds of basswood and maple would reach ecosystems several hundred meters away from a source. The long distance dispersed by ash seeds relates to their later dispersal time in winter, when wind velocities and smooth snow surfaces are great.[1711] In general, a heterogeneous mosaic can be expected to decrease dispersal distance.

Mark McDonnell, Edmund Stiles, and others provide important insight into the general pattern of dispersal across a mosaic, by focusing on the dispersal curve for seeds moved by animals. Some seeds get a free ride by hooking onto an animal's exterior. Others produced in fleshy fruits and eaten by birds, mammals, and certain other vertebrates get a ride inside. The animal gains food from the fleshy fruit tissue that may be rich in sugars or fats.[1631,1632] But the seed goes right through the gut and out undamaged.

Many of these seeds are dropped at patch or corridor edges when birds defecate and lighten their load before flying across an open area. Seeds may be dropped anywhere in the open area, but overall exhibit a decreasing density with distance. However, the open matrix commonly has a heterogeneous distribution of woody plants or small patches scattered about. These may act as 'recruitment foci', attracting birds that land and drop plenty of seeds.[376,1082,630] Consequently, the overall dispersal curve is a decreasing curvilinear trend, but punctuated with scattered peaks.[1082,145] The peaks in turn also become lower with distance. This 'decreasing curve-and-peaks' model may be a good general representation of species movement across landscapes (chapter 9). It should also be widely used in land restoration.

Multihabitat vertebrates

Probably all wide-ranging animals are multihabitat species (chapter 3). This may involve a daily use of different habitats, such as deer that in winter move between bedding areas (deer yards) in evergreen stands and feeding areas in openings.[1864,554,1443] Alternatively, different habitats may be used at different stages in the life cycle, such as insects that emerge in one, pupate in another, and use a third habitat as adults. Or multihabitat species may use different habitats at different seasons over the year. Since the animals normally return to the same habitats the following year, this is an example of migration.[322,1669] Migration is commonly thought of as cycling between two habitats in winter and summer, but

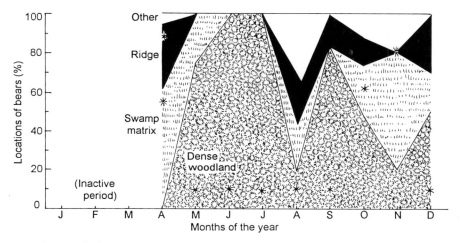

Fig. 11.8. Sequence of bear habitats and food over the year. Percentage of active, radiotracked black bears (*Ursus americanus*) in the North Carolina (USA) coastal plain is plotted by habitat. Asterisks indicate habitats where most natural food (>33%) is consumed each month. Adapted from Landers *et al.* (1979).

as in the following example, it may be a cycle through three, four, or more habitats.

In a mountainous park area of North Carolina and Tennessee (USA) black bears (*Ursus americanus*) use different habitats in summer and autumn.[549] The summer and autumn ranges overlap somewhat for some bears, but for about half of the radiotracked animals studied the ranges are completely separate. The intervening space between the seasonal ranges is rarely used. The center of summer activity moves *c.* 1.1 km (0.7 mi) from year to year. In contrast, the distance between centers of summer and autumn activity is 2.8 km for animals with overlapping seasonal ranges, and 10.0 km for bears with separate seasonal ranges. The boundaries of summer ranges often correlate with watershed divides, and are relatively fixed from year to year. The length-to-width ratio or shape of summer ranges averages approximately 2:1. Width is relatively constant compared with the more variable length of summer ranges. Seasonal ranges are commonly less than half the total area of the respective annual home ranges.

For this same multihabitat species in flatland 400 km eastward, spatial patterns provide additional insight.[917] Here, the matrix is usually swampland with hardwoods. Bears generally avoid five patch types, which cover a quarter of the area (residential, lake, most farmland, pine plantation, and sparsely vegetated sand ridge) (labelled 'Other' in Fig. 11.8). But three ecosystem types are important for bears. First, scattered ridges

389

with moderately dense vegetation provide food in late autumn (November) (Fig. 11.8). Second, dense woodlands (moist evergreen 'Carolina bays') provide both winter denning habitat and rich spring-summer foraging habitat. (Farmland corn (*Zea*) and a small amount of animal matter supplement the diet throughout the year.) However, these woodland patches used for winter dens are relatively small, and hence the bears are subject to disturbance.

To escape disturbance by high water, hunting dogs, and hunters with radio communication, some bears relocate dens to hollow trees or dry spots in the third important ecosystem type, i.e., the swampland matrix (Fig. 11.8). In short, three habitats are critical to bear survival here: dense woodland for denning and spring-summer food, ridge for autumn food, and swampland to escape disturbance. A few bears could probably survive with loss of one of the first two patch types. But the third type represents *escape cover*, an area in the neighborhood where an animal rapidly moves to avoid predation and disturbance. Escape cover, in this case the swamp matrix, is considered most critical to maintain the bear population.

The bear example is instructive in planning and managing for multi-habitat species in general.[501,688] Such species tend to be generalists, and hence can utilize a range of foods and find suitable spots for denning or nesting in different habitats. Two or more habitats must be protected. These are used at different seasons for different objectives. A minimum size of each is required. The area between the habitats may be little used. Thus, a corridor system may be sufficient to connect the habitats. Optimal corridor width varies by season (e.g., wider if bear cubs in spring must move, or if dogs or predators from adjacent areas line the corridor; see Fig. 5.7). Escape cover is essential for each of the patch and corridor steps along the route. Only by maintaining escape cover from disturbance is the plan and management apt to succeed.

Insects

Nevertheless, insects usually are much heavier than bears (measured as total biomass per square kilometer) and, indeed, heavier than all vertebrates combined.[1911] Insects also move across a mosaic. They may be blown passively, and deposited sparsely or densely in a given location according to wind patterns (chapters 6 and 10). For example, insects from a hedgerow are deposited in abundance immediately downwind of the hedgerow. In contrast, insects deposited further downwind are predominately those from other habitats across the landscape.[963,964] Alternatively insects using locomotion in flying exhibit patterns similar to mammals, which are based on behavior, sensitivity to habitat, and spatial arrangement.

In a region of southern France with several landscapes, M. Samways evaluated the effects of patch, corridor, and matrix on nine species of bush crickets (Orthoptera: Tettigoniidae – these decticine crickets are strong hoppers; seven of the species also have wings).[1484] Streams and rivers are almost complete barriers to movement for three of the species, partial barriers to one species, and minor barriers to the remaining five species. Paved tarmac roads exert a barrier effect on four species, and dirt roads on the same four species. In contrast, roadside verges (edges) and drainage ditches are not a barrier to crossing by any of the species. Verge and ditch corridors are preferred habitats for four cricket species.

Cricket use was compared in seven types of matrix, i.e., separate areas dominated by: (1) 'etangs' with natural grasses, *Bromus*, *Triticum*, and *Hordeum*; (2) 'sansouire' with *Tamarix* and *Salicornia*; (3) mixed grasses and forbs; (4) 'garrigue' with *Calycotome*, *Prunus*, *Thymus*, *Rosmarinus*, *Quercus*, and *Juniperus*; (5) vineyards; (6) wheatfields (grain); and (7) buildings. None of the nine cricket species penetrates much into etangs. All or almost all of the crickets penetrate only slightly into four other matrix types. Only the heterogeneous garrigue and mixed grass/forb matrix types are used in abundance by the bush crickets.

Most patches within a matrix are also penetrated only slightly by the crickets.[1484] In 55% of the possible cases (9 species × 8 patch types) crickets rarely penetrate beyond the edge. Indeed, edges between patches, or between patch and matrix, appear to be channels for movement of bush crickets through the landscape (Fig. 11.7a). Edges of certain patch or matrix types (wheatfields and vineyards) appear particularly suitable, and edges of the etangs matrix and the wooded patches are unsuitable. Thus, the bush cricket study points to certain barriers, as well as certain conduits, in the movement patterns of multihabitat species crossing a landscape mosaic.

Birds and mammals

Finally, ten unrelated studies are listed here to illustrate a range of patterns and concepts related to movement through mosaics.

(1) Blue jays (*Cyanocitta cristata*) consume 20% of the acorns from an oak stand (*Quercus*) in Virginia (USA).[357,812] But they transport and bury (cache) 54% of the acorns. Acorns are carried an average of 1.1 km, and 91% is buried in suburban lawns.

(2) Around Yellowstone Park in the Rocky Mountains six large winter range areas hold most of the tens of thousands of elk (*Cervus canadensis*) present.[322] The animals break up into nine smaller summer-range areas, in addition to widely dispersed tiny groups. Winter range areas are elongated in valley bottoms, and summer range areas are oval in higher-elevation meadow areas. A small percentage of the animals in a winter

(or a summer) range area moves to a different range area the following year, and hence provides genetic mixing among the separate herds.

(3) A seed-eating cockatoo, little corella (*Cacatua sanguinea*), in dry Australian terrain annually alternates two periods of roving in large flocks, with two periods of being in more sedentary small groups.[101] The large flocks wander across the mosaic during periods of scarce food, whereas the small groups remain in local habitats when seeds are abundant.

(4) A beautiful but scarce parrot-like bird, Carnaby's cockatoo (*Calyptorhynchus funereus latirostris*), essentially only moves between patches of remnant native woody vegetation in Western Australia wheat-land if the bird in a woods can visually detect a target woods.[1495]

(5) Winter ranges for white-tailed deer (*Odocoileus virginianus*) in Minnesota (USA) vary in size from year to year, but are smaller than summer ranges.[1443] Females and juveniles move directly from winter to summer ranges, whereas males tend to wander in the route. Without a dense evergreen patch (deer yard) for winter cover, deer density is low and animals are scattered in small groups.[1864]

(6) The alpine ibex (*Capra ibex*) in the Swiss Alps feeds in eight plant communities at different elevations in mountains, and causes little damage to the vegetation by grazing or browsing.[1690] Fecal droppings alter the vegetation only in local spots, and trampling damages the vegetation and accelerates soil erosion only in the alpine area.

(7) In a mosaic of fields and woods in Sweden, vegetation-feeding small mammals both enter and leave a field at a low rate along the field–woods boundary.[669] They enter and leave at a high rate in narrow wooded areas between fields, such as at field corners. Both field species and woodland species that eat vegetation exhibit this pattern. In contrast, seed- and insect-feeding small mammals move more as a broad front through various habitats of a matrix.

(8) Game reserves have been designed to produce a higher density of animals than the reserve can support, that is, to exceed carrying capacity.[943] Consequently, the animals disperse into the surrounding matrix. The optimal distance between game reserves therefore depends on the travel distance of the animals, plus hunting success rates. Corridors extending out from reserves should extend travel distances.[501] (chapter 4).

(9) Based on simulation modeling it is suggested that the rate of vertebrate movement out from a patch is dependent on edge permeability, plus the ratio of boundary to area of the patch.[1613] Greater permeability (chapter 3) and a higher boundary-to-area ratio (chapter 4) enhance emigration rate.

(10) Although vegetation structure is a primary determinant of movement route, chemicals applied to plants, soil, and air by animals doubt-

less also play an important role for many species. 'Pheromones' (intraspecific chemical messengers) are used by some species to attract mates to a particular location, or to mark plants to be avoided.[1269,1438] Pouring simulated cow urine onto patches of Colorado grassland produces greater plant growth, and a change in relative abundance of species.[373] Herbivores love the result. Urine patches that cover only 2% of the area provide 7% of the biomass and 14% of the nitrogen consumed. For those benefits, altering a route through the landscape is well worthwhile for a herbivore.

EXOTICS AND INVADERS

In a native ecosystem populations of virtually all species increase and decrease over time, often with radical fluctuations.[1317,1320] In addition, species from elsewhere invade and species go locally extinct. Human activities commonly increase the rates of invasion, population fluctuation, and extinction. With a long-term exponential growth in human population, and concomitant spread and increased activity on the land, the introduction of species to new areas is a dependable ongoing process.[749,333,769,1338] No maples were in New England (USA) 11 000 years ago, and no European rabbits (*Oryctolagus cuniculus*) were in Britain 3 000 years ago. Today, both are considered part of the native flora and fauna.

Exotics (non-natives, aliens, or introductions) are species out of their native range. Except where native, a canary in a cage and a llama in the yard are exotics. Similarly, African violets in the window and papayas in the garden are exotics. Some exotic species called *invaders* successfully colonize native ecosystems. These invading species are often of special economic or conservation importance. But concern about exotics in general is warranted, because before introduction it is hard to guess whether a species will be a non-invasive exotic or an invader.

Invaders and invasive sites

Pulses of arriving exotics, and especially invading species, have occurred in continents and regions. Several thousand years ago this occurred in the Mediterranean region, so the present biota essentially represents those species that can survive there.[1338] Scientists of this region spend little time attempting to differentiate native from exotic species, or trying to control exotic species. In contrast, Canada and the USA are currently in a four-century pulse, and New Zealand and Australia in a two-century pulse, where new invaders are recognized every year[333,624,1159] (Fig. 11.9).

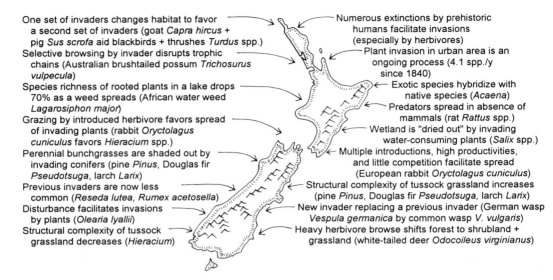

One set of invaders changes habitat to favor a second set of invaders (goat *Capra hircus* + pig *Sus scrofa* aid blackbirds + thrushes *Turdus* spp.)

Selective browsing by invader disrupts trophic chains (Australian brushtailed possum *Trichosurus vulpecula*)

Species richness of rooted plants in a lake drops 70% as a weed spreads (African water weed *Lagarosiphon major*)

Grazing by introduced herbivore favors spread of invading plants (rabbit *Oryctolagus cuniculus* favors *Hieracium* spp.)

Perennial bunchgrasses are shaded out by invading conifers (pine *Pinus*, Douglas fir *Pseudotsuga*, larch *Larix*)

Previous invaders are now less common (*Reseda lutea*, *Rumex acetosella*)

Disturbance facilitates invasions by plants (*Olearia lyallii*)

Structural complexity of tussock grassland decreases (*Hieracium*)

Numerous extinctions by prehistoric humans facilitate invasions (especially by herbivores)

Plant invasion in urban area is an ongoing process (4.1 spp./y since 1840)

Exotic species hybridize with native species (*Acaena*)

Predators spread in absence of mammals (rat *Rattus* spp.)

Wetland is "dried out" by invading water-consuming plants (*Salix* spp.)

Multiple introductions, high productivities, and little competition facilitate spread (European rabbit *Oryctolagus cuniculus*)

Structural complexity of tussock grassland increases (pine *Pinus*, Douglas fir *Pseudotsuga*, larch *Larix*)

New invader replacing a previous invader (German wasp *Vespula germanica* by common wasp *V. vulgaris*)

Heavy herbivore browse shifts forest to shrubland + grassland (white-tailed deer *Odocoileus virginianus*)

Fig. 11.9. Examples of ecological effects of invading species in New Zealand. Based on Norton (1992).

To protect economic or ecological bits of the status quo, various techniques to control certain invaders are used.[1325,351,1705,375] Pulling plants, shooting animals, and other 'fly swatter' approaches are readily used for local spots. Chemical controls are often used for larger areas. And further introductions of competitors or predators are used for biological control that usually affects still broader areas. Control by chemicals and further introductions typically are highly detrimental to certain non-target species in ecosystems.[1036,1687]

A closer look at the characteristics of invading species, followed by where they invade, is instructive.[1159,624,1764,419] Most introduced organisms die quickly due to an unsuitable environment.[439] Some survive to old age without reproducing. Some reproduce and maintain themselves as 'commensals' with humans, benefitting from continued care or activity by humans on land.[586] Garden perennials, agricultural weeds, and pet zebras are examples. These may also persist at low densities in native ecosystems. A vivid example of living with people is the house sparrow (*Passer domesticus*), probably in all cities of the world. In the 1870s a Parisian is said to have noted that the sparrows were increasing exponentially. This resulted from the increase in seeds as fresh layers of horse droppings were added from more and more horse and wagon congestion. He calculated that by 1920 the sparrows would totally blacken the sky of Paris.

394

Fig. 11.10. Water hyacinth (*Eichhornia crassipes*), a preeminent invader of slowly moving water in tropical regions. This floating exotic covers the open water surface of a bald cypress swamp (*Taxodium distichum*) with hanging Spanish moss (*Tillandsia usnioides*) in Louisiana (USA). Courtesy of USDA Forest Service.

In contrast, a small percentage of introduced species spreads widely in native ecosystems (Fig. 11.10). Such species must accomplish three steps in sequence. They must get a toehold (i.e., a spot to become established), grow (i.e., find necessary resources, and survive competition and predation), and spread (i.e., reproduce, disperse, and have subsequent generations also accomplish this sequence).

Most successful invaders exhibit the following character-istics[889, 99, 439, 1764, 1218]: (1) uncommon in its native habitat; (b) high reproductive output; (3) easily dispersed short distances; (4) phenotypically plastic (wide tolerance or generalist species); and (5) fast growth rate. Most species are carried purposely or inadvertently by humans, though

395

some arrive by natural long-distance dispersal. Some scholars suggest that most of this is like describing an invader by the definition of an invader (S. Levin[cited in 960]).

Different insight comes from examining characteristics of the environment that permit or enhance successful invaders.[451, 1264, 325] Dense forests, high mountain areas, salt marshes, and extreme deserts typically have few exotic species.[1024] Of course, where species richness is low, an invader may cause havoc.[682, 1057, 614, 1278] On the other hand, urban, suburban, and agricultural landscapes are classic centers of non-native species.[1655, 729, 565, 1492] All have a heavy and chronic human imprint, and the sheer number and density of exotic species provides a high probability that some will successfully invade surrounding native ecosystems. Tropical islands also are full of exotics because of human dispersal, similar environmental conditions around the globe, and often intense land-use activities.[587] Islands may also have 'empty niches'.[937, 1668, 721] This concept involves the relatively high isolation and extinction rates on islands, such that at a particular time there is no species fulfilling a particular functional role in the ecosystem. An arriving non-native then may readily step into that unfilled role.

A local ecosystem or a spot within it that has been disturbed is often a toehold for exotic species.[451, 264] An area with chronic human disturbance usually has a high proportion of non-native species. These locations are especially suitable as sources of potential invaders into surrounding native ecosystems. The human disturbance provides resources such as light and mineral nutrient availability, and especially space without competitors.[1264] Note that human alteration of a natural disturbance regime is also a disturbance. In a fire-prone area exotics spread if fire is controlled. In a flood-prone floodplain exotics spread if dams and levees reduce the floods. New human-created habitats such as railroad beds, waste piles from mines, and heavily polluted sites almost always have exotic species.

A species usually arrives in a new habitat without its herbivores, predators, diseases, and pests.[451, 1325] In other words, it has escaped the organisms that kept it under control 'back home'. Initially it may be considered simply a 'successful' introduction, but if it spreads, at some point it may be labelled a 'pest' or 'weed'[1026, 99, 1264] (Fig. 11.10). Initially few native species will eat it, select it as cover, or nest in it. Thus there is a lag period before it is integrated into the food webs and nutrient cycles characteristic of the surrounding native species.[1803, 1222] Aldo Leopold (1941) captured the idea as follows, 'One simply woke up one fine spring to find the range dominated by a new weed.'

To evaluate whether an exotic will become an invader in native ecosystems, it is useful to examine interactions with two types of existing species. Most ecosystems have leading dominant species, and the interactions between leading dominants and new arrivals should be a key to whether an invader will emerge. Second in the common case where the ecosystem invaded is in a successional stage, the interactions of new arrivals with later successional dominants is probably equally significant in predicting spread.[1805, 1705]

Species introductions and spatial pattern

Animals are introduced by humans in many familiar ways. See the pets in and around homes. Visit zoos, aquaria, and pet stores. See the livestock on farms. Indeed, simply go fishing in lakes and rivers, or hunting for pheasant and other game. Exotic species sources are 'everywhere', and each brings an inadvertant array of other exotics such as insects, fungi, and diseases. Chemical pesticides in agriculture are increasingly used in combination with species introductions under the rubric of biological control.[20,351,1325,375] For each introduced species that successfully controls a pest, a few to many other species are normally introduced, sometimes with little or no pretesting. How many escape and become invaders of native ecosystems? No one knows.

Plants also are introduced all around us. Humans like to surround themselves with diverse plants, and generally exotic species form the palette from which to choose. Visit a backyard or a park. Stop by a florist, a nursery, or an arboretum. Examine the crops in fields, and the forage for livestock. See the trees used in plantation forestry. All come with inadvertant fellow travellers, some of which emerge as invaders.

Simply watch a 'neophyte' and you can tell it is a new plant.[1316,724] Relative to native plants, the leaves are less chewed, the berries last longer (low-quality fruit for birds, E. W. Stiles, pers. commun.), bird nests in it are scarce, and the timing of growth and leaf fall differs. For example, European woody plants in native northeastern USA ecosystems tend to remain green longer and have dull colors in autumn, presumably adaptations to different winter severity and avian migration patterns.

An ongoing litany of cases is documented for invaders causing major economic loss, especially in Australia.[624] Similarly, epidemiology and public health have tracked a long ongoing sequence of diseases spreading through livestock, crops, forests, and humans. Exotics, and especially invaders, produce a wide range of effects that cascade through the ecosystem (Fig. 11.9). In the context of plummeting biodiversity worldwide, primarily due to habitat loss, the role of invading species in altering the species composition and richness of native ecosystems warrants monitoring. Many native populations become locally extinct with the spread of invaders in ecosystems, but limited evidence is available for species going extinct directly due to invaders other than humans.

397

Despite numerous case studies of invaders, and the search for population characteristics and for environment/ecosystem/community characteristics, the subject of exotic species has yet to gel. Few common patterns or principles and little predictive ability have emerged.[1159,624] The largely missing ingredient may be a focus on spatial pattern[1222] (C. Bull, pers. commun.).

Some disparate observations are useful. The patch-corridor-matrix principle should underlie the identification of toeholds and routes of spread. Thus, perforation of a matrix (e.g., in dispersed patch cutting[518,967]), disturbance patch size and shape[1320,501], and adjacencies (e.g., an adjoining source of chronic disturbance inputs, of changes in native species structure, and of reinfections by exotic species[1208]) are important patch characteristics. Corridors such as railroads and ditches are key conduits[336,1278] (chapters 5 and 6). And the matrix with its varying degrees of resistance (chapter 8) is a key to spread, especially determining whether the species becomes a dominant at the landscape scale. The study of exotics and invaders may crystallize when landscape spatial pattern is added to population and environmental characteristics.

The lessons from numerous case studies of exotics and invaders provide important guidelines for the homeowner, land manager, landscape architect, and policymaker. These include: (1) Never introduce or use a known invader. (2) Use almost any mix of non-invasive natives or exotics in urban or central suburban areas where exotics are rampant. (3) Rigorously limit the use of exotics near natural areas where protecting native biodiversity or rare species is important.[1764] (4) Rigorously limit the use of exotics in rural areas where natural ecosystems could be invaded.

We end this section with perhaps the best case study for an exotic plant.[1026,1027] From 1900 to 1980 cheatgrass (*Bromus tectorum*) replaced native grasses and shrubs over thousands of square kilometers, from the Rocky Mountains to the Sierra Nevadas/Cascades and from Nevada to British Columbia. The species got toeholds on small disturbed patches, and spread long distances along corridors.[501] Cheatgrass spread locally into large newly formed patches, and then became a landscape dominant. The last phase was 'like a group of coalescing leopard spots'.[1026]

LIVESTOCK

Livestock introduced into a pasture (paddock) apparently undergo a relatively distinct learning process in the way they use space. First consider some background characteristics of the animals (Fig. 11.11). In general, domesticated animals in a herd or flock are not closely related geneti-

Fig. 11.11. Herd characteristics, physiological needs, and learning the pastureland govern livestock movement patterns. Longhorn steers in Oklahoma (USA). Courtesy of US Fish and Wildlife Service.

cally, though partial or full family groups may be included. The animals are social in the sense of foraging as groups, but at least at the outset, exhibit little territorial or pecking-order behavior. Livestock learn movement patterns, and are creatures of habit. At the same time an inherent component of behavior is present, and foraging strategies still have their roots in the evolutionary history of the species.[1863]

Consider also the pasture space. It is vegetationally heterogeneous with patches and corridors of higher and lower palatability. Accessibility to the vegetation types varies. Water is usually available in one or a few distinct locations. Specific mineral salts may be limited to specific areas. Temperature regimes, both warmth and cold, are heterogeneously distributed depending on slope and wind. And constraining boundaries surround the pasture.

New livestock introduced to a pasture typically first learn the boundaries[1649] (Fig. 11.12). Next they usually locate water, the key resource, especially free standing water. Then they find suitable forage. The animals are 'central-place foragers', and move outward from water in search of food.[1820] Thus, foraging intensity tends to decrease with distance from water. For example, the optimal foraging range for cattle and sheep is generally a radius of c. 1.3 km, and the maximum range about twice that distance.[1765, 35] Severe drought reduces vegetation and may cause animals to forage further. However, the energy required to return to water is a major constraint on foraging distance.[1608, 567, 1815, 34]

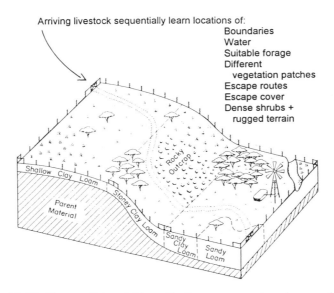

Arriving livestock sequentially learn locations of:
Boundaries
Water
Suitable forage
Different
  vegetation patches
Escape routes
Escape cover
Dense shrubs +
  rugged terrain

Fig. 11.12. Key spatial variables affecting movement and usage by live-stock. Generalized pasture pattern and learning sequence. Adapted from Stuth (1991). Illustration by L. Lyddon. Reprinted with permission from *Grazing Management: An Ecological Perspective* edited by R. K. Heitsch-midt and J. W. Stuth, copyright 1991 Timber Press.

Animals also must learn the distribution of different vegetation patches, escape routes, and escape cover (Fig. 11.12). Dense shrubs and rugged terrain tend to be avoided by most livestock while foraging.[1649] Conversely, open terrain and established path corridors are favored in foraging.

The focus on learning means that experience is important in foraging efficiency and livestock production.[1535,1534,1649] An inexperienced herd that follows established habits may remain relatively inefficient and unproductive. Simply introducing some experienced animals can increase the efficiency and production of a herd, because it 'plays follow the leader'. Therefore, harvesting a whole herd or flock, for example in a severe drought, means that the next herd will be less efficient until its animals have learned the terrain.

An ordered sequence of physiological needs is also thought to govern an animal's behavior, and hence use of space[1585 cited by 1649] (Fig. 11.11). Thirst for water is the top priority. Maintaining a heat balance between hot and cold is next. If those two are satisfied, the animal focuses on food, the caloric balance. Nighttime, when few cues for these visually oriented animals are available, requires a focus on remaining oriented

400

and avoiding predators. Finally, resting permits the conservation of calories.

Each step in this inherent series of steps affects how and when the animal moves through the heterogeneous pastureland. For example, nighttime grazing by cattle is typically concentrated in the area where the herd ended up at dusk.[1819] Goats and sheep tend to graze into the wind to conserve heat.[1585 cited by 1649] This results in more grazing pressure on the upwind portion of a pasture. When forage is limited, grazing time increases. But in near starvation times the animal often tends to give up foraging, apparently due to the considerable energy lost in searching.[295]

Thus, livestock movement through the heterogeneous pasture shows similarities to that of wildlife described earlier in this chapter.[1095,567] It is probable that many of the details well worked out for livestock also apply to wildlife movement. Just as for wildlife, many management manipulations and additions are available to increase livestock production.[1326,875,1946] The primary difference, other than the surrounding boundary, stems from the little-structured social behavior of genetically little-related animals.

Livestock management and production, however, often require a *grazing system*, i.e., using two or more pastures in alternating time periods.[1955,1590,713] Sustained production avoids overgrazing as well as heavy trampling and waste accumulation.[1818,1578,1678,713] Two basic tactics are used to increase forage quantity or quality.[714] 'High utilization grazing' (HUG) means that all plants are moderately or intensively defoliated, whereas 'high performance (or production) grazing' (HPG) means that only preferred plants are defoliated and at light to moderate levels. HUG requires animals to eat non-preferred plants, and is the only tactic for high livestock densities. Either HUG or HPG can be used for light to moderate animal densities, by moving herds among different pastures.

Three basic types of grazing system offer options in using HUG or HPG tactics for maintaining or improving pasture quality.[714] The first ('deferred rotation') is a multi-herd system using HPG, the grazing of only preferred plants[1119] (Fig. 11.13b). Livestock in moderate densities graze pastures for a long period before being rotated to other pastures. The original pastures then rest (are fallow) for a long period. The second grazing system ('high intensity/low frequency') uses the HUG tactics of defoliating all species (Fig. 11.13c). In this a single herd with high animal density grazes a pasture for a moderate period. Then the herd is rotated to another pasture, allowing the first pasture to have a long rest period.

The third system ('short duration') is similar to the second, except that HPG tactics are used (Fig. 11.13d). This is attained with high livestock density by shortening the grazing period, and also the rest period. This third grazing system can only be sustained if adequate plant regrowth

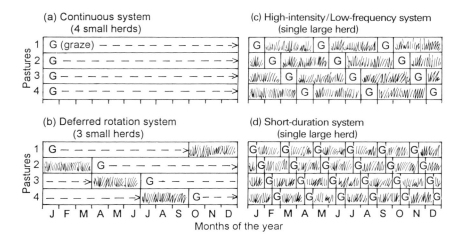

Fig. 11.13. Major grazing systems illustrated with sequence of grazed and rest periods for four pastures. G = grazing period; grass symbol = fallow or rest period for pasture. Climate and plant growth vary over the year, and lengths of grazed and rest periods for pastures normally are varied accordingly. Adapted from Heitschmidt & Taylor (1991).

takes place during the brief rest period. Unfortunately, this rarely occurs in dry landscapes.[1094,105,712,391] Various combinations of the basic approaches can also be used. These may combine HUG and HPG tactics, single and multi-herds, moderate and high densities, long and short grazing periods, and/or long and short rest periods for pastures.

In comparison with continuous grazing of a pasture (Fig. 11.13a), which of these grazing systems is best for sustained livestock production? This is a subject of lively debate worldwide.[1816,1578,1678,206] Perhaps the most valuable message in comparing the options is the importance of moderate livestock density (relative to plant growth).[714] Any increase to high density requires careful compensation, either by shortening the grazing period or lengthening the rest period or both. If not, overgrazing is certain. That means a drop in livestock production and animal protein for humans. It also means an increase in soil erosion. Indeed, most of the huge areas of desertification worldwide result from overgrazing by livestock at high density.

PART V

Changing mosaics

12

Land transformation and fragmentation

But when I consider that the nobler animals have been exterminated here — the cougar, panther, lynx, wolverine, wolf, bear, moose, deer, the beaver, the turkey, etc., etc. — I cannot but feel as if I lived in a tamed, and as it were, emasculated country . . . I listen to a concert in which so many parts are wanting . . . for instance, thinking that I have here the entire poem, and then, to my chagrin, I hear that it is but an imperfect copy that I possess and have read, that my ancestors have torn out many of the first leaves and grandest passages . . .

Henry David Thoreau, Journal, *1856*

Robert Burns' classic description of change as 'Nature's mighty law' reminds us that whatever humans do, it is set within a changing land driven by natural processes. Furthermore, change, whether natural or human induced, is the norm. Short static or equilibrium phases are common within the broad dynamics. But persistent static phases are the exception, and they never persist indefinitely.

The preceding chapters have explored the structural patterns and functional flows of mosaics during a static phase. This was useful for explanation, but in this and the following chapters we focus explicitly on the changing landscape or region. This chapter examines change driven by natural processes or by mainly unplanned human activities. The subsequent two chapters focus on change where human planning is also important.

The subject of land transformation and fragmentation is significant to all human issues that involve land. Wise forestry, economics, biodiversity conservation, agriculture, landscape architecture, sociology, wildlife biology, soil science, and so forth explicitly recognize and deal with a dynamic land. These dynamics also characterize finer and broader scales, albeit at different rates, including succession within a patch[1233,1029] and global climate change.[1413,1158]

Many of the ideas and syntheses in this chapter were developed jointly with George F. Peterken; indeed, he is almost a coauthor of chapter 12. No overall review of the subject of this chapter exists. However, individual topics are reviewed as follows: fragmentation[225,683,1793,1536,39]; spatial processes[683,501,1742,661]; and geometric modeling of mosaic sequences.[518]

Chapter 1 introduced the concepts of patch mosaic dynamics and landscape change. Chapters 2 to 8 explore changes in specific structures within a landscape or region, to wit: moving boundaries; filling in coves; tip splitting; elongating patches; shrinking patches; expanding patches; moving patches; coalescing patches; widening corridors; 'curvi-linearizing' corridors; migrating meander bands; developing rectilinear networks; and restructuring dendritic networks. The present chapter assumes familiarity with such concepts, especially those in chapters 2–6 and 11. Many ecological characteristics change as landscapes are progressively modified.[1195,582,501] In preceding chapters, change in the overall level of heterogeneity has often been of interest (chapter 9).[974,77,1748] The present chapter now focuses on spatially explicit changes in pattern, and the consequent ecological effects in land mosaics.

Thus the broad goal is to understand the spatial processes in transforming or changing land, and see if an 'ecologically-optimum' sequence for changing land can be identified. With such a sequence of mosaics, one should ultimately be able to pinpoint the best and worst location for a particular change at any stage in the sequence.

The chapter begins with the spatial processes in land transformation. This leads to a focus on one of the processes, habitat fragmentation. Next, a brief look at all types of land change is presented. The sequences of mosaics in each of the most common land changes are then described by simple geometric models, and compared ecologically in a quest for an ecologically optimum land transformation. Finally, additional frontiers in understanding land transformation are identified.

SPATIAL PROCESSES IN LAND TRANSFORMATION

In little more than a decade fragmentation, the breaking up of large habitat or land areas into smaller parcels, has become an environmental issue of worldwide proportion. Many species, including most large mammals and birds, cannot maintain viable populations in small habitat patches, which leads to extinction and loss of biodiversity.[498,991,1595,1490,678,1912] In addition, land fragmentation commonly disrupts the integrity of a stream network system, water quality of an aquifer, the natural disturbance regime in which species evolved and persist, and other ecosystem processes[1503,1303,1320,501,1742,130,730] (chapter 2).

Yet fragmentation is but a phase in the broader sequence of transforming land by natural or human causes from one type to another.[497] Other spatial processes in landscape change or transformation are equally prominent and ecologically significant.[347,1774,683,1320,1892,1307,1492] For example, a forest patch may be perforated with a clearing, or may shrink in size, but remain a single patch. Or the patch may disappear. In fact, some ecologically interesting land transformations have no fragmentation at all. Thus, it is important to examine fragmentation together with other spatial processes in the broader framework of land conversion.

Spatial processes

Perforation, the process of making holes in an object such as a habitat or land type, is probably the most common way of beginning land transformation. Thus, an extensive forest is perforated by logged or blowdown clearings, and a desert-grassland by scattered houses or clusters thereof (Fig. 12.1).

Normally, the alternative way to begin transforming land is *dissection*, the carving up or subdividing of an area using equal-width lines. A road network constructed during the nineteenth century in the USA Midwest, and roads today penetrating tropical rainforest, dissect or subdivide the landscapes into sections.

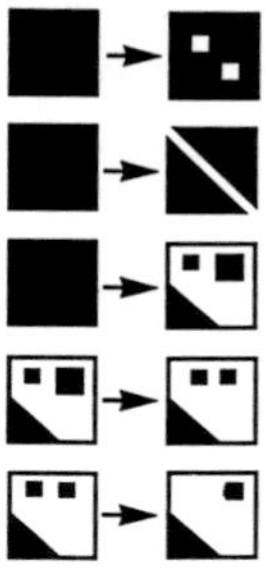

Spatial processes		Patch number	Average patch size [1]	Total interior habitat [2]	Connectivity across area [3]	Total boundary length [4]	Habitat	
							Loss	Isolation
	Perforation	0	−	−	0	+	+	+
	Dissection	+	−	−	−	+	+	+
	Fragmentation	+	−	−	−	+	+	+
	Shrinkage	0	−	−	0	−	+	+
	Attrition	−	+	−	0	−	+	+

Footnotes. [1]Perforation = "0" if size is measured as diameter rather than area; Attrition = "0" or "−" if patch lost is ≥ average patch size. [2]Shrinkage or Attrition = "0" if patch changed had no interior habitat. [3]Perforation = "−" if random straight routes are measured; Shrinkage or Attrition = "−" if measured as probability of object crossing using patches as stepping stones. [4]Shrinkage = "0" or "+" if portion lost makes no change or increases boundary of the patch.

Fig. 12.1. Major spatial processes in land transformation and their effects on spatial attributes. + = increase; − = decrease; 0 = no change. Effects are measured for the black land type or habitat. White land type surrounds the landscape. Figure prepared by Michelle Leach.

407

Fragmentation, the breaking up of a habitat or land type into smaller parcels, is here considered as similar to the dictionary sense of breaking an object into pieces.[497] It is implicit that the pieces are somewhat-widely and usually unevenly separated. Thus, breaking a plate on the floor is fragmentation, whereas carving up or subdividing an area with equal-width lines is dissection. Dissection could be considered a special case of fragmentation. The two spatial processes are differentiated in part because the separating elements typically are so different and wide-spread (roads, railroads, powerlines, windbreaks, etc. versus logged clearings, cultivated fields, housing tracts, pastures, and the like). The ecological effects of dissection and fragmentation may be similar, or highly dissimilar, depending on whether the dissecting corridor is a bar-rier to movement of the species or process considered.

Shrinkage, the decrease in size of objects, such as patches, is almost universal in land transformation (Fig. 12.1). For instance, remnant woodlots shrink as portions are removed for farming or houses. And *attrition*, the disappearance of objects such as patches and corridors, is also essentially always present in changing landscapes. Usually small patches disappear, though the occasional disappearance of large patches is apt to be especially significant ecologically.

Each of these five common spatial processes (Fig. 12.1) has distinctive spatial attributes, as well as significant effects on a range of ecological characteristics, from biodiversity to erosion and water chemistry. The first three, perforation, dissection and fragmentation, may affect either the whole area or a patch within it. The last two, shrinkage and attrition, most appropriately apply to an individual patch or corridor.

Patch number or density in the landscape increases with dissection and fragmentation, and decreases with attrition (Fig. 12.1). Average patch size decreases in the first four processes, and typically increases upon attrition, because small patches are most likely to disappear. The total amount of interior habitat normally drops with all processes. Connectivity across the area in continuous corridors or matrix typi-cally decreases with dissection and fragmentation. The total boundary length between original and new land types increases in the first three processes, and decreases with shrinkage and attrition. In short, each spatial process has a distinctive effect on the spatial attributes of a landscape, and therefore doubtless affects ecological characteristics differently.

The five spatial processes overlap through the period of land trans-formation (Fig. 12.2a). They also usually are ordered in their relative importance, beginning with perforation and dissection and ending with attrition. Perforation and dissection both peak at the outset, when a landscape is typically both perforated with patches and subdivided by

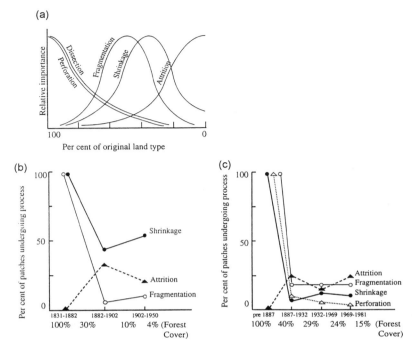

Fig. 12.2. Spatial processes changing with land transformation. (a) General spatial-process model indicating common phases of maxima and overlaps. (b) Example from Cadiz Township, Wisconsin (USA). Approx. 87.2 km² of forest in 1831; deciduous forest is mainly dominated by oak, maple, and/or ash (*Quercus* spp., *Acer saccharum*, *Fraxinus americana*); no information available on perforation or dissection. Based on Burgess & Sharpe (1981b). (c) Example from Miyako River basin, Chiba. Approx. 21.5 km²; assumed to be all forested at some point pre-1887; almost all forests are conifer plantations (mostly *Pinus densiflora*, and some *Cryptomeria japonica* and *Chamaecyparis obtusa*); no information available on dissection. Based on Ohsawa & Liang-Jun (1987). Percentage of the area in forest is given beneath each date. Figure prepared courtesy of Michelle Leach.

corridors. Fragmentation and shrinkage predominate in the middle phases of landscape change.

Land transformation examples

Two examples of land transformation illustrate several processes in the 'spatial-process model' of Fig. 12.2a. Within Cadiz Township, Wisconsin (USA), a landscape intensively studied by David Sharpe, Forest Stearns, Robert Burgess, Christopher Dunn, and Glenn Guntenspergen following earlier work by John Curtis[347,226,1541,428], the predominately forested area is mainly converted into farmland (Fig. 12.2b). On the other hand, the Miyako River basin in Chiba, Japan begins presumably nearly all forested, and is mainly converted into suburbia or built environment (Fig. 12.2c).[1242] From 1882 to 1950 Cadiz dropped from 30% to 4% forest, and from 1887 to 1981 Miyako dropped from 40% to 15% forest. Thus these two examples illustrate a quarter of the general model near its end (Fig. 12.2a).

Relatively little fragmentation takes places during the periods encompassed (after 1882 or 1887). In both landscapes fewer than 20% of the patches present at each step become fragmented (Fig. 12.2b and c). Shrinkage is low at Miyako, whereas at Cadiz about half the patches are reduced in size at each time step. Attrition is the same in both landscapes; about a quarter of the patches disappear at each step. Thus, although more long-term studies are needed to test the spatial-process model, it appears to portray important processes in real-world land transformations.

The differences in the Cadiz and Miyako curves also provide insight into the nature of deforestation. At Cadiz little fragmentation takes place after 1882, but the forest patches continue to shrink in size (Fig. 12.3). Thus, agricultural intensification[61,10,1504,1305], especially pastureland for dairy cows, leads to a progressive shrinkage in forest area and patch size. In contrast, at Miyako fragmentation continues (approximately 17% of the patches at each step), and few patches shrink in size (Fig. 12.2c). These Miyako patterns, combined with the observation that some patches increase in size, some become combined into one, and some new patches appear, strongly point to the ongoing logging and regrowth process taking place at Miyako.[1242]

Comparing differences between a particular land-transformation history, such as the Cadiz and Miyako examples, and the general spatial-process model (Fig. 12.2a) can thus be quite informative. For instance, fragmentation could peak early, if agriculture spread evenly along all roads of an initial road network, perhaps as in certain Amazon areas. Or dissection could peak twice, if later in the sequence roads were constructed within already fragmented patches. This is illustrated by the construction of 'rides' (c. 2 to 20 m wide paths) through many British woods, e.g., in the eighteenth century, for wildlife, hunting, and recreation.[1370,1306]

Similarly, all five processes in the spatial-process model (Fig. 12.2a) do not have to be present in a particular land transformation. Widening ribbon or strip development can lead directly from dissection to fragmentation, with no perforation. Logging in strips progressively from one boundary to the opposite boundary causes no fragmentation at all. Indeed, the exclusion or minimization of particular spatial processes,

Fig. 12.3. Shrinkage and attrition causing habitat loss and isolation. This 1980s photograph probably shows an area of southern Wisconsin (USA) near and similar to Cadiz Township. Perforation and fragmentation essentially ended after most of the forest cover was rapidly removed in a few decades of the mid nineteenth century. Over the subsequent century, shrinkage and attrition have been the major spatial processes affecting wooded patches. R. Forman photo.

those with negative ecological characteristics, is the key to a search later in this chapter for an ecologically optimal sequence of land transformation.

The Cadiz and Miyako examples also remind us that land transformation does not proceed at a constant rate. Spurts followed by periods of stasis are the norm. Reversals in direction occur, such as periods of reforestation within an overall deforestation pattern. And new directions appear. For example, part way through a transformation from rainforest to small-plot cultivation, ranchers arrive and redirect the change so

411

rainforest is directly replaced by pastureland. In contrast to the relative directionality of succession within a patch, diverse trajectories are possible in landscape change.

HABITAT FRAGMENTATION

The concept

The concept of fragmentation is sharpened using this spatial process analysis. Fragmentation has often been used in the general sense of land transformation that includes the breaking of a large habitat into smaller pieces.[225,683,1793,1493,1577] This concept would include perforation, shrinkage, and indeed all five spatial processes (Fig. 12.2a).

Alternatively, the fragmentation concept has been used in a narrower sense, of being essentially the combination of 'habitat loss' and 'isolation'.[1893,991,1892] However, both habitat loss and isolation increase with all five spatial processes (Fig. 12.1). Furthermore, habitat loss can take place with or without fragmentation, and likewise isolation can increase with or without fragmentation (Fig. 12.3). In essence, habitat loss and isolation are both useful concepts, but different and broader than fragmentation. Their increase may result from, rather than is, fragmentation. Indeed, many additional spatial and ecological characteristics result from or are correlated with fragmentation[225,683,983,730,967] (Fig. 12.1).

Thus, maintaining the fragmentation concept similar to that of the dictionary[497], i.e., the breaking up of a habitat or land type into smaller parcels, enhances communication among disciplines and the public. Obviously, habitats rarely move horizontally like the pieces of a dropped plate. However, aerial photos, taken say at decade intervals, often reveal a net change for a habitat type. The type may be more fragmented, less fragmented, or unchanged. Placing fragmentation in the broader framework of land transformation (Fig. 12.2), should help in designing specific tests of its ecological effects.

Fragmentation is caused by natural processes as well as human activities.[983] Fire and herbivore explosions break continuous habitat into small patches just as effectively as logging and suburbanization. Indeed, some human activities cause fragmentation, whereas others, such as the spread of irrigation or severe degradation due to overgrazing, typically change the land without fragmentation.

Furthermore, each mechanism operates at a range of scales.[1488,983,39,1493] Thus, a nesting of fragment sizes over a range of scales is probably a normal result of land fragmentation.[10] The finer scale pattern of habitat fragments is suggested to be especially detrimental to small organisms, specialist species, and ecosystem functions.[983]

412

Finally, it warrants emphasis that scattered patches result from several mechanisms (chapter 2). One is fragmentation of previously continuous habitat. Another is a patchy substrate of different soils and slopes. Alternatively, scattered patches may be due to colonization into new separate locations. Thus, climate change could convert an area of continuous woodland to grassland, while in the adjacent land woodland concurrently colonizes and forms scattered patches. No fragmentation occurs in this change. Climate change also could simply shrink the original continuous woodland, at least initially causing habitat loss without fragmentation. Or it could shrink the woodland, leaving scattered remnant patches, hence causing habitat loss with fragmentation. In short, fragmentation is one of several important spatial processes in land transformation. A multitude of spatial, species, and other effects introduced below result from this process.

Effects of fragmentation

Many of the major effects of habitat fragmentation have been separately examined in other contexts of previous chapters. These include the ecological effects of patch size and number (chapter 2), connectivity and isolation (chapters 5 and 6), and species movement (chapter 11). Nevertheless, a list here provides an overview and illustrates the wide range of fragmentation effects. Exceptions exist for some effects listed, and an increase, decrease, or no change often depends upon what the initial and new habitat types are. Most references are given in the previous chapters; a few illustrative and additional references are provided here. Fragmentation effects and their increase or decrease are listed in three categories: spatial; species; and other.

Spatial effects

Aerial photographs or maps of an area are widely available for a sequence of three or more time periods, and are an especially useful beginning for analyzing fragmentation. Many spatial patterns have been measured along this sequence, especially for forest land.[347,1541,1242,428] For instance, Fig. 12.4 represents periods of all forest, about two-thirds forest, and one-third forest in the USA Pacific Northwest. The following spatial attributes are commonly reported to increase: patch density, inter-patch distance, boundary length, stepping stones, and corridors.[1843,1875,1892,1743,1744,110] Decreases in patch size, connectivity, interior-to-edge ratio, maximum size of core, and total interior area are also common.[485,1688,273,662,1602] Patch shape and fractal dimension may increase, decrease, or not change, depending upon the pattern of fragmentation.[1542,1252,1234] These results of course depend on spatial scale, just as

413

Fig. 12.4. Coniferous forest landscape transformed over three decades. Canopy trees are Douglas-fir (*Pseudotsuga menziesii*) c. 75 m tall, and understory trees include western hemlock (*Tsuga heterophylla*). Cascade Mountains, Oregon (USA). R. Forman photos.

perforation and remnant patch sizes can be scaled to the size of a landscape.[1750]

Species effects

Species in turn respond to the changing spatial patterns caused by fragmentation.[1515] Increases in the following are common: isolation[1367]; number of generalists[110]; number of multihabitat species; number of edge species[1379,1067,1467,1941]; number of exotic species[663]; nest predation[268,186,1891,32]; and extinction rate.[1691,398,1892,1488,923] In contrast,

414

decreases are characteristic for: dispersal of interior specialists[1367,564,837,1000,1434,1065]; large-home-range species (of initial habitat type); and richness of interior species.

Other effects

Fragmentation affects nearly all ecological patterns and processes, from genes to ecosystem functions. Thus, increases in metapopulation dynamics and genetic inbreeding are common.[465,1257,1116,1359,532] On the other hand, decreases are typical in: internal habitat heterogeneity and the sizes of disturbance patches.[1016,1892] Some variables increase, decrease, or remain unchanged, such as natural disturbance[518,1602], hydrologic flows[992,1481,730], wind movement[481], nutrient cycling[730], productivity[1743,1744,1746], and gene flow.[656,481]

The preceding examples emphasize the considerable body of information available on the ecological effects of fragmentation. Although a focus has been on species extinction, fragmentation as a spatial process has effects on almost all ecological patterns and processes.

Fragmentation is often caused by natural disturbance and human activity, such as fire and logging operations. It may be due to a single event, or a sequential or continuous process, where patches are progressively fragmented into smaller pieces (Fig. 12.4). In the latter case the effect could be a curvilinear response, instead of linearly increasing or decreasing.[1309,1817,518] For example, converting large natural-vegetation patches to medium patches may have a major effect on interior species, but going from small to tiny patches may have a negligible effect.

An important research frontier is to understand the other spatial processes in land transformation, i.e., perforation, dissection, shrinkage, and attrition. The ecological effects of, for instance, perforation or attrition, appear just as great as those of fragmentation.[97,657,1306]

LAND CHANGE PATTERNS

By examining aerial photographs and surveying the literature from several disciplines the common causes and patterns of land transformation can be identified. To simplify and provide direct comparability, the heterogeneity present in a landscape is reduced here to two categories, an 'initial land type' and a 'new land type.' The scale is the landscape[1430,501], ranging from a few kilometers to several tens of kilometers across. Thus land is transformed from forest to pasture, agriculture to suburb, and desert to agriculture.

As an example, a coniferous forest landscape (*Pseudotsuga–Tsuga*) is transformed by dispersed-patch logging in the USA Pacific Northwest

415

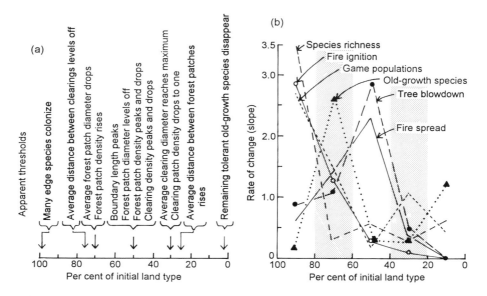

Fig. 12.5. Thresholds and periods of rapid change in a dispersed-patch deforestation process. Pacific Northwest (USA) Douglas-fir forest landscape (see Fig. 12.4). (a) Apparent thresholds of spatial attributes based on a dispersed-patch or checkerboard model.[518] (b) The rate of change of six major ecological characteristics at different phases in the deforestation process. Slopes are from Figs. 4 and 5 in Franklin & Forman (1987).

(Fig. 12.4). In perhaps the first landscape-level analysis of such a conversion process, Jerry Franklin and R. Forman evaluated its diverse ecological effects using a simple, geometric 'checkerboard model'.[518] Several apparent thresholds, where spatial attributes change rapidly in the model, are identified along the percentage-of-initial-land-type horizontal axis (Fig. 12.5a). Some of the thresholds have empirical support, some could result from simplifying a complex landscape in a model, and some await testing.[518,1602]

The transformation sequence has negative effects on many ecological factors, including windthrow, fire ignition, and interior species richness. Overall most of the rapid changes in ecological factors take place in the first 40% of transformation, and few rapid ecological changes occur in the last 40% (Fig. 12.5b). All but one of the variables change rapidly (slope >1) in the second of the five phases, i.e., when the initial land type drops from 80% to 60%. Only one variable changes rapidly in each of the last two phases.

This analysis indicates that most big ecological changes occur in the first half of a land transformation. Furthermore, the most-critical time for land planning and conservation appears to be when the landscape has 60 to 90% of its area in natural vegetation.

Several alternative possible logging sequences are compared against the checkerboard model.[518] This comparative modeling approach has been extended to evaluate the effects of changing spatial pattern on avian populations and on interactions with natural disturbance regimes.[662,1602,1662] Also a reduction and a delay in forest fragmentation are found by aggregating patch cuts, using large cuts, or including some fixed constraints, such as a nature reserve or dendritic stream network.[967,1662]

These logging and natural disturbance patterns emphasize that the 'inertia' of a landscape is a conspicuous attribute.[353,1662] At this scale spatial patterns persist for long periods, often well beyond the ecosystem or land use originally produced. Distinct spatial patterns remain from hurricanes, villages, cultivated fields, and ditches that disappeared even centuries ago.[1370,509,510,132,97,1306]

Diverse mechanisms from logging and suburbanization to wildfire and desertification transform land from one type to another. Yet a limited number of basic, changing spatial patterns in these land conversions may exist, irrespective of mechanism.[582] If so, such patterns can be analyzed for their changing spatial attributes such as patch size and boundary length.[1424,1886,1252,1745,1234] Changing patterns can also be analyzed for ecological characteristics known to correlate with these spatial attributes.[501,24,518,174,1228]

The goal of such an analysis is to identify which land-conversion sequence is 'ecologically optimum', that is, which minimizes the negative effects of fragmentation and other spatial processes. A changing spatial pattern could be evaluated for separate ecological factors such as water quality, erosion rate, butterfly species richness, or a rare predator. In contrast, to be useful in long-term societal planning, conservation, and policy, an overall concept of ecologically optimum incorporates a wide range of ecological characteristics. No pattern is optimum for every characteristic, but an ecological optimum should accommodate most major ecological characteristics.

Major land transformations

A literature survey suggests that six major causes of land transformation are widespread: deforestation, suburbanization, corridor construction, desertification, agricultural intensification, and reforestation. The six major causes are briefly introduced, and in each case the key changing patterns are encapsulated. Each changing spatial pattern is effectively a *mosaic sequence*, that is, a series of spatial patterns over time. Identifying the mosaic sequences is the key to developing spatial models that can be directly compared ecologically.

Additional examples of landscape change, such as wetland drainage or degradation due to chronic air pollution, exist and warrant spatial study. Finally, several 'instantaneous' transformations are merely mentioned at the end.

Deforestation

Cutting a forested landscape is normally a highly planned process, and consequently there is an unusually wide range of common mosaic sequences.[154,518,662,967,354,1577,1823] These include: logging expanding progressively from an edge; expansion from a few separated nuclei; dispersed patch cutting; and alternate strip cutting (Table 12.1).

Suburbanization

Individual housing clusters and areas of development are sometimes highly planned, that is, the area concerned is planned as a whole. However, more common at the landscape-scale is the unplanned or little-planned spread of suburbs.[1832,97,1809] Such unplanned patterns are highly non-random, and tend to mainly reflect geomorphic and transportation templates. Suburbanization is primarily discussed in the following chapter 13, but here a few common mosaic sequences are identified[733,436,709,1458]: concentric rings spreading outward from an adjacent city; growth along an exurban transportation corridor; and spread from satellite towns, plus infilling.

A complex mosaic sequence can include components of two or more of the preceding sequences. This is illustrated in a 20 × 10 km area extending eastward from Lyon, France that was mapped in 1900, 1950, 1970, and 1988 (J. K. Tatom, pers. commun.). Over the nine decades agriculture continued as the matrix, which remained undivided. Small patches of vegetation 'moved around' through attrition and appearance, but changed little in density. Large woods became progressively smaller and narrower. Towns grew in size and developed highly convoluted margins. Some towns became connected by strip or ribbon development. New small built areas appeared. And the city coalesced with or gobbled up nearby towns. Most of the changes occurred in spurts, and no reversals in pattern were evident.

Corridor construction

The construction of a new corridor such as a road or rail line opens up an area in a linear fashion (Fig. 12.6). Spread then proceeds outward from the corridor on opposite sides. Branch lines typically follow forming a dendritic pattern[1670,995,1577,354] (chapter 8). Transformations com-

Table 12.1. *Changing spatial patterns in land transformation. Proposed models are described in the text*

Cause	Changing spatial pattern	Proposed model
Forest cutting	Logging expands progressively from an edge	Edge
	Expansion outward from a central cut strip	Corridor
	Expansion from a new cut patch	Nucleus
	Expansion from a few separated cut patches	Nuclei
	Dispersed cuts that avoid adjacent cutting until middle of rotation	Dispersed
	Network template that avoids large cut patches until end of rotation	Network template
	Alternate strip cutting	Alternate strip
Suburbanization	Concentric rings spreading outward from adjacent city	Edge
	Growth along exurban transportation corridor	Corridor
	Spread from satellite towns, plus infilling	Nuclei
	Asynchronous bubble growth outward from city	Edge
Corridor construction	Road or railroad penetrating new area	Corridor
	Sewer line penetrating new area	Corridor
	Irrigation canal penetrating new area	Corridor
Desertification	Particulate mass spreads from adjacent area	Edge
	Spread from overgrazed spots within area	Nuclei
	Single-event heavy sediment deposition	Instantaneous
	Salinization or lowered water table over whole area	Even throughout
Settlement and agricultural spread	Scattered farmsteads throughout	Dispersed
	Villages or cooperatives without farmsteads	Nuclei
	Expansion from edge of landscape	Edge
Chronic air pollution	Vegetation damaged over whole area	Even throughout
Reforestation	Small scattered patches from abandoned fields	Dispersed
	Large planted subdivided geometric patches	Nuclei
Wetland drainage	Various human-determined patterns	Several of above
Volcanic eruption	Impact flattens, or lava flow covers	Instantaneous
Burning	Wildfire spreading from one or more sources	Instantaneous
Flooding	Dike breaks, or river rises and widens	Instantaneous
Bombing or herbiciding	Massive levels that transform landscape	Instantaneous
Nuclear explosion	Impact flattens area	Instantaneous

419

Fig. 12.6. New road corridor opening up Amazon rainforest to logging, cultivation, pastureland, and settlement. Brazil. Photo courtesy and with permission of Elizabeth L. Taylor.

monly result from construction of new roads, sewer lines, and irrigation canals (Table 12.1).

Desertification

Human-caused conversion of grasslands, savannas, and other areas into desert-like conditions of low vegetation cover and productivity is extensive worldwide.[60,214,1954,1502] However, the causes of desertification are diverse.[793,1100,1605,1930] Most important is overgrazing by livestock.[1860,420,272] But salinization and a lowered water table are also common causes. All of these usually lead to a mass movement of soil particles.[70,1570,1714] Desertified areas expanding from one side of, or from nuclei within[657], a landscape apparently are common mosaic sequences.

Agricultural intensification

Intensifying cultivation or grazing in dry areas could be an early phase of desertification. In moist areas it could be a late stage of deforestation. Yet agricultural landscapes have persisted in many regions for centuries or millennia, and their intensification is often the major change.[61,1422,1423,631,1307,531,1504,1149,82,1305] In such landscapes agriculture tends to intensify either around scattered farmsteads, or in their absence,

420

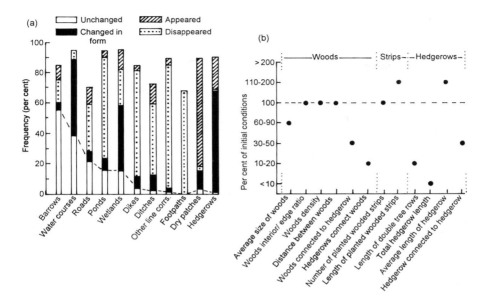

Fig. 12.7. Changes in small natural-vegetation elements during agricultural intensification. (a) The fate of 1566 elements in 11 categories over a century (1884 to 1981). Based on sampling 249 1 km² plots evenly distributed across agricultural eastern Denmark (Agger & Brandt 1988). (b) Relative changes during land consolidation in an agricultural landscape of eastern Netherlands from c. 1950 to c. 1970. Increases are above, and decreases below, the dashed line. Note that the vertical axis is nonlinear. Based on Harms et al. (1984).

around villages or cooperatives (Table 12.1). Intensification results in a progressive removal of usually small landscape elements.[1468,89,10,680,224]

Peder Agger and Jasper Brandt (1988), interested in the ecological roles and management strategies for small spatial elements (biotopes), studied the stability, appearance, and disappearance of numerous element types in an agricultural region of Denmark. Eighteen patch types and 22 corridor types are present, which are grouped into 11 categories (Fig. 12.7a). Over a century three kinds of spatial element are most likely to persist unchanged: barrows (gravel pits), water elements (water courses, ponds, and wetlands), and roads (paved and gravel). The least stable are all dry elements, i.e., those on well-drained soil: hedgerows, dry patches, footpaths, other line corridors, ditches, and dikes. More than half of the most stable elements are patches, whereas almost all the unstable elements are corridors. Hedgerows and dry patches are the types most likely to appear anew. Disappearances of small elements are frequent for many of the types. The authors pinpoint the special importance of natural vegetation elements along and attached to boundaries at various scales, e.g., between farms, along roads, and between counties.

During recent decades in several European countries, land consolidation programs have brought small-ownership parcels together to create larger, supposedly more-economic farms and fields that have large machinery.[91,679,224] Such land transformations create major changes in energy flow (Mizgajski 1990) and mosaic structure. The latter is illustrated by Bert Harms and colleagues showing changes over two decades in the eastern Netherlands.[679] Small scattered woods change little in area, shape, or number (Fig. 12.7b). Yet the connection of woods to hedgerow corridors, and via corridors to other woods, drops sharply. Planted strips of woods increase somewhat. In contrast, the total length of hedgerows, and of double tree rows along lanes and roads, drops sharply. Networks of interconnected hedgerows decrease. Such a rapid transformation resulting from agricultural intensification doubtless has numerous effects on biodiversity, species movement, mineral nutrient flows, hydrology, wind, soil characteristics, and of course aesthetics.[783]

These examples of agricultural intensification also emphasize the wide range in stability or persistence of spatial elements, especially in a fine-grained mosaic.[1422,1471,501,10,1377,1378] An environmental resource patch (chapter 2) is nearly permanent, whereas changing crops in a field represents more rapid change in resources and flows than that in most surrounding vertebrate and plant populations. The spatial arrangement of different rates of change is a key attribute of landscape change (chapter 1) and mosaic stability (chapter 14).

Reforestation

The reestablishment of a forested landscape generally is either unplanned or highly planned. Unplanned reforestation (afforestation), often on degraded soil following farm abandonment or overcutting, originates in many small irregular nuclei.[1232,1952,509] From these the forest gradually coalesces over time. Planned reforestation, on the other hand, is almost always the planting of large, highly regular tree plantations separated by road corridors.[1306] Agroforestry plantings usually have repeated farm- or field-sized units within which tree rows alternate with crops in 'alleys'.[1499,858,1945]

'Instantaneous' transformations

Varied disturbances, dramatic because of their extensiveness, transform land essentially overnight. Volcanic eruption, burning, flooding, bombing, herbiciding, and nuclear explosions are examples (ignoring cases such as burning and bombing that also may continue over months).[1320,501,1862,1419,880] These transformations doubtless have mosaic sequences, but each is compressed enough to be considered ecologically a single (mega)disturbance event.

Common and uncommon mosaic sequences

The preceding diverse causes of land transformation produce numerous changing spatial patterns or mosaic sequences (Table 12.1). However, the sequences exhibit spatial similarities that cut across geography, climate, vegetation, and cause, and can be grouped into nine categories (proposed models in Table 12.l). Five mosaic sequences are widespread[497]:

(1) *Edge*. A new land type spreads unidirectionally in more or less parallel strips from an edge. The source of the land conversion, such as a city or blowing sand, borders one side of the landscape.
(2) *Corridor*. A new corridor, such as a road or irrigation canal, bisects the initial land type at the outset, and expands outward on both sides.
(3) *Nucleus*. Spread from a single spot or nucleus within the landscape, such as a town or reforestation patch, proceeds radially and leaves a shrinking ring of initial land type.
(4) *Nuclei*. Growth from a few spots within the landscape, as for many settlement patterns or non-native species invasions, produces new areas expanding radially toward one another.
(5) *Dispersed*. Widely dispersing new patches, as in dispersed-patch logging or suburbanization with houses on large lots, rapidly eliminates large patches of the initial land type, minimizes the emergence of large patches of the new land type, and produces a temporary network of the initial land type.

The other patterns and proposed models identified in Table 12.1 are variants ('instantaneous' and 'even throughout') of the changing patterns described above, or apparently rare ('alternate strips' and 'network template'). The 'even throughout' case would produce no changing spatial patterns if the land were homogeneous, but the heterogeneity always present means that different portions of the land are differentially sensitive to the causative mechanism. The 'nuclei' and 'dispersed' models probably describe most 'even throughout' cases. 'Alternate strip' logging, and establishing a wooded 'network template' or grid that cannot be removed until the end, illustrate the rare patterns.[518]

MODELING MOSAIC SEQUENCES
Geometric models

Simple spatial models based on changing geometries are constructed to describe the five common patterns identified, plus a random pattern (Fig.

423

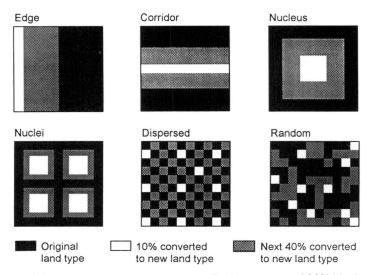

Fig. 12.8. Models of mosaic sequences. Each area starts 100% black and is progressively replaced by a white new-land-type (gray is simply used to identify locations of new land type during the 10% to 50% replacement period). White land type surrounds the landscape. Figure prepared courtesy of Kristina Hill.

12.8). The assumptions and rules for all models are: (1) only two land types are possible; (2) the model starts with 100% of initial type and ends with 100% of new type; (3) the landscape is a 100×100 unit grid; (4) each constant time interval converts 100 units to the new land type (therefore it takes, e.g., 100 years or weeks to complete the land transformation).

The *edge model* has 100-unit parallel strips of new land type progressively expanding from a boundary. The 'corridor model' has an initial 100-unit strip crossing the center of the grid, and 50-unit parallel strips synchronously expanding from it on opposite sides. The 'nucleus model' has a central 100-unit square with 'concentric boxes' progressively expanding outward. The 'nuclei model' (in this case four nuclei) begins with four 25-unit squares equi-distant from one another and from the boundary, each of which expands outward with concentric boxes of 25 units. The 'dispersed model' progressively places 100-unit patches, such that variance in inter-patch distance is minimized at each step. A 'random model', included for comparison, progressively locates 100-unit patches at random.

Spatial attributes changing

To link the models with ecological characteristics, common spatial attributes are calculated for each time step of each model. Two attri-

424

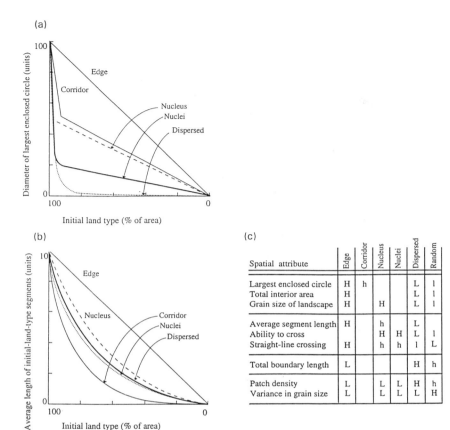

(a) — Diameter of largest enclosed circle (units); Edge; Corridor; Nucleus; Nuclei; Dispersed; Initial land type (% of area)

(b) — Average length of initial-land-type segments (units); Edge; Nucleus; Corridor; Nuclei; Dispersed; Initial land type (% of area)

(c)

Spatial attribute	Edge	Corridor	Nucleus	Nuclei	Dispersed	Random
Largest enclosed circle	H	h			L	l
Total interior area	H				L	l
Grain size of landscape	H		H		L	l
Average segment length	H		h		L	
Ability to cross			H	H	L	l
Straight-line crossing	H		h	h	l	L
Total boundary length	L				H	h
Patch density	L		L	L	H	h
Variance in grain size	L		L	L	L	H

Fig. 12.9. Comparing the land-transformation models based on their spatial attributes. (a) The five models in Fig. 12.8 are compared for a large-patch attribute, namely, the largest circle that fits within the initial land type. (b) Models compared for a connectivity attribute, i.e., how connected the initial-land-type matrix and corridors are. See below for methods. (c) Models are compared for all nine spatial attributes. The highest and lowest curves (e.g., for edge and dispersed in (a)) are indicated as H and L, respectively. Curves almost as high or low as the highest or lowest curves are indicated by h and l, respectively. 'Grain size' is the average diameter of all (initial-land-type) patches present. The three connectivity measures are based on lines starting at 40 equidistant points around the perimeter. 'Average segment length' (also see (b)) is the mean length of all initial-land-type segments encountered along 40 straight lines (perpendicular to the perimeter at the starting points). 'Ability to cross' is the percentage of the 40 lines that successfully cross along a straight or curvilinear route to the opposite side of the landscape, only using the initial-land-type. 'Straight-line crossing' is the percentage of the 40 straight perpendicular routes across that are not blocked at the beginning by the new land type. 'Total boundary length' measures the border between initial and new land types. The random model (Fig. 12.8), not plotted, would be intermediate in height among the curves in (a) and (b).

425

butes are plotted to illustrate how the models are differentiated by the shapes of curves (Fig. 12.9). When the diameter of the largest circle that fits in a patch (of initial land type) is plotted against the land transformation axis (Fig. 12.9a), the curve for the 'edge' model is noticeably higher than those for the other models. For connectivity of the matrix and corridors of the initial land type (Fig. 12.9b), again the 'edge' curve is highest.

It bears emphasis that all curves start at the same place (100%) and end at the same place (0%) (Fig. 12.9a and b). Also, no threshold is incorporated in the models beyond which further transformation is ecologically inappropriate. Therefore, the difference between models is the relative height of the curves, i.e., how long through the land transformation process a particular attribute retains a high value.

In all, nine spatial attributes are calculated for the models.[501,1745,1748] None of the five models has consistently high curves (Fig. 12.9c). The 'edge' model is highest for the three attributes reflecting the presence of large patches, namely, largest enclosed circle, total interior area, and grain size of landscape. The 'nucleus' model appears to be next highest. The 'dispersed' model is lowest for these large-patch attributes.

Connectivity of the landscape is also measured in three ways, average segment length, ability to cross the landscape in straight or curvilinear routes, and straight-line crossing (Fig. 12.9c). Overall the 'edge' and 'nucleus' models exhibit the highest levels of connectivity.

Total boundary length (Fig. 12.9c) is highest in the 'dispersed' model, and lowest in the 'edge' model. Two additional spatial attributes measured (Fig. 12.9c) appear to be expressed in different ways by the preceding attributes. The 'corridor' model produces intermediate height curves for all spatial attributes, and the 'random' or percolation model[545,546] (chapter 8) is intermediate for most attributes. In short, the spatial attributes differ markedly among the models of land transformation.

TOWARD AN ECOLOGICALLY OPTIMUM LAND TRANSFORMATION

Ecological characteristics and spatial attributes

Many ecological characteristics have been related in the literature to the preceding common spatial attributes. Three categories of attribute, patch size, connectivity, and boundary length, are considered of particular ecological significance.

Patch size

Large patches of native vegetation[991,683,501,1490,1536,678] (Fig. 12.10 Right side) are important for protecting: (a) aquifers and lakes; (b) connectivity

426

Fig. 12.10. Species sensitive to patch size, corridor connectivity, and boundary length. Left: white-faced capuchin (mono cara blanca, *Cebus capucinus*) on a leguminous tree by a corridor in Costa Rica. High corridor connectivity is a key to movement between forest patches by arboreal mammals. R. Forman photo. Right: old oaks (*Quercus*), 'shade-loving' plants, and white-tailed deer (*Odocoileus virginianus*) in a large forest patch in Arkansas (USA). The multihabitat deer use boundaries, corridors, and patches. Photo courtesy of US Fish and Wildlife Service.

of a headwaters low-order stream network; (c) habitat for patch-interior species; (d) habitat for large-home-range species; (e) sources of species dispersing through the matrix; and (f) the natural disturbance regime (chapter 2).

Small patches[943,498,225,174] are important as: (a) habitat and stepping stones for species dispersal, and for recolonization after local extinction of interior species; (b) enhancing matrix heterogeneity, thus decreasing fetch and erosion; and (c) habitat for occasional small-patch-restricted species (chapter 2).

Connectivity

High corridor connectivity[174,685,109,1114,1492,1038,1228] (Fig. 12.10 Left) is considered to: (a) enhance recolonization following frequent local extinctions; (b) enhance gene flow to combat inbreeding depression in a patch; and (c) reduce fetch and erosion in the matrix (chapters 5 and 6).

Low corridor connectivity[24,1492,1557] is suggested to: (a) decrease the spread of pests, non-native species and disturbances; and (b) decrease the sink (mortality) effect on interior species, if they do not learn to avoid unsuitable corridors (chapters 5 and 6).

427

Boundary length

More boundary and edge[943,1379,1162,685,676,503] (Fig. 12.10 Right) enhances (a) game and other species of the edge. If the increased boundary length results from convolution rather than a longer straight edge, probably (b) fewer animals move along the boundary, (c) more animals cross the boundary between habitats, and (d) erosion along the boundary typically is less (chapter 3).

Less boundary length[1224,174,503] (a) decreases edge area and (often) non-native species. If the reduced boundary length results from straightening, probably (b) more animals move along the boundary, (c) fewer cross it, and (d) erosion typically increases (chapter 3).

Comparing the geometric models

In essence, there are ecological benefits to both high and low levels of all three categories. However, the benefits vary widely in their overall significance. While it is optimum to have both large and small patches[498,1224,1309,1051], the absence of large patches is ecologically serious. The absence of small patches is generally of minimal significance, because corridors and the configuration of large patches usually provide most benefits of small patches. In short, the benefits of large patches overwhelm those of small patches.

Similarly, high connectivity is considered by most, but not all, ecologists to be ecologically much more beneficial than low connectivity.[1518,685,1114,493,1038,1228] The optimum may be to have both high and low levels separated in different portions of the landscape, but only having low connectivity is ecologically serious. Boundary length is less easily categorized, because the ecological advantages of high boundary length appear only slightly greater than those for low boundary length.

More relevant, however, is that the overall ecological advantages of large patches are greater than for any other category. For example, high corridor connectivity or many small stepping-stone patches is of limited use for movement of interior species, if large patches containing the species are absent. Overall, we conclude that the attributes providing ecological benefits are ranked in the following order: (1) large patch; (2) high corridor connectivity; (3) high density of small patches; and (4) high boundary length.

Ecological characteristics are now linked to the spatial attributes of the models. We make one operational assumption, i.e., that the initial land type is more suitable, and the new land type less suitable, ecologically. (The opposite assumption would be made, for instance, in land restoration.) Hence, the results relate to widespread land transformations such as deforestation or desertification of an initial, more-ecologically suitable land type. The results should apply to all land-scapes, though the ranking of benefits could be changed in special cases. For example, the importance of large patches might be diminished in the improbable case where there are no interior species, no large-home-range species, no aquifer, no stream network, and so forth.

Applying the ecological-ranking results to the analysis of spatial attributes (Fig. 12.9) clearly differentiates the land transformation models (Fig. 12.8). The 'edge' model is best for large-patch attributes, and is good for connectivity attributes. The 'nucleus' model is relatively good for both spatial attributes. The 'dispersed' model is the worst ecologically, because of the early loss of all large patches.

These model results suggest that the dispersed-patch logging regime, e.g., on government forest lands of the USA Pacific Northwest[518,662,967,1823], is highly detrimental ecologically. The ecological limitations of the 'corridor' model are reflected, for example, in the road construction and subsequent deforestation in Rondonia, Brazil[468,1577,354] (Fig. 12.6). The 'random' model may approximate some unplanned suburbanization patterns.[733,461,436,709,1458]

Interpretation of these land-transformation patterns is enhanced by referring to the basic spatial processes (Fig. 12.1). Perforation is high for the 'dispersed' model, and present in the 'nuclei' and 'nucleus' models (Fig. 12.8). Both dissection and fragmentation are present in the 'corridor' model. All models exhibit shrinkage, and none has attrition until the end. Only the 'random' model exhibits all five spatial processes.

In summary, the 'edge' model is considered the best of the six.[497] It has no perforation, dissection, or fragmentation. It is best for the large-patch attributes, and good for connectivity. The 'nucleus' model is second best.

'Jaws' models

Nevertheless, some ecological shortcomings of the 'edge' model are evident. In the process of changing to a new land type, there are no small patches, no corridors, minimal boundary, and an extensive patch of new land type that is subject to, e.g., wind and water erosion or heat buildup (Fig. 12.8). In addition, the initial land type becomes rectangular with an increasing length-to-width ratio, until it is only a strip near the end.

A *'jaws' model* or 'mouth' model is proposed as an ecologically-optimum pattern of land transformation (Fig. 12.11).[497] Two rules are added to those listed for the 'edge' model. The new land type spreads from two adjacent sides instead of one side. And a template of a future corridor network together with scattered small patches is established at the outset, which is not transformed until the end of the sequence.

Thus, compared with the 'edge' model, the 'jaws' model offers three ecological benefits. First, a square patch of initial habitat is always main-

(a)   (b)   (c)

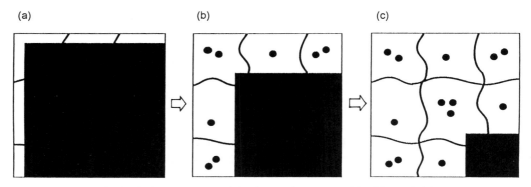

Fig. 12.11. A 'jaws' model of land transformation. (a), (b), and (c) are three stages showing 10%, 50%, and 90%, respectively, of the land transformed from black to white land types. Dots are small patches and curved lines are corridors. White land type surrounds the landscape.

tained (Fig. 12.11), an improvement in its large-patch benefit, especially in later stages of the mosaic sequence. Second, corridor connectivity is enhanced in this model. Small remnant patches act as stepping stones for species over the new-land-type area. The corridors and small patches minimize the deleterious effects of an extensive continuous area of new land type. And third, the 'jaws' model increases boundary length considerably, providing more habitat for multihabitat species, as well as edge species.

The 'jaws' model also works by spreading from four sides, producing a remnant patch in the center rather than the corner of a landscape. The 'nucleus' model, ranked second best above, is less amenable to ecological improvement. The basic problem is that the initial land type is a ring shrinking in thickness over time (Fig. 12.8). The key interior-habitat advantage of the 'edge', and especially the 'jaws', model rapidly disappears in the 'nucleus' model.

The area surrounding the landscape in the 'jaws' and other models (Figs. 12.8 and 12.11) is assumed to be and remain 'white', that is, less suitable than the 'black' landscape being transformed. Alternatively, the surrounding area could be black or a third type, or the black square being transformed could be a subset of a larger black area undergoing the same transformation sequence. The ecological effects of a mosaic sequence depend somewhat on the surrounding land type. For example, the white strips (or band of elongated patches) in Fig. 12.11a would be less desirable if the surroundings were black rather than white.

No landscape transformation mimicking the 'jaws' model is documented. However, portions of the concept apparently are beginning to be incorporated into planning of managed coniferous forest in Sweden

430

(P. Angelstam, pers. commun.). An experimental study comparing these models using micro-landscapes would be exceptionally interesting (S. Collinge, pers. commun).

The 'jaws' model can be yet further improved ecologically. Within a certain size range, a few large patches offer advantages over a single larger patch (chapters 2 and 13). The primary advantage is the ability to sustain higher total species richness for the landscape over time. If a patch exceeds the 'minimum dynamic area' (chapter 2)[1319], i.e., most major disturbances affect a small proportion of the patch area, species can persist along with the disturbances. When disturbance commonly covers a large portion or all of a patch, 'risk spreading' becomes of major importance (chapter 2). It is better to have a number of large patches. Two to a few large patches (depending on the ratio of patch species number to total species pool for the landscape) is the best current estimate for the optimum number of large patches (chapter 2).

Applying these concepts to Fig. 12.11 suggests a *jaws-and-chunks model* or metaphor for land transformation.[497] To simplify, we assume that a black square covering approximately 10% of the landscape (Fig. 12.11c) is both: (a) the area covered by most large disturbances; and (b) the minimum area to have most ecological values of large patches (see Box 2.1 in chapter 2). In this manner the size of large patches is related or scaled to the size of the whole landscape. Large-patch size is also related to disturbance and ecological benefits.

In the 'jaws' model (Fig. 12.11) the large black patches in (a) and (c) seem appropriate, but not in the middle phase represented by (b). The (a) patch is far larger than the area covered by a disturbance, and it still mimics the all-black landscape. The (c) patch is also optimum. Although it offers no risk spreading, it otherwise accomplishes most large-patch values. The black patch in (b) offers no risk spreading, e.g., against the effects of large disturbances in its central portion, and no species-richness advantage provided by two or more large patches.

A few large patches or 'chunks' of black in (b) would overcome these limitations. For instance, three squares of 10%, 10%, and 10%, and one square of approximately 20% of the landscape area would accomplish this goal. The four large patches would be separated by white strips, such as in (a), wide enough to arrest the spread of most disturbances. As the 'jaws-and-chunks' model proceeds from (b) to (c), the chunks of black sequentially decrease in number, rather than concurrently shrinking in size.

In short, the model metaphor is of open jaws moving forward, gripping a huge chunk, followed by a few large chunks, followed by a single large chunk, and leaving a final corridor network with small patches before the end. The model incorporates additional ecological benefits to those of the 'edge' and 'jaws' models. It is a hypothesis awaiting empirical evaluation. Where a less-ecologically-suitable land type replaces a more-suitable land type, the 'jaws-and-chunks' model appears to be an ecologically optimum mosaic sequence for land transformation.

REFORESTATION AND OTHER PATTERNS

Perforation, fragmentation and the other spatial processes described (Figs. 12.1 and 12.2) refer to the initial decreasing land type. They focus on remnant patches and corridors. But concurrently a new land type is increasing, with its attendant ecological characteristics. A somewhat different set of spatial processes emerge in the new land types. These may include patch appearance or proliferation, expansion, coalescence, aggregation, connection, and infilling. As habitat restoration is increasingly viewed in a whole-landscape context, models such as those in Figs. 12.2, 12.8, and 12.11 should be developed.

Reforestation ranges from huge geometric plantations of one or a few, often non-native, tree species to tiny parallel strips or patches in agroforestry designs. Generally these reforestations are highly planned and intensively managed. Alternative ecologically based designs for the large plantations are increasingly of interest.[1490,1308,1306] However, piece-by-piece land abandonment that is essentially unplanned leads to quite different reforestation patterns.[1952,1234,509]

David Foster (1992) documents a remarkable spatial and ecological history of central Massachusetts (USA) by integrating detailed data sets on land ownership, land use, forest types, and vegetation history. A 400 hectare (1000 acre) area within the Harvard Forest illustrates several dynamic patterns of ecological importance. From 1840 to 1985 fields were progressively abandoned and regrew into forest, which gradually coalesced. The rate of forest spread was greatest in the middle of the period, and forest cover reached 90% by the end (Fig. 12.12a). In spite of the scattered nature of field abandonment, the largest forest core (chapter 2) has continued to expand rather constantly since 1859. The number of forest patches increased only slightly, despite large variation in the number of regrowth patches from field abandonment (Fig. 12.12b). The boundary length of forest is greatest in the mid period, just as in the previously described deforestation process of dispersed patches.[518] Many other changing spatial variables of ecological interest can be evaluated for this mosaic sequence, including the diameter, shape, size, and perforation of forest or regrowth patches, and various connectivity measures (chapters 8 and 9).

The mid-nineteenth century period of maximum deforestation in this landscape had an intricate land-use pattern of pasture, cultivation, woodlot, and marsh.[509] The pattern is best explained by proximity to

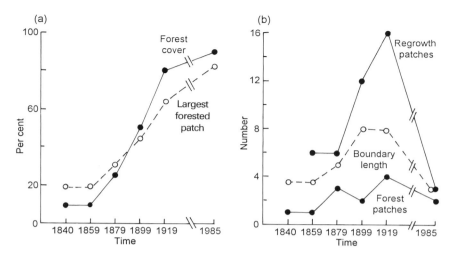

Fig. 12.12. Spatial changes during reforestation after field abandonment. (a) Forest cover and largest forested patch in 400-hectare landscape area. Relative patch size = diameter of largest circle of forest divided by diameter of largest circle in the landscape area (see similar concept for core of a patch in chapter 4). (b) Regrowth patches are of young trees on abandoned fields, and forest patches are of older trees with full canopy height. Boundary length is calculated by dividing the total field–forest border within the 400-hectare area by the diameter of the largest all-forested circle present. Harvard Forest in central Massachusetts (USA). Based on Foster (1992).

farmhouses, proximity to town roads, and soil drainage. Tree species distributions today, after 2.5 centuries of intensive human activity, strongly reflect this earlier, spatially explicit land-use pattern.

Returning to the mosaic sequence models (e.g., Fig. 12.8), one may fine-tune the overall results by adding variables to simulate the complex land transformations that are so widely observed.[1742,1743,1744,1234] For example, two original land types can be invaded by a third. One original type can be invaded by two types. Or permanent objects such as a dendritic stream network and a nature reserve can be incorporated.[967] If each new patch in the models of Fig. 12.8 is connected to the left boundary by a corridor (without forming a loop), the shapes of the curves for spatial attributes (Fig. 12.9) change somewhat, but the order of the curves from high to low remains the same.

Similarly the context of an area commonly is as ecologically important as the attributes of the area. The transformation models used here (Figs. 12.8 and 12.11) are surrounded by 'white' area, i.e., the new land type. Alternatively, the area surrounding a model landscape could begin and remain as initial land type, a third type, or a heterogeneous area. Or this model landscape could be a subset of a broader area being transformed in the same way throughout. Or surrounding landscapes could change at a different rate or in a different direction than the landscape studied. Much opportunity for significant future research exists.

The preceding results are applicable to all land transformations (Table 12.1), though they appear particularly important in forest logging strategy and suburbanization planning. Where human planning is prevalent, additional models are warranted for comparison, since human designs appear more variable than the usual unplanned or little-planned land transformations (Table 12.1). This applies to landscape restoration, where the new land type is more, rather than less, ecologically suitable than the initial land type.

The analytic approach outlined in this chapter should also aid in our determination of the optimum, or worst, location to make the next change in a landscape. Where is the ecologically best, second-best, or least-desirable place to locate the next nature reserve, shopping center, logging operation, sewage treatment facility, windbreak, or replanted woods? Part of the answer of course results from the initial heterogeneity of soil, water, and slope in the landscape. But part also emerges from the ecologically optimum trajectory or mosaic sequence. That is, we should be able to look at any existing mosaic, and evaluate each spot for its relative suitability for a particular type of change.

13

Land planning and management

A river, with its waterfalls and meadows, a lake, a hill, a cliff or individual rocks, a forest, and ancient trees standing singly . . . If the inhabitants of a town were wise, they would seek to preserve these things, though at a considerable expense; for such things educate far more than any hired teachers or preachers, or any present recognized system of school education. I do not think him fit to be the founder of a state or even of a town who does not foresee the use of these things . . .

Henry David Thoreau, Journal, *1861*

'Think globally, act locally' is a phrase much in vogue. Yet it has two problems. First, few people will ever give primacy to the globe in decision-making. Second, local considerations overwhelmingly determine actions. Indeed, these are two roots of environmental and societal problems etched widely in the land. Typically, our best soils are thinned by erosion or covered by suburbs; biodiversity and wildlife in large wooded areas are impoverished by dispersed logging; critical riparian and hedgerow corridors are cut or razed by macroagriculture; productive land is desertified by irrigation and overuse; and parks are surrounded by the tightening noose of development.

Can this unraveling of the land, as the primary capital of a nation, be reversed? Land patterns seem logical and ecological in exceptional places within, for example, Britain, Romania, Costa Rica, Australia, and the USA, as well as in certain spatial models. In these cases efforts in design, planning, conservation, and decision-making seem both visionary and practical. They are directed at agricultural landscapes, suburban landscapes, forested landscapes, or regions. At these scales the environment, economics, and society coalesce. The landscape and region are the central linkage between global and local.

The philosophic framework for this chapter therefore might be that we better *'Think globally, plan regionally, and then act locally'*.

435

The perennial challenges in planning, design, and management of an area are not only to take a broad spatial view and a long temporal view, but also to address all major environmental and human issues present. Water, transportation, biodiversity, aesthetics, sense of community, food production, and much more are the essential factors. All plans, all designs, and all management should address them. This requires an exceptional breadth of expertise. The usual result is that an individual or group within a discipline either develops a plan for a large area focused on one primary objective, or alternatively carves out a piece of the area where this objective is primary. By focusing on the preceding principles in landscape and regional ecology the approach in this chapter is broader. In addition, cultural, socioeconomic, aesthetic, and other human dimensions that are spatially explicit, i.e., used in physical planning, are schematically integrated.

Numerous books have been written in each of the areas of planning, design, conservation, and management. Many focus on a particular societal objective such as biodiversity, housing, recreation, soil erosion, industry, or fish populations. Others focus on a particular type of land such as parks, suburbs, forestry areas, irrigated areas, or pastureland. No overall review is yet available on the application of landscape ecology principles to planning, conservation, design, or management.

This chapter synthesizes and builds on the concepts and literature of chapters 1 to 12. In fact, many points here are difficult without familiarity with the preceding material. Chapter 13 focuses on the familiar traditional day-to-day and year-to-year issues faced by planners, designers, conservationists, managers, and policy makers. Chapter 14 following also focuses on planning and management, but with a sustainability perspective, that is, over human generations.

For convenience, planning is considered separately from management, though of course wise management includes planning. We begin with a proposed basic land-planning principle. Then, issues in planning based on landscape ecology are explored. Components of a generic plan applicable to any landscape are presented. Next is a focus on the suburban landscape, and its fine- and broad-scale ecological design. A case study of a special suburban town is included (courtesy of Joan D. Ferguson and Daniel H. Monahan). Finally, management, especially of natural resources, is explored.

AGGREGATE-WITH-OUTLIERS PRINCIPLE

Landscape and regional ecology offers theory and empirical evidence to understand and compare different spatial configurations. Yet to date few

436

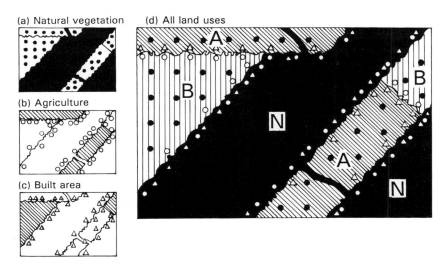

Fig. 13.1. Arrangement of land uses based on aggregate-with-outliers principle. N = natural vegetation; A = agriculture; B = built area. Outliers of natural vegetation, agriculture, and built area are represented by small black dots in (a), circles in (b), and triangles in (c), respectively. The (d) pattern integrates (a), (b) and (c). Based on Forman (1993) and Forman & Collinge (1995).

land planning or management principles based on this knowledge have been articulated.

We may pose a central question, 'What is the optimum arrangement of land uses in a landscape?' To address this, a land planning principle is proposed that links somewhat unrelated knowledge, is significant, and has widespread applicability. It is a *model*, in the sense that a complex system is simplified by filtering out apparently less-important variables. As such, it does not describe the complexity of any real landscape. It is also a 'theory' in the sense that diverse pieces of knowledge are synthesized into a coherent whole. It is called a 'principle' in the general sense. Although it has not been directly tested, considerable evidence exists for each of its foundations, i.e., the knowledge areas synthesized. Scrutiny and evaluation will identify the exceptions. They will determine whether the principle is erroneous, requires modification, or is universal.

The *aggregate-with-outliers* principle states that one should *aggregate land uses, yet maintain corridors and small patches of nature throughout developed areas, as well as outliers of human activity spatially arranged along major boundaries*. Seven mainly landscape-ecological attributes described below are incorporated into or solved by the principle.[497] A simple graphical model illustrates the concept[494] (Fig. 13.1).

437

(1) *Large patches of natural vegetation*. It bears emphasis that large patches of natural vegetation are ecologically important for at least six reasons (chapter 2). They: (a) protect aquifers; (b) protect low-order stream networks; (c) provide habitat for large-home-range species; (d) support viable population sizes of interior species; (e) permit natural disturbance regimes in which species evolved and persist; and (f) maintain a range of microhabitat proximities for multihabitat species.

(2) *Grain size*. The grain size of the landscape mosaic, i.e., the average diameter or average area of all patches in the landscape, affects many ecological factors (chapter 9). As an extreme spatial arrangement we could aggregate all natural vegetation together, all the built area together, and all agriculture together. This would be a coarse-grained landscape.[501] An alternative extreme would be to evenly intermix each small patch of natural vegetation, each house lot, and each agricultural field throughout the landscape.[1499,858,1945] This produces a fine-grained landscape.

The coarse-grained landscape provides large natural-vegetation patches for specialist interior species and protection of an aquifer, large built areas for industrial specialization, and many other characteristics requiring large patches. Except near boundaries, movement is costly (considerable distance required) for multihabitat species, including humans.

In contrast, the fine-grained landscape has predominately generalist species, since specialists requiring a large patch of one land-use type cannot survive. Species that survive need only move short distances for their requirements. In addition, some variables such as air and water quality may be low, and the overall resource base somewhat degraded.

A landscape containing a variance in grain size, especially coarse and fine grain, appears to be an important spatial configuration. No advantage to intermediate grain size is known. The landscape model (Fig. 13.1) is coarse grained, but contains fine grain near boundaries. The vicinity of intersections where three large patches converge is also fine grained.

(3) *Risk spreading*. Risk spreading is important to avoid 'placing all eggs in one basket'. For example, we want more than one large patch of natural vegetation or agriculture, in case a major windstorm or pest outbreak covers an entire patch (Fig. 13.1a and b).

(4) *Genetic variation*. Genetic variation is important to provide strains with resistance to disturbance, or to deal with environmental change. In many cases small outlier patches are required, in addition to large patches (Fig. 13.1a, b and c). For example, a pest could eliminate the major grass species, or the livestock, in a large pastureland. If some small isolated pastures also have been maintained, it is likely that the grass or livestock in them is genetically different than that in the pastureland.

Thus, using these genetic strains, the pastureland population may be reestablished following disturbance. Analogously not everyone wishes to live in a city; providing for hermits, farmers, and other homeowners is also important.

(5) *Boundary zone*. The major boundary zone between land uses, including the edge portions of each adjacent large patch, is often suitable for outliers (Fig. 13.1b, c and d). Located along a boundary, outliers do not unduly perforate and destroy the advantages of large patches.[518] The curvilinearity of boundaries reduces the apparent barrier effect of straight boundaries (chapter 3), and better mimics the results of natural processes.

(6) *Small patches of natural vegetation*. As an important exception to the preceding rule, small patches or outliers of natural vegetation are valuable throughout the developed, i.e., built and agricultural, areas (Fig. 13.1a). Small patches are significant as a supplement to, not a replacement for, large patches (chapter 2). They: (a) serve as stepping stones for species dispersal; (b) provide habitat and stepping stones for species recolonization following local extinction; (c) provide heterogeneity in the matrix that decreases wind and water erosion; (d) contain edge species with dense populations; and (e) have high species densities. A possible improvement in the arrangement of small natural-vegetation patches (Fig. 13.1a) is suggested in the following section on planning.

(7) *Corridors*. Two types of corridor are present (Fig. 13.1d). Natural vegetation corridors enhance important natural processes such as species and surface water movement. And corridors composed of diverse fine-scale land uses result in efficient human and multihabitat species movement among these land uses. Both corridor types have a concentration of weedy generalist species in a narrow area, and serve as a major filter or barrier to movement between adjacent large patches. Both corridor types concentrate movement between patches, thus minimizing unwanted movements across large patches.

The aggregate-with-outliers model for land planning thus has numerous ecological benefits.[501,518,1745,662] A range of direct benefits to humans is also suggested by this approach. These include: (1) providing an exceptionally wide range of settings; (2) locations for suburban dwellers as well as hermits; (3) fine-scale areas where jobs, homes, schools, and shops are close together; (4) the efficiency of human movements (transportation) along corridors between cities, as well as between city and suburb; (5) natural vegetation and agriculture crossing major corridors, thus preventing continuous ribbon or strip development, and enhancing town or neighborhood identity; (6) specialization within aggregated built areas (e.g., cities); (7) urban garden areas; (8) providing large patches for efficient resource extraction; (9) limiting the difficulties farms have when they are scattered; and (10) a variance in grain size providing visual diversity.

The strengths of the aggregate-with-outliers principle are its format (Fig. 13.1) and its flexibility for creative problem solving. A critical listing of pros and cons for both ecological and human dimensions is warranted for this approach, and in comparison with existing land use patterns. The principle applies to a landscape or a major portion of a landscape. The range of scales over which it applies is unknown. However, it appears applicable in any landscape, from desert to forest and from agriculture to suburb.

PLANNING

Traditionally, planners and managers have successfully planned, designed, and managed areas by explicitly integrating the human world with the biological or natural world. Familiar examples are the ancient Chinese temple areas, medieval fortress towns, notable Italian villas, Olmsted's Emerald Necklace in Boston, Letchworth in Britain, and Canberra, Australia. Yet, in some planning circles public administration and economics have been substituted for the biological component. In essence, this is an experiment, doubtless of short duration, to see if natural processes, biological patterns, and the environment can be largely ignored in planning.

The result is that planning and management themselves are now in trouble. Laws, regulations, guidelines, standard practices, building codes, and planning acts are all essentially legal 'standards' that govern transportation, mining, forestry, real estate development, grazing, environmental protection, water resources, and other human endeavors dear to our hearts (M. D. Cantwell, pers. commun.). The standards are described in manuals, law books, and the like. They were developed to protect society from human error, that is, to protect health, safety, and welfare. The unfortunate result is that the standards also protect individuals and groups who demonstrate lack of vision or make errors, as long as 'one follows the code'. We are stuck in the standards. Worse still, most standards were developed before the recent explosion in ecological understanding.

We straightened streams. We filled wetlands. We built levees along rivers. We tried to eliminate fire. We exterminated large predators. Today we are literally paying the price for wetland loss, soil erosion, massive floods, pest explosions, and 'forestlessness'. We know many of the standards are misguided, but society finds itself painted into a corner. There is no easy way out. Some standards should be rolled back or replaced,

440

but those are only important details. Nor should a cookbook of prescriptions replace the pile of outdated standards. Rather an easier and wiser path is embedded in 'vision' and an open process, and ends with a specific yet broad mandate. Read the next five paragraphs.

Chapter 7 explains why a fixed stream-corridor width makes no ecological sense, and documents in detail what does make sense. The solution is not a specific prescription to be written as a standard. Rather, it is a conceptual framework against which one evaluates, plans, and manages any stream corridor. The present chapter presents a spatial model appropriate for planning any large piece of land. It is not, and should not be, a specific legal standard. Rather, it is a framework or basis for comparing land areas for their ecological and human suitability. The model or principle is also a target toward which policy makers, planners, designers, conservationists, engineers, managers, and others can direct the use of land.

The seeds of an answer to the planning and management dilemma are in the preceding stream–corridor and spatial model examples. The proposed essentials for planning and management are knowledge, room for creative flexible solutions, and collaboration among individuals, in all three cases involving diverse fields. That sounds simple. Unfortunately, it radically differs from today's norm both for process and goals.

To implement such a proposed solution, each individual ('expert') first describes the few major, plus some minor, considerations based on his or her field of expertise. A plan then includes three parts: (1) the descriptions by each expert; (2) a synthesis of at least all major considerations; and (3) an explanation of the decisions made on the inevitable tradeoffs among major considerations, and a rationale for omitting other potentially major ones. That should sound different from today's norm. The three parts are central to such a plan produced by an interdisciplinary planning or management team. This approach could well replace the manual of inflexible and outdated standards.

But society must be protected against poor or unethical work. Thus, a code of ethics covers general standards. And specifications for detailed items such as nail size, pipe wall strength, and road width protect society, though they may be altered by plans that document better solutions. The main gain is in the area between the general ethical standards and the specifications of details. In effect, the large middle ground of planning and management must offer flexibility and new solutions.

Indeed, society gains by no longer having to consider or implement plans developed only by engineers, only by economic planners, or only by ecologists. Instead, society gets plans that incorporate and balance the range of key considerations. Wise and long-term plans, management, and decisions are more likely. This chapter uses a 'chairs-around-the-table' metaphor to illustrate portions of such planning and management.

441

Planners and managers at work

Planners and landscape architects roll up their sleeves at a table to work, meet, and communicate. This is partly because maps, models, drawings, and remotely sensed images typically are large and require a flat surface. Of course, the computer screen representing electronic data storage and display is squarely on the table. More importantly, the surrounding chairs bring experts from a range of fields together to communicate on a common subject, a specific physical space. Non-spatial public policies, market forces, and ecological principles, for example, are present. But physical design and planning lead to spatial solutions. A mosaic of land uses, with rationales, emerges. The most valuable decisions in society can be made by planners and designers in this setting.

The language of landscape ecology has added an ingredient to help make the process gel.[501,488,1745,11] The land is composed only of patches, corridors, and a background matrix (Fig. 13.2). Each has simple familiar characteristics. Patches are large and small, rounded or elongated, and smooth or lobed. Corridors are wide or narrow, straight or curvy, and connected or with gaps. The matrix is continuous or subdivided, extensive or limited, and contracting or expanding. Ecologists, hydrologists, attorneys, conservationists, elected officials, transportation engineers,

Fig. 13.2. A forest matrix surrounding agricultural patches that enclose tiny wooded patches and corridors. Planners, managers, and other disciplines communicate with the patch–corridor–matrix model for any landscape. Near Neuchatel, Switzerland. R. Forman photo.

442

foresters, geographers, and others understand and share this spatial language.

Overlays of maps, for instance of soil, vegetation, housing densities, wetlands, elevation, protected land, and industry, have long been used to identify appropriate and inappropriate locations for future land uses. Ian McHarg (1969) emphasized this approach for linking design and the environment. The availability of geographic information systems now makes the approach routine. Overlays provide direct comparison of different locations, generally based on the combination of existing characteristics within each location (or pixel).

Landscape ecology caused a rethinking of the products of this approach based on three familiar factors: structure, function, and change. First, the structure or pattern of the whole landscape, and the location of a site in that broader pattern, are more important than the internal characteristics of the site. The site location relative to large patches, adjacencies, configurations, and networks, for instance, are important. Second, the functional flows and movements over the whole landscape are central to determining any land use. Thus, corridors, stepping stones, boundary curvilinearity, sources, and sinks focus on functioning conduits and barriers. Third, landscape change places locations in the dynamic perspective of a mosaic, where each element differs in both rate and direction of change. Future land uses for a location, site, or pixel depend on habitat fragmentation, perforation, patch dynamics, spatially explicit stability mechanisms, and mosaic sequences.

Many attributes of good plans long developed at the table remain equally valid with landscape-ecological planning and design, including the following. (1) A broad spatial context extends well beyond the specific area to be planned. (2) A long temporal context includes the formative processes, human history, and natural disturbance regime. (3) Flexibility for future change is built into the design. (4) Expected changes in the area, say, at 5, 10, and 20 years, are a key part of the plan. (5) And options are included in the plan, one of which is the optimum based on the planner's judgment independent of political reality, so tradeoffs using other options can be clearly seen.

When the plan is complete the planners and designers typically leave the table to focus on other projects. The manager then takes the table, and is responsible for ongoing stewardship, anticipating and responding to disturbances and changes, and articulating and meeting the objectives of the area. This requires researchers at the table to provide useful information for management. Experts in many fields are at the table since the manager usually has a single expertise. The public and decision-makers normally visit the table. Planning must continue since the manager is responsible for both the future and the present. Periodic reevaluations

443

and changes in emphasis are required. The manager has intimate knowledge of the area over time, and is deeply committed to molding it. Good land managers provide the finest public service in society.

Landscape-ecological planning

As noted at the outset of this chapter an extensive literature on land planning exists, focused either on particular environmental or societal objectives, or on particular types of land. Before illustrating planning based on landscape ecology, it is important to pinpoint *landscape planning* which developed independently of landscape ecology. This approach usually focuses on humans, and how the land can be effectively designed for their use. Environmental or land characteristics and visual quality or cultural characteristics are carefully examined to place human activities in the landscape with the least amount of impact. Useful syntheses and reviews provide particular insight on Britain[121,609,1752,1004], Middle Europe[1506,844,859], Canada[726,848], USA[1088,1960,461,1055,1625,1624,1626], and cities.[1017,1604]

The application of landscape ecology theory has significantly affected several planning areas.[1517,501,640,585,649,639,1231] These are grouped below into three categories primarily illustrated with book-length syntheses and reviews (numerous specific applications exist in articles): rural and agricultural; natural resource areas for forestry, wildlife, and biodiversity; and corridors and greenways.

Rural and agricultural land

Some of the earliest uses of landscape ecology were in land evaluation to determine optimal land uses across a landscape.[1953,1193] This has been used in many developing nations. Soil erosion control and agricultural production are especially important in this planning process.[1800] Landscape ecological planning, particularly in rural and semi-rural agricultural areas[1490,1492], is also effective in more industrialized nations.[638,1472] These approaches usually incorporate a strong hierarchical approach and include a focus on stability. A different approach to countryside planning emphasizes the value of a 'tool kit' of landscape-ecological principles, which are applied flexibly in different parts of a landscape.[1533]

Natural resource areas for forestry, wildlife, and biodiversity

Planning of forested mosaics using landscape and regional ecology includes meshing values such as biodiversity, water, and fish with spatially explicit logging regimes. Logging that protects plant species of conservation interest must be planned around habitat types, especially wetlands and stream and river corridors, as well as disturbance regimes.[1306] Creative logging patterns can be created to mesh with key populations

444

of mobile wildlife.[683,1393,518,1986,764,403] This includes important demographic and genetic characteristics.[1595] The specific design of forested areas, and the steps in such a planning process, can be usefully outlined.[1483,490,403,583,1526]

Corridors and greenways

Landscape-ecological planning has also focused on major linear features in a landscape or region. Natural corridors along roads can be especially important for species movement and protection[1492,91] (chapter 5). Ecologists and transportation engineers jointly plan and manage them.[772] Greenways that have a major recreation emphasis, in addition to ecological objectives, are particularly important near urban fringe areas.[1028,977,759,1582]

Despite the diverse examples here and a larger literature of specific articles on landscape-ecological planning, the surface is barely scratched. Few case studies are carefully documented in print. Follow-up analyses apparently are yet to take place. Controls to see what would happen in the absence of a particular action are apparently absent. Comparisons with standard planning approaches are lacking. Sample sizes are minuscule, that is, in most cases only a single example exists for the application of a particular landscape-ecology principle. Other fields such as range management, transportation, and engineering are only beginning to consider the concepts. In short, landscape-ecological planning is a growth area.

Patterns from nature, planning, and lack of planning

Yet, what is the overall product of land planning? How detailed should designs be? Should the whole world be planned and designed? If not, what should be? There are no easy answers to these questions. Nevertheless, we at least should compare the general results of human planning with those from natural process, and those from unplanned human activities.

We focus on 'physical planning' of space or land,[1770] and use the terms planned or unplanned in this sense. Of course, much policy planning occurs independently. Roads, powerlines, sewer pipes, and school construction individually are highly designed and usually carefully constructed. Visit a suburb where this is not so, and one immediately realizes that engineers and policy makers are central to civilization.

Natural patterns and forms result from a geomorphic template on which natural flows and natural disturbances take place. Moving water, wind, and ice form the geomorphic template (chapter 9). Ice tends to form somewhat straight lines on the land plus patches with curved

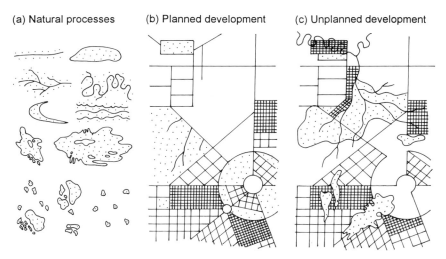

(a) Natural processes   (b) Planned development   (c) Unplanned development

Fig. 13.3. Spatial patterns produced by three groups of processes. (a) Separate patterns; (b) planned refers to the area as a whole; (c) unplanned refers to the area as a whole, though some or many pieces are highly planned and designed. Dotted = natural vegetation; white = agriculture; coarse grid = residential; fine grid = commercial.

boundaries (Fig. 13.3a). Water often forms dendritic networks, with straight lines in steep areas, and curvy boundaries from meanders in flat areas. Wind generally deposits curvy patches of sand and loess (chapter 4). Large patches and coarse-grained areas are present.[1151]

Inherently natural flows in a landscape are relatively straight due to gravity or streamline airflow. However, geomorphic, vegetational, built structures, and other heterogeneity cause turbulence. This results in irregular spatial patterns.

Natural disturbances over the template, such as hurricane patches, pest outbreaks, erosion and deposition from floods, and fire (Fig. 13.3a), primarily produce irregular boundaries and patches. The ends of burned areas tend to be especially convoluted. A river flood makes a wide corridor with bulges. In addition, because disturbances usually skip spots, a mosaic pattern commonly appears.

The dispersion of objects is also of interest. Random patterns in nature are usually difficult to find. Regular patterns are common in some areas (chapter 9). But by and large nature is aggregated[612,856] (Fig. 13.3a). It comes mainly in clusters, groves, flocks, herds, groups, and aggregations.

In short, nature produces distinctive patterns.[1252,1151] Most patches, boundaries, corridors, or matrix appear elongated, curvilinear, irregular, mosaic-like, and aggregated. Dendritic patterns are common, whereas

446

straight lines and symmetrical polygons represent a relatively uncommon geometry. The complexity of forms produced by nature also generally exceeds that created by humans.

Human *planning*, in contrast, overwhelmingly produces straight lines, rectangles, squares, and occasionally circles with radiating lines (Fig. 13.3b). Symmetrical geometric forms appear to be the goal of many planners and designers. And regularity replaces aggregation. Planted lines of exotic trees, shrubs along buildings, and lawn grasses predominate, species that could never naturally coexist (and sometimes are reduced to being called 'plant material'). They are the memorials to smothered nature, and they are the reminders of what could be.

A symmetric geometric form placed within an asymmetric natural pattern may be inspirational, unnoticed, or decried, depending in part on the provenance of the observer.[1961,1958,1626] Nevertheless, the form is unstable and generally unsustainable. Natural forces alter its symmetry (chapter 4). Such a form is costly to maintain and repair. The early square cities of northern China, and the straight roads by rivers and in mountains could retain their symmetry only with major human investment. Such geometric products of planning counteract nature, and hence are often ephemeral.

Planning and design of course do much more than produce patterns and forms. Nevertheless, the amazing richness of forms produced and maintained by nature represents a huge opportunity for planning and design.

Patterns from *unplanned* or little-planned human activities such as most suburbanization, livestock grazing, and settlement of new areas, are less easily defined.[947,1080,920] Some pieces of the puzzle may be nicely, even brilliantly, planned or designed, while other pieces appear to have been overlooked or lost. 'Lack of planning' means the puzzle as a whole is unplanned. The result is a mixture of irregular, curvy, mosaic-like, and geometric forms[1499,1945,1151] (Fig. 13.3c). Both aggregation and regularity are common.

This is conspicuous in a newly developed area. Even a cursory glance at a recent suburban landscape reveals its inefficiencies and ineffectiveness in dealing with most environmental and human issues.[755,1604,157,1458,1582] It seems arranged to both inhibit natural processes and defy planning principles.

Paradoxically, the so-called unplanned suburb is often highly planned. Roads, housing clusters, shopping areas, and infrastructure as individual items may be carefully planned, designed, and built down to fine details. The developer is frequently constrained by planners and planning acts, and by engineers and engineering standards (M. Cantwell, pers. commun.). Thus, it is not lack of planning, but rather the implementation of planning that results in the suburban form.

In effect, design and planning efforts in suburbs are focused on the wrong scale. Suburbia is highly planned at the fine scale, and nearly unplanned at the landscape and regional scale. Fire vehicles can turn around, new houses do not extend onto the street, and sewer pipes have thick walls. But from a landscape ecology, or energy efficiency, or social perspective, it is hard to call a suburb planned.

An interesting variant of lack of planning is a landscape that has been 'time tested' over human generations. Some similarity in the form of the pieces, such as farms or commercial centers, may evolve. Perhaps this suggests self-organizing principles at work[920] (A. R. Johnson, pers. commun.). One wonders whether these forms work well for humans, for nature, or for both.

The questions posed at the beginning of this section are not answered by this pattern analysis. Yet perhaps *biophilia*[1910] provides the seed of an answer. This concept indicates that humans evolved with, depend on, and have an affinity with a richness of other species in nature. Perhaps rather than the existing human planned and unplanned patterns, the answer lies in planning a whole landscape, within which both natural patterns and planned geometric patterns coexist. This does not mean regularly alternating square blocks of woodland and housing clusters. Rather, regular geometric areas are mixed with irregular, aggregated, mosaic-like, curvy patches and corridors. These individual patches and corridors in turn are distributed in irregular, aggregated, mosaic-like, and curvy arrangements over the planned landscape. Would that accomplish the goals of both biophilia and socioeconomic efficiency?

Three examples are instructive. First, in the diagram of the aggregate-with-outliers principle (Fig. 13.1a and d) the regular orchard-like distribution of small natural-vegetation patches could better mimic nature. Small irregular aggregations of patches, with lonely patches between them, would be present throughout the developed areas. These aggregations and individual patches would be located to enhance species movement over all portions of the landscape, and to all adjacent landscapes. The clusters of patches would provide flexibility for stopping in saltatory movement, and for some animals would mimic a larger patch and support a home range.

A second example comes from observing open space in cities. The most biologically impoverished places are produced by human design, whereas the biologically richest places have escaped notice or design. The grass-bench-tree ecosystem and the cemetery are depauperate compared with the vacant lot and urban streamside. Consciously eliminating biodiversity is puzzling, especially where people's links to nature are tenuous. A 'bio-rich place' for native species provides many human and ecological benefits, and is an essential component of all plans and designs.

The Third example is Canberra, Australia, a model of a precisely planned area with glorious boulevards, vistas, parks, separated suburbs, and geometric forms (see Fig. 1.7b). A leader of Aboriginal people from a remote natural landscape is reported to have described Canberra as chaotic and boring[819] (chapter 1). It is devoid of water holes, fire patterns, human-land linkages, and religious significance, and full of linear barriers separating adjacent homes and neighborhoods. Perhaps the preceding approach of integrating nature and biophilia into planning, would give meaning both to the Aboriginal leader and to the technological society in today's communities.

PRINCIPLES IN A GENERIC PLAN

Concepts, theories, empirical examples, and literature cited in the preceding chapters clearly point to the need to rethink existing planning principles and approaches. Rather than addressing the large diverse subject of planning methodology, here we pull together some of the key principles and issues based on landscape and regional ecology that could usefully appear in every plan. After summarizing these for a generic plan of any landscape, additional principles tailored to forested, arid, and agricultural landscapes are pinpointed. Suburban landscapes are highlighted in a subsequent section. The principles also apply to management.

The following planning objectives and approaches are assumed as background. Inventories and maps of key resources are available. Aerial photographs are used together with careful checking on the ground. These are supplemented by maps, GIS images, and so forth, as simplified models that provide different information, but omit key ecological information. The plan assumes ongoing major natural disturbances and changes in human activities, and thus provides appropriate flexibility. The plan shows the expected patterns at a number of future points up to at least two decades. And the plan provides at least two options, one of which is the optimum based on the planners' best judgment independent of political feasibility. Thus, all can clearly see the tradeoffs between the optimum and any particular politically-more-feasible plan.

Some of the following components appear 'indispensable', that is, they should be essential foundations of any land plan. This is because they accomplish major ecological or human objectives, and no other practical mechanism is known to accomplish them. Other components simply appear to be efficient or optimal solutions to difficult land planning problems. The components are introduced in five categories: context; whole landscape; key locations within a landscape; targeted ecological characteristics; and targeted spatial attributes.

449

Context

Planning a landscape effectively requires placing it in the spatial context of other landscapes in a region, or even a continent (see Box 13.1). Different landscapes act as sources and sinks, and thus their arrangement is important. For example, the transportation and communication flows from a major urban landscape in the region differentially affect all landscapes. In similar manner, flows of mammals and birds to and from the only forested landscape present affect all other areas, and heat from a dry agricultural landscape affects those downwind.

Box 13.1: Continental ecology

Each continent has a collection of natural resource areas, planned and managed in highly diverse ways because of the multitude of nations and government agencies involved. No continent has a system of reserves linked for flows and movements.

Yet in North America, with a few hundred million people, the idea for such a system is beginning to develop. Philip Lewis (1964) proposed a network of corridors, mainly following stream and river valleys for Wisconsin (USA) and nearby states. Larry Harris, Reed Noss, and colleagues further proposed a corridor network for Florida that focused on connecting large reserve areas.[1229,685,688] This was designed especially to protect wide-ranging species such as the black bear and Florida panther (*Ursus americanus, Felis concolor*). Corridor networks have been designed for many landscapes, often meshing natural resources and recreation.[1393,977,1492,1582] The hypothesized optimum structure of a landscape network, as well as six categories of public policy issues addressed, has been outlined.[493] It bears emphasis that a corridor system is no substitute for a number of large patches in a landscape. And corridors operate at different scales, from hedgerows on a farm to mountain chains across regions and nations.

It is logical then to extend the idea to the continental and global scales. Peter Bridgewater (1987) recognizes the continent as an important ecological scale, and identifies the connectivity of corridors in Australia as a keystone. A 'wildlands project' has been proposed for North America, primarily focused on wide-ranging species such as wolf, grizzly bear, and mountain lion (*Canis lupus, Ursus horribilis, Felis concolor*). Reed Noss describes this as a set of core reserves large enough to maintain viable populations of all native species, connected by corridors, and surrounded by buffer or multiple-use zones.[1041]

Such a design benefits several ecological dimensions, but focuses on sustaining large-home-range mammals. Instead of corridors and large reserves, could this mammal goal be accomplished with only a few huge reserves (chapter 2)?[1483] If the goal is broadened to maintain viable populations of all native species, what continental design would best accomplish that? What is the minimum design that might be successful? How would these answers differ if the objective were to prevent significant loss of species richness? Of fertile soil? Of water quantity and quality? Of ecological processes? Of all major ecological dimensions? Of both ecological and human dimensions? When the guidelines evident in each chapter are carefully examined, the spatial answers to these questions are radically different.

Spatial ecological thinking at continental and global scales is an important and exciting development. In the context of 'think globally, plan regionally, and then act locally', planning and management rarely strays far from people's actual concerns and support. Yet increasing numbers of people recognize the shortcomings of local action without a regional plan. Consequently, the core of effective ecological planning, conservation, management, and public policy ever-so-slowly moves toward the critical landscape and regional scales.

In addition, for planning a landscape or portion thereof a detailed evaluation is needed of a surrounding area whose diameter is perhaps ten times greater. This focuses on flows and movements such as ground and surface waterflow, species dispersal and migration routes, transportation, air pollution movement, and recreation use. It also focuses on change, including spreading desertification, housing, or logging, and construction of a new road, canal, or sewerline corridor. No plan can ignore movements and changes in the surroundings at this scale.

Temporal context within the landscape being planned is as important as spatial context. The original formation processes are an especially useful background. But it is also important to understand the regimes of major natural disturbances.[316,1662] How often does a wildfire, flood, pest outbreak, or cyclone hit? How large are they, where are they concentrated, and so forth? These disturbance regimes help mold the landscape, determine how rapidly it is changing, and underlie the spatial pattern of appropriate and inappropriate land uses for the planner. Similar issues require understanding the human history of the landscape.

The preceding considerations and those in the following sections provide bases to differentiate good from poor planning and design. Yet, because some considerations are well known but often ignored, we wonder about *'ethics'*, the self-imposed limitations on the freedom to act.[1390,1910,1103] For instance, in protecting land Aldo Leopold recommended a *land ethic* instead of government regulation and economic self-interest: land as a community of soils, waters, plants, and animals is to be loved and respected.[945] Unfortunately after several decades this has had little direct impact on land degradation or protection in the USA or worldwide. The solution probably lies in some combination of government, economics, and land ethics.[489] But there is a missing

451

ingredient. Something must cause people to think and plan long term. In some form that ingredient must involve landscape and regional ecology.

This leads inexorably to the *ethics of isolation*.[489] Simply stated, in land-use decisions and actions it is unethical to evaluate an area in isolation from its surroundings, or from its development over time. Ethics impel us to consider an area in its broadest spatial and temporal perspectives.

In chapter 14, on creating a sustainable environment, additional spatial and temporal contexts are pinpointed. We must spatially mesh ecological integrity and basic human needs. And it must be done to persist over human generations.

Whole landscapes

Four *indispensable patterns*, the top priority components in a plan, are recognized (Fig. 13.4).[497] No substitute for their benefits is known.

Maintaining *a few large patches of natural vegetation* is a familiar theme in this book (chapter 2). Another key component is *wide vegetation corridors along major water courses*. Almost all natural resources and human activities in a landscape depend in some measure on stream and river systems, if present in a landscape (chapter 7).

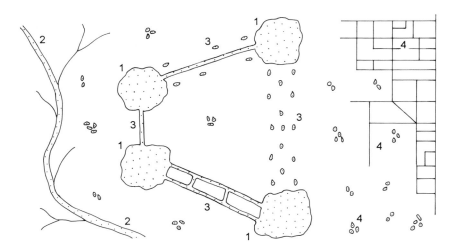

Fig. 13.4. Top-priority ecological 'indispensables' in planning a landscape. 1 = a few large patches of natural vegetation; 2 = major stream or river corridor; 3 = connectivity with corridors and stepping stones between large patches; 4 = heterogeneous bits of nature across the matrix.

A third indispensable component is maintaining *connectivity for movement of key species among the large patches* (Fig. 13.4). Wide continuous corridors forming a major green network are considered to be the best mechanism. A second best alternative apparently is a cluster of stepping stones (chapter 6), as long as the intervening matrix is not completely unsuitable for movement of key species.

A fourth such component is to maintain *heterogeneous bits of nature throughout human-developed areas*. This provides some connectivity for movement of most other species over all portions of the landscape. It also interrupts extensive areas of matrix subject to wind erosion, heat buildup, and the like. This component is effectively accomplished with a heterogeneous assemblage of small natural vegetation patches (chapter 2) and a network of line corridors (chapter 5) (Fig. 13.2). Only small patches or only the network may be almost as effective as both (Fig. 13.4). Natural heterogeneity includes the rare habitats and species scattered through the matrix. Large natural-resource areas protect many key species, but most species probably depend on the arrangement of land uses in the surroundings.

When these four principles are established in the plan, it is appropriate to construct a spatial framework for the whole landscape. This also should incorporate the following. Grain size varies widely across the landscape, especially in including some coarse-grained portions, which may be separated by fine-grained strips or areas. Disturbance regimes are essential to maintain most species, yet human activities often intervene, for instance by increasing fire frequency or decreasing flood frequency. Hence, designing to maintain disturbance regimes appropriate to the flora and fauna requires attention to disturbance initiation areas, major conduits, and major barriers or filters to spread (chapter 11).

Incorporating the irregular, curvy, mosaic-like, and aggregated nature of nature, as described in the previous section (Fig. 13.3a), is probably another indispensable pattern whose benefits cannot be otherwise replicated. However, at present the ecological and human benefits are only beginning to be explored (chapters 3 and 4). Similarly, the importance of all elements in a configuration or cluster of ecosystems is only a research frontier (chapter 9), but may become a foundation in planning.

The aggregate-with-outliers principle appears to be a robust way to fit elements together into a whole landscape pattern (Fig. 13.1). However, it cannot be considered indispensable until it is time tested.

Finally, the whole landscape pattern established must be evaluated in terms of the time trajectory of the landscape. What mosaic sequence is being followed, how far along is the existing landscape, and if

453

unchanged, what is the future sequence (chapter 12)? Normally, planning would alter this trajectory to approach the ecologically optimum mosaic sequence. At present, a modification of the 'jaws' model appears optimum (chapter 12).

Key locations within a landscape

With a spatial framework for the whole landscape established, we turn to additional principles that focus on key locations within it. These include the following.

Unusual features (chapter 1) tend to be especially important sources and sinks. Protecting elements with the highest species richness is efficient for landscape-wide biodiversity conservation. Large nodes in networks have unusual importance because of their ramifying connections. Gaps in major corridors are keys to movements along, and flows across, a corridor. Similarly flux centers (chapter 9) are critical because movements and flows through them affect many other landscape elements. Protecting elements with high sensitivity to human impact provides important stability to the landscape. And finally, focusing on strategic points (chapter 9) permits an unusually important control over flows, as well as protection from external forces.

Few if any of these locations are indispensable for control or protection. But all have unusually wide impacts on the landscape through flows and movements. Therefore they are exceptional opportunities and ready handles for the planner.

Ecological characteristics targeted

Rather than focusing on particular spatial attributes, the planner may use landscape and regional ecology to target or affect particular ecological characteristics and objectives to society. The results of targeting these characteristics of course produce spatial solutions in physical plans. The following are important examples.

(1) Maintain healthy populations of migratory fish by attention to the river mosaic, from first-order stream basins to river mouth. (2) Maintain viable populations of important species. (3) Minimize inbreeding depression. (4) Plan for the needs of multihabitat species. (5) Plan for wide-ranging species. (6) Avoid the spread of exotic species. (7) Protect keystone species whose decimation affects so many others. (8) Design to concurrently enhance interior and edge game species. (9) Channel snow accumulations to desirable locations. (10) And protect our best soils from overuse, or being covered by buildings or paved tarmac.

454

Planners and conservationists can accomplish all of these important ecological characteristics. Options are available for each objective, and consultation with other experts appears essential.

Spatial attributes targeted

Finally, a wide range of relatively detailed spatial attributes can be targeted using principles from landscape and regional ecology. Some overlap the preceding broader approaches. Options apparently are available to the planner to accomplish all of them. The attributes are listed by groups below. Detailed descriptions are given in the chapters indicated.

Patch and boundary attributes (chapters 2 to 4) include the following: small patches; minimum dynamic area; soft curvilinear and mosaic boundaries; generated edges of natural resource areas; major lobes as funnels and drift fences; and compact shapes.

Corridor attributes (chapters 5 to 7) include: gap size and number; cluster of stepping stones; major road corridors as conduits, barriers, and sources; locations such as gaps and the heads of valleys where the Venturi effect is pronounced; first-order stream control of hydrologic and nutrient flows; 2nd- to *c.* 4th-order stream controls on erosion, nutrients, and fish; meandering river controls on many conflicting environmental and human uses; floodplain ladder pattern; minimum widths along a river system; corridor connectivity; and a string-of-lights design.[501,1393]

Network, matrix, and mosaic attributes (chapters 8 to 9) generally have widespread implications: connectivity and circuitry of a network or matrix; perforation of the matrix; configurations of linked elements; a connected system rather than a collection of natural resource areas; convergency points; and interspersion of habitats.

Flows and movements over the landscape (chapters 10 to 11) include: sources and sinks for air-borne heat, gases, and materials; decreased local extinction in metapopulations; increased recolonization in metapopulations; the landscape elements learned in sequence by introduced animals; and the major routes through a mosaic.

Finally, the preceding patterns, processes, and principles emphasize how divergent the present planning process is from the rapidly developing knowledge base. Very little of the preceding has yet to be effectively incorporated into any planning for towns, parks, forests, suburbs, pastureland, natural areas, cities, game, water resources, biodiversity, soil conservation, or regions. Incorporating the most-important components above into the normal planning process now is an opportunity, a challenge, and an expectation. If the expectation is not met, a new cadre of people who can incorporate the bulk of these issues will soon step forward, and be asked to plan our land. Such planners using fresh approaches will doubtless capitalize on the

Fig. 13.5. Perforation of forest and erosion on clearcut slopes. Associated logging roads commonly are greater sources of erosion and sedimentation, too often damaging streams and fish populations below. Oregon (USA). Courtesy of USDA Soil Conservation Service.

many linkages between the preceding landscape- and regional-ecology principles, and the expectations for creating sustainable landscapes outlined in chapter 14.

FOREST, DRY AREA, AND AGRICULTURE

The components of the preceding generic plan appear to apply to all landscapes. A few additional principles seem particularly applicable to planning specific types of landscape. Nevertheless, the generic plan is the core for forested, dry, and agricultural landscapes here, and suburbia in the following section.

Forested landscapes

Most forested landscapes mesh wood harvest, nature conservation, recreation, and other major societal objectives[1700,583,1437,1004,1277,1306] (Fig. 13.5). The forested matrix is commonly perforated, dissected, and later fragmented. These spatial processes should be avoided or channeled into

456

a mosaic sequence that minimizes manifold, negative ecological effects (chapter 12). Logging patches can usefully mimic the aggregated, irregular, curvilinear, and mosaic-like patterns of nature.[1662]

Roads, including access roads for logging, perhaps have the greatest direct and indirect negative impacts (chapter 5). They open up areas to houses and built areas, whose distribution in turn is almost always best aggregated. Minimizing construction of roads and closing off existing ones decreases road density, which has beneficial effects on many large-mammal populations.[187,1099,1515] The five corridor functions, i.e., conduit, barrier, habitat, source, and sink, are ecologically significant for road, railroad, and powerline corridors in forested landscapes. In addition to minimizing road density, planning needs to focus on these functions of existing roads.

Erosion caused by water is a major concern in most forested landscapes (Fig. 13.5). Road construction and maintenance is often the major cause of erosion.[1731,229,1661] Therefore, minimizing road density has significant benefits here too. Logging techniques such as clearcutting, selective cutting, or group selection cutting differentially affect erosion.[229,309,154,1665] The shapes and locations of roads and landings on or off steep slopes do as well. The arrangement of logged patches in a mosaic sequence is equally significant. This sequence in turn determines whether a logging road can be constructed, used once, and eliminated for the length of a forest rotation, or whether it must be maintained for repeated use. The maintenance and persistence of roads and other areas of disturbed bare soil cause sediment to pour downslope.

This sediment typically accumulates in streams, rivers, and floodplains.[860,1663,1661] It smoothes stream bottoms, thus removing fish habitat and fish.[885] Mineral nutrients move in solution and with eroded particles, and pollute streams. The nutrients, especially nitrogen and phosphorus, may eutrophicate aquifers, lakes, and wetlands.

In conclusion, the principles for a generic plan provide for most planning in forested landscapes. Additional principles specifically targeted to forested landscapes focus on providing logical patterns for logging, roads, erosion, and especially forest protection.

Dry landscapes

Not surprisingly, a different set of additional principles is targeted to dry landscapes such as desert and most grassland.[732] Wind and exposed surfaces are much more significant ecologically. The various designs and arrangements of windbreaks have diverse effects on wind speed, turbulence, and vortex flows. These in turn strongly affect evapotranspiration, plant production, soil stability, and soil and snow accumulation patterns.

The effects of albedo and advection, say on oases, can be altered by the arrangement of spatial elements (chapter 10). And cold air drainage patterns are particularly pronounced in the dry night air.

Surface particles are eroded by both wind and water over large areas, and principles of soil erosion control depend heavily on spatial arrangements. Windstorm deposits of sand and loess also are altered by vegetation and windbreak patterns. Heavy rainstorms may reshape the surface dendritic network depending on these patterns. Preventing overgrazing by livestock with one of the management rotation schemes (chapter 11) is a major erosion control step.

Stream and river corridors often are the most prominent features in a dry landscape, and are central to both ecological and human uses. Many ways of ecologically planning these corridors are available. To protect oases and scattered livestock water sources, special spatial planning is required relative to trampling patterns, productive areas, and so forth. This particularly includes the immediate surroundings of, and the strips between, water holes. Irrigated agriculture often occupies bits of a dry landscape, and this requires careful thought to protect surface water areas, and guard against surface salinization and a lowered water table.

Road density and landscape dissection may require planning consideration, though roads are usually of more local significance in dry landscapes. Roadsides watered by runoff from paved tarmac surfaces often have lusher, different vegetation that attracts wildlife. And little-used dirt-road networks represent one of the best examples of a network through which species move, in this case predators at night (chapter 5).

In short, wind, moving soil, and sparse vegetation are the key players to be managed over most dry landscapes. Water is scarce, so scattered sources and stream corridors knifing through the area are primary foci for planning and management.

Agricultural landscapes

A handful of issues here require planning concepts beyond those in the generic plan above. Remnant or regenerated patches of nature, wooded strips, and water-caused erosion from fields tend to be central issues requiring planning.[680,10,531]

The arrangement of fields on slopes and relative to streams is critical to erosion and amenable to planning. And so are the cultivation practices within the fields. Wind erosion may be similarly significant in limited places and limited times. Keeping cultivation off steep slopes, and buildings off the best soils is fundamental. In agricultural intensification, erosion and mineral nutrients are especially threatening to streams, wetlands, and aquifers.

Controlling the grain size of the agricultural landscape by maintaining a fine-mesh network of hedgerows, windbreaks, and woodland corridors provides many benefits. Windbreaks provide the quiet and wake zones, and roadside natural strips (chapter 5) are promising solutions to many ecological problems.

The patches of nature usually left on poor agricultural soils can be planned for size, shape, orientation, number, and arrangement.[732] Edges can be sculpted for wildlife and may be subjected to nutrient enrichment from adjacent fields. In the mosaic sequence of an agricultural landscape, natural patches are usually particularly susceptible to shrinkage and attrition, but undergo little perforation, dissection, or fragmentation. Planners can alter the sequence to minimize shrinkage and attrition effects.

In summary, a set of principles should underlie planning, conservation, design, and management of any landscape. A few of these appear indispensable to attain important ecological objectives. Other principles address prominent issues that are distinctive in forested, arid, or agricultural landscapes.

PLANNING A SUBURB

The suburban landscape

The *suburban landscape* outside the high density of buildings in a city is distinctive in its ecological structure, functioning, and changes (Fig. 13.6). Before focusing on the unusual attributes of suburbia, its major structural and change patterns are introduced.[1604,755]

Its location between urban and rural landscapes means the suburban landscape receives strong inputs from both.[565,1342,324,1080,1081] Acting as a source, the rural area showers suburbia with seeds, animals, agricultural and wood products, and in some cases airborne soil particles or floodwater. Similarly, the urban area disperses pollutants, people, information, products, and in some cases heat, throughout suburbia. Of course, the suburban landscape is also a source of many of these objects, that then enter the urban and rural areas.

Heterogeneity within suburbs appears primarily related to geomorphology and roads[1604,1533,246] (Fig. 13.6). Geomorphic patterns provide hills, valleys, slopes, flat areas, streams, rivers, wetlands, and contrasting soils. Road networks of varying mesh size imprint the land, and correlate well with housing, commercial, wooded, and other land uses. The suburban landscape may mask portions of two or more adjacent previous landscapes, by exhibiting the same built pattern over different geomorphic areas.

Fig. 13.6. Housing development (estate) illustrating several of the unusual features of a suburban landscape. See text. Here many back lot-lines have channelized streams, and many house lots have serious backyard erosion. Morris County, New Jersey (USA). Courtesy of USDA Soil Conservation Service.

Within suburbia it is ecologically useful to differentiate the *open space* (unbuilt areas) and 'built areas'.[1435,1713,1655,565] Within open space are large and small patches that may encompass a range of land uses from nature reserves, reservoirs, and flood-control wetlands to cemeteries, golf courses, and ball fields. Both their size and shape are ecologically important.[601,1717,1597] Open space corridors include large structures such as major streams, rivers, and greenways primarily for recreation.[5,1753,1582] Small corridors such as shrublines and small streams generally are considered and managed as part of the built area.

Built areas include large and small housing, commercial, and industrial zones.[1832,1833,1826,1827] The ecological design of housing areas is examined in a later section.

Water flows relatively fast across suburbs, over roads and through pipes. Resident species such as domestic cats and vacant-lot plants readily move through this landscape.[696,565] But the movement of native rural species is much inhibited. Butterflies are most inhibited, small and large mammals intermediate, and forest birds are somewhat inhibited in crossing[877] (see Fig. 8.14). These effects are especially pronounced when forest covers less than 5% of the suburb.

460

Change in suburban structure is particularly evident.[1832,1597] This results from both internal and external forces, as well as the short average residence time of people in a suburban home. It is also because typically the urban landscape is spreading over the suburbs, and the suburbs are rolling over the rural landscape.[461,709,1458] The suburban landscape may have the same overall appearance year after year, but that appearance is moving away from the city center.

Unusual attributes

Just as the forested, arid, and agricultural landscapes exhibit their distinctive attributes, so does suburbia have its own (Fig. 13.6). Planning, conservation, and management must deal with and can capitalize on these unusual attributes, which are grouped into four categories.

1. *Species*. Generalist edge species predominate throughout suburbia.[890,6] A high species richness is present due to the abundance of exotic species.[1270,565]

2. *Patches*. The polygonal spatial arrangement of both house lots and housing clusters is almost unique.[436,246] An abundance of mowed grassy areas is usually present.[727,1655] And the density of boundaries and patches is high, reflecting a fine-grained landscape.

3. *Corridors*. Rectilinear networks of medium scale (between urban fine-scale and rural coarse scale) for human movement are the norm in suburban landscapes. Stream networks are mainly truncated and disconnected by filling, piping, road crossing, and other gaps. Eroded sediment flows tend to be rapid, due to rapid hydrologic flows as well as minimal management of erosion and sedimentation processes.[565]

4. *Suburbs*. Adjacent urban and rural lands exert a strong control on suburbs by atmospheric flows. Ecosystem eutrophication and dysfunction result from high accumulations of nutrients and toxics.[1342,565,324,1081] Natural areas are subjected to heavy human use and impact.[1717,1064] The urban landscape moves over the suburban, which expands at the expense of different rural landscapes.

Concord case study

Concord, Massachusetts (USA) was the home of Henry David Thoreau whose observations introduce each chapter.[622,1874,1511] Let us together plan the future open space of this 65 km^2 (25 mi^2) area, now in the outer edge of an extensive suburban landscape. It is an opportunity to use certain landscape-ecological principles that apply in almost any large area[473] (J. D. Ferguson and D. H. Monahan, pers. commun.).

Concord is a town with commercial and residential areas surrounded by woods and scattered fields, but is also a suburb intricately linked to exciting Boston, 25 km to the east. Approximately 37% of the town is

461

developed for buildings and roads, 27% reserved as open space, and 36% uncommitted and potentially developable. Concord attracts tourists to the origin of the American Revolution, the nineteenth-century literary center of R. W. Emerson, L. M. Alcott, H. D. Thoreau and N. Hawthorne, Walden Woods, and other special places. Yet it is primarily an active community of 15 000 people.

The number of houses has doubled in three decades, with a concurrent drop in open space.[480,473] Equally striking, however, is the fragmentation of open space into small pieces, resulting from misplaced buildings. In this respect Concord represents a typical scenario in many urban fringes.

Our broad objective is to outline an open space framework that provides for an integrated system of land and water resources, supported by a rational basis for land-use decision-making. The specific goals are to: (1) identify major town-wide or landscape-wide features, and integrate them; (2) identify all types of small special site, comparing and ranking them for open-space priority; and (3) overlay the two approaches to delineate priorities for land protection.

The approach could be considered as a way to effect 'growth management'.[381,193,538] However, because land and water provide as many and different resources as the built environment, the perspective is more akin to 'sustainability'. This means a balance between humans and nature that will continue over human generations.[1929,1380,492]

We first examine the pattern of open space in surrounding towns (Fig. 13.7a). Large open-space areas to the south and east, and a paucity of nearby areas to the west, are especially pronounced. Any plan for Concord should address the fragmentation problem within the town, as well as mesh with the surrounding pattern.

Planning at this scale is rarely integrated. Typically it focuses on the patch pattern of vegetation and land uses[1088,121,1490,775,639,1472,1201,859], or on corridors that emphasize connectivity and functioning.[1518,122,977,1492,1753,1582] Our focus is on examples of both large patches and major corridors.[277,678,11,1596] We identify actual patterns using aerial photographs plus ground checking, rather than starting with land ownership boundaries. The open space framework outlined below requires detailed knowledge only of local conditions.

Large patches and major corridors are the components that create a town-wide pattern here (Fig. 13.7b). We can identify three types of *large patches* (or areas) in Concord: built areas; natural vegetation areas; and agricultural areas. Three types of *major corridor* are also prominent: water protection corridors; wildlife corridors; and human corridors. Delineating the rationale for protecting different locations is an important value of this planning, and therefore is outlined below for most patch and corridor types.

462

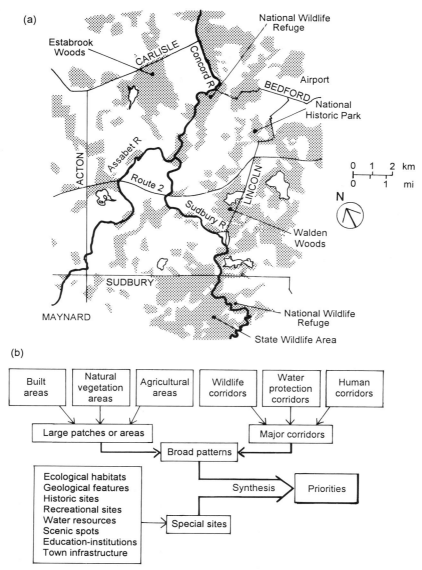

Fig. 13.7. Open space in and surrounding a town, and a framework for planning open space. (a) Existing areas with varying degrees of protection are shaded. A case-study suburban town, Concord, Massachusetts (USA) in center, and parts of seven surrounding towns. Rivers and major ponds in the glacial terrain are shown. (b) Components and sequence in the open space framework to establish land-protection priorities. Based on Ferguson *et al.* (1993); courtesy and with permission of Joan D. Ferguson.

463

In addition to large patches and major corridors, we highlight the *special sites* in town as part of the open-space framework (Fig. 13.7b). These are the small, relatively rare places that have open space value. Examples are specific ecological habitats, historic sites, water resources, and town infrastructure. Let us first outline the large patches and major corridors that create the town-wide pattern of Concord.

Town-wide pattern: large patches

1. *Built areas*. Residential, commercial, and industrial development provides structure to the town, and seven large built areas are present, including Concord Center and West Concord (B2 and B4 respectively in Fig. 13.8b). Each large built area has a continuous high density of roads and buildings, even though it may include small patches of low building density or unbuilt sites. The values of the large patches are as centers for citizens to live, work, shop, and interact.

2. *Natural-vegetation areas*. We recognize these as large, intact, and somewhat-compact patches of natural vegetation (Fig. 13.8a). Each has a considerable interior area of forest or marsh (i.e., relatively remote from obvious edge effects and direct human impacts).[991,1490,1434,672,1066] The size of a patch is sufficient to support species with large home ranges, to protect an aquifer and/or connected network of headwater streams, and for most natural disturbances to affect only a portion of the patch.[1319,683,685] Narrow lobes or fingers at the perimeter of patches enhance species dispersal to the surroundings, and also channel species into the patch (drift-fence effect)[503] (chapter 4). More than three large patches are probably required to maintain the total richness of native species in the landscape.[498] The distance between patches is less than the dispersal distance of key species, although distance apart may be greater where usable corridors are present.[1490,678,1228]

Why should the town and its citizens protect, pay or vote for large natural-vegetation patches? Water and biodiversity are the primary values, though local recreation is close behind. A large vegetated area protects surface and ground water (60% of Concord's drinking water comes from wells in town). Many wildlife and plant species are primarily restricted to the large vegetated patches. We can find large-home-range animals and healthy populations of plants and animals that essentially require patch interior conditions. And when surrounding populations disappear, the large patch acts as a ready source of colonizers (see quote at beginning of chapter 2). The large patch maintains many combi-

Fig. 13.8. Large patches and major corridors in Concord. B1–B7 = large built areas; N1–N6 = large natural-vegetation patches; A1–A5 = large agricultural areas; C1–C13 = major water and wildlife corridors. Based on Ferguson et al. (1993); courtesy and with permission of Joan D. Ferguson.

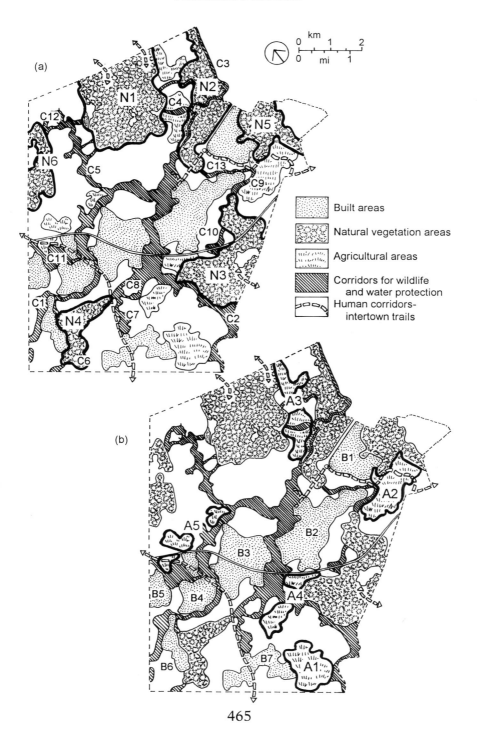

(a)

(b)

Built areas

Natural vegetation areas

Agricultural areas

Corridors for wildlife
and water protection

Human corridors-
intertown trails

465

nations of natural microhabitat conditions in proximity to one another for species requiring diverse habitats. It provides for a natural disturbance regime required for survival by many species. And by 'spreading risk' in more than three large patches, key wildlife and rare species are more likely to persist.

We find six large natural-vegetation areas in Concord, including Walden Woods and Estabrook Country (N3 and N1 respectively in Fig. 13.8a). At least one large patch with the biological and water attributes just described exists in each major section of town.

3. *Agricultural areas.* We also identify five large agricultural areas in this suburb, including Lexington Road and Nine Acre Corner (A2 and A1 respectively in Fig. 13.8b). All are predominantly open, though they include scattered buildings and wooded parcels. Fields include pastures, meadows, and cultivation. Prime farmland soil, i.e., fertile and well drained, is present over a major portion (more than a third) of each area. We include adjacent wooded parcels on prime soil, since they represent potential food-producing areas. In this glaciated terrain good farmland soil is usually subdivided, so clusters of large fields are included.

The values to the town and its citizens are exceptionally diverse. These farmlands provide many products, including vegetables, fruits, hay, and milk. Concord citizens value the historic symbolism of farmland, the active roles of farm families, the educational dimensions of farms, the availability and convenience of fresh produce for residents and metropolitan markets, and protecting prime food-producing areas in a world where hunger grows. Farmland near roads, railroads, and paths helps preserve scenic vistas and some rural character in this town. Agricultural areas enhance game populations. They also increase the biodiversity of the town and region, by providing habitat for species requiring large open areas.

With Concord's large patches identified and mapped, we next turn to major corridors. Three types are present, wildlife, water protection, and human corridors.

Town-wide pattern: major corridors

1. *Wildlife corridors.* Animal movement corridors are most effective when continuous without major gaps or narrows. With at least a high shrub layer present, moving wildlife has some protection from human activities and domestic animals (e.g., dogs and cats).[948,501,1518,685,109,1492,1228] Corridors connect large patches of natural vegetation, and together form a town-wide network with loops (Fig. 13.8a). The network extends to the town boundaries, especially in the direction of large natural-vegetation patches in surrounding towns. Water protection corridors are part of the wildlife network.

The value of wildlife corridors is to serve as conduits for the movement of species across, into and out of town. These strips counteract fragmentation effects, isolation, and the gradual species impoverishment of Concord. The network of corridors, with at least two links to each natural-vegetation area, provides optional routes for species to avoid disturbances, hunters, and predators.

2. *Water protection corridors*. These are the strips of natural vegetation around lakes and along rivers and streams (Fig. 13.8a). They typically include tree and shrub cover, and are exceptionally important to both biological diversity and ecological process.

Water protection corridors help provide a bundle of values and resources to Concord's citizens: clean water for drinking, swimming, and fishing; fallen logs and branches for fish habitat; shade to maintain cool water temperature in summer; and leaf litter as the base of most aquatic food webs.[86,130] These corridors also help absorb inputs from the surroundings: water runoff; eroded soil particles; and chemical runoff, including fertilizers, pesticides, sewage seepage, chlorine from swimming pools, radioactive nuclides, and other inorganic and organic toxic substances.[1503,999,1303] Stream and river corridors help control problems of flooding (height and frequency), sedimentation, loss of bottom fish habitat, and water quality (turbidity, BOD, algae buildup).[1503,1038,130] Water protection corridors also function as wildlife corridors.[1228,1038,1490]

We identify 13 major wildlife and water protection corridors (Fig. 13.8a). Eight are for both water and wildlife, such as the Assabet River (C1) and Sawmill Brook (C4). Five are for wildlife alone, including Walden Woods to Town Forest via a railroad underpass (C10), and Annursnac/Strawberry Hill to Estabrook Country (C12).

3. *Human corridors*. Concord has ten major types of human corridor or network, including the primary road system, secondary road system, railroad, gas and power lines, canoe routes on rivers, and inter-town walking paths. These diverse strips[961,977,1582] provide efficient transportation routes for goods, people, and energy, and are wide enough to minimize interference of moving objects with the surroundings. Corridors typically interconnect to form networks with loops and entrance/exit nodes. Most network types are spatially separated due to the incompatibility of the objects moving.

Special sites

Our next step is to identify 'special sites', the small specific locations of open space value, produced by either natural process or human activity.[473] Local experts have pinpointed more than 200 special sites in Concord, which we group into eight categories: (1) ecological habitats; (2) geological features; (3) historic sites; (4) recreational sites; (5) water resources; (6) scenic spots; (7) education and institutions; and (8) town

467

infrastructure. Each category is subdivided into its characteristic site types; for instance, ecological habitats include old forest stands, riverine meadows, and high-biodiversity sites for birds. Site types with four or more examples are considered common and therefore we remove these from the list. This helps focus protection efforts on the rare (special) sites.

Somehow we must determine the relative benefits or values of each site in this diverse assemblage.[1385,1713,473] To compare such 'apples and oranges', we evaluate each site for two attributes: (1) rarity; and (2) recovery (or replacement) time. Sites that are nationally rare are assigned greater value (e.g., 3 stars) than those at state (2 stars) or local (1 star) levels. Sites that are irreplaceable, or that would require centuries to recover to a similar form if destroyed, are assigned higher value (3 stars) than those that would recover in decades (2 stars) or in years (1 star). The overall values or benefits of sites to Concord, and also to other people, are then estimated, based primarily upon these rarity and recovery criteria.

Next, we map all high-value special sites. Those not presently protected are circled on the map. In this manner, unprotected special sites of high value, either to Concord or beyond, are pinpointed for priority protection.

Synthesis and land protection priorities

We now can overlay and link the large patch, major corridor, and special site components. As noted, the town-wide attributes fit together to form an integrated whole. It happens that 80% of the priority special sites overlap the large natural-vegetation and agricultural patches, and the water and wildlife corridors.

Thus, our analysis answers three essential questions for a plan that provides recommendations for priority action. How do the patches, corridors, and sites spatially relate to one another? Which elements are most valuable? And what are the priorities for protection?

Four priority categories emerge for open-space land protection in Concord (see Fig. 13.8) (1) *Highest*, natural vegetation patches N1 and N2, agricultural areas A1 and A4, and corridors C1, C2, C3, C4 and C5; (2) *High*, natural vegetation patches N4 and N5, agricultural areas A2, A3 and A5, and corridors C8 and C10; (3) *Medium high*, with natural vegetation patch N3, corridors C13, C9 and C12, and special sites (White Pond, Route 2, Virginia Rd., and a cluster in Concord Center); (4) *Other*, corridors C6, C7 and C11, and special sites (receiving strong but not top ranking). In essence, our open space framework provides the basis for recommendations and a five-year, priority-based action plan for Concord.[473]

General application

The analytical approach of the open space framework is widely applicable to many scales, including park, township, county, state, province, national forest, landscape, and region. The sequential categories of large intact areas or patches, major corridors, and special sites are universal. Specific types of category are readily tailored to the land uses of interest, e.g., industrial instead of agricultural patch, snowmobile rather than wildlife corridor, or oasis in lieu of historic site. The initial working-out of the town-wide patterns, and the direct comparison and ranking of diverse (apples and oranges) small sites are both reasonably objective. The framework determines the appropriateness of locations for future development. Use of this framework helps the town retain its keystone open-space features.

The present population of 15 000 residents may slowly grow in an integrated setting of open space. The town depends on major inputs or outputs of food, fuel, goods, and jobs. Yet it also supplements these by contributing its own resources, including valuable water, wildlife, aesthetics, recreation, soil, and food products. Establishing the integrated open-space pattern for these resources provides, over human generations, a more sustainable balance between people and the land.

In effect, the open space framework objectively integrates the broad-scale and fine-scale resources for decision-makers. Rather than allowing haphazard fragmentation in suburban communities, we can effectively enhance and sustain their value by distributing resources in an integrated spatial pattern. This is primarily accomplished at the scale of a landscape or regional mosaic.

ECOLOGICAL DESIGN OF HOUSE LOTS

To illustrate some of the finer-scale ecological issues in suburban landscapes, we will explore the design of an individual *house lot* based on ecology. A typical elongated fraction-of-a-hectare (or <1 acre) property with a single house toward one end that borders a road is used as a model.

Some observers will reasonably consider this to be at too fine a scale for landscape ecology. Yet house lots are included because they often cover the bulk of a suburban landscape, and their ecology is scarcely documented.

After a designer or planner demonstrates that a house lot, garden, or other space is ecologically sound, the design then may be enhanced to accomplish other objectives, such as symbolism, beauty, or meeting

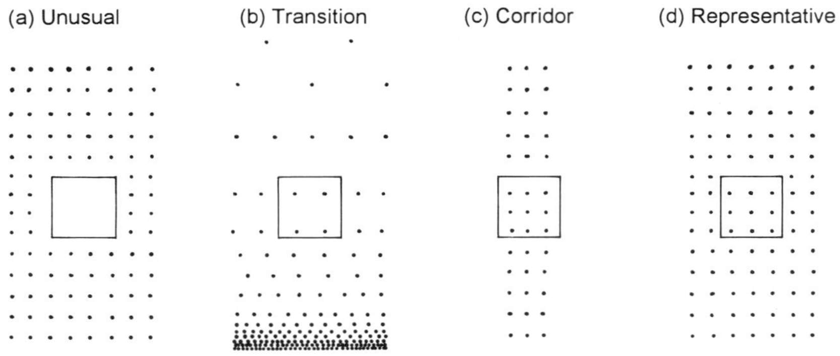

Fig. 13.9. Neighborhood context of a house lot. Square indicates house lot and dots represent a type of land use or house-lot design.

place. Vertical cypresses (*Cupressus*) 'reach to heaven', and plump orange trees symbolize fertility. Shrubs channel human movements, rushing water murmurs, and luscious scents attract romantics.

Neighborhood context

The first step in ecologically designing a house lot is to understand its context. Is the lot unusual in the neighborhood (Fig. 13.9a)? In an open area it may be oasis-like, the only lot with tree cover. Or in a wooded neighborhood it may be the only clearing. In either case, the meteorologic, soil, water, animal, and vegetational characteristics and flows are highly distinctive, much like an unusual feature in a landscape (chapter 9). All design steps thereafter can take this unusual importance into account.

Alternatively, the house lot to be designed may be in a transition zone (Fig. 13.9b). Here it is subject to strong contrasting flows from opposite sides, and perhaps a third type of flow from the remaining sides. Or if the lot interrupts a conduit such as a stream or wildlife corridor, it is likely to be inundated by floods or herds of herbivores, or patrolled by predators (Fig. 13.9c). Its design will relate strongly to the corridor. Finally, the lot may simply be representative of the surrounding lots (Fig. 13.9d). This also influences the possible range of ecological designs.

Goals and techniques

Our second step in designing a house lot is to articulate the primary goals (this of course could be the first step). Designing ecologically is analogous to designing for humans; there are many dimensions to

470

humans, and decisions must be made whether the primary goal is beauty, security, a sense of community, home energy efficiency, or what. Similarly, the design goal may maximize for one ecological component, or optimize among a few.

The goal may target high species richness. Or focus on a particular important or appealing species. Or control wind and airborne inputs. Or control soil erosion. Or emphasize recycling of materials. Or protect water or a wetland. Or design for a native ecosystem.[153] Other examples can be added, but it is clear that many goals and products of ecological design are possible.

A more detailed breakdown of goals can be combined with techniques to illustrate better the array of options on the ecological designer's palette. Four somewhat-overlapping groups are illustrated.

1. *Species*. One may focus on characteristics of species themselves, such as germination, growth rate, defense mechanisms, behavior, adaptations, coevolved mechanisms, and other species interactions.[380] Such approaches could be targeted to produce: high species richness[6]; an abundance of native species; maintenance of an unusual species; attraction of some nearby target species; 'flags' to attract migrants[1632]; a minimum of pests; or avoidance by domestic animals.

2. *Special habitats*. A related approach is to focus on habitats, whether natural or artificial, that affect particular species.[896,1696] Examples are birdhouses, walls and rock piles, logs and snags, slopes and exposures, still and moving water, wetland zonation, and distinctive exterior surfaces and structures on a house.

3. *Flows and movements*. The sense of active movement in a house lot may involve flowing water, swimming fish, flying birds or butterflies, mammal movement, moving sand or seeds, flowing scents, moving sounds, 'sensitive plants', waving branches, and fluttering leaves.

4. *Change*. Dynamics can be illustrated at several scales concurrently. These include diurnal change, lunar or tidal fluctuation, seasonal time, unidirectional or cyclic succession, paleoecological change, evolution, and coevolution.

Spatial arrangement

Design and physical planning spatially arrange objects. In the present case, how do we fit, say, trees, shrubs, grass, a house, and other objects together on a lot to accomplish the preceding ecological goals? Several issues are identified. One approach, using patches and corridors, applies landscape ecology to this fine-scale house lot.

First, the house itself, perhaps with accompanying structures such as driveway, garage, septic system, and shed, may be partially located using ecological characteristics. Solar angle, slope, soil type, wind, location

471

of stream or wetland, and existing trees have effects. Yet government regulations such as setbacks from roads and side-lot lines may have no ecological basis.

The designer may usefully consider the arrangement of small corridors and patches within the lot. Two types of corridor are prominent. One is higher than its surroundings, such as a wooded strip, and the other is lower than its surroundings, an open or trough corridor (chapter 5). Open corridors are useful as sight lines for animals and humans, and as escape routes for prey. Wooded corridors may inhibit slope erosion, enhance animal movement, be a barrier along a boundary, serve as a windbreak, provide house foundation planting, funnel people to a door, or screen the road from movement and noise. The typical house lot seems to have rather few locations used for open corridors, but several common locations of wooded corridors.

Small patches also come in open and wooded flavors. Open patches are frequent, both in front of and behind the house. Wooded patches most frequently are attached to the back line of the lot. The size, shape, and number of patches varies widely. Habitat variety usually is high when both dense dark patches and open sunny patches are present. Overall, the range of available options is much greater for small patches than for corridors in ecologically designing a house lot.

Edges and junctions generally receive considerably more attention at a fine scale than in broad-scale planning. Boundaries are what people commonly see at the finer scale. Hard boundaries and a rich variety of soft boundaries can be created (chapter 3). Boundary length may be of major interest. Interspersion can enhance the number of different adjacent habitats (chapter 9). And 'junctions' or convergency points of three or more habitat types have distinctive functions (chapter 9).[896]

Vertical height diversity also is of greater interest at fine than broad scales. That is, the number of vegetation strata or layers is not only readily seen, but has important effects on avian diversity, wind, shading, plant growth, and much more.[6] To visualize the three dimensional vegetation over a lot, one might draw horizontal slices at, say, the height of the herbaceous, shrub, and tree canopy layers. Or simply consider yourself a small arboreal vertebrate like a lizard, squirrel, or glider, and see how much of the volume over the lot is accessible. That means accessible without getting within reach of dogs, cats, and other ground predators. Such analyses may highlight the ecological importance of the shrub layer in suburbia.

We end this consideration of fine-scale ecological design by returning to the neighborhood scale, that is, the surrounding group of house lots.[584] The ecological design of a house lot depends on understanding and meshing with the three-dimensional vegetation structure of the neighborhood. In a 20-year-old residential area of Newark, Delaware (USA)

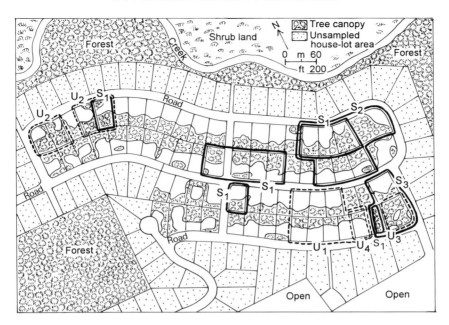

Fig. 13.10. Vegetation corridors, and habitat for a forest bird in a residential housing area. Somewhat-natural vegetation is aligned along back lot lines. S_1 to S_3 are suitable habitats (house lot areas) for wood thrush (*Hylocichla mustelina*) in suburban Delaware (USA); see text. U_1 to U_4 are the least-suitable habitats (U_1 = nested only 1 year; U_2 = no birds returned after migration; U_3 = no nests; U_4 = zero for all variables studied). Seventy house lots sampled for 4 years. Adapted from Roth (1987).

tree cover was continuous over the back line of 70 neighboring house lots studied[1450] (Fig. 13.10). This illustrates a widespread pattern in many suburbs. The *back lot line* is the least designed and manicured area. Consequently, it is commonly the most natural portion of a lot, the most likely to support native plant and animal species. Furthermore, the typical alignment of back lot lines means that a corridor of natural vegetation is often present (Fig. 13.10) These back-lot-line corridors are the most-important wildlife corridors in many suburban neighborhoods. Consequently, house lot design with connected multilayered natural vegetation plays a key role in providing or maintaining connectivity and wildlife benefits.

An additional neighborhood pattern is evident in the Newark example[1450] (Fig. 13.10). The wood thrush (*Hylocichla mustelina*), a forest bird with a hauntingly beautiful flute-like call, was sampled on the 70 house lots for four years. Five lots or groups of lots (labelled S_1 in Fig. 13.10) had wood thrush nests in at least three of the four years. One

473

group of lots (S_2) had at least three young thrushes fledged during the period, and one area (S_3) had at least two previous thrushes return after migration in a subsequent year. Fig. 13.10 shows that five small areas in the 70 residential lots appear particularly suitable for this forest bird species.

Four areas (labelled U_1 to U_4 in Fig. 13.10) in the wooded back-lot-line corridors seem especially unsuitable to wood thrushes. The remaining portions of the corridors are intermediate in suitability. It is unclear why specific lots are particularly suitable or unsuitable for this species, though more-suitable sites tend to have larger patches of continuous tree canopy. Nevertheless, the design of a house lot can be consistent with the distribution of suitable natural habitats for key native species in the neighborhood. It also can enhance movement of plants, animals, and water in neighborhood corridors.

MANAGEMENT

Most of the key principles for management at the landscape scale were presented earlier in this chapter under planning. But managers do much more than plan.[1800,631,1483,87] They are protectors and stewards of the land. They implement plans. They anticipate surprises. They deal with crises. They capitalize on change. They alter the land with subsidies and removals. They maximize and optimize for particular objectives. They determine the best place, and the undesirable places, for alterations. They implement plans. Sometimes they throw plans away, but not the land. Managers know and love the land.

Broad spatial and temporal scales

A key challenge to managers is how to keep an eye on the big picture. It is simple to manage little spots but difficult to do big mosaics. It is easy to produce something for tomorrow, but hard to create a decade-long pattern. Thus, landscape-based techniques to expand the spatial horizon are first presented, followed by a shorter list to expand the temporal perspective.[488]

Management in a broad spatial context is enhanced in many ways. When changing one landscape element, evaluate its effects on species, soil, and so forth in all elements. Effects spread, and in a reasonably predictable manner (chapter 9). Manage concurrently for two or more objectives, thus optimizing and providing stability.[1959,764,1582] Keep the 'indispensables' such as large natural-vegetation patches and river system protection, up front in priority. Observe what happens in the surroundings as carefully as what happens in the managed area.[794,501,17,39] Maintain appropriate movements and flows, not only within the area, but to and from all surrounding areas.[109,1492,583] Decide how important the configuration concept (chapter 9) is in different portions of the area. Dividing the mosaic into separately managed pieces is easy, but wise management then becomes difficult. A paradox of management is that one can have a large effect on a small area, but the probability of success in management is greater in a large area (chapter 14).

Certain management objectives make a broad spatial perspective obligatory.[1437,334,39,732] Management to maintain or enhance wide-ranging species, major species dispersal and migration routes, or many fish populations requires a broad landscape perspective.[685,403,1228] Managing for flood control, water quality, 'tight' mineral nutrient cycles, or food webs that include humans only works at a broad scale.[17] And maintaining or creating a number of large natural-vegetation patches, a natural disturbance regime[1550], a giant green network, or an aggregate-with-outliers land requires landscape ecological management.

Management to minimize *cumulative impacts*[404,815,1817,1529,1025] is especially relevant here. These impacts on a system result from the additive or synergistic effects of a number of variables, which in turn often originate in separate locations. Common examples are cumulative impacts on streams resulting from, for instance, streambank vegetation removal, sediment from roads, nitrogen and phosphorus from septic systems, and overfishing. Because cumulative impacts are ubiquitous, and involve different locations, different variables, and additive or synergistic effects, it is impossible to address them without a landscape or regional perspective.

To apply more effectively ecological principles in management, planning, and policy, a rich variety of approaches is emerging. These include sustainable development, adaptive management, ecosystem management, and new forestry. Sustainable environments may integrate all of the approaches[1929,286,1002], and are presented in detail in the next chapter.

Adaptive management stresses the need to adjust and change, as research and monitoring provide new information[745] (T. R. Crow, pers. commun.). *Ecosystem management* appears to focus on natural process, sustainability, use of models, cognizance of context, adaptability, and accountability, though usage varies[1603] (N. L. Christensen, pers. commun.). Rather than only focusing on local ecosystems (chapter 1) or ecosystem functioning[1233], it embraces population dynamics, mineral nutrient cycling, biodiversity, information flow, and other important variables.

New forestry meshes logging activities with the multiple uses of forested land.[39,516,149] It focuses on taking a landscape or regional view, a long past-and-future time perspective, and a mimicking of natural patterns when logging sites. Wood products, roads, biodiversity, erosion, and river systems are major players to be integrated.

Thinking long term is the hardest part of management. Some objectives almost force the long view: growing wood; maintaining viable populations[1595]; protecting old-growth forest; building persistent game populations[943]; preventing overgrazing and overbrowsing[713]; protecting wilderness; minimizing erosion[1798,1800,1661]; minimizing sedimentation of streams and reservoirs; preventing aquifer pollution; increasing population sizes of rare species; and landscape restoration. Since all these cases require a large area, they emphasize that if we take a broad spatial perspective, usually we automatically think long term (see Box 13.2 and Fig. 13.11).

Box 13.2: Landscape restoration

Scores of books have been written on habitat restoration and rehabilitation. Most provide rich information on small areas, such as most mines, abandoned lots, and gravel (borrow) pits, and on plant and soil treatments. They say little about restoring or rehabilitating landscapes using principles such as in the preceding chapters, though interest is growing.[821,245,1494]

Tropical dry forest in Central America has been extensively removed for grazing and other uses. Unlike nearby rainforests, it has been a low priority for protection.[795] Daniel Janzen launched a long-term project to reestablish a 75 000 hectare (280 mi²) tropical dry forest in Guanacaste National Park in Costa Rica.[796,18] The area is spatially integrated and connected with higher-elevation forest, river systems, and surrounding natural resources. And its long-term horizon is enhanced by deeply involving the regional human population in its management, as well as in natural resource education.

The initial major challenges were to decrease livestock grazing, fire regimes, and the cover of exotic grasses (D. H. Janzen, pers. commun.). After four years the area appeared as a tall-grass prairie, except for scattered pre-existing trees and an enormous number of tree seedlings. After six years, a virtually complete canopy of woody plants 2–6 m high covers the area. Almost all are deciduous species with lightweight seeds, whereas almost all trees in tiny remnant old-growth patches outside the park are evergreen species with large seeds. More than two centuries may be required for the park to reach a self-reproducing climax mosaic.

In Florida the Kissimmee River meandered 165 km (103 mi) from approximately Orlando to Lake Okeechobee. It was channelized in the twentieth century to a wide, deep canal of half the length (Fig. 13.11). The surrounding water table dropped, and most floodplain wetlands dried out.[838] The seasonal pattern of

MANAGEMENT

water flow was altered, so constant water levels degraded the remaining wetlands. The river straightening reduced or eliminated plant community complexity, habitat for waterfowl and other wetland wildlife, nursery areas for fish, and interactions between river and floodplain.

However, a recent demonstration project (using weirs) redirected a minuscule flow from the canal into selected stretches of the old river channel. This tiny bit of river burst into life. Almost overnight these channels had more forage fish, game, flushing of muck accumulation, original riparian vegetation, connections among wetland habitats, and habitat diversity.[838] Adjacent previously-dried-out wetlands had more waterfowl and wading birds, export of invertebrate food to channels, and cover by wetland plants. Perhaps these demonstration results will lead to restoration of the long, several-kilometers-wide river landscape. If so, it will be the first restored river, and the largest restored wetland worldwide.

Many other projects and approaches address landscape and regional restoration.[497] Massive planting of windbreaks in northern China has rehabilitated landscapes, and tree planting over whole mountain ranges continues. The twentieth-century reestablishment of beaver (Castor) populations in many areas of North America has rapidly restored natural stream and floodplain processes in the landscapes (chapter 7). Land evaluation techniques identifying appropriate land uses for different spatial elements have been used to improve land.[1953,1193] The reestablishment of distinct patterns of patches and corridors has been spelled out in a variety of locations.[1645,1494,1808] This includes the modeling process underlying a restoration plan for new forests in the densely-populated Netherlands.[678,877,1260] Grassland restoration has included long strips by railroads[1400,155], analogous to the wooded roadside-natural-areas in Australia (chapter 5). Agroforestry techniques offer exceptional promise for whole landscape rejuvenation.[563,1499,858,1187,1945] Restoration efforts for the 'North Woods' region (northeastern USA) emphasize large-area programs with spatially overlapping requirements, including recovery of the imperiled Atlantic salmon (Salmo salar), restoration of the eastern timber wolf (Canis lupus), creation of a 1.3-million-hectare Maine Woods National Park (see quotation at beginning of chapter 8), and renewed resource protection of national forests (M. J. Kellett, pers. commun.). Ecological restoration or rehabilitation of whole landscapes and regions is an enormously important frontier.

Some managers plan to manage an area for a long time, a good incentive to long-term thinking. Designated natural resource areas, such as parks, forestry areas, and game refuges, usually have legal mandates and are inherently persistent. Regular inventories and evaluations of objectives and management actions encourage an ongoing perspective. Management that includes the expectation of climatic change, such as linking nature reserves in a giant network, forces one to take the long view.

Fig. 13.11. Kissimmee River channelized to a canal, in the area where restoration efforts have begun. The straight route replaced meanders, the deep channel dried out floodplain wetlands on both sides, and rich biodiversity was replaced by productive cattle ranching. Former channel flow, wetlands, and wildlife populations have now been reestablished in a floodplain area behind the photographer. Central Florida. R. Forman photo.

Protecting natural resources

Natural resource areas, i.e., protected areas or land with administrative boundaries, represent an important subset of the land mosaic being managed.[943,488,1483,1513,1813] Examples are parks, forestry areas, game refuges, nature reserves, wilderness areas, and trail or greenway corridors. In general, each has a primary designated objective plus a number of additional objectives. Thus, usually the primary goal of forestry areas is to produce wood, nature reserves to protect biodiversity and natural processes, and trail corridors for recreation. Yet all three typically provide some protection to water, soil, and rare species. Furthermore, the primary objective can change. A natural resource area primarily for wood production may become more valuable as a nature reserve.[22] A park for recreation may become more valuable for flood control, or a flood control area more valuable for recreation.

Preventing overuse by humans, though, ties different types of natural resource areas together. Management to keep people out is usually unsuccessful. Rather, differentiating people according to their expected uses, and spatially separating those uses, provides a useful handle in management.

478

One of the key separations is to concentrate high-people-density high-impact locations in the edges of the protected natural resource. Such locations function like 'magnets in the edge', attracting people to certain places near the boundary of a protected area. For example, many families with small children that want to see wildlife are generally delighted with seeing some common birds or mammal tracks near the beginning of a trail. Even signs indicating what animals are nearby can accomplish this objective. In contrast, a few hearty backpackers want to head quickly to the interior of the area away from people. Hence, a trail beginning that is designed and planted to attract dense wildlife populations produces happy visitors near the edge of a natural resource area. It also helps protect interior species and other natural resources in the central portion. Based on simple geometry usually a lot of the total area is near the boundary. Edge magnets should both enhance public support and protect interior resources.

Managing a natural resource area requires good inventories. Vegetation, soil, water, elevation, and the like are usually mapped. Perhaps more important are inventories of flux centers (active flow or movement areas; chapter 9), natural disturbance regimes, and differential sensitivities to human impact.[501] Almost any factor that involves flows or movements tends to emphasize that management cannot be based on the usual static maps, with boundaries drawn as if they were barriers. No absolute barrier exists in nature, only filters.

A small natural resource area, such as a remnant wetland or woodlot, has two additional major constraints for managers[501,1514,1393,1515,1306]: species isolation and human impacts from the matrix. Connectivity for species through corridors and stepping stones in different directions is important. And developing mechanisms at the edges and in the surroundings to manage human inputs is critical for a small resource area.[1717]

Farmers manage farmland. Sometimes soil or agricultural experts help. Timber and paper companies manage private forestland. Individual landowners or residents manage woodlots. Ranchers, farmers, and many others manage pastures, paddocks, and rangeland. This includes extensive landscape areas of grassland, shrubland, and tropical savanna. Soil conservationists manage to limit erosion by wind and water. Water resource experts manage to prevent floods and maintain hydrologic water levels.

Politicians and professional civil-service personnel manage cities, suburbs, towns, and villages. Transportation engineers manage road and railroad corridors. Industrial managers manage industrial areas. Turf specialists manage golf courses. And government manages many areas.

Does anyone manage deserts? Tundra? Muskegs? Aquifers? Air quality? Water quality? How well are the other areas managed? Is a landscape and regional perspective central in land management? It should and will be.

479

14

Creating sustainable environments

Only when the city, the hamlet, or the cottage is viewed from a distance does man's life seem in harmony with the universe . . . The sunlight on cities at a distance is a deceptive beauty, but foretells the final harmony of man with Nature.

Henry David Thoreau, Journal, *1841*

Dreams of utopia have sparked the establishment of communities in every century. Social, political, or religious fervor and commitment have created such communities, especially in rural landscapes. Yet rather few of these experiments have lasted a century. At the other extreme are civilizations and societies that have continued for centuries and even millennia (Fig. 14.1). In all cases success or failure depends both on human dimensions and the land.

Two central components of sustainability are thus introduced. The time frame is over human generations, and the key characteristic is a dynamic with equal emphasis on human and environmental dimensions. The preceding chapter on land planning and management elucidated principles based on this combination of land processes and human resource use. But the time scale was years (or up to a few decades), the normal framework for planning, design, conservation, management, and policy. The difference in time scale is the primary characteristic separating sustainability from normal year-to-year planning. For planning over human generations, if we are to learn from history, the most instructive history is of human–land interactions over two or more centuries.[286]

Most sustainability literature focuses on the globe.[1414,286,1929,768] Only a few countries have begun seriously addressing the implications of sustainability.[434,1002,1431,476,778] Although this chapter touches on the built environment, the idea of sustainable urban development is not directly addressed. However, agriculture, economics, forestry, fisheries, sociology, and more recently ecology have taken a particular interest in sus-

480

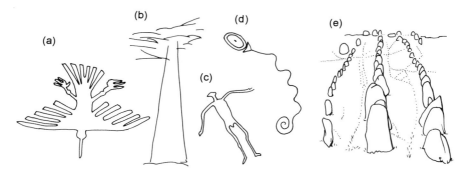

Fig. 14.1. Huge sacred land-markings by ancient cultures. (a) Bird with 145 m (450 ft) wingspan; Nazca culture in southern Peru, c. 400 BC to 800 AD. (b) Strip extending for kilometers into the foothills of the Andes; Nazca. (c) Human figure 28 m (88 ft) high, and perhaps 500 years old; Pima-Pagago people, southern California. (d) Serpent mound 400 m (1250 ft) long; Adena people c. 1000–300 BC, Ohio (USA). (e) Rows of megaliths, stones up to 4 m (12.5 ft) high, extending for >1.5 km, and constructed 2500–2000 BC; Carnac, Brittany, France. Adapted from Bridges *et al.* (1986) and Woodward & McDonald (1986).

tainability, often considering it within their own disciplines. Surprisingly, the role of spatial arrangement of land is commonly overlooked. Little literature on sustainability exists at the landscape and regional scales.[492] Yet these scales may be the most important for attaining sustainability.

This chapter assumes general familiarity with principles in most preceding chapters; indeed many points will otherwise be unclear or counterintuitive. The chapter begins with the basic concepts of sustainability. It then focuses in detail on key sustainability attributes, organized by time, space, humans, assays, and adaptability. Next, we learn from human histories, and briefly revisit regional ecology. Finally, some guidelines for making a landscape or region sustainable are presented.

CONCEPTS OF SUSTAINABILITY

Let us begin with the fascinating and important changes in the idea of agricultural sustainability, followed in turn by fisheries, forestry, economics, and other social sciences. *Sustainable agriculture* has traditionally meant maintaining maximum short-term market production. In recent years at least the following four alternative perspectives on the concept have emerged.[438,1327,858,82]

One view of agricultural sustainability is maximum yield based on locally available resources plus long-term environmental conditions. A

481

second perspective, the maintenance of agricultural production through periods of disturbance or stress, drops the word maximum and adds the concept of disturbance. A third view is an integrated system where the overall level of productivity achieved is dependent upon simultaneously maintaining soil, water, plant, and animal resources on a whole farm or larger area[1327] (J. R. Brandle, pers. commun.). And a fourth perspective is 'low-input' agriculture, where instead of increasing productivity, one raises profitability or net gain by sharply decreasing the expensive inputs of fertilizer, pesticides, machinery, and so forth. Yet in actual practice, few agriculturalists are concerned with sustainability (D. Pimentel, pers. commun.), and the absence of including soil erosion in essentially all the concepts is conspicuous.

In fisheries, *sustained yield* has referred to maintaining production over time. This has focused on the direct human dimension of attempting to prevent overfishing, though more recently an equal focus on preventing habitat loss and degradation has emerged.[1402,1444]

Similarly in forestry, sustained yield has referred to maintaining production. Preventing overcutting has been the focus, and now minimizing soil and habitat loss are also critical.[476]

Forestry is especially relevant to a sustainability discussion because trees grow at the time scale of human generations. In addition, just as agricultural practices can prevent or increase soil erosion rates, tree cover is critical to preventing soil loss. In most regions highly erodible slopes and thin soils are more likely to be covered by trees or shrubs than by farmland. Protection against soil loss on such sites is especially critical. Forest recovery or restoration is commonly at the scale of human generations, and *soil stabilization* (a steady state where soil loss equals soil recovery or replacement) is also attainable at this time scale. However, soil recovery to replace lost soil generally requires centuries.[800,740,1301] Many nations, including the USA today, have 10, 20, or more times as much soil loss as recovery.

Plato in 360 BC describes the future of such nations, based on what generations before did to Greece, 'there are remaining only the bones of the wasted body ... all the richer and softer parts of the soil having fallen away, and the mere skeleton of the land being left. ... now losing the water which flows off the bare earth into the sea ... there may be observed sacred memorials in places where fountains once existed'. A preoccupation with things other than soil makes many nations poor candidates for a sustainable future.

In economics, maintaining a *healthy economic growth rate* is the traditional objective.[1560,1478] Three interpretations or variants of this goal are to maintain (net) production, seek maximum production, or continue increasing production. Here, 'market forces' determine the value of resources, and channel assets to the economically-efficient segments of

482

society. Corrective mechanisms, usually involving government, take care of market failures, hidden costs, scarce resources, and non-economic resources. In this manner it is considered that society grows its way to sustainability and prosperity.

In contrast, a limited but growing number of economists has focused on the finiteness of the planet, the importance of the second law of thermodynamics (chapter 1), and the serious undervaluing of major non-economic resources by market forces.[355,162,1414,1716] This *steady state economics*, where inputs equal outputs, recognizes that all production ultimately depends on the raw materials of ecosystems, plus the energy transfers to make the materials useful. Thus, limiting impacts on ecological systems and the energy flows (increasing entropy) in processing, transportation, and so forth are central economic considerations. Government's role is limiting the negative impacts. Sustainability rather than growth is the goal.

Several ecologists have also highlighted the importance and conceptual opportunities for interweaving economics and ecology.[1233,1236,1235,1265,316,1575,1277] One major approach focuses on energy as a 'universal currency'. Anything economic or non-economic is converted into energetic terms, permitting the construction of highly informative models of energy flow. Another approach assigns an economic value to things that traditionally do not appear in the marketplace.[1233,1277] Thus, all objects can be directly compared in economic terms. Additional approaches build a range of bridges between the diverse conceptual areas of economics and those of ecology.

Other social scientists tend to focus on more qualitative issues, such as equal access to resources and quality of life. This especially reflects the development component of 'sustainable development'.[791] Indeed, in future development the importance of access to resources, distribution of costs and benefits across society, providing basic human needs, overall quality of life, and equity between generations is often pinpointed.[1929,778] These are only possible with a limited human population density. The steady-state economic perspective implies a significant reordering of social patterns and institutions. Generally a governmental stimulus is required to arrest the degrading resource base and energy inefficiency produced by unchecked market forces.

In 1987 a United Nations committee (World Commission on Environment and Development 1987) proposed a concept of sustainability that went beyond individual disciplines and has become the most widely used concept. *A sustainable condition for this planet is one in which there is stability for both social and physical systems, achieved through meeting the needs of the present without compromising the ability of future generations to meet their own needs.* Any specific concept of sustainability must be consistent with this generic statement of principle.[192]

In the context of this chapter on creating sustainable landscapes, the preceding general principle is recast into the following operational form: *a sustainable environment is an area in which ecological integrity and basic human needs are concurrently maintained over generations*. 'Sustainability' therefore is the condition where this is achieved or maintained. Adaptability, not constancy, is central to success. Analogously, a species, a person, and an industry only remain if they adapt or adjust to changing conditions.

Area refers to a landscape, region, continent, or planet, that is, a mosaic in which people live, extract resources, and so forth. *Ecological integrity* is the combination of near-natural levels of productivity, biodiversity, water, and soil characteristics. *Basic human needs* are considered to be adequate food, water, health, housing, energy, and cultural cohesion (these are discussed later in the chapter). The time scale of human generations is roughly a century to a millennium.

A caveat is in order. A sustainable condition or environment may be impossible. Many people strongly believe in sustainability. Some people do not, including those who consider it to be a cruel hoax. Humankind may never attain or see a sustainable environment. Therefore, it may be better to consider 'sustainability' as a direction or trajectory, rather than a specific end point.

Before exploring the attributes of a sustainable environment, a specific case study is presented that illustrates some of the patterns and issues. The human community and the environment of a forested 250 km (150 mi) wide landscape of New Brunswick, Canada are considered over a two-century period, during which woodcutting is the predominant economic activity.[1934,652,1402] This forest was discovered and exploited by Europeans as part of a world economy of timber, reaching from Sweden and Russia to Central America and The Philippines. Like almost all landscapes and regions, New Brunswick is part of a global economy.

In the early 1800s very large pines (*Pinus strobus*) were harvested and exported for ship masts (Fig. 14.2). In the mid 1800s with these large pines becoming scarce, a major sawmill industry developed using mid-sized pines and spruces (*Picea*), and wood for construction was exported. At *c.* 1900, with mid-sized trees becoming scarce, sawmills decreased, and a pulp and paper industry emerged using small spruces and firs (*Abies*). Several decades later 'whole-tree biomass' harvest of trees began. By the mid 1980s extensive areas were covered by tiny trees, with low-wood-quality hardwoods (*Populus*, *Betula*, *Prunus*, *Acer*) predominating.

Successive technological changes permitted each step to take place. Resource-based industries were developed, the regional economy was developed, and local economies were developed. However, the forest

484

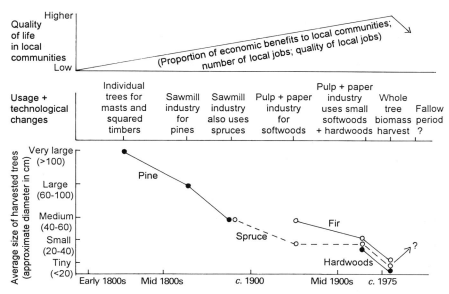

Fig. 14.2. Gradual changes in quality of life and resource base in a landscape being logged to provide wood products. New Brunswick, Canada. Based mainly on Regier & Baskerville (1986).

resource being harvested was not tended. The size of trees in stands diminished. The species composition in stands in the landscape worsened. And the age structure of stands in the landscape degraded.

At each step in the process the number of jobs, quality of jobs, and quality of life in the local communities increased somewhat (Fig. 14.2). A larger proportion of economic benefits was channeled to local communities. For nearly two centuries the forest industry was maintained, and the quality of human life is considered to have improved. But the puny trees left were no longer competitive in the global economy. Global economic interests do not care about the New Brunswick landscape, or about the long-term well being of its land or people. The forest, indeed the land, was degraded.

Tree-by-tree and stand-by-stand decisions caused the problem.[1402] Only later did the landscape or regional impacts become stunningly evident. Environment, society, and economy are indissolubly linked. The scenario could have been avoided, had a regional view been adopted and a governmental regulatory role formulated, where planning affected major decisions on cutting. And only in this way will the forest land once again be restored to a productive capacity. The immediate future of the predominant forest industry is grim. A decades-long fallow period without cutting is best for the forest, but unpalatable to the people of the region.

485

An illusion of stability and sustainability persisted during the two centuries, as human needs, both economic and quality of life, were the only important assay. The large area of forest contributed to the scenario: 'There is always more to cut.' But ecological integrity was ignored, and planning for future human generations absent. Consequently, the strategy adopted was not sustainable either in environmental terms or in satisfying basic human needs in the long run.

TIME AND SPACE

We now turn to the attributes of sustainability. How would you plan a sustainable environment, and how would you recognize one? Time and space, human dimensions, assays, and adaptability are considered in sequence.

Temporal scale

In considering sustainability, the concept of a number of human generations implies a minimum of 30–50 years for two generations, and extending to perhaps 500–2000 years. In the long term sustainability overlaps natural climate change. Few mortals care to think centuries ahead, but many think about their children and grandchildren.[1391] This latent interest in the upcoming generations is an opportunity. We can concentrate on general directions and rates of change over human generations that are apt to be propagated over centuries. In this manner we effectively plan or draw the outline for a particular sustainable environment for a particular macroclimate in the temporal range of 50–500 years.

For example, about six generations is apparently needed for a human population to develop considerable immunity to a new disease, such as measles (S. B. Warner, Jr., pers. commun.). The Iroquois Indians native to North America are said to have considered only things that last at least seven generations (c. 100 to 150 years) to be important.

Alternatively, a particular environment may be designed with a greater emphasis on adaptive mechanisms, so that it persists sustainably through climatic changes. Such a design is especially relevant in the face of rapid, human-caused climate change.

The time–space concept (chapter 1) provides a useful framework for determining the appropriate or optimum spatial dimensions of a sustainable environment, to be considered below. In addition, grain size and the arrangement of patches and corridors within the area are hypothesized to be important in sustainability.

486

Spatial scale

The planet or globe must be planned and planned sustainably, because there is no other known place for humankind to live. The planet is practically unique as an ecological system in that, although energy enters and leaves, very little matter arrives from, and leaves to, outer space.[1928,1158] In this sense the globe or biosphere is a closed system, with finite resources, both economic for production and non-economic of natural and human value.[1926,1592,1741] Material resources, however, can also be extracted from, and deposited in, rocks and other geological deposits. As ecologists and steady-state economists point out, these inputs and outputs are limited though, by the energetics of extraction and processing, as well as the degradation impacts on ecological systems at the land surface.

Flows of materials by transportation of people and goods, ocean currents, and atmospheric circulation tie the parts of the globe together. The major parts of the globe, i.e., continents and oceans, are exceedingly large and different, offering formidable obstacles to centralized planning and management for sustainability. Thus, much discussion of 'global sustainability' focuses on energy and material transport mechanisms.[1928,768,1741] The immigration and emigration of people, as well as the shipping of soil-based commodities and manufactured products, are somewhat amenable to planning and management. The transport of exotic species is less so. Ocean movement of pollutants between continents is widespread, though the effects are mainly limited to coastlines. Atmospheric circulation, also probably affected by ocean pollution, carries its own set of gaseous and particulate pollutants around the globe. Fallout onto the land is often locally significant. However, a greater effect of aerial pollutants is exerted on incoming and outgoing energy flows between outer space and the land surface.

Despite science's interest in transport mechanisms, much about climate and oceans remains difficult to predict. This unpredictability is an important backdrop or framework for any projection of sustainability.

The globe or biosphere is important in sustainability also because in a hierarchical system the conditions at a broad scale affect those at finer scales.[652,1926,283,605,1250,17] Conversely conditions in a small unit also affect the broader scale, as well as conditions in neighboring, comparable small units. Therefore, it is critical to identify the most appropriate scale (finer than the globe) to plan and manage for sustainability.

The basic options are the continent, biome, region, landscape, and local ecosystem[492] (also see drainage basin in chapter 1). The continent and biome[1096] usually have distinct boundaries, but in most cases are only loosely tied together by transportation and economics. They also encompass extremely dissimilar areas of climate, ecology, and human land use. Except for Australia no continent or biome has a single govern-ment, which is needed for planning and policy implementation for sus-tainability. As discussed in this chapter, the region and landscape are promising candidates for sustainable environments.

At a very fine scale the local ecosystem, such as a swamp, pasture, or woodlot, might also be managed for sustainability. Especially in areas remote from human activity, some individual ecosystems remain in simi-lar form for generations or centuries. However, in such areas major natu-ral disturbances significantly alter most ecosystems in the time frame of human generations.[1463,1160,1320] Furthermore, human population or activity is pronounced in most landscapes, and few local ecosystems escape major and frequent alterations in this time scale. In short, the local ecosystem in a landscape mosaic should be planned, managed, and cared for, but overall it is not a promising spatial scale for planning a sustainable environment. This conclusion is of particular importance in the conservation of natural areas.

The paradox of management

The farther we get away from an individual caring for his or her own garden, the less effective planning and management decisions are.[286,492] For instance, cumulative impacts (chapter 13) usually resulting from widely dispersed individual actions, are magnified in importance in large areas. (One may also ask whether it is good for humankind to have an individual or group of persons managing the world). A manager or plan-ner can easily affect a local ecosystem, but few will affect the planet.

Conversely, the probability of achieving sustainability decreases at finer scales. Large rapid fluctuations in individual ecosystems are normal, whereas broad-scale, natural regulatory processes provide con-siderable stability, as suggested by the Gaia hypothesis or empirical result.[993,1085,1475]

We are left with the *paradox of management*. One can more likely cause or create an effect at a fine scale, whereas success is more likely to be achieved at a broad scale. Management and planning for sustainability at an intermediate scale, the landscape or region, appears optimum. We will return to this paradox later in the chapter.

Patches and corridors

The heterogeneity provided by patches and corridors in an area plays a key role in sustainability. Thus, a mosaic structure provides at least four major 'handles' for planning. First, resources are spatially separated, pro-viding for specialization and differentiation, as well as redundancy, of human communities and activities. Second, the juxtaposition of different resources in adjacent patches provides a complementarity of resources, plus stability, for human and natural communities located near boundaries in the area (chapter 3). Third, heterogeneity is a basic cause of flows and movements, and the various types of boundary in a land mosaic mean that the amounts and rates of flows vary widely across the mosaic (chapters 1 and 9). And fourth, the presence of corridors means that channels and barriers provide a wide range of flow rates and directions in the mosaic (chapter 10). All four of these increase the adaptability of the system to inevitable small changes. Thus we hypothesize that, at least up to some level, increasing patch–corridor heterogeneity increases sustainability.

Heterogeneity *per se* appears useful to planning a sustainable environment, but more important is the actual arrangement of patches and corridors. The aggregate-with-outliers principle nicely illustrates this (see Fig. 13.1). Distributing small patches of natural vegetation evenly over the built and agricultural areas provides stepping stones and other ecological benefits (chapter 2). Distributing small patches of agriculture and houses along the boundaries of large patches, and at distinctive inter-patch distances, provides for risk spreading, transportation efficiency, different housing situations, and other ecological and human benefits. As in previous chapters, spatial arrangement matters.

In addition, corridors concentrate movement and increase its efficiency, in effect avoiding the search process and convoluted route (chapter 5). This also protects the surroundings from impacts of movement. In short, corridors and patches appear to increase sustainability, and their arrangement is a key handle for planning a sustainable environment.

Grain size

A coarse-grained landscape provides large patches for interior species, natural disturbance regimes, the integrity of stream networks, and other ecological benefits (chapters 2 and 8). Such a landscape has high 'landscape diversity', but low 'site diversity'. That is, the whole landscape tends to contain interior and edge conditions of each patch type present, but moving locally from spot to spot is monotonous (Fig. 14.3a). Each adjacent spot appears the same except when crossing an occasional boundary between large patches. Thus, a coarse-grained landscape is ideal for interior species. However, it is inefficient and costly for multi-habitat species, including humans. Except for locations near occasional boundaries, such species must move long distances among habitats to satisfy their requirements. Finally, the habitat for edge species is minimal in a coarse-grained mosaic.

489

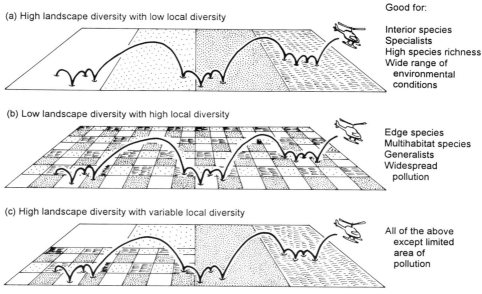

Fig. 14.3. Ecological effects of grain size. Helicopter stops for sampling at four local spots in three separated portions of each landscape.

A fine-grained landscape with small patches solves these problems, but also exhibits serious disadvantages. The area has low landscape diversity, but high site diversity (Fig. 14.3b). That is, the landscape as a whole is monotonous, with different portions having the same fine-scale heterogeneity repeated throughout (Fig. 14.4). But in moving locally, each spot or adjacent small patch is different. Thus, little movement is required to satisfy the needs of multihabitat species, including humans. Boundaries and edge species are nearly everywhere.[1474]

The problem with a fine-grained landscape is that the range of ecological conditions is limited, due to the absence of large areas with interior conditions (Fig. 14.4). Essentially only generalist species are present. Species specialized for interior conditions do not persist in small patches adjacent to other patch types. Overused land and the wastes of society are widely scattered in a fine-grained landscape, meaning that hardly any area may escape air and water pollution. Extreme or unusual conditions tend to be averaged out. In essence, both the overall range and the level of resources are lower in a fine-grained landscape.

What about insects, fungi, and other small organisms that make up >99% of the species on our planet?[743,1911,1327] The patterns and processes throughout this book primarily apply to trees, shrubs, herbs, mammals, birds, reptiles, and other conspicuous organisms. They, like landscapes

Fig. 14.4. Constructing fine- and coarse-grained environments for experiments. Patches of surface sand from a different ecosystem alternating with patches of local silt loam[1897] could be used, e.g., to compare edge and interior species distributions, animal movement patterns, changes along a land transformation axis, and so forth.[518,1887,803,662] R. Forman photos.

and regions, are at the human scale. And a reasonable amount of scientific evidence is available on their species richness, dispersal patterns, edge-versus-interior preferences, and so forth. Little such evidence exists for small organisms. Since grain size is related to the sensitivity of the species concerned, people and deer can be expected to respond to a landscape differently than spiders and earthworms. Someone someday should write a similar book on small species.

To increase sustainability, the obvious solution to the shortcomings of both coarse- and fine-grained landscapes is to vary grain size.[488] More specifically, create a coarse-grained landscape that contains fine-grained areas. This provides environmental resources and conditions for almost everything (Fig. 14.3c). Rather than maximizing for one or the other set of advantages it optimizes for both. Again, as illustrated in the aggregate-with-outliers principle (chapter 13), the arrangement of coarse- and fine-grained areas within the landscape is doubtless a key to increasing the probability of achieving a sustainable environment.

HUMAN DIMENSIONS
Cohesive forces

Since a sustainable environment by definition maintains both human and ecological characteristics, exploring the human dimensions is essential. Major conflict or war is incompatible with sustainability. Rather, people cooperating or working together appears important, so that competition for space and resources does not destroy the necessities of a sustainable environment.

What therefore are the bonding or cohesive forces that tie people together, and how do they differ in relative strength? These are impossibly broad questions, yet even brief partial answers are helpful for comparison with those from the equally broad environmental sphere. 'Culture' is a core subject of anthropology (referring, e.g., to the overall body of extrasomatic, behavioral adaptive measures of the human species). Kinship, politics, economics, ideology, religion, technology, and infrastructure are all parts of culture. In this extremely broad sense, the term 'cultural cohesion' appears not to be useful or measurable. Perhaps we could simply refer to a single overall bonding force as 'community cohesion'.

An alternative approach for this sustainability analysis uses a narrower operational concept of culture, similar to one in the dictionary. Thus, *cultural cohesion* refers to the linking of people by common intellectual, aesthetic, and moral traditions. In this manner, culture can be considered as a bonding force in its own right, separate from religion, economics, politics, and so forth.

Culture and religion appear to be the two strongest cohesive forces[494] (Fig. 14.5). The strength of a cohesive force of course depends on the depth of commitment of people in a landscape or region. If the bulk of the population shares a single culture or religion, the cohesive force may be great. If several coexist, there may be strength in diversity. Where two or three predominate and have a tradition of tolerance and mutual

Fig. 14.5. Monuments that highlight culture as a strong cohesive force. Top: The intellectual and artistic aspects of culture, perhaps also with strong religion and government, supported the construction of these Maya temples in a city that disappeared over centuries due to several ecological and human factors (Figs. 14.9 and 14.10). Tikal, Guatemala. Bottom: Temple of Confucius (K'ung Ch'iu), whose philosophy emphasizes the practical and ethical, teaching, writing, moral standards, duty to the state, and reverence to parents. Beijing. R. Forman photos.

respect, these may provide a strong cohesive force for sustainability. 'Religion' involves shared faith and worship. Ironically, sustainability itself might be considered almost a religion in which people have faith[192] (M. D. Cantwell, pers. commun.). Culture, however, with its strong educational and adaptive dimensions, and not being focused on faith, appears particularly promising as a key cohesive component of a sustainable environment.[215,536,284]

At least four social-science dimensions can also serve as strong cohesive forces. An overall order by bonding strength is hypothesized to be as follows: social structure (including values); economics; politics or government; and adversity or persecution. 'Social structure', referring to the cooperative and interdependent relationships of human society, thus ties people together through person-to-group interactions, and includes various community, housing, and service organizations. Of course, the social science approach to human cohesion and culture includes varying rates of change. 'Economics' focuses on the production, distribution, and consumption of goods and services, and links people in a particular industry or type of work. 'Politics' or government links people concerned with creating or influencing policy, attaining power or providing leadership, and exercising authority in administering policy. 'Adversity or persecution' often links people who have suffered the same hardship.[1097] All four of these cohesive forces may be strong, but are usually relatively short lived, i.e., at the years to decades scale. Except perhaps for farming, the people of a landscape today are rarely tied together by these forces over human generations.

This brief consideration suggests that culture, and to a lesser extent religion, are particularly important as cohesive forces at the region and landscape scales (Fig. 14.5). Religion generally changes very slowly, perhaps more slowly than physical changes across the land. Culture also changes slowly, though generally is more adaptable and able to change. The social science dimensions all can change overnight, e.g., as governments form, industries fail, prisoners are liberated, and communities splinter.

Cultural cohesion within landscapes and regions means 'cultural diversity' at broader scales such as continents and the globe. Analogously, cultural homogenization at the continental level, say through television and export of educational resources, means less cultural diversity, and doubtless less cultural cohesion. In essence, cohesion, especially cultural, is an independent variable essential for sustainably planning and managing environments.

Linkages with other areas

Massive landscape changes have resulted from relatively small 'innovations' (K. Deegan, pers. commun.). Some ideas have appeared within

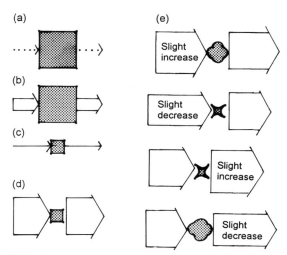

Fig. 14.6. Black-box models illustrating the effect of varying inputs and outputs. The shaded objects can be landscapes. (a) to (d) Steady state systems. (e) Four ways in which an input or output is altered, thus causing a severe stress (e.g., expansion or contraction) on the object depicted in (d).

an area, while others have come from outside. The domestication of plants and the invention of irrigation systems transformed the Middle East region from within. The introduction of the rifle and horse to Native American Indians of the North American Plains is a classic external impetus that radically altered the environment. The invention and diffusion of metal, fire control, and electricity, as relatively distinct traits, produced impacts that reverberated across landscapes and regions.

One might design a sustainable landscape to be relatively isolated, with few interactions to the outside (Fig. 14.6a). This assumes that required resources, such as food, energy, and housing materials, are sufficient within the area. Alternatively, frequent interactions with the surroundings may be a key to continuing over human generations. What types of interactions could increase the probability of sustainability?

A human community gains ideas and innovations from outside. An active trading center means that exports provide income, and imports provide otherwise scarce or unavailable resources. Frequent interactions permit a community to keep up with inevitable outside socioeconomic changes, and avoid getting 'out of date'. Immigration of people provides new ideas, plus hybrid vigor when meshed with the existing population. Emigration may decrease or increase costs to society depending on who leaves. Clearly there are also shortcomings and destabilizing aspects to each of these cases. On balance, sustainability may be favored by some intermediate level of outside interactions.

495

To evaluate better the optimum degree of openness of a landscape, it is useful to consider simple 'black box' *input-output models*.[1235] Here we can compare the amount of an object in a steady state system versus the amount flowing through it. If the amount in the object (e.g., a landscape) is large relative to the input (or output), a reasonable degree of stability is expected (Fig. 14.6b and c). If the input is large relative to the amount in the object, the object is apt to fluctuate widely over time (Fig. 14.6d). This is because small changes in inputs (or outputs) cause large changes in the amount in the object[1781,1782] (Fig. 14.6e). This applies directly to the landscape. If immigration, emigration, import, or export is large relative to the population or goods produced in the landscape, stability is unlikely. Overall, therefore to maintain the human population over generations, it appears that some interactions with the outside are better than none. However, the amount should be small relative to the population within.

Built environment

Although the built environment was examined in chapter 13, the literature identifies a few additional attributes that are particularly relevant to sustainability.[582,501] Significant problems with the basic idea of sustainability have been pinpointed in this chapter. However, the idea of sustainable urban development appears to be an oxymoron. Increasing human population and development, where the local resource base is already nearly obliterated, seems incompatible with the idea of sustainability. Street trees, grassy parks, skyscraper canyon winds, swirling pigeons, polluted wetlands, and storm-sewer streams rushing underfoot illustrate that nature is present in the center of a city. Alas, they are but shreds in a place where ecological integrity cannot exist.

Only one approach appears possible concurrently to maintain ecological integrity and basic human needs for the built environment. That is to plan and manage the urban landscape as only one of several linked landscapes considered together. The group as a whole could theoretically be sustainable. Because of the massive inputs and outputs involved, the overwhelming focus of planning and management will have to be on the non-urban landscapes to maintain their ecological integrity, and prevent land degradation. In an era of growing city numbers and sizes, this is a non-trivial step to take.

The preceding input-output model is helpful here. Some cities have a storm-drainage pipe system to expel all water from precipitation as fast as possible. Ironically such cities therefore must pipe in from elsewhere all their water needs for industry, home, and other uses. A broken major pipeline, unwanted chemicals in the pipeline, or a drought in the landscape containing a reservoir causes severe problems in the city. Similarly, many other inputs and outputs are large relative to the amount in the city (Fig. 14.6d and e). Consequently, strikes by transport lines, sewage treatment personnel, or trash collectors cause huge problems, as do bridge collapses along commuter routes and prolonged temperature inversions without air-cleaning winds.

To decrease the severity of effects of fluctuating inputs and outputs, people within the system may strive for *self reliance*. This can be enhanced by focusing on four important aspects of human settlements.[310,1725,1773,790,1827,246] (1) 'Energy efficiency and recycling of materials' lead to lower input and output levels. (2) 'Efficient transportation' is important within the built area. (3) 'Private spaces', especially providing for family bonding and services, are considered vital for people's well being. (4) And 'metropolitan food systems', from tomatoes and parsley in windows to garden plots in open areas, increase the self reliance of residents.

In addition to decreasing the ratio of inputs/outputs to the amount in the system, the expansion of the built environment in planned logical patterns is important. That is, both suburbanization and increasing density within the built area should be designed carefully to increase the four self-reliance objectives.

ASSAYS

Values

If you produced two alternative plans for making an area sustainable, how would one know which is better? More importantly, how can we get agreement among people of different traditions that a particular plan is or is not likely to be sustainable? The key is to identify attributes that are good assays of sustainability, and also are perceived in similar manner by diverse people.

No attribute or assay is value free or completely objective. Each contains some value judgment by the observer. For convenience we will use a 'relative-objectivity' scale, ranging from completely objective to completely value laden.[284] Alternatively, one could rank assays according to the degree of agreement among diverse people. For instance, reasonable agreement among people would probably occur when considering the minimum amount of water for drinking to sustain a particular community. This is because physiological requirements of the human body are fairly well known and widely applicable. On the other hand, diverse peoples doubtless would differ widely if asked to evaluate plans for their provision of adequate security or justice.

In the broad sense sustainability is highly value laden. A sustainable environment is perceived to be 'good' or 'desirable'. At the other end of the scale are quantitative, usually narrowly focused variables that are relatively value free.[286,191] These more objective attributes tend to be assays of the physical environment, which integrate ecological integrity and society's interactions with it. Is there enough fertile soil for crop production? Vegetation to protect water supplies? Trees to build houses? Biodiversity to provide food and medicine? In general such variables are easily measured, and lead to agreement on 'yes or no' questions.

Once relatively objective attributes are chosen, the difficult issue is to set priorities among them. This is more subject to value judgment and diverging views among people than selecting the individual assays. Yet 'apples and oranges' questions such as this can be decided relatively objectively, as illustrated by the Concord example (chapter 13). In summary, the focus on relatively objective attributes in planning and evaluating sustainability offers substantial benefits in generating agreement among diverse peoples.

Linking broad values to change raises another fundamental problem with sustainability. When the natural forces of climate alter a terrestrial environment, it is a natural process or change, and is considered 'OK'. These same forces, as well as the altered land, also provoke cultural evolution, which similarly is considered inevitable or 'OK'. But if human intervention as part of cultural change alters the land, it is considered unnatural, and most often 'negative'. There is an inherent value problem here. The forces often causing cultural change are positive, and the effects of this change on the land are negative, as perceived by the agents of land change (humans) who must do any sustainable planning and management.

Slowly changing attributes

Since sustainability focuses on the time scale of human generations, time provides another criterion for selecting appropriate attributes to assay. Obviously variables that fluctuate daily, such as stock market indices, the prices of goods, and the routes of birds in their territories, are inappropriate. Also inappropriate, but more difficult for people to realize, are attributes that primarily change over months and years. These are illustrated by most floods, droughts, governments, economic indices, agricultural surpluses, pest outbreaks, industrial closures, fish population levels, and so forth. In short time scales (relative to sustainability), fluctuating peaks and dips, and human and natural responses to them, become the primary focus. Underlying trends are present, but usually obscured by the noise of fluctuations.

The key is to look beyond these numerous and conspicuous fluctuating attributes, and focus on attributes that primarily vary up and down at the time scale of human generations or longer.[283,191,746] What are these *slowly changing* or *foundation variables*?[492] They are tentatively identified near the end of this chapter. Attributes such as soil depth, culture, energy, biodiversity, religion, major irrigation systems, desertification, and sedimentation of large reservoirs are examples. Both the identification of foundation variables, and how to link or separate them from short-term fluctuating variables, remain a worthy research frontier.

When the news media report a change in government, a 10% increase in the price of bread or rice, or a major lethal typhoon, it gets our attention. But who would be interested in learning that one millimeter of soil was lost last month over agricultural lands, or leaders of rainforest tribes report yet another drop in apprentices learning the cultural and medicinal uses of native plants, or one centimeter of sediment was added last month to the reservoirs of a nation? These reports should be the headlines. They foretell what our children's and grandchildren's lives will be like. They are what determine whether a landscape or region is sustainable.

Ecological integrity

The preceding two sections conclude that attributes to assay should be relatively objective and slowly changing. Here, we further consider the attributes of ecological integrity, followed by those of basic human needs, since both must be maintained in a sustainable environment.[1929,215,536,1656,434]

Ecological integrity could be measured as the single most important or sensitive attribute of an ecological system. At the other extreme it could be a long list of ecological attributes that must be maintained at a minimum level, or some optimum level among them attained. An intermediate option used here is to focus on a few key components of ecological attributes that together capture the general meaning of ecological integrity.[492,1354,840] Thus, a system with *ecological integrity* has near-natural conditions for four broad characteristics: productivity, biodiversity, soil, and water (Fig. 14.7). The ecological and human importance of each is extensively described elsewhere, and numerous feedbacks among the various attributes in Fig. 14.7 exist. Here, we consider how they can be usefully measured for a sustainable environment.

'Plant production' of course varies widely within a landscape or region from, say, negligible on an airport runway to high in a fertilized field. A near-natural level of productivity thus is an average for a landscape. It indicates a level not too far from that which would prevail if the whole landscape had native ecosystems. Quantifying near-natural more precisely is an important challenge in sustainability. Clearly, an area that has been severely eroded, salinized, covered by chemicals that inhibit

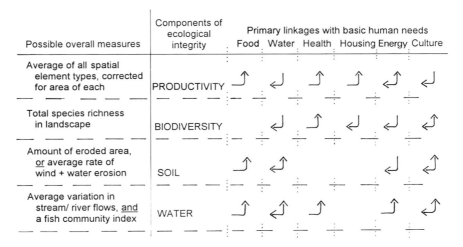

Possible overall measures	Components of ecological integrity	Primary linkages with basic human needs					
		Food	Water	Health	Housing	Energy	Culture
Average of all spatial element types, corrected for area of each	PRODUCTIVITY	↱	↵	↱	↱	↵	↵
Total species richness in landscape	BIODIVERSITY		↵	↰	↵	↵	↵
Amount of eroded area, or average rate of wind + water erosion	SOIL	↰	↵			↵	↵
Average variation in stream/ river flows, and a fish community index	WATER	↱	↵	↱		↱	↵

Fig. 14.7. Ecological integrity: key components, possible measures, and links with basic human needs. Only primary links are shown; direction of arrow indicates that increasing one attribute has a major effect on the second attribute.

plant growth, or covered with houses has a low average productivity, and is unsustainable.[1298] Several related ecological attributes, such as biomass, animal (or secondary) production, length of food chains, herbivory, and decomposition, are indirectly assayed by the productivity category (see Fig. 1.13).

The 'biodiversity' category is probably best measured as number of native species for each major group, such as mammals and trees (Fig. 14.7). A near-natural level refers to the total number in the landscape, and indicates that relatively few species have been extirpated. Thus a landscape that has lost certain habitat types, or has natural habitats so fragmented and isolated that their species number progressively drops, is not sustainable. The biodiversity category indirectly assays attributes including community types, keystone species, rare species, and genetic diversity[1592,1912] (see Fig. 1.13).

The 'soil' category perhaps is best measured by the amount of soil erosion. Large areas undergoing wind erosion or water erosion render a landscape unsustainable. The soil category indirectly measures soil moisture, soil texture, runoff of mineral nutrients such as phosphorus and nitrogen, and perhaps salinization.

The 'water' category must include quantity and quality (Fig. 14.7). Near-natural levels of water quantity integrate hydrological attributes such as floods, low-flow events, evapotranspiration, water tables, stream flow, and aquifer recharge. Water quality is an index of ecological attri-

500

butes such as turbidity, nutrient status, and fish populations. The fish community appears to be the best overall assay for at least the water quality category of sustainability, because fish are sensitive to so many land and water variables.[839]

Basic human needs

Just as in the case of ecological integrity, *basic human needs* (Fig. 14.7) can be assayed in a variety of ways. Six such needs appear central for a sustainable environment. Food, water, health, and housing (shelter) are normally considered essential needs by the United Nations. Energy for cooking and heat is commonly considered a basic need as well. In the time scale of sustainability we add cultural cohesion as a basic human need. As described earlier in this chapter some cohesive force is important in keeping people working together, and culture appears to be the most appropriate cohesive force. Other bonding forces of course operate for shorter periods, or only in certain landscapes.

It is easier to identify the categories of basic human needs than to determine quantitatively the minimum level of each for sustainability.[1104] Nevertheless, nutritional requirements and public health statistics are appropriate for the food, water, and health areas. Shelter against weather and hostilities, and energy for cooking and heating, determine levels of housing and fuel. We may look to anthropology to estimate the minimum level of cultural cohesion needed.

It should be emphasized that sustainability focuses on basic human needs. Hence good roads, education, visual quality, employment, and income level are not, or are only indirect, issues in sustainability. Each is usually valued highly. But just as we must cut through rapidly fluctuating attributes and focus our attention on the key slowly changing ones, here again we must keep our eye on the essentials. Basic human needs (Fig. 14.7) over generations are the essentials.

A more general integrative approach to evaluating the human dimension in sustainability is simply to consider 'quality of life'. This could be estimated in general classes from high to low. Or it could be subdivided and quantified, say, as infant mortality, life expectancy at birth, and literacy rate (W. C. Clark, pers. commun.).

Integrating the discussion of values, ecological integrity, and basic human needs (Fig. 14.8) provides added insight. It means that maintaining poverty is not sustainable.[791,1105] Maintaining a degraded ecological resource base is not sustainable. Maintaining adequate human needs in one landscape by degrading a second is unsustainable; in other words, strongly linked areas must be considered together. And for the same reason, maintaining ecological integrity in one area by degrading a nearby human community is not sustainable. Any of these cases might

501

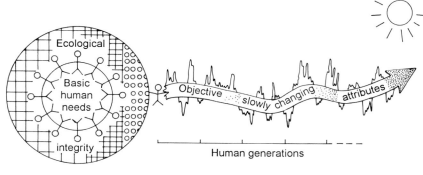

Landscape or region

Fig. 14.8. Visualizing a sustainable environment. The area is a landscape or region with varying grain size.[494] Ecological integrity and basic human needs receive equal emphasis. Cultural cohesion (symbolized by ring of people) is a key bonding force for cooperation among people. Spatial land planning is required. Time is over human generations. Relatively objective assays are selected. Slowly-changing attributes, rather than rapidly fluctuating variables, are the focus. The system adapts to change (symbolized by changes in dot density). Sustainability is considered good.

be maintained over human generations. Yet because quantitatively they do not fulfill the levels of ecological integrity and basic human needs specified, they do not fit this concept of a sustainable environment.

ADAPTABILITY AND STABILITY

We have emphasized the importance of looking beyond fluctuations to focus on long term trends. Such trends are usually upward, downward, cyclic, or level (steady state)[501] (Fig. 14.8). Any of these trends may characterize a sustainable environment. A level trend indicates that after a number of human generations the area does not differ significantly from that beforehand. The upward and downward trends indicate a progressive gradual change in one or more major foundation variables. A cyclic trend also represents slow gradual change.

Constancy is not the goal of sustainability. Constancy is impossible, and history is strewn with failed industries, fallen governments, utopian communities, and fallow fields where people attempted to prevent change. External and internal changes are the norm, and therefore a key attribute of a sustainable environment is *adaptability*, a pliable capacity permitting a system to become modified in response to disturbance (Fig. 14.8).

502

The regularity and frequency of disturbance provide a good indication of the adaptability of a system (chapter 10). If disturbance is regular (cyclic), the system develops mechanisms to deal with disturbance, and persists.[744,1219,581] The disturbances may be severe, but if frequent and predictable the system persists. However, if disturbance is irregular and rare, adaptability is apt to be low. Such a system rarely encounters a disturbance, and consequently does not develop adjustment mechanisms. The highest adaptability, however, is expected in systems where disturbances are frequent, but irregular. Such systems not only develop mechanisms to deal with disturbance, but also with the unpredictable timing of disturbance.[192]

One additional characteristic is similar to that of the input-output model (Fig. 14.6). More stability can be expected if the amplitude of fluctuation is small relative to the amount of an attribute in the system. If the amount in a system is small, for instance, fluctuations may readily lead to disappearance altogether. Amount tends to be most important when it is low, due to the higher probability of extinction. Yet, generally the rate of change is a better indicator to watch. For instance, in exponential or J-shaped curves, increases appear small for a long time, until almost overnight the curve shoots steeply upward, and the system is transformed.[944,501]

What are some key adaptability mechanisms in a sustainable environment? In the ecological realm these include natural selection and evolution, species migration, altering reproduction rates, switching diets, accumulating nutrient-conserving organic matter, and various vegetation regrowth mechanisms.

Pigeons and peregrine falcons (*Columba livia*, *Falco peregrinus*) may thrive among skyscrapers, distinctive mosses (*Bryum argenteum*, *Ceratodon purpureus*) are favored in the cracks of sidewalks, and certain plant species are most common around mines. These species have an altered behavior or tolerance in these human-created environments. They have adapted over generations to changed conditions. Herein lies an important message for planning, design, and conservation. A detailed spatial design, closely tailored to the behavior of a particular species, may become unsuitable if the species changes. Similarly, a species may change and begin using a previously unsuitable spatial arrangement. Flexibility, redundancy, and risk spreading are the watchwords. The following examples provide safety margins for the future: loops in networks; a cluster of stepping stones; several large patches; varied patch sizes; and varied grain size. Mimicking the several common shapes and spatial patterns of nature (chapter 13) may provide the greatest adaptability.

In the human domain adaptability mechanisms are also diverse. Population change takes place by altering birth, death, immigration, or emigration rates. Many socioeconomic and institutional changes can occur.[191] *Technological change* or innovation, including discovery and development, takes place. And boundaries of a system themselves sometimes are altered.

Technological change illustrates an intriguing point about disturbance and stress periods.[1449,191,192] Disturbances tend to stimulate innovations. More broadly, disturbances stimulate ecological and human adaptations. Species without the adaptations, and people that are not sufficiently adaptive, are apt to be bypassed by others in the long run.

The nature of adaptability leads to the question of stability in a sustainable environment. How are ecological integrity and basic human needs to be maintained over human generations in a landscape? 'Physical-sytem stability'[501], such as a road surface that looks about the same year after year, is not appropriate, because ecological and human systems contain organisms that change markedly. Nor is 'resistance stability', such a dense forest with considerable biomass and soil organic matter that help resist change in the face of disturbance. Few regions or landscapes retain sufficient buffering capacity against environmental changes. Nor is it 'resilience or recovery stability', such as a grassland that has little resistance to fire or cattle grazing, but recovers rapidly so that its appearance is similar year after year.[744,1219] Almost all landscapes contain small or large portions that are persistent, and if disturbed, require long periods to recover.

More likely stability in a landscape and region is effectively a *mosaic stability*, where interactions among neighboring elements dampen fluctuations from disturbance.[492] For example, a burned woodlot is rapidly recolonized by plants from adjacent patches, plus the seeds carried by birds along connecting corridors from nearby woodlots. Or, following a drought farmers from nearby farmsteads or villages quickly replow the land, and cattle growers reestablish and rotate herds among pastures on their land (chapter 11).

In Slovakia Milan Ruzicka and colleagues (1982) effectively studied mosaic stability by estimating the stability (biological response to physical-environment disturbance) of each spatial element in seven landscape areas. They conclude that as a whole hill country is more stable than lower plains and floodplains. This result is consistent with the two-century history of land use in the region. Within each landscape area, individual spatial elements vary widely in both stability and persistence. Understanding the pattern of different persistences and responses to disturbance among spatial elements in a mosaic is a wide open research area.[679,10,1377,1378] The dampening effect of interactions in a neighborhood configuration (chapter 9) is an especially important frontier.

The control or regulation mechanisms that produce stability are usually interpreted either in terms of hierarchy or cybernetic theory. In a 'hierarchy' a predominant top level to bottom level flow keeps all lower levels in control.[605,1250] Additional regulation in a hierarchy is possible by upward flow, and by interactions among different units at the same level. These interactions may be feedback loops (chapter 1). 'Negative feedback loops' in a cybernetic, ecological, or human system provide regulation. Interactions among elements in a mosaic may include both hierarchies and feedbacks. Again the nature of mosaic stability is a promising research area.

Indeed, Wolfgang Haber and colleagues (1990) suggest that hierarchy theory and cybernetics may be combined to regulate landscapes, e.g., in Germany or Middle Europe. Three hierarchical levels of regulation are recognized. The lowest level represents the normal day-by-day operation of the ecosystems. Many spatial elements are identifiable, the direction of flows and interactions are reasonably clear, processes vary in a narrow range, and processes are relatively predictable. The middle level represents regulation with normal, but irregularly occurring, events or disturbances. It is characterized by fewer recognizable spatial units, which are well linked with feedback loops, a longer time frame, generalized modelable trends, and less predictive ability. The top level represents the 'strategy of system survival'. Here, both the elements and their interactions are uncertain and unpredictable. Strong external influences exert major control, system responses may be called strategic, and long-term cycles and adaptations provide resilience.[744] Erratic events keep the system in a kind of steady 'training' that maintains its viability.

In the long term, the land and humans represent a huge feedback system at the core of a sustainable environment. The land provides resources (food, water, building materials, and so on) that affect humans (permit more to live), and humans harvest the resources, often degrading the productive land.[1298,1741] Such a negative feedback loop may look like this: more (productive) land means more people; more people, less land; less land, fewer people; fewer people, more land; more land, more people; more people, less land; and on and on. Both the human population and productive land persist, but with major cyclic fluctuations. However, humans usually dislike major fluctuations, especially of themselves.

Finally, we recognize that both cyclic and unpredictable disturbances are important inexorable events. They stimulate diverse ecological and human adaptations, which provide stabilizing interactions among elements of the landscape mosaic. Planning for a sustainable environment must expect such disturbances, and build in adequate adaptability.

LEARNING FROM HISTORY
Paleoecology

Suppose you had a meter or more of slimy bluish-gray muck that resembles toothpaste squeezed from a tube by a giant the size of a big tree. That is what gets most *paleoecologists* excited. They take these core samples from the bottoms of lakes and bogs, and can tell you history over centuries and millennia.

This is a remarkable feat of scientific detective work.[1085,131,129,1516] Like preserves in pickle juice, pollen deposits in peat are protected and provide plenty of clues to the past. Sediments sequentially deposited over long periods may be analyzed for pollen (especially trees), charcoal, clay–silt–sand, seeds–leaves–twigs, organic material, microscopic animals, and chemistry. Tree-ring histories (dendrochronology) spanning millennia have been worked out for whole regions. Glacial ice has yielded changes in dust particles and atmospheric CO_2 concentration, as well as a rare whole person carrying seeds and pollen. Other techniques include analyzing piles of rodent droppings, archaeological shell heaps, lake levels, marine plankton deposits, amphorae from Mediterranean shipwrecks, written historical records, and modern vegetation patterns.

One of the major goals of paleoecology is to understand *climatic history* over centuries and millennia, especially since the last glacial maximum *c.* 18 000 years ago, and since retreat of the ice sheets *c.* 10 000 years ago.[1084,1848,1085,1516] Some climate change results from varying atmospheric composition and from tilting of the Earth's axis over tens of millennia.[1516,1084,769] The distribution and effects of the continental ice sheets are of major importance, producing some climates and vegetation unlike those of today.[1848] From 18 000 to 6 000 years ago global temperature warmed by *c.* 6 °C (10 °F), and water was redistributed to different parts of the globe. Embedded within these long climatic cycles and changes are centuries-long cycles and changes. Written human history records many of the recent climate changes. Not surprisingly, these regional changes in macroclimate at each temporal scale cause major ecological and human effects.

Indeed, understanding *long-term vegetation change* is a second major goal of paleoecology. In a general sense, vegetation types during the glacial period moved across the land to their present distribution in Europe[767,133,132,1338], North America[117,1850,1593,382,93], and elsewhere. However, rather than a vegetation type moving as a front across the land[1850,384], species move individually at different rates.[1850,767,369] This means that preceding plant communities had different species compositions and structure than those of today.[132,767] Time lags occur, where certain species move rapidly but others slowly in response to climate change.[812,1301,369] Sharp drops or disappearances of species, such as elm (perhaps *Ulmus glabra*) in Europe and hemlock (*Tsuga canadensis*) in North America, are conspicuous in the paleoecological record.[133,1466] The changing vegetation patterns provide important clues to changes in a range of ecological processes, from soil to animals, over time.[1594,812,133] Reconstructing regional vegetation is a primary focus, though finer-scale patterns at landscape and sublandscape scales are sometimes measured. The finer-scale approach overlaps that of historical ecologists, who rely strongly on written historical accounts, plus modern vegetation and landscape analysis.[1368,1465,1370,341,132,1306]

To reconstruct regional vegetation, pollen samples are usually taken from sediments in medium to large lakes. However, some studies use small 'receiving basins' down to tiny hollows under a forest canopy. The area of vegetation represented by a pollen sample depends on the area of the receiving basin.[133,621,1516] A small hollow mainly receives pollen from plants of an area less than 100 m in diameter (though variable amounts of regional pollen sometimes complicate the picture). Also, locations where roots or earthworms have lived, or where livestock or moose mucked around, are normally avoided. Small lakes of 1 to 10 ha (2.5–25 acres) appear mainly to sample vegetation of an area perhaps 10 to 40 km (6–25 mi) in diameter, and medium-sized lakes of 10 to 100 ha sample perhaps a 40 to 100 km diameter area. By sampling small receiving basins pollen analysis may become especially useful in analyzing long-term, fine-scale ecological patterns within landscapes, an unusual combination according to the space–time principle (chapter 1).

The *long-term human impact* on land, a third focus of paleoecology, is of particular interest in sustainability. Traces of agriculture appear at the end of the last ice age, about 12 000 to 10 000 years ago.[133] Also, the mammoths (*Mammuthus*), sabre-toothed tigers (*Thylacosmilus, Smilodon*), and numerous other large mammals (megafauna) and birds became extinct about that time.[133] Indeed, a whole mammoth has since been found frozen in glacial ice of that age, so lucky scientists could see what the mammoth ate and then eat mammoth steaks.

The localized use of fire by scattered human communities in forest landscapes, followed by sharp increases when humans begin removing forest, is indicated by charcoal deposits and pollen of tree species with different fire tolerances.[1465,1466] Many other human impacts leave clear signals in the paleoecological record.[131,181,133,132,204,128,1466] The iron age brings strong plows that really transform the land for agriculture.[340] Whole forested regions can be quickly axed, such as essentially all the original forest cover of the USA Upper Midwest, removed in only three decades. Smelters for iron ore have insatiable appetites for fuel. All nearby wood is immediately burned. Wood farther away is converted to charcoal for transport to smelters. Tree species composition changes markedly in extensive charcoaling areas, leaving mainly rapidly sprouting species that are repeatedly harvested. Livestock numbers typically mushroom, and pastureland spreads widely. Erosion, sedimentation, and flooding accelerate. The human population grows exponentially. Paleoecologists see all these changes in the pages of prehistory.

Although thousands of paleoecological studies are published, a small subset is at a fine spatial scale. For example, over several millennia beech and hemlock (*Fagus, Tsuga*) may move only a few hundred kilometers.[370] A hemlock woodlot that today appears as a stable old-growth stand has changed profoundly, both within a three-century scale and a twelve-millennium scale.[509,511]

Comparing studies of one century, one millennium, and ten millennia in lake basins of tropical and subtropical America, Michael Binford and colleagues find that human activities consistently increase movement of materials from basin to lake.[129] Nitrogen and phosphorus runoff is proportional to human population size. Lake eutrophication also increases with population size, except where human activities accelerate siltation and water turbidity. Tropical ecosystems recover from severe disturbances, but the rate varies widely according to the nature of the disturbance.

What are the take-home messages from paleoecological studies for sustainability? Major changes in climate, vegetation, and human impact are the norm at the scales of both centuries and millennia. Whole landscapes, even whole regions, are typically transformed over millennia. Yet sometimes landscapes are rapidly altered, i.e., at the scale of human generations or faster. Vegetation change, deforestation, soil erosion and depth, and lake sedimentation are often major slowly changing variables. Both climate and human population density strongly affect each of these variables. Nevertheless, our predictive ability for a particular landscape is limited, partly because time lags occur in the response to climate and other changes.

The magnitude of human-caused versus climate-caused change is also instructive.[1084] The human conversion of eastern North American forest to artificial savanna in 300 years appears comparable to climate-caused vegetation change over the past 6000 years. Human-caused erosion over 900 years in the USA Southwest and over a few millennia in the Mesopotamian region may be significant compared with erosion associated with climate change. Human-caused desertification of the Sahel region in Africa is a major transformation within the twentieth century, but is minor compared with lake level, river system, and vegetation changes associated with climate there over several millennia.

Landscape ecology is a promising linkage, both in space and time, between paleoecology and the ecology of local ecosystems or communities. Most present paleoecological evidence paints clear pictures of regional change. Future studies at time scales of human generations, and at spatial scales of hundreds of meters to tens of kilometers, will be especially revealing and important. Reviews of this linkage between paleoecology and landscape and regional ecology are needed to help launch a frontier of hypotheses and understanding.

Civilizations and communities

Let us examine specific civilizations and communities as a more direct way to identify ecological and human attributes most important for sustainability. This is a preliminary analysis to illustrate the promise of the approach, and to identify any broad overall patterns. Twenty cases are selected based on four criteria: the civilization or community persisted for a century or more; a reasonable amount of relevant information is available; examples vary widely in area; and cases are distributed worldwide. The only information extracted from a case study is a list of major ecological and human variables that appear associated with less sustainability, and those associated with more sustainability. Specific attributes identified can be grouped or expressed in various ways, and should be considered as hypotheses to be evaluated or verified by anthropologists, historians, and other experts.

The communities and civilizations examined vary in size from a 2 km diameter Shaker village in Massachusetts (USA) to 2800 km across Australia. The length of persistence varies from about one century for Letchworth (United Kingdom) to 25 centuries for the Tiwanaku civilization by Lake Titicaca in the Andes.

No single attribute or group of attributes is consistently associated with 'less sustainability' (Fig. 14.9). Of the environmental variables identified, water problems and soil erosion correlate with the most cases. Of the human variables, high population growth, war, and a drop in commodity exports are most frequent. For most civilizations and communities, both environmental and human attributes are associated with less sustainability.

Different primary attributes are associated with 'more sustainability' (Fig. 14.10). No environmental attribute stands out. However, among the human variables, cultural cohesion, low population growth, and major export/import trade are associated with the most cases. Again, most civilizations and communities have both environmental and human variables important in increasing sustainability.

Overall a wide range of environmental and human attributes appears to be important in both decreasing and increasing sustainability (Figs. 14.9 and 14.10). This conclusion from our brief study is unfortunate if one were searching for a few primary attributes that could be used in

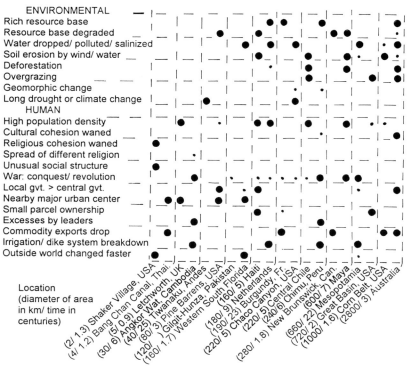

Fig. 14.9. Attributes associated with less sustainability in diverse civilizations and communities. Small dots = attributes whose importance is less clear; diameter = approximate square root of area; civilizations and communities listed left to right from smallest to largest area. Shaker village in Harvard, Massachusetts[36, 390, 752] (1790–1920 AD); Bang Chan Canal in rice-growing valley 35 km northeast of Bangkok[659, 782, 707] (1850–1970); Letchworth, a planned community 50 km north of London[7] (1900–1990); Angkor Wat, the productive wet-rice-growing center of Southeast Asia, north of Lake Tonle Sap, Kampuchea[242, 574, 1630] (800–1450); Tiwanaku, raised-field agriculture south and west of Lake Titicaca in Bolivia and Peru[129, 884, 1420] (1500 BC–1000 AD); Pine Barrens on sandy acid soils of southern New Jersey[484, 297] (1650–1980); Gilgit and Hunza valleys in rugged mountains of northern Pakistan[1349, 247]; western South Florida southwest of Lake Okechobee[415, 838] (1820–1990); Haiti[1010, 1854, 1853] (1500–1990); The Netherlands (1100–1990); Burgundy (Bourgogne) in east-central France[341, 340] (300 BC–1980 AD); Chaco Canyon in northwestern New Mexico[311,823,632] (700–1200); Central Chile, between Combarbala and Linares below 1500 m elevation[459,306,535] (1500–1990); Chimu Empire of coastal desertland[304,1266,1345] (900–1480); New Brunswick[1934,652,1402] (1800–1980); Maya civilization including western El Salvador, Belize, Yucatan, and Tabasco[1540,1335,292] (250–950); Mesopotamia around Tigris and Euphrates rivers[793,1262,160] (2350 BC–150 BC); Great Basin including eastern Oregon, western Utah, Nevada, and southeastern California[1860,420,1946] (1790–1990); Corn Belt from Ohio to Nebraska[61,457,783] (1830–1990); Australia[1922,433,1490] (1700–1990). Information was kindly contributed in part by: D. B. Abramson, E. T. Baker, M. W. Binford, K. T. Brinchmann, M. J. Buchenau, C. J. Bull, M. D. Cantwell, S. E. Costello, J. Fitzpatrick, K. E. Hill, K. S. Holden, W. J. Judge, S. Leevanijkul, L. A. Lewis, J. Meyer, F. G. Muller, K. A. Poole, D. Thaitakoo, and S. D. Weitzman.

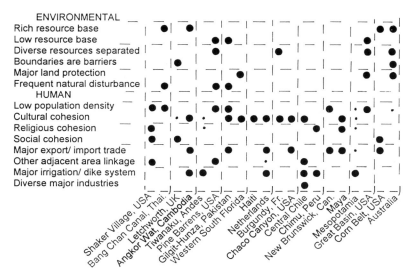

Fig. 14.10. Attributes associated with more sustainability in diverse civilizations and communities. See Fig. 14.9 caption.

planning sustainable environments. However, it is consistent with the hypothesis that there is more than one scenario or solution to creating sustainability. The results indicate that a half dozen or so major variables are critical in each case, and that these include both environmental and human variables.

Three of the attributes, soil, water, and culture, were discussed in an earlier section as particularly appropriate assays of sustainability. At least the first two are relatively objective attributes, and all three are slowly changing.

The two figures also suggest some finer-scale patterns. For decreasing sustainability, several environmental attributes appear more important in larger areas (Fig. 14.9). On the other hand, cultural cohesion appears more important in medium-sized areas (Fig. 14.10). However, due to the general nature of this analysis, such finer-scale patterns should be interpreted with caution.

For some attributes, such as high and low population density, cause and effect is unclear. Does population density primarily affect sustainability, or does sustainability affect population density?

The overall population-density curve for many of the civilizations and communities probably could be divided into three phases, increasing, somewhat level, and decreasing. It is tempting to list characteristics for the level phase as the sustainable attributes. Yet by the time population density levels off, the environment may be severely degrading. This pattern occurred at Chaco Canyon, an important pre-European-settlement community in New Mexico (USA)[823] (Figs. 14.9 and 14.10) (W. J. Judge, pers. commun.).

Technological innovation or change deserves special mention. This attribute was considered a major factor in a majority of the 20 cases. However, technological change is often the hardest factor to evaluate, and whether it increases or decreases sustainability is unclear (hence it is not included in Figs. 14.9 and 14.10). Often technological change increases the economic base, quality of life, and population growth. But it leads directly to population densities greater than the *carrying capacity* of the land (i.e., maximum population size that the land will sustain), and severe land degradation follows.[1927,1697] Such cases are usually characterized by deforestation or overgrazing and by soil erosion or water problems.

Some suggest that a strong centralized government is essential for sustainable development. The results of this preliminary comparison do not point to any particular governmental, social structure, or religious arrangement as being important for sustainability. The absence of a factor, such as gold mines or hostile neighbors, is also omitted from Figs. 14.9 and 14.10, though this absence may be important to sustainability.

More in-depth analyses of more civilizations and communities would provide additional insight (Fig. 14.1). Nevertheless, this learning from history has provided key pieces in the puzzle of attributes needed to plan a sustainable environment.

REGIONAL ECOLOGY REVISITED

The concept of a *region* involves a broad geographic area, a common macroclimate, and a common sphere of human activity and interest (chapter 1). The single macroclimate puts limits on the range of species and natural processes, though varied topography, natural disturbances, and human activities still provide a rich diversity of ecological conditions within a region.[583] The sphere of human activity and interest, commonly tied together by transportation, communication, and culture, also limits the range of human activities. Yet diversity exists within this range as humans interact with topography and ecological conditions.

A region consists of smaller highly contrasting landscapes ranging, say, from urban to forest and rice culture to mangrove swamp. These landscapes interdigitate, and interactions among them are significant. Just as for patches at a finer scale, the edges of landscapes exhibit distinctive attributes, such as small mammal dynamics and species distributions varying with distance from landscape borders.[39,1060,979] The ecological effects of the configuration of landscapes are yet another frontier awaiting questions, hypotheses, and tests.

Regional ecology has been mentioned in this volume much less than landscape ecology, mainly because much more is known about the latter. Nevertheless, *regional ecology* is included for three reasons. First, some people have loosely used the concepts of landscape and region synonymously. Second, assuming the concepts are indeed quite distinct, many of the familiar landscape-ecology concepts can be applied directly to understanding regions. Third, regional ecology is a research frontier of major significance in planning, conservation, policy, and sustainability.

To understand better how much we know about regional ecology based on the now-considerable landscape literature, a systematic look at the 80 major topics or concept areas in this book is instructive. Ninety per cent of the topics apply to regions as well as landscapes. Most of the topics that do not or hardly apply to regions concern narrow corridors, such as windbreaks, road corridors, trails and powerlines, hedgerow habitats, and rectilinear networks. Other topics of limited applicability are fine-scale suburban design, and plans tailored to forested, arid, and agricultural landscapes.

Seven-eighths of the topics that apply to regions appear widely applicable. The remaining ones apply in special cases, as the following examples suggest. Patch size when applied to landscapes within a region affects aquifers, the linkage of river systems, and habitat for a few wide-ranging species such as tigers and grizzly bears, but rarely affects species richness significantly. Internal stream and river heterogeneity is important within a narrow stream/river corridor, but is probably of limited significance when the river corridor is a landscape extending for several kilometers in width. Despite these exceptions the bulk of the landscape-ecology principles and concepts apply directly to regions composed of landscapes. This provides a solid foundation upon which a theory of regional ecology can eventually develop.

Finally in the context of sustainability, it is likely that regions are more stable than landscapes. That is, on average regions are less variable and persist longer for a particular assay or assays. Two reasons seem important. First, because the region is larger than the landscape within, it simply has more inertia. A greater disturbance is required to transform a region. Second, the region contains a wider range of ecological conditions and human activities than a landscape. Thus, we hypothesize that more regulation by hierarchies or feedbacks is present in a region. This provides stability, i.e., the capacity to resist or recover from change.

Consequently, regions offer a significant advantage in planning a sustainable environment. Over human generations they are inherently more stable. Success in attaining sustainability is more probable for a region.

513

Yet, landscapes also offer significant advantages. Their range of heterogeneity is less, with more commonality in ecological conditions and human activities.[484] Planning, conservation, and policy are more likely to make a difference, i.e., to have a visible effect. The 'management paradox' reappears. Our ability to affect the outcome is greater in a landscape, whereas the probability of success is greater in a region.

MAKING A LANDSCAPE OR REGION SUSTAINABLE

Rather than ending this book with broad generalities, we conclude this chapter on sustainable environments with specific approaches and handles for the planner, designer, conservationist, land manager, and policy maker.[488,434,476,492,778,1444] How can we convert the many degrading landscapes around us to sustainable environments? And how can we keep the scarce sustainable ones from degrading?

As a brief review of previous chapters we list a dozen broad principles of landscape and regional ecology. Then guidelines for designing a local area, followed by those for planning a larger area, are outlined. We end by incorporating sustainability into day-to-day planning and management.

Some general principles of landscape and regional ecology

Chapter 1 pointed out the importance of first principles or background theory as useful foundations for developing principles at the broad landscape and regional scales. The 'general principles' listed here integrate diverse knowledge areas, are significant, have predictive ability, have broad applicability (despite exceptions), and their foundations, and in some cases they, have a reasonable amount of supporting evidence.[1430,501,1428,627,1318] The principles are organized into categories much like this book: landscapes and regions; patches and corridors; mosaics; applications.

Landscapes and regions

(1) *Landscape and region.* A mix of local ecosystem or land-use types is repeated over the land forming a landscape, which is the basic element in a region of the next broader scale composed of a non-repetitive, high-contrast, coarse-grained pattern of landscapes. See chapters 1 and 9.

(2) *Patch–corridor–matrix.* The arrangement or structural pattern of patches, corridors, and a matrix that constitute a landscape is a major determinant of functional flows and movements through the landscape, and of changes in its pattern and process over time. See chapters 1, 9, 10, and 12.

Patches and corridors

(3) *Large natural-vegetation patches*. These are the only structures in a landscape that protect aquifers and interconnected stream networks, sustain viable populations of most interior species, provide core habitat and escape cover for most large-home-range vertebrates, and permit near-natural disturbance regimes. See chapters 2 and 11.

(4) *Patch shape*. To accomplish several key functions, an ecologically optimum patch shape usually has a large core with some curvilinear boundaries and narrow lobes, and depends on orientation angle relative to surrounding flows. See chapter 4.

(5) *Interactions among ecosystems*. All ecosystems in a landscape are interrelated, with movement or flow rate of objects dropping sharply with distance, but more gradually for species interactions between ecosystems of the same type. See chapters 9, 10, and 11.

(6) *Metapopulation dynamics*. For subpopulations on separate patches, the local extinction rate decreases with greater habitat quality or patch size, and recolonization increases with corridors, stepping stones, a suitable matrix habitat, or short inter-patch distance. See chapter 11.

Mosaics

(7) *Landscape resistance*. The arrangement of spatial elements, especially barriers, conduits, and highly heterogeneous areas, determines the resistance to flow or movement of species, energy, material, and disturbance over a landscape. See chapters 9, 10, and 11.

(8) *Grain size*. A coarse-grained landscape containing fine-grained areas is optimum to provide for large-patch ecological benefits, multi-habitat species including humans, and a breadth of environmental resources and conditions. See chapters 8 and 13.

(9) *Landscape change*. Land is transformed by several spatial processes overlapping in order, including perforation, fragmentation and attrition, which increase habitat loss and isolation, but otherwise cause very different effects on spatial pattern and ecological process. See chapter 12.

(10) *Mosaic sequence*. Land is transformed from more- to less-suitable habitat in a small number of basic mosaic sequences, the best ecologically being in progressive parallel strips from an edge, though modifications of this pattern lead to an 'ecologically optimum' sequence. See chapter 12.

Applications

(11) *Aggregate-with-outliers*. Land containing humans is best arranged ecologically by aggregating land uses, yet maintaining small patches and corridors of nature throughout developed areas, as well as outliers of human activity spatially arranged along major boundaries. See chapter 13.

(12) *Indispensable patterns.* Top-priority patterns for protection, with no known substitute for their ecological benefits, are a few large natural-vegetation patches, wide vegetated corridors protecting water courses, connectivity for movement of key species among large patches, and small patches and corridors providing heterogeneous bits of nature throughout developed areas. See chapters 2, 6, 7, 8 and 13.

Designing a local area

Suppose you were the chief planner in charge of designing a new sustainable community in a sparsely populated area. And suppose further that you could ignore political considerations. So you explore the area and ponder which should be the initial guiding principle. Find the best place to locate a known optimal spatial model of land uses? Use the existing homes and land uses around which to organize the community? Or plan it based on the arrangement of nature and natural resources?

The preexisting spatial model might fit beautifully on the land, but commonly the unique arrangement of natural resources poorly fits the idealized model. This would lead to inefficiencies and spatial incompatibilities. Similarly the scattered homes and land uses might be beautifully attuned to nature. But often, especially in an expanding community, they are already degrading the land in small ways.

A design that first understands nature and is fit to natural resources appears most likely to be sustainable (Fig. 14.11a). Such a design cannot be timid. It normally includes the removal of misplaced buildings and land uses. This may produce nasty surprises where residents have self-defined perceptions, values, and economic aspirations. Nevertheless, by following nature's unique local arrangement, your design will also be unique. It will differ in appearance from other communities. Of course, optimal spatial models and existing buildings will be secondary factors affecting the natural-resource-based design.

How detailed should the plan be? Probably the key decisions for sustainability are the general outline of major areas, and the mechanisms for adaptability or flexibility (Fig. 14.11a). Clearly the designer must include locations of specific roads, pipelines, boundaries, schools, parks, and the like. These are important in the short term, and can be easily changed if necessary. More important over human generations is the determination of major areas for water protection, building, biodiversity, livestock, and so forth. Once selected these major areas are fixed, semi-permanent. One of the central messages from this book is that sustainability absolutely depends on the sizes, shapes, and juxtaposition of these major areas, i.e., on the quality of the land mosaic.

Within these major areas adaptability is the keystone. We can predict with certainty that unpredictable changes and disturbances of both natu-

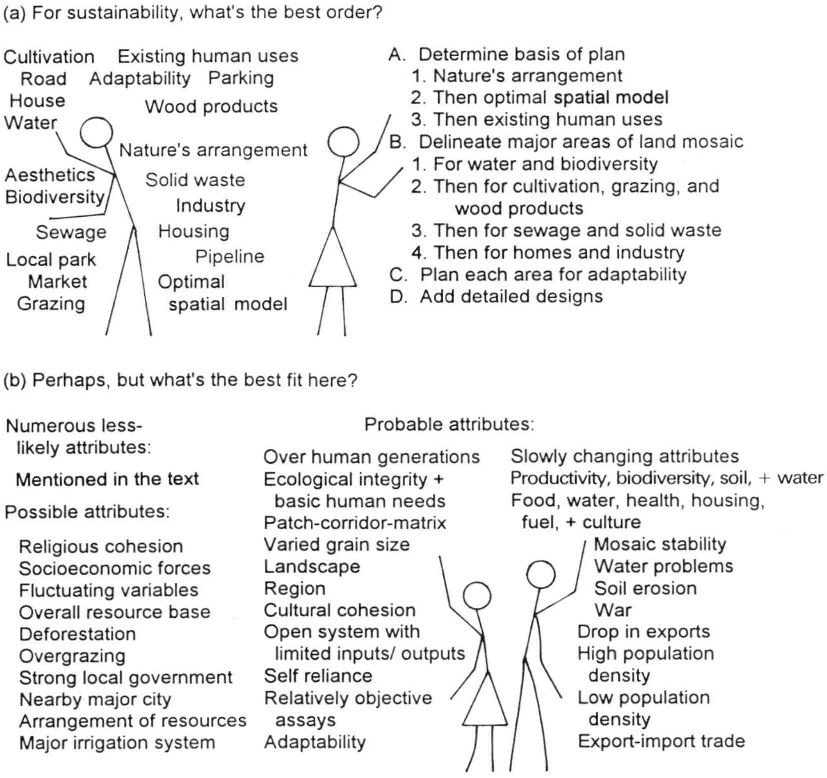

(a) For sustainability, what's the best order?

Cultivation   Existing human uses
  Road   Adaptability   Parking
House   Wood products
Water

      Nature's arrangement

Aesthetics   Solid waste
Biodiversity
          Industry
  Sewage   Housing
Local park   Pipeline
  Market   Optimal
  Grazing   spatial model

A. Determine basis of plan
   1. Nature's arrangement
   2. Then optimal spatial model
   3. Then existing human uses
B. Delineate major areas of land mosaic
   1. For water and biodiversity
   2. Then for cultivation, grazing, and
      wood products
   3. Then for sewage and solid waste
   4. Then for homes and industry
C. Plan each area for adaptability
D. Add detailed designs

(b) Perhaps, but what's the best fit here?

Numerous less-
  likely attributes:

Mentioned in the text

Possible attributes:

  Religious cohesion
  Socioeconomic forces
  Fluctuating variables
  Overall resource base
  Deforestation
  Overgrazing
  Strong local government
  Nearby major city
  Arrangement of resources
  Major irrigation system

Probable attributes:

Over human generations
Ecological integrity +
  basic human needs
Patch-corridor-matrix
Varied grain size
Landscape
Region
Cultural cohesion
Open system with
  limited inputs/ outputs
Self reliance
Relatively objective
  assays
Adaptability

Slowly changing attributes
Productivity, biodiversity, soil, + water
Food, water, health, housing,
  fuel, + culture
Mosaic stability
Water problems
Soil erosion
War
Drop in exports
High population
  density
Low population
  density
Export-import trade

Fig. 14.11. Sequence for planning, and major attributes in sustainability. (a) Unordered issues on left, and a synthesis or sequence for proceeding on right. (b) A summary of key attributes identified in this chapter. These may be considered as hypotheses and research frontiers for the future.

ral and human origin will occur. Objects that are semipermanently fixed in place are expensive to maintain, e.g., in the face of floods, earthquakes, wars, and changes in water table. Good designers have a deep scientific understanding of natural processes to identify effectively the general areas. However, at a finer scale they also can design structures and land uses that are easily adjusted, repaired, moved, and removed. More good designers are needed.

In selecting the major areas, the first question is usually where to place homes and other buildings. Alas, for sustainability that is a serious error. Rather the first areas to be laid out are those for water and biodiversity (Fig. 14.11a). Second, the areas for cultivation, grazing, and wood products are selected. This ordering means that food and fibre may not be produced on the best, most productive soils or slopes. For example, the

517

best soils may support the highest biodiversity[171,170,172] or water seepage areas, and therefore cultivation and grazing are relegated to the second best soils. Third, the areas for sewage and solid waste are selected. And finally areas for homes and industry are chosen.

This simplified sequence means that for sustainability most of the total area is already committed before buildings are located. A designer that locates buildings ahead of these more important land uses degrades those land uses. He or she is only thinking in the short term, or is betting on continuous large inputs and outputs to maintain the community. Neither leads to a sustainable environment.

A wide range of more-specific considerations described in earlier chapters can be used in your design to enhance sustainability of the community. For example, a managed greenbelt ring or string-of-lights[501,1393] can be included. Protection against human overuse of different types of natural resource areas is enhanced with various edge and boundary designs. Water supply may require protecting a whole drainage basin. Preponderant winds at different seasons transport soil and snow, and may be erosion sources. The adjacency effects among land uses[489,492] (see Fig. 1.10 and chapter 9), such as cultivation, industry, and nature reserves, are a challenge as well as an opportunity in designing for human generations. Designing ecologically is to create changing patterns based on the panoply of near and far linkages affecting species.

Key attributes for planning: a summary

The concept of a sustainable environment focuses on a time frame of human generations and the concurrent maintenance of ecological integrity and basic human needs. Here, we list or summarize the attributes presented in this chapter that appear most promising in attaining sustainability (Fig. 14.11b).

The landscape and region are the optimal spatial scales for planning and action. The patch–corridor–matrix model is useful in dealing with the heterogeneity of a mosaic. A variance in grain size offers many advantages, including flexibility.

Culture is a key cohesive force at these scales. Religion and various socioeconomic forces can also be important in enhancing cooperative effort. An open system with linkages to other areas provides important ideas, trade, immigration, and so forth. For stability, however, inputs and outputs should be small relative to the amount in a system. Self reliance, such as local food systems, private spaces, and efficiency in energy use, recycling, and transportation, is useful in a built environment.

Assays of sustainability are best if relatively objective. General trends of slowly changing attributes are a key focus, rather than conspicuous fluctuating variables.

Adaptability is essential in planning over human generations. Adaptive mechanisms are most likely to develop in response to relatively frequent but unpredictable disturbances or changes. Mosaic stability, where interactions among adjacent and nearby elements dampen fluctuation, is most important at these scales.

To maintain ecological integrity requires focusing on the categories of productivity, biodiversity, soil, and water. To maintain basic human needs entails concentrating on food, water, health, and housing, as well as energy and cultural cohesion.

The lessons from history reviewed in this chapter indicate that water problems, soil erosion, high population density, war, and a drop in exports are key attributes associated with decreased sustainability (Fig. 14.11b). Others include the overall resource base, deforestation, overgrazing, the presence of strong local governments or a major nearby city, and the breakdown of major irrigation systems.

Similarly, key attributes associated with greater sustainability are cultural cohesion, low population growth, and export–import trade. Others include the overall level and arrangement of the resource base, religious cohesion, varied linkages with adjacent areas, and a major irrigation or dike system.

In effect, the statements in this summary are guidelines or 'kilometer-posts' to creating a sustainable environment. Although relatively numerous, they represent a distillation from among a much larger set of attributes, linkages, and options. In addition to having significance, generality, and predictive ability, they appear internally consistent, i.e., to generally fit together.

Therefore, the original statement of principle constitutes a general hypothesis or theory of sustainable environments: 'a sustainable environment is an area in which ecological integrity and basic human needs are concurrently maintained over generations'. The theory is testable, the variables are subject to refinement, and the specific linkages remain to be worked out. Furthermore, the wide range of ways of combining attributes is compatible with the idea that there is more than one trajectory to, or scenario of, a sustainable environment.[286,492]

Planning for concurrent objectives

Most development plans are driven by a single major objective. Thus, maps, descriptions, options, and justifications are given primarily to locate some houses, a new road, or a nature reserve, or to protect a water source, biodiversity, soil fertility, or a city park. Such plans are usually conceived by experts in a particular discipline, are relatively easy to develop, and often are insensitive or counterproductive in the context of sustainability. The reason is simple. The plans are politically, economically, or environmentally unsound. By focusing on a single objective they tend to maximize that attribute, while other objectives receive minor consideration.

A sustainable environment, however, has several major sensitive attributes, some associated with decreasing and some with increasing sustainability. Thus, planning, conservation, and management need to focus on a number of concurrent variables rather than one. This is *optimization* rather than maximization.

Optimizing among all relevant variables would be optimum. But in practical terms the decision maker sets priorities, and focuses on the few variables perceived to be most important.

Suppose you are governor of a hilly rainforest province in the tropics (Fig. 14.12). If your planners focus on production and erosion control, their plan might have most of the area in agriculture using contour planting and perhaps hedgerows, and leaving small scattered nature reserves on the steepest slopes. If your planners instead focus on protecting biodiversity, especially interior species, most of the area would be in large nature reserves, separated by small farmland areas where species richness (and perhaps soil quality) is low.

However suppose, unlike almost any society on Earth, you seriously value both erosion control and biodiversity. Ask your planners to focus concurrently on both major objectives, and your plan will be unlike either of the preceding plans. A few large nature reserves (chapter 2) will encompass steep slopes and many moderate slopes, farmland will cover the remaining land, and hedgerows will be arranged to connect nature reserves and minimize erosion. Not only will the third design optimize and accomplish both objectives, it should gain a wider range of political support under your leadership as governor. And of course it is more sustainable, since both biodiversity and soil are less likely to degrade over human generations.

Similarly one can optimize for three variables, say, flood control, interior species richness, and cattle grazing. In landscapes dominated by ongoing agriculture, flood control is best addressed with wide bands of natural vegetation around low-order streams and gullies (chapter 7). Interior species richness requires at least a few large patches of natural vegetation. Optimizing for both variables could mean having thin corridors on some small streams, but large vegetation patches encompassing most small streams. Optimizing further for these two variables plus cattle grazing, means maintaining pastures away from streams to protect the large patches and stream corridor system, and providing water

Fig. 14.12. Moist tropical valley overcommitted to agriculture. In this area of intense rainfalls, note: absence of large forest patch on relatively steep slopes; scarcity of small stepping-stone patches; very narrow stream corridors; low connectivity of stream corridor network; no protective vegetation between pond and fields; erosion scars on far slope; and expectation of pond disappearance due to rapid sedimentation. Angel region, Puerto Rico. Courtesy of USDA Soil Conservation Service.

access points only on large streams or rivers. Concurrently providing these three major benefits to society is planning for a sustainable environment.

In this way one should consider it normal to optimize for two or three major objectives. Pairs of objectives are illustrated by clean water and crop production, wildlife movement and a sense of community for people, or game populations and industrial production. Planning and managing for three concurrent objectives is better, such as wind erosion control, wood production, and recreation, or transportation, aesthetics, and fish populations. It is instructive to compare the benefits from single, double, and triple sets of objectives, and the plans that might result. Afterward, single objective plans will doubtless be dropped from the planning repertoire.

521

Incorporating sustainability into planning and management

Large problems remain in creating sustainability. Human population growth, the expansion of megacities, linking landscape level effects with the globe, and developing economic models that give full value to natural resources and processes are poorly addressed.[1097,494,1277] Human population growth is the factor most likely to make sustainability an impossible pipe dream.

Nevertheless, we hypothesize that for any landscape, or major portion of a landscape, there exists an optimal spatial arrangement of ecosystems and land uses to maximize ecological integrity. The same is true for achieving basic human needs and for creating a sustainable environment. If so, the major but tractable challenge is to discover the arrangement.[492]

An important consequence of the principles outlined in this book is to simplify the work of planners, conservationists, managers, and others involved with land use. Inventories of all the species present, mineral nutrient cycles, soil erosion rates, and variations in water quality in a landscape or region are useful and should be developed. But planning, conservation, and management does not require them and should not await them.

Instead, *spatial solutions* now exist. These are spatial arrangements of ecosystems and land uses that make ecological sense in any landscape or region. Putting spatial solutions in place permits us to predict with some confidence that biodiversity, soil, and water will be sustainably conserved for future generations. Every species, every soil particle, and every spot in a water body will not be protected or sustained. But the spatial patterns will conserve the bulk of the attributes, as well as the most important ones. Ecological integrity (e.g., as defined and used in this chapter) will be maintained over time. And it can be done in a landscape one has never seen.

The basic spatial patterns are known, described in previous chapters, and supported with a reasonable amount of evidence. For example at the broad scale, start with the 'indispensable patterns' for which no alternative is known. Fit them together with the 'aggregate-with-outliers' design that arranges any land uses. And channel their change to mimic the 'jaws-and-chunks' transformation, an ecologically optimum pattern over time.

Then, at a finer scale, a host of spatial arrangements makes ecological sense. Incorporate wide connected corridors, clusters of stepping stones, curvilinear boundaries, networks with circuitry, and on and on. All patterns have pros and cons, and some species and soil particles will disappear, but the bulk of both will be conserved. It will be an enormous improvement for any landscape or region with a human imprint. Indeed,

522

Fig. 14.13. A land mosaic viable for both humans and nature requires broad vision, both in space and time. Left: courtesy of USDA Soil Conservation Service. Right: courtesy of US Fish and Wildlife Service.

the future is now. Why can't every landscape and region have a broad spatial-and-ecological plan with a flexible trajectory? Indeed, why can't alternative plans appear that make society and government come to grips with land mosaics?

The process of day-to-day and year-to-year planning, design, conservation, management, and policy was summarized in chapter 13. How can sustainable planning be effectively incorporated into this ongoing process? Creating a sustainable environment involves incorporating and optimizing the attributes summarized in the preceding sections.

Five specific dimensions seem especially ripe for immediate incorporation into day-to-day and year-to-year planning and management[494]: (1) a time frame of human generations; (2) an equal balance of ecological and human dimensions; (3) a focus on slowly-changing attributes; (4) a focus on relatively objective assays; and (5) the optimal spatial arrangement of elements, now rapidly emerging in the study of land mosaics. The effects of incorporating these dimensions will be quickly visible on the land around us (Fig. 14.13). And certainly our children, grandchildren, and great grandchildren will live in a more promising land. They would learn what happened at this point in history to begin transforming the globe into sustainable landscapes and regions.

A linkage between space and time has appeared in several chapters. We the people can tend our garden, maintain our yard, and farm our field. We can plan for tomorrrow, for the upcoming season, and for next year. But as individuals and as humanity we rarely can plan over human generations. The subject of this book is the key to escaping from this straightjacket. Landscapes and regions, i.e., large spatial areas, are a 'surrogate for long term.' When we plan, when we conserve, when we design, when we manage, and when we make wise decisions for landscapes, and especially for regions, we manifest sustainable thinking and act for human generations.

References and Author Index

Numbers following an asterisk are the chapters in which a reference is cited.

1 Aase, J. K. & Siddoway, F. H. 1976. Influence of tall wheatgrass wind barriers on soil drying. *Agronomy Journal* **68:** 627–31. * 6
2 Aase, J. K., Siddoway, F. H. & Black, A. L. 1985. Effectiveness of grass barriers for reducing wind erosiveness. *Journal of Soil and Water Conservation* **40:** 354–60. * 6
3 Abbott, I. 1980. Theories dealing with the ecology of landbirds on islands. *Advances in Ecological Research* **11:** 329–71. * 2
4 Adams, L. W. 1984. Small mammal use of an interstate highway median strip. *Journal of Applied Ecology* **21:** 175–8. * 5
5 Adams, L. W. & Dove, L. E. 1989. *Wildlife Reserves and Corridors in the Urban Environment*. National Institute for Urban Wildlife, Columbia, Maryland, USA. * 6, 9, 13
6 Adams, L. W. & Leedy, D. L., eds. 1991. *Wildlife Conservation in Metropolitan Environments*. National Institute for Urban Wildlife, Columbia, Maryland, USA. * 13
7 Adams, T. 1905. *Garden City and Agriculture: How to Solve the Problem of Rural Depopulation*. Simpkin, Marshall, Hamilton Kent & Co., London. * 14
8 Adamson, R. S. 1931. The plant communities of Table Mountain. II. Life-form dominance and succession. *Journal of Ecology* **19**: 304–20. * 2
9 Adler, G. H., Reich, L. M. & Tamarin, R. H. 1984. Characteristics of white-footed mice in woodland and grassland in Eastern Massachusetts. *Acta Theriologica* **29:** 57–62. * 6
10 Agger, P. & Brandt, J. 1988. Dynamics of small biotopes in Danish agricultural landscapes. *Landscape Ecology* **1:** 227–40. * 8, 12, 13, 14
11 Ahern, J. 1991. Planning for an extensive open space system: linking landscape structure and function. *Landscape and Urban Planning* **21:** 131–45. * 1, 13
12 Ahrens, C. D. 1991. *Meteorology Today: An Introduction to Weather, Climate, and the Environment*. West Publishing, St. Paul, Minnesota, USA. * 10
13 Aldridge, H. D. J. N. & Rautenbach, I. L. 1987. Morphology, echolocation, and resource partitioning in insectivorous bats. *Journal of Animal Ecology* **56**: 763–78. * 9
14 Alexander, C. 1964. *Notes on the Synthesis of Form*. Harvard University Press, Cambridge, Massachusetts. * 1
15 Alexander, R. R. 1964. Minimizing windfall around clear cuttings in Spruce–Fir forests. *Forest Science* **10**: 130–42. * 10
16 Allan, G. & Baker, L. 1990. Uluru (Ayers Rock – Mt Olga) National Park: an assessment of a fire management programme. *Proceedings of the Ecological Society of Australia* **16**: 215–20. * 1
17 Allen, T. F. H. & Hoekstra, T. W. 1992. *Toward a Unified Ecology*. Columbia University Press, New York. * 1, 8, 9, 13, 14
18 Allen, W. H. 1988. Biocultural restoration of a tropical forest: architects of Costa Rica's emerging Guanacaste National Park plan to make it an integral part of local culture. *BioScience* **38:** 156–61. * 13
19 Allendorf, F. W. & Leary, R. F. 1986. Heterozygosity and fitness in natural populations of animals. In M. E. Soulé, ed. *Conservation Biology: The Science of Scarcity and Diversity*, pp. 57–76. Sinauer Associates, Sunderland, Massachusetts, USA. * 2,11

20 Altieri, M. A. & Letourneau, D. K. 1982. Vegetation management and biological control in agroecosystems. *Crop Protection* **1**: 405–30. * 6,9,11

21 Altieri, M. A., Martin, P. B. & Lewis, W. J. 1983. A quest for ecologically based pest management systems. *Environmental Management* **7**: 91–100. * 5,9

22 Alverson, W. S., Waller, D. M. & Solheim, S. L. 1988. Forests too deer: edge effects in northern Wisconsin. *Conservation Biology* **2**: 348–58. * 3,13

23 Ambrose, J. P. 1987. *Dynamics of Ecological Boundary Phenomena Along the Borders of Great Smoky Mountains National Park.* NPS–CPSU Technical Report 34, US National Park Service, Athens, Georgia. 193 pp. * 3,4

24 Ambuel, B. & Temple, S. A. 1983. Area dependent changes in the bird communities and vegetation of southern Wisconsin forests. *Ecology* **64**: 1057–68. * 2,4,6,12

25 Ames, C. R. 1977. Wildlife conflicts in riparian management: grazing. In R. R. Johnson & D. A. Jones, eds. *Importance, Preservation and Management of Riparian Habitat: A Symposium*, pp. 44–51. General Technical Report RM-43, US Forest Service, Washington, DC * 7

26 Amor, R. L. & Stevens, P. L. 1976. Spread of weeds from a roadside into sclerophyll forests at Dartmouth, Australia. *Weed Research* **16**: 111–18. * 5

27 Anderson, S. H. 1979. Changes in forest bird species composition caused by transmission-line corridor cuts. *American Birds* **33**: 3–6. * 5

28 Anderson, S. H., Mann, K. & Shugart, H. H. 1977. The effect of transmission-line corridors on bird populations. *American Midland Naturalist* **97**: 216–21. * 2,5

29 Anderson, W. L. 1978. Waterfowl collisions with power lines at a coal-fired power plant. *Wildlife Society Bulletin* **6**: 77–83. * 5

30 Andow, D. 1983. The extent of monoculture and its effects on insect pest populations with particular reference to wheat and cotton. *Agriculture, Ecosystems and Environment* **9**: 25–35. * 6

31 Andow, D. A., Kareiva, P. M., Levin, S. A. & Okubo, A. 1990. Spread of invading organisms. *Landscape Ecology* **4**: 177–88. * 11

32 Andren, H. & Angelstam, P. 1988. Elevated predation rates as an edge effect in habitat islands: experimental evidence. *Ecology* **69**: 544–7. * 3,12

33 Andren, H. & Angelstam, P. 1993. Moose browsing on Scots pine in relation to stand size and distance to forest edge. *Journal of Applied Ecology* **30**: 133–42. * 2,3

34 Andrew, M. H. 1988. Grazing impact in relation to livestock watering points. *Trends in Ecology and Evolution* **3**: 336–9. * 11

35 Andrew, M. H. & Lange, R. T. 1986. Development of a new piosphere in arid chenopod shrubland grazed by sheep. 1. Changes to the soil surface. *Australian Journal of Ecology* **11**: 395–409. * 11

36 Andrews, E. D. 1963. *People Called Shakers: A Search for the Perfect Society.* Dover, New York. * 14

37 Andrzejewski, R. 1983. W poszukiwaniu teorii fizjocenozy. (Searching for the theory of physiocenosis.) *Wiadomosci Ekologiczne* 29, no. 2. * 4

38 Angelstam, P. 1986. Predation on ground-nesting birds' nests in relation to predator densities and habitat edge. *Oikos* **47**: 365–73. * 3

39 Angelstam, P. 1992. Conservation of communities – the importance of edges, surroundings and landscape mosaic structure. In L. Hansson, ed. *Ecological Principles of Nature Conservation: Applications in Temperate and Boreal Environments*, pp. 9–70. Elsevier, London. * 1,9,12,13,14

40 Angelstam, P., Hansson, L. & Pehrsson, S. 1987. Distribution borders of field mice *Apodemus:* the importance of seed abundance and landscape composition. *Oikos* **50**: 123–30. * 3

41 Angermeier, P. L. & Karr, J. R. 1984. Relationships between woody debris and fish habitat in a small warmwater stream. *Transactions of the American Fisheries Society* **113**: 716–26. * 7

42 Antonovics, J. & Bradshaw, A. D. 1970. Evolution in closely adjacent plant populations. VIII. Clinal patterns at a mine boundary. *Heredity* **25**: 349–62. * 8

43 Antrop, M. 1985. Teledetection et analyse du paysage. In V. Berdoulay & M. Phipps, eds. *Paysage et Systeme.* Editions de l'Universite d'Ottawa, Ottawa. * 9

44 Archbold, E. E. A. 1949. The specific character of plant communities. I. Herbaceous communities. *Journal of Ecology* **37**: 260–73. * 2

45 Aristotle. *c.* 323 BC (1988, S. Everson, ed.) *Politics*. Cambridge University Press, Cambridge. * 2

46 Armand, A. D. 1992. Sharp and gradual mountain timberlines as a result of species interaction. In A. J. Hansen & F. di Castri, eds. *Landscape Boundaries: Consequences for Biotic Diversity and Ecological Flows*, pp. 360–78. Springer-Verlag, New York. * 1,3

47 Armand, A. D. & Vediushkin. 1989. *Triggernie Geosistemi* (Trigger Geosystems). Geographical Institute AS, Moscow, Russia. * 3

48 Arner, D. & Tillman, R. E., eds. 1981. *Environmental Concerns in Rights-of-way Management*. Special Study Project WS 78–141, Electric Power Research Institute, Palo Alto, California. * 5

49 Arno, S. F. 1980. Forest fire history in the northern Rockies. *Journal of Forestry* **78**: 460–5. * 10

50 Arnold, G. W. 1983. The influence of ditch and hedgerow structure, length of hedgerows, and area of woodland and garden on bird numbers in farmland. *Journal of Applied Ecology* **20**: 731–50. * 1,9

51 Arnold, G. W., Steven, D. E. & Weeldenburg, J. R. 1993. Influences of remnant size, spacing pattern and connectivity on population boundaries and demography in euros *Macropus robustus* living in a fragmented landscape. *Biological Conservation* **64**: 219–30. * 11

52 Arnold, G. W. & Weeldenburg, J. R. 1990. Factors determining the number and species of birds in road verges in the wheatbelt of Western Australia. *Biological Conservation* **53**: 295–315. * 5,6

53 Arnold, G. W., Weeldenburg, J. & Steven, D. E. 1988. The movements of kangaroos between remnants of woodland and heath in the wheatbelt of Western Australia. In M. Ruzicka, T. Hrnciarova & L. Miklos, eds. *Proceedings of the VIIth International Symposium on Problems of Landscape Ecological Research*, pp. 59–66. Institute of Experimental Biology and Ecology, Bratislava, Czechoslovakia. * 5

54 Arnold, G. W., Weeldenburg, J. R. & Steven, D. E. 1991. Distribution and abundance of two species of kangaroo in remnants of native vegetation in the central wheatbelt of Western Australia and the role of native vegetation along road verges and fencelines as linkages. In D. A. Saunders & R. J. Hobbs, eds. *Nature Conservation 2: The Role of Corridors*, pp. 273–80. Surrey Beatty, Chipping Norton, Australia. * 5,8

55 Arrhenius, O. 1921. Species and area. *Journal of Ecology* **9**: 95–9. * 2

56 Arrhenius, O. 1923. Statistical investigations in the constitution of plant associations. *Ecology* **4**: 68–73. * 2

57 Art, H. W., Bormann, F. H., Voight, G. K. & Woodwell, G. M. 1974. Barrier island forest ecosystem: role of meteorologic nutrient inputs. *Science* **184**: 60–2. * 10

58 Ashton, P. S. 1978. Crown characteristics of tropical trees. In P. B. Tomlinson & M. H. Zimmermann, eds. *Tropical Trees as Living Systems*, pp. 591–615. Cambridge University Press, Cambridge. * 2

59 Askins, R. A., Philbrick, M. J. & Sugeno, D. S. 1987. Relationship between the regional abundance of forest and the composition of forest bird communities. *Biological Conservation* **39**: 129–52. * 11

60 Aubreville, A. 1949. *Climats, Forets et Desertification de l'Afrique Tropicale*. Societe d'Editions Geographiques, Maritimes, et Coloniales, Paris. * 12

61 Auclair, A. N. 1976. Ecological factors in the development of intensive-management ecosystems in the midwestern United States. *Ecology* **57**: 431–44. * 12,14

62 Augspurger, C. K. & Franson, S. E. 1987. Wind dispersal of artificial fruit varying in mass, area, and morphology. *Ecology* **68**: 27–42. * 11

63 Aulitzky, H. 1963. Grundlagen und Anwendung des vorlaufigen Wind-Schnee-Okogrammes. *Forstliche Bundes versuchsanstalt Mariabrunn, Mitteilungen* **60**: 763–834. * 10

64 Aulitzky, H. 1984. The microclimatic conditions in a subalpine forest as basis for the management. *GeoJournal* **8.3**: 277–81. * 10

527

65 Austin, M. P., Cunningham, R. B. & Fleming, P. M. 1984. New approaches to direct gradient analysis using environmental scalars and statistical curve-fitting procedures. *Vegetatio* **55:** 11–27. * 3,9

66 Austin, M. P. & Heyligers, P. C. 1989. Vegetation survey design for conservation: gradsect sampling of forests in north-eastern New South Wales. *Biological Conservation* **50:** 13–32. * 9

67 Austin, R. F. 1984. Measuring and comparing two-dimensional shapes. In G. L. Gaile & C. J. Willmott, eds. *Spatial Statistics and Models*, pp. 293–312. D. Reidel, Boston. * 4

68 Bache, D. H & MacAskill, I. A. 1984. *Vegetation in Civil and Landscape Engineering.* Granada, London. * 6

69 Bache, D. H. & MacAskill, I. A. 1981. Vegetation in coastal and stream-bank protection. *Landscape Planning* **8:** 363–85. * 7

70 Bagnold, R. A. 1954. *The Physics of Blown Sand and Desert Dunes*. Methuen, London. * 6,12

71 Bailey, R. G. 1983. Delineation of ecosystem regions. *Environmental Management* **7:** 365–73. * 1

72 Bailey, R. G. 1987. Suggested hierarchy of criteria for multi-scale ecosystem mapping. *Landscape and Urban Planning* **14:** 313–19. * 1,9

73 Bailey, R. G. 1988. *Ecogeographic Analysis: A Guide to the Ecological Division of Land for Resource Management*. US Department of Agriculture, Forest Service Miscellaneous Publication 1465, Washington, DC * 1,9

74 Baines, W. D. & Peterson, E. G. 1951. An investigation of flow through screens. *Transactions of the American Society of Mechanical Engineers* **73:** 467–80. * 6,10

75 Bairlein, von F. 1978. Uber die Biologie einer sudwest-deutschen Population der Monchsgrasmucke (*Sylvia atricapilla*). *Journal für Ornithologie* **119:** 14–51. * 4

76 Baker, V. R. 1977. Stream-channel response to floods with examples from central Texas. *Geological Society of America Bulletin* **88:** 1057–71. * 7

77 Baker, W. L. 1989. A review of models of landscape change. *Landscape Ecology* **2:** 111–33. * 1,12

78 Baker, W. L. 1993. Spatially heterogeneous multi-scale response of landscapes to fire suppression. *Oikos* **66:** 66–71. * 2,9,10

79 Ballard, J. T. 1979. Fluxes of water and energy through the Pine Barrens ecosystems. In R. T. T. Forman, ed. *Pine Barrens: Ecosystem and Landscape*, pp. 133–46. Academic Press, New York. * 10

80 Balser, D., Bielak, A., De Boer, G., Tobias, T., Adindu, G. & Dorney, R. S. 1981. Nature reserve designation in a cultural landscape, incorporating island biogeographic theory. *Landscape Planning* **8:** 329–47. * 2

81 Barbour, C. D. & Brown, J. H. 1974. Fish species diversity in lakes. *American Naturalist* **108:** 473–89. * 2

82 Barrett, G. W. 1992. Landscape ecology: designing sustainable agricultural landscapes. Journal of Sustainable Agriculture **2:** 83–103. * 12,14

83 Barrett, G. W. & Rosenberg, R., eds. 1981. *Stress Effects on Natural Ecosystems*. John Wiley, New York. * 10

84 Bartholomew, B. 1970. Bare zone between California shrub and grassland communities: the role of animals. *Science* **170:** 1210–12. * 3

85 Bartley, D., Bagley, M., Gall, G. & Bentley, B. 1992. Use of linkage disequilibrium data to estimate effective size of hatchery and natural fish populations. *Conservation Biology* **6:** 365–75. * 2

86 Barton, D. R., Taylor, W. D. & Biette, R. M. 1985. Dimensions of riparian buffer strips required to maintain trout habitat in southern Ontario streams. *North American Journal of Fisheries Management* **5:** 364–78. * 7,13

87 Bashore, T. L., Tzilkowski, W. M. & Bellis, E. D. 1985. Analysis of deer–vehicle collision sites in Pennsylvania. *Journal of Wildlife Management* **49:** 769–74. * 6

88 Bastian, O. 1990. Structure, function and change – three main aspects in investigation of biotic landscape components. *Ekologia* (Czechoslovakia) **9:** 405–18. * 1

89 Baudry, J. 1984. Effects of landscape structure on biological communities: the case of hedgerow network landscapes. In J. Brandt & P. Agger, eds. *Proceedings of the First International Seminar* on *Methodology in Landscape Ecological Research and Planning*, vol 1., pp. 55–65. Roskilde Universitetsforlag GeoRuc, Roskilde, Denmark. * 8,9,12

528

90 Baudry, J. 1988. Hedgerows and hedgerow networks as wildlife habitat in agricultural landscapes. In J. R. Park, ed. *Environmental Management in Agriculture: European Perspectives*, pp. 111–24. Belhaven, London. * 8

91 Baudry, J. & Burel, F. 1984. Landscape project: Remembrement: landscape consolidation in France. *Landscape Planning* **11**: 235–41. * 1,2,8,12

92 Baudry, J. & Merriam, G. 1988. Connectivity and connectedness: functional versus structural patterns in landscapes. In K–F. Schreiber, ed. *Connectivity in Landscape Ecology*, pp. 23–8. Munstersche Geographischer Arbeiten 29. Ferdinand Schoningh, Paderborn, Germany. * 8

93 Baumgartner, L. 1943. Fox squirrels in Ohio. *Journal of Wildlife Management* **7**: 193–202. * 6

94 Baur, G. N. 1964. *The Ecological Basis of Rainforest Management*. New South Wales Forestry Commission, Sydney. * 2

95 Baxter, R. M. 1977. Environmental effects of dams and impoundments. *Annual Review of Ecology and Systematics* **8**: 255–83. * 7

96 Bayfield, N. H. & Bathe, G. M. 1982. Experimental closure of footpaths in a woodland national nature reserve in Scotland. *Biological Conservation* **22**: 229–37. * 5

97 Bayliss-Smith, T. & Owens, S. 1990. *Britain's Changing Environment from the Air*. Cambridge University Press, Cambridge. * 12

98 Bazzaz, F. A. 1983. Characteristics of populations in relation to disturbance in natural and man-modified ecosystems. In H. A. Mooney & M. Godron, eds. *Disturbance and Ecosystems: Components of Response*, pp. 259–75. Springer-Verlag, New York. * 2

99 Bazzaz, F. A. 1986. Life history of colonizing plants. In H. A. Mooney & J. A. Drake, eds. *Ecology of Biological Invasions of North America and Hawaii*, pp. 96–110. Springer-Verlag, New York. * 11

100 Beasom, S. L., Wiggers, E. P. & Giardino, J. R. 1983. A technique for assessing land surface ruggedness. *Journal of Wildlife Management* **47**: 1163–6. * 9

101 Beeton, R. J. S. 1985. The little corella: a seasonally adapted species. *Proceedings of the Ecological Society of Australia* **13**: 53–63. * 11

102 Bellis, E. D. & Graves, H. B. 1971. Deer mortality on a Pennsylvania interstate highway. *Journal of Wildlife Management* **35**: 232–7. * 5,6

103 Bellis, E. D. & Graves, H. B. 1978. Highway fences as deterrents to vehicle-deer collisions. *Transportation Research Records* **674**: 53–8. * 5

104 Belovsky, G. E. 1978. Diet optimization in a generalist herbivore: the moose. *Theoretical Population Biology* **14**: 105–34. * 11

105 Belsky, A. J. 1986. Does herbivory benefit plants? A review of the evidence. *American Naturalist* **127**: 870–92. * 3,11

106 Benke, A. C., Henry R. L., III, Gillespie, D. M. & Hunter, R. J. 1985. Importance of snag habitat for animal production in southeastern streams *Fisheries* **10(5)**: 8–13. * 7

107 Bennett, A. F. 1987. Conservation of mammals within a fragmented forest environment: the contributions of insular biogeography and autecology. In D. A. Saunders, G. W. Arnold, A. A. Burbridge, & A. J. M. Hopkins, eds. *Nature Conservation: The Role of Remnants of Native Vegetation*, pp. 41–52. Surrey Beatty, Chipping Norton, Australia. * 2

108 Bennett, A. F. 1988. Roadside vegetation: a habitat for mammals at Naringal, southwestern Victoria. *Victorian Naturalist* **105**: 106–13. * 5

109 Bennett, A. F. 1990a. *Habitat Corridors: Their Role in Wildlife Management and Conservation*. Department of Conservation and Environment, Victoria, Australia. 37 pp. * 3, 5, 6, 11, 13

110 Bennett, A. F. 1990b. Habitat corridors and the conservation of small mammals in a fragmented forest environment. *Landscape Ecology* **4**: 109–22. * 5, 6, 12

111 Bennett, A. F. 1991. Roads, roadsides and wildlife conservation: a review. In D. A. Saunders & R. J. Hobbs, eds. *Nature Conservation 2: The Role of Corridors*, pp. 99–117. Surrey Beatty, Chipping Norton, Australia. * 3, 5, 6

112 Bennett, A. F., Henein, K. & Merriam, G. 1994. Corridor use and the elements of corridor quality: chipmunks and fencerows in a farmland mosaic. *Biological Conservation* **68**: 155–66. * 6

113 Benninger-Truax, M., Vankat, J. L. & Schaefer, R. L. 1992. Trail corridors as habitat and conduits for movement of plant species in Rocky Mountain National Park, Colorado, USA. *Landscape Ecology* **6**: 269–78. * 5

114 Berdoulay, V. & Phipps, M., eds. 1985. *Paysage et Systeme*. Editions de l'Universite d'Ottawa, Ottawa. * 1, 9

115 Berg, H. C. 1983. *Random Walks in Biology*. Princeton University Press, Princeton, New Jersey. * 11

116 Berger, J. 1990. Persistence of different-sized populations: an empirical assessment of rapid extinctions in bighorn sheep. Conservation Biology **4**: 91–8. * 2, 11

117 Bernabo, J. C. & Webb, T., III. 1977. Changing patterns in the Holocene pollen record of northeastern North America: a mapped summary. *Quaternary Research* **8**: 64–96. * 3, 14

118 Berndt, E. R., Cox, A. J. & Pearse, P. H. 1979. Estimation of logging costs and timber supply curves from forest inventory data. *The Forestry Chronicle* **55**: 144–7. * 2

119 Bernhardt, Von K-G. & Schreiber, K-F. 1988. Synokologische Untersuchungen eines Hecken–Feld–Waldrand–Biotopkomplexes in Westfalen (Norddeutschland). *Landschaft und Stat* **20**: 106–13. * 3

120 Beschta, R. L. & Platts, W. S. 1986. Morphological features of small streams: significance and function. *Water Resources Bulletin* **22**: 369–79. * 7

121 Best, R. H. 1981. *Land Use and Living Space*. Methuen, London. * 13

122 Bickford, W. E. & Dymon, U. J. 1990. *An Atlas of Massachusetts River Systems: Environmental Designs for the Future*. University of Massachusetts Press, Amherst, Massachusetts, USA. * 7, 8, 13

123 Bider, J. R. 1968. Animal activity in uncontrolled terrestrial communities as determined by a sand transect technique. *Ecological Monographs* **38**: 269–308. * 3, 5, 11

124 Biel, E. R. 1961. Microclimate, bioclimatology and notes on comparative dynamic climatology. *American Scientist* **49**: 326–57. * 10

125 Bilbro, J. D. & Fryrear, D. W. 1988. Annual herbaceous windbarriers for protecting crops and soils and managing snowfall. *Agriculture, Ecosystems and Environment* **22/23**: 149–61. (Reprinted 1988 in *Windbreak Technology*. Elsevier, Amsterdam.) * 6

126 Bill, R. G., Jr., Sutherland, R. A., Bartholic, J. F. & Chen, E. 1978. Observations of the convective plume of a lake under cold-air advective conditions. *Boundary-Layer Meteorology* **14**: 543–56. * 10

127 Billington, H. L. 1991. Effect of population size on genetic variation in a dioecious conifer. *Conservation Biology* **5**: 115–19. * 2, 11

128 Binford, M. W., Brenner, M. & Engstrom, D. 1992. Temporal sedimentation patterns in the nearshore littoral of Lago Huinaimarca. In C. Dejoux & A. Iltis, eds. *Lake Titicaca: A Synthesis of Limnological Knowledge*, pp. 29–37. Kluwer, Dordrecht, Netherlands. * 10, 14

129 Binford, M. W., Brenner, M., Whitmore, T. J., Higuera-Gundy, A., Deevey, E. S. & Leyden, B. 1987. Ecosystems, paleoecology and human disturbance in subtropical and tropical America. *Quaternary Science Reviews* **6**: 115–28. * 10, 14

130 Binford, M. W. & Buchenau, M. 1993. Riparian greenways and water resources. In D. S. Smith & P. C. Hellmund, eds. *Ecology of Greenways: Design and Function of Linear Conservation Areas*, pp. 69–104. University of Minnesota Press, Minneapolis, Minnesota, USA. * 5, 7, 8, 10, 12, 13

131 Binford, M. W., Deevey, E. S. & Crisman, T. L. 1983. Paleolimnology: an historical perspective on lacustrine ecosystems. *Annual Review of Ecology and Systematics* **14**: 255–86. * 9, 10, 14

132 Birks, H. H., Birks, H. J. B., Kaland, P. E. & Moe, D., eds. 1988. *The Cultural Landscape: Past, Present, and Future*. Cambridge University Press, Cambridge. * 9, 10, 12, 14

133 Birks, H. J. B. 1986. Late-Quaternary biotic changes in terrestrial and lacustrine environments, with particular reference to north-west Europe. In B. E. Berglund, ed. *Handbook of Holocene Palaeoecology and Palaeohydrology*, pp. 3–65. John Wiley, Chichester, UK. * 1, 14

134 Black, A. L., Brown, P. L., Halvorson, A. D. & Siddoway, F. H. 1981. Dryland cropping strategies for efficient water use to control saline seeps in the Northern Great Plains, U. S. A. *Agricultural Water Management* **4**: 295–311. * 6,7

135 Black, A. L. & Siddoway, F. H. 1971. Tall wheatgrass barriers for soil erosion control and water conservation. *Journal of Soil and Water Conservation* **26**: 107–11. * 6

530

136 Blake, J. G. 1983. Trophic structure of bird communities in forest patches in east-central Illinois. *Wilson Bulletin* **95:** 416–30. * 2

137 Blake, J. G. & Karr, J. R. 1987. Breeding birds of isolated woodlots: area and habitat relationships. *Ecology* **68:** 1724–34. * 2, 11

138 Blakers, M., Davies, S. J. J. F. & Reilly, P. N. 1984. *The Atlas of Australian Birds*. Melbourne University Press, Melbourne. * 6

139 Blondel, J. 1987. From biogeography to life history theory: a multithematic approach illustrated by the biogeography of vertebrates. *Journal of Biogeography* **14:** 405–22. * 2

140 Blouin, M. S. Z. & Connor, E. F. 1985. Is there a best shape for nature reserves? *Biological Conservation* **32:** 277–88. * 4

141 Boaler, S. B. & Hodge, C. A. H. 1962. Vegetation stripes in Somaliland. *Journal of Ecology* **50:** 465–74. * 4, 9

142 Boaler, S. B. & Hodge, C. A. H. 1964. Observation on vegetation arcs in the northern region, Somali Republic. *Journal of Ecology* **52:** 511–44. * 3, 4, 9

143 Boecklen, W. J. & Gotelli, N. J. 1984. Island biogeographic theory and conservation practice: species–area or specious–area relationships? *Biological Conservation* **29:** 63–80. * 2

144 Boer Leffee, W. J. 1958. De entomologische waarde van eiken-berkenbos. *De Levende Natuur* **61:** 97–102. * 5

145 Boerner, R. E. J. 1985. Alternate pathways of succession on the Lake Erie islands. *Vegetatio* **63:** 35–44. * 11

146 Boerner, R. E. J. & Forman, R. T. T. 1982. Hydrologic and mineral budgets of New Jersey Pine Barrens upland forests following two intensities of fire. *Canadian Journal of Forest Research* **12:** 503–10. * 10

147 Boerner, R. E. J. & Kooser, J. G. 1989. Leaf litter redistribution among forest patches within an Allegheny Plateau watershed. *Landscape Ecology* **2:** 81–92. * 9

148 Bollinger, E. K. & Caslick, J. W. 1985. Factors influencing blackbird damage to field corn. *Journal of Wildlife Management* **49:** 1109–15. * 6

149 Booth, D. L., Boulter, D. W. K., Neave, D. J., Rotherham, A. A. & Welsh, D. A. 1993. Natural forest landscape management: a strategy for Canada. *The Forestry Chronicle* **69:** 141–5. * 13

150 Bormann, F. H. 1953. The statistical efficiency of sample plot size and shape in forest ecology. *Ecology* **34:** 474–87. * 4, 5

151 Bormann, F. H. 1982. Air pollution stress and energy policy. In C. H. Reidel, ed. *New England Prospects: Critical Choices in a Time of Change*. University Press of New England, Hanover, New Hampshire. * 10

152 Bormann, F. H. 1987. Landscape ecology and air pollution. In M. G. Turner, ed. *Landscape Heterogeneity and Disturbance*, pp. 37–57. Springer-Verlag, New York. * 10

153 Bormann, F. H., Balmiori, D. & Geballe, G. T. 1993. *Redesigning the American Lawn: A Search for Environmental Harmony*. Yale University Press, New Haven, Connecticut. * 13

154 Bormann, F. H. & Likens, G. E. 1979. *Pattern and Process in a Forested Ecosystem: Disturbance, Development and the Steady State Based on the Hubbard Brook Ecosystem Study*. Springer-Verlag, New York. * 2, 7, 10, 12, 13

155 Borowske, J. R. & Heitlinger, M. E. 1981. Survey of native prairie on railroad rights-of-way in Minnesota. *Transportation Research Records* (Washington) **822:** 22–6. * 5, 13

156 Bosch, W. 1978. A procedure for quantifying certain geomorphological features. *Geographical Analysis* **10:** 241–7. * 4

157 Botkin, D. B. 1990. *Discordant Harmonies: A New Ecology for the Twenty-First Century*. Oxford University Press, New York. * 13

158 Botkin, D. B., Estes, J. E., MacDonald, R. M. & Wilson, M. V. 1984. Studying the earth's vegetation from space. *BioScience* **34:** 508–14. * 1

159 Botkin, D. B., Melillo, J. M. & Wu, L. S-Y. 1981. How ecosystem processes are linked to large mammal population dynamics. In C. F. Fowler & T. D. Smith, eds. *Dynamics of Large Mammal Populations*, pp. 373–87. John Wiley, New York. * 7

160 Bottero, J. 1992. *Mesopotamia: Writing, Reasoning, and the Gods*. University of Chicago Press, Chicago. * 14

161 Bottom, D. L., Howell, P. J. & Rodger, J. D. 1983. *Final Report: Fish Research Project. Oregon Salmonid Habitat Restoration.* Oregon Department of Fish and Wildlife, Portland, Oregon, USA. * 7

162 Boulding, K. E., ed. 1984. *The Economics of Human Betterment.* State University of New York Press, Albany, New York. * 14

163 Bovet, P. & Benhamou, S. 1988. Spatial analysis of animals' movements using a correlated random walk model. *Journal of Theoretical Biology* **131:** 419–33. * 11

164 Bowman, D. 1977. Stepped-bed morphology in arid gravelly channels. *Geological Society of America Bulletin* **88:** 291–8. * 7

165 Boyce, R. R. & Clark, W. A. V. 1964. The concept of shape in geography. *Geographical Review* **54:** 561–72. * 4

166 Boycott, A. E. 1936. The habitats of fresh-water Mollusca in Britain. *Journal of Animal Ecology* **5:** 116–86. * 11

167 Boyden, S. V. 1981. *The Ecology of a City and Its People: The Case of Hong Kong.* Australian National University Press, Canberra. * 1

168 Braakhekke, W. G. & Braakhekke-Ilsink, E. I. 1976. Nitrophile Saumgesellschaften im Sudosten der Niederlande. *Vegetatio* **32:** 55–60. * 3

169 Bradshaw, A. D. 1972. Some evolutionary consequences of being a plant. *Evolutionary Biology* **5:** 25–47. * 8

170 Braithwaite, L. W., Austin, M. P., Clayton, M., Turner, J. & Nicholls, A. O. 1989. On predicting the presence of birds in *Eucalyptus* forest types. *Biological Conservation* **50:** 33–50. * 9,14

171 Braithwaite, L. W., Turner, J. & Kelly, J. 1984. Studies on the arboreal marsupial fauna of eucalypt forests being harvested for woodpulp at Eden, NSW. III. Relationships between the faunal densities, eucalypt occurrence and foliage nutrients, and soil parent materials. *Australian Wildlife Research* **11:** 41–8. * 9,14

172 Braithwaite, W. 1991, Arboreal mammals of the Eden (New South Wales) woodchip forests: a regional survey. In C. R. Margules & M. P. Austin, eds. *Nature Conservation: Cost Effective Biological Surveys and Data Analysis,* pp. 127–33. CSIRO Australia, Canberra. * 9,14

173 Brakke, T. W., Verma, S. B. & Rosenberg, N. J. 1978. Local and regional components of sensible heat advection. *Journal of Applied Meteorology* **17:** 955–63. * 10

174 Brandle, J. R., Hintz, D. L. & Sturrock, J. W., eds. 1988. *Windbreak Technology.* Elsevier, Amsterdam. (Reprinted from *Agriculture, Ecosystems and Environment*, Vols. 22–23, 1988.) * 1,2,4,5,6,8,12

175 Brandle, J. R., Johnson, B. B. & Akeson, T. 1992. Field windbreaks: are they economical? *Journal of Production Agriculture* **5:** 393–8. * 6

176 Brandt, C. S. & Heck, W. W. 1968. Effects of air-pollutants on plants. In A. C. Stern, ed. *Air Pollution*, vol. 1, pp. 401–43. Academic Press, New York. * 10

177 Brandt, J. & Agger, P., eds. 1984. *Proceedings of the First International Seminar on Methodology in Landscape Ecology Research and Planning.* 5 vols. Roskilde Universitetsforlag GeoRuc, Roskilde, Denmark. * 1

178 Bratton, S. P., Hickler, M. G. & Graves, J. H. 1979. Trail erosion patterns in Great Smoky Mountains National Park. *Environmental Management* **3:** 431–45. * 5

179 Brazel, A. J. & Outcalt, S. I. 1973a. The observation and simulation of diurnal surface thermal contrast in an Alaskan alpine pass. *Archives for Meteorology, Geophysics, and Bioclimatology*, Series B **21:** 157–74. * 10

180 Brazel, A. J. & Outcalt, S. I. 1973b. The observation and simulation of diurnal evaporation contrast in an Alaskan alpine pass. *Journal of Applied Meteorology* **12:** 1134–43. * 10

181 Brenner, M. & Binford, M. W. 1986. Material transfer from water to sediment in Florida lakes. *Hydrobiologia* **143:** 55–61. * 10,14

182 Bridgeland, W. T. & Caslick, J. W. 1983. Relationships between cornfield characteristics and blackbird damage. *Journal of Wildlife Management* **47:** 824–9. * 6

183 Bridges, M., O'More, H., Reiche, M., Gallenkamp, C., Lippard, L. & Critchlow, K. 1986. *Markings: Aerial Views of Sacred Landscapes.* Phaidon, Oxford. * 14

184 Bridgewater, P. B. 1987. Connectivity: an Australian perspective. In D. A. Saunders, G. W. Arnold, A. A. Burbidge & A. J. M. Hopkins, eds. *Nature Conservation: The Role of*

Remnants of Native Vegetation, pp. 195–200. Surrey Beatty, Chipping Norton, Australia. * 5,13

185 Brinson, M. M. 1990. Riverine forests. In A. E. Lugo, M. Brinson, & S. Brown, eds. *Forested Wetlands: Ecosystems of the World*, Vol. 15, pp. 87–141. Elsevier, Amsterdam. * 7

186 Brittingham, M. C. & Temple, S. A. 1983. Have cowbirds caused forest songbirds to decline? *BioScience* **33**: 31–5. * 3,5,12

187 Brocke, R. H., O'Pezio, J. P. & Gustafson, K. A. 1990. A forest management scheme mitigating impact of road networks on sensitive wildlife species. In *Is Forest Fragmentation a Management Issue in the Northeast?*, pp. 13–17. General Technical Report NE-140, US Forest Service, Radnor, Pennsylvania. * 5,8,13

188 Brokaw, N. V. L. 1982. The definition of treefall gap and its effect on measures of forest dynamics. *Biotropica* **14**: 158–60. * 2

189 Brokaw, N. V. L. 1985. Gap-phase regeneration in a tropical forest. *Ecology* **66**: 682–7. * 2

190 Brooker, M. P. 1981. The impact of impoundments on the downstream fisheries and general ecology of rivers. In T. H. Coaker, ed. *Advances in Applied Biology*, VI, pp. 91–152. Academic Press, London. * 7

191 Brooks, H. 1986. The typology of surprises in technology, institutions, and development. In W. C. Clark & R. E. Munn, eds. *Sustainable Development of the Biosphere*, pp. 325–48. Cambridge University Press, Cambridge. * 14

192 Brooks, H. 1993. Sustainability and technology. In N. Keyfitz, ed. *Science and Sustainability: Selected Papers on IIASA's 20th Anniversary*, pp. 1–31. (Reprinted 1993 in *Environment*.) * 14

193 Brower, D. J., Godschalk, D. R. & Porter, D. R. 1989. *Understanding Growth Management: Critical Issues and a Research Agenda*. Urban Land Institute, Washington, DC. * 13

194 Brown, G. W. & Krygier, J. T. 1970. Effects of clear-cutting on stream temperature. *Water Resources Research* **6**: 1133–9. * 7

195 Brown, G. W. & Krygier, J. T. 1971. Clear-cut logging and sediment production in the Oregon Coast Range. *Water Resources Research* **7**: 1189–98. * 5

196 Brown, J. H. 1971. Mammals on mountaintops: non-equilibrium insular biogeography. *American Naturalist* **105**: 467–78. * 1,2

197 Brown, J. H. 1978. The theory of insular biogeography and the distribution of boreal birds and mammals. In K. T. Kimball & J. L. Reveal, eds. *Biogeography of the Intermountain West*, pp. 209–27. Brigham Young University Press, Provo, Utah, USA. * 2

198 Brown, J. H. & Gibson, A. C. 1983. *Biogeography*. Mosby, St. Louis, Missouri, USA. * 1,2

199 Brown, J. H. & Kodric-Brown, A. 1977. Turnover rates in insular biogeography: effect of immigration on extinction. *Ecology* **58**: 445–9. * 2,11

200 Brown, J. H. & Lieberman, G. A. 1973. Resource utilization and coexistence of seed-eating desert rodents in sand dune habitats. *Ecology* **54**: 788–97. * 11

201 Brown, K. W. & Rosenberg, N. J. 1971. Turbulent transport and energy balance as affected by a windbreak in an irrigated sugar beet (*Beta vulgaris*) field. *Agronomy Journal* **63**: 351–5. * 6

202 Brown, M. & Dinsmore, J. J. 1986. Implications of marsh size and isolation for marsh bird management. *Journal of Wildlife Management* **50**: 392–7. * 2,11

203 Brunet, P. 1976. Physionomie et signification des haies. In *Les Bocages: Histoire, Ecologie, Economie*, pp. 37–41. Institut National de la Recherche Agronomique, Centre National de la Recherche Scientifique, Universite de Rennes, Rennes, France. * 6

204 Brush, G. S. 1989. Rates and patterns of estuarine and sediment accumulation. *Limnology and Oceanography* **34**: 1235–46. * 10

205 Brush, J. E. 1953. The hierarchy of central places in southwestern Wisconsin. *Geographical Review* **43**: 380–402. * 9

206 Bryant, F. C., Dahl, B. E., Pettit, R. D. & Britton, C. M. 1989. Does short duration grazing work in arid and semiarid regions? *Journal of Soil and Water Conservation* **44**: 290–6. * 11

207 Buckley, R. 1982. The habitat-unit model of island biogeography. *Journal of Biogeography* **9**: 339–44. * 2

208 Buckley, R. C. & Knedlhans, S. B. 1986. Beachcomber biogeography: interception of dispersing propagules by islands. *Journal of Biogeography* **13**: 69–70. * 4
209 Buckner, C. A. & Shure, D. J. 1985. The response of *Peromyscus* to forest opening size in the southern Appalachian Mountains. *Journal of Mammalogy* **66**: 299–307. * 2
210 Budd, W. W., Cohen, P. L., Saunders, P. R. & Steiner, F. R. 1987. Stream corridor management in the Pacific Northwest: I: Determination of stream corridor widths. *Environmental Management* **11**: 587–97. * 7
211 Buechner, M. 1987a. A geometric model of vertebrate dispersal: tests and implications. *Ecology* **68**: 310–18. * 3,4
212 Buechner, M. 1987b. Conservation in insular parks: simulation models of factors affecting the movement of animals across park boundaries. *Biological Conservation* **41**: 57–76. * 3,5,11
213 Buechner, M. 1989. Are small-scale landscape features important factors for field studies of small mammal dispersal sinks? *Landscape Ecology* **2**: 191–9. * 4
214 Buffington, L. C. & Herbel, C. H. 1965. Vegetational change on a semidesert grassland range. *Ecological Monographs* **35**: 139–64. * 12
215 Bugnicourt, J. 1987. Culture and environment. In P. Jacobs & D. A. Munro, eds. *Conservation with Equity*: Strategies for Sustainable Development, pp. 95–106. International Union for the Conservation of Nature and Natural Resources, Gland, Switzerland. * 1,14
216 Bunce, M. 1982. *Rural Settlement in an Urban World*. St. Martin's Press, New York. * 9
217 Bunce, R. G. H., Howard, D. C., Hallam, C. J., Barr, C. J. & Benefield, C. B. 1992. *Ecological Consequences of Land Use Change*. Institute of Terrestrial Ecology, Cumbria, UK. * 1
218 Bunce, R. G. H. & Shaw, M. W. 1973. A standardized procedure for ecological survey. *Journal of Environmental Management* **1**: 239–58. * 9
219 Bunge, W. W. 1966. *Theoretical Geography*. Lund Studies in Geography, Series C, General and Mathematical Geography No. 6, Gleerup, Lund, Sweden. * 4
220 Bunnell, S. D. & Johnson, D. R. 1974. Physical factors affecting pika density and dispersal. *Journal of Mammalogy* **55**: 866–9. * 4
221 Burel, F. 1988. Biological patterns and structural patterns in agricultural landscapes. In Schreiber, K-F., ed. *Connectivity in Landscape Ecology*, pp. 107–10. Munstersche Geographische Arbeiten 29. Ferdinand Schoningh, Paderborn, Germany. * 8
222 Burel, F. 1989. Landscape structure effects on carabid beetles spatial patterns in western France. *Landscape Ecology* **2**: 215–26. * 6,11
223 Burel, F. & Baudry, J. 1990a. Hedgerow network patterns and processes in France. In I. S. Zonneveld & R. T. T. Forman, eds. *Changing Landscapes: An Ecological Perspective*, pp. 99–120. Springer-Verlag, New York. * 1,5,6,8,11
224 Burel, F. & Baudry, J. 1990b. Structural dynamic of a hedgerow network landscape in Brittany, France. *Landscape Ecology* **4**: 197–210. * 1,8,12
225 Burgess, R. L. & Sharpe, D. M., eds. 1981a. *Forest Island Dynamics in Man-dominated Landscapes*. Springer-Verlag, New York. * 1,2,12
226 Burgess, R. L. & Sharpe, D. M. 1981b. Introduction. In R. L. Burgess & D. M. Sharpe, eds. *Forest Island Dynamics in Man-dominated Landscapes*, pp. 1–5. Springer-Verlag, New York. * 12
227 Burke, I. C., Kittel, T. G. F., Lauenroth, W. K., Snook, P., Yonker, C. M. & Parton, W. J. 1991. Regional analysis of the Central Great Plains: sensitivity to climate variability. *BioScience* **41**: 685–92. * 1
228 Burke, I. C., Schimel, D. S., Yonker, C. M., Parton, W. J., Joyce, L. A. & Lauenroth, W. K. 1990. Regional modeling of grassland biogeochemistry using GIS. *Landscape Ecology* **4**: 45–54. * 1,9
229 Burns, J. W. 1972. Some effects of logging and associated road construction on Northern Californian streams. *Transactions of the American Fisheries Society* **101**: 1–17. * 5,7,13
230 Burrough, P. A. 1981. Fractal dimensions of landscapes and other environmental data. *Nature* **294**: 241–2. * 1,3,8
231 Burrough, P. A. 1986. *Principles of Geographical Information Systems for Land Resources Assessment*. Clarendon Press, Oxford. * 1,9

232 Burrough, P. A. 1989. Matching spatial databases and quantitative models in land resource assessment. *Soil Use and Management* **5**: 3–8. * 1,9

233 Busack, S. D. & Hedges, S. B. 1984. Is the peninsula effect a red herring? *American Naturalist* **123**: 266–75. * 2,4,5

234 Businger, J. A. & Frisch, A. S. 1972. Cold plumes. *Journal of Geophysical Research* **77**: 3270–1. * 10

235 Buttel, F. H., Gillespie, G. W., Larson, III, O. W. & Harris, C. K. 1981. The social bases of agrarian environmentalism: a comparative analysis of New York and Michigan farm operators. *Rural Sociology* **46**: 391–410. * 9

236 Buttel, F. H. & Larson, O. W. III. 1979. Farm size, structure, and energy intensity: an ecological analysis of U. S. agriculture. *Rural Sociology* **44**: 471–88. * 2,9

237 Bye, R. A., Jr. 1981. Quelite-ethnoecology of edible greens-past, present and future. *Journal of Ethnobiology* **1**: 109–23. * 9

238 Bynner, W. & Kiang, K-H. 1929. *The Jade Mountain: A Chinese Anthology, Being Three Hundred Poems of the T'ang Dynasty* 618–906. Alfred A. Knopf, New York. * Preface.

239 Cable, T. T. & Cook, P. S. 1990. The use of windbreaks by hunters in Kansas. *Journal of Soil and Water Conservation* **45**: 575–7. * 6

240 Caborn, J. M. 1960. The planning of shelterbelts: design and structure. *Scottish Forestry* **14**: 123–9. * 6

241 Caborn, J. M. 1965. *Shelterbelts and Windbreaks*. Faber and Faber, London. * 2,5,6

242 Cady, J. F. 1964. *Southeast Asia: Its Historical Development*. McGraw-Hill, New York. * 14

243 Cain, S. A. 1943. Sample-plot technique applied to alpine vegetation in Wyoming. *American Journal of Botany* **30**: 240–7. * 2

244 Cairns, J., Jr. 1982, *Biological Monitoring in Water Pollution*. Pergamon Press, New York. * 7

245 Cairns, J., Jr. 1988. Restoration ecology: the new frontier. In J. Cairns, Jr., ed. *Rehabilitating Damaged Ecosystems*, pp. 1–11. CRC Press, Boca Raton, Florida. * 13

246 Calthorpe, P. 1993. *The Next American Metropolis: Ecology, Communities, and the American Dream*. Princeton Architectural Press, New York, * 13, 14

247 Cameron, I. 1984. *Mountains of the Gods*. Century Publishing, London. * 14

248 Campbell, C. J. 1970. Ecological implications of riparian vegetation management. *Journal of Soil and Water Conservation* **25**: 49–52. * 7

249 Campbell, C. J. & Dick-Peddie, W. A. 1964. Comparison of phreatophyte communities on the Rio Grande in New Mexico. *Ecology* **45**: 492–502. * 7

250 Canny, M. J. 1981. A universe comes into being when a space is severed: some properties of boundaries in open systems. *Proceedings of the Ecological Society of Australia* **11**: 1–11. * 3

251 Cantwell, M. D. & Forman, R. T. T. 1994. Landscape graphs: ecological modeling with graph theory to detect configurations common to diverse landscapes. *Landscape Ecology* **8**: 239–55. * 1, 3, 4, 8, 9

252 Capel, S. W. 1988. Design of windbreaks for wildlife in the Great Plains of North America. *Agriculture, Ecosystems and Environment* **22/23**: 337–47. (Reprinted 1988 in *Windbreak Technology*. Elsevier, Amsterdam.) * 6

253 Carbaugh, B. T., Vaughan, J. P., Bellis, E. D. & Graves, H. B. 1975. Distribution and activity of white-tailed deer along an interstate highway. *Journal of Wildlife Management* **39**: 570–81. * 5

254 Carbiener, R. 1970. Un exemple de type forestier exceptionnel pour l'Europe occidentale: la foret du lit majeur du Rhin au niveau du fosse rhenan. *Vegetatio* **20**: 97–148. * 7

255 Carey, A. B., Reid, J. A. & Horton, S. P. 1990. Spotted owl home range and habitat use in southern Oregon Coast Ranges. *Journal of Wildlife Management* **54**: 11–17. * 9

256 Cargill, A. S., II. 1980. Lack of rainbow trout movement in a small stream. *Transactions of the American Fisheries Society* **109**: 484–90. * 2, 8

257 Carlson, T. N. 1986. Regional-scale estimates of surface moisture availability and thermal inertia using remote thermal measurements. *Remote Sensing Reviews* **1**: 197–247. * 10

258 Carnet, C. 1976. Role du bocage sur la distribution des sols et la circulation de l'eau dans les sols. In *Les Bocages: Histoire, Ecologie, Economie*, pp. 159–62. Institut National

de la Recherche Agronomique, Centre National de la Recherche Scientifique, Universite de Rennes, Rennes, France. * 6

259 Carpenter, E. D. 1970. Salt tolerance of ornamental plants. *American Nurseryman* **131(2):** 12, 54–71. * 5

260 Carpenter, J. R. 1935a. Forest edge birds and exposures of their habitats. *Wilson Bulletin* **47:** 106–8. * 3

261 Carpenter, J. R. 1935b. Fluctuations in biotic communities. I. Prairie–forest ecotone of central Illinois. *Ecology* **16:** 203–12. * 3

262 Case, R. M. 1975. Interstate highway road-killed animals: a data source for biologists. *Wildlife Society Bulletin* **6:** 8–13. * 5

263 Caswell, H. 1976. Community structure: a neutral model analysis. *Ecological Monographs* **46:** 327–54. * 1, 2

264 Chaloupka, M. Y. & Domm, S. B. 1986. Role of anthropochory in the invasion of coral cays by alien flora. *Ecology* **67:** 1536–47. * 11

265 Chapman, K. 1979. *People, Pattern and Process: An Introduction to Human Geography.* Edward Arnold, London. * 9

266 Charlesworth, D. & Charlesworth, B. 1987. Inbreeding depression and its evolutionary consequences. *Annual Review of Ecology and Systematics* **18:** 237–68. * 2,11

267 Charnov, E. L. 1976. Optimal foraging: attack strategy of a mantid. *American Naturalist* **110:** 141–51. * 11

268 Chasko, G. G. & Gates, J. E. 1982. Avian habitat suitability along a transmission-line corridor in an oak-hickory forest region. *Wildlife Monographs* **82:** 1–41. * 3,5,12

269 Chatwin, B. 1987. *The Songlines.* Penguin Books, New York. * 1

270 Chauvet, E. & DeCamps, H. 1989. Lateral interactions in a fluvial landscape: the river Garonne, France. *Journal of the North American Benthological Society* **8:** 9–17. * 7

271 Chen, C. 1986. On geoecology. (In Chinese.) *Acta Ecologica Sinica* **6:** 289–94. * 9

272 Chen, C. & Zhang, L. 1985. The extremely arid deserts of China. *Chinese Journal of Arid Land Research* **1:** 37–46. * 12

273 Chen, J., Franklin, J. F. & Spies, T. A. 1992. Vegetation responses to edge environments in old-growth Douglas-fir forests. *Ecological Applications* **2:** 387–96. * 3, 12

274 Chepil, W. S. & Woodruff, N. P. 1963. The physics of wind erosion and its control. *Advances in Agronomy* **15:** 211–302. * 6

275 Ching, F. D. K. 1979. *Architecture: Form, Space and Order.* Van Nostrand Reinhold, New York. * 1

276 Chisholm, M. 1962. *Rural Settlement and Land Use: An Essay in Location.* Hutchinson, London. * 9

277 Chmielewski, T. J. 1988. O Strefowo-pasmowo-wezlowej Strukturze Ukladow Ponade-kosystemowych. (On zone-stripe-node structure of overecosystemic systems). *Wiadomosci Ekologiczne* **34:** 165–85. * 13

278 Christaller, W. 1933. *Die Zentralen Orte in Suddeutschland.* Gustav Fischer, Jena. * 1,4,7,9,11

279 Christensen, N. L. 1985. Shrubland fire regimes and their evolutionary consequences. In S. T. A. Pickett & P. S. White, eds. *The Ecology of Natural Disturbance and Patch Dynamics*, pp. 85–100. Academic Press, New York. * 10

280 Christian, C. S. & Stewart, G. A. 1953. *General Report on Survey of Katherine-Darwin Region, 1946.* Land Research Series Number 1, CSIRO, Melbourne. * 9

281 Chutter, R. M. 1969. The effects of silt and sand on the invertebrate fauna of streams and rivers. *Hydrobiologia* **34:** 57–76. * 5,7

282 Clapham, A. R. 1932. The form of the observational unit in quantitative ecology. *Journal of Ecology* **20:** 192–7. * 4

283 Clark, W. C. 1985. Scales of climate impacts. *Climatic Change* **7:** 5–27. * 14

284 Clark, W. C. 1989a. The human ecology of global change. *International Social Science Journal* **121:** 315–45. * 14

286 Clark, W. C. & Munn, R. E., eds. 1986. *Sustainable Development of the Biosphere.* Cambridge University Press, Cambridge. * 13, 14

287 Clarke, K. C. 1982. Fourier synthesis of choropleth data. *Michigan Academician* **14:** 337–45. * 4

288 Cliff, A. D., Haggett, P., Ord, J. K., Bassett, K. & Davies, R. B. 1975. *Elements of Spatial Structure: A Quantitative Approach*. Cambridge University Press, Cambridge. * 9

289 Cliff, A. D. & Ord, J. K. 1981. *Spatial Autocorrelation*. Pion, London. * 9

290 Clinnick, P. F. 1985. Buffer strip management in forest operations: a review. *Australian Forestry* 48: 34–45. * 5

291 Cocking, W. D. & Forman, R. T. T. 1982. Vegetation responses of an oldfield ecosystem to single and repeated sulfur dioxide disturbances. *Hutcheson Memorial Forest Bulletin* 6: 4–19. * 10

292 Coe, M. D. 1993. *The Maya*, 5th edn. Thames and Hudson, London. * 14

293 Cohen, J. E. & Newman, C. M. 1988. Dynamic basis of food web organization. *Ecology* 69: 1655–64. * 8

294 Cole, G. A. 1983. *Textbook of Limnology*. 3rd edn. Mosby, St. Louis, Missouri, USA. * 4,7

295 Coleman, S. W., Forbes, T. D. A. & Stuth, J. W. 1989. Measurements of the plant–animal interface in grazing research. In G. C. Marten, ed. *Grazing Research: Design, Methodology and Analysis*, pp. 37–52. American Society of Agronomy, Madison, Wisconsin. * 11

296 Collinge, S. K. & Louda, S. M. 1989. Influence of plant phenology on the insect herbivore/bittercress interaction. *Oecologia* 79: 111–6. * 3

297 Collins, B. R. & Russell, E. W. B., eds. 1988. *Protecting the New Jersey Pinelands: A New Direction in Land-use Management*. Rutgers University Press, New Brunswick, New Jersey, USA. * 9,14

298 Cone, C. D., Jr. 1962. Thermal soaring of birds. *American Scientist* 50: 180–209. * 10

301 Connell, J. H. & Keough, M. J. 1985. Disturbance and patch dynamics of subtidal marine animals on hard substrata. In S. T. A. Pickett & P. S. White, eds. *The Ecology of Natural Disturbance and Patch Dynamics*, pp. 125–51. Academic Press, New York. * 2

302 Connor, E. F. & McCoy, E. D. 1979. The statistics and biology of the species–area relationship. *American Naturalist* 113: 791–833. * 2

303 Connor, E. F., McCoy, E. D. & Cosby, B. J. 1983. Model discrimination and expected slope values in species–area studies. *American Naturalist* 122: 789–96. * 2

304 Conrad, G. W. 1981. Cultural materialism, split inheritance, and the expansion of ancient Peruvian empires. *American Antiquity* 46: 3–26. * 14

305 Constant, P., Eybert, M. C. & Maheo, R. 1976. Avifaune reproductrice du bocage de l'Ouest. In *Les Bocages: Histoire, Ecologie, Economie*, pp. 327–32. Institut National de la Recherche Agronomique, Centre National de la Recherche Scientifique, et Universite de Rennes, Rennes, France. * 8

306 Contreras, D., Gasto, J. & Cosio, F., eds. 1986. *Ecosistemas Pastorales de la Zona Mediterranea de Chile: Estudio de las Comunidades Agricolas de Carquindano y Yerba Loca del Secano Costero de la Region Coquimbo*. Publicaciones UNESCO, Montevideo, Uruguay. * 14

307 Cooper, J. R., Gilliam, J. W., Daniels, R. B. & Robarge, W. P. 1987. Riparian areas as filters for agricultural sediment. *Soil Science Society of America Journal* 51: 416–20. * 7

308 Cooper, J. R., Gilliam, J. W. & Jacobs, T. C. 1986. Riparian areas as a control of nonpoint pollutants. In D. C. Correll, ed. *Watershed Research Perspectives*, pp. 166–90. Smithsonian Institution Press, Washington, DC. * 7

309 Corbett, E. S., Lynch, J. A. & Sopper, W. E. 1978. Timber harvesting practices and water quality in the eastern United States. *Journal of Forestry* 76: 484–8. * 7, 13

310 Corbett, M. N. 1981. *A Better Place to Live: New Designs for Tomorrow's Communities*. Rodale Press, Emmaus, Pennsylvania, USA. * 14

311 Cordell, L. S. & Plog, F. 1979. Escaping the confines of normative thought: a reevaluation of Puebloan prehistory. *American Antiquity* 44: 405–29. * 14

312 Cornet, A. F., Montana, C., Delhoume, J. P. & Lopez-Portillo, J. 1992. Water flows and the dynamics of desert vegetation stripes. In A. J. Hansen & F. di Castri, eds. *Landscape Boundaries: Consequences for Biotic Diversity and Ecological Flows*, pp. 327–45. Springer-Verlag, New York. * 1,3,4

313 Correll, D. L. 1991. Human impact on the functioning of landscape boundaries. In M. M. Holland, P. G. Risser & R. J. Naiman, eds. *Ecotones: The Role of Landscape Boundaries*

537

in the Management and Restoration of Changing Environments, pp. 90–109. Chapman and Hall, New York. * 3

314 Correll, D. L., Goff, N. M. & Peterjohn, W. T. 1984. Ion balances between precipitation inputs and Rhode River watershed discharges. In O. P. Bricker, ed. *Geological Aspects of Acid Deposition*, pp. 77–111. Butterworth, Boston. * 1, 10

315 Costanza, R., Sklar, F. H. & Day, J. W., Jr. 1986. Modeling spatial and temporal succession in the Atchafalaya/Terrebonne marsh estuarine complex in South Louisiana. In D. A. Wolfe, ed. *Estuarine Variability*, pp. 387–404. Academic Press, New York. * 1, 10

316 Costanza, R., Sklar, F. H. & White, M. L. 1990. Modeling coastal landscape dynamics. *BioScience* **40**: 91–107. * 1, 10, 13, 14

317 Couclelis, H. 1985. Cellular worlds: a framework for modeling micro–macro dynamics. *Environment and Planning* A **17**: 585–96. * 1

318 Coulson, G. 1982. Road-kills of macropods on a section of highway in central Victoria. *Australian Wildlife Research* **9**: 21–6. * 5

319 Coulson, G. 1989. The effect of drought on road mortality of macropods. *Australian Wildlife Research* **16**: 79–83. * 5

320 Coulson, R. N., Lovelady, C. N., Flamm, R. O., Spradling, S. L., & Saunders, M. C. 1991. Intelligent geographic information systems for natural resource management. In M. G. Turner & R. H. Gardner, eds. *Quantitative Methods in Landscape Ecology*, pp. 153–72. Springer-Verlag, New York. * 1

321 Covich, A. P. 1976. Analyzing shapes of foraging areas: some ecological and economic theories. *Annual Review of Ecology and Systematics* **7**: 235–57. * 1, 4

322 Craighead, J. J., Atwell, G. & O'Gara, B. W. 1972. Elk migration in and near Yellowstone National Park. *Wildlife Monographs* **29**: 1–48. * 8,11

323 Crapper, P. F. 1981. Geometric properties of regions with homogeneous biophysical characteristics. *Australian Geographical Studies* **19**: 117–24. * 4

324 Craul, P. J. 1992. *Urban Soil in Landscape Design*. John Wiley, New York. * 1, 13

325 Crawley, M. J. 1986. The population biology of invaders. *Philosophical Transactions of the Royal Society of London* B **314**: 711–31. * 11

326 Crist, T. O., Guertin, D. S., Wiens, J. A. & Milne, B. T. 1992. Animal movement in heterogeneous landscapes: an experiment with *Eleodes* beetles in shortgrass prairie. *Functional Ecology* **6**: 536–44. * 1, 11

327 Crome, F. H. J. 1976. Some observations on the biology of the Cassowary in northern Queensland. *Emu* **76**: 8–14. * 2

328 Crome, F. H. J. & Bentrupperbaumer, J. 1993. Special people, a special animal and a special vision: the first steps to restoring a fragmented tropical landscape. In D. A. Saunders, R. J. Hobbs & P. R. Ehrlich, eds. *Nature Conservation 3: Reconstruction of Fragmented Ecosystems*, pp. 267–79. Surrey Beatty, Chipping Norton, Australia. * 2

329 Crome, F. H. J. & Moore, L. A. 1990. Cassowaries in north-eastern Queensland: report of a survey and a review and assessment of their status and conservation and management needs. *Australian Wildlife Research* **17**: 369–85. * 2

331 Crome, F. H. J. & Richards, G. C. 1988. Bats and gaps: microchiropteran community structure in a Queensland rainforest. *Ecology* **69**: 1960–9. * 5

332 Cronon, W. 1983. *Changes in the Land: Indians, Colonists, and the Ecology of New England*. Hill & Wang, New York. * 9

333 Crosby, A. W. 1986. *Ecological Imperialism, The Biological Expansion of Europe, 900–1900*. Cambridge University Press, Cambridge. * 11

334 Crow, T. R. 1991. Landscape ecology: the big picture approach to resource management. In D. J. Decker, M. E. Krasny, G. R. Goff, C. R. Smith & D. W. Gross, eds. *Challenges in the Conservation of Biological Resources: A Practitioner's Guide*, pp. 55–65. Westview Press, Boulder, Colorado, USA. * 13

336 Crowder, A. 1983. Impact indices based on introduced plant species and litter: a study of paths in St. Lawrence Islands National Park, Ontario, Canada. *Environmental Management* **7**: 345–54. * 5, 11

337 Crowell, K. L. 1973. Experimental zoogeography: introductions of mice to small islands. *American Naturalist* **107**: 535–58. * 2

338 Crowell, K. L. 1986. A comparison of relict versus equilibrium models for insular mammals of the Gulf of Maine. *Biological Journal of the Linnean Society* **28**: 37–64. * 2

339 Crowley, P. H. 1978. Effective size and the persistence of ecosystems. *Oecologia* **35**: 185–95. * 2

340 Crumley, C. L. 1987. Historical ecology. In C. L. Crumley & W. H. Marquardt, eds. *Regional Dynamics: Burgundian Landscapes in Historical Perspective*, pp. 237–64. Academic Press, New York. * 14

341 Crumley, C. L. & Marquardt, W. H., eds. 1987. *Regional Dynamics: Burgundian Landscapes in Historical Perspective*. Academic Press, New York. * 9,14

342 Cubbage, F. 1983. *Economics of Forest Tract Size: Theory and Literature*. General Technical Report SO-41, US Forest Service, New Orleans, Louisiana. 21 pp. * 2

343 Culver, D. C. 1970. Analysis of simple cave communities. I. Caves as islands. *Evolution* **24**: 463–74. * 2

344 Culver, D., Holsinger, J. R. & Baroody, R. 1973. Toward a predictive cave biogeography: the Greenbrier Valley as a case study. *Evolution* **27**: 689–95. * 2

345 Cummins, K. W., Minshall, G. W., Sedell, J. R., Cushing, C. E. & Petersen, R. C. 1984. Stream ecosystem theory. *Verhandlungen Internationale Vereinigung Limnologie* **22**: 1818–27. * 7

346 Curatolo, J. A. & Murphy, S. M. 1986. The effects of pipelines, roads, and traffic on the movements of Caribou *Rangifer tarandus*. *Canadian Field Naturalist* **100**: 218–24. * 5

347 Curtis, J. T. 1956. The modification of mid-latitude grasslands and forests by man. In W. L. Thomas, Jr., ed. *Man's Role in Changing the Face of the Earth*, pp. 721–36. University of Chicago Press, Chicago. * 12

348 Curtis, J. T. 1959. *The Vegetation of Wisconsin: An Ordination of Plant Communities*. University of Wisconsin Press, Madison, Wisconsin, USA. * 3

350 Curtis, J. T. & McIntosh, R. P. 1951. An upland forest continuum in the prairie–forest border region of Wisconsin. *Ecology* **32**: 476–96. * 3

351 Dahlsten, D. L. 1986. Control of invaders. In H. A. Mooney & J. A. Drake, eds. *Ecology of Biological Invasions of North America and Hawaii*, pp. 275–302. Springer-Verlag, New York. * 11

352 Daines, R. H. 1968. Sulfur dioxide and plant response. *Journal of Occupational Medicine* **10**: 516–24. * 10

353 Dale, V., Hemstrom, M. & Franklin, J. 1986. Modeling the long-term effects of disturbances on forest succession, Olympic Peninsula, Washington. *Canadian Journal of Forest Research* **16**: 56–67. * 12

354 Dale, V. H., O'Neill, R. V., Pedlowski, M. & Southworth, F. 1993. Causes and effects of land-use change in Central Rondonia, Brazil. *Photogrammetric Engineering & Remote Sensing* **59**: 997–1005. * 12

355 Daly, H. E. 1991. *Steady State Economics*. Island Press, Washington, DC. * 14

356 Dambach, C. A. & Good, E. E. 1940. The effect of certain land use practices on populations of breeding birds in southwestern Ohio. *Journal of Wildlife Management* **4**: 63–76. * 6, 9

357 Darley-Hill, S. & Johnson, W. C. 1981. Acorn dispersal by the blue jay (*Cyanocitta cristata*). *Oecologia* **50**: 231–2. * 11

358 Darlington, P. J. 1957. *Zoogeography: The Geographical Distribution of Animals*. John Wiley, New York. * 2

359 Dasmann, R. F. 1964. *Wildlife Biology*. John Wiley, New York. * 9

360 Daubenmire, R. 1968. *Plant Communities*. Harper and Row, New York. * 3

361 Davies, J. A. 1963. Albedo investigations in Labrador–Ungava. *Archives for Meteorology, Geophysics, and Bioclimotology, Series B* **13**: 137–51. * 10

362 Davis, A. & Glick, T. 1978. Urban ecosystems and island biogeography. *Environmental Conservation* **5**: 299–304. * 2,6

363 Davis, C. A. 1987. A strategy to save the Chesapeake shoreline. *Journal of Soil and Water Conservation* **42**: 72–5. * 7

364 Davis, J. C. 1986. *Statistics and Data Analysis in Geology*. 2nd edn. John Wiley, New York. * 1,4,9

365 Davis, J. M., Roper, T. J. & Shepherdson, D. J. 1987. Seasonal distribution of road kills in the European badger (*Meles meles*). *Journal of Zoology (London)* **211**: 525–9. * 5

366 Davis, K. 1989. Social science approaches to international migration. In M. S. Teitelbaum & J. M. Winter, eds. *Population and Resources in Western Intellectual Traditions*, pp. 245–61. Cambridge University Press, New York. * 3, 11

367 Davis, M. B. 1981. Quaternary history and the stability of forest communities. In D. C. West, H. H. Shugart & D. B. Botkin, eds. *Forest Succession: Concepts and Application*, pp. 132–53. Springer-Verlag, New York. * 3,8,10,14

368 Davis, M. B. 1989. Lags in vegetation response to greenhouse warming. *Climatic Change* **15**: 75–82. * 3, 14

369 Davis, M. B. 1990. Insights from paleoecology on global change. *Bulletin of the Ecological Society of America* **71**: 222–8. * 14

370 Davis, M. B., Woods, K. D., Webb, S. L. & Futyma, R. P. 1986. Dispersal versus climate: expansion of *Fagus* and *Tsuga* into the upper Great Lakes region. *Vegetatio* **67**: 93–103. * 14

371 Dawson, B. L. 1991. South African road reserves: valuable conservation areas? In D. A. Saunders & R. J. Hobbs, eds. *Nature Conservation 2: The Role of Corridors*, pp. 119–30. Surrey Beatty, Chipping Norton, Australia. * 5

372 Dawson, B. L. & van der Breggen, J. P. 1991. Re-establishment and maintenance of indigenous vegetation in South African road reserves. In D. A. Saunders & R. J. Hobbs, eds. *Nature Conservation 2: The Role of Corridors*, pp. 327–32. Surrey Beatty, Chipping Norton, Australia. * 5

373 Day, T. A. & Detling, J. K. 1990. Grassland patch dynamics and herbivore grazing preference following urine deposition. *Ecology* **71**: 180–8. * 11

374 DeAngelis, D. L., Waterhouse, J. C., Post, W. M. & O'Neill, R. V. 1985. Ecological modeling and disturbance evaluation. *Ecological Modeling* **29**: 399–419.* 1

375 Debach, P. & Rosen, D. 1991. *Biological Control by Natural Enemies*, 2nd edn. Cambridge University Press, Cambridge.* 11

376 Debussche, M., Escarre, J. & Lepart, J. 1982. Ornithochory and plant succession in Mediterranean abandoned orchards. *Vegetatio* **48**: 255–66.* 1,11

377 Decamps, H. 1984. Towards a landscape ecology of river valleys. In J. H. Cooley & F. B. Golley, eds. *Trends in Ecological Research for the 1980s*, pp. 163–78. NATO Series, Plenum Press, New York.* 5,7

378 Decamps, H. & Naiman, R. J. 1989. L'ecologie des fleuves. *La Recherche* **20**: 310–19.* 7

379 DeCola, L. 1989. Fractal analysis of a classified Landsat scene. *Photogrammetric Engineering and Remote Sensing* **55**: 601–10.* 3

380 DeGraaf, R. M. & Rudis, D. D. 1987. *New England Wildlife: Habitat, Natural History, and Distribution*. General Technical Report NE–108, U.S.D.A. Forest Service, Radnor, Pennsylvania.* 11,13

381 DeGrove, J. M. 1984. *Land, Growth and Politics*. APA Planners Press, Washington, DC.* 13

382 Delcourt, H. R. & Delcourt, P. A. 1988. Quaternary landscape ecology: relevant scales in space and time. *Landscape Ecology* **2**: 23–44.* 1, 10, 14

383 Delcourt, H. R., Delcourt, P. A. & Webb, T, III. 1983. Dynamic plant ecology: the spectrum of vegetational change in space and time. *Quaternary Science Reviews* **1**: 153–75.* 1

384 Delcourt, P. A. & Delcourt, H. R. 1992. Ecotone dynamics in space and time. In A. J. Hansen & F. di Castri, eds. *Landscape Boundaries: Consequences for Biotic Diversity and Ecological Flows*, pp. 19–54. Springer-Verlag, New York.* 1, 3, 14

385 Delwiche, L. L. D. & Haith, D. A. 1983. Loading functions for predicting nutrient losses from complex watersheds. *Water Resources Bulletin* **19**: 951–9.* 7

386 Demek, J. 1977. *Tieorija Sistiem i Izuczenije Landszafta*. Progriess, Moskwa.* 4

387 den Boer, P. J. 1981. On the survival of populations in a heterogeneous and variable environment. *Oecologia* **50**: 39–53.* 11

388 Denmead, O. T. 1969. Comparative micrometeorology of a wheat field and a forest of *Pinus radiata*. *Agricultural Meteorology* **6**: 357–71.* 3,10

389 Denslow, J. S. 1980. Gap partitioning among rainforest trees. *Biotropica* 12 (Supplement to Number 2): 47–55.* 2

390 Desroche, H. 1971. *The American Shakers: From Neo-Christianity to Presocialism*. University of Massachusetts Press, Amherst, Massachusetts, USA.* 14

391 Detling, J. K. 1988. Grasslands and savannas: regulation of energy flow and nutrient cycling by herbivores. In L. R. Pomeroy & J. J. Alberts, eds. *Concepts of Ecosystem Ecology: A Comparative View*, pp. 131–48. Springer-Verlag, New York.* 11

392 Deveaux, D. 1976. Repartition et diversite des peuplements en carabiques en zone bocagere et arasee. In *Les Bocages: Histoire, Ecologie, Economie*, pp. 377–84. Institut National de la Recherche Agronomique, Centre National de la Recherche Scientifique, et Universite de Rennes, Rennes, France. * 8

393 deVos, A. 1949. Timberwolves (*Canis lupus lycaon*) killed by cars on Ontario highways. *Journal of Mammalogy* **30:** 197. * 5

394 Dhindsa, M. S., Sandhu, J. S., Sandhu, P. S. & Toor, H. S. 1988. Roadside birds in Punjab (India): relation to mortality from vehicles. *Environmental Conservation* **15:** 303–10. * 5

395 Diamond, J. M. 1972. Biogeographic kinetics: estimation of relaxation times for avifaunas of Southwest Pacific Islands. *Proceedings of the National Academy of Science (USA)* **69:** 3199–203. * 2

396 Diamond, J. M. 1973. Distributional ecology of New Guinea birds. *Science* **179:** 759–69. * 2, 3

397 Diamond, J. M. 1975. The island dilemma: lessons of modern biogeographic studies for the design of natural reserves. *Biological Conservation* **7:** 129–46. * 2

398 Diamond, J. M. 1984. 'Normal' extinctions of isolated populations. In M. H. Nitecki, ed. *Extinctions*, pp. 191–246. University of Chicago Press, Chicago. * 11, 12

399 Diamond, J. M. & Keegan, W. F. 1984. Supertramps at sea. *Nature* **311:** 704–5. * 2

400 Diamond, J. M. & May, R. M. 1981. Island biogeography and the design of natural reserves. In R. M. May, ed. *Theoretical Ecology*, 2nd edn., pp. 228–52. Blackwell, Oxford. * 1, 2

401 Diana, J. S. & Lane, E. D. 1978. The movement and distribution of Paiute Cutthroat Trout (*Salmo clarki seleniris*) in Cottonwood Creek, California. *Transactions of the American Fisheries Society* **107:** 444–8. * 8

402 Diaz, N. & Apostl, D. 1992. *Forest Landscape Analysis and Design: A Process for Developing and Implementing Land Management Objectives for Landscape Patterns*. US Forest Service, Report R6-ECO-TP-043–92, Portland, Oregon. * 1, 13

403 di Castri, F., Hansen, A. J. & Holland, M. M., eds. 1988. *A New Look at Ecotones: Emerging International Projects on Landscape Boundaries*. Biology International, Special Issue 17, International Union of Biological Sciences, Paris. 163 pp. * 3

404 Dickert, T. G. & Tuttle, A. E. 1985. Cumulative impact assessment in environmental planning: a coastal wetland watershed example. *Environmental Impact Assessment Review* **5:** 37–64. * 13

405 Dickinson, R. E. 1970. *Regional Ecology: The Study of Man's Environment*. John Wiley, New York. * 1

406 Dickstein, A. J., Erramilli, S., Goldstein, R. E., Jackson, D. P. & Langer, S. A. 1993. Labyrinthine pattern formation in magnetic fluids. *Science* **261:** 1012–15. * 4

407 Diershke, H. 1974. *Saumgesellschaften im Vegetations- und Standortsgefalle an Waldrandern*. Verlag Erich Goltze KG, Gottingen, Germany. * 2, 3

408 Dietrich, W. E. & Dunne, T. 1978. Sediment budget for a small catchment in mountainous terrain. *Zeitschrift fur Geomorphologie, Supplementbaende* **29:** 191–206. * 9

409 Dietz, D. R. & Tigner, J. R. 1968. Evaluation of two mammal repellents applied to browse species in the Black Hills. *Journal of Wildlife Management* **32:** 109–14. * 5

410 Diggle, P. J. 1983. *Statistical Analysis of Spatial Point Patterns*. Academic Press, London. * 1

411 Dix, M. E. & Leatherman, D. 1988. Insect management in windbreaks. *Agriculture, Ecosystems, and Environment* **22/23:** 513–37. (Reprinted 1988 in *Windbreak Technology*. Elsevier, Amsterdam.) * 6

412 Dmowski, K. & Kozakiewicz, M. 1990. Influence of a shrub corridor on movements of passerine birds to a lake littoral zone. *Landscape Ecology* **4:** 98–108. * 5,6

413 Dorrance, M. F., Savage, P. J. & Huff, D. E. 1975. Effects of snowmobiles on white-tailed deer. *Journal of Wildlife Management* **39:** 563–9. * 5

414 Douglas, I. 1983. *The Urban Environment*. Edward Arnold, London. * 6

415 Douglas, M. S. 1947. *The Everglades: River of Grass*. Rinehart, New York, * 14

417 Dowler, R. C. & Swanson, G. A. 1982. High mortality of Cedar Waxwings associated with highway plantings. *Wilson Bulletin* **94**: 602–3. * 5

418 Dowling, J. E. 1992. *Neurons and Networks: An Introduction to Neuroscience*. Harvard University Press, Cambridge. * 8

419 Drake, J. A., Mooney, H. A., di Castri, F., Groves, R. H., Kruger, F. J., Rejmanek, M. & Williamson, M., eds. 1989. *Biological Invasions: A Global Perspective*. John Wiley, Chichester, UK. * 6, 11

420 Dregne, H. E. 1984. North American deserts. In F. El-Baz, ed. *Deserts and Arid Lands*, pp. 147–55. Martinus Nijhoff Publishers, The Hague, Netherlands. * 12, 14

421 Dreyer, G. D. & Niering, W. A. 1986. Evaluation of two herbicide techniques on electric transmission rights-of-way: development of relatively stable shrublands. *Environmental Management* **10**: 113–18. * 5

422 Dronen, S. I. 1988. Layout and design criteria for livestock windbreaks. *Agriculture, Ecosystems and Environment* **22/23**: 231–40. (Reprinted 1988 in *Windbreak Technology*. Elsevier, Amsterdam.) * 6

423 Duchelle, S., Skelly, J. M., Sharich, T. L., Chevone, B. I., Yang, Y. S. & Nellessen, J. E. 1983. Effects of ozone on the productivity of natural vegetation in a high meadow of the Shenandoah National Park of Virginia. *Journal of Environmental Management* **17**: 299–308. * 10

424 Duda, R. O. & Hart, P. E. 1973. *Pattern Classification and Scene Analysis*. John Wiley, New York. * 4

425 Dufour, J. 1976. Un bocage tardif et ephemere: le bocage de la Champagne de Conlie (Nord de la Champagne Mancelle). In *Les Bocages: Histoire, Ecologie, Economie*, pp. 49–54. Institut National de la Recherche Agronomique, Centre National de la Recherche Scientifique, et Universite de Rennes, Rennes, France. * 8

426 Dunford, E. G. & Fletcher, P. W. 1947. Effect of removal of stream-bank vegetation upon water yield. *Transactions of the American Geophysical Union* **28**: 105–10. * 7

427 Dunn, C. P. & Loehle, C. 1988. Species–area parameter estimation: testing the null model of lack of relationship. *Journal of Biogeography* **15**: 721–8. * 2

428 Dunn, C. P., Sharpe, D. M., Guntenspergen, G. R., Stearns, F. & Yang, Z. 1991. Methods for analyzing temporal changes in landscape pattern. In M. G. Turner & R. H. Gardner, eds. *Quantitative Methods in Landscape Ecology*, pp. 173–98. Springer-Verlag, New York. * 1, 12

429 Dunne, T. & Leopold, L. B. 1978. *Water in Environmental Planning*. W. H. Freeman, San Francisco. * 7

430 Dunning, J. B., Danielson, B. J. & Pulliam, H. R. 1992. Ecological processes that affect populations in complex landscapes. *Oikos* **65**: 169–75. * 9, 11

431 Durand, J. B. 1979. Nutrient and hydrological effects of the Pine Barrens on neighboring estuaries. In R. T. T. Forman, ed. *Pine Barrens: Ecosystem and Landscape*, pp. 195–211. Academic Press, New York. * 7

432 Durner, G. M. & Gates, J. E. 1993. Spatial ecology of black rat snakes on Remington Farms, Maryland. *Journal of Wildlife Management* **57**: 812–26. * 6

433 Eckersley, R. 1989. *Regreening Australia: The Environmental, Economic and Social Benefits of Reforestation*. CSIRO, Canberra. * 14

434 Ecologically Sustainable Development: A Commonwealth Discussion Paper. 1990. Australian Government Publishing Service, Canberra. 41 pp. * 14

435 Edminster, F. C. 1938. Woody vegetation for fence rows. *Soil Conservation* **4**: 99–101. * 6

436 Edwards, A. M. 1981. *The Design of Suburbia: A Critical Study in Environmental History*. Pembridge Press, London. * 12, 13

437 Ehmann, H. & Cogger, H. 1985. Australia's endangered herpetofauna: a review of criteria and policies. In G. Grigg, R. Shine & H. Ehmann, eds. *Biology of Australasian Frogs and Reptiles*, pp. 435–47. Surrey Beatty and Royal Zoological Society of New South Wales, Sydney. * 5

438 Ehrenfeld, D. 1987. Implementing the transition to a sustainable agriculture: an opportunity for ecology. *Bulletin of the Ecological Society of America* **68**: 5–8. * 14

439 Ehrlich, P. R. 1986. Which animal will invade? In H. A. Mooney & J. A. Drake, eds. *Ecology of Biological Invasions of North America and Hawaii*, pp. 79–95. Springer-Verlag, New York. * 11

440 Ehrlich, P. R. & Ehrlich, A. H. 1981. *Extinction: The Causes and Consequences of the Disappearance of Species*. Random House, New York. * 9

441 Ehrlich, P. R. & Mooney, H. A. 1983. Extinction, substitution, and ecosystem services. *BioScience* **33:** 248–54. * 9

442 Ehrlich, R. & Weinberg, B. 1970. An exact method for characterization of grain shape. *Journal of Sedimentary Petrology* **40:** 205–12. * 4

443 Eide, S., Miller, S. & Chihuly, M. 1988. Oil pipeline crossing sites utilized in winter by moose, *Alces alces*, and caribou, *Rangifer tarandus*, in southcentral Alaska. *The Canadian Field Naturalist* **100:** 197–207. * 5

444 Eisner, T. 1991. Chemical prospecting: a proposal for action. In F. H. Bormann & S. R. Kellert, eds. *Ecology, Economics and Ethics: The Broken Circle*, pp. 196–202. Yale University Press, New Haven, Connecticut. * 2

445 Eisner, T. 1992. The hidden value of species diversity. *BioScience* **42:** 578 * 2

446 Ekern, P. C. 1964. Direct interception of cloud water on Lanaihale, Hawaii. *Soil Science Society of America Proceedings* **28:** 419–21. * 10

447 Elfring, C. 1986. Wildlife and the National Park Service. In R. L. DiSilvestro, ed. *Audubon Wildlife Report 1986*, pp. 462–94. National Audubon Society, New York. * 3

448 Elfstrom, B. A. 1974. Tree Species Diversity and Forest Island Size on the Piedmont of New Jersey. Masters thesis, Rutgers University, New Brunswick, New Jersey. 73 pp. * 2

449 Ellenberg, H. 1986. *Vegetation Ecology of Central Europe*, 4th edn. Cambridge University Press, Cambridge. * 3

450 Ellstrand, N. C. 1992. Gene flow by pollen: implications for plant conservation genetics. *Oikos* **63:** 77–86. * 2,11

451 Elton, C. S. 1958. *The Ecology of Invasions by Animals and Plants*. Methuen, London. * 6,11

452 Engstrom, D. E. & Hanssen, B. S. 1985. Post-glacial vegetational change in southeastern Labrador as inferred from pollen and chemical stratigraphy. *Canadian Journal of Botany* **63:** 543–61. * 10

453 Erickson, R. O. 1945. *The Clematis Fremontii var. Riehlii* population in the Ozarks. *Annals of the Missouri Botanical Garden* **32:** 413–60. * 1,4

454 Erickson, R. O. 1966. Relative elemental rates and anisotropy of growth in area: a computer programme. *Journal of Experimental Botany* **17:** 51, 390–403. * 4

455 Eriksson, E. 1955. Air borne salts and the chemical composition of river waters. *Tellus* **7:** 243–50. * 10

456 Erman, D. C. & Hawthorne, V. M. 1976. The quantitative importance of an intermittent stream in the spawning of rainbow trout. *Transactions of the American Fisheries Society* **105:** 675–81. * 7

457 Ervin, D. E. & Dicks, M. R. 1988. Cropland diversion for conservation and environmental improvement: an economic welfare analysis. *Land Economics* **64:** 256–68. * 14

458 Ervin, S. M. 1992. Intra-medium and inter-media constraints. In G. N. Schmitt, ed. *CAAD Futures, '91: Computer Aided Architectural Design Futures; Education, Research, Applications*, pp. 365–80. Vieweg, Braunscheig, Germany. * 1,9

459 Etienne, M., Caviedes, E. & Prado, C. 1983. *Bases Ecologiques du Developpement de la Zone Aride Mediterranee du Chili*. Centre National de la Recherche Scientifique et Centre d'Etudes Ecologiques et Phytosociologiques, Montpellier, France. * 14

460 Everest, F. H., Armantrout, N. B., Keller, S. M., Parante, W. D., Sedell, J. R., Nickelson, T. E., Johnston, J. M. & Haugen, G. N. 1982. *Salmonids Westside Forest–Wildlife Habitat Relationship Handbook*. US Forest Service, Portland, Oregon. * 7

461 Fabos, J. G. 1979. *Planning the Total Landscape: A Guide to Intelligent Land Use*. Westview Press, Boulder, Colorado, USA. * 12,13

462 Faeth, S. H. & Kane, T. C. 1978. Urban biogeography: city parks as islands for Diptera and Coleoptera. *Oecologia* **32:** 127–33. * 2,4

463 Fahrig, L. 1988. A general model of populations in patchy habitats. *Applied Mathematics and Computation* **27:** 53–66. * 11

464 Fahrig, L. 1991. Simulation methods for developing general landscape-level hypotheses of single-species dynamics. In M. G. Turner & R. H. Gardner, eds. *Quantitative Methods in Landscape Ecology*, pp. 417–42. Springer-Verlag, New York. * 11

465 Fahrig, L. & Merriam, G. 1985. Habitat patch connectivity and population survival. *Ecology* **66:** 1762–8. * 1,5,6,8,12

466 Fahrig, L. & Paloheimo, J. 1988. Effect of spatial arrangement of habitat patches on local population size. *Ecology* **69**: 468–75. * 9,11

467 Falk, N. W., Graves, H. B. & Bellis, E. D. 1978. Highway right-of-way fences as deer deterrents. *Journal of Wildlife Management* **42**: 646–50. * 5

468 Fearnside, P. M. 1986. *Human Carrying Capacity of the Brazilian Rainforest*. Columbia University Press, New York. * 12

469 Feinsinger, P., Swarm, L. A. & Wolfe, J. A. 1985. Nectar-feeding birds on Trinidad and Tobago: comparison of diverse and depauperate guilds. *Ecological Monographs* **55**: 1–28. * 9

470 Feldhamer, G. A., Gates, J. E., Harman, D. M., Loranger, A. J. & Dixon, K. R. 1986. Effects of interstate highway fencing on white-tailed deer activity. *Journal of Wildlife Management* **50**: 497–503. * 5

471 Feller, M. C. 1981. Effects of clearcutting and slashburning on stream temperature in southwestern British Columbia. *Water Resources Bulletin* **17**: 863–7. * 7

472 Fergus, C. 1991. The Florida panther verges on extinction. *Science* **251**: 1178–80. * 2,11

473 Ferguson, J. D., Connelly, M., Forman, R., Kellett, M., Mackenzie, C., Monahan, D., Schnitzer, S., Sprott, J. & Stokey, B. 1993. *Town of Concord 1992 Open Space Plan*. Concord Natural Resources Commission, Concord, Massachusetts, USA. * 13

474 Ferguson, M. A. D. & Keith, L. B. 1982. Influence of nordic skiing on distribution of moose and elk in Elk Island National Park, Alberta. *Canadian Field-Naturalist* **96**: 69–78. * 5

475 Ferris, C. R. 1979. Effects of Interstate 95 on breeding birds in northern Maine. *Journal of Wildlife Management* **43**: 421–7. * 5

476 *Final Report–Forest Use*. 1991. Ecologically Sustainable Development Working Groups, Australian Government Publishing Service, Canberra. 227 pp. * 14

477 Finean., J. B., Coleman, R. & Michell, R. H. 1984. *Membranes and Their Cellular Functions*. Blackwell, Oxford. * 3

478 Flyger, V. & Gates, J. 1982. Fox and gray squirrels. In J. A. Chapman & G. A. Feldhamer, eds. *Wild Mammals of North America*, pp. 209–29. Johns Hopkins University Press, Baltimore, Maryland, USA. * 6

479 Foin, T. C., Jr. 1976. *Ecological Systems and the Environment*. Houghton Mifflin, Boston. * 1

480 Forbes, A. M. 1992. *Concord Survey of Historic, Architectural, and Cultural Resources*. Concord Historical Commission, Concord, Massachusetts, USA. * 13

481 Fore, S. A., Hickey, R. J., Vankat, J. L., Guttman, S. I. & Schaefer, R. L. 1992. Genetic structure after forest fragmentation: a landscape ecology perspective on *Acer saccharum*. *Canadian Journal of Botany* **70**: 1659–68. * 1,11,12

482 Forman, R. T. T. 1964. Growth under controlled conditions to explain the hierarchical distributions of a moss, *Tetraphis pellucida*. *Ecological Monographs* **34**: 1–25. * 1,4

483 Forman, R. T. T. 1979a. The Pine Barrens of New Jersey: an ecological mosaic. In R. T. T. Forman, ed. *Pine Barrens: Ecosystem and Landscape*, pp. 569–85. Academic Press, New York. * 1,9,10

484 Forman, R. T. T., ed. 1979b. *Pine Barrens: Ecosystem and Landscape*. Academic Press, New York. * 1,10,14

485 Forman, R. T. T. 1981. Interaction among landscape elements: a core of landscape ecology. In S. P. Tjallingii & A. A. de Veer, eds. *Perspectives in Landscape Ecology*, pp. 35–48. PUDOC, Wageningen, Netherlands. * 2,3,4,12

486 Forman, R. T. T. 1983. Corridors in a landscape: their ecological structure and function. *Ekologia (Czechoslovakia)* **2**: 375–87. * 5,6

487 Forman, R. T. T. 1983. An ecology of the landscape. *BioScience* **33**: 535. * 1

488 Forman, R. T. T. 1986. Emerging directions in landscape ecology and applications in natural resource management. In R. Herrmann & T. Bostedt-Craig, eds. *Proceedings of the Conference on Science in the National Parks: The Plenary Sessions*, vol. 1, pp. 59–88. US National Park Service and The George Wright Society, Fort Collins, Colorado. * 1,9,13,14

489 Forman, R. T. T. 1987. The ethics of isolation, the spread of disturbance, and landscape ecology. In M. G. Turner, ed. *Landscape Heterogeneity and Disturbance*, pp. 213–29. Springer-Verlag, New York. * 9,10,13,14

544

490 Forman, R. T. T. 1989. Landscape ecology plans for managing forests. In *Proceedings of the Society of American Foresters 1988 National Convention*, pp. 131–6. Society of American Foresters, Bethesda, Maryland. (Reprinted 1990 in R. M. DeGraaf & W. M. Healy, compilers. *Is Forest Fragmentation a Management Issue in the Northeast?* pp. 27–32. General Technical Report NE-140, US Forest Service, Radnor, Pennsylvania.) * 9,13

491 Forman, R. T. T. 1990a. The beginnings of landscape ecology in America. In I. S. Zonneveld & R. T. T. Forman, eds. *Changing Landscapes: An Ecological Perspective*, pp. 35–41. Springer-Verlag, New York. * 1

492 Forman, R. T. T. 1990b. Ecologically sustainable landscapes: the role of spatial configuration. In I. S. Zonneveld & R. T. T. Forman, eds. *Changing Landscapes: An Ecological Perspective*, pp. 261–78. Springer-Verlag, New York. * 1,9,10,13,14

493 Forman, R. T. T. 1991. Landscape corridors: from theoretical foundations to public policy. In D. A. Saunders & R. J. Hobbs, eds. *Nature Conservation 2: The Role of Corridors*, pp. 71–84. Surrey Beatty, Chipping Norton, Australia. * 1,5,6,8,9,12,13

494 Forman, R. T. T. 1993. An 'aggregate-with-outliers' land planning principle, and the major attributes of a sustainable environment. In *Proceedings of the International Conference on Landscape Planning and Environmental Conservation*, pp. 71–95. University of Tokyo, Tokyo. * 13,14

495 Forman, R. T. T. & Baudry, J. 1984. Hedgerows and hedgerow networks in landscape ecology. *Environmental Management* 8: 495–510. * 6,9

496 Forman, R. T. T. & Boerner, R. E. J. 1981. Fire frequency and the Pine Barrens of New Jersey. *Bulletin of the Torrey Botanical Club* 108: 34–50. * 1,4,5,10

497 Forman, R. T. T. & Collinge, S. K. 1995. The 'spatial solution' to conserving biodiversity in landscapes and regions. In R. M. DeGraaf & R. I. Miller, eds. *Conservation of Faunal Diversity in Forested Landscapes*. Chapman & Hall, London. In press. * 11,12,13,14

498 Forman, R. T. T., Galli, A. E. & Leck, C. F. 1976. Forest size and avian diversity in New Jersey woodlots with some land use implications. *Oecologia* 26: 1–8. * 2,12,13

499 Forman, R. T. T. & Godron, M. 1981. Patches and structural components for a landscape ecology. *BioScience* 31: 733–40. * 1,2,5,6,9

500 Forman, R. T. T. & Godron, M. 1984. Landscape ecology principles and landscape function. In J. Brandt & P. Agger, eds. *Methodology in Landscape Ecological Research and Planning*, vol. 5, pp. 4–15. Roskilde Universitetsforlag GeoRuc, Roskilde, Denmark. * 5,8

501 Forman, R. T. T. & Godron, M. 1986. *Landscape Ecology*. John Wiley, New York. * Preface,1,2,3,4,5,6,8,9,10,11,12,13,14

502 Forman, R. T. T. & Hahn, D. C. 1980. Spatial patterns of trees in a Caribbean semievergreen forest. *Ecology* 61: 1267–74. * 2

503 Forman, R. T. T. & Moore, P. N. 1992. Theoretical foundations for understanding boundaries in landscape mosaics. In A. J. Hansen & F. di Castri, eds. *Landscape Boundaries: Consequences for Biotic Diversity and Ecological Flows*, pp. 236–58. Springer-Verlag, New York. * 1,3,4,5,8,9,12,13

504 Forney, K. A. & Gilpin, M. E. 1989. Spatial structure and population extinction: a study with *Drosophila* flies. *Conservation Biology* 3: 45–51. * 1,2

505 Forsman, E. D., Meslow, E. C. & Wight, H. M. 1984. Distribution and biology of the spotted owl in Oregon. *Wildlife Monographs* 87: 1–64. * 9

506 Foster, D. R. 1983. The history and pattern of fire in the boreal forest of southeastern Labrador. *Canadian Journal of Botany* 61: 2459–71. * 4,9,10

507 Foster, D. R. 1988a. Disturbance history, community organization and vegetation dynamics of the old-growth Pisgah Forest, south-western New Hampshire, U.S.A. *Journal of Ecology* 76: 105–34. * 10

508 Foster, D. R. 1988b. Species and stand response to catastrophic wind in central New England, U.S.A. *Journal of Ecology* 76: 135–51. * 10

509 Foster, D. R. 1992. Land-use history (1730–1990) and vegetation dynamics in central New England, USA. *Journal of Ecology* 80: 753–72. * 12,14

510 Foster, D. R. & Boose, E. 1992. Patterns of forest damage resulting from catastrophic wind in central New England, USA. *Journal of Ecology* 80: 79–99. * 1,10,12

511 Foster, D. R. & Zebryk, T. M. 1993. Long-term vegetation dynamics and disturbance history of a *Tsuga*-dominated forest in New England. *Ecology* 74: 982–98. * 14

545

512 Fox, H. S. A. 1976. The functioning of bocage landscapes in Devon and Cornwall between 1500 and 1800. In *Les Bocages: Histoire, Ecologie, Economie*, pp. 55–61. Institut National de la Recherche Agronomique, Centre National de la Recherche Scientifique, et Universite de Rennes, Rennes, France. * 8

513 Frank, A. B. & George, E. J. 1975. *Windbreaks for Snow Management in North Dakota*, pp. 144–54. Publication 73, Great Plains Agricultural Council, Fargo, North Dakota, USA. * 6

514 Frankel, O. H. & Soulé, M. E. 1981. *Conservation and Evolution*. Cambridge University Press, Cambridge. * 2,6,11

515 Franklin, I. R. 1980. Evolutionary change in small populations. In M. E. Soulé & B. A. Wilcox, eds. *Conservation Biology: An Evolutionary–Ecological Perspective*, pp. 135–49. Sinauer Associates, Sunderland, Massachusetts, USA. * 2,11

516 Franklin, J. F. 1993. Preserving biodiversity: species, ecosystems, or landscapes? *Ecological Applications* **3:** 202–5. * 2,13

517 Franklin, J. F. & Dyrness, C. T. 1988. *Natural Vegetation of Oregon and Washington*. Oregon State University Press, Covallis, Oregon, USA. * 9

518 Franklin, J. F. & Forman, R. T. T. 1987. Creating landscape patterns by forest cutting: ecological consequences and principles. *Landscape Ecology* **1:** 5–18. * 1, 3, 8, 10, 11, 12, 13, 14

519 Free, J. B., Gerrard, D., Stevenson, J. H. & Williams, I. H. 1975. Beneficial insects present on a motorway verge. *Biological Conservation* **8:** 61–72. * 5

520 Free, J. B. & Williams, I. H. 1980. The value of white clover *Trifolium repens* L. cultivar S100 planted on motorway verges to honeybees *Apis mellifera* L. *Biological Conservation* **18:** 89–92. * 5

521 Freeman, C. E. & Dick-Peddie, W. A. 1970. Woody riparian vegetation in the Black and Sacramento mountain ranges, southern New Mexico. *Southwestern Naturalist* **15:** 145–64. * 7

522 Freemark, K. E. 1990. Landscape ecology of forest birds in the Northeast. In R. M. DeGraaf & W. M. Healy, compilers. *Is Forest Fragmentation a Management Issue in the Northeast?* pp. 7–12. General Technical Report NE-140, US Forest Service, Radnor, Pennsylvania. * 2

523 Freemark, K. E. & Collins, B. 1989. Landscape ecology of birds breeding in temperate forest fragments. In J. M. Hagan III & D. W. Johnston, eds. *Ecology and Conservation of Neotropical Migrant Landbirds*, pp. 443–54. Smithsonian Institution Press, Washington, DC. * 2

524 Freemark, K. E. & Merriam, H. G. 1986. Importance of area and habitat heterogeneity to bird assemblages in temperate forest fragments. *Biological Conservation* **36:** 115–41. * 2

525 Fretwell, S. D. & Lucas, H. L. 1970. On territorial behaviour and other factors influencing habitat distribution in birds. *Acta Biotheoretica* **19:** 16–36. * 11

526 Frissell, C. A., Liss, W. J., Warren, C. E. & Hurley, M. C. 1986. A hierarchical framework for stream habitat classification: viewing streams in a watershed context. *Environmental Management* **10:** 199–214. * 7, 8

527 Fritschen, L. J., Balick, L. K. & Smith, J. A. 1982. Interpretation of infrared nighttime imagery of a forested canopy. *Journal of Applied Meteorology* **21:** 730–4. * 10

528 Fritz, R. S. 1979. Consequences of insular population structure: distribution and extinction of spruce grouse populations. *Oecologia* **42:** 57–65. * 11

529 Frolov, Y. S. 1975. Measuring the shape of geographical phenomena: a history of the issue. *Soviet Geography: Review and Translation* **16:** 676–87. *4

530 Froment, A. & Wildmann, B. 1987. Landscape ecology and rural restructuring in Belgium. *Landscape and Urban Planning* **14:** 415–26. * 1

531 Fry, G. L. A. 1989. Conservation in agricultural ecosystems. In I. F. Spellerberg, F. B. Goldsmith & M. G. Morris, eds. *The Scientific Management of Temperate Communities for Conservation*, pp. 415–43. Blackwell Scientific, Oxford. * 3, 5, 8, 12, 13

532 Fry, G. & Main, A. R. 1993. Restoring seemingly natural communities on agricultural land. In D. A. Saunders, R. J. Hobbs & P. R. Ehrlich, eds. *Nature Conservation 3: Reconstruction of Fragmented Ecosystems*, pp. 225–41. Surrey Beatty, Chipping Norton, Australia. * 11,12

533 Fu, B. 1989. Soil erosion and its control in the loess plateau of China. *Soil Use and Management* **5**: 76–81. * 1

534 Fu, C. 1992. Transitional climate zones and biome boundaries: a case study from China. In A. J. Hansen and F. di Castri, eds. *Landscape Boundaries: Consequences for Biotic Diversity and Ecological Flows*, pp. 394–402. Springer-Verlag, New York. * 3

535 Fuentes, E. R. 1990. Landscape change in Mediterranean-type habitats of Chile: patterns and processes. In I. S. Zonneveld & R. T. T. Forman, eds. *Changing Landscapes: An Ecological Perspective*, pp. 165–90. Springer-Verlag, New York. * 1,9,14

536 Gadgil, M. 1987. Culture, perceptions and attitudes to the environment. In P. Jacobs & D. A. Munro, eds. *Conservation with Equity: Strategies for Sustainable Development*, pp. 85–94. Columbia University Press, New York. * 14

537 Gaines, M. S. & McClenaghan, L. R., Jr. 1980. Dispersal in small mammals. *Annual Review of Ecology and Systematics* **11**: 163–96. * 11

538 Gale, D. E. 1992. Eight state-sponsored growth management programs: a comparative analysis. *Journal of the American Planning Association* **58**: 425–39. * 13

539 Galli, A. E., Leck, C. F. & Forman, R. T. T. 1976. Avian distribution patterns in forest islands of different sizes in central New Jersey. *Auk* **93**: 356–64. * 2,3

540 Game, M. 1980. Best shape for nature reserves. *Nature* **287**: 630–2. * 2,4

541 Game, M. & Peterken, G. F. 1984. Nature reserve selection strategies in the woodlands of Central Lincolnshire, England. *Biological Conservation* **29**: 157–81. * 2

542 Gandemer, J. 1979. Wind shelters. *Journal of Industrial Aerodynamics* **4**: 371–89. * 6

543 Gardner, R. H., Milne, B. T., Turner, M. G. & O'Neill, R. V. 1987. Neutral models for the analysis of broad-scale landscape pattern. *Landscape Ecology* **1**: 19–28. * 1,8,9,11

544 Gardner, R. H. & O'Neill, R. V. 1991. Pattern, process, and predictability: the use of neutral models for landscape analysis. In M. G. Turner & R. H. Gardner, eds. *Quantitative Methods in Landscape Ecology*, pp. 289–307. Springer-Verlag, New York. * 1,8,9

545 Gardner, R. H., O'Neill, R. V., Turner, M. G. & Dale, V. H. 1989. Quantifying scale-dependent effects of animal movement with simple percolation models. *Landscape Ecology* **3**: 217–27. * 1,11,12

546 Gardner, R. H., Turner, M. G., Dale, V. H. & O'Neill, R. V. 1992. A percolation model of ecological flows. In A. J. Hansen & F. di Castri, eds. *Landscape Boundaries: Consequences for Biotic Diversity and Ecological Flows*, pp. 259–69. Springer-Verlag, New York. * 3,12

547 Gardner, R. H., Turner, M. G., O'Neill, R. V. & Lavorel, S. 1991. Simulation of the scale-dependent effects of landscape boundaries on species persistence and dispersal. In M. M. Holland, P. G. Risser & R. J. Naiman, eds. *Ecotones: The Role of Landscape Boundaries in the Management and Restoration of Changing Environments*, pp. 76–89. Chapman and Hall, New York. * 3

548 Garland, T. & Bradley, W. G. 1984. Effects of a highway on Mojave Desert rodent populations. *American Midland Naturalist* **111**: 47–56. * 5

549 Garshelis, D. L. & Pelton, M. R. 1981. Movements of black bears in the Great Smoky Mountains National Park. *Journal of Wildlife Management* **45**: 912–25. * 11

550 Gates, D. M. 1962. *Energy Exchange in the Biosphere*. Harper & Row, New York. * 10

551 Gates, D. M. 1963. The energy environment in which we live. *American Scientist* **51**: 327–48. * 10

552 Gates, J. E. 1991. Powerline corridors, edge effects, and wildlife in forested landscapes of the Central Appalachians. In J. E. Rodiek & E. G. Bolen, eds. *Wildlife and Habitats in Managed Landscapes*, pp. 13–32. Island Press, Washington, DC. * 5

553 Gates, J. E. & Gysel, L. W. 1978. Avian nest dispersion and fledgling success in field–forest ecotones. *Ecology* **59**: 871–83. * 3

554 Gates, J. E. & Harman, D. M. 1980. White-tailed deer wintering area in a hemlock–northern hardwood forest. *The Canadian Field-Naturalist* **94**: 259–68. * 9,11

555 Gates, J. E. & Mosher, J. A. 1981. A functional approach to estimating habitat edge width for birds. *American Midland Naturalist* **105**: 189–92. * 3

556 Geiger, R. 1965. *The Climate Near the Ground*. Harvard University Press, Cambridge, Massachusetts. * 3,10

557 Geluso, K. N. 1971. Habitat distribution of *Peromyscus* in the Black Mesa region of Oklahoma. *Journal of Mammalogy* **52**: 605–7. * 6

558 George, E. J. 1971. *Effect of Tree Windbreaks and Slat Barriers on Wind Velocity and Crop Yields*. Product Research Report 121, US Department of Agriculture, Agricultural Research Service, Washington, DC. 23 pp. * 6

559 Gerfen, C. R. 1989. The neostriatal mosaic: striatal patch–matrix organization is related to cortical lamination. *Science* **246**: 385–8. * 1

560 Getz, L. L., Best, L. B. & Prather, M. 1977a. Lead in urban and rural song birds. *Environmental Pollution* **12**: 235–8. * 5

561 Getz, L. L., Cole, F. R. & Gates, D. L. 1978. Interstate roadsides as dispersal routes for *Microtus pennsylvanicus*. *Journal of Mammalogy* **59**: 208–12. * 5,8,11

562 Getz, L. L., Verner, L. & Prather, M. 1977b. Lead concentrations in small mammals living near highways. *Environmental Pollution* **13**: 151–7. * 5

563 Gholz, H. L. 1987. *Agroforestry: Realities, Possibilities and Potentials*. Kluwer Academic Publishers, Dordrecht, Netherlands. * 13

564 Gilbert, F. S. 1980. The equilibrium theory of island biogeography: fact or fiction? *Journal of Biogeography* **7**: 209–35. * 2,12

565 Gilbert, O. L. 1991. *The Ecology of Urban Habitats*. Chapman & Hall, London. * 2,10,11,13

566 Giles, R. H., Jr. 1978. *Wildlife Management*. W. H. Freeman, San Francisco. * 3,9

567 Giles, R. H., Jr. 1985. Planning the distribution of watering and similar developments for terrestrial wildlife. *Wildlife Society Bulletin* **13**: 411–15. * 4,11

568 Gilmore, R. M. & Gates, J. E. 1985. Habitat use by the southern flying squirrel at a hemlock – northern hardwood ecotone. *Journal of Wildlife Management* **49**: 703–10. * 3

569 Gilpin, M. E. 1981. Peninsula diversity patterns. *American Naturalist* **118**:291–6. * 3, 4

570 Gilpin, M. E. 1987. Spatial structure and population vulnerability. In M. E. Soulé, ed. *Viable Populations for Conservation*, pp. 126–39. Cambridge University Press, New York. * 11

571 Gilpin, M. E. & Diamond, J. M. 1980. Subdivision of nature reserves and the maintenance of species diversity. *Nature* **285**: 567–8. * 2

572 Gilpin, M. E. & Soulé, M. E. 1986. Minimum viable populations: processes of species extinction. In M. E. Soulé, ed. *Conservation Biology: The Science of Scarcity and Diversity*, pp. 19–34. Sinauer Associates, Sunderland, Massachusetts, USA. * 2,11

573 Girard, C. M. & Girard, M. C. 1985. Interpretation du paysage a petite echelle a partir de cliches de la chambre metrique Spacelab: approche botanique et pedologique. *Bulletin de la Societe Francaise de Photogrammetrie et Teledetection* **99**: 41–51. * 9

574 Giteau, M. 1976. *The Civilization of Angkor*. Rizzoli, New York. * 14

575 Gittins, P. 1983. Road casualties solve toad mysteries. *New Scientist* **97**: 530–1. * 5

576 Gleason, H. A. 1922. On the relation between species and area. *Ecology* **3**: 158–62. * 2

577 Gleason, H. A. 1925. Species and area. *Ecology* **6**: 66–74. * 2

578 Gliwicz, J. 1980. Island populations of rodents: their organization and functioning. *Biological Review* **55**: 109–38. * 2

579 Godron, M. 1968. Quelques applications de la notion de fréquence en écologie végétale (Recouvrement, information mutuelle entre espèces et facteurs écologiques, échantillonnage). *Oecologia Plantarum* **3**: 185–212. * 1,8,9

580 Godron, M. 1982. L'étude du 'grain' de la structure de la vegetation: application a quelques exemples mediterraneens. *Ecologia Mediterranea* **8**: 191–5. * 1

581 Godron, M. 1994. The natural hierarchy of ecological systems. In F. Klijn, ed. *Ecosystem Classification for Environmental Management*, pp. 69–83. Kluwer Academic Publishers, Netherlands. * 14

582 Godron, M. & Forman, R. T. T. 1983. Landscape modification and changing ecological characteristics. In H. A. Mooney & M. Godron, eds. *Disturbance and Ecosystems: Components of Response*, pp. 12–28. Springer-Verlag, New York. * 12, 14

583 Goldstein, B. 1992. The struggle over ecosystem management at Yellowstone. *BioScience* **42**: 183–7. * 13, 14

584 Goldstein, E. L, Gross, M. & DeGraaf, R. M. 1981. Explorations in bird-land geometry. *Urban Ecology* **5**: 113–24. * 13

585 Golley, F. B. & Bellot, J. 1991. Interactions of landscape ecology, planning and design. *Landscape and Urban Planning* **21**: 3–11. * 13

548

586 Golubev, G. & Vasiliev, O. 1978. Interregional water transfers as an interdisciplinary problem. *Water Supply and Management* **2**: 67–77. * 7

587 Good, R. 1974. *The Geography of the Flowering Plants*, 4th edn. Longmans, London. * 11

588 Good, R. E., Good, N. F. & Andresen, J. W. 1979. The Pine Barren plains. In R. T. T. Forman, ed. *Pine Barrens: Ecosystem and Landscape*, pp. 283–95. Academic Press, New York. * 10

589 Goodall, D. W. 1961. Objective methods for the classification of vegetation. IV. Pattern and minimal area. *Australian Journal of Botany* **9**: 162–96. * 2

590 Goodall, D. W. 1974. A new method for analysis of spatial pattern by random pairing of quadrats. *Vegetatio* **29**: 135–46. * 1,9

591 Goodfellow, S. & Peterken, G. F. 1981. A method for survey and assessment of woodlands for nature conservation using maps and species lists: the example of Norfolk woodlands. *Biological Conservation* **21**: 177–95. * 2

592 Gooselink, J. G., Shaffer, G. P., Lee, L. C., Burdick, D. M., Childers, D. L., Leibowitz, N. C., Hamilton, S. C., Boumans, R., Cushman, D., Fields, S., Koch, M. & Visser, J. M. 1990. Landscape conservation in a forested wetland watershed. *BioScience* **40**: 588–600. * 9

593 Gorham, E., Vitousek, P. M. & Reiners, W. A. 1979. The regulation of chemical budgets over the course of terrestrial ecosystem succession. *Annual Review of Ecology and Systematics* **10**: 53–84. * 1,7,10

594 Gorman, O. T. & Karr, J. R. 1977. Habitat structure and stream fish communities. *Ecology* **59**: 507–15. * 7

595 Gosz, J. R. 1991. Ecological functions in a biome transition zone: translating local responses to broad-scale dynamics. In A. J. Hansen & F. di Castri, eds. *Landscape Boundaries: Consequences for Biotic Diversity and Ecological Flows*, pp. 55–75. Springer-Verlag, New York. * 3

596 Gosz, J. R., Brookins, D. G. & Moore, D. I. 1983. Using strontium isotope ratios to estimate inputs to ecosystems. *BioScience* **33**: 23–30. * 1,10

597 Gosz, J. R., Dahm, C. N. & Risser, P. G. 1988. Long-path FTIR measurement of atmospheric trace gas concentrations. *Ecology* **69**: 1326–30. * 1

598 Gosz, J. R., Likens, G. E. & Bormann, F. H. 1972. Nutrient content of litter fall on the Hubbard Brook Experimental Forest, New Hampshire. *Ecology* **53**: 769–84. * 7

600 Gosz, J. R. & Sharpe, P. J. H. 1989. Broad-scale concepts for interactions of climate, topography, and biota at biome transitions. *Landscape Ecology* **3**: 229–43. * 3

601 Gotfryd, A. & Hansell, R. I. C. 1986. Predictions of bird-community metrics in urban woodlots. In J. Verner, M. L. Morrison & C. J. Ralph, eds. *Wildlife 2000: Modeling Habitat Relationships of Terrestrial Vertebrates*, pp. 321–6. University of Wisconsin Press, Madison, Wisconsin, USA. * 4,13

602 Gottfried, B. M. 1979. Small mammal populations in woodlot islands. *American Midland Naturalist* **102**: 105–12. * 2

603 Goudie, A. 1990. *The Human Impact on the Natural Environment*, 3rd edn. MIT Press, Cambridge, Massachusetts. * 7

604 Gould, S. J. 1979. An allometric interpretation of species–area curves: the meaning of the coefficient. *American Naturalist* **114**: 335–43. * 2

605 Gould, S. J. 1985. The paradox of the first tier: an agenda for paleobiology. *Paleobiology* **11**: 2–12. * 1, 8,14

606 Gouyon, P. H., Lumaret, R., Valdeyron, G. & Vernet, P. 1983. Reproductive strategies and disturbance by man. In H. A. Mooney & M. Godron, eds. *Disturbance and Ecosystems: Components of Response*, pp. 214–25. Springer-Verlag, New York. * 2

607 Grace, J. 1988. Plant response to wind. *Agriculture, Ecosystems and Environment* **22/23**: 71–88. (Reprinted 1988 in *Windbreak Technology*. Elsevier, Amsterdam.) * 6

608 Graetz, R. D. & Cowan, I. 1979. Microclimate and evaporation. In D. W. Goodall, R. A. Perry & K. M. W. Howes, eds. *Arid-land Ecosystems: Structure, Functioning and Management*. Cambridge University Press, Cambridge. * 10

609 Green, B. 1985. *Countryside Conservation: The Protection and Management of Amenity Ecosystems*. George Allen & Unwin, London. * 13

610 Gregory, K. J. & Walling, D. E. 1973. *Drainage Basin Form and Process: A Geomorphological Approach*. John Wiley, New York. * 4, 5, 7, 8, 9

549

611 Gregory, S. V., Swanson, F. J., McKee, W. A. & Cummins, K. W. 1991. An ecosystem perspective of riparian zones. *BioScience* **41**: 540–51. * 7

612 Greig-Smith, P. 1983. *Quantitative Plant Ecology*, 3rd edn. University of California Press, Berkeley. * 1, 2, 9, 13

613 Griffin, G. F., Price, N. F. & Portlock, H. F. 1983. Wildfires in the central Australian rangelands, 1970–1980. *Journal of Environmental Management* **17**: 311–23. * 10

614 Griffin, G. F., Stafford Smith, D. M., Morton, S. R., Allan, G. E. Masters, K. A. & Preece, N. 1989. Status and implications of the invasion of tamarisk (*Tamarix aphylla*) on the Finke River, Northern Territory, Australia. *Journal of Environmental Management* **29**: 297–315. * 7, 11

615 Griffith, D. A. 1982. Geometry and spatial interaction. *Annals of the Association of American Geographers* **72**: 332–46. * 4

616 Griffith, D. A., O'Neill, M. P., O'Neill, W. A., Leifer, L. A. & Mooney, R. G. 1986. Shape indices: useful measures or red herrings? *Professional Geographer* **38**: 263–70. * 4

617 Griffiths, J. C. 1967. *Scientific Method in Analysis of Sediments*. McGraw-Hill, New York. * 4

618 Griggs, R. F. 1914. Observations on the behavior of some species at the edges of their ranges. *Bulletin of the Torrey Botanical Club* **41**: 25–49. * 3

619 Griggs, R. F. 1937. Timberlines as indicators of climatic trends. *Science* **85**: 251–5. * 3, 4

620 Griggs, R. F. 1942. Indications as to climatic changes from the timberline of Mount Washington. *Science* **95**: 515–19. * 3, 4

621 Grimm, E. C. 1988. Data analysis and display. In B. Huntley & T. Webb III, eds. *Vegetation History*, pp. 43–76. Kluwer, Dordrecht, Netherlands. * 14

622 Griscom, L. 1949. *The Birds of Concord*. Harvard University Press, Cambridge, Massachusetts. * 13

623 Gromadzki, M. 1970. Breeding communities of birds in mid-field afforested areas. *Ekologia Polska* **18**: 307–50. * 2, 6

624 Groves, R. H. & Burdon, J. J. 1986. *Ecology of Biological Invasions*. Cambridge University Press, Cambridge. * 11

625 Grubb, P. J. 1977. The maintenance of species-richness in plant communities: the importance of the regeneration niche. *Biological Review (Cambridge Philosophical Society)* **52**: 107–45. * 2

626 Grubb, P. J. 1988. The uncoupling of disturbance and recruitment, two kinds of seed bank, and persistence of plant populations at the regional and local scales. *Annales Zoologici Fennici* **25**: 23–36. * 3

627 Grubb, P. J. 1989. Toward a more exact ecology: a personal view of the issues. In P. J. Grubb & J. B. Whittaker, eds. *Toward a More Exact Ecology*, pp. 3–29. Blackwell Scientific, Oxford. * 1, 14

628 Grunbaum, B. & Shephard, G. C. 1987. *Tilings and Patterns*. W. H. Freeman, San Francisco. * 4

629 Grunert, F., Benndorf, D. & Klingbeil, K. 1984. Neuere Ergebnisse zum Aufbau von Schutzpflanzungen. *Beitraege fuer die Forstwirtschaft* **18**: 108–15. * 6

630 Guevara, S., Purata, S. E. & van der Maarel, E. 1986. The role of remnant forest trees in tropical secondary succession. *Vegetatio* **66**: 77–84. * 1, 3, 11

631 Gulinck, H. 1986. Landscape ecological aspects of agro-ecosystems. *Agriculture, Ecosystems and Environment* **16**: 79–86. * 12, 13

632 Gumerman, G. J., ed. 1988. *The Anasazi in a Changing Environment*. Cambridge University Press, Cambridge. * 14

633 Gutzwiller, K. J. & Anderson, S. H. 1987. Multiscale associations between cavity-nesting birds and features of Wyoming streamside woodlands. *Condor* **89**: 534–48. * 4, 11

634 Gutzwiller, K. J. & Anderson, S. H. 1992. Interception of moving organisms: influences of patch shape, size, and orientation on community structure. *Landscape Ecology* **6**: 293–303. * 4, 11

635 Haase, G. 1991. Naturraumkartierung und Bewertung des Naturraumpotentials. *Der Schriftenreihe des Deutschen Rates für Landespflege* **59**: 923–40. * 9

636 Haase, G. & Aurada, K. D. 1984. Natural conditions and problems of their complex utilization in the German Democratic Republic. *GeoJournal* **8.1**: 53–65. * 9

637 Haase, G. & Richter, H. 1983. Current trends in landscape research. *GeoJournal* **7.2:** 107–19. * 1

638 Haber, W. 1990. Using landscape ecology in planning and management. In I. S. Zonneveld & R. T. T. Forman, eds. *Changing Landscapes: An Ecological Perspective*, pp. 217–32. Springer-Verlag, New York. * 1, 13, 14

639 Haber, W. 1992. Erfahrungen und Erkenntnisse aus 25 Jahren der Lehre und Forschung in Landschaftsokologie: Kann man okologisch planen? In F. Duhme, R. Lenz & L. Spandau, eds. *Lehrstuhl für Landschaftsokologie in Weihenstephan mit Prof. Dr. Dr. h. c. W. Haber*, pp. 1–28. 25 Jahre Landschaftsokologie, Weihenstephan, Germany. * 1, 13

640 Haber, W. & Burkhardt, I. 1988. Landschaftsokologie in der Landespflege in Weihenstephan. *Berichte zur deutschen Landeskunde* **62:** 155–73. * 13

641 Hack, J. T. 1960. Interpretation of erosional topography in humid temperate regions. *American Journal of Science* **258A:** 80–97. * 7, 9, 10

642 Hack, J. T. & Goodlett, J. C. 1960. *Geomorphology and Forest Ecology of a Mountain Region in the Central Appalachians*. Geological Survey Professional Paper 347, US Government Printing Office, Washington, DC. * 9, 10

643 Hagan, J. M., III & Johnston, D. W., eds. 1989. *Ecology and Conservation of Neotropical Migrant Landbirds*. Smithsonian Institution Press, Washington, DC. * 2,3

644 Hagen, L. J. & Skidmore, E. L. 1971. Turbulent velocity fluctuations and vertical flow as affected by windbreak porosity. *Transactions of the American Society of Agricultural Engineers* **14:** 634–7. * 6

645 Hagen, L. J., Skidmore, E. L., Miller, P. L. & Kipp, J. E. 1981. Simulation of effect of wind barriers on airflow. *Transactions of the American Society of Agricultural Engineers* **24:** 1002–8. * 6

646 Haggett, P., Cliff, A. D. & Frey, A. 1977. *Locational Analysis in Human Geography*, 2nd edn. John Wiley, New York. * 1,4,5,8,9,11

647 Haines-Young, R., Green, D. R. & Cousins, S., eds. 1993. *Landscape Ecology and Geographic Information Systems*. Taylor & Francis, London. * 1

648 Hall, C. A. S. 1988. An assessment of several of the historically most influential theoretical models used in ecology and of the data provided in their support. *Ecological Modelling* **43:** 5–31. * 11

649 Hall, D. L. 1991. Landscape planning: functionalism as a motivating concept from landscape ecology and human ecology. *Landscape and Urban Planning* **21:** 13–19. * 13

650 Hall, F. G., Strebel, D. E. & Sellers, P. J. 1988. Linking knowledge among spatial and temporal scales: vegetation, atmosphere, climate and remote sensing. Landscape Ecology **2:** 3–22. * 1

651 Hall, P. 1966. *Von Thunen's Isolated State*. Pergamon, New York. * 9

652 Hall, T. H. 1981. Forest management decision-making: art or science? *The Forestry Chronicle* **57:** 233–8. * 14

653 Halle, F., Oldeman, R. A. A. & Tomlinson, P. B. 1978. *Tropical Trees and Forests: An Architectural Analysis*. Springer-Verlag, Berlin. * 2

654 Hammett, J. E. 1992. The shapes of adaptation: historical ecology of anthropogenic landscapes in the southeastern United States. *Landscape Ecology* **7:** 121–35. * 4

655 Hammond, E. H. 1954. Small-scale continental landform maps. *Annals of the Association of American Geographers* **44:** 33–42. * 9

656 Hamrick, J. L., Linhart, Y. B. & Mitton, J. B. 1979. Relationships between life history characteristics and electrophoretically detectable genetic variation in plants. *Annual Review of Ecology and Systematics* **10:** 173–200. * 12

657 Hanan, N. P., Prevost, Y., Diouf, A. & Diallo, O. 1991. Assessment of desertification around deep wells in the Sahel using satellite imagery. *Journal of Applied Ecology* **28:** 173–86. * 1,12

658 Hanes, T. L. 1971. Succession after fire in the chaparral of Southern California. *Ecological Monographs* **41:** 27–52. * 10

659 Hanks, L. M. 1972. *Rice and Man: Agricultural Ecology in Southeast Asia*. University of California Press, Berkeley. * 14

660 Hanley, T. A. 1983. Black-tailed deer, elk, and forest edge in a western Cascades watershed. *Journal of Wildlife Management* **47:** 237–42. * 3

661 Hansen, A. J. & di Castri, F., eds. 1992. *Landscape Boundaries: Consequences for Biotic Diversity and Ecological Flows*. Springer-Verlag, New York. * 1,3,12

551

662 Hansen, A. J., Urban, D. L. & Marks, B. 1992. Avian community dynamics: the interplay of landscape trajectories and species life histories. In A. J. Hansen & F. di Castri, eds. *Landscape Boundaries: Consequences for Biotic Diversity and Ecological Flows*, pp. 170–95. Springer-Verlag, New York. * 1,12,13,14

663 Hanski, I. 1982. Distributional ecology of anthropochorous plants in villages surrounded by forest. *Annales Botanici Fennici* **19**: 1–15. * 11,12

664 Hanski, I. 1991. Single species metapopulation dynamics: concepts, models and observations. *Biological Journal of the Linnean Society* **42**: 17–38. * 11

665 Hansson, L. 1977. Landscape ecology and stability of populations. *Landscape Planning* **4**: 85–93. * 9

666 Hansson, L. 1979. On the importance of landscape heterogeneity in northern regions for the breeding population densities of homeotherms: a general hypothesis. *Oikos* **33**: 182–9. * 9

667 Hansson, L. 1982. Daggdjurens vinterutnyttjande av kanter mellan hyggen och barrskog. *Fauna och flora* **77**: 301–8. * 3

668 Hansson, L. 1983. Bird numbers across edges between mature conifer forest and clear-cuts in Central Sweden. *Ornis Scandinavica* **14**: 79–103. * 3

669 Hansson, L. 1987. Dispersal routes of small mammals at an abandoned field in central Sweden. *Holarctic Ecology* **10**: 153–9. * 11

670 Hansson, L. 1995. Landscape ecology of boreal forests. *Trends in Ecology and Evolution*. In press. * 1

671 Hansson, L. & Angelstam, P. 1988. New ecological processes in small reserves. In L. Svensson, ed. *Nature Conservation*, pp. 25–32. National Swedish Environmental Protection Board, Research Secretariat, Solna, Sweden. * 2

672 Hansson, L. & Angelstam, P. 1991. Landscape ecology as a theoretical basis for nature conservation. *Landscape Ecology* **5**: 191–201. * 1,2,13

673 Haralick, R. M. 1979. Statistical and structural approaches to texture. *Proceedings of the Institute of Electrical and Electronic Engineers* **67**: 786–804. * 1,9

674 Harary, F. 1969. *Graph Theory*. Addison-Wesley, Reading, Massachusetts, USA. * 4,9

675 Hardin, G. 1968. The tragedy of the commons. *Science* **162**: 1243–8. * 1

676 Hardt, R. A. & Forman, R. T. T. 1989. Boundary form effects on woody colonization of reclaimed surface mines. *Ecology* **70**: 1252–60. * 1,3,4,12

677 Hargrove, W. W. & Pickering, J. 1992. Pseudoreplication: a *sine qua non* for regional ecology. *Landscape Ecology* **6**: 251–8. * 1

678 Harms, B. & Opdam, P. 1990. Woods as habitat patches for birds: application in landscape planning in The Netherlands. In I. S. Zonneveld & R. T. T. Forman, eds. *Changing Landscapes: An Ecological Perspective*, pp. 73–97. Springer-Verlag, New York. * 1,8,9,12,13

679 Harms, W. B., Stortelder, A. H. F. & Vos, W. 1984. Effects of intensification of agriculture on nature and landscape in The Netherlands. *Ekologia* (Czechoslovakia) **3**: 281–304. * 12,14

680 Harms, W. B., Stortelder, A. H. F. & Vos, W. 1987. Effects of intensification of agriculture on nature and landscape in The Netherlands. In M. G. Wolman & F. G. A. Fournier, eds. *Land Transformation in Agriculture*, pp. 357–80. SCOPE Publication 32. John Wiley, New York. * 8, 12, 13

681 Harper, J. L. 1977. *Population Biology of Plants*. Academic Press, New York. * 8, 11

682 Harris, D. R. 1966. Recent plant invasions in the arid and semi-arid Southwest of the United States. *Annals of the Association of American Geographers* **56**: 408–22. * 7, 11

683 Harris, L. D. 1984. *The Fragmented Forest: Island Biogeography Theory and the Preservation of Biotic Diversity*. University of Chicago Press, Chicago. * 1,2,3,7,8,9,12,13

684 Harris, L. D. 1988. Edge effects and conservation of biotic diversity. *Conservation Biology* **2**: 330–2. * 3

685 Harris, L. D. & Gallagher, P. B. 1989. New initiatives for wildlife conservation: the need for movement corridors. In G. Mackintosh, ed. *Preserving Communities and Corridors*, pp. 11–34. Defenders of Wildlife, Washington, DC. * 1,2,5,6,7,12,13

686 Harris, L. D. & Kangas, P. 1979. Designing future landscapes from principles of form and function. In G. H. Pilsner & R. C. Smardon, eds. *Our National Landscape: Applied*

Techniques for Analysis and Management of the Visual Resource, pp. 725–9. General Technical Report PSW-34, US Forest Service, Washington, DC. * 1,4

687 Harris, L. D. & McElveen, J. D. 1981. *Effect of Forest Edges on North Florida Breeding Birds*. IMPAC Report 6(4). School of Forest Resources and Conservation, University of Florida, Gainesville, Florida. 24 pp. * 3

688 Harris, L. D. & Scheck, J. 1991. From implications to applications: the dispersal corridor principle applied to the conservation of biological diversity. In D. A. Saunders & R. J. Hobbs, eds. *Nature Conservation 2: The Role of Corridors*, pp. 189–220. Surrey Beatty, Chipping Norton, Australia. * 5,6,8,11,13

689 Harrison, R. & Lunt, G. G. 1980. *Biological Membranes: Their Structure and Function*, 2nd edn. John Wiley, New York. * 3

690 Harrje, D. T., Buckley, C. E. & Heisler, G. M. 1982. Building energy reductions: windbreak optimization. *Journal of the Energy Division, Proceedings of the American Society of Civil Engineering* **108** (EY3): 143–54. * 6

691 Hart, J. F. 1975. *The Look of the Land*. Prentice-Hall, Englewood Cliffs, New Jersey, USA. * 4

692 Hartigan, J. P., Douglas, B., Biggers, D. J., Wessel, T. J. & Stroh, D. 1979. Areawide and local frameworks for urban nonpoint pollution management in northern Virginia. In *Proceedings of National Conference on Stormwater Management Alternatives*. Wilmington, Delaware, USA. * 7

693 Hartshorn, G. S. 1978. Treefalls and tropical forest dynamics. In P. B. Tomlinson & M. H. Zimmermann, eds. *Tropical Trees as Living Systems*, pp. 617–38. Cambridge University Press, Cambridge. * 2

694 Hasel, A. A. 1938. Sampling error in timber surveys. *Journal of Agricultural Research* **57**: 713–36. * 4

695 Haskova, J. 1992. The role of corridors for plant dispersal in the landscape. In *Ecological Stability of Landscape; Ecological Infrastructure; Ecological Management*, pp. 88–99. Institute of Applied Ecology, Kostelec, Czechoslovakia. * 5,6,8

696 Haspel, C. & Calhoon, R. E. 1991. Ecology and behavior of free-ranging cats in Brooklyn, New York. In L. W. Adams & D. L. Leedy, eds. *Wildlife Conservation in Metropolitan Environments*, pp. 27–30. National Institute for Urban Wildlife, Columbia, Maryland, USA. * 13

697 Hastings, A. 1988. Food web theory and stability. *Ecology* **69**: 1665–8. * 8,11

698 Hastings, H. M., Pekelney, R., Monticciolo, R., Vun Kannon, D. & Del Monte, D. 1982. Time scales, persistence and patchiness. *BioSystems* **15**: 281–9. * 8,9

699 Haupt, H. F. & Kidd, W. J., Jr. 1965. Good logging practices reduce sedimentation in Central Idaho. *Journal of Forestry* **63**: 664–70. * 5

700 Hay, K. G. 1991. Greenways and biodiversity. In W. E. Hudson, ed. *Landscape Linkages and Biodiversity*, pp. 162–75. Island Press, Washington, DC. * 6

701 Hayes, T. D., Riskind, D. H. & Pace, W. L., III. 1987. Patch-within-patch restoration of man-modified landscapes within Texas state parks. In M. G. Turner, ed. *Landscape Heterogeneity and Disturbance*, pp. 173–98. Springer-Verlag, New York. * 1,4

702 Haynes, K. E. & Rube, M. I. 1973. Directional bias in urban population density. *Annals of the Association of American Geographers* **63**: 40–7. * 4

703 He, H. & Xiao, D. 1990. Landscape ecology – the development of a holistic concept. (In Chinese.) *Journal of Applied Ecology* (China) **1**: 264–9. * 1

704 Heady, E. O. 1952. *Economics of Agricultural Production and Resource Use*. Prentice-Hall, New York. * 2

705 Healy, R. G. 1985. *Competition for Land in the American South*. Conservation Foundation, Washington, DC. * 1

706 Heaney, L. R. & Patterson, B. D., eds. 1986. *Island Biogeography of Mammals*. Academic Press, Orlando, Florida. * 2

707 Heckman, C. W. 1979. *Rice Field Ecology in Northeastern Thailand*. W. Junk Publishers, The Hague. * 14

708 Heede, B. H. 1972. Influences of a forest on the hydraulic geometry of two mountain streams. *Water Resources Bulletin* **8**: 523–30. * 7

709 Heinritz, G. & Lichtenberger, E., eds. 1986. *The Take-off of Suburbia and the Crisis of the Central City*. Franz Steiner Verlag, Wiesbaden, Germany. * 13

710 Heinselman, M. L. 1973. Fire in the virgin forests of the Boundary Waters Canoe Area, Minnesota. *Quaternary Research* **3**: 329–82. * 4,10

711 Heisler, G. M. & DeWalle, D. R. 1988. Effects of windbreak structure on wind flow. In J. R. Brandle, D. L. Hintz & J. W. Sturrock, eds. *Windbreak Technology*, pp. 41–69. Elsevier, Amsterdam. (Reprinted from *Agriculture, Ecosystems and Environment* vol. **22/ 23**, 1988.) * 3,6,9

712 Heitschmidt, R. K., Dowhower, S. L. & Walker, J. W. 1987. 14- vs 42-paddock rotational grazing: aboveground biomass dynamics, forage production, and harvest efficiency. *Journal of Range Management* **40**: 216–23. * 11

713 Heitschmidt, R. K. & Stuth, J. W., eds. 1991. *Grazing Management: An Ecological Perspective*. Timber Press, Portland, Oregon. * 7,8,11,13

714 Heitschmidt, R. K. & Taylor, C. A., Jr. 1991. Livestock production. In R. K. Heitschmidt & J. W. Stuth, eds. *Grazing Management: An Ecological Perspective*, pp. 161–77. Timber Press, Portland, Oregon. * 11

715 Helle, E. & Helle, P. 1982. Edge effect on forest bird densities on offshore islands in the northern Gulf of Bothnia. *Annales Zoologici Fennici* **19**: 165–9. * 3

716 Helle, P. & Muona, J. 1985. Invertebrate numbers in edges between clear-fellings and mature forests in northern Finland. *Silva Fennica* **19**: 281–94. * 3

717 Helliwell, D. R. 1976. The effects of size and isolation on the conservation value of wooded sites in Britain. *Journal of Biogeography* **3**: 407–16. * 2

718 Henderson, M. T., Merriam, G. & Wegner, J. 1985. Patchy environments and species survival: chipmunks in an agricultural mosaic. *Biological Conservation* **31**: 95–105. * 1,5,6,8,11

719 Hendl, M. 1963. *Einfuhrung in die Physikalische Klimatologie: Band II, Systematische Klimatologie*. VEB Deutscher Verlag der Wissenschaften, Berlin. * 1

720 Henein, K. M. & Merriam, G. 1990. The elements of connectivity where corridor quality is variable. *Landscape Ecology* **4**: 157–70. * 5,6,8,9

721 Herbold, B. & Moyle, P. B. 1986. Introduced species and vacant niches. *American Naturalist* **128**: 751–60. * 11

722 Herrera, C. M. 1984. A study of avian frugivores, bird-dispersed plants, and their interaction in Mediterranean scrublands. *Ecological Monographs* **54**: 1–23. * 1

723 Hessing, M. B., Johnson, C. D. & Balda, R. P. 1981. Secondary succession of desert grassland in north-central Arizona powerline corridors. *Journal of Environmental Management* **13**: 59–73. * 5

724 Heywood, V. H. 1989. Patterns, extents and modes of invasions by terrestrial plants. In J. A. Drake & H. A. Mooney, eds. *Biological Invasions: A Global Perspective*, pp. 31–60. John Wiley, Chichester, UK. * 11

725 Hibberd, J. K. & Soutberg, T. L. 1991. Roadside reserve condition 1977–89 in the Southern Tablelands of New South Wales. In D. A. Saunders & R. J. Hobbs, eds. *Nature Conservation 2: The Role of Corridors*, pp. 177–86. Surrey Beatty, Chipping Norton, Australia. * 5,8

726 Hills, G. A. 1974. A philosophical approach to landscape planning. *Landscape Planning* **1**: 339–71. * 13

727 Hobbs, E. R. 1988. Species richness of urban forest patches and implications for urban landscape diversity. *Landscape Ecology* **1**: 141–52. * 13

728 Hobbs, R. J. 1987. Disturbance regimes in remnants of natural vegetation. In D. A. Saunders, G. W. Arnold, A. A. Burbidge & A. J. M. Hopkins, eds. *Nature Conservation: The Role of Remnants of Native Vegetation*, pp. 233–40. Surrey Beatty, Chipping Norton, Australia. * 2

729 Hobbs, R. J. 1989. The nature and effects of disturbance relative to invasions. In J. A. Drake, H. A. Mooney, F. di Castri, R. H. Groves, F. J. Kruger, M. Rejmanek, & M. Williamson, eds. *Biological Invasions: A Global Perspective*, pp. 389–406. John Wiley, Chichester, UK. * 11

730 Hobbs, R. J. 1993. Effects of landscape fragmentation on ecosystem processes in the Western Australian wheatbelt. *Biological Conservation* **64**: 193–201. * 2, 12

731 Hobbs, R. J. & Hopkins, A. J. M. 1991. The role of conservation corridors in a changing climate. In D. A. Saunders & R. J. Hobbs, eds. *Nature Conservation 2: The Role of Corridors*, pp. 281–90. Surrey Beatty, Chipping Norton, Australia. * 6

732 Hobbs, R. J., Saunders, D. A. & Arnold, G. W. 1993. Integrated landscape ecology: a Western Australian perspective. *Biological Conservation* **64**: 231–8. * 1, 13

733 Hodge, C. O. & Hodges, C. N., eds. 1974. *Urbanization in the Arid Lands: A Symposium*. International Center for Arid and Semi-arid Land Studies, Publication 75–1, Lubbock, Texas. * 12

734 Hodson, N. L. 1962. Some notes on the causes of bird road casualties. *Bird Study* **9:** 168–73. * 5

735 Hodson, N. L. 1966. A survey of road mortality in mammals (and including data for the Grass snake and Common frog). *Journal of Zoology (London)* **148:** 576–9. * 5

736 Hodson, N. L. & Snow, D. W. 1965. The road deaths enquiry, 1960–61. *Bird Study* **12:** 90–9. * 5

737 Hofstra, G. & Hall, R. 1971. Injury on roadside trees: leaf injury on pine and white cedar in relation to foliar levels of sodium and chloride. *Canadian Journal of Botany* **49:** 613–22. * 5.

738 Hogarth, P. & Clery, V. 1979. *Dragons*. Penguin Books, New York. * 11

739 Holbo, H. R. & Luvall, J. C. 1989. Modeling surface temperature distributions in forest landscapes. *Remote Sensing of Environment* **27:** 11–24. * 10

740 Hole, F. D. & Campbell, J. B. 1985. *Soil Landscape Analysis*. Routledge & Kegan Paul, London. * 1, 3, 9, 14

741 Holland, M. M. 1988. SCOPE/MAB technical consultations on landscape boundaries: report of a SCOPE/MAB workshop on ecotones. In F. di Castri, A. J. Hansen & M. M. Holland, eds. *A New Look at Ecotones: Emerging International Projects on Landscape Boundaries*, pp. 47–104. Biology International, Special Issue 17, International Union of Biological Sciences, Paris. * 3

742 Holland, M. M., Risser, P. G. & Naiman, R. J., eds. 1991. *Ecotones: The Role of Landscape Boundaries in the Management and Restoration of Changing Environments*. Chapman and Hall, New York. * 3

743 Holldobler, B. & Wilson, E. O. 1990. *The Ants*. Belknap Press of Harvard University Press, Cambridge, Massachusetts. * 11, 14

744 Holling, C. S. 1973. Resilience and stability of ecological systems. *Annual Review of Ecology and Systematics* **4:** 1–23. * 14

745 Holling, C. S., ed. 1978. *Adaptive Environmental Assessment and Management*. John Wiley, New York. * 13

746 Holling, C. S. 1986. The resilience of terrestrial ecosystems: local surprise and global change. In W. C. Clark & R. E. Munn, eds. *Sustainable Development of the Biosphere*, pp. 292–320. Cambridge University Press, Cambridge. * 14

747 Holmes, R. M. 1970. Meso-scale effects of agriculture and a large prairie lake on the atmospheric boundary layer. *Agronomy Journal* **62:** 546–9. * 10

748 Holt, R. D. 1985. Population dynamics in two-patch environments: some anomalous consequences of an optimal habitat distribution. *Theoretical Population Biology* **28:** 181–208. * 11

749 Holzner, W., Werger, M. J. A. & Ikusima, I., eds. 1983. *Man's Impact on Vegetation*. W. Junk Publishers, The Hague. * 11

750 Hooper, M. D. 1976. Historical and biological studies on English hedges. In *Les Bocages: Histoire, Ecologie, Economie*, pp. 225–7. Institut National de la Recherche Agronomique, Centre National de la Recherche Scientifique, et Universite de Rennes, Rennes, France. * 6

751 Hopkins, B. 1955. The species–area relations of plant communities. *Journal of Ecology* **43:** 409–26. * 2

752 Horgan, E. R. 1987. *Shaker Holy Land: A Community Portrait*. Harvard Common Press, Harvard, Massachusetts, USA. * 14

753 Hori, T. 1953. *Studies on Fogs in Relation to Fog-preventing Forest*. Tanne, Sapporo, Japan. (Summary in D. H. Miller. 1957. Coastal fogs and clouds. *Geographical Review* **47:** 591–4.) * 10

754 Horton, R. E. 1945. Erosional development of streams and their drainage basins; hydrophysical approach to quantitative morphology. *Bulletin of the Geological Society of America* **56:** 275–370. * 4,9

755 Hough, M. 1984. *City Form and Natural Processes: Towards an Urban Vernacular*. Van Nostrand Reinhold, New York. * 2,13

756 Howard, R. A. & Baynton, H. W. 1969. The ecology of an elfin forest in Puerto Rico, 3. Hilltop and forest influences on the microclimate of Pico del Oeste. *Journal of the Arnold Arboretum* **50:** 80–92. * 10

757 Howe, H. F. 1984. Implications of seed dispersal by animals for tropical reserve management. *Biological Conservation* **30**: 261–81. * 2,11

758 Howe, R. W. 1984. Local dynamics of bird assemblages in small forest habitat islands in Australia and North America. *Ecology* **65**: 1585–601. * 9,11

759 Hudson, W. E., ed. 1991. *Landscape Linkages and Biodiversity*. Island Press, Washington, DC. * 13

760 Huet, M. 1954. Biologie, profils en long et en travers des eaux courantes. *Bulletin Francais de Pisciculture* **175**: 41–53. * 7

761 Huey, L. M. 1941. Mammalian invasion via the highway. *Journal of Mammalogy* **22**: 383–5. * 5

762 Hughes, F. M. R. 1988. The ecology of African floodplain forests in semi-arid and arid zones: a review. *Journal of Biogeography* **15**: 127–40. * 7

763 Hunt, A., Dickens, H. J. & Whelan, R. J. 1987. Movement of mammals through tunnels under railway lines. *Australian Zoology* **24**: 89–93. * 5

764 Hunter, M. L., Jr. 1990. *Wildlife, Forests, and Forestry: Principles of Managing Forests for Biological Diversity*. Prentice-Hall, Englewood Cliffs, New Jersey. * 2,9,13

765 Hunter, M. L., Jr. 1993. Natural fire regimes as spatial models for managing boreal forests. *Biological Conservation* **65**: 115–20. * 10

766 Hunter, M. L., Jr. & Yonzon, P. 1993. Altitudinal distributions of birds, mammals, people, forests, and parks in Nepal. *Conservation Biology* **7**: 420–3, * 4, 9

767 Huntley, B. & Birks, H. J. B. 1983. *An Atlas of Past and Present Pollen Maps for Europe: 0–13 000 Years Ago*. Cambridge University Press, Cambridge. * 1, 3, 8, 14

768 Huntley, B. J., Ezcurra, E., Fuentes, E. R., Fujii, K., Grubb, P. J., Haber, W., Harger, J. R. E., Holland, M. M., Levin, S. A., Lubchenco, J., Mooney, H. A., Neronov, V., Noble, I., Pulliam, H. R., Ramakrishnan, P. S., Risser, P. G., Sala, O., Sarukhan, J. & Sombroek, W. G. 1991. A sustainable biosphere: the global imperative. *Ecology International* **20**: 5–14. (Reprinted 1992 in *Bulletin of the Ecological Society of America* **73**: 7–14). * 13, 14

769 Huntley, B. & Webb, T., III. 1989. Migration: species' response to climatic variations caused by changes in the earth's orbit. *Journal of Biogeography* **16**: 5–19. * 11, 14

770 Hupp, C. R. 1982. Stream-grade variation and riparian-forest ecology along Passage Creek, Virginia. *Bulletin of the Torrey Botanical Club* **109**: 488–99. * 7

771 Hussey, B. M. J. 1991. The flora roads survey – volunteer recording of roadside vegetation in Western Australia. In D. A. Saunders & R. J. Hobbs, eds. *Nature Conservation 2: The Role of Corridors*, pp. 41–8. Surrey Beatty, Chipping Norton, Australia. * 5

772 Hussey, B. M. J., Hobbs, R. J. & Saunders, D. A. 1991. *Guidelines for Bush Corridors*. CSIRO Division of Wildlife and Ecology, Perth, Western Australia. 21 pp. * 5, 13

773 Hutton, M. 1980. Metal contamination of feral pigeons *Columba livia* from the London area: part 2 – biological effects of lead exposure. *Environmental Pollution Series A* **22**: 281–93. * 5

774 Hylgaard, T. 1980. Recovery of plant communities on coastal sand dunes disturbed by human trampling. *Biological Conservation* **19**: 15–25. * 5

775 Ide, H. & Takeuchi, K. 1987. Landuse planning based on potential natural vegetation. In A. Miyawaki, A. Bogenrieder, S. Okuda & J. White, eds. *Vegetation Ecology and Creation of New Environments*, pp. 277–82. Tokai University Press, Tokyo. * 13

776 Illies, J. 1961. Versuch einer allgemeinen biozonotischen Gliederung der Fliessgewasser. *Internationale Revue der gesamten Hydrobiologie* **46**: 205–13. * 7

777 Ims, R. A. & Stenseth, N. C. 1989. Divided the fruit flies fall. *Nature* **342**: 21–2. * 1, 9

778 *Intersectoral Issues Report*. 1992. Ecologically Sustainable Development Working Group Chairs, Australian Government Publishing Service, Canberra. 248 pp. * 14

779 Isachenko, A. G. 1973. *Principles of Landscape Science and Physical-geographic Regionalization*. (Translated from Russian.) Melbourne University Press, Carlton, Victoria, Australia. * 9

780 Isard, W. 1972. *Ecologic-Economic Analysis for Regional Development*. Free Press, New York. * 1

781 Isard, W. 1975. *Introduction to Regional Science*. Prentice-Hall, Englewood Cliffs, New Jersey, USA. * 1

782 Ishii, Y., ed. 1978. *Thailand: A Rice Growing Society*, translated by P. Hawkes & S. Hawkes. University of Hawaii Press, Honolulu. * 14

783 Iverson, L. R. 1988. Land-use changes in Illinois, USA: the influence of landscape attributes on current and historic land use. *Landscape Ecology* **2**: 45–61. * 1,12,14

784 Ives, R. L. 1942. The beaver–meadow complex. *Journal of Geomorphology* **5**: 191–203. * 7

785 Iwaszutina, L. I. & Nikolajew, W. A. 1971. *Kontrastnost Landszaftnoj Struktury i Niekotoryje Aspiekty Jejo Izuczenija*. Wiesstnik MU, Ser. 5, Gieografia nr 5, Moscow. * 4

786 Jaccard, P. 1908. Nouvelles recherches sur la distribution florale. *Bulletin de la Societe Vaudoise des Sciences Naturelles* **44**: 223–70. * 2

787 Jackson, L. W. R. 1959. Relation of pine forest overstory opening diameter to growth of pine reproduction. *Ecology* **40**: 478–80. * 2

788 Jacobs, A. F. G. 1984. Wind reduction near the surface behind a thin solid fence. *Agricultural and Forest Meteorology* **33**: 157–62. * 6

789 Jacobs, A. F. G. 1985. Turbulence around a thin solid fence. *Agricultural and Forest Meteorology* **34**: 315–21. * 6

790 Jacobs, P. 1991. Sustainable urban development. *International Journal of Sustainable Development* **1**: 48–54. * 14

791 Jacobs, P. & Munro, D. A., eds. 1987. *Conservation with Equity: Strategies for Sustainable Development*. International Union for the Conservation of Nature and Natural Resources, Gland, Switzerland. * 14

792 Jacobs, T. C. & Gilliam, J. W. 1985. Riparian losses of nitrate from agricultural drainage waters. *Journal of Environmental Quality* **14**: 472–8. * 7

793 Jacobsen, T. & Adams, R. M. 1958. Salt and silt in ancient Mesopotamian agriculture: progressive changes in soil salinity and sedimentation contributed to the breakup of past civilizations. *Science* **128**: 1251–8. * 12,14

794 Janzen, D. H. 1983. No park is an island: increase in interference from outside as park size decreases. *Oikos* **41**: 402–10. * 3,8,13

795 Janzen, D. H. 1988. Tropical dry forests: the most endangered major tropical ecosystem. In E. O. Wilson, ed. *Biodiversity*, pp. 130–7. National Academy Press, Washington, DC. * 13

796 Janzen, D. H. 1988. Tropical ecological and biocultural restoration. *Science* **239**: 243–4. * 13

797 Jarvinen, O. 1982. Conservation of endangered plant populations: single large or several small reserves? *Oikos* **38**: 301–7. * 2

798 Jenkins, S. H. 1980. A size-distance relation in food selection by beavers. *Ecology* **61**: 740–6. * 7

799 Jennrich, R. I. & Turner, F. B. 1969. Measurement of non-circular home range. *Journal of Theoretical Biology* **22**: 227–37. * 4

800 Jenny, H. 1980. *The Soil Resource: Origin and Behavior*. Springer-Verlag, New York. * 7,9,14

801 Jensen, J. R. 1986. *Introductory Digital Image Processing*. Prentice-Hall, Englewood Cliffs, New Jersey, USA. * 1

802 Jensen, M. 1954. *Shelter Effect: Investigations into the Aerodynamics of Shelter and Its Effects on Climate and Crops*. The Danish Technical Press, Copenhagen. * 6,8

803 Johnson, A. R., Milne, B. T. & Wiens, J. A. 1992. Diffusion in fractal landscapes: simulations and experimental studies of tenebrionid beetle movements. *Ecology* **73**: 1968–83. * 1,9,11,14

804 Johnson, A. R., Wiens, J. A., Milne, B. T. & Crist, T. O. 1992. Animal movements and population dynamics in heterogeneous landscapes. *Landscape Ecology* **7**: 63–75. * 1, 11

805 Johnson, A. S. 1989. The thin green line: riparian corridors and endangered species in Arizona and New Mexico. In G. Mackintosh, ed. *Preserving Communities and Corridors*, pp. 35–46. Defenders of Wildlife, Washington, DC. * 7

806 Johnson, H. B. 1976. *Order Upon the Land: The U. S. Rectangular Land Survey and the Upper Mississippi Country*. Oxford University Press, New York. * 8

807 Johnson, L. 1990. Analyzing spatial and temporal phenomena using geographical information systems: a review of ecological applications. *Landscape Ecology* **4**: 31–43. * 1

808 Johnson, R. J. & Beck, M. M. 1988. Influences of shelterbelts on wildlife management and biology. *Agriculture, Ecosystems and Environment* **22/23**: 301–35. (Reprinted 1988 in *Windbreak Technology*. Elsevier, Amsterdam.) * 6

809 Johnson, R. J., Brandle, J. R., Fitzmaurice, R. L. & Poague, K. L. 1992. Vertebrates for biological control of insects in agroforestry systems. In *Biological Control of Forest Pests in the Great Plains: Status and Needs – A Symposium*, pp. 77–84. Great Plains Agricultural Council Publication 145, Bismark, North Dakota, USA. * 6

810 Johnson, R. R. & Jones, D. A., eds. 1977. *Importance, Preservation and Management of Riparian Habitat: A Symposium*. General Technical Report RM-43, US Forest Service, Fort Collins, Colorado, USA. * 7

811 Johnson, W. C. 1988. Estimating dispersability of *Acer*, *Fraxinus* and *Tilia* in fragmented landscapes from patterns of seedling establishment. *Landscape Ecology* **1**: 175–87. * 11

812 Johnson, W. C. & Atkisson, C. S. 1985. Dispersal of beech nuts by blue jays in fragmented landscapes. *American Midland Naturalist* **113**: 319–24. * 6,7,9,11,14

813 Johnson, W. C., Schreiber, R. K. & Burgess, R. L. 1979. Diversity of small mammals in a powerline right-of-way and adjacent forest in East Tennessee. *American Midland Naturalist* **101**: 231–5. * 5

814 Johnston, C. A. & Bonde, J. 1989. Quantitative analysis of ecotones using a geographic information system. *Photogrammetric Engineering and Remote Sensing* **55**: 1643–7. * 1,3

815 Johnston, C. A., Datenbeck, N. E., Bonde, J. P. & Niemi, G. J. 1988. Geographic information systems for cumulative impact assessment. *Photogrammetric Engineering and Remote Sensing* **54**: 1609–15. * 1,13

816 Johnston, C. A. & Naiman, R. J. 1987. Boundary dynamics at the aquatic-terrestrial interface: the influence of beaver and geomorphology. *Landscape Ecology* **1**: 45–57. * 3, 7

817 Johnston, C. A. & Naiman, R. J. 1990. The use of a geographic information system to analyze long-term landscape alteration by beaver. *Landscape Ecology* **4**: 5–19. * 1, 7

818 Jones, R. 1985. Ordering the landscape. In I. Donaldson & T. Donaldson, eds. *Seeing the First Australians*, pp. 181–209. George Allen and Unwin, Sydney. * 1, 13

819 Jongman, R. G. H., ter Braak, C. J. F. & van Tongeren, O. F. R. 1987. *Data Analysis in Community and Landscape Ecology*. PUDOC, Wageningen, Netherlands. * 1, 9

820 Jonkers, D. A. & de Vries, G. W. 1977. *Verkeerssalachtoffers onder der fauna*. Zeist, Nederlandsa Vereniging tot Bescherming van Vogels. * 5

821 Jordan, W. R. III, Gilpin, M. E. & Aber, J. D., eds. 1987. *Restoration Ecology: A Synthetic Approach to Ecological Research*. Cambridge University Press, Cambridge. * 13

822 Journel, A. G. & Huijbregts, C. J. 1978. *Mining Geostatistics*. Academic Press, London. * 1

823 Judge, W. J. 1984. New light on Chaco Canyon. In D. G. Noble, ed. *New Light on Chaco Canyon*, pp. 1–12. School of American Research Press, Santa Fe, New Mexico, USA. * 14

824 Justesen, S. H. 1932. Influence of size and shape of plots on the precision of field experiments with potatoes. *Journal of Agricultural Science* **22**: 365–72. * 4

825 Kahn, F. & Mejia, K. 1990. Palm communities in wetland forest ecosystems of Peruvian Amazonia. *Forest Ecology and Management* **33/34**: 169–79. * 7

826 Kaiser, H. 1959. Die Stromung an Windschutzstreifen. *Berichte Deutsch. Wetterd.* **7**: 1–36. * 6

827 Kalamkar, R. J. 1932. Experimental error and the field plot technique with potatoes. *Journal of Agricultural Science* **22**: 373–83. * 4

828 Kalma, J. D. & Badham, R. 1972. The radiation balance of a tropical pasture, I. The reflection of short-wave radiation. *Agricultural Meteorology* **10**: 251–9. * 10

829 Kandinsky, W. 1979. *Point and Line to Plane*. Dover Publications, New York. * 1

830 Kangas, P. C. 1989. An energy theory of landscape for classifying wetlands. In A. Lugo, M. Brinson & S. Brown, eds. *Forested Wetlands of the World*, pp. 15–23. Elsevier, Amsterdam. * 4, 10, 11

831 Kapos, V. 1989. Effects of isolation on the water status of forest patches in the Brazilian Amazon. *Journal of Tropical Ecology* **5**: 173–85. * 2, 3

832 Kareiva, P. M. 1983. Local movement in herbivorous insects: applying a passive diffusion model to mark–recapture field experiments. *Oecologia* (Berlin) **57**: 322–7. * 11

833 Kareiva. P. M. 1985. Finding and losing host plants by Phyllotreta: patch size and surrounding habitat. *Ecology* **66**: 1809–16. * 1, 11

558

834 Kareiva, P. M. 1987. Habitat fragmentation and the stability of predator–prey interactions. *Nature* **326:** 388–90. * 8

835 Kareiva, P. M. 1990. Population dynamics in spatially complex environments: theory and data. *Philosophical Transactions of the Royal Society* (London) B **330:** 175–90. * 11

836 Kareiva, P. M. & Shigesada, N. 1983. Analyzing insect movement as a correlated random walk. *Oecologia* (Berlin) **56:** 234–8. * 11

837 Karr, J. R. 1982. Population variability and extinction in the avifauna of a tropical land bridge island. *Ecology* **63:** 1975–8. * 11,12

838 Karr, J. R. 1989. Kissimmee River: restoration of degraded resources. In *Proceedings Kissimmee River Restoration Symposium*. South Florida Water Management District, West Palm Beach, Florida. * 13,14

839 Karr, J. R. 1991. Biological integrity: a long-neglected aspect of water resource management. *Ecological Applications* **1:** 66–84. *8,14

840 Karr, J. R. 1993. Protecting ecological integrity: an urgent societal goal. *Yale Journal of International Law* **18:** 297–306. * 14

841 Karr, J. R. & Freemark, K. E. 1983. Habitat selection and environmental gradients: dynamics in the 'stable' tropics. *Ecology* **64:** 1481–94. * 9

842 Karr, J. R. & Schlosser, I. J. 1977. *Impact of Nearstream Vegetation and Stream Morphology on Water Quality and Stream Biota*. Ecological Research Series, EPA-600/3–77–097, US Environmental Protection Agency, Athens, Georgia. 103 pp. * 7

843 Karr, J. R. & Schlosser, I. J. 1978. Water resources and the land–water interface. *Science* **201:** 229–34. * 7

844 Kaule, G. 1991. *Arten- und Biotopschutz, Zweite uberarbeitete und erweiterte Auflage*. Stuttgart, Germany. * 13

845 Kay, A. M. & Keough, M. J. 1981. Occupation of patches in the epifaunal communities on pier pilings and the bivalve *Pinna bicolor* at Edithburgh, South Australia. *Oecologia* **48:** 123–30. * 2

846 Keals, N. & Majer, J. D. 1991. The conservation status of ant communities along the Wubin–Perenjori corridor. In D. A. Saunders & R. J. Hobbs, eds. *Nature Conservation 2: The Role of Corridors*, pp. 387–93. Surrey Beatty, Chipping Norton, Australia. * 5

847 Keegan, W. F. & Diamond, J. M. 1987. Colonization of islands by humans: a biogeographical perspective. *Advances in Archaeological Method and Theory* **10:** 49–92. * 2

848 Kehm, W. 1993. Toward the recovery of nature: the Toronto Waterfront Regeneration Trust Agency case study. In *Proceedings of the International Conference on Landscape Planning and Environmental Conservation*, pp. 224–52. University of Tokyo, Tokyo. * 13

849 Keller, E. A. 1971. Pools, riffles and meanders: discussion. *Geological Society of America Bulletin* **82:** 279–80. * 7

850 Keller, E. A. 1980. The fluvial system: selected observations. In A. Sands, ed. *Riparian Forests in California: Their Ecology and Conservation*, pp. 39–46. Agricultural Sciences Publications 4101, University of California, Berkeley, California. * 7,13

851 Keller, E. A. & Swanson, F. J. 1979. Effects of large organic material on channel form and fluvial processes. *Earth Surface Processes* **4:** 361–80. * 7

852 Kemp, J. C. & Barrett, G. W. 1989. Spatial patterning: impact of uncultivated corridors on arthropod populations within soybean agroecosystems. *Ecology* **70:** 114–28. * 5,6,8

853 Kendeigh, S. C. 1944. Measurement of bird populations. *Ecological Monographs* **14:** 67–106. * 3

854 Kendeigh, S. C. 1982. *Bird Populations in East Central Illinois: Fluctuations, Variations, and Development Over a Half-century*. Illinois Biological Monograph 52, University of Illinois Press, Champaign, Illinois. * 2

855 Kenny, D. & Loehle, C. 1991. Are food webs randomly connected? *Ecology* **72:** 1794–9. * 8

856 Kershaw, K. A. & Looney, J. H. H. 1985. *Quantitative and Dynamic Plant Ecology*. 3rd edn. Edward Arnold, London. * 1,9,13

857 Kessell, S. R. 1979. *Gradient Modeling*. Springer-Verlag, New York. * 1,3

858 Kidd, C. V. & Pimentel, D., eds. 1992. *Integrated Resource Management: Agroforestry for Development*. Academic Press, San Diego, California. * 12,13,14

859 Kiemstedt, H. 1993. Landscape planning in Germany – a systematic ecological approach for environmental protection. In *Proceedings of the International Conference on*

Landscape Planning and Environmental Conservation, pp. 35–57. University of Tokyo, Tokyo. * 13

860 Kikkawa, J. 1964. Movement, activity and distribution of the small rodents *Clethrionomys glareolos* and *Apodemus sylvaticus* in woodland. *Journal of Animal Ecology* **33:** 259–99. * 4,11

861 Kilgore, B. M. & Taylor, D. 1979. Fire history of a sequoia-mixed conifer forest. *Ecology* **60:** 129–42. * 10

862 Kimball, J. W., Kozicky, E. L. & Nelson, B. A. 1956. Pheasants of the plains and prairies. In D. L. Allen, ed. *Pheasants in North America*, pp. 204–63. The Stockpole Co., Harrisburg, Pennsylvania, and the Wildlife Management Institute, Washington, DC. * 6

863 Kimura, M. & Crow, J. F. 1963. The measurement of effective population number. *Evolution* **17:** 279–88. * 2

864 King, A. W. 1991. Translating models across scales in the landscape. In M. G. Turner & R. H. Gardner, eds. *Quantitative Methods in Landscape Ecology*, pp. 479–517. Springer-Verlag, New York. * 1

865 Kinnaird, M. F. & O'Brien, T. G. 1991. Viable populations for an endangered forest primate, the Tana River Crested Mangabey (*Cercocebus galeritus galeritus*). *Conservation Biology* **5:** 203–13. * 2,11

866 Kirkpatrick, J. B. 1983. An iterative method for establishing priorities for the selection of nature reserves: an example from Tasmania. *Biological Conservation* **25:** 127–34. * 2

867 Kitchener, D. J., Chapman, A., Dell, J., Muir, B. G. & Palmer, M. 1980. Lizard assemblage and reserve size and structure in the Western Australian wheatbelt – some implications for conservation. *Biological Conservation* **17:** 25–62. * 2

868 Kitchener, D. J., Chapman, A., Muir, B. J. & Palmer, M. 1980. The conservation value for mammals of reserves in the Western Australian wheatbelt. *Biological Conservation* **18:** 179–207. * 2

869 Kitchener, D. J., Dell, J. & Muir, B. G. 1982. Birds in Western Australian wheatbelt reserves – implications for conservation. *Biological Conservation* **22:** 127–63. * 2

870 Klee, P. 1964. *The Thinking Eye: The Notebooks of Paul Klee.* 2nd edn. G. Wittenborn, New York. * 1,3

871 Klein, B. C. 1989. Effects of forest fragmentation on dung and carrion beetle communities in central Amazonia. *Ecology* **70:** 1715–25. * 2

872 Klein, D. R. 1971. Reaction of reindeer to obstructions and disturbances. *Science* **173:** 393–8. * 5

873 Klinge, H., Junk, W. J. & Revilla, C. J. 1990. Status and distribution of forested wetlands in tropical South America. *Forest Ecology and Management* **33/34:** 81–101. * 7

874 Klokk, T. 1981. Classification and ordination of river bank vegetation from middle and upper parts of the River Gaula, central Norway. *Kongelig Novske Videnskaps Selskaps Skrifter* **2:** 1–43. * 7

875 Klopatek, J. M. & Risser, P. G. 1982. Energy analysis of Oklahoma rangelands and improved pastures. *Journal of Range Management* **35:** 637–43. * 11

876 Klotzli, F. 1978. Ufersicherung-eine Kontaktzone zwischen Naturschutz und Wasserbau. *Berichte der ANL* **2:** 81–9. * 7

877 Knaapen, J. P., Scheffer, M. & Harms, B. 1992. Estimating habitat isolation in landscape planning. *Landscape and Urban Planning* **23:** 1–16. * 2,5,7,8,9,11,13

878 Knight, D. H. 1987. Parasites, lightning, and the vegetation mosaic in wilderness landscapes. In M. G. Turner, ed. *Landscape Heterogeneity and Disturbance*, pp. 59–83. Springer-Verlag, New York. * 1,10.

879 Knight, D. H. 1991. Congressional incentives for landscape research. *Bulletin of the Ecological Society of America* **72:** 195–203. * 1

880 Knight, D. H. & Wallace, L. L. 1989. The Yellowstone fires – issues in landscape ecology. *BioScience* **39:** 700–6. * 10, 12

881 Knutson, R. M. 1987. *Flattened Fauna: A Field Guide to Common Animals of Roads, Streets and Highways.* Ten Speed Press, Berkeley, California. * 5

882 Koeppl, J. W., Slade, N. A. & Hoffman, R. S. 1975. A bivariate home range model with possible application to ethological data analysis. *Journal of Mammalogy* **56:** 81–90. * 4

883 Kolasa, J. & Pickett, S. T. A., eds. 1991. *Ecological Heterogeneity.* Springer-Verlag, New York. * 1,8

884 Kolata, A. L. 1993. *The Tiwanaku: Portrait of an Andean Civilization*. Blackwell, Cambridge, Massachusetts, USA. * 14

885 Kondolf, G. M. & Wolman, M. G. 1993. The sizes of salmonid spawning gravels. *Water Resources Research* **29:** 2275–85. * 7,13

886 Kopecky, K. 1988. Einfluss der Strassen auf die Synanthropisierung der Flora und Vegetation nach Beobachtungen in der Tschechoslowakei. *Folia Geobotanica et Phytotaxonomica* (Praha) **23:** 145–71. * 5

887 Kopecky, K. & Hejny, S. 1973. Neue syntaxonomische Auffassung der Gesellschaften ein- bis zweijahriger Pflanzen der *Galio-Urticetea* in Bohmen. *Folio Geobotanica et Phytotaxonomica* (Praha) **8:** 49–66. * 3

888 Koppen, W. P. 1931. *Grundriss der Klimakunde*. Walter de Gruyter, Berlin. * 1,9

889 Kornas, J. 1983. Man's impact upon the flora and vegetation in Central Europe. In W. Holzner, M. J. A. Werger & I. Ikusima, eds. *Man's Impact on Vegetation*, pp. 277–86. W. Junk Publishers, The Hague. * 11

890 Kot, H. 1988. The effect of suburban landscape structure on communities of breeding birds. *Polish Ecological Studies* **14:** 235–61. * 13

891 Kotliar, N. B. & Wiens, J. A. 1990. Multiple scales of patchiness and patch structure: a hierarchical framework for the study of heterogeneity. *Oikos* **59:** 253–60. * 1

892 Kovar, P., Brabec, E. & Holubova, J. 1982. Particle deposition in Prague grasslands. *Ekologia* (Czechoslovakia) **1:** 251–6. * 10

893 Kozlowski, T. T. & Ahlgren, C. E., eds. 1974. *Fire and Ecosystems*. Academic Press, New York. * 10

894 Kozova, M., Smitalova, K. & Vizyova, A. 1986. Use of measures of network connectivity in the evaluation of ecological landscape stability. *Ekologia* (Czechoslovakia) **5:** 187–202. * 1,5,8

895 Kratz, T. K., Benson, B. J., Blood, E. R., Cunningham, G. L. & Dahlgren, R. A. 1991. The influence of landscape position on temporal variability in four North American ecosystems. *The American Naturalist* **138:** 355–78. * 9

896 Kress, S. W. 1985. *The Audubon Society Guide to Attracting Birds*. Scribner's, New York. * 13

897 Kroodsma, R. L. 1982. Bird community ecology on power-line corridors in East Tennessee. *Biological Conservation* **23:** 79–94. * 5

898 Kroodsma, R. L. 1984. Effect of edge on breeding forest bird species. *Wilson Bulletin* **96:** 426–36. * 3

899 Krummel, J. R., Gardner, R. H., Sugihara, G., O'Neill, R. V. & Coleman, P. R. 1987. Landscape patterns in a disturbed environment. *Oikos* **48:** 321–4. * 1,3,10

900 Kucera, C. L., ed. 1983. *Proceedings of the Seventh Annual North American Prairie Conference*. Southwest Missouri State University, Springfield, Missouri, USA. * 2

901 Kuchler, A. W. & Zonneveld, I. S. 1988. *Vegetation Mapping*. Kluwer Academic Publishers, Dordrecht, Netherlands. * 3,9

902 Kuennen, T. 1989. New Jersey's I-78 preserves mountain habitat. In T. Kuennen, ed. *Roads and Bridges* (February 1989), pp. 69–73. * 5

903 Kullman, L. 1986. Recent tree-limit history of *Picea abies* in the southern Swedish Scandes. *Canadian Journal of Forest Research* **16:** 761–71. * 3

904 Kung, E. C., Bryson, R. A. & Lenschow, D. H. 1964. Study of a continental surface albedo on the basis of flight measurements and structure of the earth's surface cover over North America. *Monthly Weather Review* **92:** 543–64. * 10

905 Kupfer, J. A. & Malanson, G. P. 1993. Structure and composition of a riparian forest edge. *Physical Geography* **14:** 154–70. * 7

906 Kushlan, J. A. 1979. Design and management of continental wildlife reserves: lessons from the Everglades. *Biological Conservation* **15:** 281–90. * 9

907 Lack, D. L. 1969. The number of bird species on islands. *Bird Study* **16:** 193–209. * 2

908 Lack, D. L. 1976. *Island Biology: Illustrated by the Land Birds of Jamaica*. University of California Press, Berkeley, California. * 2

909 Lack, P. C. 1988. Hedge intersections and breeding bird distribution in farmland. *Bird Study* **35:** 133–6. * 8

910 Lacy, R. C. 1987. Loss of genetic diversity from managed populations: interacting effects of drift, mutation, immigration, selection, and population subdivision. *Conservation Biology* **1:** 143–58. * 2, 11

561

911 Ladino, A. G. & Gates, J. E. 1981. Responses of animals to transmission-line corridor management practices. In D. Arner & R. E. Tillman, eds. *Environmental Concerns in Rights-of-way Management*, pp. 53–1 to 53–10. Special Study Project WS 78–141. Electric Power Research Institute, Palo Alto, California. * 5

912 Lagerwerff, J. V. & Specht, A. W. 1970. Contamination of roadside soil and vegetation with cadmium, nickel, lead, and zinc. *Environmental Science and Technology* **4**: 583–6. * 5

913 Laikhtman, D. L. 1964. *Physics of the Boundary Layer of the Atmosphere*. Israel Program for Scientific Translations, Jerusalem. * 10

914 Lalo, J. 1987. The problem of road kill. *American Forests* (September–October): 50–2, 72. * 5

915 Lande, R. 1988. Genetics and demography in biological conservation. *Science* **241**: 1455–60. * 2, 11

916 Lande, R. & Barrowclough, G. F. 1987. Effective population size, genetic variation, and their use in population management. In M. E. Soulé, ed. *Viable Populations for Conservation*, pp. 87–123. Cambridge University Press, New York. * 2, 11

917 Landers, J. L., Hamilton, R. J., Johnson, A. S. & Marchington, R. L. 1979. Foods and habitat of black bears in southeastern North Carolina. *Journal of Wildlife Management* **43**: 143–53. * 2, 9, 11

918 Langton, T. E. S., ed. 1989. *Amphibians and Roads*. ACO Polymer Products Ltd, Shefford, Bedfordshire, England. * 5

919 Lankester, K., van Apeldoorn, R., Meelis, E. & Verboom, J. 1991. Management perspectives for populations of the Eurasian badger (*Meles meles*) in a fragmented landscape. *Journal of Applied Ecology* **28**: 561–73. * 11

920 Lansing, J. S. & Kremer, J. N. 1994. Emergent properties of Balinese water temple networks: coadaptation on a rugged fitness landscape. In C. G. Langton, ed. *Artificial Life III*, pp. 201–23. Addison-Wesley, Reading, Massachusetts, USA. * 9, 13, 14

921 Larsen, J. A. 1988. *The Northern Forest Border in Canada and Alaska*. Springer-Verlag, New York. * 3

922 Laudenslayer, W. F., Jr. & Balda, R. P. 1976. Breeding bird use of a pinyon–juniper–ponderosa pine ecotone. *Auk* **93**: 571–86. * 3

923 Laurance, W. F. 1991. Ecological correlates of extinction proneness in Australian tropical rain forest mammals. *Conservation Biology* **5**: 79–89. * 12

924 Laurance, W. F. & Yensen, E. 1991. Predicting the impacts of edge effects in fragmented habitats. *Biological Conservation* **55**: 77–92. * 2, 3, 4

925 Lawley, B. J. 1957. The discovery, investigation and control of scrub typhus in Singapore. *Transactions of the Royal Society of Tropical Medicine and Hygiene* **51**: 56–61. * 5

926 Lawton, R. M. 1967. The conservation and management of the riparian evergreen forests of Zambia. *Commonwealth Forestry Review* **46**: 223–32. * 7

927 Lawton, R. O. 1990. Community gaps and light penetration into a wind-exposed tropical lower montane rainforest. *Canadian Journal of Forest Research* **20**: 659–67. * 2,3

928 Lay, D. 1938. How valuable are woodland clearings to birdlife. *Wilson Bulletin* **50**: 254–6. * 2,3

929 Learner, M. A., Bowker, D. W. & Halewood, J. 1990. An assessment of bank slope as a predictor of conservation status in river corridors. *Biological Conservation* **54**: 1–13. * 7

930 Leberg, P. L. 1991. Influence of fragmentation and bottlenecks on genetic divergence of wild turkey populations. *Conservation Biology* **5**: 522–30. * 11

931 Leck, C. F. 1979. Avian extinctions in an isolated tropical wet-forest preserve, Ecuador. *Auk* **96**: 343–52. * 2

932 Ledig, F. T. 1986. Heterozygosity, heterosis, and fitness in outbreeding plants. In M. E. Soulé, ed. *Conservation Biology: The Science of Scarcity and Diversity*, pp. 77–104. Sinauer Associates, Sunderland, Massachusetts, USA. * 2

933 Ledig, F. T. & Little, S. 1979. Pitch pine (*Pinus rigida* Mill.): ecology, physiology, and genetics. In R. T. T. Forman, ed. *Pine Barrens: Ecosystem and Landscape*, pp. 347–71. Academic Press, New York. * 2

934 Leduc, J. P. 1979. Le role du bocage. *Recherche et Nature* **20**: 4–9. * 8

935 Lee, D. R. & Sallee, G. T. 1970. A method of measuring shape. *Geographical Review* **60**: 555–63. * 4

937 Lee, M. A. B. 1974. Distribution of native and invader plant species on the island of Guam. *Biotropica* **6**: 158–64. * 11

938 Lees, G. 1964. A new method for determining the angularity of particles. *Sedimentology* **3**: 2–21. * 4

939 Lefkovitch, L. P. & Fahrig, L. 1985. Spatial characteristics of habitat patches and population survival. *Ecological Modelling* **30**: 297–308. * 8

940 LeFranc, M. N., Moss, M. B., Patnode, K. A. & Sugg, W. C., III, eds. 1987. *Grizzly Bear Compendium*. National Wildlife Federation and Interagency Grizzly Bear Committee, Washington, DC. * 2,7

941 Lemly, A. D. 1982. Modification of benthic insect communities in polluted streams: effects of sedimentation and nutrient enrichment. *Hydrobiologia* **87**: 229–45. * 5,7

942 Leonard, P. L. & Cobham, R. O. 1977. The farming landscape of England and Wales: a changing scene. *Landscape Planning* **4**: 205–36. * 8

943 Leopold, A. 1933. *Game Management*. Scribners, New York. * 3,4,8,9,11,12,13

944 Leopold, A. 1941. Cheat takes over. *The Land* **1**: 310–13. * 11,14

945 Leopold, A. 1949. *A Sand County Almanac, and Sketches Here and There*. Oxford University Press, New York. * 13,14

946 Leopold, L. B., Wolman, M. G. & Miller, J. P. 1964. *Fluvial Processes in Geomorphology*. W. H. Freeman, San Francisco. * 7

947 Lepart, J. & Debussche, M. 1992. Human impact on landscape patterning: Mediterranean examples. In A. J. Hansen & F. di Castri, eds. *Landscape Boundaries: Ecological Consequences on Biotic Diversity and Ecological Flows*, pp. 76–106. Springer-Verlag, New York. * 1, 13

948 *Les Bocages: Histoire, Ecologie, Economie*. 1976. Institut National de la Recherche Agronomique, Centre National de la Recherche Scientifique, et Universite de Rennes, Rennes, France. 586 pp. * 1,5,6,8,13

950 Leser, H. 1978. *Landschaftsokologie*. 2nd edn. Ulmer, Stuttgart. * 1

951 Leser, H. 1991. *Landschaftsokologie: Ansatz, Modelle, Methodik, Anwendung*. Verlag Eugen Ulmer, Stuttgart. * 1

952 Leung, Y. 1987. On the imprecision of boundaries. *Geographic Analysis* **19**: 125–51. * 3

953 Levenson, J. B. 1981. Woodlots as biogeographic islands in southeastern Wisconsin. In R. L. Burgess & D. M. Sharpe, eds. *Forest Island Dynamics in Man-dominated Landscapes*, pp. 13–39. Springer-Verlag, New York. * 2

954 Levin, S. A. 1976. Population dynamic models in heterogeneous environments. *Annual Review of Ecology and Systematics* **7**: 287–310. * 1, 11

955 Levin, S. A. 1981. Models of population dispersal. In S. N. Busenberg & K. L. Cooke, eds. *Differential Equations and Applications in Ecology, Epidemics, and Population Problems*, pp. 1–18. Academic Press, New York. * 11

956 Levin, S. A. 1989. Challenges in the development of a theory of community and ecosystem structure and function. In J. Roughgarden, R. M. May & S. A. Levin, eds. *Perspectives in Ecological Theory*, pp. 242–55. Princeton University Press, Princeton, New Jersey. * 8

957 Levin, S. A. 1992. The problem of pattern and scale in ecology. *Ecology* **73**: 1943–67. * 1

958 Levins, R. 1968. *Evolution in Changing Environments*. Princeton University Press, Princeton, New Jersey. * 1

959 Levins, R. 1970. Extinction. *Some Mathematical Questions in Biology*, vol. 2, pp. 75–107. American Mathematical Society, Providence, Rhode Island, USA. * 11

960 Lewin, R. 1987. Ecological invasions offer opportunities. *Science* **238**: 752–3. * 11

961 Lewis, P. H., Jr. 1964. Quality corridors in Wisconsin. *Landscape Architecture* **54**: 100–7. * 1,6,13

962 Lewis, S. A. 1991. The conservation and management of roadside vegetation in South Australia. In D. A. Saunders & R. J. Hobbs, eds. *Nature Conservation 2: The Role of Corridors*, pp. 313–18. Surrey Beatty, Chipping Norton, Australia. * 5,6

963 Lewis, T. 1969a. The distribution of flying insects near a low hedgerow. *Journal of Applied Ecology* **6**: 443–52. * 8,11

964 Lewis, T. 1969b. The diversity of the insect fauna in a hedgerow and neighboring fields. *Journal of Applied Ecology* **6:** 453–8. * 6,8,11

965 Lewontin, R. C. 1974. *The Genetic Basis of Evolutionary Change*. Columbia University Press, New York. * 11

966 Li, H. & Franklin, J. F. 1988. Landscape ecology: a new conceptual framework in ecology. (In Chinese.) *Advances in Ecology* (China) **5:** 23–33. * 1

967 Li, H., Franklin, J. F., Swanson, F. J. & Spies, T. A. 1993. Developing alternative forest cutting patterns: a simulation approach. *Landscape Ecology* **8:** 63–75. * 11,12

968 Liddle, M. J. 1975. A selective review of the ecological effects of human trampling on natural ecosystems. *Biological Conservation* **7:** 11–36. * 5

969 Lidicker, W. Z., Jr., Wolff, J. O., Lidicker, L. N. & Smith, M. H. 1992. Utilization of a habitat mosaic by cotton rats during a population decline. *Landscape Ecology* **6:** 259–68. * 9,11

970 Lienenbecker, H. & Raabe, U. 1981. Veg auf Bahnhofen des Ost-Munsterlandes. Berichte naturw. ver. Bielefeld **25:** 129–41. * 5

971 Liew, T. C. & Wong, F. O. 1973. Density, recruitment, mortality and growth of dipterocarp seedlings in virgin and logged-over forest in Sabah. *The Malaysian Forester* **36:** 3–15. * 2

972 Likens, G. E. & Bormann, F. H. 1974. Linkages between terrestrial and aquatic ecosystems. *BioScience* **24:** 447–56. * 7

973 Likens, G. E., Bormann, F. H., Johnson, N. M., Fisher, D. W. & Pierce, R. S. 1970. Effects of forest cutting and herbicide treatment on nutrient budgets in the Hubbard Brook watershed-ecosystem. *Ecological Monographs* **40:** 23–47. * 7

974 Likens, G. E., Bormann, F. H., Pierce, R. S., Eaton, J. S. & Johnson, N. M. 1977. *Biogeochemistry of a Forested Ecosystem*. Springer-Verlag, New York. * 7,10,12

975 Likens, G. E. & Butler, T. J. 1981. Recent acidification of precipitation in North America. *Atmospheric Environment* **15:** 1103–9. * 10

976 Lindenmayer, D. B. & Nix, H. A. 1993. Ecological principles for the design of wildlife corridors. *Conservation Biology* **7:** 627–30. * 6,7

977 Little, C. E. 1990. *Greenways for America*. Johns Hopkins University Press, Baltimore, Maryland. * 5,6,13

978 Little, S. 1979. Fire and plant succession in the New Jersey Pine Barrens. In R. T. T. Forman, ed. *Pine Barrens: Ecosystem and Landscape*, pp. 297–314. Academic Press, New York. * 10

979 Liu, J., Cubbage, F. W. & Pulliam, H. R. 1994. Ecological and economic effects of forest landscape structure and rotation length: simulation studies using ECOLECON. *Ecological Economics*. **10:** 249–263. * 1,9

980 Lo, C. P. 1980. Changes in the shape of Chinese cities, 1934–1974. *Professional Geographer* **32:** 173–83. * 4

981 Loman, J. 1991. Small mammal and raptor densities in habitat islands; area effects in a south Swedish agricultural landscape. *Landscape Ecology* **5:** 183–9. * 2

982 Lomolino, M. V., Brown, J. H. & Davis, R. 1989. Island biogeography of montane forest mammals in the American Southwest. *Ecology* **70:** 180–94. * 2

983 Lord, J. M. & Norton, D. A. 1990. Scale and the spatial concept of fragmentation. *Conservation Biology* **4:** 197–202. * 12

984 Lorimer, C. G. 1977. The presettlement forest and natural disturbance cycle of northeastern Maine. *Ecology* **58:** 139–48. * 10

985 Losch, A. 1944. *Die Raumliche Ordnung der Wirtschaft*, 2nd edn. Gustav Fischer, Jena. * 4

986 Loucks, O. L. 1970. Evolution of diversity, efficiency, and community stability. *American Zoologist* **10:** 17–25. * 2, 10

987 Loucks, O. L., Plumb-Mentjes, M. L. & Rogers, D. 1985. Gap processes and large-scale disturbance in sand prairies. In S. T. A. Pickett & P. S. White, eds. *The Ecology of Natural Disturbance and Patch Dynamics*, pp. 71–83. Academic Press, New York. * 2

988 Love, D., Grzybowski, J. A. & Knopf, F. L. 1985. Influence of various land uses on windbreak selection by nesting Mississippi Kites. *Wilson Bulletin* **97:** 561–5. * 6

989 Lovejoy, T. E. 1987. National parks: how big is big enough? In R. Herrmann & T. B. Craig, eds. *Conference on Science in National Parks: The Fourth Triennial Conference on Research in the National Parks and Equivalent Reserves*, pp. 49–58. George Wright Society and US National Park Service, Denver, Colorado. * 2

564

990 Lovejoy, T. E., Bierregaard, R. O., Jr., Rylands, A. B., Malcolm, J. R., Quintela, C. E., Harper, L. H., Brown, K. S., Jr., Powell, A. H., Powell, G. V. H., Schubart, H. O. R. & Hays, M. B. 1986. Edge and other effects of isolation on Amazon forest fragments. In M. Soulé, ed. *Conservation Biology: The Science of Scarcity and Diversity*, pp. 251–85. Sinauer Associates, Sunderland, Massachusetts, USA. * 1, 2, 3

991 Lovejoy, T. E., Rankin, J. M., Bierregaard, R. O., Jr., Brown, K. S., Jr., Emmons, L. H. & Van der Voort, M. 1984. Ecosystem decay of Amazon forest remnants. In M. H. Niteki, ed. *Extinctions*, pp. 295–325. University of Chicago Press, Chicago. * 1, 2, 12, 13

992 Lovejoy, T. E. & Salati, E. 1983. Precipitating change in Amazonia. In E. F. Moran, ed. *The Dilemma of Amazonian Development*, pp. 211–20. Westview Press, Boulder, Colorado, USA. * 12

993 Lovelock, J. E. & Margulis, L. 1974. Atmospheric homeostasis by and for the biosphere: the Gaia hypothesis. *Tellus* **26**: 1–10. * 14

994 Lovett, G. M., Reiners, W. A. & Olson, R. K. 1982. Cloud droplet deposition in subalpine balsam fir forests: hydrological and chemical inputs. *Science* **218**: 1303–4. * 10

995 Lowe, J. C. & Moryadas, S. 1975. *The Geography of Movement*. Houghton Mifflin, Boston. * 1, 5, 8, 9, 11, 12

996 Lowrance, R., Leonard, R. & Sheridan, J. M. 1985. Managing riparian ecosystems to control nonpoint pollution. *Journal of Soil and Water Conservation* **40**: 87–91. * 7

997 Lowrance, R., McIntyre, S. & Lance, C. 1988. Erosion and deposition in a field/forest system estimated using cesium-137 activity. *Journal of Soil and Water Conservation* **43**: 195–9. * 7, 10

998 Lowrance, R., Sharpe, J. K. & Sheridan, J. M. 1986. Long-term sediment deposition in the riparian zone of a coastal plain watershed. *Journal of Soil and Water Conservation* **41**: 266–71. * 7, 10

999 Lowrance, R., Todd, R., Fail, J., Jr., Hendrickson, O., Jr., Leonard, R. & Asmussen, L. 1984. Riparian forests as nutrient filters in agricultural watersheds. *BioScience* **34**: 374–7. * 1, 5, 7, 9, 13

1000 Loyn, R. H. 1987. Effects of patch area and habitat on bird abundances, species numbers and tree health in fragmented Victorian forests. In D. A. Saunders, G. W. Arnold, A. A. Burbidge & A. J. M. Hopkins, eds. *Nature Conservation: The Role of Remnants of Native Vegetation*, pp. 65–77. Surrey Beatty, Chipping Norton, Australia. * 2, 12

1001 Loyn, R. H., Runnalls, R. G., Forward, G. Y. & Tyers, J. 1983. Territorial bell miners and other birds affecting populations of insect prey. *Science* **221**: 1411–13. * 2

1002 Lubchenko, J., Olson, A. M., Brubaker, L. B., Carpenter, S. R., Holland, M. M., Hubbell, S. P., Levin, S. A., MacMahon, J. A., Matson, P. A., Melillo, J. M., Mooney, H. A., Peterson, C. H., Pulliam, H. R., Real, L. A., Regal, P. J. & Risser, P. G. 1991. The Sustainable Biosphere Initiative: an ecological research agenda. *Ecology* **72**: 371–412. (Reprinted 1991 in *Ecology International* **20**: 17–58.) * 13, 14

1003 Lubina, J. A. & Levin, S. A. 1988. The spread of a reinvading species: range expansion in the California sea otter. *American Naturalist* **131**: 526–43. * 11

1004 Lucas, O. W. R. 1991. *The Design of Forest Landscapes*. Oxford University Press, Oxford. * 13

1005 Ludwig, J. A., Whitford, W. G., Rodney, A. B. & Grieve, R. E. 1977. An evaluation of transmission line construction on piñon–juniper woodland and grassland communities in New Mexico. *Journal of Environmental Management* **5**: 127–37. * 5

1006 Luey, J. E. & Adelman, I. R. 1980. Downstream natural areas as refuges for fish in drainage development watersheds. *Transactions of the American Fisheries Society* **109**: 332–5. * 8

1007 Lull, H. W. 1951. Forest fire smoke of September 1950. *Journal of Forestry* **49**: 286. * 10

1008 Lulla, K. & Mausel, P. 1983. Ecological applications of remotely sensed multispectral data. In B. F. Richason, Jr., ed. *Introduction to Remote Sensing of the Environment*, 2nd edn., pp. 354–77. Kendall/Hunt, Dubuque, Iowa, USA. * 1

1009 Lunan, J. S. & Habeck, J. R. 1973. The effects of fire exclusion on ponderosa pine communities in Glacier National Park, Montana. *Canadian Journal of Forest Research* **3**: 574–9. * 10

1010 Lundahl, M. 1979. *Peasants and Poverty: A Study of Haiti*. St. Martin's Press, New York. * 14

1011 Luvall, J. C. & Holbo, H. R. 1991. Thermal remote sensing methods in landscape ecology. In M. G. Turner & R. H. Gardner, eds. *Quantitative Methods in Landscape Ecology*, pp. 127–52. Springer-Verlag, New York. * 1, 10

1012 Lyles, L. 1988. Basic wind erosion processes. *Agriculture, Ecosystems and Environment* **22/23:** 91–101. (Reprinted 1988 in *Windbreak Technology*. Elsevier, Amsterdam.) * 6

1013 Lynch, J. F. 1987. Responses of breeding bird communities to forest fragmentation. In D. A. Saunders, G. W. Arnold, A. A. Burbidge & A. J. M. Hopkins, eds. *Nature Conservation: The Role of Remnants of Native Vegetation*, pp. 123–40. Surrey Beatty, Chipping Norton, Australia. * 2

1014 Lynch, J. F. & Johnson, N. K. 1974. Turnover and equilibria in insular avifaunas, with special reference to the California Channel Islands. *Condor* **76:** 370–84. * 2

1015 Lynch, J. F. & Saunders, D. A. 1991. Responses of bird species to habitat fragmentation in the wheatbelt of Western Australia: interiors, edges and corridors. In D. A. Saunders & R. J. Hobbs, eds. *Nature Conservation 2: The Role of Corridors*, pp. 143–58. Surrey Beatty, Chipping Norton, Australia.* 3, 5, 6

1016 Lynch, J. F. & Whigham, D. F. 1984. Effects of forest fragmentation on breeding bird communities in Maryland, USA. *Biological Conservation* **28:** 287–324.* 2, 11, 12

1017 Lynch, K. 1960. *The Image of the City*. MIT Press & Harvard University Press, Cambridge, Massachusetts.* 1,13

1018 Lyon, L. J. 1959. An evaluation of woody cover plantings as pheasant winter cover. *Transactions of the North American Wildlife Conference* **24:** 277–89.* 6

1019 Lyon, L. J. 1983. Road density models describing habitat effectiveness for elk. *Journal of Forestry* **81:** 592–5.* 8

1020 Mabbutt, J. A. & Fanning, P. C. 1987. Vegetation banding in arid western Australia. *Journal of Arid Environments* **12:** 41–59.* 4

1021 Mabelis, A. 1990. Natuurwaarden in cultuurlandschappen. *Landschap* **7:** 253–68.* 11

1022 MacArthur, R. H. & Levins, R. 1964. Competition, habitat selection and character displacement in a patchy environment. *Proceedings of the National Academy of Science* (USA) **51:** 1207–10.* 1

1023 MacArthur, R. H. & Wilson, E. O. 1967. *The Theory of Island Biogeography*. Princeton University Press, Princeton, New Jersey.* 1,2

1024 Macdonald, I. A. W., Loope, L. L., Usher, M. B. & Hamann, O. 1989. Wildlife conservation and the invasion of nature reserves by introduced species: a global perspective. In J. A. Drake, H. A. Mooney, F. di Castri, R. H. Groves, F. J. Kruger, M. Rejmanek & M. Williamson, eds. *Biological Invasions: A Global Perspective*, pp. 215–56. John Wiley, Chichester, UK.* 11

1025 MacDonald, L. H., Smart, A. W. & Wissmar, R. C. 1991. *Monitoring Guidelines to Evaluate Effects of Forestry Activities on Streams in the Pacific Northwest and Alaska*. US Environmental Protection Agency, Corvallis, Oregon.* 13

1026 Mack, R. N. 1981. Invasion of *Bromus tectorum* L. into western North America: an ecological chronicle. *Agro-Ecosystems* **7:** 145–65.* 11

1027 Mack, R. N. 1986. Alien plant invasion into the Intermountain West: a case history. In H. A. Mooney & J. A. Drake, eds. *Ecology of Biological Invasions of North America and Hawaii*, pp. 191–213. Springer-Verlag, New York.* 11

1028 Mackintosh, G., ed. 1989. *Preserving Communities & Corridors*. Defenders of Wildlife, Washington, DC.* 13

1029 MacMahon, J. A. 1980. Ecosystems over time: succession and other types of change. In R. H. Waring, ed. *Forests: Fresh Perspectives from Ecosystem Analyses*, pp. 27–58. Oregon State University Press, Corvallis, Oregon, USA. * 2,12

1030 Madden, J. P. 1967. *Economies of Size in Farming, Theory, Analytical Procedures, and Review of Selected Studies*. Economic Research Service, Report 107, US Department of Agriculture, Washington, DC. 83 pp.* 2

1031 Mader, H-J. 1981. Untersuchungen zum Einfluss der Flachengrosse von Inselbiotopen auf deren Funktion als Trittstein oder Refugium. *Natur und Landschaft* **56:** 235–42.* 2

566

1032 Mader, H-J. 1984. Animal habitat isolation by roads and agricultural fields. *Biological Conservation* **29**: 81–96. * 1,4,5

1033 Mader, H-J. 1988. The significance of paved agricultural roads as barriers to ground dwelling arthropods. In K-F. Schreiber, ed. *Connectivity in Landscape Ecology*, pp. 97–100. Munstersche Geographische Arbeiten 29, Ferdinand Schoningh, Paderborn, Germany. * 1,5

1034 Mader, H-J. & Pauritsch, G. 1981. Nachweis des Barriere-Effektes von verkehrsarmen Strassen und Forstwegen auf Kleinsauger der Waldbiozonose durch Markierungs- und Umsetzungsversuche. *Natur und Landschaft* **56**: 451–4. *1,5

1035 Maehr, D. 1990. The Florida panther and private lands. *Conservation Biology* **4**: 167–70. *2,5

1036 Magnuson, J. J. 1976. Managing with exotics – a game of chance. *Transactions of the American Fisheries Society* **105**: 1–9. * 2,11

1037 Maitland, P. S. 1974. The conservation of freshwater fishes in the British Isles. *Biological Conservation* **6**: 7–14. * 8

1038 Malanson, G. P. 1993. *Riparian Landscapes*. Cambridge University Press, Cambridge. * 1,5,7,8,10,12,13

1039 Mandelbrot, B. B. 1983. *The Fractal Geometry of Nature*. W. H. Freeman, New York. * 1,3,4,9

1040 Mander, U., Jagomaegi, J. & Kuelvik, M. 1988. Network of compensative areas as an ecological infrastructure of territories. In K-F. Schreiber, ed. *Connectivity in Landscape Ecology*, pp. 35–8. Munstersche Geographische Arbeiten 29. Ferdinand Schoningh, Paderborn, Germany. * 9

1041 Mann, C. C. & Plummer, M. L. 1993. The high cost of biodiversity. *Science* **260**: 1868–71. * 13

1042 Mansergh, I. M. & Scotts, D. J. 1989. Habitat continuity and social organisation of the mountain pygmy-possum restored by tunnel. *Journal of Wildlife Management* **53**: 701–7. * 5

1043 Marcot, B. G. & Meretsky, V. J. 1983. Shaping stands to enhance habitat diversity. *Journal of Forestry* **81**: 527–8. * 4

1044 Margalef, R. 1963. On certain unifying principles in ecology. *American Naturalist* **97**: 357–74. * 3

1045 Margalef, R. 1979. The organization of space. *Oikos* **33**: 152–9. * 3

1046 Margules, C. R. 1989a. Introduction to some Australian developments in conservation evaluation. *Biological Conservation* **50**: 1–11. * 9

1047 Margules, C. R. 1989b. Selecting nature reserves in South Australia. In J. C. Noble & R. A. Bradstock, eds. *Mediterranean Landscapes in Australia: Mallee Ecosystems and Their Management*, pp. 498–505. CSIRO, Melbourne. * 2, 9

1048 Margules, C. R. 1992. The Wog Wog habitat fragmentation experiment. *Environmental Conservation* **19**: 316–25. * 1,2

1049 Margules, C. R. & Austin, M. P., eds. 1991. *Nature Conservation: Cost Effective Biological Surveys and Data Analysis*. CSIRO Australia, Canberra. * 2,9

1050 Margules, C. R. & Nicholls, A. O. 1987. Assessing the conservation value of remnant habitat 'islands': mallee patches on the western Eyre Peninsula, South Australia. In D. R. Saunders, G. W. Arnold, A. A. Burbidge & A. J. M. Hopkins, eds. *Nature Conservation: The Role of Remnants of Native Vegetation*, pp. 89–102. Surrey Beatty, Chipping Norton, Australia. * 2

1051 Margules, C. R., Nicholls, A. O. & Pressey, R. L. 1988. Selecting networks of reserves to maximise biological diversity. *Biological Conservation* **43**: 63–76. * 2,9,12

1052 Margules, C. & Usher, M. B. 1981. Criteria used in assessing wildlife conservation potential: a review. *Biological Conservation* **21**: 79–109. * 2,9

1053 Marr, J. W. 1977. The development and movement of tree islands near the upper limit of tree growth in the southern Rocky Mountains. *Ecology* **58**: 1159–64. * 4

1054 Marsh, P. C. & Luey, J. E. 1982. Oases for aquatic life within agricultural watersheds. *Fisheries* **7(6)**: 16–19,24. * 8

1055 Marsh, W. M. 1991. *Landscape Planning: Environmental Applications*. John Wiley, New York. * 13

567

1056 Marshall, J. K. 1967. The effect of shelter on the productivity of grasslands and field crops. *Field Crop Abstracts* **20:** 1–14. *2,3,6

1057 Martin, A. A. & Tyler, M. J. 1978. The introduction into Western Australia of the frog *Limnodynastes tasmaniensis. Australian Zoologist* **19:** 320–44. * 6,11

1058 Martin, T. E. 1981. Limitation in small habitat islands: chance or competition? *Auk* **98:** 715–34. * 6

1059 Martin, T. E. & Karr, J. R. 1986. Patch utilization by migrating birds: resource oriented? *Ornis Scandinavica* **17:** 165–74. * 6

1060 Martinsson, B., Hansson, L. & Angelstam, P. 1993. Small mammal dynamics in adjacent landscapes with varying predator communities. *Annales Zoologici Fennici* **30:** 31–42. * 14

1061 Maruyama, T. & Kimura, M. 1980. Genetic variability and effective population size when local extinction and recolonization of subpopulations are frequent. *Proceedings of the National Academy of Science* (USA) **77:** 6710–14. * 2,11

1062 Massam, B. H. 1975. *Location and Space in Social Administration.* John Wiley, New York. * 4

1063 Matheron, G. 1965. *Les Variables Regionalisees et Leur Estimation.* Masson, Paris. * 4

1064 Matlack, G. R. 1993. Sociological edge effects: spatial distribution of human impact in suburban forest fragments. *Environmental Management* **17:** 829–35. * 3,13

1065 Matlack, G. R. 1994a. Plant species migration in a mixed-history forest landscape in eastern North America. *Ecology* **75:** 1491–1502. * 12

1066 Matlack, G. R. 1994b. Vegetation dynamics of the forest edge: trends in space and successional time. *Journal of Ecology.* **82:** 113–23. * 3

1067 Matthiae, P. E. & Stearns, F. 1981. Mammals in forest islands in southeastern Wisconsin. In R. L. Burgess & D. M. Sharpe, eds. *Forest Island Dynamics in Man-dominated Landscapes,* pp. 55–66. Springer-Verlag, New York. * 2, 12

1068 Matthysen, E. 1987. A longterm study on a population of the European nuthatch, *Sitta europaea caesia* Wolf. *Sitta* **1:** 2–17. * 11

1069 May, R. M. 1975. Island biogeography and the design of wildlife preserves. *Nature* **254:** 177–8. * 2

1070 Mayr, E. 1963. *Animal Species and Evolution.* Belknap Press of Harvard University Press, Cambridge, Massachusetts. * 2

1071 Mayr, E. 1982. Speciation and macroevolution. *Evolution* **36:** 1119–32. * 2, 3

1072 McArthur, D. & Erlich, R. 1977. An efficiency evaluation of four drainage basin shape ratios. *Professional Geographer* **24:** 290–5. * 4

1073 McArthur, J. V., Kovacic, D. A. & Smith, M. H. 1988. Genetic diversity in natural populations of a soil bacterium across a landscape gradient. *Proceedings of the National Academy of Science* (USA) **85:** 9621–4. * 11

1074 McClanahan, T. R. 1986. Seed dispersal from vegetation islands. *Ecological Modeling* **32:** 301–9. * 11

1075 McColl, J. G. 1978. Ionic composition of forest soil solutions and effects of clearcutting. *Soil Science Society of America Journal* **42:** 358–63. * 10

1076 McColl, R. H. S. 1978. Chemical runoff from pastures: the influence of fertilizer and riparian zones. *New Zealand Journal of Marine and Freshwater Research* **12:** 371–80. * 7

1077 McCoy, E. D. 1983. The application of island-biogeographic theory to patches of habitat: how much land is enough? *Biological Conservation* **25:** 53–61. * 2

1078 McCullouch, C. E. & Cain, M. L. 1989. Analyzing discrete movement data as a correlated random walk. *Ecology* **70:** 383–8. * 11

1079 McDonnell, M. J. 1981. Trampling effects on coastal dune vegetation in the Parker River National Wildlife Refuge, Massachusetts, USA. *Biological Conservation* **21:** 289–301. * 5

1080 McDonnell, M. J. & Pickett, S. T. A., eds. 1993. *Humans as Components of Ecosystems: The Ecology of Subtle Human Effects and Populated Areas.* Springer-Verlag, New York. * 13

1081 McDonnell, M. J., Pickett, S. T. A. & Pouyat, R. V. 1993. The application of the ecological gradient paradigm to the study of urban effects. In M. J. McDonnell & S. T. A.

Pickett, eds. *Humans as Components of Ecosystems: The Ecology of Subtle Human Effects and Populated Areas*, pp. 1–20. Springer-Verlag, New York. * 13

1082 McDonnell, M. J. & Stiles, E. W. 1983. The structural complexity of old field vegetation and the recruitment of bird-dispersed plant species. *Oecologia* **56:** 109–16. * 1, 11

1083 McDowell, C. R., Low, A. B. & McKenzie, B. 1991. Natural remnants and corridors in Greater Cape Town: their role in threatened plant conservation. In D. A. Saunders & R. J. Hobbs, eds. *Nature Conservation 2: The Role of Corridors*, pp. 27–39. Surrey Beatty, Chipping Norton, Australia. * 6

1084 McDowell, P. F., Webb, T. III & Bartlein, P. J. 1990. Long-term environmental change. In B. L. Turner II, W. C. Clark, R. W. Kates, J. F. Richards, J. T. Mathews & W. B. Meyer. 1990. *The Earth as Transformed by Human Action*, pp. 143–62. Cambridge University Press, Cambridge. * 1, 10, 14

1085 McElroy, M. B. 1986. Change in the natural environment of the Earth: the historical record. In W. C. Clark & R. E. Munn, eds. *Sustainable Development of the Biosphere*, pp. 199–211. Cambridge University Press, Cambridge. * 14

1086 McGinley, M. A. & Whitham, T. G. 1985. Central place foraging by beavers (*Castor canadensis*): test of foraging predictions and the impact of selective feeding on the growth form of cottonwoods (*Populus fremontii*). *Oecologia* **66:** 558–62. * 7

1087 McGuinness, K. A. 1984. Equations and explanations in the study of species–area curves. *Biological Review* **59:** 423–40. * 2

1088 McHarg, I. L. 1969. *Design with Nature*. Doubleday, Garden City, New York. * 13

1089 McIntosh, R. P. 1985. *The Background of Ecology: Concept and Theory*. Cambridge University Press, Cambridge. * 1

1090 McKendry, J. E. & Machlis, G. E. 1993. The role of geography in extending biodiversity gap analysis. *Applied Geography* **13:** 135–52. * 9

1091 McKenzie, N. L., Belbin, L., Margules, C. R. & Keighery, G. J. 1989. Selecting representative reserve systems in remote areas: a case study in the Nullarbor region, Australia. *Biological Conservation* **50:** 239–61. * 9

1092 McNaughton, K. G. 1983. The direct effect of shelter on evaporation rates: theory and an experimental test. *Agricultural Meteorology* **29:** 125–36. * 6

1093 McNaughton, K. G. 1988. Effects of windbreaks on turbulent transport and microclimate. *Agriculture, Ecosystems and Environment* **22/23:** 17–39. (Reprinted 1988 in *Windbreak Technology*. Elsevier, Amsterdam.) * 6

1094 McNaughton, S. J. 1978. Serengeti ungulates: feeding selectivity influences the effectiveness of plant defense guilds. *Science* **199:** 806–7. * 11

1095 McNaughton, S. J. 1985. Ecology of a grazing ecosystem: the Serengeti. *Ecological Monographs* **55:** 259–94. * 7, 11

1096 McNeely, J. A. 1987. Learning lessons from nature: the biome approach to sustainable development. In P. Jacobs & D. A. Munro, eds. *Conservation with Equity: Strategies for Sustainable Development*, pp. 251–72. International Union for the Conservation of Nature and Natural Resources, Gland, Switzerland. * 1, 14

1097 McNeill, W. H. 1977. *Plagues and Peoples*. Basil Blackwell, Oxford. * 14

1098 Mech, L. D. 1970. *The Wolf: The Ecology and Behavior of an Endangered Species*. Natural History Press, Garden City, New York. * 11

1099 Mech, L. D., Fritts, S. H., Raddle, G. L. & Paul, W. J. 1988. Wolf distribution and road density in Minnesota. *Wildlife Society Bulletin* **16:** 85–7. * 5, 8, 13

1100 Meckelein, W., ed. 1980. *Desertification in Extremely Arid Environments*. Geographisches Institut der Universitat Stuttgart, Stuttgart. * 12

1101 Meentemeyer, V. 1984. The geography of organic decomposition rates. *Annals of the Association of American Geographers* **74:** 551–60. * 2

1102 Meentemeyer, V. & Box, E. O. 1987. Scale effects in landscape studies. In M. G. Turner, ed. *Landscape Heterogeneity and Disturbance*, pp. 15–36. Springer-Verlag, New York. * 1, 3

1103 Meffe, G. K. & Carroll, C. R. 1994. *Principles of Conservation Biology*. Sinauer Associates, Sunderland, Massachusetts, USA. * 1,2,6,13

1104 Mellor, J. W. & Gavian, S. 1987. Famine: causes, prevention, and relief. *Science* **235:** 539–45. * 14

1105 Mellor, J. W. 1988. The intertwining of environmental problems and poverty. *Environment* **30**: 8–13, 28–30. * 14

1106 Melman, P. J. M., Verkaar, H. J. & Heemsbergen, H. 1988. The maintenance of road verges as possible 'ecological corridors' of grassland plants. In K. -F. Schreiber, ed. *Connectivity in Landscape Ecology*, pp. 131–3. Munstersche Geographische Arbeiten 29, Ferdinand Schoningh, Paderborn, Germany. * 5

1107 Melton, M. A. 1957. *An Analysis of the Relation Among Elements of Climate, Surface Properties, and Geomorphology*. Department of Geology Technical Report 11, Columbia University, New York. * 9

1108 Menges, E. S. 1991. The application of minimum viable population theory to plants. In D. A. Falk & K. E. Holsinger, eds. *Genetics and Conservation of Rare Plants*, pp. 45–61. Oxford University Press, New York. * 2

1109 Menges, E. S. 1992. Stochastic modeling of extinction in plant populations. In P. L. Fiedler & S. K. Jain, eds. *Conservation Biology: The Theory and Practice of Nature Conservation, Preservation and Management*, pp. 253–75. Chapman & Hall, New York. * 2

1110 Mergen, F. 1954. Mechanical aspects of wind-breakage and windfirmness. *Journal of Forestry* **52**: 119–25. * 10

1111 Merriam, G. 1984. Connectivity: a fundamental ecological characteristic of landscape pattern. In J. Brandt & P. Agger, eds. *Methodology in Landscape Ecological Research and Planning*, vol. 1, pp. 5–15. Roskilde Universitetsforlag GeoRuc, Roskilde, Denmark. * 5, 8, 9

1112 Merriam, H. G. 1988a. Landscape dynamics in farmland. *Trends in Ecology and Evolution* **3**: 16–20. * 1, 6, 8, 11

1113 Merriam, H. G. 1988b. Modelling woodland species adapting to an agricultural landscape. In K-F. Schreiber, ed. Connectivity in Landscape Ecology, pp. 67–8. Munstersche Geographische Arbeiten 29. Ferdinand Schoningh, Paderborn, Germany. * 8

1114 Merriam, G. 1990. Ecological processes in the time and space of farmland mosaics. In I. S. Zonneveld & R. T. T. Forman, eds. *Changing Landscapes: An Ecological Perspective*, pp. 121–33. Springer-Verlag, New York. * 1, 6, 7, 8, 9, 11, 12

1115 Merriam, G. 1991. Corridors and connectivity: animal populations in heterogeneous environments. In D. A. Saunders & R. J. Hobbs, eds. *Nature Conservation 2: The Role of Corridors*, pp. 133–42. Surrey Beatty, Chipping Norton, Australia. * 8

1116 Merriam, G., Henein, K. & Stuart-Smith, K. 1991. Landscape dynamics models. In M. G. Turner & R. H. Gardner, eds. *Quantitative Methods in Landscape Ecology*, pp. 399–416. Springer-Verlag, New York. * 1, 8, 9, 11, 12

1117 Merriam, G., Kozakiewicz, M., Tsuchiya, E. & Hawley, K. 1989. Barriers as boundaries for metapopulations and demes of *Peromyscus leucopus* in farm landscapes. *Landscape Ecology* **2**: 227–35. * 3, 5, 6

1118 Merriam, G. & Lanoue, A. 1990. Corridor use by small mammals: field measurement for three experimental types of *Peromyscus leucopus*. *Landscape Ecology* **4**: 123–31. * 5,6,7,8

1119 Merrill, L. B. 1954. A variation of deferred rotation grazing for use under Southwest range conditions. *Journal of Range Management* **7**: 152–4. * 11

1120 Messenger, K. G. 1968. A railway flora of Rutland. *Proceedings of the Botanical Society of the British Isles* **7**: 325–44. * 5

1121 Meyer, J. L. & Edwards, R. T. 1990. Ecosystem metabolism and turnover of organic carbon along a blackwater river continuum. *Ecology* **71**: 668–77. * 7

1122 Michener, G. R. & Michener, D. R. 1977. Population structure and dispersal in Richardson's ground squirrels. *Ecology* **58**: 359–68. * 4

1123 Middleton, J. & Merriam, G. 1981. Woodland mice in a farmland mosaic. *Journal of Applied Ecology* **18**: 703–10. * 6,8

1124 Middleton, J. & Merriam, G. 1983. Distribution of woodland species in farmland woods. *Journal of Applied Ecology* **20**: 625–44. * 1,2,8,9

1125 Mierau, G. W. & Favara, B. E. 1975. Lead poisoning in roadside populations of deer mice. *Environmental Pollution* **8**: 55–64. * 5

1126 Miller, B. A., Daniel, T. C. & Berkowitz, S. J. 1979. Computer programs for calculating soil loss on a watershed basis. *Environmental Management* **3**: 237–70. * 7,10

1127 Miller, D. H. 1956. The influence of open pine forest on daytime temperature in the Sierra Nevada. *Geographical Review* **46**: 209–18. * 10

570

1128 Miller, D. H. 1957. Coastal fogs and clouds. *Geographical Review* **47**: 591–4. * 10

1129 Miller, D. H. 1977. *Water at the Surface of the Earth: An Introduction to Ecosystem Hydrodynamics*. Academic Press, New York. * 1,10

1130 Miller, D. H. 1978. The factor of scale: ecosystem, landscape mosaic, and region. In K. A. Hammond, G. Macinko & W. B. Fairchild, eds. *Sourcebook on the Environment: A Guide to the Literature*, pp. 63–88. University of Chicago Press, Chicago. * 1,9

1131 Miller, D. H. 1981. *Energy at the Surface of the Earth: An Introduction to the Energetics of Ecosystems*. Academic Press, New York. * 10

1132 Miller, D. H. 1984. Ecosystem contrasts in interaction with the planetary boundary layer. *GeoJournal* **8.3**: 211–19. * 10

1133 Miller, D. R. 1975. Microclimate at an oak forest–asphalt parking lot interface. In *Proceedings of the 12th Conference on Agricultural and Forest Meteorology*, pp. 35–6. American Meteorological Society, Boston, Massachusetts. * 3

1134 Miller, D. R., Rosenberg, N. J. & Bagley, W. T. 1973. Soybean water use in the shelter of a slat-fence windbreak. *Agricultural Meteorology* **11**: 405–18. * 6

1135 Miller, D. R., Rosenberg, N. J. & Bagley, W. T. 1975. Wind reduction by a highly permeable tree shelterbelt. *Agricultural Meteorology* **14**: 321–33. * 6

1136 Miller, R. I. 1978. Applying island biogeographic theory to an East African reserve. *Environmental Conservation* **5**: 191–5. * 2

1137 Milne, B. T. 1988. Measuring the fractal geometry of landscapes. *Applied Mathematics and Computation* **27**: 67–79. * 1,3,8,9

1138 Milne, B. T. 1991a. Lessons from applying fractal models to landscape patterns. In M. G. Turner & R. H. Gardner, eds. *Quantitative Methods in Landscape Ecology*, pp. 199–235. Springer-Verlag, New York. * 1,3,8,9,11

1139 Milne, B. T. 1991b. The utility of fractal geometry in landscape design. *Landscape and Urban Planning* **21**: 81–90. * 4

1140 Milne, B. T. 1992. Spatial aggregation and neutral models in fractal landscapes. *American Naturalist* **139**: 32–57. * 1,9

1141 Milne, B. T. & Forman, R. T. T. 1986. Peninsulas in Maine: woody plant diversity, distance, and environmental patterns. *Ecology* **67**: 967–74. * 2,3,4,9

1142 Milne, B. T., Johnston, K. M. & Forman, R. T. T. 1989. Scale-dependent proximity of wildlife habitat in a spatially-neutral Bayesian model. *Landscape Ecology* **2**: 101–10. * 1,9

1143 Mimura, M. & Murray, J. D. 1978. On a diffusive prey–predator model which exhibits patchiness. *Journal of Theoretical Biology* **75**: 249–62. * 11

1144 Minckler, L. S. & Woerheide, J. D. 1965. Reproduction of hardwoods: 10 years after cutting as affected by site and opening size. *Journal of Forestry* **63**: 103–7. * 2

1145 Mineau, P. & Madison, D. 1977. Radiotracking of *Peromyscus leucopus*. *Canadian Journal of Zoology* **55**: 465–8. * 6

1146 Minshall, G. W., Cummins, K. W., Petersen, R. C., Cushing, C. E., Bruns, D. A., Sedell, J. R. & Vannote, R. L. 1985. Developments in stream ecosystem theory. *Canadian Journal of Fisheries and Aquatic Sciences* **42**: 1045–55. * 7

1147 Mitchell, W. J. & McCullough, M. 1991. *Digital Design Media*. Van Nostrand Reinhold, New York. * 1

1148 Mitsch, W. J. & Gosselink, J. G. 1986. *Wetlands*. Van Nostrand Reinhold, New York. * 7

1149 Mizgajski, A. 1990. *Entwicklung von Agrarlandschaften im Mitteleuropasichen Tiefland seit dem 19. Jahrhundert in energetischer Sicht*. Munstersche Geographische Arbeiten 33. Ferdinand Schoningh, Paderborn, Germany. 110 pp. * 12

1150 Mladenoff, D. J., White, M. A., Crow, T. R. & Pastor, J. 1994. Applying principles of landscape design and management to integrate old-growth forest enhancement and commodity use. *Conservation Biology* **8**: 752–62. * 9

1151 Mladenoff, D. J., White, M. A., Pastor, J. & Crow, T. R. 1993. Comparing spatial pattern in unaltered old-growth and disturbed forest landscapes. *Ecological Applications* **3**: 294–306. * 9

1152 Mlot, C. 1990. Restoring the prairie. *BioScience* **40**: 804–9. * 2

1153 Moellering, H. & Rayner, J. N. 1979. *Measurement of Shape in Geography and Cartography*. Department of Geography, Ohio State University, Columbus, Ohio, USA. * 4

1154 Moellering, H. & Rayner, J. N. 1981. The harmonic analysis of spatial shapes using dual axis Fourier shape analysis (DAFSA). *Geographical Analysis* **13**: 64–77. * 4

1155 Moellering, H. & Rayner, J. N. 1982. The dual axis Fourier shape analysis of closed cartographic forms. *The Cartographic Journal* **19**: 53–9. * 4

1156 Montegut, J. 1976. Le bocage et les commensales des cultures. In *Les Bocages: Histoire, Ecologie, Economie*, pp. 229–238. Institut National de la Recherche Agronomique, Centre National de la Recherche Scientifique, et Universite de Rennes, Rennes, France. * 8

1157 Montgomery, D. R. & Dietrich, W. E. 1992. Channel initiation and the problem of landscape scale. *Science* **255**: 826–30. * 9

1158 Mooney, H. A. 1991. Toward the study of the earth's metabolism. *Bulletin of the Ecological Society of America* **72**: 221–8. * 12,14

1159 Mooney, H. A. & Drake, J. A., eds. 1986. *Ecology of Biological Invasions of North America and Hawaii*. Springer-Verlag, New York. * 6,11

1160 Mooney, H. A. & Godron, M., eds. 1983. *Disturbance and Ecosystems: Components of Response*. Springer-Verlag, New York. * 10,14

1161 Moore, N. W. & Hooper, M. D. 1975. On the number of bird species in British woods. *Biological Conservation* **8**: 239–50. * 2

1162 Moran, J. M. & Morgan, M. D. 1994. *Meteorology: The Atmosphere and the Science of Weather*. Macmillan, New York. * 10

1163 Morgan, K. A. & Gates, J. E. 1982. Bird population patterns in forest edge and strip vegetation at Remington Farms, Maryland. *Journal of Wildlife Management* **46**: 933–44. * 3,5,12

1164 Morisawa, M. 1985. *Rivers: Form and Process*. Longman, London. * 7

1165 Morrison, S. W. 1988. The Percival Creek Corridor Plan. *Journal of Soil and Water Conservation* **43**: 465–7. * 7

1166 Moss, M. R. 1983. Landscape synthesis, landscape processes and land classification: some theoretical and methodological issues. *Geojournal* **7**: 145–53. * 9

1167 Muehlenbach, V. 1979. Contributions to the synanthropic (adventive) flora of the railroads in St. Louis, Missouri, USA. *Annals of the Missouri Botanical Garden* **66**: 1–108. * 5

1168 Muehrcke, P. C. & Muehrcke, J. O. 1980. *Map Use: Reading, Analysis, and Interpretation*. JP Publications, Madison, Wisconsin, USA. * 4

1169 Mueller-Dombois, D. & Ellenberg, H. 1974. *Aims and Methods of Vegetation Ecology*. John Wiley, New York. * 3,9

1170 Mulhearn, P. J. & Bradley, E. F. 1977. Secondary flows in the lee of porous shelterbelts. *Boundary-layer Meteorology* **12**: 75–92. * 6

1171 Mulholland, P. J., Newbold, J. D., Elwood, J. W. & Webster, J. R. 1985. Phosphorus spiralling in a woodland stream: seasonal variations. *Ecology* **66**: 1012–23. * 7

1172 Munn, R. E. 1966. *Descriptive Micrometeorology*. Academic Press, New York. * 10

1173 Murphy, D. D. 1989. Conservation and confusion: wrong species, wrong scale, wrong conclusions. *Conservation Biology* **3**: 82–4. * 2

1174 Musick, H. B. & Grover, H. D. 1991. Image textural measures as indices of landscape pattern. In M. G. Turner & R. H. Gardner, eds. *Quantitative Methods in Landscape Ecology*, pp. 77–103. Springer-Verlag, New York. * 1,9

1175 Mwalyosi, R. B. B. 1991. Ecological evaluation for wildlife corridors and buffer zones for Lake Manyara National Park, Tanzania, and its immediate environment. *Biological Conservation* **57**: 171–86. * 6

1176 Myers, K., Margules, C. R. & Musto, I., eds. 1984. *Survey Methods for Nature Conservation*. CSIRO, Division of Water and Land Resources, Canberra. * 9

1177 Myster, R. W. & Pickett, S. T. A. 1992. Effects of palatability and dispersal mode on spatial patterns of trees in oldfields. *Bulletin of the Torrey Botanical Club* **119**: 145–51. * 3,11

1178 Naegeli, W. 1946. Weitere Untersuchungen uber die Windverhaltnisse im Bereich von Windschutzanlagen. *Mitteilungen Schweizerische Anstalt für das Forstliche Versuchswesen* **24**: 660–737. * 6

1179 Naegeli, W. 1953. Untersuchungen uber die Windverhaltnisse im Bereich von Schilfrohrwanden. *Mitteilungen Schweizerische Anstalt fur das Forstliche Versuchswesen* **29**: 213–66. * 6

1180 Nagel, E. 1961. *The Structure of Science: Problems in the Logic of Scientific Explanation*. Harcourt, Brace and World, New York. * 1

572

1181 Naiman, R. J. & Decamps, H., eds. 1990. *The Ecology and Management of Aquatic–Terrestrial Ecotones*. Man and the Biosphere Series, vol. 4., UNESCO, Paris. * 7

1182 Naiman, R. J., Decamps, H., Pastor, J. & Johnston, C. A. 1988. The potential importance of boundaries to fluvial ecosystems. *Journal of the North American Benthological Society* **7**: 289–306. * 3,7

1183 Naiman, R. J., Johnston, C. A. & Kelley, J. C. 1988. Alteration of North American streams by beaver. *BioScience* **38**: 753–62. * 3,7,9

1184 Naiman, R. J., Melillo, J. M. & Hobbie, J. E. 1986. Ecosystem alteration of boreal forest streams by beaver (*Castor canadensis*). *Ecology* **67**: 1254–69. * 7

1185 Naiman, R. J., Melillo, J. M., Lock, M. A., Ford, T. E. & Reice, S. R. 1987. Longitudinal patterns of ecosystem processes and community structure in a subarctic river continuum. *Ecology* **68**: 1138–56. * 7

1186 Naiman, R. J., Pinay, G., Johnston, C. A. & Pastor, J. 1994. Beaver influences on the long-term biogeochemical characteristics of boreal forest drainage networks. *Ecology* **75**: 905–21. * 7

1187 Nair, P. K. R. 1993. *An Introduction to Agroforestry*. Kluwer Academic Publishers, Dordrecht, Netherlands. * 13

1188 Nankinov, D. N. & Todorov, N. M. 1983. Bird casualties on highways. *The Soviet Journal of Ecology* **14**: 288–93. * 5

1189 Nassauer, J. I. 1992. The appearance of ecological systems as a matter of policy. *Landscape Ecology* **6**: 239–50. * 9

1190 Nassauer, J. I. & Westmacott, R. 1987. Progressiveness among farmers as a factor in heterogeneity of farmed landscapes. In M. G. Turner, ed. *Landscape Heterogeneity and Disturbance*, pp. 199–210. Springer-Verlag, New York. * 9

1191 National Research Council. 1986. *Global Change in the Geosphere-Biosphere: Initial Priorities for an IGBP*. National Academy Press, Washington, D. C. * 3,10

1192 Naveh, Z. 1982. Landscape ecology as an emerging branch of human ecosystem science. *Advances in Ecological Research* **12**: 189–237. * 1

1193 Naveh, Z. & Lieberman, A. S. 1993. *Landscape Ecology: Theory and Application*. Springer-Verlag, New York. * 1,13

1194 Neef, E. 1967. *Die theoretischen Grundlagen der Landschaftslehre*. Verlag Hermann Haack, Gotha/Leipzig, Germany. * 1

1195 Neef, E. 1979. *Analyse und prognose von Nebenwirkungen Gesellschaftlicher Aktivitaten im Naturraum*. Abhandlungen der Sachsischen Akademie der Wissenschaften zu Leipzig, Mathematisch-naturwassenschaftliche Klasse, Band 54, Heft 1. Akademie-Verlag, Berlin. * 12

1196 Neef, E. 1981. Stages in the development of landscape ecology. In S. P. Tjallingii & A. A. de Veer, eds. *Perspectives in Landscape Ecology*, pp. 19–27. PUDOC, Wageningen, Netherlands. * 1

1197 Neilson, R. P. 1991. Climatic constraints and issues of scale controlling regional biomes. In M. M. Holland, P. G. Risser & R. J. Naiman, eds. *Ecotones: The Role of Landscape Boundaries in the Management and Restoration of Changing Environments*, pp. 31–51. Chapman and Hall, New York. * 10

1198 Neilson, R. P., King, G. A., DeVelice, R. L. & Lenihan, J. M. 1992. Regional and local vegetation patterns: the responses of vegetation diversity to subcontinental air masses. In A. J. Hansen & F. di Castri, eds. *Landscape Boundaries: Consequences for Biotic Diversity and Ecological Flows*, pp. 129–49. Springer-Verlag, New York. * 1,3,10

1200 Nellis, M. D. & Briggs, J. M. 1989. The effect of spatial scale on Konza landscape classification using textural analysis. *Landscape Ecology* **2**: 93–100. * 1,9

1201 Nelson, A. C. 1992. Preserving prime farmland in the face of urbanization: lessons from Oregon. *Journal of the American Planning Association* **58**: 467–88. * 13

1202 Neumeister, H. 1972. Raumtypisierung und Faktorenanalyse. *Wissenschaftliche Abhandlungen der Geographischen Gessellschaft der Deutschen Demokratischen Republik* **9**: 000. * 4

1203 Neustein, S. A. 1971. Damage to forests in relation to topography, soil and crops. In *Windthrow of Scottish Forests in January 1968*, pp. 42–8. Forestry Commission Bulletin **45**, United Kingdom. * 10

1204 Newbey, B. J. & Newbey, K. R. 1987. Bird dynamics of Foster Road Reserve, near Ongerup, Western Australia. In D. A. Saunders, G. W. Arnold, A. A. Burbidge & A. J. M.

573

Hopkins, eds. *Nature Conservation: The Role of Remnants of Native Vegetation*, pp. 341–3. Surrey Beatty, Chipping Norton, Australia. * 5

1205 Newbold, J. D., Erman, D. C. & Roby, K. B. 1980. Effects of logging on macroinvertebrates in streams with and without buffer strips. *Canadian Journal of Fisheries and Aquatic Science* **37**: 1076–85. * 7

1206 Newmark, W. D. 1985. Legal and biotic boundaries of western North American national parks: a problem of congruence. *Biological Conservation* **33**: 197–208. * 3,9

1207 Newmark, W. D. 1986. Species–area relationship and its determinants for mammals in western North American parks. *Biological Journal of the Linnean Society* **28**: 65–82. * 2

1208 Newsome, A. E. & Noble, I. R. 1986. Ecological and physiological characters of invading species. In R. H. Groves & J. J. Burdon, eds. *Ecology of Biological Invasions*, pp. 1–21. Cambridge University Press, Cambridge. * 11

1209 Nicholls, A. O. & Margules, C. R. 1991. The design of studies to demonstrate the biological importance of corridors. In D. A. Saunders & R. J. Hobbs, eds. *Nature Conservation 2: The Role of Corridors*, pp. 49–61. Surrey Beatty, Chipping Norton, Australia. * 6

1210 Niemi, A. 1969. On the railway vegetation and flora between Esbo and Inga, southern Finland. *Acta Botanica Fennica* **83**: 1–28. * 5

1211 Niering, W. A. & Goodwin, R. H. 1974. Creation of relatively stable shrublands with herbicides: arresting 'succession' on rights-of-way and pasture land. *Ecology* **55**: 784–95. * 5

1212 Nilsson, C. 1986. Methods of selecting lake shorelines as nature reserves. *Biological Conservation* **35**: 269–91. * 7

1213 Nilsson, S. G. 1986. Are bird communities in small biotope patches random samples from communities in large patches? *Biological Conservation* **38**: 179–204. * 2,9

1214 Nip-Van der Voort, J., Hengeveld, R. & Haeck, J. 1979. Immigration rates of plant species in three Dutch polders. *Journal of Biogeography* **6**: 301–8. * 5

1215 Nittmann, J. & Stanley, H. E. 1986. Tip splitting without interfacial tension and dendritic growth patterns arising from molecular anisotropy. *Nature* **321**: 663–8. * 4

1216 Nixon, C., McClain, M. & Donohoe, R. 1980. Effects of clear-cutting on gray squirrels. *Journal of Wildlife Management* **44**: 403–12. * 6

1217 Nixon, P. R. & Lawless, G. P. 1968. Advective influences on the reduction of evapotranspiration in a coastal environment. *Water Resources Research* **4**: 39–46. * 10

1218 Noble, I. R. 1989. Attributes of invaders and the invading process: terrestrial and vascular plants. In J. A. Drake, H. A. Mooney, F. di Castri, R. H. Groves, F. J. Kruger, M. Rejmanek & M. Williamson, eds. *Biological Invasions: A Global Perspective*, pp. 301–13. John Wiley, Chichester, UK. * 11

1219 Noble, I. R. & Slatyer, R. O. 1980. The use of vital attributes to predict successional changes in plant communities subject to recurrent disturbances. *Vegetatio* **43**: 5–21. * 14

1220 Noer, H. 1975. Okonomisk vurdering av fordelene med bedre arronderte skogeiendommer. (An evaluation of economic disadvantages associated with fragmentation of forest holdings.) *Tidsskrift for Skogbruk* **83**: 233–48. * 2

1221 Norin, B. N. 1961. Chto takoe lesotundra? (Forest–tundra transition zone, what is it?) *Botanicheskij Journal* **46**: 21–36. * 3

1222 Norton, D. A. 1992. Disruption of natural ecosystems by biological invasion. In J. H. M. Willison, S. Bondrup-Nielsen, C. Drysdale, T. B. Herman, N. W. P. Munro & T. L. Pollock, eds. *Science and Management of Protected Areas*, pp. 309–19. Elsevier, Amsterdam. * 11

1223 Norton, D. A. & Lord, J. M. 1990. On the use of 'grain size' in ecology. *Functional Ecology* **4**: 719. * 1

1224 Noss, R. F. 1983. A regional landscape approach to maintain diversity. *BioScience* **33**: 700–6. * 1,3,9,12

1225 Noss, R. F. 1987a. From plant communities to landscapes in conservation inventories: a look at The Nature Conservancy (USA). *Biological Conservation* **41**: 11–37. * 7,9

1226 Noss, R. F. 1987b. Corridors in real landscapes: a reply to Simberloff and Cox. *Conservation Biology* **1**: 159–64. * 6

1227 Noss, R. F. 1991. Effects of edge and internal patchiness on avian habitat use in an old-growth Florida hammock. *Natural Areas Journal* **11**: 34–47. * 3

574

1228 Noss, R. 1993. Wildlife corridors. In D. S. Smith & P. C. Hellmund, eds. *Ecology of Greenways: Design and Function of Linear Conservation Areas*, pp. 43–68. University of Minnesota Press, Minneapolis, Minnesota, USA. * 5,6,7,8,11,12,13

1229 Noss, R. F. & Harris, L. D. 1986. Nodes, networks and MUMS: preserving diversity at all scales. *Environmental Management* 10: 299–309. * 5,6,8,9,13

1230 Novitzki, R. P. 1979. Hydrologic characteristics of Wisconsin's wetlands and their influence on floods, stream flow, and sediment. In P. E. Greeson, J. R. Clark & J. E. Clark, eds. *Wetland Functions and Values: The State of Our Understanding*, pp. 377–88. American Water Resources Associates, Minneapolis, Minnesota, USA. *7,10

1231 Numata, M. 1993. Basic concepts and methods of landscape ecology. In *Proceedings of the International Conference on Landscape Planning and Environmental Conservation*, pp. 96–110. University of Tokyo, Tokyo. * 1,13

1232 Nyland, R. D., Zipperer, W. C. & Hill, D. B. 1986. The development of forest islands in exurban central New York State. *Landscape and Urban Planning* 13: 111–23. * 12

1233 Odum, E. P. 1971. *Fundamentals of Ecology*. 3rd edn. Saunders, Philadelphia. * 3,12,14

1234 Odum, E. P. & Turner, M. G. 1990. The Georgia landscape: a changing resource. In I. S. Zonneveld & R. T. T. Forman, eds. *Changing Landscapes: An Ecological Perspective*, pp. 137–64. Springer-Verlag, New York. * 1,9,12

1235 Odum, H. T. 1983. *Systems Ecology: An Introduction*. John Wiley, New York. * 3, 8, 9, 14

1236 Odum, H. T. & Odum, E. C. 1981. *Energy Basis for Man and Nature*. McGraw-Hill, New York. * 3, 14

1237 Odum, H. W. 1951. The promise of regionalism. In M. Jensen, ed. *Regionalism in America*, pp. 395–425. University of Wisconsin Press, Madison, Wisconsin. * 1

1238 Oetting, R. B. & Cassel, J. F. 1971. Waterfowl nesting on interstate highway right-of-way in North Dakota. *Journal of Wildlife Management* 35: 774–81. * 5

1239 Ogilvie, R. T. & Furman, T. 1959. Effect of vegetational cover of fence rows on small mammal populations. *Ecology* 40: 140–1. * 6

1240 Ogle, C. C. 1987. The incidence and conservation of animal and plant species in remnants of native vegetation within New Zealand. In D. A. Saunders, G. W. Arnold, A. A. Burbidge & A. J. M. Hopkins, eds. *Nature Conservation: The Role of Remnants of Native Vegetation*, pp. 79–87. Surrey Beatty, Chipping Norton, Australia. * 2

1241 Ogle, C. C. & Wilson, P. R. 1985. Where have all the mistletoes gone? *Forest and Bird* 16: 10–13. * 2

1242 Ohsawa, M. & Liang-Jun, D. 1987. Urbanization and landscape dynamics in a watershed of the Miyako River, Chiba, Japan. In H. Obara, ed. *Integrated Studies in Urban Ecosystems as the Basis of Urban Planning (II)*, pp. 187–97. Special Research Project on Environmental Science B334-R15-3, Ministry of Education, Culture and Science, Tokyo, Japan. * 12

1243 Oke, T. R. 1987. *Boundary Layer Climates*. Routledge, London. * 10

1244 Okubo, A. 1980. *Diffusion and Ecological Problems: Mathematical Models*. Springer-Verlag, Berlin. * 6,11

1245 Okubo, A. & Levin, S. A. 1989. A theoretical framework for the analysis of data on the wind dispersal of seeds and pollen. *Ecology*, 70: 329–38. * 11

1246 Oldeman, R. A. A. 1978. Architecture and energy exchange of dicotyledonous trees in the forest. In P. B. Tomlinson & M. H. Zimmermann, eds. *Tropical Trees as Living Systems*, pp. 535–60. Cambridge University Press, London. * 2, 8

1247 O'Meara, T. E., Monkler, J. R., Stelter, H. & Nagy, J. G. 1981. Nongame wildlife responses to chaining of pinyon-juniper woodlands. *Journal of Wildlife Management* 45: 381–9. * 3

1248 Omernik, J. M., Abernathy, A. R. & Male, L. M. 1981. Stream nutrient levels and proximity of agricultural and forest land to streams: some relationships. *Journal of Soil and Water Conservation* 36: 227–31. * 7

1250 O'Neill, R. V., DeAngelis, D. L., Waide, J. B. & Allen, T. F. H. 1986. *A Hierarchical Concept of Ecosystems*. Princeton University Press, Princeton, New Jersey. * 1, 3, 4, 8, 14

1251 O'Neill, R. V., Johnson, A. R. & King, A. W. 1989. A hierarchical framework for the analysis of scale. *Landscape Ecology* 3: 193–205. * 1

1252 O'Neill, R. V., Krummel, J. R., Gardner, R. H., Sugihara, G., Jackson, B., DeAngelis, D. L., Milne, B. T., Turner, M. G., Zygmunt, B., Christensen, S. W., Dale, V. H. & Graham, R. L. 1988a. Indices of landscape pattern. *Landscape Ecology* **1:** 153–62. * 1, 3, 4, 9, 12, 13

1253 O'Neill, R. V., Milne, B. T., Turner, M. G. & Gardner, R. H. 1988b. Resource utilization scales and landscape pattern. *Landscape Ecology* **2:** 63–9. * 1

1254 Opdam, P. 1988. Populations in fragmented landscape. In K-F. Schreiber, ed. *Connectivity in Landscape Ecology*, pp. 75–7. Munstersche Geographische Arbeiten 29, Ferdinand Schoningh, Paderborn, Germany. * 11

1255 Opdam, P. 1990a. Dispersal in fragmented populations: the key to survival. In R. G. H. Bunce & D. C. Howard, eds. *Species Dispersal in Agricultural Habitats*, pp. 3–17. Belhaven Press, London. * 11

1256 Opdam, P. 1990b. Understanding the ecology of populations in fragmented landscapes. In S. Myrberget, ed. *Transactions of the 19th IUGB Congress*, pp. 373–80. Norwegian Institute for Nature Research, Trondheim, Norway. * 11

1257 Opdam, P. 1991. Metapopulation theory and habitat fragmentation: a review of holarctic breeding bird studies. *Landscape Ecology* **5:** 93–106. * 1, 2, 11, 12

1258 Opdam, P., Rijsdijk, G. & Hustings, F. 1985. Bird communities in small woods in an agricultural landscape: effect of area and isolation. *Biological Conservation* **34:** 333–52. * 1, 6, 11

1259 Opdam, P. & Schotman, A. 1987. Small woods in rural landscape as habitat islands for woodland birds. *Acta Oecologia/Oecologia Generalis* **8:** 269–74. * 11

1260 Opdam, P., van Apeldoorn, R., Schotman, A. & Kalkhoven, J. 1992. Population responses to landscape fragmentation. In C. C. Vos & P. Opdam, eds. *Landscape Ecology of a Stressed Environment*, pp. 147–71. Chapman & Hall, London. * 1, 11, 13

1261 Opdam, P., van Dorp, D. & ter Braak, C. J. F. 1984. The effect of isolation on the number of woodland birds of small woods in The Netherlands. *Journal of Biogeography* **11:** 473–8. * 11

1262 Oppenheim, A. L. 1964. *Ancient Mesopotamia: Portrait of a Dead Civilization*. University of Chicago Press, Chicago. * 14

1263 Orians, G. H. 1982. The influence of tree falls in tropical forests on tree species richness. *Tropical Ecology* **23:** 255–79. * 2

1264 Orians, G. H. 1986. Site characteristics favoring invasions. In H. A. Mooney & J. A. Drake, eds. *Ecology of Biological Invasions of North America and Hawaii*, pp. 133–48. Springer-Verlag, New York. * 11

1265 Orians, G. H. 1990. Ecological concepts of sustainability. *Environment* **32:** 10–15, 34–9. * 14

1266 Ortloff, C. R., Moseley, M. E. & Feldman, R. A. 1982. Hydraulic engineering aspects of the Chimu Chicama–Moche intervalley canal. *American Antiquity* **47:** 572–95. * 14

1267 Osawa, R. 1989. Road-kills of the swamp wallaby, *Wallabia bicolor*, on North Stradbroke Island, south-east Queensland. *Australian Wildlife Research* **16:** 95–104. * 5

1268 Osborne, L. L. & Wiley, M. J. 1988. Empirical relationships between landuse/cover and stream water quality in an agricultural watershed. *Journal of Environmental Management* **26:** 9–27. * 1, 7

1269 Oshima, K., Honda, H. & Yamamoto, I. 1973. Isolation of an oviposition marker from Azuki bean weevil, *Callosobruchus chinensis*. *Agricultural and Biological Chemistry* **37:** 2679–80. * 11

1270 Owen, J. 1991. *The Ecology of a Garden: The First Fifteen Years*. Cambridge University Press, Cambridge. * 13

1271 Oxley, D. J. & Fenton, M. B. 1976. The harm our roads do to nature and wildlife. *Canadian Geographical Journal* **92(3):** 40–5. * 5

1272 Oxley, D. J., Fenton, M. B. & Carmody, G. R. 1974. The effects of roads on populations of small mammals. *Journal of Applied Ecology* **11:** 51–9. * 5

1273 Packer, C., Pusey, A. E., Rowley, H., Gilbert, D. A., Martenson, J. & O'Brien, S. J. 1991. Case study of a population bottleneck: lions of the Ngorongoro Crater. *Conservation Biology* **5:** 219–30. * 2, 11

1274 Paine, R. T. 1966. Food web complexity and species diversity. *American Naturalist* **100:** 65–75. * 2, 8, 9

1275 Paine, R. T. 1988. Food webs: road maps of interactions or grist for theoretical development? *Ecology* **69:** 1648–54. * 8

1276 Paine, R. T. & Levin, S. A. 1981. Intertidal landscapes: disturbance and the dynamics of pattern. *Ecological Monographs* **51:** 145–78. * 2, 8

1277 Panayotou, T. & Ashton, P. S. 1992. *Not by Timber Alone: Economics and Ecology for Sustaining Tropical Forests*. Island Press, New York. * 14

1278 Panetta, F. D. & Hopkins, A. J. M. 1992. Weeds in corridors: invasion and management. In D. A. Saunders & R. J. Hobbs, eds. *Nature Conservation 2: The Role of Corridors*, pp. 341–51. Surrey Beatty, Chipping Norton, Australia. * 5, 6, 11

1279 Parks, P. J. 1991. Models of forested and agricultural landscapes: integrating economics. In M. G. Turner & R. H. Gardner, eds. *Quantitative Methods in Landscape Ecology*, pp. 309–22. Springer-Verlag, New York. * 1

1280 Parks, P. J. & Alig, R. J. 1988. Land base models for forest resource supply analysis: a critical review. *Canadian Journal of Forest Research* **18:** 965–73. * 1

1281 Parsons, J. J. 1960. 'Fog drip' from coastal stratus, with special reference to California. *Weather* **15:** 58–62. * 10

1282 Pasek, J. E. 1988. Influence of wind and windbreaks on local dispersal of insects. *Agriculture, Ecosystems and Environment* **22/23:** 539–54. (Reprinted 1988 in *Windbreak Technology*. Elsevier, Amsterdam.) * 6

1283 Pastor, J. & Broschart, M. 1990. The spatial pattern of a northern conifer–hardwood landscape. *Landscape Ecology* **4:** 55–68. * 1,9

1284 Patrick, R., Matson, B. & Anderson, L. 1979. Streams and lakes in the Pine Barrens. In R. T. T. Forman, ed. *Pine Barrens: Ecosystem and Landscape*, pp. 169–93. Academic Press, New York. * 7,8

1285 Pattee, H. H. 1973. *Hierarchy Theory: The Challenge of Complex Systems*. Braziller, New York. * 1

1286 Patterson, B. D. & Atmar, W. 1986. Nested subsets and the structure of insular mammalian faunas and archipelagos. *Biological Journal of the Linnean Society* **28:** 65–82. * 2

1287 Patton, D. R. 1975. A diversity index for quantifying habitat edge. *Wildlife Society Bulletin* **394:** 171–3. * 3,4

1288 Pavlidis, T. 1978. A review of algorithms for shape analysis. *Computer Graphic and Image Processing* **7:** 243–58. * 4

1289 Pavlidis, T. & Feng, H-Y. F. 1977. Shape discrimination. In K. S. Fu, ed. *Syntactic Pattern Recognition and Applications*, pp. 125–45. Springer-Verlag, Berlin. * 4

1290 Payette, S. 1983. The forest tundra and present tree-lines of the northern Quebec–Labrador peninsula. In P. Morisset & S. Payette, eds. *Tree-line Ecology: Proceedings of the Northern Quebec Tree-line Conference*, pp. 3–23. Centre d'Etudes Nordiques, Universite Laval, Quebec. * 3,9

1291 Payette, S., Morneau, C., Sirois, L. & Desponts, M. 1989. Recent fire history of the northern Quebec biomes. *Ecology* **70:** 656–73. * 10

1292 Paynter, R. 1982. *Models of Spatial Inequality: Settlement Patterns in Historical Archeology*. Academic Press, New York. * 1

1293 Pearsall, W. H. 1924. The statistical analysis of vegetation: a criticism of the concepts and methods of the Upsala school. *Journal of Ecology* **12:** 135–9. * 2

1294 Pearson, J. E. 1993. Complex patterns in a simple system. *Science* **261:** 189–92. * 4

1295 Pechanec, J. F. & Stewart, G. 1940. Sagebrush–grass range sampling studies: size and structure of sampling unit. *Journal of the American Society of Agronomy* **32:** 669–82. * 4

1296 Pedevillano, C. & Wright, R. G. 1987. The influence of visitors on mountain goat activities in Glacier National Park, Montana. *Biological Conservation* **39:** 1–11. * 5

1297 Pedroli, G. B. M. & Borger, G. J. 1990. Historical land use and hydrology: a case study from eastern Noord-Brabant. *Landscape Ecology* **4:** 237–48. * 7,10

1298 Peet, R. K., Glenn-Lewin, D. C. & Wolf, J. W. 1983. Prediction of man's impact on vegetation. In W. Holzner, M. J. A. Werger & I. Ikusima, eds. *Man's Impact on Plant Species Diversity: A Challenge for Vegetation Science*, pp. 41–54. W. Junk Publishers, The Hague. * 14

577

1300 Pelikan, J. 1986. Small mammals in windbreaks and adjacent fields. *Acta Scientarium Naturalium Academiae Scientarium Bohemoslovacae Brno* **20(4):** 1–38. * 6

1301 Pennington, W. 1986. Lags in adjustment of vegetation to climate caused by the pace of soil development: evidence from Britain. *Vegetatio* **67:** 105–18. * 3,10,14

1302 Pennycuick, C. J. 1973. The soaring flight of vultures. *Scientific American* **229(12):** 102–9. * 10

1303 Peterjohn, W. T. & Correll, D. L. 1984. Nutrient dynamics in an agricultural watershed: observations on the role of a riparian forest. *Ecology* **65:** 1466–75. * 7,12,13

1304 Peterken, G. F. 1968. International selection of areas for reserves. *Biological Conservation* **1:** 55–61. * 9

1305 Peterken, G. F. 1992. Coppices in the lowland landscape. In G. P. Buckley, ed. *Ecology and Management of Coppice Woodlands*, pp. 3–17. Chapman & Hall, London. * 12

1306 Peterken, G. F. 1993. *Woodland Conservation and Management*, 2nd edn. Chapman and Hall, London. * 11,12,13,14

1307 Peterken, G. F. & Allison, H. 1989. *Woods, Trees and Hedges: A Review of Changes in the British Countryside*. Nature Conservancy Council, Peterborough, UK. * 12

1308 Peterken, G. F., Ausherman, D., Buchenau, M. & Forman, R. T. T. 1992. Old-growth conservation within British upland conifer plantations. *Forestry* **65:** 127–44. * 4,12

1309 Peterken, G. F. & Game, M. 1984. Historical factors affecting the number and distribution of vascular plant species in the woodlands of central Lincolnshire. *Journal of Ecology* **72:** 155–82. * 1,2,12

1310 Peters, R. H. 1991. *A Critique for Ecology*. Cambridge University Press, Cambridge. * 2

1311 Peterson, G. W. 1988. Disease management in windbreaks. *Agriculture, Ecosystems and Environment* **22/23:** 501–11. (Reprinted 1988 in *Windbreak Technology*. Elsevier, Amsterdam.) * 6

1312 Petrides, G. A. 1942. Relation of hedgerows to wildlife in central New York. *Journal of Wildlife Management* **6:** 261–80. * 6

1313 Petts, G. E. 1984. *Impounded Rivers: Perspectives for Ecological Management*. John Wiley, Chichester, UK. * 7

1314 Phillips, D. L. & Shure, D. J. 1990. Patch-size effects on early succession in southern Appalachian forests. *Ecology* **71:** 204–12. * 2

1315 Phipps, M. 1981. Entropy and community pattern analysis. *Journal of Theoretical Biology* **93:** 253–73. * 1

1316 Pickard, J. 1984. Exotic plants on Lord Howe Island: distribution in space and time – 1853–1981. *Journal of Biogeography* **11:** 181–208. * 11

1317 Pickett, S. T. A. 1980. Non-equilibrium coexistence of plants. *Bulletin of the Torrey Botanical Club* **107:** 238–48. * 2,11

1318 Pickett, S. T. A. & Kolasa, J. 1989. Structure of theory in vegetation science. *Vegetatio* **83:** 7–15. * 1, 14

1319 Pickett, S. T. A. & Thompson, J. N. 1978. Patch dynamics and the design of nature reserves. *Biological Conservation* **13:** 27–37. * 2,12,13

1320 Pickett, S. T. A. & White, P. S., eds. 1985. *The Ecology of Natural Disturbance and Patch Dynamics*. Academic Press, New York. * 1,2,3,10,11,12,14

1321 Picton, H. D. 1979. The application of insular biogeographic theory to the conservation of large mammals in the Northern Rocky Mountains. *Biological Conservation* **15:** 73–9. * 2

1322 Pielou, E. C. 1977. *Mathematical Ecology*. John Wiley, New York. * 1,8,9

1323 Pielou, E. C. 1979. *Biogeography*. John Wiley, New York. * 2,11

1324 Pietrzak, M. 1989. *Problemy I Metody Badania Struktury Geokompleksu*. Uniwersytet Im. Adama Mickiewicza W Poznaniu, Seria Geografia Nr 45, Poznan, Poland. 125 pp. * 4

1325 Pimentel, D. 1986. Biological invasions of plants and animals in agriculture and forestry. In H. A. Mooney & J. A. Drake, eds. *Ecology of Biological Invasions of North America and Hawaii*, pp. 149–62. Springer-Verlag, New York. * 11

1326 Pimentel, D., Oltenacu, P. A., Nesheim, M. C., Krummel, J., Allen, M. S. & Chick, S. 1980. The potential for grass-fed livestock: resource constraints. *Science* **207:** 843–8. * 11

1327 Pimentel, D., Stachow, U., Takacs, D. A., Brubaker, H. W., Dumas, A. R., Meaney, J. J., O'Neil, J. A. S., Onsi, D. E. & Corzilius, D. B. 1992. Conserving biological diversity in agricultural/forestry systems. *BioScience* **42:** 354–62. * 11,14

1328 Pimm, S. L., Jones, H. L. & Diamond, J. M. 1989. On the risk of extinction. *American Naturalist* **132:** 757–85. * 1, 11

1329 Pinchot, G. 1899. *A Study of Forest Fires and Wood Production in Southern New Jersey. Appendix: Annual Report of the State Geologist for 1898*. Geological Survey of New Jersey, Trenton, New Jersey. * 1

1330 Pjavchenko, N. I. 1980. O sovremennom zabolachivanii severnyh lesov. (Modern expansion of bogs in the northern forests). In *Strukturnofunkcionalnaja Organizacija Biogeocenozov*, pp. 144–63. Nauka, Moscow. * 3

1331 Platts, W. S. & Rinne, J. N. 1985. Riparian and stream enhancement management and research in the Rocky Mountains. *North American Journal of Fisheries Management* **5:** 115–25. * 7

1332 Plotnick, R. E. & McKinney, M. L. 1993. Ecosystem organization and extinction dynamics. *Palaios* **8:** 202–12. * 10

1333 Podoll, E. B. 1979. Utilization of windbreaks by wildlife. In *Windbreak Management*. Publication 92, Great Plains Agricultural Council, Lincoln, Nebraska. * 6

1334 Poff, N. L. & Ward, J. V. 1989. Implications of streamflow variability and predictability for lotic community structure: a regional analysis of streamflow patterns. *Canadian Journal of Fisheries and Aquatic Sciences* **46:** 1805–18. * 7

1335 Pohl, M. 1985. *Prehistoric Lowland Maya Environment and Subsistence Economy*. Papers of the Peabody Museum, vol. 77. Harvard University, Cambridge, Massachusetts. * 14

1336 Pojar, T. M., Prosence, R. A., Reed, D. F. & Woodard, T. N. 1975. Effectiveness of a lighted, animated deer crossing sign. *Journal of Wildlife Management* **39:** 87–91. * 5

1337 Pollard, E., Hooper, M. D. & Moore, N. W. 1974. *Hedges*. W. Collins, London. * 1,3,5,6,8

1338 Pons, A., Couteaux, M., de Beaulieu, J. L. & Reille, M. 1990. Plant invasions in southern Europe from the paleoecological point of view. In F. di Castri, A. J. Hansen & M. Debussche, eds. *Biological Invasions in Europe and the Mediterranean Basin*, pp. 169–77. Kluwer Academic Publishers, Dordrecht, Netherlands. * 10,11,14

1339 Port, G. R. & Thompson, J. R. 1980. Outbreaks of insect herbivores on plants along motorways in the United Kingdom. *Journal of Applied Ecology* **17:** 649–56. * 5

1340 Portmann, A. 1967. *Animal Forms and Patterns*. (Translated by H. Czech.) Schocken Books, New York. * 1,4

1341 Poston, T. & Stewart, I. 1978. *Catastrophe Theory and Its Applications*. Pitman, San Francisco. * 3

1342 Pouyat, R. V. & McDonnell, M. J. 1991. Heavy metal accumulations in forest soils along an urban–rural gradient in southeastern New York, USA. *Water, Air and Soil Pollution* **57–58:** 797–807. * 13

1343 Powell, A. H. & Powell, G. V. N. 1987. Population dynamics of male euglossine bees in Amazonian forest fragments. *Biotropica* **19:** 176–9. * 2

1344 Poynton, J. C. & Roberts, D. C. 1985. Urban open space planning in South Africa: a biogeographical perspective. *South Africa Journal of Science* **81:** 33–7. * 6

1345 Pozorski, T. & Pozorski, S. 1982. Reassessing the Chicama–Moche intervalley canal: comments on 'Hydraulic engineering aspects of the Chimu Chicama–Moche intervalley canal'. *American Antiquity* **47:** 851–68. * 14

1346 Pressey, R. L. & Nicholls, A. O. 1989. Application of a numerical algorithm to the selection of reserves in semi-arid New South Wales. *Biological Conservation* **50:** 263–78. * 9

1347 Preston, F. W. 1962a. The canonical distribution of commonness and rarity: part I. *Ecology* **43:** 185–215. * 2

1348 Preston, F. W. 1962b. The canonical distribution of commonness and rarity: part II. *Ecology* **43:** 410–32. * 2

1349 Price, L. W. 1981. *Mountains and Man: A Study of Process and Environment*. University of California Press, Berkeley, California. * 14

1350 Price, P. W. 1976. Colonization of crops by arthropods: nonequilibrium communities in soybean fields. *Environmental Entomology* **5:** 605–11. * 3,6,9

1351 Price, W. 1961. The effects of the characteristics of snow fences on the quantity and shape of deposited snow. In *Publication 54*, pp. 89–98. Association Internationale d'Hydrologie Scientifique, Oxford, UK. * 6

1352 Primack, R. B. 1993. *Essentials of Conservation Biology*. Sinauer Associates, Sunderland, Massachusetts, USA. * 2,11

1353 Primack, R. B. & Miao, S. L. 1992. Dispersal can limit local plant distribution. *Conservation Biology* **6:** 513–19. * 2

1354 Probst, J. R. & Crow, T. R. 1991. Integrating biological diversity and resource management: an essential approach to productive, sustainable ecosystems. *Journal of Forestry* **89:** 12–17. * 14

1355 Pugh, H. D. & Price, W. I. J. 1954. Snow drifting and the use of snow fences. *Polar Record* **7:** 4–23. * 6

1356 Puglisi, M. J., Lindzey, J. S. & Bellis, E. D. 1974. Factors associated with highway mortality of white-tailed deer. *Journal of Wildlife Management* **38:** 799–807. * 5

1357 Pulliam, H. R. 1988. Sources, sinks, and population regulation. *American Naturalist* **132:** 652–61. * 3,6,11

1358 Pulliam, H. R. & Danielson, B. J. 1991. Sources, sinks, and habitat selections: a landscape perspective on population dynamics. *American Naturalist* **137** (Supplements): 50–66. * 6,11

1359 Pulliam, H. R., Dunning, J. B., Jr. & Liu, J. 1992. Population dynamics in complex landscapes: a case study. *Ecological Applications* **2:** 165–77. * 11, 12

1360 Purdie, R. W. 1987. Selection of key area networks for regional nature conservation – the revised Bolton and Specht method. *Proceedings of the Royal Society of Queensland* **98:** 59–71. * 9

1361 Putz, F. E., Coley, P. D., Lu, K., Montalvo, A. & Aiello, A. 1983. Uprooting and snapping of trees: structural determinants and ecological consequences. *Canadian Journal of Forest Research* **13:** 1011–20. * 10

1362 Pyne, S. J. 1982. *Fire in America*. Princeton University Press, Princeton, New Jersey. * 9

1363 Quarles, H. D. III, Hanawalt, R. B. & Odum, W. E. 1974. Lead in small mammals, plants and soil at varying distances from a highway. *Journal of Applied Ecology* **11:** 937–49. * 5

1364 Quattrochi, D. A. & Pelletier, R. E. 1991. Remote sensing for analysis of landscapes: an introduction. In M. G. Turner & R. H. Gardner, eds. Quantitative Methods in Landscape Ecology, pp. 51–76. Springer-Verlag, New York. * 1

1365 Quinby, P. A. 1988. The contribution of ecological science to the development of landscape ecology: a brief history. *Landscape Research* **13:** 9–11. * 1

1366 Quinn, J. F. & Robinson, G. R. 1987. The effects of experimental subdivision and flowering plant diversity in a California annual grassland. *Journal of Ecology* **75:** 837–56. * 1,2

1367 Rackham, O. 1975. *Hayley Wood: Its History and Ecology*. Cambridgeshire & Isle of Ely Naturalists' Trust, Cambridge, UK. * 11,12

1368 Rackham, O. 1976. *Trees and Woodland in the British Landscape*. J. M. Dent, London. * 6,14

1369 Rackham, O. 1980. *Ancient Woodland*. Edward Arnold, London. * 9,11

1370 Rackham, O. 1986. *The History of the Countryside*. J. M. Dent, London. * 12,14

1371 Radforth, N. W. 1952. Suggested classification of muskeg for the engineer. *Engineering Journal* **35:** 1199–210. * 10

1372 Radforth, N. W. 1977. Muskeg hydrology. In N. W. Radforth & C. O. Brawner, eds. *Muskeg and the Northern Environment in Canada*, pp. 130–47. University of Toronto Press, Toronto. * 10

1373 Rafe, R. W., Usher, M. B. & Jefferson, R. G. 1985. Birds on reserves: the influence of area and habitat on species richness. *Journal of Applied Ecology* **22:** 327–35. * 2

1374 Rahamimoff, A. & Bornstein, N. 1981. Edge conditions – climatic considerations in the design of buildings and settlements. *Energy and Buildings* **4:** 43–9. * 3

1375 Raine, J. K. & Stevenson, D. C. 1977. Wind protection by model fences in a simulated atmospheric boundary layer. *Journal of Industrial Aerodynamics* **2:** 159–80. * 6

1376 Ralls, K., Harvey, P. H. & Lyles, A. M. 1986. Inbreeding in natural populations of birds and mammals. In M. E. Soulé, ed. *Conservation Biology: The Science of Scarcity and Diversity*, pp. 35–56. Sinauer Associates, Sunderland, Massachusetts, USA. * 2

1377 Rambouskova, H. 1988. Comments on the ecostabilizing functions of small-scale landscape structures – I. Part. *Ekologia* (Czechoslovakia) **7**: 397–412. * 3,12,14

1378 Rambouskova, H. 1989. Comments on the ecostabilizing functions of small-scale landscape structures – II. Part. *Ekologia* (Czechoslovakia) **8**: 35–48. * 3,12,14

1379 Ranney, J. W., Bruner, M. C. & Levenson, J. B. 1981. The importance of edge in the structure and dynamics of forest islands. In R. L. Burgess & D. M. Sharpe, eds. *Forest Island Dynamics in Man-dominated Landscapes*, pp. 67–96. Springer-Verlag, New York. * 2,3,6,12

1380 Rapoport, A. 1990. *History and Precedent in Environmental Design*. Plenum Press, New York. * 13

1381 Rapoport, E. H. 1982. Areography: Geographical Strategies of Species. Fundacion Bariloche Series Number 1. Pergamon Press, New York. * 4,11

1382 Rappaport, R. A. 1968. *Pigs for the Ancestors*. Yale University Press, New Haven, Connecticut. * 9

1383 Rappaport, R. A. 1971. The flow of energy in an agricultural society. *Scientific American* **225**: 116–32. * 9

1384 Rasid, H. & Pramanik, M. A. H. 1990. Visual interpretation of satellite imagery for monitoring floods in Bangladesh. *Environmental Management* **14**: 815–21. * 1

1385 Ratcliffe, D., ed. 1977. *A Nature Conservation Review*. 2 vols. Cambridge University Press, Cambridge. * 2,9,13

1386 Ratcliffe, E. J. 1983. *Through the Badger Gate: The Story of Badgers, Their Persecution and Protection, and of a Cub Reared and Returned to the Wild*. Dalesman Books, Clapham, North Yorkshire, UK. * 5

1387 Ratti, J. T. & Reese, K. P. 1988. Preliminary test of the ecological trap hypothesis. *Journal of Wildlife Management* **52**: 484–91. * 3

1388 Raty, M. 1982. Effects of highway traffic on tetraonid densities. *Ornis Fennica* **56**: 169–70. * 5

1389 Rauner, I. I. 1963. Izmerenie teplo- i vlagoobmena mezhdu lesom i atmosferoi pod vliianiem okruzhaiushchikh territorii. *Izvestiia Akademii Nauk SSSR, Seriia Geografii* **1**: 15–28. * 10

1390 Raven, P. H. 1976. Ethics and attitudes. In J. B. Simmons, R. I. Beyer, P. E. Brandham, G. Ll. Lucas & V. T. H. Parry, eds. *Conservation of Threatened Plants*, pp. 155–79. Plenum Press, New York. * 13

1391 Raven, P. H. & Wilson, E. O. 1992. A fifty-year plan for biodiversity surveys. *Science* **258**: 1099–100. * 14

1392 Ray, G. C. & Hayden, B. P. 1992. Coastal zone ecotones. In A. J. Hansen & F. di Castri, eds. *Landscape Boundaries: Consequences for Biotic Diversity and Ecological Flows*, pp. 403–20. Springer-Verlag, New York. * 3

1393 Recher, H. F., Shields, J., Kavanagh, R. & Webb, G. 1987. Retaining remnant mature forest for nature conservation at Eden, New South Wales: a review of theory and practice. In D. A. Saunders, G. W. Arnold, A. A. Burbidge & A. J. M. Hopkins. *Nature Conservation: The Role of Remnants of Native Vegetation*, pp. 177–94. Surrey Beatty, Chipping Norton, Australia. * 1, 2, 5, 6, 7, 11, 13, 14

1394 Redford, K. & da Fonseca, G. 1986. The role of gallery forests in the zoogeography of the Cerrado's non-volant mammalian fauna. *Biotropica* **18**: 126–35. * 7

1395 Reed, D. F. 1981. Mule deer behavior at a highway underpass exit. *Journal of Wildlife Management* **45**: 542–3. * 5

1396 Reed, D. F., Pojar, T. M. & Woodard, T. N. 1974. Use of one-way gates by mule deer. *Journal of Wildlife Management* **38**: 9–15. * 5

1397 Reed, D. F. & Woodard, T. N. 1981. Effectiveness of highway lighting in reducing deer–vehicle accidents. *Journal of Wildlife Management* **45**: 721–6. * 5

1398 Reed, D. F., Woodard, T. N. & Pojar, T. M. 1975. Behavioral response of mule deer to a highway underpass. *Journal of Wildlife Management* **39**: 361–7. * 5

1400 Reed, D. M. & Schwarzmeier, J. A. 1978. The prairie corridor concept: possibilities for planning large scale preservation and restoration. In Lewin & Landers, eds. *Proceedings of the Fifth Midwest Prairie Conference*, pp. 158–65. Iowa State University, Ames, Iowa, USA. * 5, 13

581

1401 Reed, L. E. 1976. The long range transport of air pollutants. *Ambio* **5**: 202. * 10
1402 Regier, H. A. & Baskerville, G. L. 1986. Sustainable redevelopment of regional ecosystems degraded by exploitive development. In W. C. Clark & R. E. Munn, eds. *Sustainable Development of the Biosphere*, pp. 75–101. Cambridge University Press, Cambridge. * 14
1403 Reh, W. 1989. Investigations into the influence of roads on the genetic structure of populations of the common frog *Rana temporaria*. In T. E. S. Langton, ed. *Amphibians and Roads*, pp. 101–3. ACO Polymer Products, Shefford, Bedfordshire, England. * 5,11
1404 Reichholf, von J. & Esser, J. 1981. Daten zur mortalitat des Igels (*Erinaceus europaeus*) verursacht durch den Strassenverkehr. *Zeitschrift fuer Saugeterkunde* **46**: 216–22. * 5
1405 Reifsnyder, W. E. & Lull, H. W. 1965. *Radiant Energy in Relation to Forests*. Technical Bulletin 1344, US Forest Service, Washington, D. C. * 3, 10
1406 Reijnen, M. J. S. M., Thissen, J. & Bekker, G. J. 1987. Effects of road traffic on woodland breeding bird populations. *Acta Oecologia Generalis* **8**: 312–13. * 5
1407 Reiner, R. & Griggs, T. 1989. TNC undertakes riparian restoration projects in California. *Restoration and Management Notes* **7**: 3–8. * 5,7
1408 Reiners, W. A. 1979. Ecological reseach opportunities in the New Jersey Pine Barrens. In R. T. T. Forman, ed. *Pine Barrens: Ecosystem and Landscape*, pp. 557–67. Academic Press, New York. * 10
1409 Reiners, W. A. 1983. Transport processes in the biogeochemical cycles of carbon, nitrogen, phosphorus, and sulphur. In B. Bolin & R. B. Cook, eds. *The Major Biogeochemical Cycles and Their Interactions*, pp. 143–76. John Wiley, Chichester, UK. * 1,10
1410 Reiners, W. A. & Lang, G. E. 1979. Vegetational patterns and processes in the Balsam fir zone, White Mountains, New Hampshire. *Ecology* **60**: 403–17. * 1,4,10
1411 Reiners, W. A., Marks, R. H. & Vitousek, P. M. 1975. Heavy metals in sub-alpine and alpine soils of New Hampshire. *Oikos* **26**: 264–75. * 10
1412 Remillard, M. M., Gruendling, G. K. & Bogucki, D. J. 1987. Disturbance by beaver (*Castor canadensis* Kuhl) and increased landscape heterogeneity. In M. G. Turner, ed. *Landscape Heterogeneity and Disturbance*, pp. 103–22. Springer-Verlag, New York. * 7
1413 *Remote Sensing of the Biosphere*. 1986. National Academy Press, Washington, DC. * 10,12
1414 Repetto, R., ed. 1985. *The Global Possible: Resources, Development and the New Century*. Yale University Press, New Haven, Connecticut, USA. * 14
1415 Repetto, R. & Gillis, M., eds. 1988. *Public Policies and the Misuse of Forest Resources*. Cambridge University Press, New York. * 1
1416 Rex, K. D. & Malanson, G. P. 1990. The fractal shape of riparian forest patches. *Landscape Ecology* **4**: 249–58. * 4
1417 Rey, J. R. 1981. Ecological biogeography of arthropods on Spartina islands in Northwest Florida. *Ecological Monographs* **51**: 237–65. * 2,11
1418 Rey, J. R. 1984. Experimental tests of island biogeographic theory. In D. R. Strong, Jr. et al., eds. *Ecological Communities: Conceptual Issues and the Evidence*, pp. 101–12. Princeton University Press, Princeton, New Jersey. * 2
1419 Richards, P. W. 1984. The forests of South Viet Nam in 1971–72: a personal account. *Environmental Conservation* **11**: 147–53. * 12
1420 Richerson, P. J. 1993. Humans as a component of the Lake Titicaca ecosystem: a model system for the study of environmental deterioration. In M. J. McDonnell & S. T. A. Pickett, eds. *Humans as Components of Ecosystems*, pp. 125–40. Springer-Verlag, New York. * 14
1421 Richling, A. 1976. *Analiza i Struktura: Srodowiska Geograficznego i Nowa Metoda Regionalizacji Fizycznogeograficznej (Na Przykadzie Wojewodztwa Biaostockiego)*. Rozprawy UW. nr 104. Wydawnictwa Uniwersytetu Warszawskiego, Warszawa. * 4
1422 Richter, H. 1981. Die inhaltliche Konzeption der Karte 'Flachennutzung und naturraumliche Ausstattung' 1: 750 000 im 'Atlas DDR'. *Petermanns Geographische Mitteilungen* **3**: 207–12. * 12
1423 Richter, H. 1984a. Land use and land transformation. *GeoJournal* **8**: 67–74. * 12
1424 Richter, H. 1984b. Geographische Landschaftsprognose. *Geographische Berichte* **111**: 91–102. * 4,12

1425 Ripley, B. D. 1981. *Spatial Statistics*. John Wiley, New York. * 1,9
1426 Ripple, W. J., Johnson, D. H., Hershey, K. T. & Meslow, E. C. 1991. Old-growth and mature forests near spotted owl nests in western Oregon. *Journal of Wildlife Management* **55**: 316–18. * 8,9
1427 Risch, S. J., Andow, D. & Altieri, M. A. 1983. Agroecosystem diversity and pest control: data, tentative conclusions, and new research directions. *Environmental Entomology* **12**: 625–9. * 3,9
1428 Risser, P. G. 1987. Landscape ecology: state-of-the-art. In M. G. Turner, ed. *Landscape Heterogeneity and Disturbance*, pp. 3–14. Springer-Verlag, New York. * 1,14
1429 Risser, P. G., Birney, E. C., Blocker, H. D., May, S. W. D., Parton, W. J. & Wiens, J. A. 1981. *The True Prairie Ecosystem*. Hutchinson & Ross, Stroudsburg, Pennsylvania. * 2
1430 Risser, P. G., Karr, J. R. & Forman, R. T. T. 1984. *Landscape Ecology: Directions and Approaches*. Special Publication 2, Illinois Natural History Survey, Champaign, Illinois. * 1,12,14
1431 Risser, P. G., Lubchenco, J. & Levin, S. A. 1991. Biological research priorities: a sustainable biosphere. *BioScience* **41**: 625–7. * 13,14
1432 Ritchie, J. C. 1987. *Postglacial Vegetation of Canada*. Cambridge University Press, Cambridge. * 3
1433 Robbins, C. S., Dowell, B. A., Dawson, D. K., Colon, J., Espinoza, F., Rodriguez, J., Sutton, R. & Vargas, T. 1987. Comparison of neotropical winter bird populations in isolated patches versus extensive forest. *Acta Oecologia Generalis* **8**: 285–92. * 2
1434 Robbins, C. S., Dawson, D. K. & Dowell, B. A. 1989. Habitat area requirements of breeding forest birds of the Middle Atlantic States. *Wildlife Monographs* **103**: 1–34. * 2,12,13
1435 Robinette, G. O. 1972. *Plants, People and Environmental Quality*. US Department of Interior, National Park Service, Washington, DC.
1436 Robinson, G. R. & Quinn, J. F. 1988. Extinction, turnover, and species diversity in an experimentally fragmented California annual grassland. *Oecologia* **76**: 71–82. * 2,11
1437 Rodiek, J. E. & Bolen, E. G., eds. 1991. *Wildlife and Habitats in Managed Landscapes*. Island Press, Washington, DC. * 13
1438 Roitberg, B. D. & Prokopy, R. J. 1987. Insects that mark host plants. *BioScience* **37**: 400–6. * 11
1439 Romme, W. H. 1982. Fire and landscape diversity in subalpine forests of Yellowstone National Park. *Ecological Monographs* **52**: 199–221. * 9,10
1440 Romme, W. H. & Knight, D. H. 1981. Fire frequency and subalpine forest succession along a topographic gradient in Wyoming. *Ecology* **62**: 319–26. * 10
1441 Romme, W. H. & Knight, D. H. 1982. Landscape diversity: the concept applied to Yellowstone Park. *BioScience* **32**: 664–70. * 1,9
1442 Ronco, F. & Noble, D. L. 1971. Englemann spruce regeneration in clearcut openings not insured by record seed crop. *Journal of Forestry* **69**: 578–9. * 2
1443 Rongstad, O. J. & Tester, J. R. 1969. Movements and habitat use of white-tailed deer in Minnesota. *Journal of Wildlife Management* **33**: 366–79. * 11
1444 Rosenberg, A. A., Fogarty, M. J., Sissenwine, M. P., Beddington, J. R. & Shepherd, J. G. 1993. Achieving sustainable use of renewable resources. *Science* **262**: 828–9. * 14
1445 Rosenberg, N. J. 1979. Windbreaks for reducing moisture stress. In B. J. Barfield & J. F. Gerber, eds. *Modification of the Aerial Environment of Plants*, pp. 394–408. American Society of Agricultural Engineers, St. Joseph, Michigan, USA. * 6
1446 Rosenberg, N. J., Blad, B. L. & Verma, S. B. 1983. *Microclimate: The Biological Environment*. John Wiley, New York. * 6, 10
1447 Rosenfeld, A. 1969. *Picture Processing by Computer*. Academic Press, New York. * 4
1448 Rost, G. R. & Bailey, J. A. 1979. Distribution of mule deer and elk in relation to roads. *Journal of Wildlife Management* **43**: 634–41. * 5, 8
1449 Rostow, W. W. 1978. *The World Economy: History & Prospect*. University of Texas Press, Austin, Texas. * 14
1450 Roth, R. R. 1987. Assessment of habitat quality for wood thrush in a residential area. In L. W. Adams & D. L. Leedy, eds. *Integrating Man and Nature in the Metropolitan Environment*, pp. 139–49. National Institute for Urban Wildlife, Columbia, Maryland, USA. * 13

1451 Rouse, W. R. 1976. Microclimatic changes accompanying burning in subarctic lichen woodland. *Arctic and Alpine Research* **8:** 357–76. * 10

1452 Rouse, W. R. 1984a. Microclimate at arctic tree line. 1. Radiation balance of tundra and forest. *Water Resources Research* **20:** 57–66. * 3, 10

1453 Rouse, W. R. 1984b. Microclimate of arctic tree line. 2. Soil microclimate of tundra and forest. *Water Resources Research* **20:** 67–73. * 3, 10

1454 Rouse, W. R. 1984c. Microclimate at arctic tree line. 3. The effects of regional advection on the surface energy balance of upland tundra. *Water Resources Research* **20:** 74–8. * 10

1455 Row, C. 1978. Economies of tract size in timber growing. *Journal of Forestry* **76:** 576–82. * 2

1456 Rowe, J. S. & Sheard, J. W. 1981. Ecological land classification: a survey approach. *Environmental Management* **5:** 451–64. * 9

1457 Rowe, P. B. 1963. Streamflow increases after removing woodland riparian vegetation from a southern California watershed. *Journal of Forestry* **61:** 365–70. * 7

1458 Rowe, P. G. 1991. *Making a Middle Landscape*. MIT Press, Cambridge, Massachusetts. * 12, 13

1459 Rubec, C. D. A. 1983. Applications of remote sensing in ecological land survey in Canada. *Canadian Journal of Remote Sensing* **9:** 19–30. * 1

1460 Rudnicky, T. C. & Hunter, M. L., Jr. 1993a. Reversing the fragmentation perspective: effects of clearcut size on bird species richness in Maine. *Ecological Applications* **3:** 357–66. * 2

1461 Rudnicky, T. C. & Hunter, M. L., Jr. 1993b. Avian nest predation in clearcuts, forests, and edges in a forest-dominated landscape. *Journal of Wildlife Management* **57:** 358–64. * 3

1462 Rumney, G. R. 1968. *Climatology and the World's Climates*. Macmillan, London. * 10

1463 Runkle, J. R. 1985. Disturbance regimes in temperate forest. In S. T. A. Pickett & P. S. White, eds. *The Ecology of Natural Disturbance and Patch Dynamics*, pp. 17–34. Academic Press, New York. * 2, 10

1464 Runkle, J. R. & Yetter, T. C. 1987. Treefalls revisited: gap dynamics in the southern Appalachians. *Ecology* **68:** 417–24. * 2

1465 Russell, E. W. B. 1983. Indian set fires in the forests of the northeastern United States. *Ecology* **64:** 78–88. * 9, 14

1466 Russell, E. W. B., Davis, R. B., Anderson, R. S., Rhodes, T. E. & Anderson, D. S. 1993. Recent centuries of vegetational change in the glaciated north-eastern United States. *Journal of Ecology* **81:** 647–64. * 9, 10, 14

1467 Ruthsatz, B. 1984. Kleinstrukturen im Raum Ingolstadt: Schutz-und Zeigerwert Teil II: Waldsaume. *Tuexenia* **4:** 227–49. * 12

1468 Ruthsatz, B. & Haber, W. 1981. The significance of small-scale landscape elements in rural areas as refuges for endangered plant species. In S. P. Tjallingii & A. A. de Veer, eds. *Perspectives in Landscape Ecology*, pp. 117–24. PUDOC, Wageningen, Netherlands. * 12

1469 Ruthsatz, B. & Otte, A. 1987. Kleinstrukturen im Raum Ingolstadt: Schutz und Zeigerwert. Teil III. Feldwegrander und Ackerraine. *Tuexenia* **7:** 139–63. * 5

1470 Ruzicka, M., ed. 1982. *Proceedings of the VIth International Symposium on Problems in Landscape Ecological Research*. Institute for Experimental Biology and Ecology, Bratislava, Czechoslovakia. * 1

1471 Ruzicka, M., Jurke, A., Kozova, M., Zigrai, F. & Svetlosanov, V. 1982. Evaluation methods of landscape stability on agricultural territories in Slovakia. In M. Ruzicka, ed. *Proceedings of the VIth International Symposium on Problems in Landscape Ecological Research*, pp. 1–31. Institute for Experimental Biology and Ecology, Bratislava, Czechoslovakia. * 12, 14

1472 Ruzicka, M. & Miklos, L. 1990. Basic premises and methods in landscape ecological planning and optimization. In I. S. Zonneveld & R. T. T. Forman, eds. *Changing Landscapes: An Ecological Perspective*, pp. 233–60. Springer-Verlag, New York. * 1, 13

1473 Ryman, N. & Laikre, L. 1991. Effects of supportive breeding on the genetically effective population size. *Conservation Biology* **5:** 325–9. * 2

1474 Ryszkowski, L. 1992. Energy and material flows across boundaries in agricultural landscapes. In A. J. Hansen & F. di Castri, eds. *Landscape Boundaries: Consequences for Biotic Diversity and Ecological Flows*, pp. 270–84. Springer-Verlag, New York. * 1,8,14

1475 Ryszkowski, L. & Kedziora, A. 1987. Impact of agricultural landscape structure on energy flow and water cycling. *Landscape Ecology* **1:** 85–94. * 10,14

1476 Saarenmaa, H., Stone, N. D., Folse, L. J., Packard, J. M., Grant, W. E., Makela, M. E. & Coulson, R. N. 1988. An artificial intelligence modelling approach to simulating animal/habitat interactions. *Ecological Modeling* **44:** 125–41. * 1

1477 Sabins, F. F., Jr. 1987. *Remote Sensing: Principles and Interpretation*. W. H. Freeman, New York. * 1,10

1478 Sagoff, M. 1988. *The Economy of the Earth: Philosophy, Laws and the Environment*. Cambridge University Press, Cambridge. * 14

1479 Saint Girons, M. C. 1981. Notes sur les mammiferes de France. XV. Les pipistrelles et la circulation routiere. *Mammalia* **45:** 131. * 5

1480 Salati, E., Dall'Olio, A., Matsui, E. & Gat, J. R. 1979. Recycling of water in the Amazon Basin: an isotope study. *Water Resources Research* **15:** 1250–8. * 10

1481 Salati, E., Lovejoy, T. E. & Vose, P. B. 1983. Precipitation and water recycling in tropical rain forests with special reference to the Amazon Basin. *The Environmentalist* **3:** 67–72. * 10,12

1482 Salo, J., Kalliola, R., Hakkinen, I., Makinen, Y., Niemela, P., Puhakka, M. & Coley, P. D. 1986. River dynamics and the diversity of Amazon lowland forest. *Nature* **322:** 254–8. * 7

1483 Salwasser, H., Schonewald-Cox, C. & Baker, R. 1987. The role of interagency cooperation in managing for viable populations. In M. E. Soulé, ed. *Viable Populations for Conservation*, pp. 159–73. Cambridge University Press, Cambridge. * 2,13

1484 Samways, M. J. 1989. Insect conservation and landscape ecology: a case-history of bush crickets (Tettigoniidae) in southern France. *Environmental Conservation* **16:** 217–26. * 1,11

1485 Sapoval, B., Gobron, T. & Margolina, A. 1991. Vibrations of fractal drums. *Physical Review Letters* **November 18:** 2974–7. * 3

1486 Sargeant, A. B. 1981. Road casualties of prairie nesting ducks. *Wildlife Society Bulletin* **9:** 65–9. * 5

1487 Saunders, D. A. 1980. Food and movements of the short-billed form of the white-tailed black cockatoo. *Australian Wildlife Research* **7:** 257–69. * 5

1488 Saunders, D. A. 1989. Changes in the avifauna of a region, district and remnant as a result of fragmentation of native vegetation: the wheatbelt of Western Australia. A case study. *Biological Conservation* **50:** 99–135. * 1,12

1489 Saunders, D. A. 1990. Problems of survival in an extensively cultivated landscape: the case of Carnaby's cockatoo, *Calyptorhynchus funereus latirostris*. *Biological Conservation* **54:** 277–90. * 5

1490 Saunders, D. A., Arnold, G. W., Burbidge, A. A. & Hopkins, A. J. M., eds. 1987. *Nature Conservation: The Role of Remnants of Native Vegetation*. Surrey Beatty, Chipping Norton, Australia. * 1,2,5,9,11,12,13,14

1491 Saunders, D. A. & de Rebeira, C. P. 1991. Values of corridors to avian populations in a fragmented landscape. In D. A. Saunders & R. J. Hobbs, eds. *Nature Conservation 2: The Role of Corridors*, pp. 221–40. Surrey Beatty, Chipping Norton, Australia. * 5,6

1492 Saunders, D. A. & Hobbs, R. J., eds. 1991. *Nature Conservation 2: The Role of Corridors*. Surrey Beatty, Chipping Norton, Australia. * 1,5,6,8,11,12,13

1493 Saunders, D. A., Hobbs, R. J. & Arnold, G. W. 1993. The Kellerberrin project on fragmented landscapes: a review of current information. *Biological Conservation* **64:** 231–8. * 1,9,12

1494 Saunders, D. A., Hobbs, R. J. & Ehrlich, P. R., eds. 1993. *Nature Conservation 3: The Reconstruction of Fragmented Ecosystems: Global and Regional Perspectives*. Surrey Beatty, Chipping Norton, Australia. * 1,13

1495 Saunders, D. A. & Ingram, J. A. 1987. Factors affecting survival of breeding populations of Carnaby's cockatoo *Calyptorhynchus funereus latirostris* in remnants of native vegetation. In D. A. Saunders, G. W. Arnold, A. A. Burbidge & A. J. M. Hopkins, eds.

Nature Conservation: The Role of Remnants of Native Vegetation, pp. 249–58. Surrey Beatty, Chipping Norton, Australia. * 5,9,11

1496 Savill, P. S. 1983. Silviculture in windy climates. *Forestry Abstracts* **44:** 473–88. * 10

1497 Schafer, J. A. & Penland, S. T. 1985. Effectiveness of Swareflex reflectors in reducing deer–vehicle accidents. *Journal of Wildlife Management* **49:** 774–6. * 5

1498 Scherer, F. M. 1973. The determinants of industrial plant sizes in six nations. *Review of Economic Statistics* **55:** 135–45. * 2

1499 Scherr, S. J., Roger, J. H. & Oduol, P. A. 1990. Surveying farmers' agroforestry plots; experiences in evaluating alley-cropping and tree border technologies in western Kenya. *Agroforestry Systems* **11:** 141–73. * 12,13

1500 Schimel, D. S., Parton, W. J., Adamsen, F. J., Woodmansee, R. G., Senft, R. L. & Stillwell, M. A. 1986. The role of cattle in the volatile loss of nitrogen from a shortgrass steppe. *Biogeochemistry* **2:** 39–52. * 9

1501 Schimel, D., Stillwell, M. A. & Woodmansee, R. G. 1985. Biogeochemistry of C, N, and P in a soil catena of the shortgrass steppe. *Ecology* **66:** 276–82. * 2,9

1502 Schlesinger, W. H., Reynolds, J. F., Cunningham, G. L., Huenneke, L. F., Jarrell, W. M., Virginia, R. A. & Whitford, W. G. 1990. Biological feedbacks in global desertification. *Science* **247:** 1043–8. * 12

1503 Schlosser, I. J. & Karr, J. R. 1981. Riparian vegetation and channel morphology impact on spatial patterns of water quality in agricultural watersheds. *Environmental Management* **5:** 233–43. * 1,7,12,13

1504 Schluter, H., Bottcher, W. & Bastian, O. 1990. Vegetation change caused by land-use intensification – examples from the hilly country of Saxony. *GeoJournal* **22:** 167–74. * 12

1505 Schmandt, J. & Clarkson, J., eds. 1992. *The Regions and Global Warming: Impacts and Response Strategies*. Oxford University Press, New York. * 1, 10

1506 Schmid, W. A. & Jacsman, J. 1987. *Grundlagen der Landschaftsplanung, Die Landschaft und ihre Nutzung, Lehrmittel fur Orts-, Regional- und Landesplanung*. Eidgenossische Technische Hochschule, Zurich. * 13

1507 Schoener, A. & Schoener, T. W. 1981. The dynamics of the species–area relation in marine fouling systems: I. Biological correlates of changes in the species–area slope. *American Naturalist* **118:** 339–60. * 2

1508 Schoener, T. W. 1983. Simple models of optimal-feeding-territory size: a reconciliation. American Naturalist **121:** 608–29. * 4, 11

1509 Schoener, T. W. 1987. The geographical distribution of rarity. *Oecologia* (Berlin) **74:** 161–73. * 8

1510 Schoener, T. W. & Spiller, D. A. 1987. High population persistence in a system with high turnover. *Nature* **330:** 474–7. * 11

1511 Schofield, E. A. & Baron, R. C., eds. 1993. *Thoreau's World and Ours*. North American Press, Golden, Colorado, USA. * 13

1512 Scholten, H. 1988. Snow distribution on crop fields. *Agriculture, Ecosystems and Environment* **22/23:** 363–80. (Reprinted 1988 in *Windbreak Technology*. Elsevier, Amsterdam.) * 6

1513 Schonewald-Cox, C. M. 1988. Boundaries in the protection of nature reserves. *BioScience* **38:** 480–6. * 3, 5, 9, 13

1514 Schonewald-Cox, C. M. & Bayless, J. W. 1986. The boundary model: a geographic analysis of design and conservation of nature reserves. *Biological Conservation* **38:** 305–22. * 1, 2, 3, 4, 9, 13

1515 Schonewald-Cox, C. & Buechner, M. 1990. Park protection and public roads. In P. L. Fiedler & S. K. Jain, eds. *Conservation Biology: The Theory and Practice of Nature Conservation, Preservation and Management*, pp. 373–95. Chapman & Hall, New York. * 5, 8, 12, 13

1516 Schoonmaker, P. K. & Foster, D. R. 1991. Some implications of paleoecology for contemporary ecology. *Botanical Review* **57:** 204–45. * 10, 14

1517 Schreiber, K-F. 1987. Beitrage der Landschaftsokologie zur Okosystemforschung und ihre Anwendung. *Verhandlungen des Deutschen Geographentages* **45:** 134–45. * 13

1518 Schreiber, K-F., 1988. *Connectivity in Landscape Ecology*. Munstersche Geographische Arbeiten 29, Ferdinand Schoningh, Paderborn, Germany. * 1, 5, 6, 8, 12, 13

586

1519 Schreiber, K-F. 1990. The history of landscape ecology in Europe. In I. S. Zonneveld & R. T. T. Forman, eds. *Changing Landscapes: An Ecological Perspective*, pp. 21–34. Springer-Verlag, New York. * 1

1520 Schreiber, K-F. & Kias, U. 1983. A concept for environmental impact assessment of new roads. *Applied Geography and Development* 21: 95–107. *5

1521 Schreiber, R. K. & Graves, J. H. 1977. Powerline corridors as possible barriers to the movements of small mammals. *American Midland Naturalist* 97: 504–8. * 5

1522 Schroeder, R. L., Cable, T. T. & Haire, S. L. 1992. Wildlife species richness in shelterbelts: test of a habitat model. *Wildlife Society Bulletin* 20: 264–73, * 6

1523 Schulz, J. P. 1960. Ecological studies on rain forest in northern Suriname. *Verhandelingen der Koninklijke nederlandse akademie van wetenschappen, Afdeeling natuurkunde (sectie 2)* 53: 1–267. * 2

1524 Schumm, S. A. 1956. Evolution of drainage basins and slopes in badlands at Perth Amboy, New Jersey. *Bulletin of the Geological Society of America* 67: 597–646. * 9, 10

1525 Schwartz, C. C. & Ellis, J. E. 1981. Feeding ecology and niche separation in some ungulates on the shortgrass prairie. *Journal of Applied Ecology* 18: 343–53. * 9

1526 Scott, J. M., Davis, F., Csuti, B., Noss, R., Butterfield, B., Groves, C., Anderson, H., Caicco, S., D'Erchia, F., Edwards, T. C., Jr., Ulliman, J. & Wright, R. G. 1993. Gap analysis: a geographic approach to protection of biological diversity. *Wildlife Monographs* 123: 1–41. * 9

1527 Scott, R. E., Roberts, L. J. & Cadbury, C. J. 1972. Bird deaths from power lines at Dungeness. *British Birds* 65: 273–86. * 5

1528 Seal, U. S. 1985. The realities of preserving species in captivity. In R. J. Hoage, ed. *Animal Extinctions: What Everyone Should Know*, pp. 71–95. Smithsonian Institution Press, Washington, DC. * 2

1529 Sebastiani, M., Sambrano, A., Villamizar, A. & Villalba, C. 1989. Cumulative impact and sequential geographic analysis as tools for land use planning. A case study: Laguna La Reina, Miranda State, Venezuela. *Journal of Environmental Management* 29: 237–48. * 13

1530 Sedell, J. R. & Froggatt, J. L. 1984. Importance of streamside forests to large rivers: the isolation of the Williamette River, Oregon, USA, from its floodplain by snagging and streamside forest removal. *Vehandlungen – Internationale Vereinigung fur theoretische und Angewandte Limnologie* 22: 1828–34. * 7

1531 Seginer, I. 1975. Flow around a windbreak in oblique wind. *Boundary-layer Meteorology* 9: 133–41. * 6

1532 Selander, R. K. 1983. Evolutionary consequences of inbreeding. In C. M. Schonewald-Cox, S. M. Chambers, B. MacBryde & L. Thomas, eds. *Genetics and Conservation: A Reference for Managing Wild Animal and Plant Populations*, pp. 201–15. Benjamin/Cummings, Menlo Park, California. * 2

1533 Selman, P. 1993. Landscape ecology and countryside planning: vision, theory and practice. *Journal of Rural Studies* 9: 1–21. * 13

1534 Senft, R. L. 1989. Hierarchical foraging models: effects of stocking and landscape composition on simulated resource use by cattle. *Ecological Modeling* 46: 283–303. * 2, 11

1535 Senft, R. L, Coughenour, M. B., Bailey, D. W., Rittenhouse, L. R., Sala, O. E. & Swift, D. M. 1987. Large herbivore foraging and ecological hierarchies. *BioScience* 37: 789–99. * 1, 11

1536 Shafer, C. L. 1990. *Nature Reserves: Island Theory and Conservation Practice*. Smithsonian Institution Press, Washington, DC. * 2, 11, 12

1537 Shaffer, M. L. 1981. Minimum population sizes for species conservation. *BioScience* 31: 131–4. * 2

1538 Shalaway, S. D. 1985. Fencerow management for nesting birds in Michigan. *Wildlife Society Bulletin* 13: 302–6. * 6

1539 Shantz, H. L. 1917. Plant succession on abandoned roads in eastern Colorado. *Journal of Ecology* 5: 19–42. * 5

1540 Sharer, R. J., ed. 1980. The Quirigua Project 1974–1979. *Expedition* 23: 5–10. * 14

1541 Sharpe, D. M., Guntenspergen, G. R., Dunn, C. P., Leitner, L. A. & Stearns, F. 1987. Vegetation dynamics in a southern Wisconsin agricultural landscape. In M. G. Turner, ed. *Landscape Heterogeneity and Disturbance*, pp. 137–55. Springer-Verlag, New York. * 1, 12

1542 Sharpe, D. M., Stearns, F. W., Burgess, R. L. & Johnson, W. C. 1981. Spatio-temporal patterns of forest ecosystems in man-dominated landscapes of the eastern United States. In S. P. Tjallingii & A. A. de Veer, eds. *Perspectives in Landscape Ecology*, pp. 109–16. PUDOC, Wageningen, Netherlands. * 12

1543 Shaver, G. R. & Chapin, F. S., III. 1986. Effect of NPK fertilization on production and biomass of Alaskan tussock tundra. *Arctic and Alpine Research* **18:** 261–8. * 1, 9

1544 Shaver, G. R., Nadelhoffer, K. J. & Giblin, A. E. 1991. Biogeochemical diversity and element transport in a heterogeneous landscape, the North Slope of Alaska. In M. G. Turner & R. H. Gardner, eds. *Quantitative Methods in Landscape Ecology*, pp. 105–25. Springer-Verlag, New York. * 1, 9

1545 Shaw, D. L. 1988. The design and use of living snowfences in North America. *Agriculture, Ecosystems and Environment* **22/23:** 351–62. (Reprinted 1988 in *Windbreak Technology*. Elsevier, Amsterdam.) * 6

1546 Shelley, R. C. 1982. *An Introduction to Sedimentology*, 2nd edn. Academic Press, New York. * 9

1547 Shmida, A. & Wilson, M. V. 1985. Biological determinants of species diversity. *Journal of Biogeography* **12:** 1–20. * 1

1548 Shreeve, T. G. & Mason, C. F. 1980. The number of butterfly species in woodlands. *Oecologia* **45:** 414–18. * 2

1550 Shugart, H. H. & Seagle, S. W. 1985. Modeling forest landscapes and the role of disturbance in ecosystems and communities. In S. T. A. Pickett & P. S. White, eds. *The Ecology of Natural Disturbance and Patch Dynamics*, pp. 353–68. Academic Press, New York. * 1, 13

1551 Shuldiner, P. W., Cope, D. F. & Newton, R. B. 1979. *Ecological Effects of Highway Fills on Wetlands; User's Manual*. Transportation Research Board, National Research Council, Washington, DC. * 5

1552 Shure, D. J. & Phillips, D. L. 1991. Patch size of forest openings and arthropod populations. *Oecologia* **86:** 325–34. * 2

1553 Shure, D. J. & Wilson, L. A. 1993. Patch-size effects on plant phenolics in successional openings of the Southern Appalachians. *Ecology* **74:** 55–67. * 2

1554 Siddle, D. J. 1970. Location theory and the subsistence economy: the spacing of rural settlements in Sierra Leone. *The Journal of Tropical Geography* **31:** 79–90. * 9

1555 Siderits, K. & Radtke, R. E. 1977. Enhancing forest wildlife habitat through diversity. *Transactions of the North American Wildlife and Natural Resources Conference* **42:** 425–34. * 9

1556 Simberloff, D. S. 1976. Experimental zoogeography of islands: effects of island size. *Ecology* **57:** 629–48. * 2

1557 Simberloff, D., Farr, J. A., Cox, J. & Mehlman, D. W. 1992. Movement corridors: conservation bargains or poor investments? *Conservation Biology* **6:** 493–504. * 6, 12

1558 Simberloff, D. S. & Gotelli, N. 1984. Effects of insularization on plant species richness in the prairie–forest ecotone. *Biological Conservation* **29:** 27–46. * 2

1559 Simberloff, D. S. & Wilson, E. O. 1969. Experimental zoogeography of islands: the colonization of empty islands. *Ecology* **50:** 278–96. * 2

1560 Simon, J. L. 1982. the Ultimate Resource. Princeton University Press, Princeton, New Jersey. * 14

1561 Simpson, G. G. 1940. Mammals and landbridges. *Journal of the Washington Academy of Sciences* **30:** 137–63. * 2,6

1562 Simpson, G. G. 1964. Species density of North American recent mammals. *Systematic Zoology* **13:** 57–73. * 3

1563 Simpson, R. H. & Riehl, H. 1981. *The Hurricane and Its Impact*. Louisiana State University Press, Baton Rouge, Louisiana, USA. * 10

1564 Sinclair, N. R., Getz, L. L. & Bock, F. S. 1967. Influence of stone walls on the local distribution of small mammals. *The University of Connecticut Occasional Papers, Biological Science Series* **1:** 43–62. * 6

1565 Singer, F. J. 1978. Behavior of mountain goats in relation to US Highway 2, Glacier National Park, Montana. *Journal of Wildlife Management* **42:** 591–7. * 5

1566 Singer, F. J. & Doherty, J. L. 1985. Movements and habitat use in an unhunted population of mountain goats (*Oreamnos americanus*). *Canadian Field Naturalist* **99:** 205–17. * 5

588

1567 Singer, F. J., Langlitz, W. L. & Samuelson, E. C. 1985. Design and construction of highway underpasses used by mountain goats. *Transportation Research Records* **1016:** 6–10. * 5

1568 Singh, R. L., ed. 1969. *Readings in Settlement Geography*. National Geographical Society of India, Varanasi. * 9

1569 Sisk, T. D. & Margules, C. R. 1993. Habitat edges and restoration: methods for quantifying edge effects and predicting the results of restoration efforts. In D. A. Saunders, R. J. Hobbs & P. R. Ehrlich, eds. *Nature Conservation 3: Reconstruction of Fragmented Ecosystems*, pp. 57–69. Surrey Beatty, Chipping Norton, Australia. * 4

1570 Skidmore, E. L. 1987. Wind-erosion direction factors as influenced by field shape and wind preponderance. *Soil Science Society of America Journal* **51:** 198–202. * 2,4,12

1571 Skidmore, E. L, Fisher, P. S. & Woodruff, N. P. 1970. Wind erosion equation: computer solution and application. *Soil Science Society of America Proceedings* **34:** 931–5. * 4

1572 Skidmore, E. L. & Hagen, L. J. 1977. Reducing wind erosion with barriers. *Transactions of the American Society of Agricultural Engineers* **20:** 911–15. * 6

1573 Skidmore, E. L., Jacobs, H. S. & Hagen, L. J. 1972. Microclimate modification by slat-fence windbreaks. *Agronomy Journal* **64:** 160–2. * 6

1574 Skinner, M. & Iffrig, G. F. 1982. Boardwalk construction in a Missouri Fen – special precautions to protect natural integrity. *Natural Areas Journal* **2:** 10–11. * 5

1575 Sklar, F. H. & Costanza, R. 1991. The development of dynamic spatial models for landscape ecology: a review and prognosis. In M. G. Turner & R. H. Gardner, eds. *Quantitative Methods in Landscape Ecology*, pp. 239–88. Springer-Verlag, New York. * 1, 10, 14

1576 Sklar, F. H., Costanza, R. & Day, J. W., Jr. 1985. Dynamic spatial simulation modeling of coastal wetland habitat succession. *Ecological Modeling* **29:** 261–81. * 1,10

1577 Skoke, D. & Tucker, C. 1993. Tropical deforestation and habitat fragmentation in the Amazon: satellite data from 1978 to 1988. *Science* **260:** 1905–10. * 12

1578 Skovlin, J. 1987. Southern Africa's experience with intensive short duration grazing. *Rangelands* **9:** 162–7. * 11

1579 Slatyer, R. O. 1965. Measurements of precipitation interception by an arid zone plant community (*Acacia aneura* F. Muell). *Arid Zone Research* **25:** 181–92. * 10

1580 Slatyer, R. O. & Noble, I. R. 1992. Dynamics of montane treelines. In A. J. Hansen & F. di Castri, eds. *Landscape Boundaries: Consequences for Biotic Diversity and Ecological Flows*, pp. 346–59. Springer-Verlag, New York. * 1,3,4

1581 Smith, D. G. 1976. Effect of vegetation on lateral migration of anastomosed channels of a glacier meltwater river. *Geological Society of America Bulletin* **87:** 857–60. * 7

1582 Smith, D. S. & Hellmund, P. C., eds. 1993. *Ecology of Greenways: Design and Function of Linear Conservation Areas*. University of Minnesota Press, Minneapolis, Minnesota, USA. * 1,3,5,6,13

1583 Smith, F. E. 1972. Spatial heterogeneity, stability, and diversity in ecosystems. In E. S. Deevey, ed. *Growth by Intussusception: Ecological Essays in Honor of G. Evelyn Hutchinson*. Connecticut Academy of Arts & Sciences, New Haven, Connecticut. (Reprinted in *Transactions of the Connecticut Academy of Science* **44:** 309–35.) * 8

1584 Smith, I. K. & Vankat, J. L. 1992. Dry evergreen forest (coppice) communities of North Andros Island, Bahamas. *Bulletin of the Torrey Botanical Club* **119:** 181–91. * 2,3

1585 Smith, M. S. 1988. *Modeling: Three Approaches to Predicting How Herbivore Impact is Distributed in Rangelands*. Regional Research Report 628, New Mexico Agricultural Experiment Station, Las Cruces, New Mexico, USA. * 11

1586 Smith, R. A., Alexander, R. B. & Wolman, M. G. 1987. Water-quality trends in the nation's rivers. *Science* **235:** 1607–15. * 7

1587 Smith, W. H. & Siccama, T. G. 1981. The Hubbard Brook Ecosystem Study: biogeochemistry of lead in the northern hardwood forest. *Journal of Environmental Quality* **10:** 323–33. * 10

1588 Snow, D. W. & Mayer-Gross, H. 1967. Farmland as nesting habitat. *Bird Study* **14:** 43–52. * 6

1589 Snyder, W. D. 1985. Survival of radio-marked hen ring-necked pheasants in Colorado. *Journal of Wildlife Management* **49:** 1044–50. * 6

1590 Society for Range Management. 1989. *A Glossary of Terms Used in Range Management*, 3rd edn. Society for Range Management, Denver, Colorado, USA. * 11

1591 Sokal, R. & Oden, N. L. 1978. Spatial autocorrelation in biology. 2. Some biological implications and four applications of evolutionary and ecological interest. *Biological Journal of the Linnean Society* **10:** 229–49. * 1

1592 Solbrig, O. T., van Emden, H. M. & van Oordt, P. G. W. J., eds. 1992. *Biodiversity and Global Change*. International Union of Biological Sciences, Paris. * 14

1593 Solomon, A. M. 1986. Transient response of forests to CO₂ induced climate change: simulation modeling experiments in eastern North America. *Oecologia* **68:** 567–79. * 3, 10, 14

1594 Solomon, A. M. & Webb, T., III. 1985. Computer-aided reconstruction of late-Quaternary landscape dynamics. *Annual Review of Ecology and Systematics* **16:** 63–84. * 14

1595 Soulé, M. E., ed. 1987. *Viable Populations for Conservation*. Cambridge University Press, Cambridge. * 1, 2, 11, 12, 13

1596 Soulé, M. E. 1991. Land use planning and wildlife maintenance: guidelines for conserving wildlife in an urban landscape. *Journal of the American Planning Association* **57:** 313–23. * 2, 6, 13

1597 Soulé, M. E., Bolger, D. T. & Alberts, A. C. 1988. Reconstructed dynamics of rapid extinctions of chaparral-requiring birds in urban habitat islands. *Conservation Biology* **2:** 75–92. * 6, 11, 13

1598 Soulé, M. E. & Gilpin, M. E. 1991. The theory of wildlife corridor capability. In D. A. Saunders & R. J. Hobbs, eds. *Nature Conservation 2: The Role of Corridors*, pp. 3–8. Surrey Beatty, Chipping Norton, Australia. * 5, 6

1599 Soulé, M. E., Wilcox, B. A. & Holtby, C. 1979. Benign neglect: a model of faunal collapse in the game reserves of East Africa. *Biological Conservation* **15:** 259–72. * 2

1600 Sousa, W. P. 1979. Experimental investigations of disturbance and ecological succession in a rocky intertidal algal community. *Ecological Monographs* **49:** 227–54. * 2

1601 Sousa, W. P. 1985. Disturbance and patch dynamics on rocky intertidal shores. In S. T. A. Pickett & P. S. White, eds. *The Ecology of Natural Disturbance and Patch Dynamics*, pp. 101–24. Academic Press, New York. * 2

1602 Spies, T. A., Ripple, W. J. & Bradshaw, G. A. 1994. Dynamics and pattern of a managed coniferous forest landscape in Oregon. *Ecological Applications* **4:** 555–68. * 12

1603 Spies, T. A., Tappeiner, J., Pojar, J. & Coates, D. 1991. Trends in ecosystem management at the stand level. *Transactions of the North American Wildlife and Natural Resources Conference* **56:** 628–39. * 13

1604 Spirn, A. W. 1984. *The Granite Garden: Urban Nature and Human Design*. Basic Books, New York. * 2, 13

1605 Spooner, B. & Mann, H. S., eds. 1982. *Desertification and Development: Dryland Ecology in Social Perspective*. Academic Press, London. * 12

1606 Sprugel, D. G. 1976. Dynamic structure of wave-generated *Abies balsamea* forests in the northeastern United States. *Journal of Ecology* **64:** 889–91. * 3, 4

1607 Spurr, S. H. 1957. Local climate in the Harvard Forest. *Ecology* **38:** 37–46. * 10

1608 Squires, V. R. 1982. Behaviour of free-ranging livestock on native grasslands and shrublands. *Tropical Grasslands* **16:** 161–70. * 11

1609 Stabler, D. F. 1985. Increasing summer flow in small streams through management of riparian areas and adjacent vegetation: a synthesis. In R. R. Johnson, D. D. Ziebell, D. R. Patton, P. F. Folliott & R. H. Hamre, eds. *Riparian Ecosystems and Their Management: Reconciling Conflicting Uses*, pp. 206–10. General Technical Report RM-120, US Forest Service, Fort Collins, Colorado, USA. * 7

1610 Stafford, C. R. 1994. Structural changes in Archaic landscape use in the dissected uplands of southwestern Indiana. *American Antiquity* **59:** 219–39. * 9

1611 Stafford, C. R. & Hajic, E. R. 1992. Landscape scale: geoenvironmental approaches to prehistoric settlement strategies. In J. Rossignol & L. Wandsnider, eds. *Space, Time, and Archaeological Landscapes*, pp. 137–65. Plenum Press, New York. * 9

1612 Stafford Smith, D. M. & Pickup, G. 1990. Pattern and production in arid lands. *Proceedings of the Ecological Society of Australia* **16:** 195–200. * 8,9

1613 Stamps, J. A., Buechner, M. & Krishnan, V. V. 1987a. The effects of edge permeability and habitat geometry on emigration from patches of habitat. *American Naturalist* **129:** 533–52. * 3,4,11

1614 Stamps, J. A., Buechner, M. & Krishnan, V. V. 1987b. The effects of habitat geometry on territorial defense costs: intruder pressure in bounded habitats. *American Zoologist* **27:** 307–25. * 3,4

1615 Stapleton, J. & Kiviat, E. 1979. Rights of birds and rights of way. *American Birds* **33:** 7–10. * 5

1616 Statzner, B. & Higler, B. 1985. Questions and comments on the river continuum concept. *Canadian Journal of Fisheries and Aquatic Sciences* **42:** 1038–44. * 7

1617 Stauffer, D. 1985. *Introduction to Percolation Theory*. Taylor and Francis, London. * 1,8

1618 Stauffer, D. F. & Best, L. B. 1980. Habitat selection by birds of riparian communities: evaluating effects of habitat alterations. *Journal of Wildlife Management* **44:** 1–15. * 7

1619 Stein, B. A. 1974. *Size, Efficiency, and Community Enterprise*. Center for Community Economic Development, Cambridge, Massachusetts, USA. * 2

1620 Stein, W. D. 1986. *Transport and Diffusion Across Cell Membranes*. Academic Press, Orlando, Florida. * 3

1621 Steinbeck, J. 1945. *Cannery Row*. Viking, New York. * 5

1622 Steinbeck, J. 1989. *Grapes of Wrath*. Viking, New York. * 5

1623 Steinblums, I. J., Froehlich, H. A. & Lyons, J. K. 1984. Designing stable buffer strips for stream protection. *Journal of Forestry* **82:** 49–52. * 7

1624 Steiner, F. 1991. *The Living Landscape: An Ecological Approach to Landscape Planning*. McGraw-Hill, New York. * 13

1625 Steiner, F., Young, G. & Zube, E. 1988. Ecological planning: retrospect and prospect. *Landscape Journal* **7:** 31–9. * 13

1626 Steinitz, C. 1990. Toward a sustainable landscape with high visual preference and high ecological integrity: the loop road in Acadia National Park, USA. *Landscape and Urban Planning* **19:** 213–50. * 13

1627 Stenseth, N. C. 1979. Where have all the species gone? On the nature of extinction and the Red Queen hypothesis. *Oikos* **33:** 196–227. * 2

1628 Steward, J. H. 1955. *Theory of Culture Change*. University of Illinois Press, Urbana, Illinois, USA. * 9

1629 Stickel, L. F. 1968. Home range and travels. In J. A. King, ed. *Biology of* Peromyscus (*Rodentia*), pp. 373–411. Special Publication 2, American Society of Mammalogists, Stillwater, Oklahoma, USA. * 6

1630 Stierlin, H. 1984. *The Cultural History of Angkor*. Aurum Press, London. * 4, 14

1631 Stiles, E. W. 1980. Patterns of fruit presentation and seed dispersal in bird-disseminated woody plants in the eastern deciduous forest. *American Naturalist* **116:** 670–88. * 1, 3, 11

1632 Stiles, E. W. 1982. Fruit flags: two hypotheses. *American Naturalist* **120:** 500–9. * 11, 13

1633 Stilgoe, J. R. 1980. The wildering of rural New England 1850 to 1950. *New England and St. Lawrence Valley Geographical Society, Proceedings* **10:** 1–6. * 9

1634 Stilgoe, J. R. 1989. Everyday rural landscapes and Thoreau's wild apples. *New England Landscape* **1:** 5–11. * 1,9

1635 Stocker, G. C. & Irvine, A. K. 1983. Seed dispersal by Cassowaries (*Casuarius casuarius*) in North Queensland rainforests. *Biotropica* **15:** 170–6. * 2, 11

1636 Stoddart, D. R. 1965. The shape of atolls. *Marine Geology* **3:** 369–83. * 4

1637 Stoeckeler, J. H. 1962. *Shelterbelt Influence on Great Plains Field Environment and Crops: A Guide for Determining Design and Orientation*. US Forest Service, Production Research Report 62, Washington, DC. * 6

1638 Stoeckeler, J. H., Strothman, R. O. & Krefting, L. W. 1957. Effect of deer browsing on reproduction in the northern hardwood-hemlock type in northeastern Wisconsin. *Journal of Wildlife Management* **21:** 75–80. * 3

1639 Storm, G. L., Andrews, R. D., Phillips, R. L., Bishop, R. A., Siniff, D. B. & Tester, J. R. 1976. Morphology, reproduction, dispersal and mortality of midwestern red fox populations. *Wildlife Monographs* **49:** 5–82. * 5, 7, 11

1640 Stout, B. B. 1952. *Species Distribution and Soils in the Harvard Forest*. Bulletin 24, Harvard Forest, Petersham, Massachusetts, USA. * 9

1641 Stout, I. J. & Corwell, G. W. 1976. Nonhunting mortality of fledged North American waterfowl. *Journal of Wildlife Management* **40:** 681–93. * 5

1642 Strahler, A. N. 1952. Hypsometric (area–altitude) analysis of erosional topography. *Bulletin of the Geological Society of America* **63:** 1117–42. * 9

1643 Strahler, A. N. 1964. Quantitative geomorphology of drainage basins and channel networks. In V. T. Chow, ed. *Handbook of Applied Hydrology*, pp. (4) 39–76. McGraw-Hill, New York. * 7, 10

1644 Strelke, W. K. & Dickson, J. G. 1980. Effect of forest clear-cut edge on breeding birds in east Texas. *Journal of Wildlife Management* **44:** 559–67. * 3

1645 Stritch, L. R. 1990. Landscape-scale restoration of barrens-woodland within the oak–hickory forest mosaic. *Restoration and Management Notes* **8:** 73–7. * 13

1646 Strong, D. R., Jr. 1979. Biogeographic dynamics of insect–host plant communities. *Annual Review of Entomology* **24:** 89–119. * 2

1647 Sturges, D. 1984. *The Geometry of Snowdrifts Cast by Shelterbelts as Determined by Small-scale Outdoor Modeling*. US Forest Service, Laramie, Wyoming. 13 pp. * 6

1648 Sturrock, J. W. 1972. Aerodynamic studies of shelterbelts in New Zealand. 2. Medium height to tall shelterbelts in mid-Canterbury. *New Zealand Journal of Science* **15:** 113–40. * 6

1649 Stuth, J. W. 1991. Foraging behavior. In R. K. Heitschmidt & J. W. Stuth, eds. *Grazing Management: An Ecological Perspective*, pp. 65–83. Timber Press, Portland, Oregon. * 11

1650 Suckling, G. C. 1984. Population ecology of the sugar glider *Petaurus breviceps* in a system of fragmented habitats. *Australian Wildlife Research* **11:** 49–75. * 5, 6

1651 Suffling, R., Lihou, C., & Morand, Y. 1988. Control of landscape diversity by catastrophic disturbance: a theory and a case study of fire in a Canadian boreal forest. *Environmental Management* **12:** 73–8. * 10

1652 Sugihara, G. & May, R. M. 1990. Applications of fractals in ecology. *Trends in Ecology and Evolution* **5:** 79–86. * 11

1653 Sugihara, G., Schoenly, K. & Trombla, A. 1989. Scale invariance in food web properties. *Science* **245:** 48–52. * 8

1654 Sukachev, V. N. & Dylis, N. 1964. *Fundamentals of Forest Biogeocoenology*. Oliver and Boyd, Edinburgh. * 1, 3

1655 Sukopp, H., Hejny, S. & Kowarik, I., eds. 1990. *Urban Ecology: Plants and Plant Communities in Urban Environments*. SPB Academic Publishing, The Hague, Netherlands. * 11, 13

1656 Sunkel, O. 1987. Beyond the World Conservation Strategy: integrating development and the environment in Latin America and the Caribbean. In P. Jacobs & D. A. Munro, eds. *Conservation with Equity: Strategies for Sustainable Development*, pp. 35–54. International Union for the Conservation of Nature and Natural Resources, Gland, Switzerland. * 14

1657 Sutton, R. K. 1992. Landscape ecology of hedgerows and fencerows in Panama Township, Lancaster County, Nebraska. *Great Plains Research* **2:** 223–54. * 6

1658 Sutton, W. R. J. 1973. The importance of size and scale in forestry and forest industries. *New Zealand Journal of Forestry Science* **18:** 63–80. * 2

1659 Swank, W. T. & Caskey, W. H. 1982. Nitrate depletion in a second-order mountain stream. *Journal of Environmental Quality* **11:** 581–4. * 7

1660 Swank, W. T. & Douglass, J. E. 1974. Streamflow greatly reduced by converting deciduous hardwood stands to pine. *Science* **185:** 857–9. * 10

1661 Swanson, F. J., Franklin, J. F. & Sedell, J. R. 1990. Landscape patterns, disturbance, and management in the Pacific Northwest, USA. In I. S. Zonneveld & R. T. T. Forman, eds. *Changing Landscapes: An Ecological Perspective*, pp. 191–213. Springer-Verlag, New York. * 1, 7, 10, 13

1662 Swanson, F. J., Jones, J. A., Wallin, D. O. & Cissel, J. H. 1994. Natural variability – implications for ecosystem management. In *East Side Forest Ecosystem Health Assessment. Volume 2. Ecosystem Management: Principles and Applications*, pp. 89–103. Report GTR, US Forest Service, Portland, Oregon. *12,13

1663 Swanson, F. J., Kratz, T. K., Caine, N. & Woodmansee, R. G. 1988. Landform effects on ecosystem patterns and processes. *BioScience* **38:** 92–8. * 7,9,10,13

1664 Swanson, F. J. & Swanston, D. N. 1977. Complex mass-movement terrains in the western Cascade Range, Oregon. In D. R. Coates, ed. *Reviews in Engineering Geology*, vol. III, pp. 113–24. Geological Society of America, Boulder, Colorado. * 7,10

1665 Swanson, F. J., Wondzell, S. M. & Grant, G. E. 1992. Landforms, disturbance, and ecotones. In A. J. Hansen & F. di Castri, eds. *Landscape Boundaries: Consequences for Biotic Diversity and Ecological Flows*, pp. 304–23. Springer-Verlag, New York. * 4,13

1666 Swihart, R. K. & Slade, N. A. 1984. Road crossing in *Sigmodon hispidus* and *Microtus ochrogaster*. *Journal of Mammalogy* **65**: 357–60. * 5

1667 Swihart, R. K., Slade, N. A. & Bergstrom, B. J. 1988. Relating body size to the rate of home range use in mammals. *Ecology* **69**: 393–9. * 11

1668 Swincer, D. E. 1986. Physical characteristics of sites in relation to invasions. In R. H. Groves & J. J. Burdon, eds. *Ecology of Biological Invasions*, pp. 67–76. Cambridge University Press, Cambridge. * 11

1669 Swingland, I. R. & Greenwood, P. J., eds. 1983, *The Ecology of Animal Movement*. Clarendon Press, Oxford. * 8,11

1670 Taaffe, E. J. & Gauthier, H. L., Jr. 1973. *Geography of Transportation*. Prentice-Hall, Englewood Cliffs, New Jersey, USA. * 1,5,8,9,11,12

1671 Tabler, R. D. 1974. New engineering criteria for snow fence systems. *Transportation Research Record* (Washington) **506**: 65–84. * 6

1672 Takeuchi, K. 1991. *Regional (Landscape) Ecology*. (In Japanese.) Asakura Publishing, Tokyo. * 1

1673 Tamm, C. O. & Troedsson, T. 1955. An example of the amounts of plant nutrients supplied to the ground in road dust. *Oikos* **6**: 61–70. * 5

1674 Tanfiljev, G. I. 1894. *Predely Lesov na Juge Rossii (The Forest Limits in the South of Russia)*. Sankt Peterburg, Russia. * 3

1675 Tanner, C. B. 1957. Factors affecting evaporation from plants and soils. *Journal of Soil and Water Conservation* **12**: 221–7. * 10

1676 Taub, F. B. 1974. Closed ecological systems. *Annual Review of Ecology and Systematics* **5**: 139–60. * 1

1677 Taylor, C. 1975. *Fields in the English Landscape*. J. M. Dent, London. * 4

1678 Taylor, C. A., Jr. 1989. Short duration grazing: experiences from the Edwards Plateau region in Texas. *Journal of Soil and Water Conservation* **44**: 297–302. * 11

1679 Taylor, M. E. 1971. Bone disease and fractures in east African viverrids. *Canadian Journal of Zoology* **49**: 1035–42. * 5

1680 Taylor, M. W. 1977. A comparison of three edge indexes. *Wildlife Society Bulletin* **5**: 192–3. * 3,4

1681 Taylor, R. J. 1987a. The geometry of colonization: 1. Islands. *Oikos* **48**: 225–31. * 4

1682 Taylor, R. J. 1987b. The geometry of colonization: 2. Peninsulas. *Oikos* **48**: 232–7. * 3,4

1683 Taylor, R. J. & Regal, P. J. 1978. The peninsular effect on species diversity and the biogeography of Baja California. *American Naturalist* **112**: 583–93. * 2,4

1684 Tchou, Y-T. 1951. Etudes ecologiques et phytosociologiques sur les forets riveraines du Bas-Languedoc. *Vegetatio* **1**: 2–28, 93–128, 217–57, 347–83. * 7

1685 Temple, S. A. 1986a. Predicting impacts of habitat fragmentation on forest birds: a comparison of two models. In J. Verner, M. L. Morrison & C. J. Ralph, eds. *Wildlife 2000: Modeling Habitat Relationships of Terrestrial Vertebrates*, pp. 301–4. University of Wisconsin Press, Madison, Wisconsin, USA. * 2,3,4

1686 Temple, S. A. 1986b. Recovery of the endangered Mauritius kestrel from an extreme population bottleneck. *Auk* **103**: 632–3. * 2

1687 Temple, S. A. 1990. The nasty necessity: eradicating toxics. *Conservation Biology* **4**: 113–15. * 11

1688 Temple, S. A. & Cary, J. R. 1988. Modeling dynamics of habitat-interior bird populations in fragmented landscapes. *Conservation Biology* **2**: 340–7. * 2,12

1689 Templeton, A. R. 1986. Coadaptation and outbreeding depression. In M. E. Soulé, ed. *Conservation Biology: The Science of Scarcity and Diversity*, pp. 105–16. Sinauer Associates, Sunderland, Massachusetts. * 2,11

1690 ten Houte de Lange, S. M. 1978. Zur futterwahl des Alpensteinbockes (*Capra ibex* L.): Eine Untersuchung an der Steinbockkolonie am Piz Albris bei Pontresina. *Zeitschrift für Jagdwissenschaft* **24**: 113–38. * 9,11

1691 Terborgh, J. 1974. Preservation of natural diversity: the problem of extinction prone species. *BioScience* **24**: 715–22. * 2,12

1692 Terborgh, J. 1976. Island biogeography and conservation: strategy and limitations. *Science* **193**: 1029–30. * 2

1693 Terborgh, J. 1985. The role of ecotones in the distribution of Andean birds. *Ecology* **66**: 1237–46. * 3

1694 Terborgh, J. & Winter, B. 1980. Some causes of extinction. In M. E. Soulé & B. A. Wilcox, eds. *Conservation Biology: An Evolutionary–Ecological Perspective*, pp. 119–33. Sinauer Associates, Sunderland, Massachusetts, USA. * 2

1695 Terborgh, J. & Winter, B. 1983. A method for siting parks and reserves with special reference to Colombia and Ecuador. *Biological Conservation* **27**: 45–58. * 9

1696 Terres, J. K. 1987. *Songbirds in Your Garden: How to Attract, Feed, and Enjoy Birds in Your Garden and Backyard*. Harper & Row, New York. * 13

1697 Thayer, R. L., Jr. 1990. Pragmatism in paradise, technology and the American landscape. *Landscape* **30**: 1–11. * 14

1698 Thebaud, C. & Debussche, M. 1991. Rapid invasion of *Fraxinus ornus* L. along the Herault River system in southern France: the importance of seed dispersal by water. *Journal of Biogeography* **18**: 7–12. * 7

1699 Theberge, J. B. 1989. Guidelines to drawing ecologically sound boundaries for national parks and nature reserves. *Environmental Management* **13**: 695–702. * 3,9

1700 Thomas, J. W., ed. 1979. *Wildlife Habitats in Managed Forests: The Blue Mountains of Oregon and Washington*. Agriculture Handbook 553, US Forest Service, Portland, Oregon. * 3,8,11,13

1701 Thomas, J. W., Black, H., Jr., Scherzinger, R. J. & Pederson, R. J. 1979. Deer and elk. In J. W. Thomas, ed. *Wildlife Habitats in Managed Forests: The Blue Mountains of Oregon and Washington*, pp. 104–27. Agriculture Handbook 553, US Forest Service, Portland, Oregon. * 9

1702 Thomas, J. W., Maser, C. & Rodiek, J. E. 1979a. Edges. In J. W. Thomas, ed. *Wildlife Habitats in Managed Forests: The Blue Mountains of Oregon and Washington*, pp. 48–59. Agriculture Handbook 553, US Forest Service, Portland, Oregon. * 3,9

1703 Thomas, J. W., Maser, C. & Rodiek, J. E. 1979b. Riparian zones. In J. W. Thomas, ed. *Wildlife Habitats in Managed Forests: The Blue Mountains of Oregon and Washington*, pp. 40–7. Agriculture Handbook 553, US Forest Service, Portland, Oregon. * 7

1704 Thomas, J. W., Toweill, D. E. & Metz, D. P. 1982. *Elk of North America: Ecology and Management*. Wildlife Management Institute, Stackpole, Harrisburg, Pennsylvania, USA. * 8,9,11

1705 Thomas, L. K., Jr. 1988. Some principles of exotic species ecology and management and their interrelationships. In L. K. Thomas, Jr., ed. *Management of Exotic Species in Natural Communities*, pp. 96–110. US National Park Service and The George Wright Society, Fort Collins, Colorado. * 11

1706 Thomas, W. L., ed. 1956. *Man's Role in Changing the Face of the Earth*. University of Chicago Press, Chicago. * 8,9

1707 Thompson, D. W. 1961. *On Growth and Form*. (Abridged edition, J. T. Bonner, ed.) Cambridge University Press, Cambridge. * 1,4

1708 Thompson, J. N. & Willson, M. F. 1978. Disturbance and the dispersal of fleshy fruits. *Science* **200**: 1161–3. * 3, 11

1709 Thompson, W. A., Vertinsky, I. & Krebs, J. R. 1974. The survival value of flocking in birds: a simulation model. *Journal of Animal Ecology* **43**: 785–820. * 9

1710 Thomson, J. W., Jr. 1940. Relic prairie areas in central Wisconsin. *Ecological Monographs* **10**: 685–717. * 5

1711 Thoreau, H. D. 1993. The dispersion of seeds. In B. P. Dean, ed. *Faith in a Seed*, pp. 23–173. Island Press, Washington, DC. * 9,11

1712 Thorne, J. F. & Miller, R. W. 1984. Landscape-scale geomorphic control over nitrogen mineralization in a subalpine spruce–fir forest, Adirondack Mountains, New York. *Landscape Ecology*. In press. * 9

1713 Thrall, G. I., Swanson, B. & Nozzi, D. 1988. Green-space acquisition ranking program (GARP): a computer-assisted decision strategy. *Computers, Environment and Urban Systems* **12**: 161–84. * 13

1714 Tibke, G. 1988. Basic principles of wind erosion control. *Agriculture, Ecosystems and Environment* **22/23**: 103–22. (Reprinted 1988 in *Windbreak Technology*. Elsevier, Amsterdam.). * 6, 12

594

1715 Ticknor, K. A. 1988. Design and use of field windbreaks in wind erosion control systems. *Agriculture, Ecosystems and Environment* **22/23:** 123–32. (Reprinted 1988 in *Windbreak Technology*. Elsevier, Amsterdam.) * 6

1716 Tietenberg, T. H. 1988. *Environmental and Natural Resource Economics*. Scott, Foresman & Co., Glenview, Illinois, USA. * 14

1717 Tilghman, N. 1987. Characteristics of urban woodlands affecting breeding bird diversity and abundance. *Landscape and Urban Planning* **14:** 481–95. * 2,3,5,13

1719 Tilman, D. 1985. The resource-ratio hypothesis of plant succession. *American Naturalist* **125:** 827–52. * 2

1720 Tilzer, M. M. & Serruya, C., eds. 1990. *Large Lakes: Ecological Structure and Function*. Springer-Verlag, New York. * 2

1721 Timm, R. M. 1988. Vertebrate pest management in windbreak systems. *Agriculture, Ecosystems and Environment* **22/23:** 555–70. (Reprinted 1988 in *Windbreak Technology*. Elsevier, Amsterdam). * 3,6

1722 Tjallingii, S. P. & de Veer, A. A., eds. 1981. *Perspectives in Landscape Ecology*. PUDOC Centre for Agricultural Publishing and Documentation, Wageningen, Netherlands. * 1

1723 *Toads on Roads*. No date (approximately 1986). Fauna and Flora Preservation Society, Zoological Society of London, London. * 5

1724 Tobler, W. R. 1978. Comparison of plane forms. *Geographical Analysis* **10:** 154–62. * 4

1725 Todd, N. & Todd, J. 1984. *Bioshelters, Ocean Arks, City Farming: Ecology as the Basis of Design*. Sierra Club Books, San Francisco. * 14

1726 Tomlin, C. D. 1990. *Geographic Information Systems and Cartographic Modeling*. Prentice-Hall, Englewood Cliffs, New Jersey, USA. * 1,9

1727 Tosi, J. A., Jr. 1964. Climatic control of terrestrial ecosystems: a report on the Holdridge model. *Economic Geography* **40:** 173–81. * 9

1728 Toth, R. E. 1988. Theory and language in landscape analysis, planning, and evaluation. *Landscape Ecology* **1:** 193–201. * 1

1729 Trabaud, L. 1970. Quelques valeurs et observations sur la phyto-dynamique des surfaces incendiees dans le Bas-Languedoc. *Naturalia Monspeliensia Serie Botanique* **21:** 231–42. * 10

1730 Trabaud, L. & Lepart, J. 1980. Diversity and stability in garrigue ecosystems after fire. *Vegetatio* **43:** 49–57. * 10

1731 Trimble, G. R., Jr. & Sartz, R. S. 1957. How far from a stream should a logging road be located? *Journal of Forestry* **55:** 339–41. * 7,13

1732 Trinci, A. P. J. 1971. Influence of the width of the peripheral growth zone on the radial growth rate of fungal colonies on solid media. *Journal of General Microbiology* **67:** 325–44. * 4

1733 Trinci, A. P. J. 1974. A study of the kinetics of hyphal extension and branch initiation of fungal mycelia. *Journal of General Microbiology* **81:** 225–36. * 4

1734 Troll, C. 1939. Luftbildplan und okologische Bodenforschung. *Zeitschraft der Gesellschaft fur Erdkunde Zu Berlin*, pp. 241–98. * 1

1735 Troll, C. 1950. Die geografische Landschaft und ihre Erforschung. *Studium Generale* **3:** 163–81. Springer-Verlag, Berlin. * 1

1737 Troll, C. 1968. Landschaftsokologie. In R. Tuxen, ed. *Pflanzensoziologie und Landschaftsokologie*, pp. 1–21. Dr. W. Junk Publishers, The Hague, Netherlands. * 1

1738 Troll, C. 1971. Landscape ecology (geo-ecology) and biogeocenology – a terminological study. (Translated by E. M. Yates.) *GeoForum* **8:** 43–6. * 1

1739 Tucker, C. J., Gatlin, J. A. & Schneider, S. R. 1984. Monitoring vegetation in the Nile delta with NOAA-6 and NOAA-7 AVHRR imagery. *Photogrammetric Engineering and Remote Sensing* **50:** 53–61. * 1

1740 Tucker, C. J., Townshend, J. R. G. & Goff, T. E. 1985. African land-cover classification using satellite data. *Science* **227:** 369–75. * 9

1741 Turner, B. L. II, Clark, W. C., Kates, R. W., Richards, J. F., Mathews, J. T. & Meyer, W. B., eds. 1990. *The Earth as Transformed by Human Action: Global and Regional Changes in the Biosphere Over the Past 300 Years*. Cambridge University Press, Cambridge. * 14

1742 Turner, M. G., ed. 1987a. *Landscape Heterogeneity and Disturbance*. Springer-Verlag, New York. * 1,4,9,10,12

595

1743 Turner, M. G. 1987b. Spatial simulation of landscape changes in Georgia: a comparison of three transition models. *Landscape Ecology* **1**: 29–36. * 1,12

1744 Turner, M. G. 1988. A spatial simulation model of land use changes in a piedmont county of Georgia. *Applied Mathematics and Computation* **27**: 39–51. * 1,4,9,12

1745 Turner, M. G. 1989. Landscape ecology: the effect of pattern on process. *Annual Review of Ecology and Systematics* **20**: 171–97. * 1,5,9,11,12,13

1746 Turner, M. G. 1990. Spatial and temporal analysis of landscape patterns. *Landscape Ecology* **4**: 21–30. * 4,12

1747 Turner, M. G. & Bratton, S. P. 1987. Fire, grazing and the landscape heterogeneity of a Georgia barrier island. In M. G. Turner, ed. *Landscape Heterogeneity and Disturbance*, pp. 85–101. Springer-Verlag, New York. * 10

1748 Turner, M. G. & Gardner, R. H., eds. 1991. *Quantitative Methods in Landscape Ecology: The Analysis and Interpretation of Landscape Heterogeneity*. Springer-Verlag, New York. * 1,3,8,9,10,12

1749 Turner, M. G., Gardner, R. H. & O'Neill, R. V. 1991. Potential responses of landscape boundaries to global environmental change. In M. M. Holland, P. G. Risser & R. J. Naiman, eds. *Ecotones: The Role of Landscape Boundaries in the Management and Restoration of Changing Environments*, pp. 52–75. Chapman and Hall, New York. * 3

1750 Turner, M. G., Romme, W. H., Gardner, R. H., O'Neill, R. V. & Kratz, T. K. 1993. A revised concept of landscape equilibrium: disturbance and stability on scaled landscapes. *Landscape Ecology* **8**: 213–27. * 2, 9, 12

1751 Turner, S. J., O'Neill, R. V., Conley, W., Conley, M. R. & Humphries, H. C. 1991. Pattern and scale: statistics for landscape ecology. In M. G. Turner & R. H. Gardner, eds. *Quantitative Methods in Landscape Ecology*, pp. 17–49. Springer-Verlag, New York. * 1,9

1752 Turner, T. 1987. *Landscape Planning*. Hutchinson Education, London. * 5

1753 Turner, T. 1992. Open space planning in London: from standards per 1000 to green strategy. *Town Planning Review* **63**: 365–86.* 5,6,13

1754 Tuxen, R. 1952. Hecken und Gebusche. *Mitteilungen der Geographischen Gesellschaft in Hamburg* **50**: 85–117.* 3

1755 Txyning, T. 1989. Amherst's tunneling amphibians. *Defenders* (Washington, DC) (September/October): 20–3.* 5

1756 Tyler, G. & Wells, D. I. 1971. The new gerrymander threat. *The American Federationist* **78(2)**: 1–7.* 4

1757 Tyson, R. M. 1980. Road killed Platypus. *The Tasmanian Naturalist* **60**: 8.* 5

1758 Udo, R. K. 1965. Disintegration of nucleated settlement in eastern Nigeria. *Geographical Review* **55**: 53–67.* 9

1759 Uhl, C., Jordan, C., Clark, K., Clark, H. & Herrera, R. 1982. Ecosystem recovery in Amazon caatinga forest after cutting, cutting and burning, and bulldozer clearing treatments. *Oikos* **38**: 313–20.* 2

1760 Ulanowicz, R. E. 1988. On the importance of higher-level models in ecology. *Ecological Modelling* **43**: 45–56.* 11

1761 Unwin, D. J. 1981. *Introductory Spatial Analysis*. Methuen, London.* 4

1762 Urban, D. L., O'Neill, R. V. & Shugart, H. H., Jr. 1987. Landscape ecology: a hierarchical perspective can help scientists understand spatial patterns. *BioScience* **37**: 119–27.* 1,3

1763 Usher, M. B., ed. 1986. *Wildlife Conservation Evaluation*. Chapman and Hall, London.* 2,9

1764 Usher, M. B. 1988. Biological invasions of nature reserves: a search for generalisations. *Biological Conservation* **44**: 119–35.* 11

1765 Valentine, K. A. 1947. Distance from water as a factor in grazing capacity of rangeland. *Journal of Forestry* **45**: 749–54.* 11,13

1766 Valentine, S. & Dolan, R. 1979. Footstep-induced sediment displacement in the Grand Canyon. *Environmental Management* **3**: 531–3.* 5

1767 van Apeldoorn, R. C., Oostenbrink, W. T., van Winden, A. & van der Zee, F. F. 1992. Effects of habitat fragmentation on the bank vole, *Clethrionomys glareolus*, in agricultural landscape. *Oikos* **65**: 265–74.* 11

1768 van Arsdel, E. P. 1967. The noctural diffusion and transport of spores. *Phytopathology* **57**: 1221–9.* 10

1769 van der Maarel, E. 1976. On the establishment of plant community boundaries. *Berichte der Deutschen Botanischen Gesellschaft* **89**: 415–43.* 3
1770 van der Maarel, E. 1978. Ecological principles for physical planning. In M. W. Holdgate & M. J. Woodman, eds. *The Breakdown and Restoration of Ecosystems*, pp. 413–50, Plenum Press, New York.* 1, 13
1771 van der Meijden, E., De Jong, T. J., Klinkhamer, P. G. L. & Kooi, R. E. 1985. Temporal and spatial dynamics in populations of biennial plants. In J. Haeck & J. W. Woldendorp, eds. *Structure and Functioning of Plant Populations 2. Phenotypic and Genotypic Variation in Plant Populations*, pp. 91–103. North-Holland Publishing, Amsterdam.* 11
1772 van der Pijl, L. 1969. *Principles of Dispersal in Higher Plants*. Springer-Verlag, Berlin.* 11
1773 van der Ryn, S. & Calthorpe, P. 1986. *Sustainable Communities: A New Design Synthesis for Cities, Suburbs, and Towns*. Sierra Club Books, San Francisco.* 14
1774 van der Zande, A. N., Ter Keurs, W. J. & van der Weidjen, W. J. 1980. The impact of roads on the densities of four bird species in an open field habitat – evidence of a long distance effect. *Biological Conservation* **18**: 299–321. * 3,5,12
1775 van Dorp, D. & Opdam, P. F. M. 1987. Effects of patch size, isolation and regional abundance on forest bird communities. *Landscape Ecology* **1**: 59–73. * 2,6,11
1776 van Eimern, J., Karschon, R., Razumova, L. A. & Robertson, G. W. 1964. *Windbreaks and Shelterbelts*. Technical Note 59, World Meteorological Organization, Geneva, Switzerland. 191 pp. * 6
1777 van Emden, H. F. 1965. The role of uncultivated land in the biology of crop pests and beneficial insects. *Scientific Horticulture* **17**: 121–36. * 6
1778 van Gelder, J. J. 1973. A quantitative approach to the mortality resulting from traffic in a population of *Bufo bufo* L. *Oecologia* **13**: 93–5. * 5
1779 van Horne, B. 1983. Density as a misleading indicator of habitat quality. *Journal of Wildlife Management* **47**: 893–901. * 3,11
1780 van Leeuwen, B. H. 1982. Protection of migrating common toad *Bufo bufo* against car traffic in The Netherlands. *Environmental Conservation* **9**: 34–41. * 5
1781 van Leeuwen, C. G. 1966. A relation theoretical approach to pattern and process in vegetation. *Wentia* **15**: 25–46. * 3,14
1782 van Leeuwen, C. G. 1981. From ecosystem to ecodevice. In S. P. Tjallingii & A. A. de Veer, eds. *Perspectives in Landscape Ecology*, pp. 29–34. Pudoc, Wageningen, Netherlands. * 3,14
1783 van Noorden, B. 1986. *Dynamiek en Dichtheid van Bosvogels in Geisoleerde Loofbosfragmenten*. Report 86/19, Research Institute for Nature Management, Leersum, The Netherlands. * 11
1784 van Noorden, B., Opdam, P. & Schotman, A. 1988. Dichtheid van bosvogels in geisoleerde loofbosjes. *Limosa* **61**: 19–25. * 11
1785 Vannote, R. L., Minshall, G. W., Cummins, K. W., Sedell, J. R. & Cushing, C. E. 1980. The river continuum concept. *Canadian Journal of Fisheries and Aquatic Sciences* **37**: 130–7. * 7
1786 van Selm, A. J. 1988. Ecological infrastructure: a conceptual framework for designing habitat networks. In K-F. Schreiber, ed. *Connectivity in Landscape Ecology*, pp. 63–6. Munstersche Geographische Arbeiten 29, Ferdinand Schoningh, Paderborn, Germany. * 9
1787 Verboom, J., Opdam, P., & Schotman, A. 1991. Kerngebieden en kleinschalig landschap. *Landschap* **8**: 1–14. * 11
1788 Verboom, J., Schotman, A., Opdam, P. & Metz, J. A. J. 1991. European nuthatch metapopulations in a fragmented agricultural landscape. *Oikos* **61**: 149–56. * 1, 11
1789 Verboom, J. & van Apeldoorn, R. 1990. Effects of habitat fragmentation on the red squirrel, *Sciurus vulgaris* L. *Landscape Ecology* **4**: 171–6. * 11
1790 Verkaar, H. J. P. A. 1988a. Wegbermen en rivier-dijken als mogelijke migratiebanen voor planten. *Landschap* **5**: 72–82. * 5
1791 Verkaar, H. J. 1988b. The possible role of road verges and river dykes as corridors for the exchange of plant species between natural habitats. In K-F. Schreiber, ed. *Connectivity in Landscape Ecology*, pp. 79–84. Munstersche Geographische Arbeiten 29, Ferdinand Schoningh, Paderborn, Germany. * 5

597

1792 Vermeule, C. C. 1900. Forests and climate. In *Report on Forests: Annual Report of the State Geologist for the year 1899*, pp. 167–72. Geological Survey of New Jersey, Trenton, New Jersey, USA. * 1,10

1793 Verner, J., Morrison, M. L. & Ralph, C. J., eds. 1986. *Wildlife 2000: Modeling Habitat Relationships of Terrestrial Vertebrates*. University of Wisconsin Press, Madison, Wisconsin, USA. * 2,12

1794 Verry, E. S. & Timmons, D. R. 1982. Waterborne nutrient flow through an upland-peatland watershed in Minnesota. *Ecology* **63**: 1456–67. * 7

1795 Vestal, A. G. 1949. *Minimum Areas for Different Vegetations: Their Determination from Species–Area Curves*. University of Illinois Press, Urbana, Illinois, USA. * 2

1796 Vestjens, W. J. M. 1973. Wildlife mortality on a road in New South Wales. *Emu* **73**: 107–12. * 5

1797 Villard, M-A., Freemark, K. & Merriam, G. 1989. Metapopulation theory and neotropical migrant birds in temperate forests: an empirical investigation. In J. M. Hagen, III & D. W. Johnston, eds. *Ecology and Conservation of Neotropical Migrant Landbirds*, pp. 474–82. Smithsonian Institution Press, Washington, DC. * 11

1798 Vink, A. P. A. 1975. *Land Use in Advancing Agriculture*. Springer-Verlag, Berlin. * 1,13

1800 Vink, A. P. A. 1980. *Landschapsecologie en Landgebruik*. Bohn, Scheltema and Holkema, Utrecht, Netherlands. (1983 translation. *Landscape Ecology and Land Use*. Longman, London.) * 1,13

1801 Vitousek, P. M. 1983. Mechanisms of ion leaching in natural and managed ecosystems. In H. A. Mooney & M. Godron, eds. *Disturbance and Ecosystems: Components of Response*, pp. 129–44. Springer-Verlag, New York. * 10

1802 Vitousek, P. M. 1985. Community turnover and ecosystem nutrient dynamics. In S. T. A. Pickett & P. S. White, eds. *The Ecology of Natural Disturbance and Patch Dynamics*, pp. 325–33. Academic Press, New York. * 2

1803 Vitousek, P. M. 1986. Biological invasions and ecosystem properties: can species make a difference? In H. A. Mooney & J. A. Drake, eds. *Ecology of Biological Invasions in North America and Hawaii*, pp. 163–76. Springer-Verlag, New York. * 11

1804 Vitousek, P. M. & Reiners, W. A. 1975. Ecosystem succession and nutrient retention: a hypothesis. *BioScience* **25**: 376–81. * 7,10

1805 Voigt, J. W. & Weaver, J. E. 1951. Range condition classes of native midwestern pasture: an ecological analysis. *Ecological Monographs* **21**: 39–60. * 11

1806 Von Neumann, J. & Morgenstern, O. 1947. *Theory of Games and Economic Behavior*. Princeton University Press, Princeton, New Jersey. * 9

1807 Voorhees, L. D. & Cassell, J. F. 1980. Highway right-of-way: mowing versus succession as related to duck nesting. *Journal of Wildlife Management* **44**: 155–63. * 5

1808 Vos, C. C. & Opdam, P., eds. 1992. *Landscape Ecology of a Stressed Environment*. Chapman & Hall, London. * 1,13

1809 Vos, W. & Stortelder, A. 1992. *Vanishing Tuscan Landscapes: Landscape Ecology of a Submediterranean-Montane Area* (Solano Basin, Tuscany, Italy). PUDOC, Wageningen, Netherlands. * 9,12

1810 Vrijenhoek, R. C. 1985. Animal population genetics and disturbance: the effects of local extinctions and recolonizations on heterozygosity and fitness. In S. T. A. Pickett & P. S. White, eds. 1985. *The Ecology of Natural Disturbance and Patch Dynamics*, pp. 265–85. Academic Press, New York. * 11

1811 Vuillemieur, F. 1970. Insular biogeography in continental regions. I. The Northern Andes of South America. *American Naturalist* **104**: 373–88. * 2

1812 Wace, N. M. 1977. Assessment of dispersal of plant species – the carborne flora in Canberra. *Proceedings of the Ecological Society of Australia* **10**: 167–86. * 5

1813 Wagner, F. H. & Kay, C. E. 1993. 'Natural' or 'healthy' ecosystems: are U. S. national parks providing them? In M. J. McDonnell & S. T. A. Pickett, eds. *Humans as Components of Ecosystems: The Ecology of Subtle Human Effects and Populated Areas*, pp. 257–70. Springer-Verlag, New York. * 13

1814 Wales, B. A. 1972. Vegetation analysis of northern and southern edges in a mature oak–hickory forest. *Ecological Monographs* **42**: 451–71. * 2,3

1815 Walker, B. H., Emslie, R. H., Owen-Smith, R. N. & Scholes, R. J. 1987. To cull or not to cull: lessons from a southern African drought. *Journal of Applied Ecology* **24**: 381–401. * 11

1816 Walker, B. H., Ludwig, D., Holling, C. S. & Peterman, R. M. 1981. Stability of semi-arid savannah grazing systems. *Journal of Ecology* **69**: 473–98. * 11

1817 Walker, D. A., Webber, P. J., Binnian, E. F., Everett, K. R., Lederer, N. D., Nordstrand, E. A. & Walker, M. D. 1987. Cumulative impacts of oil fields on northern Alaskan landscapes. *Science* **238**: 757–61. * 13

1818 Walker, J. W. & Heitschmidt, R. K. 1986. Effect of various grazing systems on type and density of cattle trails. *Journal of Range Management* **39**: 428–31. * 5,11

1819 Walker, J. W., Heitschmidt, R. K., de Moraes, E. A., Kothman, M. M. & Dowhower, S. L. 1989. Quality and botanical composition of cattle diets under rotational and continuous grazing treatments. *Journal of Range Management* **42**: 239–42. * 11

1820 Walker, J. W., Stuth, J. W. & Heitschmidt, R. K. 1989. A simulation approach for evaluating field data from grazing trials. *Agricultural Systems* **30**: 301–16. * 11

1821 Wallace, T. P. & Wintz, P. A. 1980. An efficient three-dimensional aircraft recognition algorithm using normalized Fourier descriptors. *Computer Graphics and Image Processing* **13**: 99–126. * 4

1822 Waller, D. M., O'Malley, D. M. & Gawler, S. C. 1988. Genetic variation in the extreme endemic *Pedicularis furbishiae* (Scrophulariaceae). *Conservation Biology* **1**: 335–40. * 2

1823 Wallin, D. O., Swanson, F. J. & Marks, B. 1994. Landscape pattern response to changes in pattern-generation rules: land-use legacies in forestry. *Ecological Applications* **4**: 569–80. * 4,12

1824 Walling, E. 1985. *Country Roads: The Australian Roadside.* 2nd edn. Pioneer Design Studio, Victoria, Australia. * 5

1825 Walsh, F. J. 1990. An ecological study of traditional Aboriginal use of 'country': Martu in the Great and Little Sandy Deserts, Western Australia. *Proceedings of the Ecological Society of Australia* **16**: 23–37. * 1

1826 Wang, R., Zhao, J. & Ouyang, Z. 1990. *Human Systems Ecology.* China Science and Technology Press, Beijing. * 13

1827 Wang, R., Zhao, Q. & Ouyang, Z. 1992. *Ecopolis Planning in China: Principles and Practices of Urban Ecological Regulation.* Department of Systems Ecology, RCEES, Academia Sinica, Beijing. * 1, 13, 14

1828 Ward, A. L. 1982. Mule deer behavior in relation to fencing and underpasses on Interstate 80 in Wyoming. *Transportation Research Records* **859**: 8–13. * 5

1829 Ward, J. V. & Stanford, J. A. 1979. *The Ecology of Regulated Streams.* Plenum, New York. * 7

1830 Ward, N. I., Brooks, R. R. & Reeves, R. D. 1974. Effect of lead from motor-vehicle exhausts on trees along a major thoroughfare in Palmerston North, New Zealand. *Environmental Pollution* **6**: 149–58. * 5

1831 Warner, R. E. & David, L. M. 1982. Woody habitat and severe mortality of ring-necked pheasants in central Illinois. *Journal of Wildlife Management* **46**: 923–32. * 6

1832 Warner, S. B., Jr. 1978. *Streetcar Suburbs: The Process of Growth in Boston, 1870–1900.* Harvard University Press, Cambridge, Massachusetts. * 12, 13

1833 Warner, S. B., Jr. 1989. Introduction: when suburbs are the city. In B. M. Kelly, ed. *Suburbia Re-examined*, pp. 1–9. Greenwood Press, New York. * 13

1834 Waser, N. M. & Price, M. V. 1989. Optimal outcrossing in *Ipomopsis aggregata*: seed set and offspring fitness. *Evolution* **43**: 1097–109. * 2

1835 Watt, A. S. 1947. Pattern and process in the plant community. *Journal of Ecology* **35**: 1–22. * 4

1836 Way, D. S. 1978. *Terrain Analysis: A Guide to Site Selection Using Aerial Photographic Interpretation*, 2nd edn. Dowden, Hutchinson & Ross, Stroudsburg, Pennsylvania, USA. * 8,9

1837 Way, J. M. 1977. Roadside verges and conservation in Britain: a review. *Biological Conservation* **12**: 65–74. * 5

1838 Weathers, K. C., Likens, G. E., Bormann, F. H., Eaton, J. S., Bowden, W. B., Andersen, J. L., Cass, D. A., Galloway, J. N., Keene, W. C., Kimball, K. D., Huth, P. & Smiley,

599

D. 1986. A regional acidic cloud/fog water event in the eastern United States. *Nature* **319:** 657–8. * 10

1839 Weaver, J. E. 1954. *North American Prairie*. Johnsen Publishing, Lincoln, Nebraska, USA. * 3

1840 Weaver, J. E., Hanson, H. C. & Aikman, J. M. 1925. Transect method of studying woodland vegetation along streams. *Botanical Gazette* **80:** 168–87. * 7,10

1841 Weaver, M. & Kellman, M. 1981. The effects of forest fragmentation on woodlot tree biotas in southern Ontario. *Journal of Biogeography* **8:** 199–210. * 2

1842 Weaver, T. & Dale, D. 1978. Trampling effects of hikers, motorcycles and horses in meadows and forests. *Journal of Applied Ecology* **15:** 451–7. * 5,9

1843 Webb, N. R. 1985. Habitat island or habitat mosaic? A case study of heathlands in southern England. In W. Zielonkowski & H-J. Mader, ed. *Inselokologie – Anwendung in der Planung des landlichen Raums*, pp. 62–8. Akademie fur Naturschutz und Landschaftspflege, Laufen/Salzach, Germany. * 12

1844 Webb, N. R., Clarke, R. T. & Nicholas, J. T. 1984. Invertebrate diversity on fragmented *Calluna*-heathland: effects of surrounding vegetation. *Journal of Biogeography* **11:** 41–6. * 2,3,9

1845 Webb, N. R. & Hopkins, P. J. 1984. Invertebrate diversity on fragmented *Calluna* heathland. *Journal of Applied Ecology* **21:** 921–33. * 2,3

1846 Webb, N. R. & Vermaat, A. H. 1990. Changes in vegetational diversity on remnant heathland fragments. *Biological Conservation* **53:** 253–64. * 2

1847 Webb, T., III. 1986. Is vegetation in equilibrium with climate? How to interpret late-Quaternary pollen data. *Vegetatio* **67:** 75–91. * 1,3,10

1848 Webb, T., III., Bartlein, P. J. & Kutzbach, J. E. 1987. Climatic change in eastern North America during the past 18,000 years; comparisons of pollen data with model results. In W. F. Ruddiman & H. E. Wright, Jr., eds. *North America and Adjacent Oceans During the Last Deglaciation, Geology of North America*, vol. K-3, pp. 447–62. Geological Society of America, Boulder, Colorado, USA. * 3,8,10,14

1850 Webb, T., III, Cushing, E. J. & Wright, H. E., Jr. 1983. Holocene changes in the vegetation of the Midwest. In H. E. Wright, Jr., ed. *Late-Quaternary Environments of the United States*, pp. 142–65. University of Minnesota Press, Minneapolis, Minnesota, USA. * 8,14

1851 Webster, R. 1977. *Quantitative and Numerical Methods in Soil Classification and Survey*. Clarendon Press, New York. * 1

1852 Wegner, J. & Merriam, G. 1979. Movements by birds and small mammals between a wood and adjoining farmland habitats. *Journal of Applied Ecology* **16:** 349–58. * 4,9

1853 Weil, T. E. *et al.* 1982. *Haiti: A Country Study*. Area Handbook Series, US Army, Washington, DC. * 14

1854 Weinstein, B. & Segal, A. 1984. *Haiti: Political Failures, Cultural Successes*. Praeger, New York. * 14

1855 Weissman, G. & Claiborne, R., eds. 1975. *Cell Membranes: Biochemistry, Cell Biology, and Pathology*. HP Publishing, New York. * 3

1856 Welcomme, R. L. 1979. *Fisheries Ecology of Floodplain Rivers*. Longman, New York. * 7

1857 Wells, P. V. 1961. Succession in desert vegetation on streets of a Nevada ghost town. *Science* **134:** 670–1. * 5

1858 Wenger, E. L., Zinke, A. & Gutzweiler, K-A. 1990. Present situation of the European floodplain forests. *Forest Ecology and Management* **33/34:** 5–12. * 7

1859 Wertz, J. B. 1966. The flood cycle of ephemeral mountain streams in the southwestern United States. *Annals of the Association of American Geographers* **56:** 598–633. * 7

1860 West, N. E. 1983. North American temperate deserts and semi-deserts. In N. E. West, ed. *Temperate Deserts and Semi-deserts (Ecosystems of the World*, vol. 5), pp. 321–421. Elsevier, Amsterdam. * 12,14

1861 Westing, A. H. 1969. Plants and salt in the roadside environment. *Phytopathology* **59:** 1174–81. * 5

1862 Westing, A. H. 1980. *Warfare in a Fragile World: Military Impact on the Human Environment*. Taylor & Francis, London. * 12

1863 Westoby, M. 1974. An analysis of diet selection by large generalist herbivores. *American Naturalist* **108**: 290–304. * 11

1864 Wetzel, J. F., Wambaugh, J. R. & Peek, J. M. 1975. Appraisal of white-tailed deer winter habitats in northeastern Minnesota. *Journal of Wildlife Management* **39**: 59–66. * 11

1865 Wheeler, G. L. & Rolfe, G. L. 1979. The relationship between daily traffic volume and the distribution of lead in roadside soil and vegetation. *Environmental Pollution* **18**: 265–74. * 5

1866 Whitcomb, B. L., Whitcomb, R. F. & Bystrak, D. 1977. Long-term turnover and effects of selective logging on the avifauna of forest fragments. *American Birds* **31**: 17–23. * 2

1867 Whitcomb, R. F., Lynch, J. F., Opler, P. A. & Robbins, C. S. 1976. Island biogeography and conservation: strategy and limitations. *Science* **193**: 1030–2. * 2

1868 White, E. J. & Turner, F. 1970. A method of estimating income of nutrients in catch of air borne particles by a woodland canopy. *Journal of Applied Ecology* **7**: 441–61. * 10

1869 White, J. 1979. The plant as a metapopulation. *Annual Review of Ecology and Systematics* **10**: 109–45. * 2, 11

1870 White, L. P. 1970. 'Brousse tigree' patterns in southern Niger. *Journal of Ecology* **58**: 549–53. * 4

1871 White, P. S. 1979. Pattern, process, and natural disturbance in vegetation. *The Botanical Review* **45**: 229–99. * 8

1872 Whitmore, T. C. 1978. Gaps in the forest canopy. In P. B. Tomlinson & M. H. Zimmermann, eds. *Tropical Trees as Living Systems*, pp. 639–55. Cambridge University Press, Cambridge. * 2

1873 Whitmore, T. C. 1982. On pattern and process in forests. In E. I. Newman, ed. *The Plant Community as a Working Mechanism*, pp. 45–57. Blackwell, Oxford. * 2

1874 Whitney, G. G. & Davis, W. C. 1986. From primitive woods to cultivated woodlots: Thoreau and the forest history of Concord, Massachusetts. *Journal of Forest History* **30**: 70–81. * 13

1875 Whitney, G. G. & Somerlot, W. J. 1985. A case study of woodland continuity and change in the American Midwest. *Biological Conservation* **31**: 265–87. * 12

1876 Whittaker, R. H. 1967. Gradient analysis of vegetation. *Biological Review* **42**: 207–64. * 3

1877 Whittaker, R. H. & Levin, S. A. 1977. The role of mosaic phenomena in natural communities. *Theoretical Population Biology* **12**: 117–39. * 8

1878 Whyte, W. H. 1959. *Securing Open Space for Urban America: Conservation Easements.* Urban Land Institute, Washington, DC. * 6

1879 Widacki, W. 1979. Typologia granic geokompleksow w karpatach (Typology of geocomplex boundaries). *Zeszyty Naukowe Uniwersytetu Jagiellonskiego, Prace Geograficzne* **47**: 7–16. * 3

1880 Widacki, W. 1981. Klasyfikacja granic geokompleksow (Classification of the boundaries of geocomplexes). *Zeszyty Naukowe Uniwersytetu Jagiellonskiego, Prace Geograficzne* **53**: 19–26. * 3

1881 Wiens, J. A. 1976. Population responses to patchy environments. *Annual Review of Ecology and Systematics* **7**: 81–120. * 8

1882 Wiens, J. A. 1985. Vertebrate responses to environmental patchiness in arid and semi-arid ecosystems. In S. T. A. Pickett & P. S. White, eds. *The Ecology of Natural Disturbance and Patch Dynamics*, pp. 169–93. Academic Press, New York. * 2

1883 Wiens, J. A. 1989. Spatial scaling in ecology. *Functional Ecology* **3**: 383–97. * 1,3,11

1884 Wiens, J. A. 1990. On the use of 'grain' and 'grain size' in ecology. *Functional Ecology* **4**: 720. * 1

1885 Wiens, J. A. 1991. Ecological flows across landscape boundaries: a conceptual overview. In A. J. Hansen & F. di Castri, eds. *Landscape Boundaries: Consequences for Biotic Diversity and Ecological Flows*, pp. 217–35. Springer-Verlag, New York. * 3

1886 Wiens, J. A., Crawford, C. S. & Gosz, J. R. 1986. Boundary dynamics: a conceptual framework for studying landscape ecosystems. *Oikos* **45**: 421–7. * 1,3,5,12

1887 Wiens, J. A. & Milne, B. T. 1989. Scaling of 'landscapes' in landscape ecology, or, landscape ecology from a beetle's perspective. *Landscape Ecology* **3**: 87–96. * 1,3,9,11,14

601

1888 Wiens, J. A., Stenseth, N. C., Van Horne, B. & Ims, R. A. 1993. Ecological mechanisms and landscape ecology. *Oikos* **66:** 369–80. * 1

1889 Wight, B. C. 1988. Farmstead windbreaks. *Agriculture, Ecosystems and Envrionment* **22/23:** 261–80. (Reprinted 1988 in *Windbreak Technology*. Elsevier, Amsterdam.) * 6

1890 Wiken, E. B. & Ironside, G. 1977. The development of ecological (biophysical) land classification in Canada. *Landscape Planning* **4:** 273–5. * 9

1891 Wilcove, D. S. 1985. Nest predation in forest tracts and the decline of migratory songbirds. *Ecology* **66:** 1211–14. * 3,5,12

1892 Wilcove, D. S., McLellan, C. H. & Dobson, A. P. 1986. Habitat fragmentation in the temperate zone. In M. E. Soulé, ed. *Conservation Biology*, pp. 879–87. Sinauer Associates, Sunderland, Massachusetts, USA. * 4,12

1893 Wilcox, B. A. 1980. Insular ecology and conservation. In M. E. Soulé & B. A. Wilcox, eds. *Conservation Biology: An Evolutionary–Ecological Perspective*, pp. 95–117. Sinauer Associates, Sunderland, Massachusetts, USA. * 12

1894 Wilcox, B. A. 1986. Extinction models and conservation. *Trends in Ecology and Evolution* **1:** 46–8. * 2

1895 Wilcox, B. A. & Murphy, D. D. 1985. Conservation strategy: the effects of fragmentation on extinction. *American Naturalist* **125:** 879–87. * 2

1896 Wilcox, B. A. & Murphy, D. D. 1989. Migration and control of purple loosestrife (*Lythrium salicaria* L.) along highway corridors. *Environmental Management* **13:** 365–70. * 5

1897 Wildman, H. D. 1977. Environmental patch size in a first year plant community. MS thesis, Rutgers University, New Brunswick, New Jersey, USA. * 14

1898 Williams, C. B. 1943. Area and number of species. *Nature* **152:** 264–7. * 2

1899 Williams, M. 1989. *Americans and Their Forests*. Cambridge University Press, Cambridge. * 9

1900 Williamson, M. H. 1975. The design of wildlife reserves. *Nature* **256:** 519. * 2

1901 Williamson, M. H. 1981. *Island Populations*. Oxford University Press, Oxford. * 2, 11

1902 Williamson, M. H. & Lawton, J. H. 1991. Fractal geometry of ecological habitats. In S. S. Bell, E. D. McCoy & H. R. Mushinsky, eds. *Habitat Structure: The Physical Arrangement of Objects in Space*, pp. 69–86. Chapman & Hall, London. * 9

1903 Willis, E. O. 1973. The behavior of Ocellated Antbirds. *Smithsonian Contributions to Zoology* **144:** 1–57. * 4

1904 Willis, E. O. 1974. Populations and local extinctions of birds on Barro Colorado Island, Panama. *Ecological Monographs* **44:** 153–69. * 2, 6

1905 Willis, E. O. 1984. Conservation, subdivision of reserves, and the anti-dismemberment hypothesis. *Oikos* **42:** 396–8. * 2

1906 Willis, J. C. 1922. *Age and Area: A Study in Geographical Distribution and Origin of Species*. Cambridge University Press, Cambridge. * 2

1907 Willson, M. F. 1974. Avian community structure and habitat structure. *Ecology* **55:** 1017–29. * 3

1908 Wilmanns, O. & Brun-Hool, J. 1982. Irish mantel and saum vegetation. *Journal of Life Sciences, Royal Dublin Society* **3:** 165–74. * 3

1909 Wilson, E. O. 1975. *Sociobiology*. Belknap Press, Cambridge, Massachusetts, USA. * 11

1910 Wilson, E. O. 1984. *Biophilia*. Harvard University Press, Cambridge, Massachusetts. * 2, 13

1911 Wilson, E. O. 1987. The little things that run the world: the importance and conservation of invertebrates. *Conservation Biology* **1:** 344–6. * 11, 14

1912 Wilson, E. O. 1992. *The Diversity of Life*. Harvard University Press, Cambridge, Massachusetts. * 2, 9, 12, 14

1913 Wilson, E. O. & Willis, E. O. 1975. Applied biogeography. In M. L. Cody & J. M. Diamond, eds. *Ecology and Evolution of Communities*, pp. 522–34. Belknap Press, Cambridge, Massachusetts, USA. * 2

1914 Wilson, J. D. 1985. Numerical studies of flow through a windbreak. *Journal of Wind Engineering and Industrial Aerodynamics* **21:** 119–54. * 6

1915 Wilson, R. J. 1979. *Introduction to Graph Theory*. Academic Press, New York. * 9

1916 Winkworth, R. E. 1967. The composition of several arid spinifex grasslands of central Australia in relation to rainfall, soil water relations, and nutrients. *Australian Journal of Botany* **15:** 107–30. * 9, 10

1917 Wischmeier, W. H. 1976. Use and misuse of the universal soil loss equation. *Journal of Soil and Water Conservation* **31**: 5–9. * 7, 10

1918 Wood, J. W. 1987. The genetic demography of the Gainj of Papua New Guinea. 2. Determinants of effective population size. *American Naturalist* **129**: 165–87. * 2

1919 Woodmansee, R. G. 1990. Biogeochemical cycles and ecological hierarchies. In I. S. Zonneveld & R. T. T. Forman, eds. *Changing Landscapes: An Ecological Perspective*, pp. 57–71. Springer-Verlag, New York. * 1, 9, 10

1920 Woodruff, N. P. & Siddoway, F. H. 1965. A wind erosion equation. *Soil Science Society of America Proceedings* **29**: 602–8. * 6

1921 Woodruff, N. P. & Zingg, A. W. 1953. Wind tunnel studies of shelterbelt models. *Journal of Forestry* **51**: 173–8. * 6

1922 Woods, L. E. 1983. *Land Degradation in Australia*. Australian Government Printing Service, Canberra. * 14

1923 Woodward, F. I. 1987. *Climate and Plant Distribution*. Cambridge University Press, Cambridge. * 3, 10

1924 Woodward, S. L. & McDonald, J. N. 1986. *Indian Mounds of the Middle Ohio Valley: A Guide to Adena and Ohio Hopewell Sites*. McDonald & Woodward Publishing, Newark, Ohio, USA. * 14

1925 Woodwell, G. M. 1967. Radiation and the patterns of nature. *Science* **156**: 461–70. * 10

1926 Woodwell, G. M. 1983. The blue planet: of wholes and parts and man. In H. A. Mooney & M. Godron, eds. *Disturbance and Ecosystems: Components of Response*, pp. 2–10. Springer-Verlag, New York. * 14

1927 Woodwell, G. M. 1985. On the limits of nature. In R. Repetto, ed. *The Global Possible: Resources, Development, and the New Century*, pp. 47–65. Yale University Press, New Haven, Connecticut, USA. * 14

1928 Woodwell, G. M. 1990. *The Earth in Transition: Patterns and Processes of Biotic Impoverishment*. Cambridge University Press, Cambridge. * 10,14

1929 World Commission on Environment and Development. 1987. *Our Common Future*. Oxford University Press, Oxford. * 13,14

1930 Worster, D. 1979. *Dust Bowl*. Oxford University Press, Oxford. * 6,9,12

1931 Worster, D. 1985. *Nature's Economy: A History of Ecological Ideas*. Cambridge University Press, Cambridge. * 1

1932 Wright, H. E., Jr. 1974. Landscape development, forest fires, and wilderness management. *Science* **186**: 487–95. * 10

1933 Wunderle, J. M., Diaz, A., Velazquez, I. & Schawon, R. 1987. Forest openings and the distribution of understory birds in a Puerto Rican rainforest. *Wilson Bulletin* **99**: 22–37. * 2, 3

1934 Wynn, G. 1981. *Timber Colony: A Historical Geography of Early Nineteenth Century New Brunswick*. University of Toronto Press, Toronto. * 14

1935 Yahner, R. H. 1981a. Avian winter abundance patterns in farmstead shelterbelts: weather and temporal effects. *Journal of Field Ornithology* **52**: 157–68. * 6

1936 Yahner, R. H. 1981b. Winter bird populations in Minnesota shelterbelts. *American Birds* **35**: 39. * 6

1937 Yahner, R. H. 1982a. Avian nest densities and nest-site selection in farmstead shelterbelts. *Wilson Bulletin* **94**: 156–75. * 6

1938 Yahner, R. H. 1982b. Avian use of vertical strata and plantings in farmstead shelterbelts. *Journal of Wildlife Management* **46**: 50–60. * 5, 6

1939 Yahner, R. H. 1983. Seasonal dynamics, habitat relationships, and management of avifauna in farmstead shelterbelts. *Journal of Wildlife Management* **47**: 85–104. * 6

1940 Yahner, R. H. 1987. Use of even-aged stands by winter and spring bird communities. *Wilson Bulletin* **99**: 218–32. * 3

1941 Yahner, R. H. 1988. Changes in wildlife communities near edges. *Conservation Biology* **2**: 333–9. * 3, 5, 12

1942 Yapp, W. B. 1973. Ecological evaluation of a linear landscape. *Biological Conservation* **5**: 45–7. * 5

1943 Yates, P. & Sheridan, J. M. 1983. Estimating the effectiveness of vegetated floodplains/wetlands as nitrate–nitrite and orthophosphorus filters. *Agriculture, Ecosystems and Environment* **9**: 303–14. * 7

603

1944 Yon, D. & Tendron, G. 1981. *Alluvial Forests of Europe*. Nature and Environment Series, No. 22., Council of Europe, Strasbourg, France. * 7

1945 Young, A. 1989. *Agroforestry for Soil Conservation*. International Council for Research in Agroforestry and C.A.B. International, Nairobi, Kenya. * 13

1946 Young, J. A. & Sparks, B. A. 1985. *Cattle in the Cold Desert*. Utah State University Press, Logan, Utah, USA. * 11,14

1947 Young, S. P. 1946. History, life habits, economic status, and control. In S. P. Young & E. A. Goldman, eds. *The Puma, Mysterious American Cat*, Part I, pp. 1–173. American Wildlife Institute, Washington, DC. * 7

1948 Yu, K. 1990. Basin experience of Chinese agriculture and ecological prudence. In R. Wang, J. Zhao & Z. Ouyang, eds. *Human Systems Ecology*, pp. 63–71. China Science and Technology Press, Beijing. * 1

1949 Yu, K-j. 1992. Experience of basin landscapes in Chinese agriculture has led to ecologically prudent engineering. In L. O. Hansson & B. Jungen, eds. *Human Responsibility and Global Change*, pp. 289–99. University of Goteborg, Goteborg, Sweden. * 1

1950 Zachar, D. 1982. *Soil Erosion*. Elsevier, Amsterdam. * 6

1951 Zeigler, B. P. 1976. *Theory of Modeling and Simulation*. John Wiley, New York. * 1

1952 Zipperer, W. C., Burgess, R. L. & Nyland, R. D. 1990. Patterns of deforestation and reforestation in different landscape types in central New York. *Forest Ecology and Management* **36:** 103–17. * 12

1953 Zonneveld, I. S. 1979. *Land Evaluation and Land(scape) Science*. 2nd edn. ITC Textbook VII.4. International Institute for Aerial Survey and Earth Sciences, Enschede, Netherlands. * 1,9,13

1954 Zonneveld, I. S. 1980. Some consequences of the mutual relationship between climate and vegetation in the Sahel and Sudan. *ITC Journal* 1980–2: 255–96. * 12

1955 Zonneveld, I. S. 1988. Basic principles of land evaluation using vegetation and other attributes. In A. W. Kuchler & I. S. Zonneveld, eds. *Vegetation Mapping*, pp. 499–517. Kluwer Academic Publishers, Dordrecht, Netherlands. * 8,11

1956 Zonneveld, I. S. 1990. Scope and concepts of landscape ecology as an emerging science. In I. S. Zonneveld & R. T. T. Forman, eds. *Changing Landscapes: An Ecological Perspective*, pp. 1–20. Springer-Verlag, New York. * 1

1957 Zonneveld, I. S. & Forman, R. T. T., eds. 1990. *Changing Landscapes: An Ecological Perspective*. Springer-Verlag, New York. * 1

1958 Zube, E. H. 1987. Perceived land use patterns and landscape values. *Landscape Ecology* **1:** 37–45. * 13

1959 Zube, E. H. & Simcox, D. E. 1987. Arid lands, riparian landscapes, and management conflicts. *Environmental Management* **11:** 529–35. * 13

1960 Zube, E. H. & Zube, M. J. 1977. *Changing Rural Landscapes*. University of Massachusetts Press, Amherst, Massachusetts, USA. * 9,13

1961 Zusne, L. 1970. *Visual Perception of Form*. Academic Press, New York. * 4,13

604

Index

Aboriginal people
  land use, 17–18, 301
  painting, 18–19
accessibility of a patch, 320
acidity, 358
adaptability in sustainability,
    502–5, 516–17
adaptation
  disturbance and, 352–3,
    357–8
  genetic variation and, 67,
    384–5
adaptive management, 475
Adena people, 481
Adirondack (USA) region, 24,
    267
adjacency, 103, 285–6
  in interspersion of habitats,
    314–15
  number of, 103, 122–3
  as species source, 57
adjacency analysis, 319
adjacency arrangement, 289
adjustment period, 56–7
administrative boundary, 92–3
advection, 333–7
  as major link between
    ecosystems, 325, 335–6
advocacy, 37
aeolian landscape or terrain,
    124, 307–8
afforestation, see reforestation
age
  of corridor, 157
  of island or patch, 56, 80
age structure of population, 78
Agger, Peder, 421–2
aggregate-with-outliers
    principle, 436–40, 453,
    515
  human benefits of, 439–40
aggregation, 256
  of ecosystems, 291
  of gaps, 148, 155–6
  in nature, 301, 446
agricultural intensification
  examples of: in Denmark,
    421–2; in The
    Netherlands, 421–2; in
    Wisconsin, 410–11
  land consolidation and, 422

spatial pattern and, 419–22
  stability of spatial elements
    in, 421–2
agricultural land planning,
    458–9
agricultural landscape, 521
  albedo of, 329, 331
  birds in, 378–9
  generic plan for, 458–9
  heat flows above, 335–6
agricultural pest, 197
agricultural soil, 308
agriculture
  low input, 225, 482
  in river corridor, 225–6,
    250
  targeted on high
    biodiversity sites, 311–12
  in tropics, 521
  see also field
agroforestry, 477
air
  cold, drainage, 347
  heat given off to, 330–3
  pollution in mosaic, 357–8
airflow, see wind
albedo, 328
  of different ecosystems,
    328–31
  of edge, 88
  of landscapes, 329
Alberta, 6, 335–6
alien, see exotic species
all-patch pattern measure,
    318, 321
alluvial fan, 211–12, 307
alluvium, 307–8
alpha index, 261
Amazon rainforest
  fragmentation, 32
  road in, 420, 429
amoeboid patch, 116–17, 137–
    7
amphibians, 62, 152, 205
  tunnel for, 166
anadromous fish, 236
analogy, learn from, 30
  membrane, 95–7, 109
  nutrition, 290
analytic approach, 33–5
ancient woodland, 374–5

Angkor Wat, Cambodia
    (Kampuchea), 122,
    510–11
angle
  acute, 141
  intersection, 259, 264,
    269
  of orientation, 125–7, 181,
    186–7, 192–3, 197
animal
  in configuration, 294–8
  in stream and river
    corridor, 236–43
animal dispersal, see dispersal
animal movement, see
    movement
animal need for movement,
    203
animal perception of land
    pattern, 303–5
ants, 63, 377, 380
Antarctica, carbon dioxide in,
    325–6, 337
anthropogenic, 31, 177
  see also human
anthropology, see archaeology;
    culture
apples-and-oranges
    comparison, 498
aquatic biomass, 76
aquatic habitat for fish,
    232–6
aquatic system, size effect on,
    51–3
aquifer, 47, 52
archaeology, 290, 481
  Chaco Canyon, 510–11
area effect, see size effect
area-per-se, 57, 59–60
Arnold, Graham W., 294–5
arrangement of objects, 39
array of islands, 58–9
arrival model
  disturbances in, 359–60;
    single or continued, 359–
    60
  populations in, 368–9
arthropods, 50, 62, 377
  see also insects
ash (Fraxinus), 222, 270, 305,
    338, 388, 409

assays
  for landscape and regional
    ecology, 37
  relatively objective, 523
  for sustainability: basic
    human needs, 501–2;
    ecological integrity, 499–
    501; slowly changing
    attributes, 498–9; values,
    497–8
associated ecosystems, 292
asymmetrization, 120
atmosphere–organism cycle,
  77
atmosphere–organism–soil
  cycle, 77–8
atmosphere, temperature of,
  112
atmospheric gases
  carbon dioxide and
    methane, 77, 325–8, 341–2
  greenhouse, 330, 341–2
  nitrous oxide, 327–8
  oxygen and ozone, 327–8
  water vapor, 327–8
atmospheric pressure, 323–5
atmospheric window, 328
attrition, 407–11, 429
Austin, Mike P., 312–13
Australia, 17–19, 165, 312–13,
  393, 397, 450, 509–11
  see also Canberra;
    Queensland; southeastern
    Australia; Western
    Australia
autocorrelation, 319
avian, see birds

back lot line, 473–4
bacteria in nitrogen cycle, 77
badger (Meles), 377, 380
barrier
  corridor as, 18, 148–9, 152,
    163–4, 236
  edge as, 92–3, 97, 100–1
  in network, 271–4
  river as, 236
  and spread of disturbance,
    360–2
barrow, 421
Barton, D. R., 251
basic human need, 484, 500–2
  culture as, 500–1
  energy as, 500–1
  food, health, housing
    (shelter), and water as,
    500–1
  linkage with ecological
    integrity, 500–2
basin, receiving, 507
  see also drainage basin
Baskerville, G. L., 485
basswood (Tilia), 374, 388
bear (Ursus)
  black, 241, 267, 389–90,
    405, 450

grizzly, 105, 241, 292, 450;
    Alaska brown and, 69
beauty strip, see roadside
    natural strip
beaver (Castor), 220–3, 248–9,
    297, 405, 477
beech (Fagus), 24, 195, 305,
    357, 508
beetle (Coleoptera), 32–3, 268–
    70, 371, 374, 377–80
behavioral caution, 100, 279
behavioral connectivity, 38,
    157, 296
behavioral perception of land
    pattern, 303–5
behavioral science, 287
Bennett, Andrew, 156
beta index, 250
bicycle path, see trail corridor
Binford, Michael W., 508
biocoenose, 20
biodiversity, 37
  in ecological integrity, 500
  concept of, 38, 54
  geological substrate and,
    311–12
  mineral nutrients and, 311–
    12
  optimization for erosion
    control and, 520–1
  planning natural resource
    area for, 444–5
  size effect on, 54–66
  values of, 54
  see also species; wildlife
biodiversity areas, selection
    of, 310–13
biogeochemistry, see mineral
    nutrients
biogeocoenose, 20
biogeography, see island
    biography
biological community, see
    biological; community
biological conservation, see
    conservation
biological magnification, 77
biomass, 75, 393
  aquatic, 76
  distribution of: pyramid, 76;
    top-shaped, 76
  patch size effect on, 48–51
biome, 14
biophilia, 448
bio-rich place, 448
biosphere, 12–13
  see also global
biotope, 20, 38
  see also spatial element
birds
  in English farmland
    configuration, 294–5
  metapopulation of, 376–9
  in New Jersey woods, 59–
    62, 72–3
  road killed, 165

seed-eating and
    insect-eating, 60
  size and isolation effect on,
    59–65, 375–9
  in small woods, 125–6, 381
  soaring, 346–7
  winter and summer, 295
bird movement
  across landscape, 279–80
  among non-adjacent
    elements, 391–2
birth rate, 78
bison (Bison), 271
black-box model, 495
bleed flow, 181
blowdown, 355, 416
  at head of valley, 347–8, 355
  see also hurricane; tornado
bluff object, 181
Boerner, Ralph E. J., 354
border, 85–6
  administrative, of natural
    area, 92–3
  as filter, 92–3
  management of, 101–2
Bormann, F. Herbert, 358
boundary, 81–111, 439
  abruptness of, 82–7
  climate change and, 112
  concave and convex, 107–
    11, 140–1, 281–2
  concept of, 7, 38, 85–6
  and corridor compared, 158
  cove in, 107–9
  at different scales, 86–7
  edges in, 85–6
  fire, 84
  of landscape, 21–2
  lobe in, 107–10, 140
  management of, 101–2
  measures of, 318, 320
  mechanisms causing, 83, 90
  nature reserve centred on,
    313
  receptor in, 96–8, 152
  species movement across,
    100–1; animal, 83–4; 106–
    7; human, 92–3; plant,
    83–4
  species movement along,
    106–7
boundary-crossing frequency,
    279–81, 362, 368, 387
boundary density, see
    boundary length
boundary function or process,
    95–101
  barrier as, 511
  conduit as, 96, 386–7
  habitat as, 96–100
  sink as, 96
  smoothing as, 141
  source as, 96
boundary length, 98, 314,
    407–8, 433
  high, 428

low, 428
boundary permeability, 100–1
boundary roughness, 130–2
boundary segmentation, 93
boundary surfaces, 107–8
boundary theory, 82–5
boundary type, 82–7
  administrative, 92–3
  concave, 107–10, 140
  convex, 107–10, 140–1
  curvilinear (curvy), 83, 106–
    9, 116
  fractal, 98–9, 107–8
  hard, 83–5
  induced, 92
  irregular, 117, 132
  moving, 109–11, 122–3,
    138–41; advancing and
    retreating, 111, 139–41;
    contracting, 110;
    expanding, 110, 139–41
  soft, 83–5
  straight, 83, 107–10, 116,
    140; erosion and, 132
  tiny-patch, 83, 94
boundary zone, 85–6, 439
braided stream, 211, 214–15
  network, 215, 256
Braithwaite, Wayne, 311–12
Brandt, Jasper, 421–2
Bratton, Susan, 353
Brazil, 32, 420, 429
bridge for animal crossing,
  165
bridge line, 257
Bridgewater, Peter, 450
Britain, 59, 99, 159, 177, 313,
  374–5, 393, 410, 444
  see also England
browse
  rabbit and deer, 31, 91, 394
  of shrub layer, 195
browse line, 31, 104
  see also overgrazing or
    overbrowsing
budworm caterpillar
  (Choristoneura), 242–3
buffalo, 225, 271
buffer, 382, 450
  concept of, 292
  edge as, 92
  against extinction, 47
buildings in design, location
  of, 516–18
built area
  in aggregate-with-outliers
    principle of, 436–40
  movement of species across,
    279–80
  self-reliance in, 497
  in sustainability, 496–7
built objects in river corridor,
  226–7
Burgess, Robert L., 409–12
butterflies, 63, 279–80, 374

Cadiz Township, Wisconsin,
  409–12
California, 21, 67, 481
Canada, 6, 23–4, 27, 242–3,
  393, 444, 484–6, 510–11
canal, 214
  movement of species across,
    279–80
Canberra, Australia, 18–19,
  301, 449
candelabra from graph theory,
  294, 298
Cantwell, Margot, 293–4
carbon cycle, 77
carbon dioxide, 77
  as absorbing gas, 327–8
  atmospheric buildup of,
    112, 325–6, 336–7
  and greenhouse effect, 341–
    2
  in photosynthesis, 76, 341
caribou (Rangifer), 118, 164
Carnac, Brittany, 481
carrying capacity, 78, 392
  human, 512
cartography, 317–19
  cascade effect, species, 63
cassowary, 63
catadromous fish, 236
catastrophe theory, 111
catchment, see drainage basin
catena, 287–8
cattle, 399–401
  see also livestock
cell, see macroclimate
cellular automata, 34
centipede, 377, 380
central place theory, 137, 371
  foraging and, 249, 399
  hexagon model and, 30, 119
Chaco Canyon, New Mexico,
  510–12
channel
  intermittent, 211–12, 232,
    244–6, 351
  for spread of disturbance,
    360–2
  stream, 210–11, 214
  see also stream
channelization
  of river, 476–8
  of stream, 214–16, 224–5,
    460
chaos theory, 111
cheatgrass (Bromus), 398
checkerboard landscape, 309–
  10
  model of, 416–17
Chile, 510–11
China, 225, 301, 364, 447, 477
  built structures in:
    Confucius Temple, 492–3;
    Great Wall, 172
  culture and philosophy
    underlying landscape

structure in, 15–16, 162,
  302, 492–3
restoration in, 477
chipmunk, 296–7, 380
circuitry, 261, 266
  see also network
city, 444, 480
  albedo of, 329
  Canberra, 18–19, 301, 449
  open space in, 448
  outside of, 301
  shape of, 137
  see also urban; built area
civilization, examples for
  sustainability, 509–12
clay
  as component of soil, 308
  holding mineral nutrients,
    229–31
clearcut or clearing, see
  opening
climate and landform, 300
climate change, 336
  boundaries as indicators of,
    112
  corridor and, 151
  effect on landscape and
    region, 336–7, 413, 510
  landscape and region effect
    on, 336–7
  paleoecology analysis of,
    337
cluster
  of ecosystems: 286–94;
    flowweb in, 288; repeated,
    13, 21, 288–91
  of islands, 58–9
  of small patches, 201–2
  of stepping stones, 201–2
  see also configuration
coalescence phase of
  landscape ecology, 29
coastal plain, 307–8, 389
cockatoo, 392
cognitive map, 18, 241
cohesive force in
  sustainability, 492–4
  culture as, 492–4, 509–11
  religion as, 492–4, 510–11
  social science dimensions
    as, 494, 510–11
cold air drainage, 347
collective, see cooperative
  community
Collinge, Sharon K., 32, 106–7
colonization
  human, 58–9
  of island, 55–9
  of patches by birds, 125–6
  see also recolonization
commodity exports, 510
community, biological, 75–8
  equilibrium and
    non-equilibrium, 58
  function of, 76–8
  moving, 110

community, biological,
  continued
  plot shape in sampling of,
    125
  principles of, 54–5
  structure of, 75–6
  tree, as link between
    geology and mammal, 311
  windbreak effect on, 194–5
community, human
  Concord case study as, 461–
    9
  cooperative, 17
  examples of, for
    sustainability, 509–12
competition
  diffuse, 79
  environmental conditions
    in, 79
  interspecific and
    intraspecific, 79
  in natural selection, 79
  resources for, 79; limiting,
    79
competitive exclusion
  principle, 79
concave boundary surface
  indicating contracting
    boundary, 110
  of patch, 107–10, 140
  of matrix, 281–2
concave–convex reversal, 140–
  1
concave depression, large and
  small, 350
concave soil surface
  temperature and wind on,
    348–51
  trees on, 304–5
Concord, Massachusetts case
  study, *see* suburb
concurrent objectives,
  planning for, 519–21
conduit, 125
  edge as, 96
  road corridor as, 160–3
  roadside as, 161
  roadside natural strip as,
    161–3, 205
  in stream and river
    corridor, 228–9, 239–45
  woodland corridor as
    wildlife, 198–207
configuration, 14
  animals and usage of, 294–8
  boundary of, 289
  concept of, 38, 289
  in conservation, 289–90, 300
  familiar spatial patterns in,
    290–2
  interaction and
    interdependence within,
    298
  patterns common to all
    landscapes, detecting,
    292–4

size of, 294–8
stability and change of,
  298–300
variable ecosystems in, 298
see also cluster; mosaic;
  neighborhood
Confucius, Temple of, 492–3
connectedness, *see*
  connectivity
connectivity
  behavioral, 38, 157, 296
  concept of, 38, 155
  and connectedness, 156
  of corridor, 148, 153–7, 201,
    452–3
  of forest, 294, 296
  functional, 38, 157, 276, 296
  and gaps, 155–6, 265
  independent of context, 294
  low and high, 427
  of matrix, 277–8, 407–8
  of rectilinear network, 261,
    266; for wildlife, 274
  stepping stones and, 156,
    452–3
  of stream corridor, 249–52,
    261
  of stream network, 47, 52–3
  of trough corridor, 160
conservation, 36–7
  configuration in, 300
  of egrets and herons in
    configuration, 289–90
  genetics in, 67–8, 386
  minimum number point in,
    74
  minimum viable population
    in, 70–1
  network analysis of
    neighborhood in, 272–4
  and patch shape, 124–5
  safety factor, 70–1
  soil, 36–7
  see also management;
    planning
conservation biology, *see*
  biodiversity; conservation
construction in river corridor,
  226–7
contagion, 321
  see also aggregation
content and context, 290, 378
context
  adjacency in, 285–6
  and content, 290, 378
  in generic plan, 450–2
  location in landscape in,
    285–6
  neighborhood in, 285–6
continent, 12–13
  ecology of, 450–1
  linkage between, 325, 487
contrast, 11, 21, 81–2
control
  over dynamics, 277–8
  in experiment, 30, 32–3

convergency point, 314–15,
  472
convex boundary surface
  concave–convex reversal
    and, 140–1
  indicator of expanding, 140
  of patch, 107–10, 140–1
  of matrix, 281–2
convex soil surface
  temperature and wind on,
    348–50
  trees on, 304–5
conveyor belt, 337–45
  gases in, 340–2
  material transport in, 337–
    45
  particles in, 338–40
  water in, 342–3
cooperative community, 17
core, 129–33
  area, 376
  of patch, 129–33
corner, 107–8, 129–30
Correll, David L., 230
corridor, 32, 439
  advantage and disadvantage
    of, 203
  age of, 157
  attributes of, 6–7, 145–59; in
    generic plan, 455
  and boundary compared,
    158
  changing, 157–8
  concept of, 38, 145, 147–8
  connectivity of, 148, 153–7,
    201, 452–3
  construction in, 226–7
  construction of, 418–20
  curvilinearity of, 148, 202,
    258
  external structure of, 146–8
  gap in, 30–1, 148, 202, 250–
    1
  generalist species in, 204–5
  gradient along, 148, 201
  home range within, 164,
    241, 249, 274
  in house lot design, 472
  interior in, 150, 245, 250
  internal entity in, 146–7,
    196
  internal structure of, 146–8
  intersection in, 148
  large patches connected by,
    452–3
  length of, 148, 151, 421
  linear boundaries and, 158–
    9
  matrix effect on, 203
  matrix protected by, 147–8
  metapopulation affected by,
    203
  narrows in, 148
  node in, 148
  origin of, 157
  patchiness in, 148, 201

planning of, 445
quality of, 204, 270, 274–6
resident species in, 200–1
species movement in: along,
147–8, 198–200; dispersal,
163, 201, 203; edge
species, 200; enhancing,
200–3; exotic species,
204–6; inhibiting, 200–3;
interior species, 200; pest,
204–5; plant, 199; routes
in, 200; specialist species,
205; wildlife, 198–207
in sustainability, 488–9
corridor density, 260, 266,
376–9
corridor functions and
processes, 147–53
access for human impact as,
169–70
avoidance by species as,
164, 169–70
barrier as, 18, 148–9, 152,
236, 163–4
conduit as, 148, 150–2, 239–
43, 386, 450
filter as, 148–9, 152, 163–4
habitat as, 148–50, 195–8,
236–9
sink as, 148–9, 153, 164–8,
276
source as, 148–9, 153
corridor mosaic sequence
model, 419, 423–9
corridor network, *see* network
corridor type
disturbance, 157–9
environmental, 157
greenway, 199, 445
introduced, 157
lakeshore, 214, 216
line, 150, 154
regenerated, 157
remnant, 157
strip, 150, 154, 195
trough, 158–9
see also hedgerow;
powerline; river; road;
stream; trail; windbreak;
woodland
corridor width, 153–5, 206
for assemblage of species,
155
characteristics of, 146–8
hedgerow, 195, 302
for individual animal, 155
line and strip, 150, 154
narrows and, 148
powerline, 65, 174–5
for a species, 65, 155, 174–5
stream, 155, 232, 243–9,
251–2
variability in, 148, 270
Costa Rica, 89–90, 151, 427,
476

counter current principle, 30,
150–1
fish in, 150–1
in network, 271
cove
in boundary, 107–9
depth (length) of, and
animal use, 109
functions of, 108–9, 129
mine revegetation affected
by, 109, 141
width of, and animal use,
109, 131
cove concentration effect, 109,
141
covert, 314–15, 472
cow, *see* livestock
coyote, 107, 271
cricket (Orthoptera), 391
cultivation, *see* field
cultural change, 494, 498
cultural cohesion
role of corridor in, 152
in sustainability, 492–4,
509–11
cultural diversity, 494
culture 481, 492
as basic human need, 500–1
reflected in a mosaic, 14–19
cumulative impact, 475
Curtis, John, 410
curve-and-peaks model,
decreasing, 388
curvilinearity
of boundary, 83, 106–9, 116
of corridor, 148, 202, 258
and movement across
boundary, 106–7
and movement along
boundary, 106–7
and wildlife usage of
boundary, 106–7
cutting, *see* deforestation;
forestry
cybernetics, 9
cycle
atmosphere–organism,
atmosphere–organism–
soil, and organism–soil,
77–8
nitrogen and phosphorus,
77
predator–prey, 80
see also mineral nutrients
Czech Republic, *see*
Czechoslovakia
Czechoslovakia, 16–17, 89–90,
151, 272–4

dam
beaver, 220–3
in dry climate, 224
mini-, in mountain, 218
in river corridor, 223–4
upstream and downstream
effect of, 224

death rate, 78
decomposer, 76
decreasing curve-and-peaks
model, 388
deer, 10–11, 405, 427
browsing by, 91, 394
in corridors, edges, and
openings, 105–7, 131, 165,
349
migration and movement,
262–3, 388, 392
deer yard, 388, 392
deferred-rotation grazing
system, 401–2
deforestation, 418–19, 507–8
disturbance in: blowdown,
416; fire ignition, 416; fire
spread, 416
game populations in, 416
in Japan and Wisconsin,
409–12
metapopulation dynamics
and, 374
species change in, 416
and sustainability, 510
Delaware, 472–4
delta, 307
dendritic
and fractal, 119
landscape, 309
network, 256; stream system
as, 52–3, 141
pattern reversed, 124
ridge network, 304, 306
Denmark, 271–2, 421–2
density
boundary, 98
corridor, 260, 266, 376–9,
421
network 261; drainage, 260–
1; linkage, 259
road, 260, 266–7
species, 47, 61–2, 150
of woods, 377–9
density-independent and
density-dependent factors,
78–9
deposition of sediment, 169–
70, 224–7, 234
depression
inbreeding, 66–8, 203, 385–6
outbreeding, 66–8, 385;
see also concave depression
depth-of-view, 305
desert, 457–8
desertification, 419–20, 508
design
biophilia in, 448
bio-rich place in, 448
of configuration, 300
continent with ecological,
450–1
disturbance spread
minimized by, 361–3
ecological goals and
techniques in, 470–1, 518;

design, *continued*
change as, 471; flows and movements as, 471; special habitats as, 471; species as, 471
exotic species in, 398
forms often produced in, 447; stability of, 447
of gap in corridor, 202
house lot, 469–74
of landscape or region as surrogate for long term, 523–4
of local area, 516–18; degree of detail in, 516; sequence of decisions in, 516–18
location of buildings in, 516–18
nature reserve, 57, 137, 313
neighborhood context in: corridor as, 470; representative as, 470; transition as, 470; unusual as, 470
of pastureland, paddock, or ranch, 272
spatial arrangement in, 471–4; edges and junctions in, 472; small corridors in, 472; small patches in, 472; vertical height diversity in, 472
see also planning
designers at work, 442–3
detritus food chain, 76
Diershke, Hartmut, 95
diffuse sky radiation, 328
diffusion, 101, 367
dike, 421, 510–11
as trough corridor, 159
dispersal, 47
animal, 365, 388, 392
in corridor, 163, 201, 203
decreasing curve-and-peaks model of, 388
long-distance, 366–7, 388
plant, 366–7, 387–8
seed, *see* seed dispersal
short-distance, 366–7, 388
stepping stones for, 47
dispersal-funnel effect, 128, 132
dispersed mosaic sequence model, 419, 423–9
dispersed-patch cutting, 354, 414–19
see also forestry
dispersion of individuals, 78
dispersion of patches, 321
dissection of landscape, 407–11, 429
dissolved substances in stream corridor, 228–32, 245–8
disturbance, 338, 390
adaptability and, 502–5
adaptation and, 352–3

barrier or channel for spread of, 360–2
concept of, 38, 351–2
and exotics, 394
and heterogeneity, 359–66
human, 396
minimum dynamic area and, 57, 65–6
in mosaic, 351–63
opening caused by, 63
pattern produced by, 445–8
resistant or susceptible ecosystem to, 360–1
risk spreading and, 53
space–time principle and, 8
species richness affected by, 57, 65
spread of, 359–63; design to minimize, 361–3
in stream corridor, 354
and wooded strip, 180
disturbance corridor, 157–9
disturbance line by edge, 94
disturbance patch, 44–5, 64, 111, 351–63
size effect on, 63–5
disturbance regime, 65–6, 352–3
disturbance type
air pollution as, 357–9
blowdown as, 416
chronic, 45
fire as, 353–4; ignition of, 416; spread of, 416
frequent, 352–3, 502–5, 511
human, 396
instantaneous, of whole landscape, 352–3, 418–19, 422–3
near-natural, 47
repeated, 45
volcanic impact as, 352–3
water as, 354–5
windstorm as, 355–7
ditch, 196, 391, 421
diversion of stream, 224–5
diversity measure, 318–20
dog's mercury (*Mercurialis*), 374–5
dominance, 318, 320
doubling time, 78
douglas-fir (*Pseudotsuga*), 352–3, 394, 414–17
dragon, 364
drainage basin, 14, 219, 297
shape of, 137
drainage, cold air, 347
drainage density, 260–1
drift
–fence effect, 128–9, 132
genetic, 68
drip line, 94
drought, 354–5, 382, 399–400, 510
drumlin, 308

dry area or landscape, planning for, 457–8
dune, 291, 307–8
Barchan, 307
parallel linear, 307
Dunn, Christopher P., 409–12
dust, 340
dust bowl, 288
Dylis, N., 20
dynamic
area, minimum, 57, 65–6, 431
equilibrium, stream as, 216

East Anglia, England, 12–13, 31, 91, 374–5
Eastern Europe, 16–17, 301
see also Czechoslovakia; Poland; Romania
ecological change in land transformation, phase of, 416–17
ecological characteristics
in generic plan, 454–5
in geometric models, 426–8
ecological design of house lot, 469–74
ecological flows and shape attributes, 124–33
ecological and human dimensions, balance of, 523
ecological integrity, 37, 484, 499–501
biodiversity in, 500
linkage with basic human needs in, 500–2
productivity in, 499–500
soil in, 500
water in, 500–1
ecological process in space–time principle, 8
ecological trap, 91
ecologically optimum land transformation, 417, 426–32
ecology, 15, 19, 37
and economics, 483
of landscape, *see* landscape ecology
of region, *see* regional ecology
ecomosaic, 20, 300
economic
land-use model, 34
product in hedgerow, 299
economics
and ecology, 483
of forestry and farming, 50–1, 53
global and local, 484–6
maintain healthy growth rate in, 482
steady state, 483
ecoregion, 14

ecosystem, 75–8
  associated, 292
  cluster of, 13, 21, 286–94
  concept of, 12, 14, 38, 75
  configuration of, *see*
    configuration
  fragmentation effect on, 48–
    54
  function(ing) in, 76–8, 287
  incompatible, 292
  interactions: among, 286–
    90, 296–7, 515; between
    adjacent, 101–3, 292
  in landscape, 12–14, 20–1
  local, 12, 14
  resistant and susceptible,
    360–1
ecosystem management, 475
ecosystem structure, 75–6
ecotone, 85
ecotope, 20, 38
Ecuador, 310
eddy, 182–4
edge, 81–111
  anatomy and structure of,
    93–5; disturbance line of,
    94; drip line of, 94; in
    Germany, 95; mantel in,
    93–5, 98, 104; saum in,
    93–5, 98; three
    dimensions of, 86; veil in,
    94, 98; vertical, 85
  in a boundary, 85–6
  concept of, 38–9
  controlling mechanisms on,
    87–93
  curvilinearity of, 83, 106–9,
    116
  in design, 472
  development in, 87–95
  on different sides of patch,
    104
  human effects on, 92–3
  of landscape, 21–2
  length along, 85
  magnets in, 479
  management of, 101–2
  of matrix, 102
  microclimate of, 87–90;
    light in field, 88; light in
    forest, 87–8, 95; shade in,
    90–1; temperature in, 89;
    turbulence in, 89; wind in,
    87–90
  nest predators in, 106
  number of, 320
  plants in: nutrient-rich, 90,
    95; as survivors of
    herbivory, 91
  soil in, 90–1; nitrogen in,
    90–1
  type of: field, 48–9; forest,
    49–50; generated, 92–3;
    hard, 83–5, 106; induced,
    92; inner, 89–90, 97; soft,
    83–5

  see also boundary
edge effect, 96–100
  concept of, 38–9, 85
  on island, 55, 57
  and mantel, 94
  measurement of, 85
  in wildlife biology, 85, 313
  see also boundary
edge functions and processes
  conduit as, 96
  filter as, 92–3, 97, 100–1,
    382
  herbivory and predation as,
    91–2, 106
  impaction and particle
    deposition as, 88, 90–1
  Venturi effect as, 100
  wildlife usage as, 106–7
edge mosaic-sequence model,
  419, 423–31
edge species, 61, 97
  density and population size
    of, 47
  generalist and multihabitat,
    96–7
  in hard and soft boundaries,
    84, 132
  movement in corridor, 200
  nitrophile and phosphatile
    as, 90
  in powerline corridor, 174–5
  wildlife as, 106–7
  in woods, 61, 379
edge-to-interior ratio, *see*
  interior-to-edge ratio
edge width, 85, 104–6
  based on: different factors,
    105; light, 88, 130;
    microclimate, 87–90;
    microclimate and
    vegetation, 105; most
    sensitive species, 105–6;
    wind, 89, 130
  browsing, human, and
    external effects on,
    on different sides, 104, 130
  in field and forest, 104
  mantel effect on, 104, 106
  measurement of, 105
  variability in along
    boundary, 106, 129–30,
    140
  see also boundary
effective breeding population,
  366–7
egret, 289–90
element, 39
  see also landscape element;
    spatial element
elk (*Cervus*), 106–7, 262–3,
  266, 391–2
elm (*Ulmus*), 193, 222, 507
enclosed patch, 257
energy, 4
  advection, 325, 333–7
  albedo, 88, 328–31

  as basic human need, 500–1
  chemical and kinetic, 76,
    100, 104
  heat: given off to air, 330–3;
    for evapotranspiration,
    330–3; in food chain, 76;
    for soil, 330–3
  latent and sensible, 330
  light, 76, 87–8, 90–1, 95
  mosaic connected by, 327–
    37
  one way flow of, 76
  solar, 87–8, 327–30
  vertical flows of, 327–33
energy budget, 332
energy efficiency, 76, 497
England, 103, 156, 162, 197,
  258, 294–5, 374–5, 509–11
  see also Britain
entropy, 4, 34, 120
environmental conditions
  affecting individuals, 79
environmental corridor, 157
environmental degradation,
  508, 509–12
environmental gradient, *see*
  gradient
environmental patch, 44–5,
  111
  size effect on, 59–63
environmental resistance in
  population growth, 78
equation
  for erosion: water, 252;
    wind, 190, 207
  logistic, 78
  for mosaic pattern, 319–21
  for network, 282
  for patch: pattern, 319–21;
    shape, 141–2
  population growth, 78
  species–area: for islands,
    58, 80; for patches in
    landscape, 80; for
    samples in community,
    55, 80
  for water crossing through
    stream corridor, 246, 252
  for wind, 207
equilibrium
  community, 58
  island biogeography, theory
    of, 56
  non-, 58
  patch shape, 115–16, 139
  stream as dynamic, 216
erosion
  equations for, 190, 207, 252
  in field, 51, 125–6, 191
  in forestry, 308, 457
  landscape or terrain created
    by, 124
  patch size and, 51
  regional, 508
  in suburb, 460
  in sustainability, 509–10

erosion, *continued*
  by water, 244–9, 521; sheet
    flow in mountains, 351,
    521; valley, 305
  by wind, 189–92, 457–8;
    distances paticles move,
    189–90, 288; processes,
    189–90; results of
    selective particle
    movement, 190
  windbreak effect on, 186,
    189–92
erosion control
  optimization for
    biodiversity and, 520–1
  principles for wind, 190–2,
    207
escape cover, 47, 390
esker, 308
ethic, 451
  of isolation, 452
  land, 451–2
eucalypt (*Eucalyptus*) forest,
    50, 65, 311–12
eutrophication, 77, 508
evaporation, *see*
    evapotranspiration
evapotranspiration, 77, 330
  in edges, 87–90
  heat for, 330–3
  and water: budget, 51–2; in
    oasis, 342
  windbreak effect on, 187–9
evenness, 320
evolution, 384–5
evolutionary flexibility, 67–8
exclosure experiment, 91
exotic species, 393–8
  in corridor and edge, 203–6,
    241–2, 314
  disturbance and, 394
  guidelines for, 398
  invaders as, 394–8
  on island, 59
  and spatial pattern:
    introduction of, 161;
    invasive sites for, 393–7
experiment, 30–3
  in bottles, 46
  exclosure, 91
  in grassland, 32, 46–7
  landscape-scale, 32–3, 50,
    62
  micro-scale, 32–3, 163–4,
    431
  sulfur dioxide, 358
exponential population
    growth, 78, 393
exports, 510–11
exposure of edge, 104
extinction
  buffer against, 47
  in island biogeography, 55–
    8
  local, 70–1, 274–6, 397
  in metapopulation

dynamics, 70–1, 372–5,
    380–2

Fahrig, Lenore, 371
farming, economics of, 53
  see also field
farmstead, 191
feedback
  combined with hierarchy,
    505
  negative and positive, 9–10,
    80, 505
  in predator–prey cycle, 80
  in regulation and stability,
    10, 80, 505
  structure and function, 4
feeder line, 257
fence
  drift, 128–9, 132
  network in dry area, 272
fencerow, *see* hedgerow
Feng-shui, 15–16, 301
Ferguson, Joan D., 436, 461
fertilizer, 77, 380
fetch, 47, 49
field
  albedo of, 329
  connectivity of, in
    landscape, 294
  cultivation, 179, 191, 225
  edge and interior of, 48–9,
    104, 391–2
  erosion in, 51, 125–6, 191
  heat energy flows in, 331–2
  mineral nutrient runoff in,
    230–1
  orientation of, 125–6
  productivity of, 48–9, 194,
    231
  shape of, 122, 125–6
  size of, 48–52
  strip cropping in, 191–2
  see also agricultural;
    agriculture
fifth–*c*. 10th-order river, 245,
    249
filter
  boundary or edge as, 92–3,
    96–7, 100–1, 382
  corridor as, 148–9, 152,
    163–4
fir (*Abies*), 24, 338, 484–5
fire, 64
  boundary of, 84
  and deforestation, 416, 507
  ignition and spread, 416
  instead of tilling, 18
  in mosaic, 353–4
first-order stream, 52–3, 244–7
first principles, 29–30
fish, 62
  in counter current, 150–1
  food and habitat for, 228,
    233–5
  migratory: anadromous,
    236; catadromous, 236

and network connectivity,
    47, 52
  patch size effect on, 52, 62
  in stream and river system,
    233–6; refuge for, 265
flood, flooding, and
    floodwater, 214, 217–20,
    238–9, 244–5
  that escapes levee, 354
floodplain, 210–11, 214, 225
  habitat in, 240–2
  restoration, 476–8
  soil in, 307–8
floodplain ladder, 245, 250
Florida, USA, 131, 450
  mountain lion (panther)
    (*Felis*), 69–70, 312, 450
  river floodplain restoration,
    476–8
  South, 510–11
flow between adjacent
    ecosystems, 101–3
flowweb, 288
flux center, 316
fog-removing forest, 342
foliage
  nutrient-rich, 90, 95
  vertical stratification of, 76;
    in edge, 85; in hedgerow,
    195–6; in suburb, 472–3;
    in windbreak, 178–80,
    183–6, 193–4
food chain, 76
  detritus and grazing, 76
  species high in, 63
  in stream, 228, 233–5
food web, 77, 262
footpath, *see* trail corridor
foraging, 249, 365, 399
forest, value of, 14–15
forested landscape, planning
    for, 456–7
forestry, 36–7, 178, 456–7
  dispersed patch cutting,
    354, 414–19
  economics, 50–1
  planning natural resource
    area for, 444–5
  in river corridor, 225–6,
    250
  and roads, 456–7
  soil in, 308; erosion, 457
  two-century example of,
    484–6
form-and-function principle,
    30
  and patch shape, 124–5
Forman, Richard T. T., xvi,
    60–1, 73, 106–7, 354, 416–
    17
Foster, David R., 353, 355–7,
    432–3
foundation variables, 498–9
foundations of landscape
    ecology, 3–40
founder effect, 68, 385

Fourier analysis of patch shape, *see* harmonic analysis
fox, 10, 152, 236
fractal, 34, 99, 320
  boundary, 98–9, 107–8
  network: corridor density as, 260–1; dendritic pattern as, 119; stream dissection as, 305
  patch shape, 119
  self-similarity property and similarity ratio, 99, 120
  space-filling property, 99
fragmentation, 382, 405–34
  concept of, 39, 408, 412–13
  effects of: on ecosystem process, 48–54; other, 415; spatial, 413–14; species, 296, 380, 414–15
  experimental studies of, 32, 50
  in land transformations, 17, 405–34; as phase of major ecological change, 416–17
  in model of spatial processes, 407–11
  in mosaic-sequence models, 429
  natural process of, 412
  neighborhood analysis of, 272–4
  and scale, 412
  in suburb, 462
  of tropical rainforest, 32, 62–3
  see also land transformation
framework for open space planning, 463
France, 46, 197, 394
  cricket movement in, 391
  hedgerows and mesh size in landscape of, 256, 267–70
  human effects on land in, 418, 481, 510–11
Franklin, Jerry F., 354, 416–17
function, structure and, 4, 75
functional connectivity, 38, 157, 276, 296
functional response of predator, 79
funnel effect, 108, 128, 132
future predicted by shape, 138–9
Gaia hypothesis, 488
game, 39
  in deforestation, 416
  reserve, 128, 392
  see also wildlife
Game, Margaret, 73
gaming strategy, 317
gamma index, 261
gap, 63, 66, 76
  aggregation of, 148, 155–6
  in corridor, 30–1, 148, 187,

202; hedgerow, 31, 196; stream, 250–1, 265
  design of, 202
  size of: in corridor, 148; in patch, 63–4
  suitability of area in, 148
  Venturi effect in, 30, 187
  in windbreak, 187
  see also opening
gap analysis, 312
gap dynamics, 63, 76
Gardner, Robert H., 254–5
gas(es)
  absorbing: carbon dioxide, 327–8; methane, 327–8; nitrous oxide, 327–8; oxygen, 327–8; ozone, 327–8; water vapor, 327–8
  in conveyor belt, 340–4
  greenhouse, 341–2
  measurement of, in landscape, 36
  nitrogen, 77
gas line, 159
Gates, J. Edward, 175
gateway, 257
gene flow, 67, 383–6
  corridor role in, 203
  founder effect and, 68, 385
  genetic variation affected by, 384–5
  patch shape effect on, 127–8
  swamping with, 67
gene forms (alleles), 66, 68
general principles of landscape and regional ecology, 514–16
generalist species, *see* species
generated edge, 92–3
generic plan, 449–56
  agricultural landscapes and, 458–9
  background objectives and approaches in, 449
  context in, 450–2
  corridor attributes in, 455
  dry landscapes and, 457–8
  ecological characteristics targeted in, 454–5
  flows and movements over the landscape in, 455
  forested landscapes and, 456–7
  indispensable components in, 449
  key locations identified in, 454
  network, matrix, and mosaic attributes in, 455
  patch and boundary attributes in, 455
  spatial attributes targeted in, 455–6
  whole landscape in, 452–4
  see also plan, planning

genetic bottleneck, 385
genetic differentiation, 152, 236
genetic diversity, *see* genetic variation
genetic drift, 68
genetic mixing, 392, 394
genetic strain, 67, 354, 385
  see also subpopulation
genetic swamping, 67, 384–5
genetic variation, 66–71, 127–8, 132, 438
  gene flow and, 384–5
  minimum viable population and, 70–1, 384
  in network, 272
genetics
  in conservation, 67–8, 386
  evolutionary flexibility in, 67–8
  founder effect in, 68, 385
  inbreeding and outbreeding depression in, 66–8, 203, 385–6
  in a patch, 66–8
  techniques for population, 36
geocomplex, 20
geographic information system, 20, 34, 317–18, 443
  flexibility and cautions for use of, 34
  rastor and vector, 34
geographical range, 78
geography, 20, 36–7
  first law in, 286–7
  as source of spatial measures, 319
geological substrate and biodiversity, 311–12
geology reflected in a mosaic, 300–2
geomorphic process, 300–2, 510
geometric form as model for mountains, 303–4
geometric model, 26–7, 34, 419, 423–32
geometric patch shape, 116, 120
geomorphic process, 300–2, 510
geostatistics, 35, 319
Germany, 163–4, 344, 378, 383, 505
  edge structure in, 95
  temperature along transect in, 333–4
giant conveyor belt, *see* conveyor belt
giant green network, *see* network
giraffe, 523
GIS, *see* geographic information system
glacier, 8; terrain or

glacier, *continued*
  landscape created by,
    124, 308–9, 463–4
glider, 311
global air circulation
  high pressure regions in,
    323–5
  low pressure regions in,
    323–5
  zones, 323–6
global economy, 484–6
global framework, 323–6
global sustainability, 480
glossary, 38–40
Go, 317
goat, 394, 401
  see also livestock; mountain
    goat
Godron, Michel, xvi, 277–8
golden oriole (*Oriolus*), 376,
    378
government, 494, 510–12
gradient, 4, 76, 82–5; along
    corridor, 148, 201
grain, 10–11, 39
  fine and coarse, 10, 17, 39,
    438, 489–91
  response, 10–11, 490–1
  size, 319, 515: of landscape,
    268–9, 354, 425, 438, 515;
    in region, 23; in
    sustainability, 489–91
  variable-sized, 438, 491, 515
graph cell from graph theory,
    293–4
graph theory
  landscape graph or
    graph-theoretic graph in,
    293; candelabra as, 294;
    298; graph cell as, 293–4;
    necklace as, 293, 386;
    rigid polygon as, 294;
    spider as, 293
  model, 34, 293–4
grassland, 394
  exotic species spread in,
    398
  experiment in, 32, 46–7
  size of, 65
grassland landscape, 21, 291,
    336, 476–7
  planning in, 457–8
grassland reserve by railroad,
    172, 477
gravel (barrow) pit, 421
gravity, 30
gravity model, 260, 371
grazing food chain, 76
grazing system
  deferred rotation, 401–2
  high intensity/low
    frequency, 401–2
  high performance grazing
    (HPG), 401–2
  high utilization grazing
    (HUG), 401–2

short duration, 401–2
great blue heron (*Ardea*), 366
Greece, 482
greenhouse gas, 341–2
  see also gas(es)
greenlining, 199
  see also greenway; network
greenway, 199
  planning of, 445
growth management, 462
growth rate
  economic, 482
  population, 78
Guatemala, 492–3
Guntenspergen, Glenn R.,
    409–12

Haber, Wolfgang, 505
habitat, 39
  arrangement of, 310–15
  corridor as, 148–50, 195–8,
    236–9
  for fish, 233–5
  in floodplain, 240–2
  interspersion of, 314–15; for
    wildlife, 313–15
  microhabitat as, 47
  multihabitat species and,
    see multihabitat species
  shape of, 137
  as sink, 381–2
  small patch as, 47
  as source, 381–2, 376
  special, in house lot design,
    471
habitat fragmentation, *see*
  fragmentation; land
  transformation
habitat isolation, *see* isolation
habitat loss, 407, 412
habitat quality
  of corridor, 201, 204
  of matrix, 373, 375, 383
  metapopulation affected by,
    373, 375, 381, 383
  of patch, 373, 375, 378,
    381–3
  see also habitat suitability
habitat suitability, 148, 279,
    369, 387
  see also habitat quality
hailstorm, 355
harmonic analysis of patch
  shape, 134–7
  amplitude and frequency,
    136
  sine wave, 136
Harms, Bert, 279–80, 421–2
Harris, Larry, 154–5, 450
Hawaii, carbon dioxide
  increase at, 325–6, 337
hawthorn (*Crataegus*), 91, 97,
    178, 258
Hayley Wood, England, 374–5
heat energy
  for air, 330–3

for evapotranspiration, 330–
    3
  in field, 329, 331–2
  in food chain, 76
  latent and sensible, 330
  for soil, 330–3
hedge, *see* hedgerow
hedgerow, 31, 179, 258
  concept of, 39, 177
  economic product in, 299
  external form of, 196–7;
    curvilinear, 258; double,
    along lane, 421–2
  gap in, 31, 196
  as habitat, 195–8
  internal structure of, 195–6;
    ditch and soil bank as,
    196; internal entity as,
    196
  persistence of, 421–2
  species around: controlling
    agricultural pests, 197;
    insects as, 390; in
    movement to and from
    surroundings, 197–8;
    predator as, 198; shrub
    layer overbrowsed by, 195
  width of, 195, 302
hedgerow density, 376–9, 421
hedgerow network, 258, 261
Hellmund, Paul C., 106–7
herbaceous species in woods,
    374–5
herbivore and herbivory, 75,
    91, 392–4
  see also browse;
    overbrowsing
heron, 289–90, 366
heterogeneity, 4, 39
  and disturbance, 359–66
  and flow, 84
  of matrix, 47, 279–81
  macro- and micro-, 281
  measurement of, 318–20
  substrate, 44–5
heterogeneous bits of nature,
    452–3
hexagons
  modeling as, 30, 33, 119,
    297
  as patch shape, 119
  see also central place theory
hierarchy
  combined with feedbacks,
    505
  on land, 11–14;
    configuration in, 287–9,
    299–300
  in network, 259–60
  patch-within-patch, 11, 123
  in regulation and stability,
    505
hierarchy theory, 9, 34
high intensity/low frequency
  grazing system, 401–2

high performance grazing (HPG), 401–2
high utilization grazing, (HUG), 401–2
highway, *see* road corridor
hiker, 305
  see also trail corridor
hill, *see* mountains
hillslope, 210–11, 214, 235, 239
historical development of field, 28–9
history, learning from
  civilizations and communities, 509–12
  paleoecology, 506–9
Hobbs, Richard J., 48
Holmes, R. M., 336
home range, 365
  and configuration, 296
  corridor within, 164; road, 164
  large, 25, 47, 312, 388
  seasonal change in, 389–92
  shape of, 130, 137, 389, 391–2
  two or more patches constituting, 198, 389–90
  of wildlife, 137, 389–90
  within corridor, 241, 249; versus within patch, 274
homogeneity, 39, 84
  and matrix, 281
horizontal arrival of disturbance, 359
horizontal transport, advection as, 325, 335–6
horse, 353, 394, 494
  see also livestock; trail corridor
house clearings in landscape, connectivity of, 294
house lot, ecological design of, 469–74
  back lot line in, 473–4
  goals and techniques in, 470–1; change as, 471; flows and movements as, 471; special habitats as, 471; species as, 471
  neighborhood context in: corridor as, 470; representative as, 470; transition as, 470; unusual as, 470
  spatial arrangement in, 471–4; edges and junctions as, 472; small corridors as, 474; small patches as, 472; vertical height diversity as, 472
housing as basic human need, 500–1
housing development (estate), 297, 460; shape of, 121
HPG, 401–2

HUG, 401–2
human
  aggregate-with-outliers principle benefit to, 439–50
  colonization of islands by, 58–9
  and ecological dimensions, balance of, 523
  edge affected by, 92–3
  and landscape change, 301–2
  as multihabitat species, 490
  regular pattern produced by, 301
  river corridor altered by, 209, 223–7
  sustainability, dimensions or attributes in, 492–7, 509–12; built environment, 496–7; cohesive forces, 492–4; linkages with other areas, 494–6
human activity, 38
  boundary created by, 23
  and corridor, 158–9, 169–70
  lake sediment indicating, 508
  long-term effect of, 508–9
  and natural process, 14–19
  and patch shape, 115–17, 121–3
  region created by, 23
human adaptability, 502–5
human carrying capacity, 512
human community, *see* community
human disturbance, 396
human generations in sustainability, 486, 523–4
human impact
  long-term, 507–8
  from road corridor access, 169–70, 420
human needs, basic, 484
human population density, 392–3, 511–12
  environmental effects of, 508, 511–12
human scale, 4, 37, 490–1
humidity
  over floodplain, 217
  windbreak effect on, 188
Hunter, Malcolm, Jr., 354
hunting, 151, 196, 392, 410
hurricane, 8
  hole, 129
  in New England, 355–7; damage intensity, 356–7; damage locations, 356; patch size, 356–7; species composition, 356; vegetation height, 356
hydrology and hydrologic, 77 budget, 217

flows and fluvial, 217–19
sponge effect, 53, 213, 244–7
  see also water

ibex (*Capra*), 392
Idaho, USA, 84, 131, 232
immigration, *see* colonization
impaction, 88, 90–1, 340, 342–6
inbreeding depression, 66–8, 203, 385–6
incompatible ecosystems, 292
Index of Biotic Integrity, 234
India, 335
indifferent species, 97
indispensable patterns, 449, 452–4, 516
  in generic plan, 449
industry, 511
inertia, 417, 513
influence field, 297
information theory, xvii, 318–20
infrared (IR) radiation, 327–8
  near-infrared radiation, 328
innovation, 494–5, 504, 512
input–output model, 495–6
insects, 165, 380, 490
  birds that feed on, 60
  movement: along edge, 390; among non-adjacent elements, 388–91; in matrix, 390; in patch, 390
  in scattered woods, 377, 380
instantaneous land transformation, 352–3, 418–19, 422–3
intensification of agriculture, 410–11, 419–22
interactions
  between adjacent ecosystems, 101–3, 292
  among ecosystems, 286–90, 296–7, 386–93, 515
  water flow, 350–1
interdigitated landscape, 309–10
interior
  of corridor, 150, 245, 250
  in lobe, 129–30
  in matrix, 407–8
  of patch, 129–34, 245, 250
interior species, 47, 61–2, 97
  in corridor, 150, 174–5, 200
  in patch, 47, 125–30
  in woods, 61, 377
interior–edge borderline, 90, 105
interior-to-edge ratio
  of field and forest, 49–50, 61, 421
  of matrix, 278–9
  of patch and island, 57, 392

intermittent channel, 211–12, 232, 244–6, 351
internal entity in corridor, 146–7, 196
intersection angle, 259, 264, 269
intersection in corridor, 148
intersection effect, 266
intersection node in network, 259
  L, T, or + as, 258–9, 266
interspersion of habitats, 313–15
intrinsic rate of increase, 78
introduced
  patch and corridor, 45, 111, 157
  species, 397–8
invaders
  characteristics of, 395–7
  control of, 394
  guidelines for, 398
  invasive sites and, 393–7
  spatial pattern and, 397–8
inversion, see temperature inversion
Iran, 304
irregular
  boundary and patch shape, 117–18, 132
  network, 256
irrigation or irrigated landscape, 21, 336, 510–11
island, 113
  cluster of, 58–9; array or screen as, 58–9
  edge of, 55, 57
  human colonization of, 58–9
  land-bridge, 56–7
  tropical, 396
  width in stream or river, 211
island biogeography, 55–6, 274
  application of, 57–8
  area and age effect in, 55–7
  colonization and extinction in, 55–9
  equations, 80
  isolation (near and far) in, 55–7
  methodological problems of, 56–8
  mosaic and patch characteristics and, 57
  planning and, 57–8
  plants and, 58
isolation, 55–7, 62, 275–6, 407, 412
  as distance from woods, 375–9, 421
  ethics of, 452
isolation of a patch measure, 320

isopod, 377, 380
Italy, 299

Janzen, Daniel H., 476
Japan, 342, 409–12
jaws model, 429–32
jaws-and-chunks model, 431–2
jay, 81, 296, 378–9, 391
Johnson, Alan, 32–3, 371
junction of habitats, 314–15, 472

kame, 308–9
kangaroo, 17, 269
Kangas, Patrick D., 359
Kareiva, Peter M., 372
Karr, James R., 234
karst-limestone terrain, 309
Kedziora, A., 329, 331–2
kettle hole, 309
keystone species, 63, 77, 296
Kissimmee River, Florida, 476–7
Knaapen, J. P., 280
Knight, Dennis H., 318, 353
Kozova, M., 272–4
kriging, 319

Labrador, 353–4
labyrinthine form, 125
lack of planning, patterns from, 445–9
ladder, floodplain, 245, 250
lake, 253
  albedo of and air temperature above, 329, 335–6
  sediment in, 308; indicating human activities, 508
  shape of, 122, 137
  as sink, 335–6, 507–8; size of, as receiving basin, 507
  water in: quality of, 47, 52–3, 77; entering from submerged spring, 351
lakeshore, 214, 216
land-bridge island, 56–7
land change, see land transformation
land consolidation, 421–2
land ethic, 451–2
land evaluation, 477
land, hierarchy on, 11–14, 287–9, 299–300
land management, see management
land-markings, sacred, 481
land mosaic, 4, 300, 523
  see also mosaic
land mosaic phase of landscape ecology, 29
land planning, see planning
land protection, 511
  see also conservation;

natural resource protection
land restoration, see landscape restoration
land transformation, 44, 405–34
  ecological characteristics in, 426–8
  ecologically optimum, 417, 426–32
  fragmentation in, 17, 405–34; concept of, 39, 408, 412–13; other effects of, 415; as phase of major ecological change, 416–17; spatial effects of, 413–14; species effects of, 296, 380, 414–15
  geometric models of, 26–7, 34, 419, 423–32
  jaws models of, 429–32
  mosaic sequence in: common and uncommon, 423; modeling, 419–26; optimum, 417, 426–32
  patterns, 415–23
  spatial attributes in, 424–9
  spatial processes in, 406–12; attrition as, 407–11; dissection as, 404–10; fragmentation as, 405–34; perforation as, 407–11; shrinkage as, 404–11
land transformation, type of agricultural intensification as, 410–11, 419–22
  corridor construction as, 418–20
  deforestation as, 418–19, 507–8
  desertification as, 419–20, 508
  instantaneous as, 352–3, 418–19, 422–3
  reforestation as, 419, 422, 432–4
  suburbanization as, 410, 418–19, 429
Landers, J. L., 389
landform, 300–1, 306–9
landings, 51
landscape
  boundary of, 21–2
  concept of, 12–14, 39, 514
  as corridor, 25
  ecosystems repeated in, 13, 21
  flows linking, 335–6; in generic plan, 452–5
  gases measurement in, 36
  grain size of, 268–9, 354, 425, 438, 515
  inertia of, 417
  as living system, 5
  patch–corridor–matrix

model of, 6–7, 34, 274, 398, 514
portion of, 22, 39
position within, 316, 378–9
sources and sinks in, 340, 343–5
spatial arrangement of, within region, 20, 23, 378–9, 430
as surrogate for long-term, 523–4
see also land
landscape architecture, 36–7
see also design; planning
landscape change, 44, 301–2, 508, 515
see also land transformation
landscape detective, 4
landscape diversity, 489–90
landscape ecological planning, 444–9
see also planning
landscape ecology, 19–22, 37
alternative perspectives on, 20
foundations of, 3–40
historical phases of, 28–9
principles of, 20, 514–16; applications in, 37, 515–16; landscapes and regions in, 514; mosaics in, 515; patches and corridors in, 515
spatial language of, 7, 442–3
landscape element, 7, 20, 38–9
see also spatial element
landscape graph, 293
landscape planning, 444, 523–4
see also planning
landscape resistance, 272, 279–81, 515
boundary-crossing frequency and, 279–81, 362, 368, 387
movement of species and, 279–81
forest cover effect on, 279–81
landscape restoration
diverse examples of, 477
of North Woods, 477
of river floodplain, 476–8
of tropical dry forest, 476
landscape-scale experiment, 32–3, 50, 62
landscape structure, 381
landscape type, 306–10
aeolian, 124
agricultural, *see* agricultural landscape
dry, desert, and grassland, 21, 291, 336, 457–8, 476–7
forested, 456–7

glacial terrain, 124, 308–9, 463–4
irrigated, 21, 336, 510–11
patch–corridor–matrix
model of: checkerboard as, 309–10; dendritic as, 309; interdigitated as, 309–10; rectilinear as, 309–10; large patch as, 309; small patch as, 309
suburban, 22, 300–1
water-created terrain, 124
wind-created terrain, 124, 307–8
lane with double row of trees, 421–2
large-patch landscape, 309
large woody debris
as habitat in forest, 60
in stream, 215, 235, 245, 248–52
lawn, 391
lead pollution, 168–9
leaf, 113, 137
temperature, 194
learning, 367–70, 398–401
length-to-width ratio, 116–17, 133–5, 138, 353, 389
four-to-one, 122, 126
Leopold, Aldo, 396, 451
levee, 214, 227, 238
flood escapes, 354
Levins, Richard, 380
Lewis, Philip, 450
lichen, 358
Lieberman, Arthur, 20
life table, 78
light
edge width affected by, 88, 130
in forest and field edge, 87–8, 90–1, 95
intensity, 76
see also solar energy
Likens, Gene E., 358
lime (*Tilia*), 374, 388
line
corridor, 150, 154
disturbance and drip, 94
trunk and feeder, 257, 271
linear distribution of ecosystems, 291
linkage
between continents, 325, 487
in a network, 258–60, 266–8, 275–6
with other area in sustainability, 494–6
between regions, 27, 325, 344–5
linkage density, 259
livestock, 145, 225, 311–12, 398–402
design of pasture, paddock, or ranch for, 272

grazing system for, 401–2
herd size and number of, 401–2
in learning new pasture, 398–401; inexperienced herd of, 400
overgrazing or overbrowsing by, 195, 420, 510
riverbank eroded by, 223
routes of, 159, 172–3, 265, 393
sequence of behaviors and physiological needs of, 399–401
transport of mineral nutrients by, 51, 288, 393
livestock movement, 398–402
livestock production, 400–2, 478
living system, 4–5, 298
loam, 308, 491
lobe, 107–10, 137
drift-fence effect of, 128–9, 132
functions of, 108–9
funnel effect of, 108, 128, 132; dispersal funnel, 128, 132
interior in, 129–30
length of, 133–4
mass flow and, 128–9
number of, 116–19, 122, 133–4
peninsula effect in, 108, 127–8
tip of, 127–8
tip splitting of, 120, 140
local
area: design of, 516–18; economics of, 484–6
extinction, 70–1, 274–6, 397
local ecosystem, 12, 14
see also ecosystem
locomotion, 30, 287
loess, 307–8
logarithm or logarithmic
plot, 56
series in population growth, 78
variation smoothed out by, 57–8
log-normal distribution, 80
logging, *see* forestry, deforestation
logistic equation, 78
longhorn steer, 399
see also livestock
loop, feedback, *see* feedback
loop in network, 255, 270
see also circuitry
LOS effect on ecosystem, 45–8, 53
Lovejoy, Thomas E., 32, 62
Lynch, Kevin, 7
Lyon, France, 418

MacArthur, Robert H., 56
macroclimate, 13, 22–3, 325, 512
macroheterogeneity, 281
Mader, H-J., 33, 163
magnets in the edge, 479
Maine, USA, 65, 99, 170, 358, 477
mainland, 56, 127
mammal
  movement, 388–93
  patch size effect on, 60, 62, 64
  road crossing by, 163–5
management, 474–9
  approaches in: adaptive, 475; cumulative impact, 475; ecosystem, 475; island biogeography, 57; landscape restoration, 432–4, 474–7; natural resource protection, 478–9; new forestry, 475–6; park and recreation, 36; sustainability, 522–4
  current difficulty in, 440–1; proposed plan or solution for, 441
  ecosystem, 475
  ethics in, 451–2
  genetic issues in, 67–8
  growth, 462
  indispensable patterns in, 449, 452–4, 516
  land, 474–9; broad spatial scale in, 474–7
  paradox of, 488
  spatial issues in: broad scale, 474–7; ecosystems to consider as, 287; indispensables as, 449, 452–4; network analysis of neighborhood as, 272–4; patch shape as, 124–5; road closing as, 267
  species-centered: with beaver, 222–3; exotic species, 398; metapopulations in, 382–3; minimum viable population, 71–2; multihabitat species, 390; for road crossing by wildlife, 163–8, 176; wildlife, 11
  as surrogate for long-term, 523–4; broad temporal scale in, 475–7
  sustainability incorporated into, 522–4
  temporal scale in, 476–7
  type of habitat or area for: border as, 92–3, 101–2; boundary as, 101–2; configuration as, 294, 300; corridor to arrest

spread as, 206; edge as, 101–2; park and recreation as, 36; powerline as, 175–6; roadside natural strip, 154, 159–63, 171; windbreak as, 193
  see also planning
managers at work, 442–4, 474
mantel, 93–5, 98
  edge width determined by, 104, 106
  main-mantel and pre-mantel, 94
  species in Germany, 95
map overlay, 443
  shortcomings of, 443
maple (Acer), 24, 195, 305, 357, 388, 393, 409, 484–5
Margules, Chris R., 50, 312–13
marine substrate, 65
marsh, 65, 237–8
  area and isolation of, 65, 290, 380
  see also wetland
marsh tit (Parus), 376–9
mass flow, 30; patch shape and, 128–9
Massachusetts
  forest and hurricane in, 305, 355–7, 432–3
  human community, 461–9, 509–11
materials
  conveyor belt and, 337–45
  cycles of, 77–8
  transport source to sink of, 343–5
matrix, 277–82
  concept of, 39, 277
  connectivity of, 277–8, 407–8
  contiguity of, 319
  corridor affected by, 203
  dynamics of, 102, 279–82
  edge and interior of, 102, 407–8
  habitat quality of, 373, 375, 383
  heterogeneity and homogeneity of, 39, 84, 279–81; benefit of small patch in, 47
  how to identify the: area in, 277–8; connectivity in, 277–8; control over dynamics in, 277–8, 298
  interior-to-edge ratio in, 278–9
  measurement of, 80, 279, 317–21
  movement in, 387, 391; resistance against, 279–81
  network enclosing, 260, 272; sections of, 387

protected by corridor, 147–8
  road corridor effect on, 168–70, 420
  shape of, 278–9; changing, 281–2
  structure of, 7, 277–9
  subdivided, 278–9
  type of, 26–7; porous, 278–9; swampland as, 389–90
Maya, 492–3, 510–11
McDonnell, Mark J., 388
McHarg, Ian, 443
meander band, 211, 220–1, 307
meandering channel, 220–1
median vegetated strip in highway, 152, 160
membrane analogy with edge, 95–7, 109
membrane physiology, 30, 96
  receptor function in, 96–8, 152
  semipermeable function in, 152
membrane theory, 95–6, 109
Merriam, Gray, 36, 157, 274–6, 296–7
mesh size, 255, 267–70
  variable, 121–2, 269–70
Mesopotamia, 508, 510–11
metapopulation, 372
  corridor role in, 203
  examples of, 379–80; avian, in woods, 376–9
  in management, 382–3
  source–sink habitats of, 380–2
  spatial pattern in, 375–6
  subpopulation in, 372–5, 380–2
metapopulation dynamics, 372–83, 515
  extinction in, 70–1, 372–5, 380–2; habitat quality effect on, 373, 375, 381, 383; patch size effect on, 373, 375, 381, 383
  recolonization in, 372–5, 380–3; corridor or stepping stone effect on, 373, 375, 383; inter-patch distance effect on, 373, 375, 383; inter-patch habitat quality effect on, 373, 375, 383
  winking in, 380–5
methane, 327–8
metropolitan food system, 497
mice, 196, 200, 249, 274–6, 296–7, 380
microclimate
  of an edge, 87–90, 105
  in opening, 89–90, 333–4, 348–50
  by a windbreak, 187–9

microhabitat proximity, 47
microheterogeneity, 281
micro-scale experiment, 32–3, 163–4, 431
migration, 365, 388–9
  large mammal, 263–4
  long- versus short-distance, 62
  patch shape, orientation, and, 125–6
  of stream and river, 211, 219–21, 238–9
migratory fish, 236
Miller, David H., 12–13, 21
millipede, 377, 380
Milne, Bruce T., 32–3, 371
mine revegetation, 109, 141
mineral nutrients
  biodiversity and, 311–12
  lateral flow of, from matrix to channel, 229–32; associated with eroded particles, 231
  livestock transport of, 51, 288, 393
  patch size and, 51
  -rich plants in edge, 90, 95
  runoff in field, 230–1
  spiraling in stream, 229
  see also nitrogen; phosphorus
minimum area point, 54–5, 65–6
  for different groups, 59–60
minimum dynamic area, 57, 65–6, 431
minimum number point, 73–4
minimum viable population, 68–72
  conservation safety factor for, 70–1
  effective breeding population of, 71, 384
  extinctions in, 71–2, 384
  50/500 and few hundred/few thousand, 70–1, 383–4
  genetic variation in, 70–1, 384
  subpopulation and, 71
Miyako River basin, Japan, 409–12
modeling, 33–5
moisture
  in edge and air, 89, 187–9
  soil, 192–5
Monahan, Daniel, H., 436, 461
monkey, 427
Montana, USA, 6, 84, 172–3, 191, 221
moraine, 308–9
mortality sink, see sink
mosaic, 57
  cluster of ecosystems in, 13, 21, 286–94
  concept of, 3–7, 39, 83

configuration as, see configuration
culture reflected in, 14–19, 300–2
disturbance in, 351–63; air pollution, 357–9; fire, 353–4; water, 354–5; windstorm, 355–7
energy flow connecting, 327–37
fire in, 353–4
geology reflected in, 300–2
grain size of, see grain
land, 300
measurement of, 317–321; equations for, 319–21
neighborhood, and network analysis, 272–4; see also neighborhood
shifting, 44
species in: biodiversity areas selection, 310–13; habitat arrangements, 310–17; movement, 364–402; wildlife habitats, 313–15
strategic points in, 315–17
third dimension in, 302–6
types of landscape and region in, 306–10; glacial terrain, 124, 308–9; patch–corridor–matrix, 309–10; water-created terrain, 124, 307; wind-created terrain, 127, 307–8
mosaic patterns, 285–321
  aggregated, 291
  associated, 292
  common to all landscapes, 292–4
  linear or parallel, 291
  regular, 290–1
mosaic sequence, 417, 515
  common and uncommon, 423
  ecologically optimum, 417, 426–32
  in land transformation, 419
mosaic-sequence models
  boundary length in, 425–30
  comparing, 428–9
  connectivity in, 425–30
  ecological characteristics in, 426–9
  fragmentation in, 429
  patch size in, 425–31
  spatial attributes in, 424–9
  surrounding area effect on, 433–4
  type: alternate strips, 419, 423; corridor, 419, 423–9; dispersed, 419, 423–6; edge, 419, 423–31; even throughout, 419, 423; geometric, 423–32;

instantaneous, 419, 423;
jaws, 429–32;
jaws-and-chunks, 431–2;
mouth, 429–32; network template, 419, 423;
nuclei, 419, 423–6;
nucleus, 419, 423–6;
optimum, 417, 426–32;
random, 424–6
  variables added to, 433
mosaic stability, 298–9, 504
moss, 60, 503
motorbike, see trail corridor
mountain goat (Oreamnos), 304
mountain lion (Felis), 241, 92, 405
  in Florida, 69–70, 312, 450
mountains
  balance between vegetational and topographic height in, 302–4
  conical form of, 306
  horizontal movement and vegetation zone or belt in, 306
  ridge network in, 304, 306
  ruggedness in, 304
  sheet flow in, 350–1, 521
  snow accumulation in, 346
  vertical dimension, 302–6: flow in, 25, 306, 345–8; species movement route in, 306; wedge-shaped valleys in, 306
  wind pattern with steep slope upwind or downwind in, 181
  see also terrain
mountainous region, 21, 25, 310
  landscape mosaics in, 21
mouse, 196, 200, 249, 274–6, 296–7, 380
mouth model, 429–32
movement
  across: boundary, 83–4, 92–3, 100–1, 106–7; built area, 279–80; curvilinearity effect on, 106–7; road and canal, 279–80
  along: boundary, 106–7; corridor, 147–8, 198–200; curvilinearity effect on, 106–7; edge, 390
  animal's need for, 203
  dispersal as, see dispersal
  enhancing and inhibiting, 200–3
  gene flow, 67–8, 383–6
  landscape resistance effect on, 279–81
  location of: between ecosystems, 101–3; among

movement, *continued*
ecosystems, 296–7; in
matrix, 387, 390–1; in
mosaic, 57, 364–402; in
mountains, 306; among
non-adjacent elements,
386–93; in patch, 390
in matrix, 279–81, 387, 391
metapopulation dynamics,
372–83
models of, 365–72
effect of mosaic attributes
on, 369–70
population models of, 370–2
routes, *see* routes
source of, *see* source
species: animal, 365–6; bird,
376–9, 391–3; edge, 200;
exotic, 204–6, 393–8;
insect, 390–1; interior,
200; invader, 393–7;
livestock, 398–402;
mammal, 388–93;
multihabitat vertebrate,
388–90; pest, 204–5;
plant, 199, 366–7;
specialist, 205; wildlife,
198–207
type of, 365–9; foraging as,
249, 365, 399; home
range as, 365; migration
as, *see* migration;
navigating as, 367–8;
nocturnal as, 161;
population arrival as,
368–9; saltatory as, 367;
searching as, 367–8; seed
dispersal as, *see* seed
dispersal; wandering as,
366, 392; wind and water
flow as, 323–63
see also species; woodland
corridor
moving patch
advancing and retreating,
139–41
predicting the future from,
138–9
multihabitat species, 39, 427
mosaic for, 304, 368, 490
movement of, among
non-adjacent elements,
388–93
in patch and edge, 47, 97
vertebrate as, 388–90;
human, 490
mushroom, 60, 338
mutation, 70–1, 384
MVP, *see* minimum viable
population

narrows, 148
natural disturbance, *see*
disturbance
natural-history-and-physical-

environment phase of
landscape ecology, 28
natural patterns, 445, 516
measurement of, 30–1
patch shape and irregularity
in, 116–17, 132
natural process and human
activity, 14–19
natural resource area, 478
for biodiversity, 444–5
and boundary management,
101–5
for forestry, 444–5
human overuse prevention
in, 478–9
magnets in the edge of, 479
management, *see*
management; park
planning, *see* planning
small, 479
for wildlife, 444–5
see also nature reserve;
park
natural resource protection,
478–9
natural selection, 79
overpopulation, variation,
and competition in, 79
survival of the fittest in, 79
natural strip, roadside, *see*
roadside natural strip
natural vegetation, 39
nature, 3
nature, scattered bits of, 452–
3
see also natural patterns
nature conservation, *see*
conservation
nature reserve, 450
design of, 57, 137; centered
on boundary, 313
survey to locate, 312–13
system or network rather
than collection of, 177,
314
Naveh, Zev, 20
navigating, animals, 367–8
Nazca culture, 481
near-infrared radiation, 328
nearest neighbor probabilities,
321
necklace from graph theory,
293, 386
needs, basic human, *see* basic
human needs
neighborhood, 382
concept, 103, 285–6, 289
in context, 285–6, 378; of
house lot, 470
escape cover in, 390
model, 34
residential, 472–4; wildlife
corridor in, 473
see also configuration
neophyte, 397
Nepal, 311

nest, 126
predation, 106
predicting location of, 294–
5
Netherlands, The, 170
agricultural intensification
and sustainability in,
421–2, 510–11
birds and metapopulations
in, 289–90, 376–83
locating new forest in, 312,
477
movement across landscape
in, 279–80
network, 39, 253–76
in action, 272–6
barrier effect in, 271–4
counter current principle
in, 271
formation, development,
and change of, 254–7
genetic variation in, 272
as habitat, 269–70
linkages in, 258–60, 266–8,
275–6; location of, 269–
70; movement and flows
along, 262, 270–1, 386–7;
number of, 275–6; per
node, 259; quality of,
275–6
matrix sections enclosed by,
260, 272, 387
nodes in, 258–62, 265–7;
large, 264–5, 275–6;
number of, 275–6
optimum form of, 276, 450
ridge, 304, 306
structural attribute of, 254–
62; hierarchy as, 259–60;
intersections as, 258–9,
266; loops as, 255, 270;
shape as, 125
wind velocity in, 271
network circuitry and
complexity, 261, 266
network connectivity, 261,
266; fish and, 47, 52
network density, 261
network measures and
equations
alpha, beta, and gamma
index as, 274, 282
mesh size as, 255, 267–70
network resistance, 270, 272
network template model, 419,
423
network theory and model,
30, 34
network type
braided, 215, 256
dendritic, 52–3, 141, 256,
262–5; reversed, 124
giant green, 154, 171, 276,
450
hedgerow, 258, 261
irregular, 256

rectilinear, 258–9, 263, 265–6, 274
ridge, 304, 306
strip of network as, 156
wavy net as, 258, 269, 271
neutral model, 30, 34, 254–5
New Brunswick, Canada, 242–3, 484–6, 510–11
New England, USA, 122–3, 302, 393
  landscapes of, 23–4
  hurricane in, 355–7
new forestry, 475–6
New Jersey, USA, 270, 358, 460
  forest patch size in, 59–62, 72–3
  hedgerows, 195
  Pine Barrens, 27, 354, 510–11
New Mexico, USA, 106–7, 271, 288, 510–12
  see also southwestern USA
New South Wales, see southeastern Australia
New Zealand, 70, 288, 393–4
niche, empty, 396
Niering, William A., 176
nitrate, 358
nitrogen, 77, 287–8
  edge and plants affected by, 90–1, 393
  eutrophication by, 77, 508
  in stream corridor, 228–31
nitrogen cycle, 77
nitrophile, 90
nitrous oxide, 327–8
nocturnal movement, 161
node
  attached and intersection, 148, 257–9
  in corridor and network, 7, 148, 256–62, 266–8, 273–6
  size of, 273–4; gravity model and, 260, 371
  in string of lights pattern, 201, 259
non-adjacent elements, movement among, 386–93
non-equilibrium, 58
non-native, see exotic species
North Carolina, USA, 389
North Dakota, USA, 198
Noss, Reed F., 450
nuclei mosaic-sequence model, 419, 423–9
nucleus mosaic-sequence model, 419, 423–9
number of large patches adjacent, 103, 122–3
  bird evidence for, 72–3
  plant evidence for, 73–5
nuthatch (Sitta), 279, 376–8
nutrient, see mineral nutrients

Oak (Quercus), 24, 222, 296, 305, 357, 391, 409, 427
oasis effect, 53, 335
  evapotranspiration and, 342
oasis, linear, 236
observations in landscape ecology, 30–2
ocean, 14, 113
  current, 324–5
Oklahoma, USA, 356, 399
Omar Khayyam, 179–80
O'Neill, Robert V., 99
Ontario, Canada, 194, 250–1, 289, 354
  road crossing in, 163–4
  species movement in agricultural landscape of, 274–6, 296–7, 380–1
Opdam, Paul, 36, 279–80, 376–9
open system, see system
opening, 63–4, 66, 349, 355, 456
  microclimate in, 89–90, 333–4, 348–50
  size and shape of, 50–1, 65, 136–7
  see also gap
openness, 305
opossum (Didelphis), 366
optimization for two and three variables, 520–1
optimum
  mosaic sequence of land transformation, 417, 426–32, 434
  network form, 276
  patch shape, 131–3
  spatial arrangement: in landscape, 437, 516, 522–3; of landscapes, 25–7
organic compound, 76
organic matter, soil, 229–34
organism–soil cycle, 77–8
organizing force, 297–8
orientation angle, 125–7, 150
  field erosion and, 125–7
  of patch for birds, 125–7
  of windbreak, 181, 186–7, 192–3, 197
outbreeding depression, 66–8, 385
overgrazing or overbrowsing, 195, 420, 510
overpass for animal crossing road, 165
overpopulation in natural selection, 79
owl, 60, 198, 268
  nest location of, 296
ox intestine and belly, 15–16
oxbow, 211, 219, 221
Oxley, D., 164
oxslip (Primula), 374–5
oxygen, 327–8
ozone, 327–8, 358

Pacific Northwest, USA, 291, 296, 352, 354, 413–17, 429, 456
paddock, see pasture
paleoecology, 506–9
  climate change and, 337, 506
  landscape ecology and fine-scale, 507–9
  lessons for sustainability, 506–9
  long-term human impact in, 507–8
  long-term vegetational change in, 110, 506–7
  pollen analysis in, 278, 507
panther, see mountain lion
paradox of management, 488
parallel pattern of ecosystems, 291–2; linear dunes as, 307
park
  boundary and trails in, 92–3, 174, 312
  location of, for biodiversity, 310–11
  see also natural resource area; nature reserve
park planning and management, see planning; management
particle
  in conveyor belt, 338–40
  deposition in impaction, 88, 90–1, 340, 342–6
past predicted by shape, 138–9
pasture, 398–402
  deforestation and, 410–11
  design in dry area, 272
  rotation and grazing system, 401–2
patch, 39, 43
  area, see patch size; size effect on
  characteristics of, 6–7, 57; core as, 129–33; gap dynamics in, 63–4, 76; habitat quality of, 373, 375, 378, 381–3; interior-to-edge ratio, 57; interior as, 129–34, 245, 250; location as, 52–3; see also edge; patch shape
  cluster of small, 201–2
  dispersion of, 321
  genetics in, see genetics
  home range in corridor versus, 274
  in house lot design, 472
  interior species of, see interior species
  moving, 139–41; predicting the future from, 138–9
  origin or cause of, 44–5, 111

patch, *continued*
  persistence, 44–5, 298; *see also* stability
  in sustainability, 488–9
patch-centered measure, 318, 320
patch–corridor–matrix
  model, 6–7, 34, 274, 398, 514
  types: checkerboard, 309–10; dendritic, 309; interdigitated, 309–10; large-patch, 309; rectangular, 309; small-patch, 309
patch dynamics, 34, 44, 65–6, 238
patch–matrix interaction, adjacent and distant, 132
patch number, 43–80, 407–8
  constituting home range, 198, 389–90
  of large patches, 72–5, 431; in England, 73; in New Jersey, 72–3
  minimum number point and species pool determining, 73–4
  risk spreading and, 53, 75, 431, 438
  SLOSS and, 45–8, 53–4
patch shape, 57, 113–14, 515
  classification of, 123–4
  in conservation and wildlife management, 124–5
  convolution in, 113, 116–17; ecological effects of, 124–9; measurement of, 133–8, 142; number of lobes as, 116–19, 122, 133–4; ecological effects of interior in, 132
  elongation in, 114–17, 120, 122, 355–6; ecological effects of, 125–7; measurement of, 133–8, 141–2
  gene flow affected by, 127–8
  human activity and, 115–17, 121–3
  measurement of, 133–8, 141–2; additional methods, 135–8; convolution, 133–8; elongation, 133–8; equations for, 141–2; harmonic analysis, 134–7; index for, 135–8, 141–2; interior, 134; perimeter, 134–8
  of moving patch, *see* moving patch
  orientation angle and, 125–7
  origins and causes of, 115–23, 138–41; fire, 353;

  human activity, 115–17, 121–3
  plains, slopes, and terrain effect on, 114–16; as equilibrium, 115–16, 139
  to predict the future, past, and present, 138–9
  variability in, 114–15
patch shape, functions and processes of, 124–33
  drift-fence effect as, 128–9, 132
  form-and-function principle in, 30, 124–5
  funnel and dispersal-funnel effect as, 128, 132
  inerior species and migration as, 125–30
  perimeter effect as, 130–2; measurement of, 134–8, 142
  turbulence as, 127–8
patch shape, types of
  common and uncommon, 114–19
  expanding and contracting, 110
  geometric (commonly symmetrical), 116, 120; compact, 114–16, 124, 135, 141–2; fractal, 119; hexagon, 119; hourglass (dumbbell), 119; rectangular, 122–3, *see also* length-to-width ratio; ring, 25, 119, 306, 430; rounded, 116–17, 132; square, 117, 119, 122–3; star-shaped, 119, 123
  irregular (commonly asymmetrical), 117–18; amoeboid, 116–17, 136–7; ephemeral, 118, 120; indented, 118–19; mixed, 122; natural, 116–17, 132; perforated, 119, 123; spaceship, 132; sponge-like, 119, 123
  network, 125
  optimum, 131–3
  streamline, 117–18
patch size, 43–80
  see also size effect
patch type
  amoeboid, 116–17, 136–7
  cause of: disturbance as, 44–5, 63–5, 111, 351–63; environmental as, 44–5, 111; introduced as, 45, 111; regenerated as, 44–5; remnant as, 44–5, 50, 59–63, 111, 458–9
  enclosed, 257
  moving as, *see* moving patch
  source as, 376, 380–5

patch-within-patch, 11, 123
patchiness or patchy, 76, 148, 201, 320
path, *see* trail corridor
pattern, spatial
  causes of, 4; disturbance, 445–8
  familiar or common to all landscapes, 290–4
  from nature, planning, and lack of planning, 445–9, 453
peninsula effect
  and island biogeography, 57
  in lobe, 108, 127–8
people, *see* human
perception
  of openness and landscape, 301–2, 305
  of vegetational and topographic structure, 303–6
percolation theory, 30, 34, 254–5, 319
perforated matrix, 278–9
perforation
  of landscape, 407–11, 429, 456
  in patch shape, 119, 123
perimeter, 130–8, 142
  see also interior-to-edge ratio
permeability of membrane and boundary, 100–1, 392
persistence, 44–5, 298
  of hedgerow, 421–2
  of road corridor, 421
  see also stability
Peru, 509–11
pest, 46, 49, 197, 203–5, 396
  in stream corridor, 242–3
Peterjohn, William T., 230
Peterken, George F., 73, 374, 406
pheromone, 392–3
phosphatile, 90
phosphorus
  in edge, 90–1
  eutrophication by, 77, 508
  movement between ecosystems, 228–31, 287–8
phosphorus cycle, 77
photosynthesis, 76, 341
pied flycatcher (*Ficedula*), 379
Pima-Pagago people, 481
Pinchot, Gifford, 178
pine, 24, 46, 137, 296, 305, 338, 354, 357, 394, 409, 484–5
Pine Barrens of New Jersey, 27, 354, 510–11
plan
  degree of detail in, 516
  for every landscape and region, 523

generic, see generic plan
knowledge, flexibility, and
    collaboration based, 441,
    443
overlay based, 443;
    shortcomings of, 443
spatial and temporal
    context of, 443
see also planning
planet, 12–13
see also global
planetary boundary layer,
    325–6
planners at work, 442–4
planning
    aggregate-with-outliers
        principle in, 436–40, 453,
        515
    agricultural land, 444
    apples-and-oranges
        comparison of special
        sites in, 468
    biophilia in, 448
    bio-rich place in, 448
    Concord case study, 461–9;
        general application of,
        469; large patches in,
        462–6; major corridors in,
        465–7; special sites in,
        463–8; synthesis and
        protection priority in,
        468; town-wide pattern
        in, 464–7
    for concurrent objectives,
        519–21
    of configuration, 289–90,
        300
    and continental ecology,
        450–1
    corridors, 445; stream and
        river, 452, 457–8
    current difficulty in, 440–1
    diameter of surrounding
        area for, 451
    dry, desert, or grassland
        landscape, 457–8
    ecological design of house
        lot in, 469–74
    ecologically optimum
        mosaic sequence in, 417,
        426–32
    ecosystems to consider in,
        287
    ethics in, 451–2
    forested landscape, 456–7
    framework for open space,
        463
    general application of, 469
    greenways, 445
    history needed for, 451
    house lot, 469–74; goals and
        techniques, 470–1;
        neighborhood context,
        470; spatial arrangement,
        471–4

indispensable patterns in,
    452–3
and island biogeography,
    57–8
lack of, 445–9
landscape, 444
landscape ecological, 444–9
mosaic-sequence models in,
    453–4
for multihabitat species,
    390
natural resource area for
    biodiversity, forestry, and
    wildlife, 444–5
network analysis of
    neighborhood in, 272–4
optimum location of change
    in, 434
overlay approach in, 443,
    468
patterns: from lack of
    planning, 446–9; from
    nature, 445–7, 453; from
    planning, 446–9
physical, 445
policy, 445
principles of generic plan,
    see generic plan
proposed plan or solution
    for, 441
protecting society in, 441
rarity and recovery time of
    special sites in, 468
of region, 36–7
restoration, see restoration
rural land, 444
scale in, 448
standards in, 440–1
of suburb, 459–69; Concord,
    461–9; house lot, 469–74
sustainability incorporation
    into, 522–4
synthesis and land
    protection priorities in,
    468
see also generic plan; plan
plant
    in corridor: hedgerow
        attached to woods, 151;
        stream or river, 236–43
    in edge, nutrient rich, 90,
        95
    mesh size and, 268
    movement, 366–7; along
        corridor, 199; dispersal
        as, 366–7, 387–8
    patch size effect on, 43, 60–
        2, 64
    as survivor of herbivory,
        91
    windbreak effect on, 194–5
Plato, 482
plosion, 256
plume of smoke or moisture,
    339–40, 342
point, line, and plane, 7

Poland, 380
    albedo and heat energy
        flows in, 329, 331–2
    patch shape in, 114–16
policy, 6, 445
pollen analysis, 278, 506–7
pollution
    air, 340–2, 357–8
    soil and water, 77, 168–9,
        508
pond, 421, 521
    beaver, 221–2
    see also lake
pool and riffle, 69, 210–11,
    214, 219–21, 234
pool, species, 55, 73–4
poplar (Populus), 125–6, 184,
    222, 237, 302, 357, 381,
    484–5
population, 66, 78–9
    arrival and spread of, 368–
        70
    breeding and effective
        breeding, 71, 366–7
    human, 392–3, 508, 511–12
    local, see subpopulation
    minimum viable, 68–72
    patch size effect on, 59–60
    road kill effect on, 167–8
population density, 392–3,
    511–12
    environmental effects of,
        508, 511–12
population growth, 78–9
    age structure, life table, and
        doubling time in, 78
    declining and exponential,
        68–70, 78, 393
    density-dependent and
        -independent, 78
    of mouse, 274–5
    rate equation, 78: birth rate
        in, 78; carrying capacity
        in, 78; death rate in, 78;
        environmental resistance
        in, 78; intrinsic rate of
        increase in, 78;
        population size in, 78,
        275–6
    regulation of, 10, 80
population ecology, 78–80
population genetics
    see genetics
population models, 370–2
    diffusion: homogeneity and
        heterogeneity in, 370–1;
        -reaction, 370
    gravity model and central
        place theory as, 371
    porosity of windbreak, 183–5,
        193–4
    porous matrix, 278–9
    portion of landscape, 22,
        39
    potassium, 311
    power function, 80

powerline corridor, 65, 159, 174–6
  edge species in, 174–5
  interior habitat and barrier functions of, 174–6
  management of, 175–6
  width and birds, 65, 174–5
predation, 79–80
  in edge, 91–2, 106
  on small road at night, 161
predator, 75–7, 79, 394
  in edge and hedgerow, 91–2, 107, 198
predator–prey cycle, 80
present functioning predicted by shape, 138–9
prey, 79
principles, 437
  first, 29–30
  of landscape and regional ecology, 514–16
private spaces, 497
producer, 75–6
production or productivity, 37, 76, 341
  in ecological integrity, 499–500
  of field, 48–9, 194, 231
  of forest, 49–51
  livestock, 400–2, 478
  size effect on, 48–51
  windbreak and, 194
protected area
  see natural resource
protecting society, 441
pyramid of biomass, 76

Qi, 15–16
quality of life, 485–6, 501
Quebec, Canada, 354
Queensland, 63, 98
quiet zone, see windbreak

rabbit, 10, 31, 197, 393–4
Rackham, Oliver, 374–5
radiation, see sky radiation; solar energy
radiotracking, 35–6, 274, 285, 389
railroad corridor
  grassland reserve and, 172, 477
  as trough corridor, 159, 171–2
rain, 90, 342
rain shadow, 346
raindrop energy, 244
rainforest, see tropical
ranch, see pasture
random, 35
  grid pattern and percolation theory, 254–5
  mosaic-sequence model, 422–6, 429
  non-, 255
  patch shapes, 120

species distribution and population dispersion, 58, 78
random walk, 367
rangeland, see pasture
rare species in corridor, 149–50, 236, 246
rastor in geographic information system, 34
receptor, 96–8, 152
recolonization
  lobe and corridor role in, 128–9, 203
  in metapopulation dynamics, 372–5, 380–3
  of patch, 47, 274–6
recruitment foci, 388
rectangular shape, 122–3
  see also length-to-width ratio
rectilinear landscape, 309
rectilinear network, 256, 263, 265–9, 274
  nodes and linkages in, 263
rectilinearity, 265
recycling of materials, 497
redundancy, 299
reforestation, 419, 422, 432–4
  in Massachusetts, 432–3
refuge for fish, 265
regenerated patch and corridor, 44–5, 157
Regier, H. A., 485
region
  concept of, 12–13, 22–4, 39, 512, 514
  flows within, 25, 335–6, 344–5
  grain size of, 23
  human policy or activity as cause of, 6, 23
  landscapes within, 20, 23, 430; linkage of, 335–6, 344–5; as spatial elements, 20, 23; spatial arrangement of, 25–7, 378–9, 430; biodiversity and spatial arrangement of, 378–9
  as linkage between local and global, 435
  linked with other region, 27, 325, 344–5
  macroclimate of, 13, 22–3, 325
  policy as cause of, 6
  shape of, 137
  as surrogate for long-term, 523–4
  sustainable, making a, 514–24
  types of, 306–10; glacial terrain, 308–9, 463–4; eco-, 14; high atmospheric pressure, 323–5; low atmospheric

pressure, 323–5; mountainous, 21, 25, 310; patch–corridor–matrix, urban, 310; water-created terrain, 124; wind-created terrain, 124, 307–8
regional change, 508
regional ecology, 22–8, 37, 512–14
  landscape ecology principles applicable for, 513
  and sustainability, 513–14
  see also landscape ecology
regional planning, 36–7, 523–4
  see also planning
regional trend in biodiversity, 378–9
regional vegetation and erosion, 507–8
regular pattern of ecosystems, 290–1, 299, 304
  humans produce, 301
regulation of population, 10, 80
rehabilitation, see restoration
relation theory, 103
relative evenness, 319
relative patchiness, 320
relative richness, 319
relaxation period, 56–7
religion, 492–4, 510
remembrement, 421–2
remnant corridor, 157
remnant patch, 44–5, 50, 458–9
  size effect on, 59–63
remote sensing, 35–6, 317–18, 328
replicate, 30
reptiles, 62
rescue effect, 374
research methods, 29–36
reserve, see nature reserve
reservoir
  of carbon, phosphorus, and nitrogen, 77
  as source, 39
  of water, 53
residential neighborhood, 470, 472–4
resistance
  corridor and network, 270, 272
  landscape, 272, 279–81, 515
resistant ecosystem, 360, 369
resolution, 10
resource, limiting and competition for, 79
resource base, 510–11
restoration
  landscape, 432–4, 476–7; diverse examples of, 477
  of mine surface, 109, 141
  of North Woods, 477
  river floodplain, 476–8

of tropical dry forest, 476
rheotaxis, 30, 150, 239
rice culture, 225
ride in European woods, 172, 410
see also trail corridor
ridge
  network, 304, 306
  rain shadow and heat flow on, 346–7
riffle, 69, 210–11, 214
  see also pool and riffle; stream corridor
rights of way, 159
rigid polygon from graph theory, 294
ring shape, 25, 119, 306, 430
riparian corridor, 208
  see also river corridor; stream corridor
risk spreading
  in number of large patches, 53, 75, 431, 438
  and patch shape, 132
Risser, Paul G., 20
river, 208
  in stream order system, 245, 249
  structure of, *see* stream
river continuum, 209–10, 228
river corridor, 40, 158, 208–52
  as barrier, 236
  as conduit, 228–9
  cross-sectional structure of, 213–16
  fish in, 232–6; food for, 233–5; habitat for, 233–5; system, 235–6
  as habitat: riverbank, 210–11, 237; floodplain, 210–11, 214, 235–9, 245–9; hillslope, 210–11, 214, 235, 239; upland, 210–11, 214, 235–6, 239
  human activity in, 209, 223–7; agriculture as, 225–6; built objects as, 226–7; channelization as, 476–8; construction as, 226–7; dam as, 223–4; forestry as, 225–6, 250
  lateral flow from matrix to channel, 229–32
  linear structure of, 210–13
  natural process in, 216–27; deposition as, 219–20; erosion as, 219–20; hydrologic flow as, 217–19
  restoration of, 476–8
  terrestrial species in: animals, 220–3, 236–43; plants, 236–43
  types of, 209–16
  width of, 243–9
river mosaic, 209–10, 218, 247

river otter (*Lutra*), 223, 237
riverbank
  animals on, 223
  vegetation, 233–5
road, 22, 164, 420
  closing, 163–8, 176
  paved (tarmac) and unpaved, 163–4
road corridor, 39, 146, 159–72
  and badger, 377
  in forestry, 456–7
  material from: dust as, 169–70; particle erosion as, 456–7; lead pollution as, 168–9; salt as, 169
  matrix affected by, 168–70, 420
  median vegetated strip in, 152, 160
  persistence of, 421
  species avoidance of, 164, 169–70
  stream sedimentation downstream of, 169–70
  vehicular traffic effect in, 161–6
  width effect on crossing, 163–4; for birds, 164; for invertebrates, 163–4; for large mammals, 161–4; for mid-sized mammals, 163–4; for small mammals, 163–4
road corridor, function or process in
  access and human impact as, 169–70, 420
  barrier as, 279–80, 391
  conduit as, 160–3; hazards along, 161–2; nocturnal and diurnal, 161; for predators, 161
  filter as, 163–4
  habitat as, 154, 160
  movement of species across as, 279–80
  other, 170–2
  sink as, 164–8
  source as, 168–70
road density, 260, 266–7
road kill, 165
  animals avoid, by crossing: bridge over sunken road, 165; overpass arched over road, 165; in tunnel, 165–6, 176; in underpass, 165–7, 176
  and driver behavior, 165, 168
  food and attractants in corridor cause, 166–7
  population affected by, 167–8
  on roads of different type, 168
road network, 121, 154, 171

road reserve, *see* roadside natural strip
roadside (verge), 159
  as conduit and barrier, 161, 391
  food in, 166
roadside natural strip, 159–60, 171
  beauty strip as, 171
  as conduit, 161–3, 205
  in giant green network, 154, 171
  grassland reserve as, 172
  width of, 154, 162, 171
Rocky Mountains, 126, 266
  species movement in, 391–2, 398
Romania, 17
Romme, William H., 318, 353
rotation
  forest, 354
  pasture, 401–2
Roth, R. R., 473
rough terrain, *see* terrain
route
  across landscape, 254–5, 386–7
  in corridor, 200, 271; gap, 202
  curvilinearity of, 255, 271
  livestock, 159, 172–3, 265, 393
  in mountains, 306
  in neighborhood mosaic, 272–4; loop as alternative, 270
  patch shape and animal movement, 127
  in stream corridor, 239–42, 245
ruggedness, 304
run, *see* fetch
rural
  area, species movement across, 279–80
  land planning, 444
Ruzicka, Milan, 504
Ryzkowski, L., 329, 331–2

S-shaped curve, 78
sacred
  land-markings, 481
  site, 18
safety factor, conservation, 70–1
Sahel, 508
salinization, 118, 169
saltatory movement, 367
sampling
  patch size effect, 46, 58
  plot size and shape in, 54–5, 125
Samways, M., 391
sand, 308, 491
  dune, 307–8
sandbox, 253

saum, 93–5, 98
Saunders, Denis A., 36
scale, 8–10, 39
    boundary at different, 86–7
    broad (coarse) and fine
        versus large and small, 8–
        9, 39
    domains of, 11–12
    fine, 9, 18, 39; vegetational
        change and landscape
        ecology, 507–9
    fragmentation at different,
        412
    human, 4, 37, 490–1
    independent, 304–5
    on map, 8
    series of, 78; used in nest
        location, 294–6
    spatial, 7–14; in
        management, 474–7; in
        sustainability, 486–91
    species and, 366, 378
    temporal, 323; in
        management, 476–7; in
        sustainability, 486, 488
scholarship, 15
Schotman, Alex, 376–9
science, 15, 30
sea breeze without a sea, 89,
    333
searching, 367–8
seascape, 14
second law of
    thermodynamics, 4
second–c.4th-order stream,
    52–3
    connectivity of, 249–52, 261,
        265
    width of, 245–9
section of matrix, 278, 387
sedimentation, 169–70, 224–7,
    234
seed dispersal, 63, 65, 322,
    366–7, 387–8
    by animals, 388, 391
    by conveyor belt, 338
    distance, 387–8
    by road, 160
    by railroad, 172
    up valley, 387
seepage (seep), 212, 244–7,
    351
    width of protected area
        around, 245
segmentation
    of boundary, 93
    in stream corridor, 212
self-organizing principle, 290,
    448
self-reliance, 497
semivariance, 319
shade
    in field and forest edge, 90–
        1
    from streambank and
        riverbank, 245, 248–52

shape, 113–14, 137
    attributes and ecological
        flow, 124–33
    in the matrix, 278–82
    of patch, see patch shape
    to predict future, past, and
        present, 138–9
    of, as type of object: city,
        137; clearing, 136–7;
        drainage basin, 137; field,
        122, 125–6; habitat, 137;
        home range, 130, 137,
        389, 391–2; housing
        development (estate), 121;
        lake, 122, 137; landscape,
        25; network, 125;
        opening, 136–7; other
        objects, 137; region, 137
Sharpe, David M., 409–12
Shaver, Gus R., 287–8
sheep, 401
    see also livestock
sheet flow, 350–1, 521
shelterbelt, see hedgerow;
    windbreak
shifting mosaic, 44
short-duration grazing system,
    401–2
shrinkage, 407–11, 429
silt, 291, 308, 491
sink, 39
    corridor as, 148–9, 153, 276;
        road, 164–8
    edge as, 96, 103
    habitat, 381–2;
        metapopulation and 380–
        2; mortality, 203–4
    lake as, 335–6, 507–8
    in landscape, 340, 343–5;
        material transport to, 340,
        343–5; as particular type
        of area, 344, 358
    learning to avoid, 204, 276
sinuosity ratio, 137, 220
site, 20
    diversity, 489–90
    special, 463–4, 467–8
size (area)
    area-per-se as, 46–7
    of field and forest, 48–52
    forestry, economics, and,
        49–51
    and height of surroundings,
        65
    of hurricane created patch,
        356–7
    in island biogeography, 54–
        9, 80
    of landscape mosaic, 59–66,
        80, 407–8
    large patch: ecological value
        of, 47–8, 426–7, 438, 515;
        location of, 75; number
        of, 72–5; in planning of
        matrix sections, 278
    mid-, 47, 49

    in mosaic-sequence models,
        425–31
    of opening, 50–1, 65
    sampling of, 46–7
    small patch: ecological value
        of, 47–8, 427, 438;
        -restricted species, 47, 60
    see also patch size
size effect on ecosystem, 45–
    54
    process, 48–54; erosion, 51;
        herbivory, 50, 64;
        hydrology, 51–3;
        productivity, 48–51;
        structure: biomass, 48–51;
        mineral nutrients, 51
    water: quality, 52–3;
        quantity, 51–3
size effect on habitat
    patch: disturbance, 63–5;
        environmental, 59–63; in
        landscape, 59–66;
        remnant, 59–63
    type: aquatic system, 51;
        field, 48–9, 51–2;
        grassland, 65; island, 54–
        9; marsh, 65, 290; woods,
        59–62
size effect on species
    biodiversity, 54–66
    extinction of subpopulation,
        372–93
    invertebrate: ants, 63;
        anthropods, 50, 62;
        butterflies; snails, 60
    on islands, 55–7
    metapopulation, 373, 375,
        381, 383
    plant, 60–2, 64; mosses, 60;
        mushrooms, 60; tree
        reproduction, 64; tree
        species, 60–2
    population size, 59–60
    samples within a
        community, 54–5, 80
    vertebrates: amphibians, 62;
        birds, 59–65, 375–9; fish,
        62; large mammals, 62;
        reptiles, 62; small
        mammals, 60, 62, 64
size-independent species, 59
size-restricted species, 59–60
ski, see trail corridor
sky radiation, 328
slope, north- versus
    south-facing, 328, 346–7
SLOSS effect
    on biodiversity, 54–65
    on ecosystem, 45–48,
        53–4
Slovakia, see Czechoslovakia
slowly changing attributes in
    sustainability, 498–9, 523
small-patch landscape, 309
Smith, Daniel S., 106–7
snails, 60

snow
  accumulation, 355; in
    mountains, 346
  albedo of, 329
  windbreak effect on, 185,
    192–4
snowdrift, 288
snowflake, transport of, 343
snowmobile, *see* trail corridor
soaring birds, 346–7
soil, 37
  albedo of, 329, 330
  bank and ditch in
    hedgerow, 196
  catena, 287–8
  in ecological integrity, 500
  in edge, 90–1; mineral
    nutrients of, 90–1, 95
  in floodplain, 307–8
  in forestry, 308, 457
  heat, 330–3
  particle lifted by conveyor
    belt, 338
  sand, silt, and clay in, 246,
    308
  windbreak effect on, 186,
    189–95
soil conservation, 36–7
  see also conservation
soil erosion, *see* erosion
soil moisture, 192–5
soil organic matter, 229–34
soil stabilization, 482
soil surface, concave and
    convex, 304–5, 348–51
soil texture triangle, 308
solar energy
  in edge, 87–8
  longwave and shortwave
    radiation in, 327–30;
    visible wave lengths in,
    327–9
  produces structure, 4
  reaching Earth's:
    atmosphere, 327; surface,
    327–30
  as vertical input, 327–33
solar radiation, *see* solar
    energy
song thrush (*Turdus*), 378–9
songlines, 17–18
source, 39
  in landscape, 340, 343–5
  material transport from,
    340, 343–5
  in network, 256
  patch, 376, 381–2; gene flow
    and, 384–5; in
    metapopulation, 380–2
  seepage and spring as, 212,
    244–7, 351
  species, 18, 47; in island
    biogeography, 56–7; in
    mosaic, 273
  type of, 343; boundary as,
    96; corridor as, 148–9,

153; edge as, 96, 103; road
    as, 168–70
source–sink habitats, 376,
    381–2
southeastern Australia, 50, 62,
    74, 162, 205, 258, 265,
    311–12
southwestern USA, 22, 25, 70,
    508
  see also New Mexico
shaceship shape, 132
space–time principle, 8
  disturbance and ecological
    process in, 8
  example of, 36, 337
sparrow, 394
spatial arrangement, optimum
  in landscape, 437, 516, 522–
    3
  of landscapes, 25–7
spatial attribute
  in generic plan, 455–6
  in mosaic-sequence model,
    424–9
spatial element, 6, 38–9
  in landscape, 20; repeated
    cluster of, 13, 21
  in region, 20, 23
  see also landscape element
spatial-flow principle, 286–7,
    345
spatial language, 7, 442–3
spatial process
  general model, 409–12
  in land transformation,
    406–12
  type of: attrition as, 407–11;
    dissection as, 407–10;
    fragmentation as, 405–34;
    perforation as, 407–11;
    shrinkage as, 407–11, 429
spatial scale in sustainability,
    486–91
spatial solution, xix, 522
spatial statistics, 33–5, 319
special sites, 463–4, 367–8
specialist species, *see* species
species
  avoidance of corridor by,
    164, 169–70
  edge, *see* edge species
  exotic, *see* exotic species
  generalist and specialist,
    96–7, 149–50, 174, 204–
    5
  indifferent, 97
  interior, *see* interior species
  keystone, 63, 77, 296
  multihabitat, *see*
    multihabitat species
  rare, *see* rare species in
    corridor
  supersaturated with, 56
  wide-ranging and
    large-home-range, 25, 47,
    312, 388

species–area curve, 59–62
  equations for, 55, 58, 80
  minimum area point of, 54
  for samples, 54–5
species cascade effect, 63
species coexistence, 79
species colonization, *see*
    colonization
species density (packing)
  in small patches and
    corridors, 47, 62, 150
  tree and bird, 61–2
species dispersal, *see* disperal
species extinction, *see*
    extinction
species interaction, 79
  see also competition;
    predation
species introduction, 385,
    397–8
  see also exotic species
species movement, *see*
    movement
species pool, 55, 73–4
species richness, 40
  see also size effect on
    species
species source, 18, 47, 56–7,
    273
  see also source
species turnover, 56, 58, 63,
    381
spider, 33
  from graph theory, 293
  road effect on, 163–4
sponge effect, 53, 213, 244–5
sponge-like shape, 119, 123
spreading risk, *see* risk
    spreading
spring, 212, 244–6, 351
spruce (*Picea*), 24, 222, 243,
    484–5
spruce budworm
    (*Choristoneura*), 242–3
square shape, 117, 119, 122–3
squirrel, 60, 165, 296, 377, 379
stability
  cause of, 9–10
  of configuration, 290
  of convergency point
    (covert), 315
  conveyor belt providing
    resilience, 338
  feedbacks and, 10, 80, 505
  in food web, 77
  of forms common in design,
    447
  hierarchy and, 505
  input–output model and,
    495–6
  network connectivity and,
    274–6
  persistence and, 44–5, 298
  space–time principle and, 8,
    337

stability, *continued*
  in sustainability, 502–5,
    516–17
  type: mosaic, 298–9, 504;
    physical-system, 504;
    resilience or recovery,
    504; resistance, 504
statistics, 33–5
  geostatistics and spatial, 35,
    319
steady state economics, 483
steady state system, 495
Stearns, Forest, 409–12
stepped bed stream, 218
stepping stones, 40, 430, 452–
    3
  cluster of and row of, 201–
    2, 386–7
  for dispersal, 47
  gaps and, 155–6
  island as, 56
  small patch as, 47, 201–2
Stiles, Edmund W., 388
strain, genetic, 67, 354, 385
  see also subpopulation
strategic point
  based on: access, control, or
    protection, 316; evidence
    in book, 316; gaming,
    317; Go, 317; position in
    landscape, 316
  in landscape and patch,
    315–17
stratification, 76
  see also vertical structure or
    stratification
stratosphere, 325
stream
  diversion of, 224–5
  as dynamic equilibrium,
    216
  eutrophication of, 77
  flow rate of, 218
  meander band of, 211, 220–
    1, 307
  migration of, 211, 219–21,
    238–9
  with reversed dendritic
    pattern, 124
  structure in: hole caused by
    log or rock as, 214; large
    woody debris as, 215,
    235, 245, 248–52; narrow
    or wide island as, 211;
    open canopy over as, 214;
    particle size on bottom
    as, 218; pool and riffle,
    69, 210–11, 214, 219–21,
    234
  type of: braided, 211, 214–
    15; constrained and
    unconstrained, 211;
    ephemeral, *see*
    intermittent channel;
    intermittent, 211–12, 232,

244–6, 351; stepped bed,
    in mountains, 218
stream capture, 354
stream channel, 210–11
  flow, 210, 214
stream channelization, 214–
    16, 224–5, 460
  river and, 476–8
stream corridor, 146, 208–52
  canopy in, open or closed,
    211, 245, 248
  concept of, 40, 158, 208
  connectivity of, 249–52,
    261; gap and, 250–1, 265
  cross-sectional structure of,
    213–16
  dam in, *see* dam
  dissolved substances in,
    228–32, 245–8
  disturbance in, 354
  habitat in: aquatic, 232–6;
    floodplain, 210–11, 214,
    235–9, 245–9; hillslope,
    210–11, 214, 235, 239;
    streambank, 210–11, 237;
    upland, 210–11, 214, 235–
    6, 239
  linear structure of, 210–13
  in planning, 452, 457–8
  river mosaic concept in,
    209–10
  stream order and:
    first-order, 52–3, 244–7;
    2nd–*c*.4th-order, 52–3,
    245–52, 261, 265
  types of, 209–16
  water in: equation for
    lateral flow of, 252;
    quality of, 228–32
  width of, 155, 232, 243–9,
    251–2
  wind and, 241
  see also river corridor
stream corridor function or
    process
  functioning as: barrier, 391;
    conduit, 228–9, 239–45
  habitat as, 236–9;
    streambank, 237;
    floodplain, 237–9;
    hillslope, 239; upland,
    239
  internal, 216–27; conflicting
    ecological, 242–3;
    deposition as, 219–20;
    erosion as, 219–20;
    hydrologic flow as, 217–
    19; lateral flow from
    matrix to channel as,
    229–32; lengthwise
    movement as, 228–9
stream corridor species
  birds moving along as, 241
  routes of, 239–42, 245
  type of: animal as, 220–3;
    beaver as, 220–3; bird as,

241; fish as, 232–6; plant
    as, 236–43; terrestrial
    animal as, 236–43
stream discharge, 214–15
stream network
  connectivity and
    development of, 52–3, 141
  fish refuge in, 265
  fractal, 119, 260–1, 305
  headwaters protection of,
    47, 52–3
  large node in, 264–5
stream order
  concept, 40, 212, 218
  in dry area, 213
  type of: first-order, 52–53,
    244–7; 2nd–*c*.4th-order,
    52–53, 245–9, 261, 265;
    c.5th–10th order river, as,
    245, 249
stream recharge, 214–15
streambank, 210–11, 214, 237
  as habitat, 210–11, 237
  vegetation, 233–5
streamline
  patch shape, 117–18
  in wind, 180–1
  windbreak, 181, 185, 193
string-of-lights pattern, 201,
    259
strip corridor, 150, 154, 195
strip cropping in field, 191–2
structure and function, 4, 75
subdivided matrix, 278–9
subpopulation, 71
  genetic differentiation of,
    152, 272
  genetic variation within, 67,
    71
  in metapopulation
    dynamics, 372–5, 380–2;
    local extinction of, 70–1,
    372–5, 380–2
  minimum viable population
    and, 71
  and patch: within network,
    272; shape, 127–8
substrate heterogeneity, 44–5
subsurface flow
  of mineral nutrient, 229–31
  of water, 77, 212, 252, 351
suburb
  back lot lines in, 473
  Concord case study of, 461–
    9; general application of,
    469; synthesis and
    protection priority in, 468
  fragmentation in, 462
  framework for open space
    planning in, 463
  house lot in, 469–74: goals
    and techniques for, 470–
    1; neighborhood context
    of, 470; spatial
    arrangement of, 471–4
  large patches in, 47, 462–6;

agricultural areas, 465–6;
built areas as, 464–5;
natural-vegetation areas
as, 464–5
major corridors in, 462–7;
human corridors as, 465,
467; water protection
corridors as, 465, 467;
wildlife corridors as, 465–
6
planning of, 459–69
residential neighborhood in,
470, 472–4
special site in, 463–4, 467–
8; ecological habitat as,
463, 467; education and
institution as, 463, 467;
geological feature as, 463,
467; historic site as, 463,
467; recreational site as,
463, 467; scenic spot as,
463, 467; town
infrastructure as, 463,
467–8; water resource as,
463, 467
surrounding area effect on,
463
town-wide pattern in, 464–7
values of large patches and
major corridors in, 464–7
suburban landscape, 22, 300–
1
change in, 461
forms in, 446–7
heterogeneity in and
movement across, 459–60
open space and built area
in, 460
between rural and urban,
459
unusual attributes of:
corridors, 461; patches,
461; species, 461;
suburbs, 461
suburban lawn, 391
suburbanization
in Japan and France, 410,
418
mosaic sequences in, 418–
19, 429
succession, 44, 63
Suffling, Roger, 354
suitability, see habitat quality;
habitat suitability
Sukachev, V., 20
sulfate, 358
sulfur dioxide, 340–1
experiment, 358
sun, see solar energy
sunfleck, 8
supersaturated with species,
56
surface
boundary, of patch, 107–10,
140–1

concave and convex soil,
304–5, 348–51
surface runoff
of mineral nutrient, 229–31
of water, 77, 252, 351
survey to locate nature
reserves, 312–13
survival of the fittest, 79
susceptible ecosystem, 360–1,
369
sustainability
concept, 40, 462, 481–6,
502; in agriculture, 481–2;
in this book, 483–4; in
economics, 482–3; in
fisheries, 482; in forestry,
482; in social science,
483; of World
Commission on
Environment and
Development, 483
incorporated into planning
and management, 522–4
less, 509–10
more, 509–11
planning and management,
incorporation into, 522–4
of several linked
landscapes, 496
see also sustainable
environment
sustainable development, 483
see also sustainable
environment
sustainable environment, 480–
524, 519
adaptability and stability in,
502–5, 516–17
assays of, 497–502; values
as, 497–8; slowly
changing attributes as,
498–9
basic human needs in, 501–
2
as coarse grain with fine
grain areas, 438, 491
cohesive forces in, 492–4,
509–11
ecological integrity in, 499–
501
environmental variables in,
509–11; erosion and, 509–
10
grain size of, 489–91
human dimensions or
attributes of, 492–7, 509–
12; built environment as,
496–7; cohesive forces as,
492–4; generations as,
486, 523–4; linkages with
other areas as, 494–6
key attributes of, 517–19
learning from civilizations
and communities for,
509–12; New Brunswick
case study as, 484–6

learning from history and
paleoecology for, 506–9
making a, 514–24; general
principles for, 514–16;
designing a local area as,
516–18
patches and corridors in,
488–9
planning and management
based on, 518–19, 522–4;
concurrent objectives in,
519–21
principles of landscape and
regional ecology for, 514–
16; applications in, 515–
16; landscapes and
regions in, 514; patches
and corridors in, 515;
mosaics in, 515
as region or landscape,
513–14
regional ecology for, 27,
512–14
self reliance in, 497
slowly changing or
foundation variables in,
498–9
spatial scale in, 486–8, 518;
grain size and, 489–91;
paradox of management
and, 488; patches,
corridors, and, 488–9
technological change in,
485, 504, 512, 523
temporal scale in, 486, 488
urban, 496
sustained yield in fisheries
and forestry, 482
swamp, 238, 389–90, 395
see also wetland
swamping, genetic, 67, 384–5
Sweden, 62, 103, 392, 430–1
Switzerland, 392, 442
system
living, 4; configuration as,
298; landscape as, 5
of nature reserves, 314
open and closed, 4, 30, 84,
486–7
resilience of ecological, 338
steady state and changing,
495

target, 369
technological change or
innovation, 485, 504, 512,
523
temperature, 89
atmospheric, 112
in edge, 89
in forest versus opening in
Germany, 333–4
and particle deposition
from smokestack, 339–40
streambank vegetation and
water, 233, 250–1

temperature, *continued*
surface, 330; concave and
convex, 348–50
in valley, 347
windbreak effect on, 187–9;
of leaf, 194
temperature inversion, 339–
40, 342
temporal scale
in management, 476–7
in sustainability, 486, 488
Tennessee, USA, 174–5, 312,
389
terrace, 214, 238
terrain
and patch shape, 114–16
rough, 345–51; airflow in
ridges and valleys as,
345–8; convex and
concave surfaces in, 348–
50; sheet flows of water
in, 350–1
ruggedness of, 304
type, 306–9; glacial, 124,
308–9, 463–4;
karst-limestone, 309;
tidal-flat, 309;
valley-alluvium, 309;
water-created, 124, 307;
wind-created, 124, 307–8
territory, 365
Texas, USA, 11, 186, 288, 349
texture analysis, 319
soil triangle, 308
theory, 437, 519
thermal, 325
over landscapes, 335–6;
tilted, 335
soaring birds using, 335;
think globally, act locally, 435
think globally, plan regionally,
and then act locally, 435,
451
third dimension
of edge, 86
of land, 302–6, 345–6
see also mountains; terrain
Thoreau, Henry David, 3, 43,
81, 113, 145, 177, 208,
253, 285, 290, 296, 322,
364, 405, 435, 461–2, 480
threshold
in deforestation, 416, 426
in forest size effect on
stream network, 52
Thunen band, 297
tidal-flat terrain, 309
time-distance, 368
tip of lobe, 127–8
tip-splitting, 120, 140
topography, 302–4
see also mountains; terrain
topological space, 368
tornado, 355–6
town, 435, 461–9
traffic, vehicular, 161–6

tragedy-of-the-commons, 15–
16
trail corridor, 172–4
damage going up or down,
173–4
density, 260
as trough corridor, 159
type: animal, 172, 271;
bicycle, 172;
cross-country (touring)
ski, 172–3; horseback,
159, 172–4; livestock,
172–3; motorbike, 172–4;
ride in European woods
as, 172; snowmobile, 172–
3; walking or hiking, 172–
4, 421; wildlife, in edge,
94–5
trampling, 172–4, 392
see also trail corridor
transformation, *see* land
transformation
transportation
efficient, in urban, 497
network theory from, 257,
261–2
traplining, 368
travel lane, 386
see also wildlife; corridor
trees, patch size effect on, 60–
2, 64
Troll, Carl, 20
trophic level, 63
tropical
dry forest, 476
island, 396
landscape, human effect on,
508, 521
rainforest, 4, 32, 62–3, 93,
98, 151, 342; road in, 420,
429
troposphere, 325–6, 337–8
delivery processes in, 343
trough corridor, 158–9
connectivity of, 160
type: dike, 159; powerline,
159, 172–4; railroad, 159,
171–2; road, 146, 159–72;
trail, 159, 172–4
trout, 232–3, 351
stream corridor vegetation
protecting, 250–1
trunk line, 257, 271
tundra, 118, 164, 287–8, 303
tunnel
for amphibians, 166
for animal crossing, 165–6;
design of, 166, 176
turbulence, 118
in edge, 89
patch shape and, 127–8
planetary boundary layer
and, 325, 344
of wind, 90, 180–3; in edge,
90; windbreak effect on,
182–7

Turner, Monica G., 353
turnover, species, 56, 58, 63,
381

ultraviolet (UV) radiation,
327–8
underpass
on alligator alley, 166
for animal crossing, 165–7;
design of, 166, 176
understanding as goal, 15
unplanned forms, 445–9
unpredictability, 487, 505
unusual attributes of
suburban landscape, 461
unusual feature, 7, 23
upland habitat in stream and
river corridor, 210–11,
214, 235–6, 239
urban center nearby, 510
urban region, 310; exotics in,
394
urbanization, *see*
suburbanization
USA, 6, 17, 27, 159, 174, 393,
444, 482

valley
air flow in: along slope,
347–8; diurnal change in,
347–8; longitudinal
valley-bottom, 347–8;
up-and-down transport,
347–8
blowdown at head of, 347–
8, 355
temperature in, 347
wedge-shaped, V-shaped,
and U-shaped, 306–9
valley-alluvium terrain, 309
valley erosion, 305
values in sustainability, 497–8
van Apeldoorn, Rob, 376–9
van Dorp, D., 379
van Noorden, Boena, 376–9
variables
added to mosaic-sequence
models, 433
optimization for two or
three, 520–1
variation (variability)
in connectivity, 294
in ecosystem over time, 298
in grain size, 438, 491, 515
in land and its perception,
305
in mesh size, 121–2, 269–70
in natural selection, 79
in patch shape, 114–15
smoothed out by logarithm,
57–8
in width of: corridor, 148,
270; edge, 106, 129–30,
140; stream corridor,
245–7, 251–2
see also genetic variation

vector, 100–1, 366
in geographic information
system, 34
locomotion or mass flow of,
100
vegetation, natural, 39, 319;
zone or belt in
mountains, 306
vegetational change,
long-term, 506–7
vegetational and topographic
height, balance between,
302–4
vehicular traffic, 161–6
veil, 94, 98
Venezuela, 172, 232
Venturi effect
in corridor gap, 30, 187
in edge, 100
Verboom, Jana, 376–9
verge, see roadside
vertical arrival of disturbance,
359
vertical dimension in
mountains, 302–6
flow in, 25, 306, 345–8
vertical energy flows, 327–33
vertical structure or
stratification, 76
in edge, 85
hurricane susceptibility
and, 356
in suburb, 472–3
in windbreak and
hedgerow, 178–80, 183–6,
193–6
village
in China, 15–16
in Eastern Europe, 16–17
vineyard, 46, 391
Vink, A. P. A., 20
visible wave lengths, 327–8
vision, 523
volcanic impact, 352–3
vole, 377–80
vortex, 181, 187, 350, 355

wake zone, see windbreak
wandering movement, 366,
392
war, 492, 509–10
water, 37
area, movement of species
across, 279–80
in conveyor belt, 342–3
in ecological integrity, 500–
1
evapotranspiration and, 51–
2, 342
flow in mosaic, 350–1; as
disturbance, 354–5
patch size and, 51–3
surface runoff, subsurface
flow, and sheet runoff of,
77
water budget, 51–2

water erosion, see erosion
water hyacinth (Eichhornia),
395
water quality
of lake, 47, 52–3, 77
large patch and, 47
in stream corridor, 228–32
water quantity, see hydrology
and hydrologic
water resources degraded,
509–11
water table, 215
water vapor, 327–8
watershed, see drainage basin
wave length shift, 330
wavy net, 258, 269, 271
weather, 325
weaving phase of landscape
ecology, 28
weed, 396
Wegner, John, 296–7
Western Australia, 195, 248,
269, 392
road corridor in, 154, 160–
3, 171
wetland
etangs and sansouire as,
391
in floodplain, 214, 237–8
marsh as, 65, 237–8; area
and isolation of, 290, 380
and mineral nutrient, 231
persistence of, 421
soil, 308
species conservation in, 74
swamp as, 238, 389–90, 395
Wiens, John A., 32–3, 371
wildlands project, 450
wildlife, 40
biodiversity, 37–8, 54
in edge, 106–7;
curvilinearity effect on,
106–7
game as, 39, 128, 392, 416
home range of, 137
movement in corridor, 198–
207
near ceremonial center, 17
planning natural resource
area for, 444–5
see also species
wildlife biology, see wildlife
management
wildlife corridor
advantages and
disadvantages of, 203
enhancing and inhibiting
movement in, 200–3
in residential neighborhood,
473
see also corridor
wildlife habitat pattern, 137,
313–15
wildlife management, 11, 36–
7, 401
browse line in, 31, 104; see

also overgrazing or
overbrowsing
convergency point or covert
in, 314–15, 472
edge and edge effect in, 85,
98, 106–9, 313; path in,
94–5
interspersion in, 313–15
maintaining both edge and
interior in, 313
and patch shape, 124–5
species-centred: with
beaver, 222–3; exotic
species, 398;
metapopulations in, 382–
3; minimum viable
population, 71–2;
multihabitat species, 390;
for road crossing by
animals, 163–8, 176
see also management
Wilson, Edward O., xiii–xiv,
56, 380
wind
on concave and convex
surfaces, 348–50
flows of, in mosaic, 322–50
forest and: downwind side
of, 186; penetration in
edge of, 87–90, 130
orientation of object and:
oblique, 126, 181, 186–7,
192; parallel, 187, 192
principles and equations
for, 180–2, 207; erosion
control, 190–2, 207
seed dispersal and, 338,
387–8
with steep slope upwind or
downwind, 181
stream corridor and, 241
and streamlined or bluff
object, 181
turbulence in, 90, 180–1;
approaching, 183; eddy
in, 182–4
type of: bleed flow as, 181;
streamline, 180–1;
turbulent, 90, 180–1;
vortex, 181, 187
Venturi effect: in corridor
gap, 30, 187; in edge, 100
see also windbreak
wind directions, 126, 181,
186–7, 192
global, in January and July,
324
and ocean current, 324–5
wind erosion, see erosion
wind fetch or run, 47, 49
wind rose, 126
windbreak, 177
albedo of, 329, 331
distance from, 182–4
end portions of, 187, 192
length of, 185, 187, 193

windbreak, *continued*
management of, 193
orientation: angle, 181, 186–7, 192–3, 197; parallel to wind, 292
plant spacing and rows in, 193, 196
resistance coefficient of, 182
structure of: cross-sectional, 181, 185–6; gap in, 187; height, 181, 183–5; richness in, 178–80, 184–5; streamlined, 181, 185, 193; tall narrow, 185–6
type of 178–80: herbaceous, 192; protecting farmstead, 191; protecting pasture, 196–7; streamlined, 181, 185, 193
water affected by, 331–2
see also wind; hedgerow; corridor
windbreak capacity, 192–4
windbreak effect on
communities, 194–5
evaporation or evapotranspiration, 187–9; dependent on dry or moist overhead airflow, 188
microclimate, 187–9; humidity, 188; moisture, 187–9; temperature, 187–9
plant, 194–5; productivity, 194
snow, 185, 192–4

soil, 189–92; erosion, 186, 189–92; moisture, 192–5
wind, 180–7; erosion force, 192; speed, 182–9; turbulence, 182–7
windbreak porosity, 183–5, 193–4
high, medium or low, 183–5
measurement of, 182
in upper or lower portion, 185, 193–4
windbreak zone
quiet, 187–9
reduced windspeed, 182–7
upwind, 182–3
wake, 182–3, 186–9
windspeed
factors determining, 181
in field or edge, 48–9, 89–90
in network, 271
windbreak effect on, 182–9
windstorm, 355–7
see also hurricane; tornado
windthrow, *see* blowdown
winking, 380–5
Wisconsin, USA, 73, 82, 388, 409–12, 450
wolf (*Canis*), 298, 366, 405, 450, 477
wood thrush (*Hylocichla*), 380–1, 473–4
wooded strip, 31, 158, 177
disturbance and, 180
structural richness of, 178–80
woodland, ancient, 374–5
woodland corridor, 158, 177–80, 198–207

advantage and disadvantage of, 203
connectivity and curvilinearity of, 201–2
habitat quality of, 201, 204
matrix effect on, 203
movement in: edge species, 200; enhancing wildlife, 200–3; exotic species, 204–6; inhibiting wildlife, 200–3; interior species, 200; plant, 199; pest, 204–5; specialist species, 205
single row and multiple row, 178–9
species in: generalist, 204–5; resident, 200–1
structure of, 178–80; vertical stratification as, 178–80
as wildlife conduit, 198–207
woodland herb, 73, 374–5
Woodmansee, Robert G., 287–8
woods, *see* patch
woody debris, *see* large woody debris
woody strip, *see* wooded strip
Wyoming, USA, 125–6, 136–7, 262–4, 353, 381, 391–2
Yellowstone Park, USA, 353, 391–2

zone
global air circulation, 323–6
by windbreak, *see* windbreak zone
Zonneveld, I. S., 20